Hemp Diseases and Pests

Management and Biological Control

Hemp Diseases and Pests

Management and Biological Control

•

An Advanced Treatise

J.M. McPartland
Assistant Professor, University of Vermont

Robert Connell Clarke
Projects Manager, International Hemp Organization
Director, Pharmtex Consultants, Inc., San Diego, California

and

David Paul Watson
Director, HortaPharm B.V.
Chairman, International Hemp Organization, Amsterdam

CABI *Publishing*

CABI is a trading name of CAB International

CABI Head Office
Nosworthy Way
Wallingford
Oxfordshire OX10 8DE
UK

Tel: +44 (0)1491 832111
Fax: +44 (0)1491 833508
Email: cabi@cabi.org
Web site: www.cabi.org

CABI North American Office
875 Massachusetts Avenue
7th Floor
Cambridge, MA 02139
USA

Tel: +1 617 395 4056
Fax: +1 617 354 6875
Email: cabi-nao@cabi.org

A catalogue record for this book is available from the British Library, London, UK

Library of Congress Cataloging-in-Publication Data

McPartland, J. M.
Hemp diseases and pests : management and biological control / J.M. McPartland, R.C. Clarke, D.P. Watson
p. cm.
Includes bibliographic references (p.)
ISBN 0-85199-454-7 (alk. paper)
1. Hemp—diseases and pests. I. Clarke, Robert Connell, 1953- II. Watson, David Paul. III. Title

SB608.C28 M37 2000
633.5'399—dc21

00-023738

ISBN-13: 978-085199-454-3
ISBN-10: 0-85199-454-7

First published 2000
Reprinted 2002 (twice), 2004, 2006, 2009, 2010

Printed and bound in the UK by CPI Antony Rowe, Chippenham and Eastbourne
from copy supplied by the authors.

Contents

Preface vii
Acknowledgements viii
Forewords ix
How to use this book xi

Section One: Prerequisite Information about Crop Biology
1. PRINCIPLES OF PLANT PROTECTION 1
   Introduction to IPM concepts and keys for identifying problems
2. REQUIREMENTS FOR GROWTH 9
   Something about soil, moisture, temperature, and light
3. TAXONOMY AND ECOLOGY 13
   Classification of organisms associated with *Cannabis*, their ecological relations, and symptoms of parasitic interactions

Section Two: Diseases and Pests of *Cannabis*
4. INSECTS AND MITES 25
   Cannabis is not insect-pollinated, so these organisms primarily act as parasites, often causing major destruction
5. FUNGAL DISEASES 93
   Some fungi become *Cannabis* symbionts, but most cause disease, sometimes death
6. OTHER CANNABIS PESTS AND PATHOGENS 137
   A ménagerie of viruses, bacteria, nematodes, other plants, protozoa, slugs, non-insect arthropods, and vertebrate pests
7. ABIOTIC DISEASES 155
   Descriptions of mineral deficiencies and toxicities, sunscorch, other unsuitable environmental conditions, and herbicide damage
8. POST-HARVEST PROBLEMS 167
   Featuring hemp retters and rotters, seed eaters, and marijuana contaminants

Section Three: Control of Diseases and Pests
9. CULTURAL & MECHANICAL METHODS OF CONTROLLING DISEASES AND PESTS... 175
   Old and new techniques for reducing damage from pests and disease
10. BIOLOGICAL CONTROL 181
   Useful predators, parasites, microbial pesticides, and companion plants
11. BIORATIONAL CHEMICAL CONTROL 189
   The use and abuse of organic pesticides

Appendix 1: Synthetic chemicals 205
Appendix 2: A dichotomous key of diseases and pests 211
Appendix 3: Conversion factors 219
References 221
Index 245

Preface

As the 21st century begins, we are favoured with a worldwide resurgence in hemp cultivation. As global forests dwindle to rampikes and timber limits, we see a glimmer of hope for our future, thanks to a renewed interest in this ancient and humble source of food and fibre.

This book is our contribution to hemp's revival. Much of the literature regarding hemp diseases and pests dates back 50 years or more. Further, these publications are frequently buried in obscure agronomy journals. Cultivators of illicit *Cannabis* have published high-calibre research in the last 25 years, but they published in semi-clandestine "grey journals" such as *Sinsemilla Tips.* Our primary effort was to collect this scattered bibliography and assimilate it into a comprehensive and readable format.

Our second effort was to manoeuvre the control of diseases and pests into the 21st century. Most hemp research dates to the days when DDT was considered a glamorous panacea. We must find new control methods for sustainable hemp cultivation. Many "new" pesticides are old, such as pyrethrum, a popular insecticide before the days of DDT. Biological control also is old; the use of biocontrol against hemp pests began around 1886, in France. Être une tête à Papineau, we see biocontrol and hemp resurging together.

This book was written by amateurs, in the sense that none of the authors earns a living primarily as an entomologist or plant pathologist. Several Canadian and European professionals reviewed our work and honoured us with unequalled advice. For this book to have a second edition, we ask you, the reader, to share the results of your experience. Please send us feedback, via our research centre—the International Hemp Association, Postbus 75007, 1070 AA Amsterdam, The Netherlands.

The format and organization of this text benefited from George Lucas's *Diseases of Tobacco,* Cynthia Wescott's *Plant Disease Handbook* and *Gardener's Bug Book,* the APS *Compendium* series, and the many-authored *Diseases and Pests of Vegetable Crops in Canada,* edited by Ronald Howard, J. Allan Garland, and W. Lloyd Seaman. We hope our efforts to catalogue the *Cannabis* pest-control literature will reward the reader, and serve as a platform for launching further investigations. We mind J.D. Bernal's remark, "It is easier to make a scientific discovery than to learn whether it has already been made."

— J.M. McPartland, R.C. Clarke, D.P. Watson

Acknowledgements

McPartland: to Tom Alexander, editor of the first American journal dedicated to *Cannabis* agronomy. Thank you to Patty for her patience and support, to my brother for *Roget's Thesaurus*, and to my folks for growing me in the Garden State. Best regards to Ralph VonQualen and Don Schoeneweiss for plant pathology.

Clarke: to my parents for their unwaivering support and Mother Cannabis for her continuous supply of academic challenges.

Watson: to my mother, Evelyn Watson, for interesting me in gardening, Allen Chadwick for teaching me biodynamic farming, and the SSE for its conservation work.

Many individuals graciously reviewed parts of our manuscript (or the entire text – noted by *): J. Bissett, M. Cubeta, J. Dvorak*, W. Elmer, J. Goodwin, K. Hillig, S.J. Hughes, D. Jackson*, D. Lawn, J. Leland, P.J. Maddox, P. Mahlberg, S. Miller, S.J. Miller*, J. Moes*, G. O'Toole*, D. Pimentel, P.L. Pruitt*, W. Ravensberg, M. Robiner*, D.P. Rogers, G. Scheifele*, R.E. Schultes, K. Seifert, E.G. Simmons, J.B. Sinclair*, E. Small, B. Sutton, F. Uecker, D. West*, D. Wirtshafter*. Our 1991 edition was greatly improved by R. Rodale; we miss his equivocal wit.

We are pleased to acknowledge the financial sponsorship provided by three companies in producing the colour illustrations for this book. All three companies have a practical interest in the production of organically grown *Cannabis* species.

Koppert Biological Systems have produced clear nuclear populations of beneficial organisms in convenient application systems, which can be placed on affected plants. Beneficial insects can be positioned close to the focus of pest infestation, and the beneficials conveniently disappear when their food supply is exhausted, leaving no residues to compromise quality. Contact details: tel. +31 (0)10 5140444; +31 (0)10 5115203; e-mail info@koppert.nl; http://www.koppert.com/

GW Pharmaceuticals Ltd grows *Cannabis* under license from the Home Office in commercial quantities for medicinal purposes. Organic methods of growing are used and biological pest control is a feature of the production of medicinal *Cannabis*. The methods of organic growing ensure high yields and elimination of pesticide residues. Contact details: tel. +44 (0)1980 619000; fax: +44 (0)1980 619111; e-mail info@medicinal-cannabis.org; http://www.medicinal-cannabis.org/

Hortapharm BV breeds unique *Cannabis* varieties for medicinal or industrial applications. Hortapharm has cultivated *Cannabis* under IPM conditions for a decade, and is responsible for breeding the medical varieties utilized by GW Pharmaceuticals. Contact details: tel. +31 (0)20 6185591; fax +31 (0)20 6185726; e-mail Horta@euronet.nl

We donate 25% of our royalties to the International Hemp Association, Postbus 75007, 1070 AA Amsterdam, the Netherlands. Tel./fax: +31 (0)20 6188758; e-mail iha@euronet.nl

KOPPERT
BIOLOGICAL SYSTEMS

UK Medicinal Cannabis Project

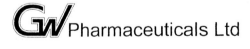

GW Pharmaceuticals Ltd

HortaPharm B.V.

Forewords

Hemp is grown for a variety of uses and there are records of its production and conversion into useful products for at least 5000 years. Important raw materials are still available from different types of plant in the genus *Cannabis*. It is a renewable starting material for many industries, including textiles, paper, rope, as a foodstuff and fuel, used from ancient times to the present. More recently it has been used in cosmetics and is being investigated as the starting point for a new generation of biodegradable plastics. Depletion of feed-stocks from fossil fuels and the need to access alternative sources of cellulose and fixed oils has refocused attention on renewable sources of industrial starting materials obtainable from *Cannabis*.

The use of *Cannabis* in medicine has had a chequered career throughout history. Until the early part of the twentieth century it was regarded as a valuable medicine in general practice. At present, *Cannabis* is classified as a Schedule 1 ("Class A" in the UK) drug and its use is proscribed under legislation following the UN Single Convention 1961. Recently, approval has been given by the UK Home Office for *Cannabis* to be grown for medicinal purposes so that the safety and efficacy of *Cannabis* as a medicine can be critically re-evaluated, with a view to its reinstatement as a prescription medicine.

Cannabis has been grown from antiquity and methods of producing it have been developed in many geographically different communities. Control of diseases and pests is a necessary part of good husbandry, and it is useful to have this information critically evaluated and brought together in one volume.

The resurgence of interest in medical and non-medical uses of *Cannabis* further underlines the need for an authoritative and comprehensive book on diseases of this valuable crop. The emphasis on biological methods of control is a recurrent theme throughout the book and reflects the move towards organic methods of horticulture for crops grown in the field and under glass. The authors have met this need by providing a comprehensive guide based on practical experience and academic expertise.

— Brian A. Whittle
Scientific Director, GW Pharmaceuticals Ltd

This comprehensive volume is an important resource for researchers, crop management consultants, and growers who wish to increase their knowledge of the pest management of hemp. Many of the pests and diseases profiled in this book are widely known to crops throughout the world. This book focuses on their damage to hemp and the specific solutions for this crop. Particularly useful is the clear, systematic approach to pest and disease descriptions. This is accompanied by an extensive overview of biological, natural, and chemical solutions to the pests and diseases of hemp, with details on products, application methods, and dosages. The emphasis on realistic biocontrol options is an outstanding feature . It is rare that such extensive information on disease and arthropod control has been combined in one volume. Moreover, it will be very useful for anyone working with pest management in other high-value crops.

Biological control of pests and diseases has evolved rapidly in the past 25 years, and many of the techniques described here have been adapted from tried-and-true strategies that have proved successful for growers of other crops. In fact, Koppert Biological Systems has been collaborating with one of the authors and we have successfully controlled many different pests on hemp with natural enemies routinely used in greenhouse vegetables.

Just as the science surrounding hemp continues to evolve, so does the science of biological control. The authors have succeeded in creating a complete, up-to-date collection of pest management information for hemp. Ultimately, the techniques described in this volume will be adapted and improved by growers themselves and I am confident that many growers of hemp will benefit greatly from the knowledge contained in this volume.

— Willem J. Ravensberg
Head, R & D Microbials, Koppert Biological Systems

How to Use This Book

You need not begin this book at page one. If you tentatively identified a pest or pathogen infesting your plants, look its name up in the index. See if the problem in hand matches its description in the book. Proceed to read all about the problem and its control measures. Control measures for common pests and pathogens are highlighted with charts, illustrations, and explicit instructions.

If you face an unknown assailant, turn to the identification keys in Chapter 1 and Appendix 2. If you dislike keys, most common pests can be diagnosed by the "picture-book method"—flip through the illustrations until you recognize your problem.

This book has no separate glossary; technical terms used in this text are defined within it. If a technical term is unfamiliar to you, use the index to locate the term's definition. Terms in **bold print** indicate they are being defined in that paragraph. We define technical terms in Chapters 1, 2, and 3, to build the precise language needed for pest identification using pest **morphology** (form and structure). We are unapologetic morphologists, and agree with Wheeler (1997), "no credible theoretical reason exists why molecular data should be more informative or less homoplastic than morphological data."

To control diseases and pests, we consider chemicals a last resort. We prefer cultural techniques, mechanical methods, and biological controls. **Cultural techniques** alter the farmscape, making it less favourable for pests and disease organisms (see Chapter 9). **Mechanical methods** utilize traps, barriers, and other ingenious techniques (see Chapter 9). **Biological controls** employ beneficial organisms to subdue pests and pathogens (see Chapter 10). Although we discourage the use of chemicals, technical information regarding their use is presented in Chapter 11 and Appendix 1. This information is presented in the spirit of harm-reduction, not as a green light for chemical abuse. Use as little as possible.

For current availability of specific biocontrol organisms and biorational chemical controls, please obtain the *Annual Directory* published by BIRC (Bio-Intergral Resource Centre), P.O. Box 7414, Berkeley, California 94707, telephone (510) 524-2567.

"In order to control mosquitoes, one must learn to think like a mosquito."
—Samuel Taylor Darling

Chapter 1: Principles of Plant Protection

INTRODUCTION

Diseases and pests cost farmers many millions of dollars in losses every year. Agrios (1997) estimated that 13% of fibre crops were lost to insects, 11% were lost to diseases, and 7% were lost to weeds and other organisms. In addition to these losses in the field, Pimentel *et al.* (1991) added another 9% in postharvest storage losses. These percentages would soar if crops were not managed for diseases and pests.

Our challenge is to sustain crop yields as we shift from chemical control to biological control. We want products made from healthy *Cannabis*—no weakened or discoloured fibre, rancid seed oil, or aflatoxin-laden medical marijuana. But we don't want pesticide-tainted products, either. Everyone benefits from nontoxic pest control—consumers, cultivators, *Cannabis,* and our environment. Nontoxic pest control is the keystone for sustainable agriculture. Sustainability requires a shift in societal consciousness, as well as an agronomic transformation, and we feel *Cannabis* cultivation can inspire the shift.

Some experts claim that hemp suffers no pests or diseases (Herer 1985, Conrad 1994). Their view may be skewed. We have grown hemp (McPartland under DEA permit, Clarke and Watson under Dutch and British licenses), and we can say with confidence that hemp suffers diseases and pests. As long as we insist on cultivating large acreages of intensive, high-output, monocropped plants, our crops will attract problems.

DISEASES AND PESTS

Do insects cause disease? No. According to Whetzel, quoted in Westcott (1990), **disease** is damage caused by the *continued or persistent* irritation of a causal factor (e.g., a nutrient deficiency or a pathogen). **Pathogens** are organisms that cause *continued* irritation, such as viruses, bacteria, fungi, parasitic plants, and nematodes.

Injury, in contrast, is caused by *transient* irritations. Insects are transient; they cause feeding injury, not disease. Hail stones cause injury, not disease. Injury, however, may lead to *secondary disease.* Hail stones injure plants, and injured tissues may become infected by disease organisms, such as the grey mould fungus (*Botrytis cinerea*). Aphids injure plants, and they often transmit viruses as they hop from diseased plants to healthy plants.

THE CROP DAMAGE TRIANGLE

Any discussion of plant protection should begin with the "crop damage triangle" (Fig 1.1). Crop damage requires the presence of: *Side 1*) a susceptible host, *Side 2*) a pest or pathogen, and *Side 3*) an environment conducive to crop damage. The length of each side of the triangle represents a condition favouring crop damage. The longer each side, the larger the triangle, the greater the crop damage. Damage is limited by the triangle's shortest side—either an absence of parasites, a healthy environment, or a resistant host. All our control methods manipulate one or more sides of the triangle.

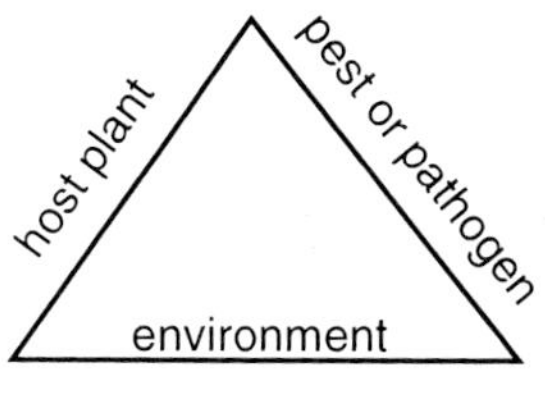

Figure 1.1: The crop damage triangle.

Host is manipulated by breeding crop plants for resistance to diseases and pests. Breeding is simply the mixing of genes and the selection of hybrids with resistant traits. The larger a crop's gene pool, the better chance of finding resistant genes. *Cannabis* still has a large gene pool. It has been bred for aeons all over Asia, and each geographic pocket of plants developed resistance to its local diseases and pests.

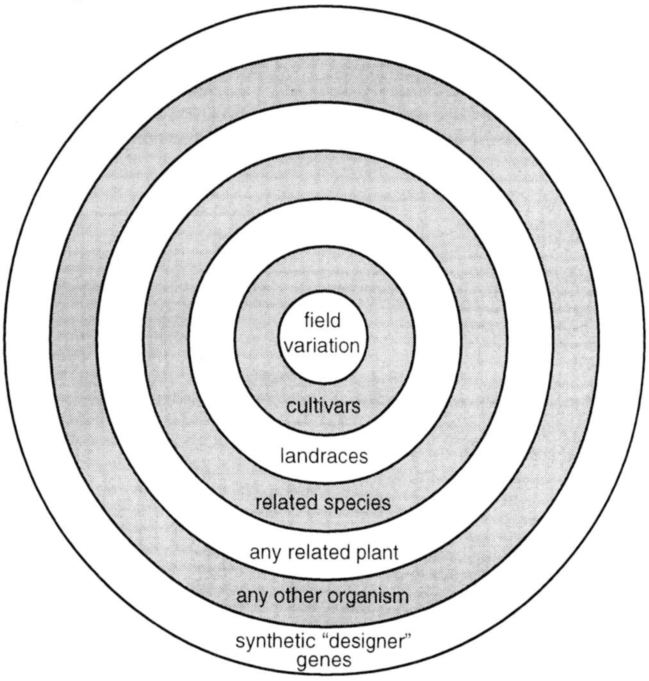

Figure 1.2: The expanding gene pool: sources for genetic resistance to parasites and pests (adapted from Cook & Qualset 1996).

When we relocate plants, they are attacked by new pests. We must breed new resistance. To breed *significant* resistance requires casting into a *larger* gene pool—perhaps crossing two different **cultivars**, such as the French cultivar 'Fibrimon 21' and the Hungarian cultivar 'Kompolti' (Fig 1.2). Larger yet, we can breed different **landraces**, such as dwarf Northern Russian hemp and giant Southern Chinese hemp. We can even breed different **species**, such as *Cannabis sativa* and *Cannabis afghanica.* Beyond species lies the breeding frontier: inter*generic* crosses between *Cannabis sativa* and *Humulus lupulus* (hops). *Cannabis* and *Humulus* are easily grafted together (Crombie & Crombie 1975), but cross-pollination does not produce viable seeds. Cereal scientists have created intergeneric crosses since the 1950s (Sears 1956), combining X-ray irradiation with embryo rescue on artificial growth media. But X-rays are now passé. Genetic engineering is the latest way to move genes from one plant to another. Engineers have used a bacterium, *Agrobacterium tumefaciens*, to splice DNA between plants (for a description, see *A. tumefaciens* in Chapter 6). More recently, *A. tumefaciens* has been replaced by microprojectile bombardment devices known as "bioblasters" or "gene guns" (Christou 1996).

Bioengineers can introduce genes from closely related genera (i.e., *Humulus*), or distantly related plants (e.g., *Sequoiadendron giganteum*). Genetic engineering can even produce inter*phyletic* crosses, splicing genes from bacteria, fungi, or animals. In the future, engineers will synthesize "designer genes"—the ultimate, infinite gene pool, with unknown benefits and dangers lurking in its depths.

Despite this "lurking unknown," we see bioengineered crops being planted across the USA and Canada, so North America has become a large-scale transgenic field study. Unanticipated catastrophes have already arisen, both ecological (Losey *et al.* 1999) and medical (Nordlee *et al.* 1996, Ewen & Pusztai 1999, Fenton *et al.* 1999).

Resistance against insects and diseases does not last forever. Pests and pathogens are moving targets; they mutate and undo the work of plant breeders. Pests can adapt within three years of a new cultivar's introduction (Gould 1991). Paradoxically, crops with *partial* resistance may last longer than strongly-resistant varieties. For more information on breeding resistance, see Chapter 7 (Genetics) and Chapter 9 (Method 5).

Environment, the base of the crop damage triangle, is manipulated by cultural and mechanical methods. Glasshouse environments are easily manipulated. To control grey mould, lower the humidity and increase the temperature. To control spider mites, raise the humidity and lower the temperature. Outdoor environments may also be manipulated. Careful site selection influences the immediate microenvironment around plants. Choosing a shady site protects against sun-loving flea beetles but exacerbates grey mould. Prevent root rot by avoiding low-lying, heavy soil (or lightening the soil and planting in raised beds).

Field locations should be selected with an eye toward neighbouring crops—if European corn borers are a problem, do not plant near maize. The previous season's crop also affects a site—expect white root grubs if rotating after sod. Lastly, host density affects the microenvironment and plays a significant role in crop protection (De Meijer *et al.* 1994).

Pests and pathogens, the third side of the triangle, are controlled by cultural, mechanical, biological, and chemical methods. Traditionally, agronomists often cite "the two E's" of pest and pathogen control: *exclusion* and *eradication*.

Exclusion is why the USA government established the Plant Quarantine Act in 1912—to exclude foreign parasites. Farmers exclude parasites in three ways: 1) cleaning all equipment before entering a field, glasshouse, or growroom, 2) sterilizing or pasteurizing soil before bringing it into a glasshouse or growroom, and 3) using certified seed which is disease- and pest-free.

Eradication requires the elimination of parasites once they have arrived. Some problems can be eradicated by cultural and mechanical methods (e.g., pruning diseased branches, starving pests via crop rotation, heat treatment of seeds). *Total eradication* usually requires the use of heavy pesticides. But spraying one pest may increase the population of another pest, resulting in more crop damage. *"Eradication" becomes an unrealistic and self-defeating goal.* Eradication is not a concept embraced by practitioners of IPM.

INTEGRATED PEST MANAGEMENT (IPM)

IPM is a holistic approach to controlling pests (and disease pathogens and weeds) that integrates all sides of the crop damage triangle, and utilizes cultural, biological, and chemical control methods (Stern *et al.* 1959). Researchers in Canada and Europe have used terms such as "complementary"and "coordinated," but the Australian concept of "integrated management" has won the acronym battle.

IPM began as a solution to *ecological* and *economic* problems associated with pesticides. It replaces the concept of *eradication* with the concept of *coexistence* (McEno 1990). IPM practitioners (IPMers) coexist with pests as long as pests remain below economically-damaging levels. What constitutes "economically-damaging" differs from pest to pest and plant to plant. A gardener growing for a flower contest may consider damage by a single budworm intolerable. Fibre crops, in contrast, endure *many* budworms before economic thresholds are reached. Biocontrol alone can keep most pest populations under economic thresholds.

IPM integrates ideas from conventional agriculture with ideas from **organic farming**. Organic farming is defined by the National Organic Standards Board as "an ecological farm management system that promotes and enhances biodiversity, biological cycles, and soil biological activity. It is based on minimal use of off-farm inputs and on management practices that restore, maintain and enhance ecological harmony." From an economic and political standpoint, the minimal use of products from commodity corporations removes organic farmers from the corporate food chain. This places organic farmers at odds with multinational giants; this also places organic farmers at odds with the USDA, which depends upon research money from the Monsantos of the world.

The organic process was postulated by Goethe, and elaborated by Rudolf Steiner, the founder of **bio-dynamic farming**. Bio-dynamic farmers control diseases and pests by stimulating natural processes and enhancing healthy ecological relationships (Steiner 1924, AGÖL 1998). The same year Steiner founded bio-dynamic farming, Sir Albert Howard began developing his "Indore Process" for maintaining soil fertility (Howard 1943). In the USA, J.I. Rodale began using methods that paralleled the research of Howard and Steiner.

Organic farmers focus on soil fertility much more than IPMers. Natural soil fertility has many benefits. Crops growing in organically-managed soils suffer less pests than crops growing in conventionally-managed soils (Phelan *et al.* 1996). Organic crops cause less pharyngitis and laryngitis in marijuana smokers than crops cultivated with chemical fertilizers (Clarke, unpublished research 1996). Marshmann *et al.* (1976) compared organically-managed crops to crops grown with chemical fertilizers, and found the former contained more Δ^9-tetrahydrocannabinol (THC). Similarly, studies with *Papaver* showed that poppies sprayed with pesticides produced less opiates than controls (Wu *et al.* 1978).

Unfortunately, the term "organic farming" has degraded into a marketing tool. The new USDA certification standards are weaker than standards previously established by individual states in the USA. The term organic farming is also confused with another USDA buzzword, **sustainable agriculture**. Sustainable agriculture shares goals with organic farming, but is difficult to define in absolute terms. Sustainable agriculture does not prescribe a concrete set of regenerative technologies, practices, or policies. It is more a process of education (Röling & Wagemakers 1998).

Organic farmers and IPMers differ in their approaches to pesticides. Organic farmers eschew synthetic pesticides, as well as many natural pesticides (see "The National List" in Chapter 11). IPMers, in contrast, use any pesticide that works. But IPMers reject the *conventional* approach of spraying pesticides, which is rigidly based on the calendar date. Instead, IPMers spray on a schedule determined by three aspects: 1) pest monitoring, 2) climate monitoring, and 3) the presence of beneficial organisms. *With IPM, careful observation replaces the brute force of conventional chemical warfare. Farmers using IPM must closely monitor crop conditions, biocontrol organisms, the weather, and all pests in the area, not just single target species. IPM is pest management for the information age.*

IPMers must understand a complex web of ecological relationships, such as parasitism, mutualism, and competition. The centre of this web is a crop. In our case, *Cannabis*. Animals and other non-photosynthetic organisms have a **parasitic** relationship with *Cannabis*. There are exceptions, such as *Homo sapiens*, who maintains a **mutualistic** relationship with *Cannabis*. Mutualism occurs when *both* species benefit from their interaction. *Homo sapiens* nurtures the plant's growth (cultivation) and disperses seeds (zoochory); the plant provides us with fibre, food, oil, and medicaments. *Cannabis* has a **competitive** relationship with most other plants, competition for the raw materials of photosynthesis—sunlight, water, and soil nutrients.

Relationships change with space and time. When two insects meet on a plant, the competition becomes intense in a space-limited environment such as the hollow stalk of a hemp plant. But insects don't have to meet to compete. Plants damaged by leaf-chewing insects are avoided by leafminers for many months (Faeth 1986). Chewed plants produce more defence chemicals, such as THC, making the remaining leaves less desirable to leafminers. Also, leafminers feeding on previously-damaged leaves suffer greater parasitism than leafminers feeding on unchewed leaves. This is because some **parasitoids** (parasites of pests) use chewed leaves as clues to locate their hosts. Amazingly, leaves damaged by wind or other mechanisms do *not* attract parasitoids unless oral secretions from pests are added (Turlings *et al.* 1990). Furthermore, parasitoids can distinguish between leaf damage caused by their hosts and leaf damage caused by other herbivores (DeMoraes *et al.* 1998). Plants emit different volatile chemicals in response to different pests, and parasitoids clue into the differences. The communication between plants and parasitoids is more sophisticated than previously realized (DeMoraes *et al.* 1998).

The final outcome of any interaction is often arbitrated by the environment, or the microclimate. A pest may flourish on a plant's lower leaves, shaded and protected, but not survive in the harsh environment near a plant's apex. This is especially true in *Cannabis*, where flowering tops accumulate THC—a chemical with pesticidal activity.

These few paragraphs indicate how complex IPM can become. In practice, IPM methods are arranged in a hierarchy depending on pest populations, crop density, and environmental concerns. The primary IPM strategy is *selectivity*. A control method should *selectively* kill pests and not beneficial organisms. *Selective* timing and *selective* treatment applied to *selective* infested plants (not the entire field) minimize collateral damage. *Selectivity requires careful identification of pests and pathogens.* Know thy enemy.

CANNABIS

Also know thy host. Unfortunately, the taxonomy of *Cannabis* remains in flux. The genus may be monotypic (consisting of one species according to Small & Cronquist 1976), or polytypic (with two or more species according to Schultes *et al.* 1974 and Emboden 1974).

Knowing your host becomes particularly important when dealing with pathogens and pests. Many parasites coevolve with their hosts, eventually becoming dependent on a single host species ("Fahrenholz's Rule"). This is why many pests attack one species and disregard others. This happens with *Cannabis* pests. Some pests attack hemp plants but cannot feed on marijuana plants (McPartland 1992, 1997a). Are these hosts different species? Rothschild & Fairbairn (1980) found the insect pest *Pieris brassicae* could distinguish between Turkish and Mexican strains of marijuana. Are these hosts different species?

Human taxonomists differentiate between plants by genetic, chemical, and morphological characteristics. The genetic characteristics of *Cannabis* are currently under close scrutiny (see "Genetics" section in Chapter 7). The chemical taxonomy of *Cannabis* is complex; *Cannabis* is a veritable chemical factory. *Cannabis* uniquely produces the **cannabinoids**, a family of C_{21} terpenophenolic compounds, including THC, cannabidiol (CBD), cannabinol (CBN), cannabichromene (CBC), cannabigerol (CBG), and at least 60 other cannabinoids (Turner *et al.* 1980). The unique smell of *Cannabis*, however, is not from cannabinoids, but from **terpenoids**. Terpenoids are polymers of a C_5 isoprene precursor, such as the monoterpenoids (with C_{10} skeletons), sesquiterpenoids (C_{15}), diterpenoids (C_{20}), and triterpenoids (C_{30}). *Cannabis* produces over 150 terpenoids, including caryophyllene, myrcere, humulene, limonene, and several pinenes (Hood *et al.* 1973, Hendriks *et al.* 1975, Ross & ElSohly 1996, Mediavilla & Steinemann 1997). Collectively, terpenoids are called the **essential oil** or **volatile oil** of the plant. One terpenoid, caryophyllene oxide, is the primary volatile sniffed by narcotic dogs (Brenneisen & ElSohly 1988). Interestingly, hemp varieties produce more caryophyllene oxide than drug varieties (Mediavilla & Steinemann 1997). Cannabinoids and terpenoids have pesticidal and repellent properties (McPartland 1997b).

This book will follow a polytypic approach to *Cannabis* taxonomy. The key below describes four prominent *Cannabis* segregates that we can tell apart on a morphological basis. The morphological key is adapted from work by Schultes *et al.* (1974), Emboden (1974), Small & Cronquist (1976), and Clarke (1987):

1. ***Cannabis sativa*** (=*C. sativa* var. *sativa*):
Plants tall (up to 6 m), stems smooth and hollow, laxly branched with long internodes; petioles short, usually 5–9 leaflets per leaf, leaflets lanceolate, largest leaflets averaging 136 mm long (length/width ratio = 7.5); racemes have long internodes, and achenes are partially exposed; achenes (seeds) usually >3.7 mm long, somewhat lens-shaped with a blunt base, surface dull light-to-dark green and usually unmarbled, seeds *usually* adherent to plants at maturity. Cultivated for fibre, oil, and sometimes for drugs.

2. ***Cannabis indica*** (=*C. sativa* var. *indica*):
Plants shorter (under 3 m), stems smooth and nearly solid, densely branched with shorter internodes; petioles shorter, usually 7–11 leaflets per leaf; leaflets narrow lanceolate, largest leaflets averaging 92 mm long (l/w ratio = 10); achenes averaging 3.7 mm long, less lens-shaped, with a more rounded base, surface green-brown and marbled or unmarbled, with or without an abscission layer. Cultivated primarily for drugs but also used for fibre and oil.

3. ***Cannabis ruderalis*** (=*C. sativa* var. *spontanea*):
Plants small (usually under 0.5 m), stems smooth and hollow, occasionally unbranched; petioles short,

usually 5–7 leaflets per leaf, leaflets elliptic, largest leaflets averaging 60 mm long (l/w ratio = 6); achenes small with a pronounced abscission structure at the base, surface dull green and marbled, abscission layer fleshy with oil-producing cells, seeds readily shed from plant. Not cultivated.

4. ***Cannabis afghanica*** (=*C. sativa* var. *afghanica*): Plants short (under 1.5 m), stems ribbed and nearly solid, densely branched with short internodes; petioles long, usually 7–11 leaflets per leaf, leaflets dark green and broadly oblanceolate, largest leaflets averaging 130 mm (l/w ratio =5); racemes have short internodes, and achenes are not exposed; nested, compound bracts sometimes produced; achenes usually <3.0 mm long, nearly round with a blunt base, surface shiny grey and marbled. Cultivated exclusively for drugs, primarily hashish.

In our opinion, researchers frequently misname these *Cannabis* segregates. *C. indica* is frequently misnamed *C. sativa*, and *C. afghanica* is frequently misnamed *C. indica*. Clarke (1987) attempted to correct the confusion by elevating *C. afghanica* Vavilov from its original subspecies level (=*C. sativa* f. *afghanica* Vavilov 1926). Clarke noted that Schultes *et al.* (1974) lumped *C. afghanica* with *C. indica*. Unfortunately, *Cannabis* from Afghanistan has come to typify *C. indica*, especially in the eyes of marijuana breeders. This is incorrect; Lamarck (the botanist who named *C. indica*) was entirely unfamiliar with Afghan *Cannabis*. His taxon refers to the biotype from India (*indica*). Marijuana breeders' use of the name "*indica*" for the *afghanica* biotype has become entrenched, causing extensive confusion. Some breeders (e.g., Schoenmakers 1986) double the confusion by calling *afghanica* plants "*ruderalis* species".

The cannabinoid content within each *Cannabis* segregate varies greatly. Reducing THC has occupied hemp breeders for years (Bredemann *et al.* 1956). *Cannabis* segregates can interbreed and hybridize, exemplified by X*C. intersita* Sojak and by hybrids illustrated by Schoenmakers (1986) and Kees (1988). *C. sativa* and *C. indica* escape cultivation and grow wild, like *C. ruderalis* (Small & Cronquist 1976).

IPM STEPS

IPM consists of five steps: 1) identifying and monitoring diseases and pests, 2) monitoring the environment, 3) deciding the proper IPM intervention, 4) implementing the intervention, 5) post-intervention reassessment.

Monitoring methods vary from casual hearsay between neighbours to daily quantitative trap sampling. Your monitoring effort should match the severity of your problem. Somewhere along the line, keeping a logbook becomes essential.

Monitoring requires the regular inspection of plants, insect traps, or soil samples. The larger the crop, the greater number of samples. Be sure to monitor hard-to-see spots, like centres of crop fields or back corners of glasshouses. Also monitor glasshouse "hotspots" located near doors and window vents. In your logbook, record the date, time, and location of any crop damage observed. If pests are present, estimate their numbers (qualitatively—"many," or quantitatively—"average of 5 aphids per leaflet"). Recording the temperature, humidity, and time of day is helpful. Mark infested plants with a bright-coloured pole or flag so they can be relocated.

A seasoned entomologist or plant pathologist can identify a problem at first glance. For the rest of us, several items of diagnostic equipment come into play, and are described below. These items can be purchased from sources listed in the annual directory published by BIRC (Bio-Integral Resource Centre), USA telephone 1-510-524-2567. For difficult-to-find items like aspirators and beat sheets, contact Gempler's for their agricultural catalogue (USA 1-800-382-8473, web site http://www.gemplers.com).

Insects and other arthropods

Most insects can be collected with tweezers or a hand trowel, plus a flashlight (many are nocturnal). An aspirator (venturi suction trap) is useful for collecting small, mobile insects. A penknife may be needed to extract recalcitrant individuals from protected places. Since smashed insects are difficult to identify, a collection jar keeps them incarcerated for closer scrutiny. Knock insects out of foliage with standard muslin sweep nets or beating sheets. Insect traps baited with foods or pheromones allow you to monitor pests 24 hours a day. Insect traps suitable for IPM monitoring are discussed in Chapter 9 ("method 12").

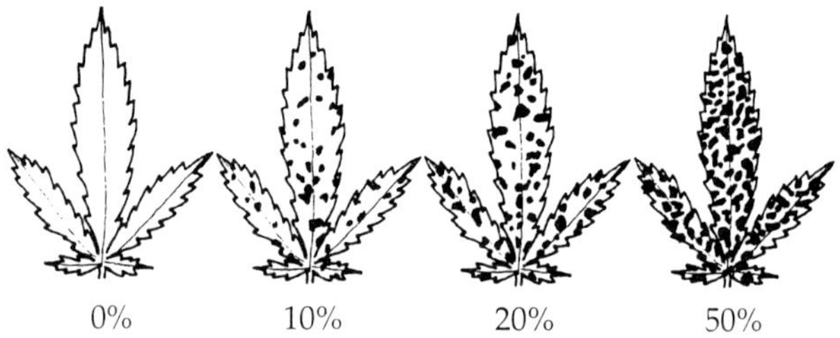

0%   10%   20%   50%

Figure 1.3: Visual scale for damage assessment by fungi or leaf-eating insects on a *Cannabis* leaf (McPartland).

A magnifying lens (10x to 16x) may be needed to identify small insects (aphids, thrips, mites). Some species require microscopic examination of their genitalia, usually after chemical clearance with potassium hydroxide. Immature larvae (caterpillars, grubs, maggots, etc.) may prove impossible to identify. With care, captured larvae can be nurtured into adulthood for proper identification.

Many *Cannabis* pests can be identified by the damage they cause. A *population* of insects can also be monitored, either directly (by counting them) or indirectly (by assessing their damage). Degree of damage can be estimated with **visual scales** (Fig 1.3). The American plant pathologist Nathan Cobb first devised visual scales. Cobb worked with disease damage, but his scales can be used for estimating insect damage. Tehon & Stout (1930) illustrated a variety of Cobb scales.

Fungi and bacteria

Few fungi and bacteria are readily identified in the field. Many must be identified with a microscope. "Immature" (nonsporulating) fungi, like immature insects, may defy identification. Specimens should be kept moist for a few days to promote spore development, or they may need to be isolated and raised on artificial media in petri plates. Hundreds of different agar-based artificial media are commercially available. Many **selective media** only allow growth of specific pathogens.

Advances in biotechnology may move petri plates to the basement (Miller 1995). Two biotechnical approaches are available. *Immunodiagnostic methodologies* include immunofluorescence, dot immunobinding, and enzyme-linked immunosorbent assays (ELISA). *Nucleic acid-based methodologies* include nucleic acid probes, restrictive fragment length polymorphism analysis, and polymerase chain reaction techniques. Immunodiagnostic **ELISA tests** are the

Identification keys

Keys are indispensable tools for diagnosing crop problems. The best keys are crop-specific. Some diseases and pests prevail in fibre crops, other problems predominate in drug crops (McPartland 1996a,b). Keys should be site-specific, because diseases and pests change between indoor and outdoor crops. Problems also change as plants grow from seedlings to flowering adults. Diseases and pests vary geographically—many virus diseases, for instance, are limited by the range of their insect vectors. Different keys can be constructed for all these different scenarios.

There are basically 2 types of identification keys: synoptic and dichotomous. Synoptic keys rely on pattern recognition, while dichotomous keys are structured decision trees. Watson prefers a synoptic "Simpleton's Key." Clarke prefers the six synoptic "Top 10" lists presented below. McPartland prefers a dichotomous key, relegated to Appendix 2.

Simpleton's Key:

Most common symptoms with their most common causes:

no seedlings—old seeds, cold soil, damping off fungi, eaten by pests

wilting—too little or too much H_2O, leaf-sucking insects, root insects, wilt fungi, nematodes

mould on buds or leaves—grey mould, brown blight, downy mildew, pink rot, powdery mildew (don't confuse mould with webbing)

webbing—spider mites, budworms, hemp borers, leaf-eating caterpillars

spots on leaves—leaf-sucking insects, leaf fungi, leafminers, too much fertilizer

brown and curling leaf margins—too much fertilizer, dry air, too little K, brown blight

holes in leaves—caterpillars, beetles, grasshoppers, bacteria

pale green or yellow leaves—not enough N, poor pH, nematodes, soil fungi, leaf-sucking insects

lumpy stems—European corn borers, hemp borers, beetle grubs, canker fungi, stem nematodes

spindly stems—not enough light, too much yellow light, temperature too hot, soil too wet, not enough N or K or Ca

disfigured roots—soil fungi, nematodes, broomrape, grubs, maggots, rodents

tips of limbs or tops missing—rodents, rabbits, deer, cattle, humans

Six "Top 10" lists of common disease & pest problems, indexed with page numbers:

Seed & seedling problems	Page
Damping-off fungi	97
Overwatering	164
Cutworms	54
Crickets	87
Rodents	153
Slugs and snails	151
Flea beetles	65
Birds	152
Old seed	169
Insufficient light, water, or temperature	164

Flower & leaf problems, outdoors	Page
Grey mould	93
Aphids	31
Yellow and brown leaf spots	101,104
Nutritional diseases	155
Flea beetles	65
Budworms & leaf-eating caterpillars	51, 57
Downy mildrew	106
Plant bugs	73
Brown blight	114
Bacterial leaf diseases	144
Deer	153

Stem & branch problems	Page
European corn borers	44
Grey mould	93
Hemp borers	48
Hemp canker	96
Beetle & weevil grubs	70, 72
Fusarium canker	107
Rhizoctonia sore shin	102
Anthracnose	121
Striatura ulcerosa (bacteria)	145
Stem nematodes	139

Root problems	Page
Root knot nematodes	137
Rhizoctonia root rot	102
Beetle grubs	65
Fusarium root rot	108
Broomrape	150
Cyst nematodes	138
Texas root rot	125
Root maggots	85
Rodents	153
White root grubs	68

Whole-plant problems	Page
Fusarium wilt	109
Charcoal rot	112
Verticillium wilt	122
Nutritional problems	155
Overwatering	164
Bacterial wilt	146
Virus diseases	142
Dodder	148
Armyworms	54
Thieves with guns	23

Indoor problems (glasshouses & grow rooms)	Page
Nutritional diseases and overwatering	155, 164
Spider mites	25
Aphids	31
Whiteflies	39
Grey mould	93
Thrips	60
Powdery mildrew	111
Fungus gnats	89
Leafhoppers	79
Virus diseases	142

most popular. They are based on monoclonal antibodies and provide portable, on-site testing with rapid results. Commercially available ELISA kits can instantly identify fungi that cause disease in *Cannabis* (e.g., *Botrytis cinerea, Rhizoctonia solani, Sclerotinia sclerotiorum, Fusarium graminearum*). These kits detect microscopic fungi *before* symptoms appear. But they are expensive—single tests cost between US$4 to $50 (Sutula 1996).

Nematodes

Some nematodes, such as female root knot nematodes, can be identified by simply inspecting roots. Other nematodes may require special techniques, such as boiling infested roots in lactophenol mixed with cotton blue stain, then decolourizing roots in plain lactophenol and viewing roots under a microscope. Agrios (1997) described three methods for separating nematodes from soil or plant tissues: the Baermann funnel method, sieving method, and centrifugal or sugar flotation method. The Baermann is simple and produces a high yield of nematodes, but takes two to five days. Centrifugal flotation can be more expensive and may yield fewer nematodes than the Baermann method, but takes much less time (only ten minutes).

Viruses

Identifying viruses is not easy. Their symptoms can be confusing. To see them requires an electron microscope. Cross-inoculation studies with other plants are time-intensive. Serological tests and fluorescent antibody techniques are expensive and not available for HSV (hemp streak virus) or HMV (hemp mosaic virus)—the most common viruses on *Cannabis*. Viral inclusion bodies can be detected with a light microscope by using special dyes (Christie *et al.* 1995), techniques described in Chapter 6.

Nutritional diseases

Mineral deficiencies are usually diagnosed by observing characteristic symptoms in foliage. Soil and plants can be chemically tested for mineral deficiencies. Agricultural agents and extension services often provide inexpensive soil and water tests. Cheap kits for testing soil are available, but may provide inaccurate readings.

The ultimate aid for identifying pests and diseases is an expert. The USA government has employed public consultants since 1854, when Asa Fitch was hired as an entomologist. The State of New York hired the first professional plant pathologist, Joseph C. Arthur, in 1882. Today, the USDA maintains a network of county extension agents across the country; find your local office in the yellow pages. Of course, until *Cannabis* cultivation is legalized in the USA, asking extension agents for assistance may be hazardous.

ENVIRONMENTAL MONITORING

While monitoring diseases and pests, also keep an eye on the weather. Weather affects the severity of diseases and pests, and the efficacy of control methods. For instance, if cutworms are a problem, cool, wet weather causes them to proliferate. Wet weather hampers biocontrol of cutworms with *Trichogramma* wasps, but enhances biocontrol with *Steinernema* nematodes.

Heat controls the development of pests and pathogens. Development cannot begin until the environment reaches a certain temperature (the low threshold), and pest development stops if the temperature climbs too high (the high threshold). The amount of heat required for development of a pest or pathogen varies very little—it always equals the sum of temperature (between thresholds) and time. Thus, we can predict the development of problems by summing the growing season's accumulated heat. This is measured in **degree days** (°Cd or °Fd, for centigrade or fahrenheit). Methods of measuring °Cd vary in complexity. A daily maximum-minimum thermometer can estimate the approximate °Cd at your location. More accurate determinations of heat require a calculus of minute-by-minute measures of air *and* soil temperatures.

Starting in early spring, record the average temperature for the day (maximum plus minimum divided by 2). From the average temperature you then subtract the low threshold ("base") temperature. For hemp, Van der Werf (1994) used a base temperature of 0°C, and began measuring on the day he sowed seed. For example, the day he sowed seed, the maximum temperature reached 15°C, the minimum was 5°C, equalling an average temperature of 10°C. Subtract 0°C as the base temperature. By this calculation, the seeds accumulated ten degree days (10 °Cd) the first 24 hours after planting. Daily degree days are continuously added to calculate the accumulated °Cd for the season. Van der Werf (1994) determined that hemp seedlings required 88.3 °Cd before they emerged from the soil. Slembrouck (1994) also used a base of 0°C, and began measuring on the day seeds were sown. She calculated that 'Fedora' plants began flowering at 1350 °Cd, whereas 'Futura' plants required 1400 °Cd before flowering.

For pests and pathogens of temperate crops, many experts use 10°C as the base temperature. In Vermont (USA), using 10°C as the base temperature and beginning measurements on March 1st, adult flea beetle emerge from soil to chew on seedlings at 90–110 °Cd (data is less accurate for soil insects, because we measure air temperature, not soil temperature). The first generation of European corn borer moths lays eggs around 250–275 °Cd. This year Vermont suffered record-breaking temperatures all spring, so pests developed early. Some farmers were caught by surprise when egg-laying bollworm moths appeared a month before usual, but the moths were right on time by our °Cd estimations, so we were ready with our *Trichogramma* wasps and Bt sprays.

IMPLEMENTING IPM STRATEGIES

The person in charge of monitoring pests should also be the decision-maker who implements control strategies. If not, then the monitor and decision-maker must keep in close communication. Similarly, if the decision-maker and the implementor are separate, communication is key. Good IPM decisions frequently require outside support, as this book hopes to provide.

LAST STEP: POST-INTERVENTION MONITORING

During the 1940s and 1950s, farmers sprayed DDT and were done, knowing their pests were dead. Not any more. Today, monitoring pests and pathogens must continue after you have intervened with control methods. Post-intervention monitoring provides feedback for evaluating the effectiveness of the IPM programme. Biocontrol methods require very careful feedback.

POST SCRIPT: SHOULDERS OF GIANTS

Our current work on *Cannabis* builds on earlier efforts by hundreds of men and women. Allow us to briefly highlight some previous researchers and their work. For a more complete history see the two-part series by McEno (1987, 1988).

Surprisingly, the first published description of a *Cannabis* disease did not appear until 1832, and it was described by an American, Lewis David von Schweinitz (Fig 1.4). Schweinitz discovered a fungus infesting hemp stalks near Salem, North Carolina. He named it *Sphaeria cannabis*. Schweinitz was an interesting character—a cigar-smoking minister, and the first American to earn a Ph.D. (Rogers 1977). He described the first fields of wild hemp growing in North America (Schweinitz 1836).

In Europe, the first *Cannabis* problems to receive attention were leaf diseases: Lasch described *Ascochyta cannabis* in 1846; Westendorp described powdery mildew in 1854; Kirchner described *Depazea cannabis* in 1856; and two researchers independently described *Septoria cannabis* in 1857. Our modern-day plague, grey mould (caused by *Botrytis cinerea*), was originally described on hemp by Hazslinszky in 1877. In the entomology world, the first *Cannabis*-specific insects were described in 1860—*Aphis cannabis* by Passerini in Italy, and *Psyche cannabinella* by Doumère in France.

One of the great workers emerging from this period was Oskar Kirchner (Fig 1.4). A "Renaissance man" from Germany, Kirchner wrote about all kinds of *Cannabis* problems—fungi and insects, as well as nematodes, bacteria, and parasitic plants (Kirchner 1906). His artwork was admired by many, and frequently imitated (see comments in the section on yellow leaf spot, Chapter 5).

Figure 1.4: Outstanding *Cannabis* researchers: *top row, left to right:* Vavilov; Schweinitz; Charles; *middle left:* Kirchner; *lower left:* Röder; *lower right:* Dewey holding a male (staminate) plant growing next to a female (pistillate) plant.

America's "Renaissance man" was Lyster Hoxie Dewey (Fig 1.4). Dewey's career at the USDA began with critical work on the Gramineae. He wrote about medicinal herbs, deploring the overharvesting of wild plants such as goldenseal and ginseng. His ecological views were unique in the USDA. Dewey explained how destroying our native prairie enabled tumbleweed (Russian thistle) to spread across Midwestern rangelands. Dewey's contemporaries, in contrast, believed tumbleweed was a Russian plot to destroy American agriculture [see *Scientific American* 264:84].

From 1899 to 1935 Dewey led fibre-plant investigations at the USDA. He became a champion of hemp, dedicating his energies and talents to the advancement of *Cannabis*. Dewey imported seeds from all over the world, from fibre *and* drug plants, and evaluated them on American soil. Bòcsa (1999) called Dewey "the first hemp breeder," inaugurating a distinguished lineage that includes Fleischmann, Grishko, Bredemann, von Sengbusch, Hoffmann, Allavena, Virovets, Mathieu, and Bòcsa himself.

Dewey (1914) called Coates Bull and Fritz Knorr the first hemp breeders. This pair from St. Paul bred 'Minnesota No. 8' from the best Chinese landraces they could buy in Kentucky. Dewey subjected 'Minnesota No. 8' to a decade of inbreeding and half-sib family selection to create his first successful variety, 'Kymington' (Dewey 1928). At the same time, Dewey imported seed directly from China, and developed 'Chington.' He subsequently bred the first inter-varietal cultivar, 'Arlington,' by crossing 'Kymington' with 'Chington.' Dewey also crossed 'Kymington' with the Italian variety 'Ferrara,' to create the celebrated hybrid 'Ferramington' (Dewey 1928). Dr. Fleischmann, in Hungary, bred a similar cultivar by crossing Dewey's 'Kymington,' with 'F-hemp,' a variety of north Italian ancestry. Fleischmann's stock sired many of the Hungarian cultivars available today (De Meijer 1995).

Dewey was an ecologist as well as a plant breeder. He worked on ways to make paper out of hemp hurds. Long ago he lamented, "There seems to be little doubt that the present wood supply cannot withstand indefinitely the demands placed upon it... Our forests are being cut three times as fast as they grow" (Dewey & Merrill 1916). Unfortunately, Dewey lived to see his hemp efforts undone by Harry Anslinger's anti-marijuana propaganda. Dewey died several years after passage of the Marihuana Tax Act of 1937.

Dewey was not expert in pests and pathogens. For these problems he collaborated with Vera Charles (Fig 1.4). Vera Charles was another USDA researcher based near Washington, D.C., but she collected *Cannabis* specimens from across the country. She also lived to see hemp cultivation outlawed, and saw her old hemp research plots on Potomac Flats replaced by the Pentagon.

Dewey's counterpart in Russia was Nikolai Vavilov (Fig 1.4). Among the great scientists of our century, Vavilov has been lionized as an international statesman of agriculture and plant genetics (Menvedev 1969). Vavilov collected *Cannabis* from around the globe, often journeying to central Asia, which he considered the centre of *Cannabis* diversity. His *Cannabis* germplasm collection is preserved at the newly-renamed Vavilov Research Institute (VIR) in St. Petersburg. Preservation of *Cannabis* germplasm by the VIR has been supported by the International Hemp Association (Lemeshev *et al.* 1994).

Vavilov's research with drug plants elicited criticism from other Soviet agronomists. His research was eventually terminated by political action, as was Dewey's research. But Vavilov also lost his life. Shortly after his publication of *The Origin of the Cultivation of our Primary Crops, in Particular of Cultivated Hemp,* Vavilov locked horns with T.D. Lysenko. Lysenko was famous for fabricating genetic theories based on Marxist doctrine. He became a powerful toady of Stalin. Lysenko had Vavilov arrested. Shortly before Vavilov died in one of Stalin's gulags, he wrote for the ages, "We shall go to the pyre, we shall burn, but we shall not renounce our convictions."

Besides Dewey and Vavilov, many other *Cannabis* researchers died during World War II, including Kirchner and Klebahn in Germany, Curzi in Italy, Guilliermond in France, Lange in Denmark, and Komarov and Tranzschel in the USSR. One researcher that survived the carnage was Kurt Röder (Fig 1.4). Röder published a half-dozen papers on *Cannabis* viruses, fungi, and insect vectors, before his laboratory in Berlin was destroyed.

In the 1950s, Italy became a centre for hemp research thanks to the tireless efforts of Ferri and Goidànich at Bologna, and Noviello at Naples. Eastern Europe got busy, with publications from Poland, Romania, Bulgaria, the former Yugoslavia, the former USSR, and especially Hungary. Hungarian research has been headed by Bòsca (an agronomist) and Nagy (an entomologist).

The study of *Cannabis* parasites became divided during the 1970s with the rise of anti-marijuana biocontrol research. While Europeans tried to control pests and pathogens of hemp, USA researchers experimented with the same pests and pathogens to control marijuana! Marijuana growers acquired their own disease and pest experts—Frank, Rosenthal, Rob Clarke, Sam Selgnij, the Bush Doctor, and Chief Seven Turtles, who published in journals such as *High Times* and *Sinsemilla Tips.* Arthur McCain, a biocontrol researcher at UC-Berkeley, says the anti-marijuana biocontrol era ended when research was cancelled by the Carter administration (Zubrin 1981). Unfortunately, the USA government may have caught the biocontrol bug again (McPartland & West 1999).

The 1980s and 1990s saw a renewal of hemp cultivation in Western Europe. This encouraged new publications concerning diseases and pests (Spaar *et al.* 1990, Gutherlet & Karus 1995). Much recent phytopathological research has come from Holland (DeMeijer 1993, 1995; Hennink *et al.* 1993, Kok *et al.* 1994, Van der Werf 1994). McPartland (1981 *et al.*) revived phytopathological research at the University of Illinois, which has a long history of *Cannabis* research (e.g., Tehon & Boewe 1939, Adams 1942, Hackleman & Domingo 1943, Tehon 1951, Boewe 1963, Haney & Bazzaz 1970, Haney & Kutscheid 1973, Haney & Kutscheid 1975). Biocontrol research continues in India, where *Cannabis* is utilized to control pests of other crops (Pandey 1982, Mojumder *et al.* 1989, Kaushal & Paul 1989, Upadhyaya & Gupta 1990, Bajpai & Sharma 1992, Kashyap *et al.* 1992, Jalees *et al.* 1993, Vijai *et al.* 1993, Sharma *et al.* 1997). The future looks promising.

"You never know what is enough unless you know what is more than enough."

—William Blake

Chapter 2: Requirements for Growth

Plants have 17 requirements for growth—moisture, warmth, light, air, and 13 nutrients found in the soil:

MOISTURE

Thanks to its extensive root system, *Cannabis* tolerates dry conditions (although it does not *thrive* in dry conditions). Lisson & Mendham (1998) detected water extraction by roots 140 cm deep in soil. On the other hand, *Cannabis* grows poorly in wetlands or saturated soil. Hemp growth peaks when soil moisture is at 80% of soil field capacity (Slonov & Petinov 1980). Duke (1982) summarized data from 50 reports and found *Cannabis* does best in areas receiving 970 mm rainfall *per year* (range, 310–4030). During the *growing season*, Lisson & Mendham (1998) measured maximum fibre yields in hemp receiving 535 mm water (rain + irrigation). They calculated a hemp "water use efficiency" equalling 3 g stem (dry weight) per kg water.

Plant hydration is expressed as **water potential** (ψ), gauged in MegaPascals (MPa) or bars. Older literature measures hydration as a percentage of total leaf saturation (TLS). Hemp growth peaks at a TLS of 85–93% (Slonov & Petinov 1980). This TLS approximates a ψ value of -0.3 MPa (= -3 bars). During dry summer months, ψ routinely drops to -1.2 MPa (= -12 bars). When ψ drops below -1.5 MPa, photosynthesis shuts down in 75% of maize plants. Note that photosynthesis stops *before* wilt symptoms are seen. *Cannabis* probably shuts down below -1.5 MPa, but the exact number awaits measurement. Inexpensive instruments for measuring ψ are becoming available.

Besides *soil* water and *plant* hydration, careful cultivators must account for *atmospheric* water. *Cannabis* grows best at a relative humidity (RH) between 40-80% (Frank 1988), but RH over 60% promotes gray mould in *afghanica* biotypes and their hybrids. So a RH between 40-60% is optimal during flowering, to avoid gray mould.

TEMPERATURE

In a meta-analysis of 50 studies, Duke (1982) determined *Cannabis* growth peaks at a temperature of 14.3°C (range 5.6-27.5°C). For CO_2-enriched plants in a glasshouse or growroom, the ideal temperature is higher—21–27°C during the day and 13–21°C at night (Frank 1988).

LIGHT

Light may be measured two ways—by quantity and energy. **Light quantity** is measured by the *brightness* cast by a candle onto a square foot of surface one foot away (1 footcandle or 1 **Lumen**). **Light brightness** is what the "exposure meter" in your camera measures. In the metric world, light imparted by a candle upon a square metre one metre away equals 1 **Lux** (1 Lux = 0.093 Lumen).

Ordinary indoor light averages 150 Lux, too dim for *Cannabis*, a plant that requires a lot of light. Researchers have grown *Cannabis* in growrooms under as little as 600 Lux (Sáringer & Nagy 1971). Paris *et al.* (1975) used fluorescent and incandescent lamps emitting 14,000–18,000 Lux. The brightest sunlight yet measured is 100,000 Lux atop Mauna Loa in Hawai'i.

You can estimate brightness with the light meter in your camera. Set the film-speed dial to ASA 200. Aim the camera meter at a sheet of matt white paper placed near plants, and orient the paper to receive maximum light. Position the camera so the meter sees only the paper, and the camera does not shadow the paper. Set the shutter speed at 1/500 second. The f-stop setting for a correct exposure at 1/500 can be converted to lumens using Table 2.1.

Table 2.1: Converting f-stops to Lumens.

f-stop:	Lumens:
f2.8	125 L
f4	250 L
f5.6	500 L
f8	1000 L
f11	2000 L
f16	4000 L
f22	8000 L

The other way to measure light is *energy*, its ability to do work (e.g., drive photosynthesis). Engineers measure light in kilocalories per hour imparted per square metre, or Watts per square metre (W m^{-2}). Sunlight entering the Earth's upper atmosphere imparts 1350 W m^{-2}. By the time sunlight reaches the Earth's surface, its energy has dropped to 1000 W m^{-2}, at noon on a clear summer day near the equator. On a cloudy day sunlight dissipates to 100 W m^{-2}. Moonlight exerts only 0.01 W m^{-2}. *Cannabis* researchers have grown seedlings under as little as 96 W m^{-2} of mixed incandescent and fluorescent lighting (McPartland 1984). Frank & Rosenthal (1978) suggested flowering *Cannabis* under a minimum of 215 W m^{-2} (=20 watts ft^{-2}). For CO_2-enriched growth chambers, Rosenthal (1990) recommended up to 320 W m^{-2} to saturate *Cannabis* photosynthesis.

Some *Cannabis* scientists have measure light energy in terms of emitted photons, as μE/m^2/second, where 1000 Lux = 19.5 μE m^{-2} s^{-1} (Balduzzi & Gigliano 1985). Bush Doctor (1993b) described the energy emitted by different lights (fluorescent, metal halide, high pressure sodium bulbs), and how they translate to watts m^{-2} and watts per dollar.

Light energy depends on colour. Colour is a function of wavelength, measured in nanometers (nm). Energy increases in proportion to wavelength. Short-wave light has less energy than long-wave light. For instance, purple light (short wavelength 420 nm) requires 130 milliWatts to generate 1 Lumen of brightness, whereas yellow (long wavelength 570 nm) needs only 1.4 milliWatts to generate the same brightness. Converting from light brightness (Lux) to light energy (W m^{-2}) is not simple, since most light represents a mixture of wavelengths. Rosenthal (1998) provides brightness-to-energy conversion factors for many different types of bulbs.

Plants prefer certain wavelengths. Plants reject (reflect) green-yellow light (500–600 nm). This is why they look green to us. Photosynthesis works best with red and blue wavelengths. Bush Doctor (1993b) described the wavelengths emitted by commercial light bulbs.

Ultraviolet (UV) radiation consists of wavelengths shorter than deep purple light. **UV-A** contains wavelengths from 420 to 315 nm, **UV-B** ranges from 315 to 280 nm, and **UV-C** ranges from 280 to 100 nm. UV radiation damages nucleic acids and proteins in plants and people, especially UV-C. Ozone in the Earth's upper atmosphere absorbs all UV-C, about 95% of UV-B, and about 50% of UV-A. It has been suggested that *Cannabis* biosynthesizes THC as a UV protectant (Pate 1983, 1994). Indeed, under conditions of high UV-B exposure, *Cannabis* produces more THC (Lydon *et al.* 1987).

Table 2.2: Soil nutrient extraction of different crops during one growing season.[1]

CROP	N ($kg\ ha^{-1}$)	P_2O_5 ($kg\ ha^{-1}$)	K_2O ($kg\ ha^{-1}$)	CaO ($kg\ ha^{-1}$)	MgO ($kg\ ha^{-1}$)	S ($kg\ ha^{-1}$)
Maize (*Zea mays*) 12,200 kg grain ha^{-1}	302	130	302	93	123	37
Wheat (*Triticum* sp.) 5200 kg grain ha^{-1}	152	61	184	34	45	23
Oats (*Avena sativa*) 3600 kg grain ha^{-1}	131	43	165	21	37	22
Hemp-whole plant ≈200,000 dry kg ha^{-1}	177	53	184	199	35	18
Hemp-stems only 6000 kg ha^{-1}	52	12	99	68	12	8
Hemp-seeds only 700 kg ha^{-1}	33	18	8	3	6	9
Hemp-flowers only 1200 kg ha^{-1}	56	30	15	6	10	9

1. Data for rows 1-3 converted from Wolf (1999), rows 4-6 from Berger (1969), row 7 from McEno (1991).

Most experts describe *Cannabis* as a *short-day plant.* It flowers in the autumn, when the photoperiod drops below 12–13 hours per day, depending on the variety and its geographical origin. Actually, *Cannabis* is best described as a *long-night plant*—interruption of dark periods by a short light period will completely prevent flowering, while an interruption of the light period by even a long dark period will not prevent flowering.

ATMOSPHERE

Plants, like all living things, require oxygen to survive. But unlike all other creatures, plants provide their own O_2 as a by-product of photosynthesis. Atmospheric carbon dioxide (CO_2) is often the limiting factor for photosynthesis. Frank (1988) reported peak growth at CO_2 levels of 1500 to 2000 ppm (=1.5–2.0%), five or six times greater than current atmospheric concentrations.

SOIL

Soil science is an interdisciplinary field, the most complex feature a farmer must manipulate. In the USA about 20,000 types of soil are recognized (Brady & Weil 1999). Soil series are named by their sites of discovery—my garden is dense Vergennes clay; up the hill, the soil lightens to a Covington silty clay loam. *Cannabis* grows best in a nutrient-rich, well-drained, well-structured, high organic matter, silty loam soil. To create this hypothetical substrate, you have to evaluate the soil's nutrient content, pH, type, and texture.

Macronutrients are elements required by plants in relatively large amounts. Organic materials in soil provide three of the six macronutrients—nitrogen (N), phosphorus (P), and sulphur (S). Minerals in soil provide the other three macronutrients—potassium (K), magnesium (Mg), and calcium (Ca).

Micronutrients (formally called **trace elements**) are also essential for plant growth, but in relatively small amounts. Most micronutrients become toxic to plants if they exceed trace amounts. Minerals in soil provide all seven micronutrients—iron (Fe), zinc (Zn), boron (B), copper (Cu), manganese (Mn), chlorine (Cl), and molybdenum (Mo). Some nutrients are needed in extremely tiny amounts. For instance, a plant needs a million N atoms for every Mo atom (Jones 1998). Some researchers argue that nickel (Ni), cobalt (Co), sodium (Na), vanadium (V), titanium (Ti), and silicon (Si) are also plant micronutrients (Jones 1998).

Cannabis places greater nutrient demands upon the soil than other crops. See Table 2.2. Fibre crops require high soil N, high K, then in descending order: Ca, P, Mg, and micronutrients. Seed crops, compared to fibre crops, extract less K and more P from the soil. The nutrient extraction of drug crops has not been measured, but we present estimates in Table 2.2. Drug crops have a high P requirement (Frank & Rosenthal 1978, Frank 1988), and Mg, Fe, and Mn may play a role in the enzyme regulation of THC synthesis (Kaneshima *et al.* 1973, Latta & Eaton 1975).

Storm (1987) described the function of *Cannabis* plant nutrients in detail. A summary is found in Table 2.3. Plants lacking nutrients produce telltale symptoms. For deficiency symptoms and their correction see Chapter 7.

Soil acidity, measured as **pH**, directly affects the availability of nutrients in the soil. See Fig 2.1 for an illustration of this relationship in organic soils. In soils with insufficient organic materials, pH has a greater influence on nutrient availability (Wolf 1999). Duke (1982) summarized pH data from 44 reports and suggested a soil pH of 6.5 is best. Test the pH of a tablespoon of wet soil by adding a pinch of baking soda. If it fizzes, then pH < 5.0 (too acid). Then test a tablespoon of dry soil by adding a few drops of vinegar—if it fizzes, then pH > 7.5 (too alkaline). Meters to measure pH are relatively inexpensive and accurate. Frank & Rosenthal (1978) provided charts and tables for adjusting different soils to a proper pH.

Understanding soil, however, is more than measuring pH and nutrients. Digging up soil for chemical tests is like grinding up your finger and conducting the same tests—you learn a lot about pH and chemistry, but nothing about structure and function of the soil.

Soil structure and function is determined by mineral particles, organic material, and microbiology. The particle size of **minerals** determines the three major soil types—sand, silt, and clay. **Sand** consists of relatively large particles, from 2.0 to 0.05 mm in diameter (these sizes are USDA standards—the British standard is 2.0-0.06 mm). Sandy soil feels gritty when rubbed between the fingers. **Silt** consists of particles from 0.05 to 0.002 mm in diameter, with a floury feel. **Clay** particles are smaller than 0.002 mm, invisible under light microscopes. Wet clay soil "slicks out," developing a continuous ribbon when pressed between the thumb and finger. Mixtures of the three soil types are called **loam soils**.

Organic material, "**humus**," consists of decomposing plant and animal matter, mixed with microorganisms. Microscopic bacteria and fungi convert the nutrients in humus into forms absorbable by plants. Bacteria and fungi also form the base of an underground food chain, including protozoans, arthropods, nematodes, and earthworms. Thus, soil organisms do more than *make* soil—they *are* the soil. The microorganisms and bugs hold nutrients in their bodies, keeping the nutrients from leaching away. As soil organisms die, they release nutrients for plants. Soil with more than 10% organic material is termed **muck**; if over 25% it becomes a bog or peat field.

Optimally, soil consists of 5–10% organic material and 40–45% mineral material. The other 50% is *space*. The lattice of space should contain equal parts air and water. The lattice allows nutrients to flow, roots to grow, bugs to move around, and everything to breath.

Coarse-textured soil is mostly sand. Sand maintains its lattice of space; it does not compact. It drains well, sometimes too well. To improve the water-holding capacity of sand, add organic material (e.g., peat moss) or synthetic soil conditioners (e.g., vermiculite). **Fine-textured** soil is mostly clay. The space in clay compacts easily. Compacted clay drains poorly, becomes puddled, and dries into large hard clods. To improve clay soil, add organic material or sand. "No-textured soil" is really **hydroponics**—soil science without the soil solids—100% space. See Storm (1987) for an introduction to *Cannabis* hydroponics. See Chapter 7 in this book for more information on soils.

ASSORTED ECOLOGICAL CHARACTERISTICS

Bush Doctor (1987a) found stressed plants growing at 3050 m above sea level. Sharma (1983) reported plants surviving up to 3700 m in altitude. *Cannabis* grows from 0° (the equator) to approximately 63° latitude (nearly the arctic circle, in Norway). Many hemp cultivars have been developed for 40–55° latitude; they perform poorly at semitropical and equatorial latitudes, due to early flowering. Drug varieties were traditionally developed at semitropical and equatorial locations, and perform poorly at high latitudes (above 45°) due to late flowering (frost kills them before flowering). Other ecological parameters described in the "triangular ordination" of Grime *et al.* (1988)—such as aspect, slope, bare soil index, and associated floristic diversity—have not yet been collected for *Cannabis*.

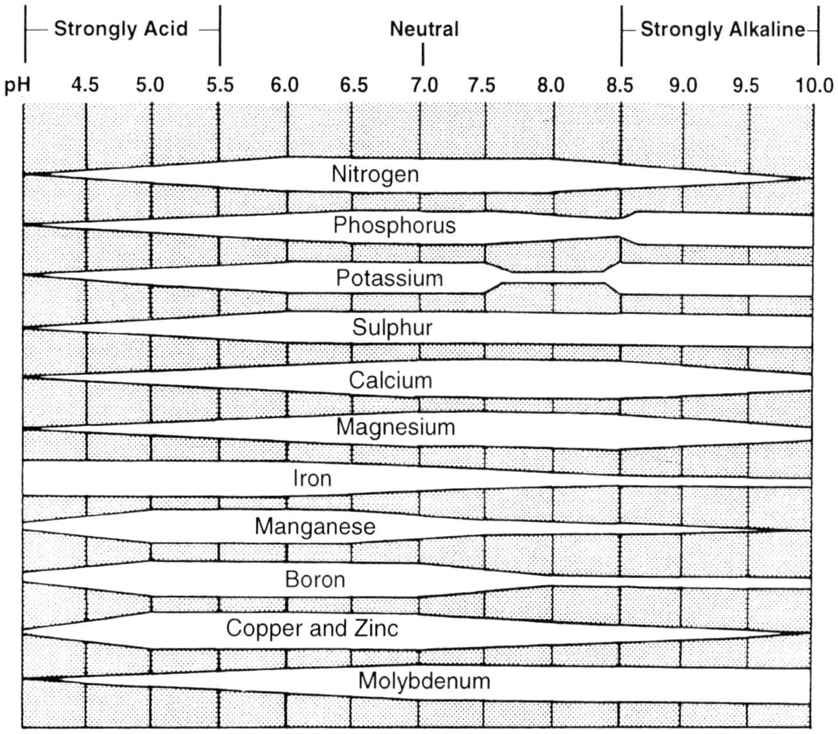

Figure 2.1:
Effect of soil pH on availability of plant nutrients. Maximum availability is indicated by the widest part of each bar. (McPartland redrawn from Thompson & Troeh 1973)

Table 2.3: Soil nutrients required for plant growth, and their function.

Nutrient	Forms taken up by plant	Nutrient Information
Nitrogen N	NH_4^+ NO_3^-	part of amino acids (proteins), nucleic acids (DNA and RNA), enzymes, coenzymes, cell membranes, and chlorophyll
Phosphorus P	$H_2PO_4^-$ HPO_4^{2-}	a component of sugar phosphates (ATP), nucleic acids, lipids, and coenzymes, promotes root formation and flowering
Potassium K	K^+	The primary intracellular cation and a major enzyme catalyst, fuels the "hydrogen pump" and drives stomatal movement
Calcium Ca	Ca^{2+}	cements the middle lamella in cell walls and regulates N metabolism, promotes healthy root and stem development
Sulphur S	SO_4^{2-}	required constituent of amino acids, enzymes, and coenzymes, involved in the formation of vitamins
Magnesium Mg	Mg^{2+}	the core of chlorophyll, regulates P metabolism, and may activate enzymes that synthesize THC
Iron Fe	Fe^{2+} Fe^{3+}	occurs in respiratory enzymes and a catalyst of chlorophyll formation and perhaps THC synthesis
Boron B	$B(OH)_4^-$ BO_3^{3-}	appears in enzymes and regulates K and Ca metabolism, promotes root development and prevents tissue necrosis due to excess oxygen
Manganese Mn	Mn^{2+} Mn^{3+}	a component of photosynthetic enzymes, perhaps THC-synthesis enzymes, involved with N and Fe metabolism
Copper Cu	Cu^{2+}	part of respiratory and photosynthetic enzymes, involved in cell wall formation and lignification
Molybdenum Mo	MoO_4^{2-}	serves as a metal component of enzymes required for utilization of N
Chlorine Cl	Cl^-	major intracellular anion, activates photosynthesis, involved with K in the regulation of osmotic pressure
Zinc Zn	Zn^{2+} $Zn(OH)_2$	required for DNA and protein synthesis and formation of auxin and other growth hormones
Cobalt Co	Co^{2+}	enhances the growth of organisms involved in symbiotic N fixture, constituent of vitamin B_{12}
Vanadium V	V^+	promotes chlorophyll synthesis, functions in oxidation-reduction reactions
Silicon Si	$Si(OH)_4$	forms enzyme complexes that act as photosynthesis regulators, plays a role in the structural rigidity of cell walls
Sodium Na	Na^+	involved in regulation of osmotic pressure

"Only when there is classification can there be analysis."
—William James

Chapter 3: Taxonomy and Ecology

Most people, especially ecologists, find taxonomy tedious. Ecologists and taxonomists examine the organisms in a hemp field from different perspectives. Ecologists examine all species to study the way organisms interact with each other, at one site. Taxonomists examine all ecosystems to study the way a species is related to other organisms, around the world (Wheeler 1997). Obviously, taxonomists and ecologists need each other.

When it comes to pest identification, taxonomists have the upper hand. Once you know the name of a pest or pathogen (especially the scientific or Latin name), you can research ways to control it. Exact identification becomes crucial if you use biocontrol, which only works against specific pests. Applying biocontrol against a misidentified pest can be a waste of time and money. Sloppy identification is permissible if you use nonspecific pesticides.

Most of us learned a two-kingdom taxonomy in school. Everything was jammed into the Plant Kingdom or the Animal Kingdom. In 1969 R. H. Whittaker described a five-kingdom taxonomy—Monera, Protista (now called Protoctista), Fungi, Plantae, and Animalia. We describe six kingdoms: **Vira,** little bits of bad news wrapped in a protein coat; **Monera,** the prokaryotes, including bacteria, phytoplasmas, and actinomycetes; **Protoctista,** unicellular eukaryotes, including protozoans, algae, slime moulds, and oömycetes; **Fungi,** moulds, mildews, smuts, etc.; **Plantae,** *Cannabis;* and **Animalia,** including nematodes, molluscs, insects, and vertebrates. See Fig 3.1 for an illustration of *Cannabis* parasites representing most of these kingdoms. Each kingdom is subdivided into Phyla, then Classes, Orders, Families, Genera, and Species ("**K**ing **P**hillip **C**ame **O**ver **F**or **G**old **S**overeigns").

VIRA

Conceptually, the viral kingdom resides within the other kingdoms. Viruses only replicate as true obligate parasites, in connection with a living host. Viruses cannot "grow," they do not eat, they do not have sex. They are complex molecules, entities between chemicals and life. Viruses contain DNA or RNA, which encode information for the re-production of identical chemicals. Viruses cause disease by reprogramming their host's metabolic machinery, causing host cells to produce foreign (viral) proteins.

A Russian botanist, Dmitri Iwanowsky, discovered viruses in 1892 while studying diseased tobacco plants. He found that the juice from diseased tobacco plants could pass through a bacterial filter and still be infective. Viruses were not actually seen until the advent of electron microscopy.

Viruses are transmitted (**vectored**) by aphids as the aphids move from diseased plants to healthy plants. To a lesser degree, viruses are vectored by leafhoppers, whiteflies, mites, mealybugs, thrips, and other insects with sucking mouthparts. Viruses and their vectors work in a symbiotic relationship. Viruses induce a change in plant metabolism—a kind of premature senescence—which makes plant sap more nutritious for sap-sucking insects (Kennedy 1951, Kennedy *et al.* 1959). The insects, in turn, transmit the viruses to new hosts.

Nematodes, fungi and parasitic plants occasionally vector viruses as they move (grow) from plant to plant. Viruses commonly spread through vegetative propagation (cloning) of infected "mother plants." Viruses may be transmitted through seeds. Viruses also spread plant-to-plant via root grafts, or by leaves rubbing in the wind, or by workers moving among diseased and healthy plants. Tobacco mosaic virus (the virus originally discovered by Iwanowsky) can even be transmitted by smoking infected cigarettes near uninfected plants.

Viruses cause symptoms of stunting, chlorosis, and overgrowth. **Stunting**, also known as **dwarfing**, indicates a slowing or cessation of plant growth. When a shoot becomes stunted, the internodes are shorter, which results in the crowding of foliage, known as a **rosette**. Viral destruction of chlorophyll results in **chlorosis**—the yellowing of normally green plant tissue. Chlorosis over an entire leaf is called **virus yellows**. Chlorosis may form in circular patterns as **ring spot**, in stripes as **hemp streak**, or appear randomly over the leaf as **virus mosaic**.

Symptoms of overgrowth caused by viruses include **hypertrophy** and **hyperplasia**. These appear as distorted enlargements and thickenings of leaves and flowers, sometimes called **witch's brooms**. The lamina and leaf margins of virus-infected plants may become distorted and these symptoms are called **wrinkle leaf**.

Luckily, few viruses attack hemp. Why? *Cannabis* extracts and purified cannabinoids inhibit the replication of viruses (Blevins & Dumic 1980, Braut-Boucher *et al.* 1985, Lancz *et al.* 1990, Lancz *et al.* 1991). We know no plant viruses causing disease in people; likewise, plants cannot catch the flu.

Many viruses infect insects. Insect viruses may contain DNA or RNA. Garrett (1994) suggested insect-borne RNA viruses were originally plant viruses that, millions of years ago, infected insects as insects fed on plant nectar. One such insect RNA virus is sold as a biocontrol agent—the *Agrotis segetum* cytoplasmic polyhedrosis virus (abbreviated AsCPV). DNA viruses are more common biocontrol agents, such as the Nuclear Polyhedrosis Virus (NPV) and its many strains (MbNPV, AcNPV, HzNPV, SeNPV, HcNPV, TnNPV, etc.), the *Agrotis segetum* granulosis virus (AsGV), and the *Melanopus sanguinipes* entomopoxvirus (MsEPV).

All the aforementioned biocontrol viruses produce **occlusion bodies** (OBs) in their hosts. OBs are viruses embedded within a proteinaceous capsule. Viruses in OBs persist longer in the environment than non-occluded viruses, which makes OB viruses more useful as biocontrol agents. OBs formed by NPV viruses contain hundreds or thousands of virus particles and grow to 20 μm in diameter (Hunter-Fujita *et al.* 1998). OBs of GVs are much smaller and only contain one virus particle. When insects ingest OBs, the proteinaceous capsule is dissolved by enzymes in the insect midgut. This releases the virus particles, allowing viruses to infect the host and replicate. Eventually the insect dies and its carcass disintegrates, releasing more OBs onto leaf surfaces. **Viroids** are similar to viruses, but have no protein coat. Plant viroids were discovered in the early 1970s. About a dozen viroid diseases have been described, none on *Cannabis.*

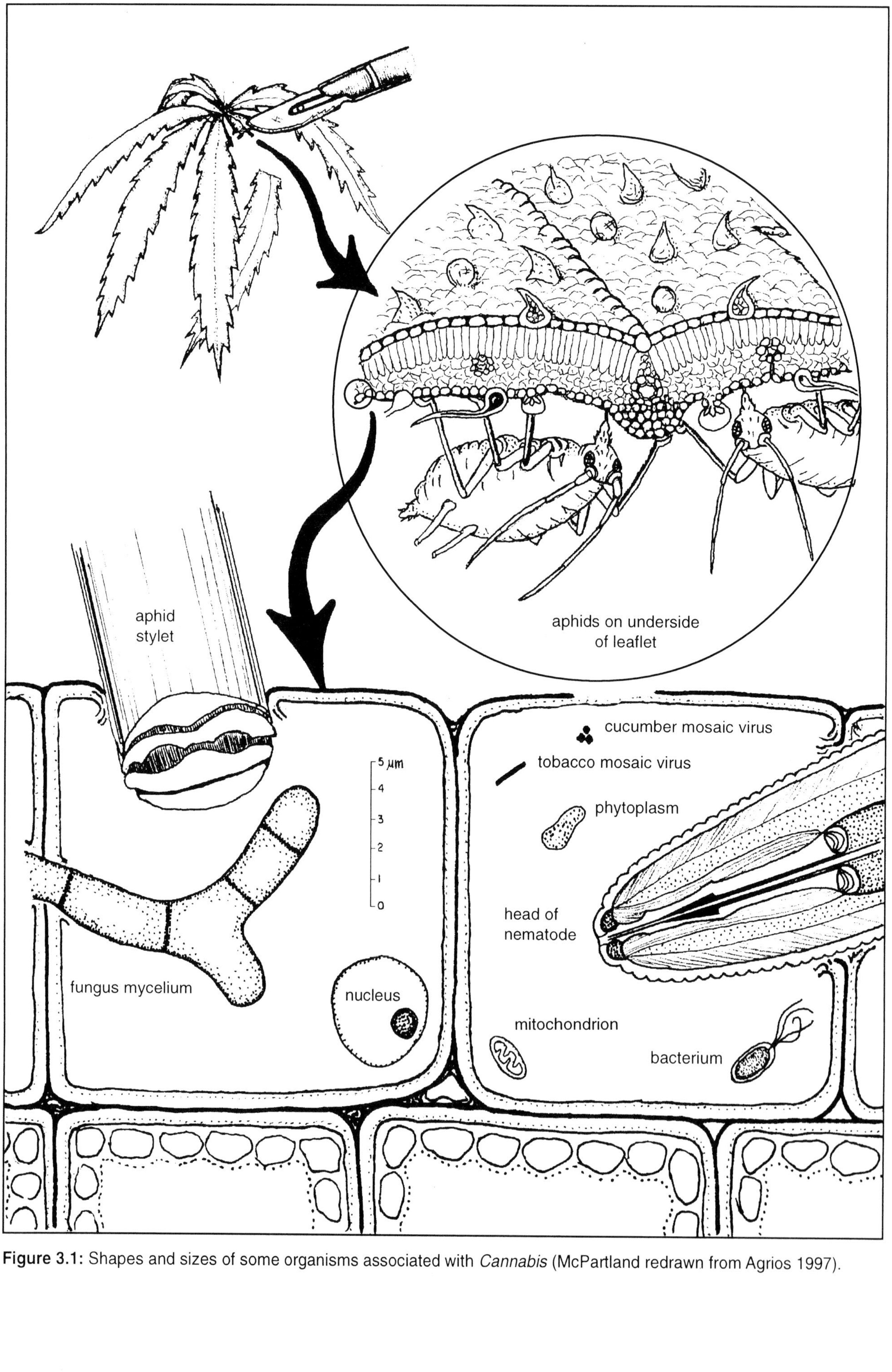

Figure 3.1: Shapes and sizes of some organisms associated with *Cannabis* (McPartland redrawn from Agrios 1997).

MONERA

This kingdom contains two phyla, the Cyanophyta (blue-green algae) and the Schizomycetes (bacteria, actinomycetes, and phytoplasmas). Antoni van Leeuwenhoek discovered bacteria in 1683, shortly after he started making microscopes. Leeuwenhoek was visually-oriented (Jan Vermeer was his best friend), and he was quite a curious fellow. Leeuwenhoek temporarily blinded himself by observing the ignition of gunpowder magnified x160. After that he turned to more pedestrian materials such as the scum on his teeth, where he discovered bacteria.

Thomas Burrill first proposed that bacteria cause disease. He studied "fire blight" of pear trees at the University of Illinois in 1878. Burrill predated Pasteur, but Pasteur studied bacteria in *people*, so he got all the glory. Thus we don't speak of the *Burrillization* of milk.

Bacteria are far smaller than plant cells or animal cells. About 1000 bacteria laid side-by-side measure a millimetre. Exceptionally small bacteria approximate the size of large viruses, but bacteria are more complex than viruses. Bacteria are enclosed by a membrane composed of proteins, carbohydrates, and lipids, similar to plant cells and our cells. Unlike us (but like plant cells), bacteria have a cell wall. Bacteria can grow, eat, and have sex. Unlike organisms in all other kingdoms, they only have one chromosome, and it is circular, without a nuclear membrane. Bacteria communicate with each other by passing genes between themselves.

Bacteria are extremely prolific. Some double their numbers every 20 minutes. If unchecked, one such bacterium could multiply into a colony covering the Pentagon in 16.50 hours. Bacteria are very hardy. One plant-pathogenic species, *Bacillus cereus*, can revert to a spore stage and survive in boiling water, frozen water, or no water at all.

Here are some generalities about *Cannabis* pathogenic bacteria: they are usually rod-shaped **bacilli**, gram-negative, aerobic, motile (moving with propeller-like **flagellae**), enter plants through small wounds or other openings, generally have a narrow host range, and rarely harm humans.

About ten species of bacteria attack *Cannabis*. Their symptoms often begin as *chlorosis*—the same as symptoms from viruses. But the chlorotic tissue subsequently dies (becomes **necrotic**), dries out, and turns brown. **Leaf spots** are localized lesions of necrotic leaf tissue, more or less circular. If lesions enlarge and become irregular in shape, they become **blotches.** Blotches have indistinct chlorotic margins, and may be accompanied by wilting. **Wilting** is a drooping of leaves or shoots, a loss of plant turgor, indicating bacteria have obstructed plant xylem. **Blights** are symptoms of wilting and necrosis which involve whole shoots or branches. If wilting becomes permanent, whole branches suffer **dieback**—the wilted leaves turn brown and dry out. Dieback begins at the tip of a branch or shoot and advances backwards towards the base. **Cankers** are localized, sunken, necrotic lesions on stalks or branches; cankers may cause wilting and dieback. **Rot** is a brown liquefying necrosis, indicating complete tissue destruction. A **root rot** indicates necrosis and collapse of part or all of the root system. **Crown rot** involves the crown (the transition zone between the root and stalk). One species of bacteria causes hyperplasia and hypertrophy of plant cells, called a **crown gall** (a cancer-like growth).

Cannabis protects itself by producing many antibacterial compounds. Extracts of *Cannabis* inhibit or kill plant-pathogenic bacteria (Bel'tyukova 1962, Vijai *et al.* 1993). An aqueous extract of hemp or wild hemp, called "cansantine" or "konsatin," was sprayed on potatoes and tomatoes to kill bacteria (Zelepukha 1960, Zelepukha 1963). *Cannabis* extracts also kill human pathogens (Krejcí 1950, Ferenczy 1956, Ferenczy *et al.* 1958, Kabelík *et al.* 1960, Radosevic *et al.* 1962, Gal *et al.* 1969, Veliky & Genest 1972, Veliky & Latta 1974, Klingeren & Ham 1976, Braut-Boucher *et al.* 1985). Cannabinoids are bactericidal or bacteriostatic, including THC and CBD (Schultz & Haffner 1959, Klingeren & Ham 1976), cannabidiolic acid (Kabelík *et al.* 1960, Gal *et al.* 1969, Farkas & Andrássy 1976), cannabigerol (Mechoulam & Gaoni 1965, Elsohly *et al.* 1982), and cannabichromene (Turner & Elsohly 1981). The non-cannabinoid essential oil of *Cannabis* is also bacteriostatic, and the essential oil derived from hashish is more bacteriostatic than the essential oil derived from fibre cultivars (Fournier *et al.* 1978).

Some bacteria act *symbiotically* with plants, not parasitically. *Rhizobium* species live within legume roots and trap atmospheric nitrogen for plants. *Azobacter, Azospirillum,* and *Klebsiella* species live on the surface of roots and also fix nitrogen. Some have been isolated from *Cannabis* (Kosslak & Bohlool 1983). Researchers have sprayed these bacteria on roots as nitrogen "biofertilizers" (Fokkema & Van Heuvel 1986).

Many bacteria colonize the surface of plants, forming resident populations. Most of these **epiphytic** bacteria are gram-negative, such as *Erwinia, Pseudomonas, Xanthomonas,* and *Flavobacterium* species. Gram-positive species occur less frequently (*Bacillus, Lactobacillus,* and *Corynebacterium* species). Epiphytes live on leaf surfaces (the **phylloplane**) more frequently than root surfaces (the **rhizoplane**). Most epiphytes colonize plants in a commensal relationship, living off cellular leakage. Others are mutualistic—in exchange for plant nutrients, the epiphytes protect plants from pathogenic organisms. Redmond *et al.* (1987) applied an *Erwinia* epiphyte as a biocontrol against *Botrytis,* the fungal pathogen that causes grey mould disease.

Some bacteria aid plants by killing insects. One such species, *Bacillus thuringiensis* (Bt) is a well-known insect killer. Spray Bt on plants and insects will die after they eat sprayed leaves. Another insect-killing bacterium, *Xenorhabdus nematophilus,* lives within soil nematodes. You can purchase nematodes that contain *X. nematophilus* and mix them into insect-infested soil. The nematodes find insects, penetrate them, then release the bacteria. The bacteria kill the insects and the nematodes feed off the cadavers. Quite a delivery system.

Phytoplasmas are essentially small bacteria without cell walls. They were discovered in a photo darkroom shared by plant and animal scientists. A plant scientist studying diseased plants was surprised by a lack of viruses appearing in his electronmicrographs. His veterinary colleague, who studied mycoplasmal pneumonia of swine, said, "But look at all those mycoplasmas." The new microorganisms were subsequently called mycoplasma-like organisms (MLOs). Genetic studies have determined these organisms are related to mycoplasmas, so now they are called Phytoplasmas (Agrios 1997). They spread by sap-sucking insects, which also become infected by them. A *Cannabis* phytoplasma has been reported in India, causing rosette symptoms with hypertrophy and leaf distortion (Phatak *et al.* 1975).

PROTOCTISTA

This kingdom contains green, plant-like organisms, as well as organisms that move about like animals. Thus, some researchers split Protoctista into two kingdoms—Chromista and Protozoa (Hawksworth *et al.* 1995). Protoctistans are unicellular or multicellular. They differ from bacteria by having several chromosomes (surrounded by a nuclear membrane), plus membrane-bound cell organelles (such as mitochondria, chloroplasts, etc.).

The green algae (Chromista phylum **Chlorophyta**) deserve mention, since they produce most of the Earth's atmospheric oxygen. Chrysophytids should always be mentioned. They lived and died 20 million years ago. For aeons their calcium microskeletons piled up on ocean bottoms. We now call these deposits **diatomaceous earth**, and dust them on plants as an organic control against aphids.

Protoctistans of the phylum **Protozoa** are animal-like—such as the classic *Amoeba* and the bacteria-eating *Paramecium*. One genus of flagellated protozoans, *Phytomonas*, causes plant disease. It has not been found in *Cannabis*. Extracts of *Cannabis* and purified cannabinoids kill protozoans (McClean & Zimmerman 1976, Pringle *et al.* 1979, Nok *et al.* 1994).

Members of the phylum **Oömycota** (also called Oömycetes, algal fungi, or phycomycetes) are difficult to differentiate from true fungi. They have cellulose in their cell walls (like plants), instead of the chitin found in fungi. Two genera, *Pseudoperonospora* and *Pythium*, commonly infect *Cannabis*. They give rise to unique symptoms. *Pseudoperonospora* species cause **downy mildew**, a blue-white felt that forms on the undersides of leaves. *Pythium* species cause **damping off**, a rapid collapse of small seedlings.

Harvey (1925) invented a technique for isolating Oömycetes, using steam-sterilized *Cannabis* seeds. He floated the seeds in pond water as Oömycete "bait." Thanks to Harvey's technique, the scientific literature is filled with reports of aquatic Oömycetes infesting *Cannabis* seeds. None of these "baited" Oömycetes cause problems unless you store seeds in pond water.

Phylum **Myxomycophyta** includes the fungus-like slime moulds—a truly curious bunch. Some "individuals" can exist as *either* one multicellular organism *or* a collection of single-celled organisms. If there is enough food around, the single cells go about their business, growing and dividing like amoebae. But if starved, they aggregate into clumps and crawl off like a slug. Finding better conditions, the slug erects a tall stalk topped by spores. The spores blow off, revert to amoebae, and go their separate ways. Some slime moulds are brightly coloured and visibly pulsate. They often make neighbourhood news when found crawling up someone's house. Gzebenyuk (1984) found one species, *Didymium clavus*, climbing hemp stems. THC and CBN are toxic to other slime moulds (Bram & Brachet 1976).

FUNGI

Classification and taxonomy get complicated here. Even Linnaeus, the genius taxonomist who first coined "*Cannabis sativa*," found the fungi frustrating. He lumped many fungi in his genus *Chaos*.

Fungi can be unicellular or multicellular. They contain one, two, or many nuclei per cell. Fungi have cell walls composed of chitin. Some are mobile. Most are sexual. They are everywhere we look. In this aspect they are more successful than insects—fungi have conquered the seas, but we find no insects in marine environments. Experts estimate there are at least 1.5 million species of fungi, but only 10% have been identified to date (Hawksworth *et al.* 1995).

Fungi have no digestive system, so they absorb (not ingest) nutrients. Fungi grow into food, exude enzymes which cause digestion to occur around them, then they absorb the nutrients. Multicellular fungi grow in threadlike tubular filaments termed **hyphae**. Hyphae may or may not contain **septa**, which are incomplete cross-walls. The body of a fungus, its collection of hyphae, is called a **mycelium**.

Fungi cause more *Cannabis* disease than the rest of earth's organisms combined. Some symptoms are similar to those caused by viruses or bacteria, such as **chlorosis, necrosis, rosettes, wilting, leaf spots, blotches, blights, diebacks, cankers, root rots**, and **crown rots**. **Powdery mildew** arises as a thin covering of white fungus upon the upper surfaces of leaves. **Black mildew** and **sooty mould** appear as black growths on leaves; the latter is associated with aphid droppings. **Rust** is distinguished by rust-coloured pustules of fungal spores.

The vast majority of fungi are **saprophytes**, living off already-dead material. They benefit us by decomposing organic matter and releasing nutrients back to the soil. Fungi also ruin our food and overrun leather, cotton, and paper in damp places. A small group of fungi "go both ways" as saprophytes of dead plants *and* parasites of living plants. Termed **Facultative Parasites** (FPs), they normally live as saprophytes but can attack living hosts. Many of the rot fungi and damping-off fungi fall into this category. Some FPs also attack us. Look in an old pint of cottage cheese. The fuzzy fungus on the lid causes lung disease (geotrichosis), the black mould causes mucormycosis, and the orange slime on the bottom causes meningitis and endocarditis in people with AIDS. **Facultative Saprophytes** normally attack living hosts, but can feed on recently-dead ones when times are tough. **Obligate Saprophytes** only eat the dead. **Obligate Parasites** (OPs) only feed on the living, which poses a problem for scientists who study them, since OPs cannot grow on agar in petri plates.

Some FPs live as symbionts within the nooks and crannies of leaves, feeding on cellular leakage, aphid honeydew, pollen grains, and other airborne debris. These phylloplane fungi live above the leaf epidermis (**epiphytes**) or in spaces below the epidermis (**endophytes**). Phylloplane fungi protect their plant hosts by repelling pathogenic fungi and herbivorous animals (Fokkema & Van den Heuvel 1986).

Mycorrhizae are symbionic fungi that live within plant roots. They extend hyphae into the deep soil, drawing water and minerals (mostly phosphorus) back to their host's roots. In return, the host supplies the mycorrhizae with photosynthetic products. Plants enjoying this fungus partnership grow faster than their nonmycorrhizal neighbours (see "Mycorrhizae" in Chapter 5).

Other "friendly" fungi act as **hyperparasites** and feed on other fungi, strangle soil nematodes in nooses of hyphae, and infest insects. Over 700 species of fungi cause diseases in insects (Roberts & Hajek 1992). One such fungus, *Cordyceps sinensis*, produces long, thin, black sclerotia, known as "dead man's fingers" (highly esteemed in Chinese medicine). Many *Cordyceps* anamorphs are sold for biocontrol of insects, including *Hirsutella, Nomuraea, Paecilomyces*, and *Verticillium* species.

Cannabis produces antifungal chemicals. In this capacity, hemp has been planted with potatoes to deter the potato blight fungus, *Phytophthora infestans* (Israel 1981). Concentrated *Cannabis* extracts are lethal to fungi (Vysots'kyi 1962, Misra & Dixit 1979, Pandey 1982, Gupta & Singh 1983, Singh & Pathak 1984, Grewal 1989, Kaushal & Paul 1989). Pure THC and CBD inhibit fungal growth (Dahiya & Jain 1977, Elsohly *et al.* 1982, McPartland 1984), as does cannabichromene (Turner & Elsohly 1981), and cannabigerol (Elsohly *et al.* 1982). Terpenoids and phenols are antifungal, such as linalool, citronellol, geraniol, eugenol (Kurita *et al.* 1981), limonene, cineole, β-myrcene, α- and β-pinene (DeGroot 1972, Wilson 1997), and these are components of *Cannabis* essential oil (Turner *et al.* 1980).

But fungi fight back. They can replicate. Fungi reproduce by **budding** (like bacteria) or by **spores**. There are several kinds of spores. **Conidia** are spores produced by **mitosis**. They are genetically haploid, $1n$ (like our sperm and ova). But unlike our sperm and ova, conidia can germinate by themselves, directly into whole $1n$ organisms. Hyphae of two $1n$ organisms can intertwine and fuse, forming **diploid** ($2n$) organisms. This is fungal sex. In some cases when hyphae fuse, their nuclei remain separate—forming **dikaryotic** ($1+1n$) organisms. The hyphae of all these organisms—haploid, diploid and dikaryotic—are identical in external appearance. Only their nuclei know for sure.

Diploids ($2n$) can produce spores by **meiosis**. These spores may form at the site of haploid fusion (at the zygote, hence "**zygospores**"). Spores also arise distally in a sac (the sac is called an **ascus** and spores arising in the ascus are called **ascospores**) or spores arise distally on a club (the club is called a **basidium** and spores arising on the basidium are called **basidiospores**). Most *Cannabis*-attacking fungi produce millions of microscopic spores. Spores come in assorted sizes and shapes (Fig 3.2).

Some spores, asci, and basidia form within reproductive structures called **fruiting bodies**. Magic mushrooms, for instance, are fruiting bodies of the basidiomycete *Psilocybe cubensis.* The rest of *P. cubensis* is an underground network of hyphae. Fruiting bodies of *Cannabis* pathogens are less spectacular. Most look like the period at the end of this sentence. Figure 3.3 illustrates some fruiting bodies found on hemp—**sporangia, apothecia, pycnidia, acervuli, synnema (coremia), cleistothecia, perithecia, spermigonia (=pycnia), aecia, uredia,** and **telia**. Fruiting bodies and spores serve as the most useful means of identifying fungi, by their size, shape, colour, and arrangement.

Spores may bud directly off specialized hyphae, without the protection of a fruiting body. Many fungi also produce **chlamydospores**, which are hyphae with hard, thickened walls that survive inhospitable conditions. After bad conditions pass, these survival units germinate and regenerate hyphae. A mass of these survival units, frequently rounded into a ball with a rind-like covering, is called a **sclerotium**.

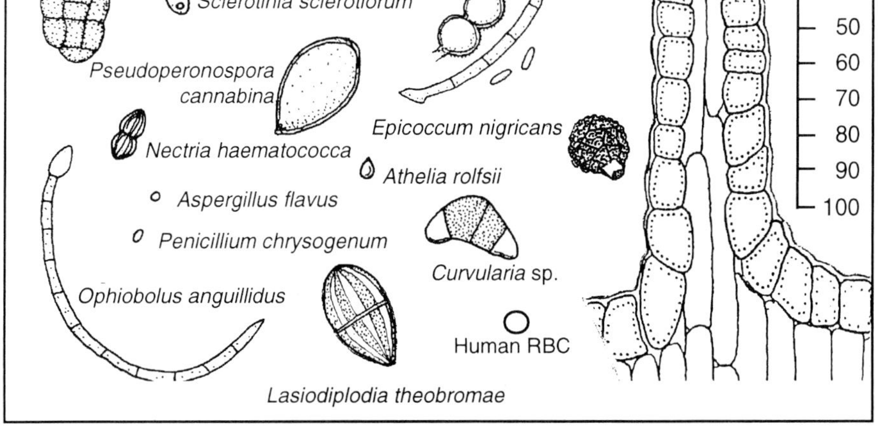

Figure 3.2: Spores of several *Cannabis*-pathogenic fungi, drawn to scale with a glandular leaf hair (McPartland).

Most plant-pathogenic fungi produce *two* types of spores. The *meiotically*-derived *sexual* spores are called the **teleomorph** stage; these include zygospores, ascospores, and basidiospores. *Mitotically*-derived *asexual* spores are called the **anamorph** stage, including conidia, chlamydospores, and sclerotia. The ability to produce two spore stages confuses taxonomists: each spore stage may have its own name! The grey mould fungus, for instance, usually produces conidia and is called *Botrytis cinerea.* But the fungus sometimes produces ascospores, and this stage is named *Botryotinia fuckeliana.* Two names, one species. How could this happen? The two spore stages were discovered by different scientists who did not recognize the stages were related to each other.

Fungi are classified by their *teleomorph* (sexual) stage, and placed the **Chytridiomycota, Zygomycota, Ascomycota** and **Basidiomycota**. The teleomorph stage of some fungi, however, has not yet been discovered; these fungi are only known by their *anamorph* (asexual) stage. Two Frenchmen called these nonsexual organisms "imperfect fungi," and the term stuck. Imperfects are placed in their own phylum, the **Deuteromycota**. Hawksworth *et al.* (1995) abandoned the Deuteromycota as "an artificial assemblage of fungi." He does not classify members of this large group of fungi, but merely calls them "mitosporic fungi." We disagree with Hawksworth and retain the phylum for its three useful classes (see Table 3.1).

Few areas in biology evoke more controversy than the classification of fungi. Within the phylum Ascomycota, for instance, different researchers recognize three to six classes, whereas Hawksworth *et al.* (1995) recognize none. We follow Ainsworth *et al.* (1973), who recognize six classes. Naming fungi (nomenclature) is governed by the International Code of Botanical Nomenclature (Greuter *et al.* 1994). The code is periodically updated and has become rather complicated. As von Arx once said, "nonspecialists have difficulties in understanding the code and adhering to its provisions."

In this text, we are primarily concerned with plant-pathogenic fungi. Thus we will bypass many otherwise-important fungi, such as truffles, the Eurotiales (human pathogens), as well as one whole phylum—the Chytridiomycota (chytrids). Table 3.1 contains a hierarchical outline of plant-pathogenic fungi, arranged by phylum, class, and order.

Some fungi attack only *Cannabis,* while other species plague a wide variety of plants. Some fungi switch hosts during their life cycles, producing different fruiting bodies on different hosts. For instance, rust fungi form up to five different fruiting bodies on different hosts. None resembles another. At one time they were considered five different organisms, rather than five different forms of one organism. *Chaos.*

PLANTAE

We split the plant kingdom into two phyla. The **Bryophyta** consists of primitive nonvascular plants with flagellated sperm cells, such as liverworts and

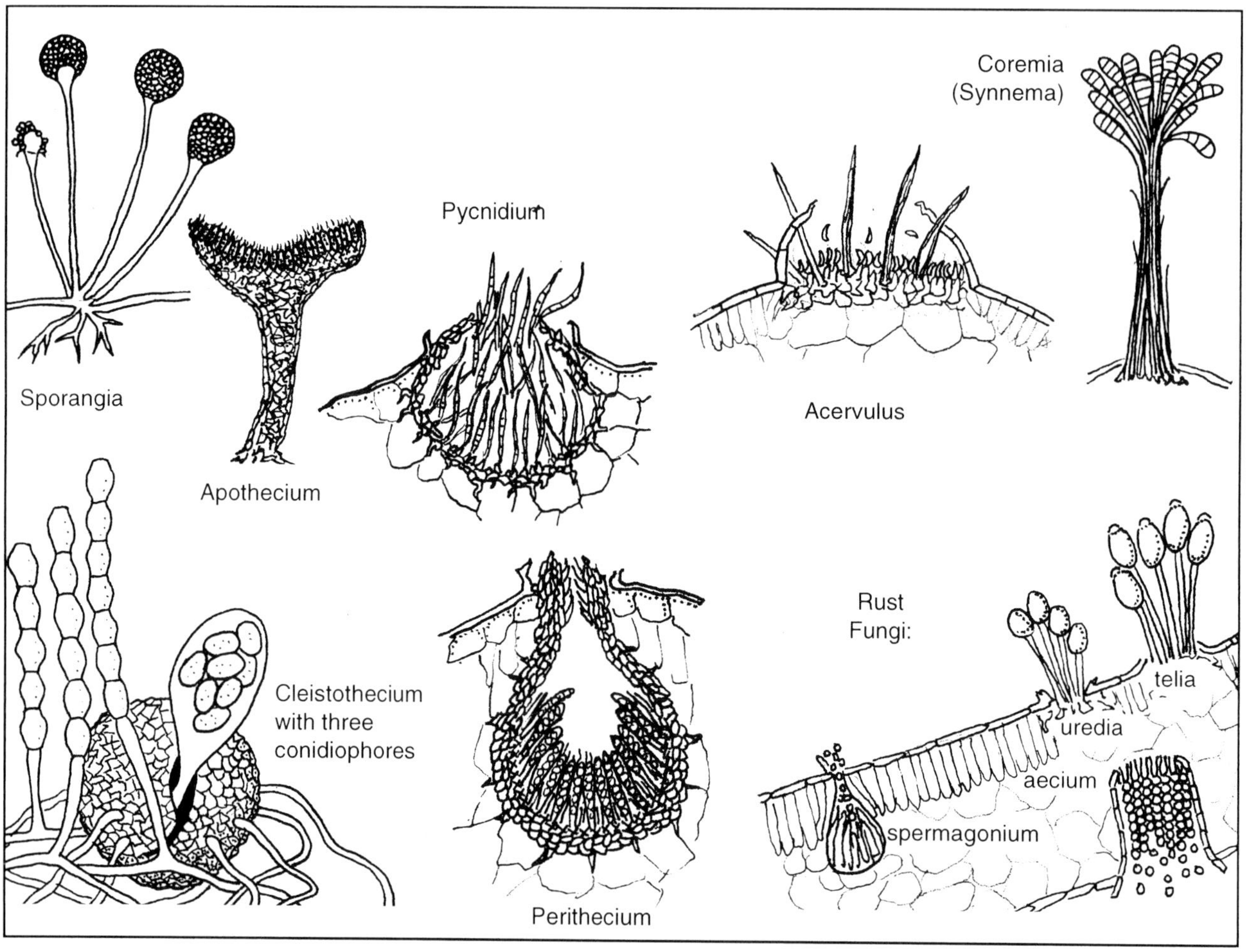

Figure 3.3: Assorted reproductive fruiting structures of *Cannabis*-pathogenic fungi (McPartland).

mosses. **Tracheophyta** plants are vascular (containing xylem and phloem) with roots, stems, and leaves. Their sperm cells (pollen) have no flagella. The Tracheophyta includes five classes: psilopsids, club mosses, horsetails, ferns, and seed plants. Seed plants are split into two orders, the **Gymnospermae** (cyads, ginkos, conifers) and the **Angiospermae** (monocots and dicots). Among the latter we find *Cannabis,* as well as some of its antagonists.

Antagonist plants that compete with crops for light, water, and nutrients are called **weeds**. For a competitive edge, evolution has armed some plants with arsenals of poisons and repellants (termed **allelochemicals**). Many weeds inhibit *Cannabis* seed germination (Muminovic 1990). According to Good (1953), *Cannabis* grows poorly near spinach (*Spinacia oleracea*), rye (*Secale cereale*), and garden cress (*Lepidium sativum*). In retaliation, *Cannabis* suppresses neighbouring plants, whether they are weeds such as purple nutsedge (Srivastava & Das 1974), quackgrass (Muminovic 1991), and chickweed (Stupnicka-Rodzynkiewicz 1970), or crop plants such as maize (Pandey & Mishra 1982), rice (Vismal & Shukla 1970), lupine, beets, brassicas (Good 1953), wheat, rye, and oats (Schwär 1972).

Cannabis has **weed mimics**. Weed mimics imitate crop plants, so farmers overlook them while weeding. Seedlings of hemp nettle (*Galeopsis* species), for instance, are very difficult to distinguish from *Cannabis* seedlings. Male plants of the dye plant *Datisca cannabina* are remarkable mimics of *Cannabis,* fooling even professional plant taxonomists (Small 1975).

The most pernicious antagonists of *Cannabis* are parasitic plants. These plants contain little or no chlorophyll. They leech off other plants by sending modified roots (**haustoria**) into roots or stems of their hosts. More than 2500 species of parasitic plants are known around the world. Luckily, less than a dozen species leech off *Cannabis,* collectively they are known as dodder and broomrape. To add injury to insult, parasitic plants may infect their hosts with viruses.

ANIMALIA

The animal kingdom is divided into 33 phyla. Four herbivorous (plant-eating) phyla are discussed below: the *Nematoda* (nematodes), *Mollusca* (snails and slugs), *Arthropoda* (insects and their ilk), and *Chordata* (e.g., birds and mammals). Herbivorous animals are nitrogen-challenged. Plants consist mainly of carbohydrates, whereas animals consist of protein. The difference is nitrogen. Animals are 7–14% nitrogen by dry weight (dw). Plant rarely contain more than 6% nitrogen (dry weight), except for actively growing tissues and reproductive parts (see Table 3.2). Thus young shoots and seeds become very attractive to nitrogen-starved herbivores (Mattson 1980).

Phylum Nematoda (Aschelminthes)

Nematodes (roundworms, eelworms) that feed on plants are nearly microscopic. Nematodes are not related to earthworms. Built on a much simpler scale, nematodes have no respiratory or circulatory systems. They have a few muscle cells, which enable nematodes to wiggle out of a predator's grip, but the muscles cannot coordinate move-

Table 3.1: Taxonomy of fungi associated with *Cannabis*

PHYLUM ZYGOMYCOTA (hyphae rarely have septa; teleomorph produces zygospores)

- **Class Zygomycetes**: produce a profuse mycelium, much of which is immersed in the host; they reproduce asexually by sporangia
  - **Order Mucorales:** saprophytic (as storage moulds) or weakly pathogenic on *Cannabis: Mucor* and *Rhizopus* species
  - **Order Endogonales**: soil fungi forming mycorrhizal associations with plants, *Cannabis* symbionts: *Glomus* species
  - **Order Entomophthorales**: parasitic on insects, biocontrol agents against pests: *Erynia, Conidiobolus, Entomophthora* species

PHYLUM ASCOMYCOTA (hyphae septate; teleomorph produces ascospores)

- **Class Plectomycetes**: asci unitunicate (walls have one layer), scattered within a closed cleistothecium or gymnothecium, with single-celled ascospores
  - **Order Erysiphales**: produce a profuse mycelium, mostly on leaf surfaces; anamorph spores formed in chains—the powdery mildews; *Cannabis* pathogens: *Leveillula, Sphaerotheca* species
- **Class Pyrenomycetes**: asci unitunicate, asci borne in a single layer within an ostiolated perithecium, with single or multicelled ascospores
  - **Order Sphaeriales**: perithecia dark, carbonaceous, with or without a stroma. Anamorph spores bud directly off hyphae or form within pycnidia or sporodochia; *Cannabis* pathogens: *Chaetomium, Gibberella, Diaporthe, Hypomyces, Melanospora, Nectria, Phyllachora* species
- **Class Discomycetes**: asci unitunicate, borne in a single layer within a cup-like apothecium, with paraphyses and single- or multi-celled ascospores
  - **Order Helotiales**: asci open with a pore or tear (inoperculate) to release ascospores; *Cannabis* pathogens: *Sclerotinia, Orbilla, Hymenoscyphus* species
- **Class Loculoascomycetes**: asci bitunicate (have 2 layers), developing in unwalled locules within a stroma; ascospores usually multicellular; anamorph stage produces conidia free or within pycnidia
  - **Order Pleosporales**: pseudothecia usually uniloculate, asci clavate (long cylindrical) in shape and usually arranged in a single layer; pseudoparaphyses present; *Cannabis* pathogens: *Botryosphaeria, Didymella, Leptospora, Leptosphaeria, Ophiobolus, Pleospora* species
  - **Order Dothideales**: pseudothecia are uni- or multiloculate, asci ovate in shape and usually scattered within locules; *Cannabis* pathogens: *Mycosphaerella, Leptosphaerulina, Schiffnerula* species

PHYLUM BASIDIOMYCOTA (hyphae septate with clamp connections; teleomorph produces basidiospores)

- **Class Teliomycetes**: simple septa present; basidiospores borne on promycelia and teliospores
  - **Order Uredinales**: obligate parasites with complicated life cycles spanning several spore types; known as the Rust fungi; *Cannabis* pathogens: *Aecidium, Uromyces, Uredo* species
- **Class Hymenomycetes**: dolipore septa present; basidia and basidiospores borne on a hymenium (fertile layer lining a fruiting body)
  - **Order Agaricales**: fruiting bodies are monomitic (one type of thin-walled hypha), hymenium often hidden by a veil—the Mushrooms.
  - **Order Aphyllophorales**: fruiting bodies are monomitic to trimitic, hymenium exposed; *Cannabis* pathogens: *Athelia, Thanatephorus*

FORM-PHYLUM DEUTEROMYCOTA (hyphae aseptate or septate; anamorph stage)

- **Form-Class Hyphomycetes**: mycelium bears conidia directly on special hyphae (conidiophores), conidiophores free or bound in tufts (coremia) or cushion-like masses (sporodochia); *Cannabis* pathogens: *Alternaria, Aspergillus, Botrytis, Cercospora, Cephalosporium, Cladosporium, Curvularia, Cylindrosporium, Epicoccum, Fusarium, Myrothecium, Penicillium, Periconia, Phymatotrichopsis, Pithomyces, Pseudocercospora, Ramularia, Sarcinella, Stemphylium, Thyrospora, Torula, Trichothecium, Ulocladium, Verticillium* species
- **Form-Class Coelomycetes**: conidia borne on conidiophores enclosed in pycnidia or acervuli; *Cannabis* pathogens: *Ascochyta, Botryodiplodia, Colletotrichum, Coniothyrium, Diplodina, Macrophomina, Microdiplodia, Phoma, Phomopsis, Phyllosticta, Rhabdospora, Septoria, Sphaeropsis* species
- **Form-Class Agonomycetes**: "Mycelia sterilia," mycelium with no reproductive structures; *Cannabis* pathogens: *Rhizoctonia* species

ment in a specific direction. Their nervous system is so simple it can be described at the level of individual cells: *Caenorhabditis elegans,* for instance, has exactly 302 neurons. A complete wiring diagram of its nervous system has been compiled. The entire DNA sequence needed to build *C. elegans* has been described—19,000 genes, a 97-megabase genomic sequence. The physical characteristics of typical plant-pathogenic nematodes are illustrated in Fig 3.4.

Nematodes are extremely abundant—a fistful of soil may contain thousands of them. They occupy every earthly niche from mountain top to sea bottom. Nematode crop losses tend to be underestimated because of the nematodes' small size and their unseen (mostly underground) damage. Indeed, the first nematode to be discovered was an odd species that attacks plants *above-ground.* In 1743 Turberville Needham extracted a white fibrous material from stunted wheat in England. To his amazement, the fibrous material began to wiggle when soaked in water. The fibres were matted larvae of *Anguina tritici,* the wheat gall nematode.

Most nematodes are dioecious, with males and females required for reproduction. Males are usually smaller than females. Males of one species, *Trichosomoides crassicauda,* are so small they live in the female's uterus! Some nematodes are **hermaphroditic** (females with additional male gonads). Others eliminate males altogether, and reproduce **parthenogenetically**.

At least seven nematode species attack *Cannabis.* All are **polyphagous**, which means they attack many different crops. They feed with a hypodermic-like **stylet**, which resembles a hollow spear. Nematodes thrust stylets into

Table 3.2: Parts of plants and the nutrients they contain[1]

Plant part	Nitrogen[2]	Carbohydrate[2]	Water[3]
leaves	1–5	5–17	(40)[4]–90
stalks	<1	50[5]	30–50
seeds	5–10	30	30–50
pollen	5–50	very low	30–70
phloem sap	0.004–0.6	5–20	80–95
xylem sap	0.0002–0.1	<0.06	>98

[1]adapted from Mattson (1980), and Young (1997); [2]Nitrogen and Carbohydrate concentrations measured as % dry weight, except for sap which is measured as % weight/volume; [3]Water expressed as % wet weight; [4]droughted leaves may have very low water levels; [5]hemp fibres contain high levels of cellulose, an indigestible carbohydrate.

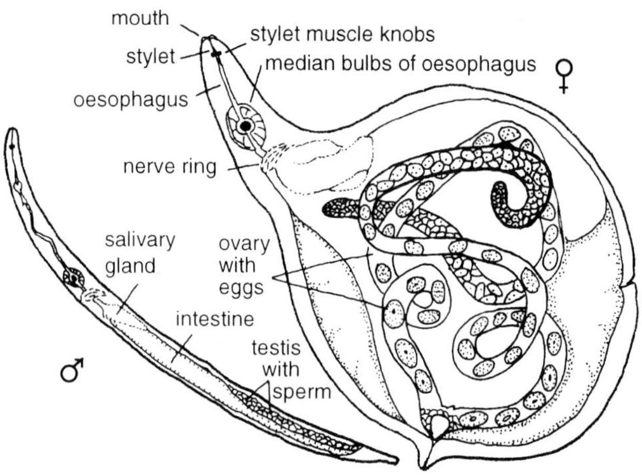

Figure 3.4: Male and female nematodes showing important parts (McPartland).

plant cells, then suck out cell cytoplasm. Stylets vary in size and shape between species and are useful for identifying nematodes. Nematodes can feed from outside roots as **ectoparasites** (e.g., *Paralongidorus maximus*) or enter plants to feed as **endoparasites**. Endoparasites are **sedentary**, remaining embedded in roots (e.g., *Meloidogyne, Heterodera* species) or **migratory**, moving root-to-root (e.g., *Pratylenchus penetrans*) or migrating to above-ground plant parts (e.g., *Ditylenchus dipsaci*).

Above-ground symptoms from nematodes mimic symptoms of root injury—stunting, chlorosis, and **insipient wilting** (drooping of leaves during midday with recovery at night). *Below-ground* symptoms are more distinctive, including **root knots** or **root galls**.

Nematode populations are naturally biocontrolled by viruses, bacteria, protozoans, other nematodes, and killer fungi (which strangle nematodes in constricting hyphal loops). Some plants, including *Cannabis*, ooze metabolites that repel or kill many nematodes (Kir'yanova & Krall 1971, Haseeb *et al.* 1978, Vijayalakshmi *et al.* 1979, Goswami & Vijayalakshmi 1986, Mojumder *et al.* 1989, Kok *et al.* 1994, Mateeva 1995). A worm related to nematodes, *Dugesia tigrina*, is killed by THC and CBD (Lenicque *et al.* 1972).

Phylum Mollusca

Here we have the snails and slugs. They are familiar creatures, seemingly benign. But garden slugs (*Limax, Agriolimax* and *Arion* species) can be nasty pests. They kill seedlings and can whack older plants. Cannabinoid receptors and anandamide have been found in several molluscs, including *Mytilus edulis*, an edible marine bivalve (Stefano *et al.* 1998).

Phylum Annelida

This group deserves a mention for our nightcrawler friend, *Lumbricus terrestris*. We owe earthworms an incalculable debt for serving as "intestines of the earth," to quote Aristotle. Earthworms digest organic material, their tunnels aerate soil, and their manure castings constitute one of the finest fertilizers available. Earthworms can be quite prolific—in prairie soil they weigh up to 6738 kg ha^{-1} (6000 lbs/acre), but in adjacent corn fields, their numbers drop to an average of 88 kg ha^{-1} (78 lbs/acre)—earthworms do not tolerate ploughing and other soil disturbances (Zaborski 1998). Pesticides kill earthworms, particularly carbamates (benomyl, carbaryl, zineb); over 200 pesticides have been tested for acute toxicity (Edwards & Bohlen 1992).

Earthworms are usually 180 mm long and 4 mm wide, but can grow up to 300 mm long. Earthworms should not be confused with nematodes, although 19th century experts erroneously urged farmers to destroy both. Parkinson (1640) described how an aqueous extract of macerated hemp leaves, "powred into the holes of earthwormes, will draw them forth, and fishermen and anglers have used this feate to get wormes to baite their hookes." This feat is still practised in eastern Europe (Kabelik *et al.* 1960). Cannabinoid receptors and anandamide have been found in a related annelid species, *Hirudo medicinalis*, the leech (Stefano *et al.* 1998).

Phylum Arthropoda

Over 800,000 arthropods have been described, with dozens of new species named every week. Arthropods account for 80% of the animal species on earth, and some wreak havoc on agriculture. Arthropods can be distinguished by their jointed chitinous exoskeletons and segmented bodies. Six classes are hemp herbivores: the **Crustacea** (including pillbugs, with five to seven pairs of legs), **Symphyla** ("garden centipedes," with 12 pairs of legs), **Chilopoda** (true centipedes, with one pair of legs per segment), **Diplopoda** (millipedes, thousand-leggers, with two pairs of legs per segment and many, many segments), **Arachnida** (spiders, ticks, and mites, with four pairs of legs), and the **Insecta**, with three pairs of legs.

Insects are the largest class. Here's some entomology terminology: the body of an insect is segmented into the **head, thorax**, and **abdomen**. *Externally*, the head may contain one pair of compound eyes, one or more pairs of simple eyes (**ocelli**), one pair of segmented **antennae**, breathing tubes (**tracheae**), and mouthparts. The thorax may sport one or two pairs of wings, and three pairs of segmented legs. Legs are jointed, with a hip (**trochanter**), upper leg (**femur**), lower leg (**tibia**), and a foot (**tarsus**). Insect abdomens may exhibit vestigial legs (**prolegs**, the fleshy unjointed stubs you see on caterpillars), **tympana** (thinned sections of abdomen which serve as "ears"), **genitalia** (modified as *stingers* in bees), and **cerci** (caudal appendages serving olfactory or tactile functions). *Internally*, the head houses brains, a blood vessel, and the oesophagus. The thorax contains nerve ganglia ("sub-brains"), the lower oesophagus (including the **crop**), two pairs of **spiracles** (breathing apparati), the aorta, and muscles for locomotion. Insect abdomens contain insect hearts, digestive organs, more nerve ganglia, eight more pair of spiracles, excretory organs, and reproductive systems (Fig 3.5).

Moths, butterflies, flies, beetles, wasps, bees, and ants have four-stage life cycles—eggs, immature larvae (caterpillars, maggots, grubs), pupae (cocoons, chrysalids), and adults. This is called **complete metamorphosis** because larvae undergo a dramatic change during the pupal state. Larvae of moths and butterflies (caterpillars) may have up to 22 legs—six in front are true legs, the rest are prolegs and disappear during metamorphosis. Beetle larvae (grubs) have six legs. Fly larvae (maggots) have no legs. Larvae generally do not feed on the same plants as adults.

Aphids, leafhoppers, thrips, plant bugs, grasshoppers, and related insects only pass through three stages—eggs, immature nymphs, and adults. This is called **incomplete metamorphosis**, without any dramatic change in appearance. Nymphs and adults often feed on the same host plants.

To grow through their hard exoskeletons, immature insects undergo several **moults**, when they shed their old exoskeletons. Stages between moults are termed **instars** (eggs hatch into first instars, then moult into second instars, moult into third instars, etc.). Insects may be **univoltine**, producing one generation per year, or **multivoltine**, producing several generations per year. Insects are cold-blooded, so their growth rate and reproduction is partially dependent on the temperature. Cold temperatures may induce **winter dormancy** or **hibernation** in insects. Generally only one stage hibernates (either eggs, larvae, pupae, or adults, depending on the species). Some insects go dormant before temperatures become unfavourable. This is called **diapause**, and is frequently triggered by short photoperiods in autumn.

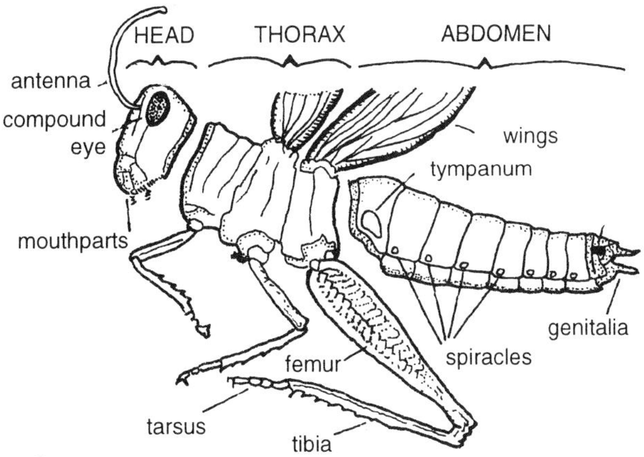

Figure 3.5:
Typical insect showing important parts (McPartland).

Insects feed with sucking or chewing mouthparts. Sucking insects insert syringe-like mouthparts into plants to suck sap. They mostly feed on phloem sap, although some also feed on xylem sap. Sap-suckers must absorb huge amounts of sap to obtain their required nitrogen (see Table 3.2); some suck fluids weighing 100–300 times their own body weight per day. Excess water and carbohydrates are excreted as sticky **honeydew**.

Insects with chewing mouthparts may selectively feed on the delicate leaf tissue between veins, leaving

Table 3.3: A synopsis of the insect orders associated with *Cannabis.*

Order	Examples	Characteristics
Collembola	springtails	simple metamorphosis, no wings, chewing mouthparts
Orthoptera	crickets, grasshoppers	incomplete metamorphosis, 4 wings, chewing parts
Dermaptera	earwigs	incomplete metamorphosis, 4 wings, chewing mouthparts
Isoptera	termites	incomplete metamorphosis, 4 wings, chewing mouthparts
Thysanoptera	thrips	incomplete metamorphosis, 4 wings, rasping-sucking mouthparts
Hemiptera	true bugs	incomplete metamorphosis, "half-wings" (part chitinous, part membranous), piercing-sucking mouthparts arising from front part of the head.
Homoptera	aphids, scales, whiteflies, leafhoppers	incomplete metamorphosis, wings either chitinous or membranous (but uniform), piercing-sucking mouthparts arising from posterior part of the head
Neuroptera	lacewings	incomplete metamorphosis, 4 wings, chewing mouthparts in larvae and adults (larvae carnivorous, eating other insects)
Lepidoptera	butterflies and moths	complete metamorphosis, 4 wings, chewing mouthparts in larvae, siphoning mouthparts in adults; larvae lack compound eyes
Coleoptera	beetles and weevils	complete metamorphosis, 4 wings (the front pair hardened into a sheath), chewing mouthparts in larvae and adults; larvae lack compound eyes
Hymemoptera	bees, wasps, ants, sawflies	complete metamorphosis, 4 wings, chewing or reduced mouthparts in larvae, and chewing-lapping in adults; larvae lack compound eyes
Diptera	flies	complete metamorphosis, 2 wings, chewing or reduced mouthparts in larvae, and sucking-sponging in adults; larvae lack compound eyes

behind a skeleton of leaf veins (**skeletonizers**). Other chewers are less selective and gnaw large holes in leaves or notch leaf edges. Chewing insects also bore into stems (**stem borers**), roots (**root borers**), or even into the narrow spaces inside leaves (**leafminers**). Seed eaters may puncture seeds and suck out the contents, or break open seeds and shell out the contents, or eat seeds whole.

Twenty-seven orders of insects are currently recognized by entomologists. About a third of them include plant pests. Table 3.3 outlines the orders involved in *Cannabis* ecology. Of course, many insects are our friends, like ladybeetles. Some insects are even domesticated, such as silk moths (*Bombyx mori*), which now exist only in captivity. More than 35 centuries of selective breeding have turned *B. mori* caterpillars into fat, sedentary, silk-spewing machines, and the adults have lost the ability to fly (Young 1997).

Cannabis is primarily wind-pollinated, so it need not attract symbiotic, pollen-collecting insects. Vavilov (1926) described a symbiotic exception—he found a red bug (*Pyrrhocoris apterus*, Fig 4.39) practising **zoochory**—the animal dispersal of plant seeds. *P. apterus* is attracted to a fat pad at the base of *Cannabis ruderalis* seeds, not the seeds themselves. In the process of fat pad feeding, the bug carries seeds "far distances" and facilitates the spread of *Cannabis*.

Herbivorous insects destroy as much *Cannabis* as bacteria, protozoans, and nematodes combined. Wounds caused by insects serve as portals for fungal infections. Some insects actually carry plant-pathogenic viruses, bacteria, and fungi from plant to plant. These pathogens often cause more damage than the insects themselves.

Plants repel insects by producing mechanical and chemical deterrents. *Cannabis* leaves are covered by **cystolith trichomes** (Fig 3.6). These microscopic hairs resemble pointed needles of glass. They are in fact heavily silicified, embedded with calcium carbonate crystals. Cystoliths impede the mobility of smaller herbivores, and damage the mouthparts of larger ones. Small insects actually impale themselves on cystolith spikes—they may remain impaled and die, or suffer a morbid series of impale-and-escape episodes (Levin 1973).

Cannabis is also covered by **glandular trichomes** (see Fig 3.2). Glandular trichomes secrete many chemicals, including terpenoids, ketones, and cannabinoids. Glandular trichomes may rupture and release their fluid contents when damaged by insects. The fluid oxidizes into a dark, gummy substance. This gummy substance accumulates on mouthparts and limbs of insects and eventually immobilizes them.

The contents of glandular trichomes may also *poison* insects. Glandular terpenoids produced by other plants are powerful insect poisons (e.g., ryania, azadirachtin, and pyrethroids). Two terpenoids produced by *Cannabis*—limonene and pinene—are almost as potent. Limonene is sold as an insecticide (Demize®). High-THC drug plants produce three to six times more limonene and pinene than most hemp varieties (Mediavilla & Steinemann 1997). The terpenoids humulene and caryophyllene also poison insects (Messer *et al.* 1990). Methyl ketones are synthesized by *Cannabis* (Turner *et al.* 1980) and repel leaf-eating insects (Kashyap *et al.* 1991). Cannabinoids, such as THC, possess insecticidal and repellent properties (Rothschild *et al.* 1977, Rothschild & Fairbairn 1980, McPartland 1997b).

Levin (1973) asked why these chemicals are secreted atop glandular trichomes rather than sequestered in the interior of the leaf. He hypothesized that trichomes serve as an "early warning system." Many insects have chemoreceptors located on their feet (their feet smell). These insects are repelled the moment their feet touch a chemical-laden trichome. The rest of the leaf avoids damage.

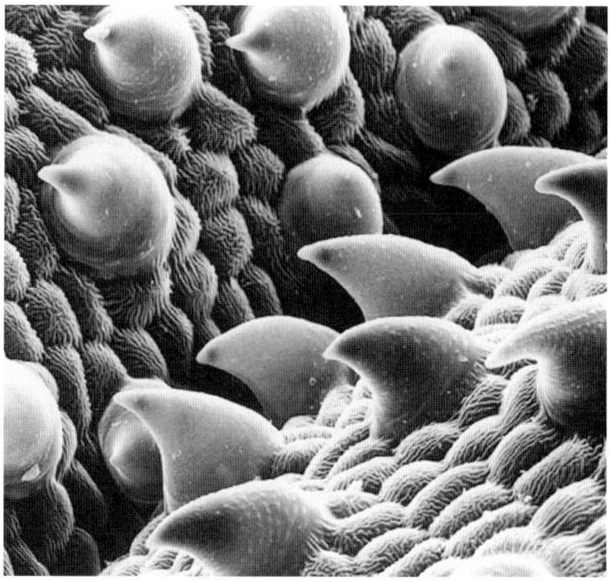

Figure 3.6: Cystolith trichomes on the surface of a *Cannabis* leaf, seen with a scanning electron microscope (SEMx2000, McPartland).

Volatile chemicals may repel pests from neighbouring plants. Riley (1885) noted that *Cannabis sativa* growing near cotton "exerted a protective influence" against cottonworms (*Alabama argillacea*, he called them *Aletia xylina*). Similarly, hemp grown around vegetable fields safeguarded the fields from attack by a cabbage caterpillar, *Pieris brassicae* (Beling 1932); potato fields were protected against the potato beetle, *Leptinotarsa decemlineata* (Stratii 1976); and wheat suffered less damage by the root maggot, *Delia coarctata* (Pakhomov & Potushanskii 1977).

Dried *Cannabis* leaves repel weevils from harvested grain (Riley & Howard 1892, MacIndoo & Stevers 1924, Khare *et al.* 1974). Scattering a 2 cm layer of leaves over piles of potatoes protected them from the tuber moth, *Phthorimaea operculella*, for up to 120 days (Kashyap *et al.* 1992). Prakash *et al.* (1987) mixed dried leaves into rice, 2% w/w, to control *Sitophilus oryza* weevils in the laboratory; but this dose failed to provide adequate protection in natural storage conditions (Prakash *et al.* 1982). Dried leaves kill ticks (Reznik & Imbs 1965), *Varroa* mites (Surina & Stolbov 1981), and drive off bedbugs when placed under mattresses (King 1854, Chopra *et al.* 1941). Sprays made from *Cannabis* leaves kill many insect pests (Bouquet 1950, Abrol & Chopra 1963, Reznik & Imbs 1965, Stratii 1976, Fenili & Pegazzano 1974, Bajpai & Sharma 1992, Jalees *et al.* 1993, Sharma *et al.* 1997). Juice squeezed from leaves or seeds has removed vermin from the scalp and ears (Pliny 1950 reprint, Culpepper 1814, Indian Hemp Drugs Commission 1894)

If *Cannabis* is such a good insecticide, how do *Cannabis*-eating insects survive? Sap-sucking insects, such as aphids and mites, use their long stylets to bypass chemicals secreted on the surface of plants. Leaf-chewing insects, however, cannot avoid cannabinoids and terpenoids. Perhaps they intersperse marijuana meals with less-toxic lunches on other plants. *Spilosoma obliqua* caterpillars, for instance, often eat *Cannabis* in India (Nair & Ponnappa 1974). But when Deshmukh *et al.* (1979) force-fed *S. obliqua* a pure *Cannabis* diet, the caterpillars died after 20 days.

Insects protect themselves from consumed toxins by rapid excretion, enzymatic detoxification, and sequestration. Sequestration involves the bioaccumulation of toxins into impervious tissues or glands, such as fat deposits or the exoskeleton. Monarch butterfly caterpillars are famous for feeding on milkweeds and sequestering poisonous pyrrolizidine alkaloids. The stored poisons are distasteful to birds and rodents and serve as predator deterrents.

Rothschild *et al.* (1977) studied sequestration of cannabinoids in caterpillars of the tiger moth, *Arctia caja*, and nymphs of the stink grasshopper, *Zonocerus elegans*. Caterpillars eating *Cannabis* stored a significant amount of THC and CBD in their exoskeleton. Rothschild estimated a single caterpillar containing 0.07 mg THC could elicit pharmacological activity in a predatory mouse. Grasshoppers stored little THC and no CBD in their exoskeletons, they excreted most cannabinoids in their frass.

Some insects metabolize poisons and convert them to other uses. The following examples are cited by Duffey (1980): Beetles blend plant oils (e.g., myrcene) with pheromones to aid the diffusion of their mating signals. Spider mites (*Tetranychus urticae*) convert farnesol to pheromones. Limonene, α-pinene, and eugenol accelerate the reproductive maturation of locusts. Aphids use *β*-farnesene as an alarm pheromone (Howse *et al.* 1998). *Cannabis* produces all these volatile chemicals (Turner *et al.* 1980).

Perhaps cannabinoids signal insects as behavioural pheromones. Some pheromones are heterocyclic structures (Howse *et al.* 1998). Rothschild *et al.* (1977) observed paradoxical behaviour in *Arctia caja* exposed to THC. Rothschild gave caterpillars a feeding choice of two *Cannabis* plants—one plant rich in THC, the other plant with little THC but with high levels of cannabidiol (CBD). The caterpillars showed a definite preference for high-THC plants, even though caterpillars feeding on high-THC plants grew stunted and died sooner than caterpillars feeding on high-CBD plants. Rothschild noted, "...should these compounds [cannabinoids] exert a fatal fascination for tiger caterpillars, it suggests another subtle system of insect control by plants."

Subsequently, Rothschild & Fairbairn (1980) studied egg-laying females of the large white butterfly (*Pieris brassicae*). Females of this species normally oviposit on cabbage leaves. Spraying cabbage leaves with extracts of THC-rich plants reduced egg laying by *P. brassicae*, compared to cabbage leaves sprayed with extracts of CBD-plants. Furthermore, spraying leaves with a 1% THC solution reduced egg laying by *P. brassicae*, compared to leaves sprayed with a 1% CBD solution. Both cannabinoids signalled females as oviposit deterrents.

Rothschild conducted her research before the discovery of cannabinoid receptors. Does THC exert behavioural changes in insects via cannabinoid receptors? Preliminary results say no, cannabinoid receptors (CB1) could not be detected in brain tissue of a locust, *Schistocerca gregaria* (Egertová *et al.* 1998).

The production of cannabinoids increases in plants under stress (Pate 1999). Other plants under stress increase their production of terpenoids. For instance, tobacco plants damaged by budworms (*Heliothis virescens*) release much more *β*-caryophyllene and slightly more *β*-farnesene than undamaged plants. Plants damaged by bollworms (*Helicoverpa zea*), produce slightly more *β*-caryophyllene and much more *β*-farnesenea—a reversed ratio. These different ratios can be distinguished by the parasitic wasp *Cardiochiles nigriceps*, which only attacks *H. virescens* (DeMoraes *et al.* 1998). Essentially, the tobacco plants utilize volatile terpenoids as "smoke signals" to alert the enemy of their enemy.

Terpenoids also have their bad side, from our perspective. Terpenoids attract certain pests (Howse *et al.* 1998), and they repel or harm some beneficial insects. See "Method 5" in Chapter 9 for more information.

Phylum Chordata

Chordates contain internal skeletons and dorsal nerve chords. Most of the 15 chordate classes ignore *Cannabis*, including Class **Reptilia** and Class **Amphibia**. Concerning Class **Osteichthyes** (fish), we have two reports: Fairbain (1976) described anglers using *Cannabis* seeds for fish bait. Anglers germinated the seeds and used them when root tips emerged from seedcoats. At this stage the seeds resemble water snails. Clarke (pers. commun. 1997) reported European fishermen mixing *Cannabis* seeds in dough balls as carp bait.

Members of only two chordate classes regularly interact with *Cannabis*. They act as herbivores, sometimes aiding plants by distributing seeds which pass *per anum:*

Class **Aves** includes four orders of *Cannabis* seed-eating birds: Galliformes (e.g., quail and pheasant), Columbiformes (doves), Pickformes (woodpeckers), and Passeriformes (linnets, starlings, grackles, sparrows, nuthatches, goldfinches, etc.). As a historical postscript, early reports from Kentucky described the now-extinct passenger pigeon (*Ectopistes migratorius*) feeding on hemp seed (Allen 1908). Seed from feral hemp is an important food for several midwestern American game birds. This has pitted wildlife agencies against police who eradicate feral hemp (Vance 1971).

Members of the class **Mammalia** are frequent *Cannabis* pests, including the order Rodentia (mice, moles, field voles, gophers, and groundhogs/woodchucks), order Lagomorpha (rabbits and hares), and order Artiodactyla (deer). Some humans (*Homo sapiens*) can be destructive. Cannabinoids are not very toxic to mammals. The oral LD_{50} of THC in mice is >21,600 mg kg^{-1} (Loewe 1946). Mixtures of cannabinoids are even less toxic than pure THC (Thompson *et al.* 1973). THC may harm ruminants (cows, sheep, deer) because of its antibacterial activity. Deer rumen microorganisms are inhibited by two terpenes, cineol and camphor (Nagy *et al.* 1964). These terpenes are also biosynthesized by *Cannabis* (Turner *et al.* 1980).

"A single man cannot help his time, he can only express its collapse."
—Søren Kierkegaard

Chapter 4: Insects and Mites

Mostafa & Messenger (1972) listed 272 species of insects and mites associated with *Cannabis*. Here we describe fewer organisms, about 150. This chapter represents a critical review of the existing literature—we believe many publications describing "*Cannabis*-insect associations" actually report insects and entomologists meeting accidently in hemp fields. Insects, after all, wander.

Mites and insects are presented here in their approximate order of economic impact. Common names are standards accepted by the Entomological Society of America (Bosik 1997). These standards may differ from common names used in Europe (Wood 1989) or Australia (Naumann 1993). Scientific (=Latin) names change occasionally; we have included synonyms or earlier names for the sake of continuity and reference to earlier literature.

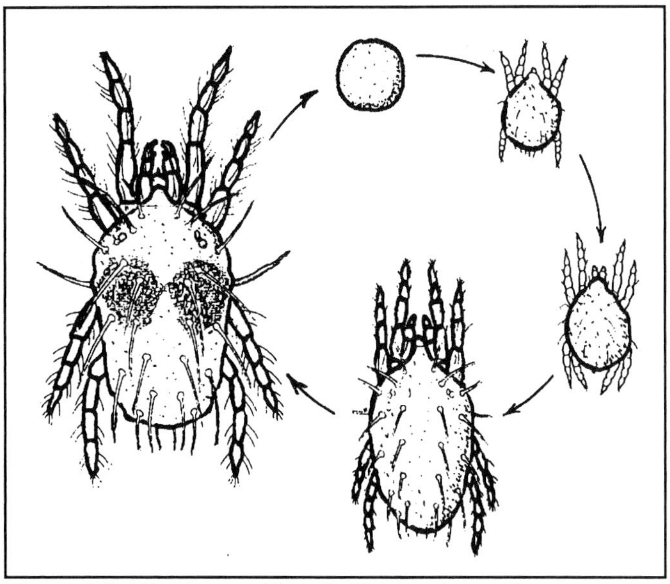

Figure 4.1: Life stages of female *Tetranychus urticae* (McPartland modified from Malais & Ravensberg 1992).

SPIDER MITES

These insidious arachnids are the most destructive pests of glasshouse and growroom *Cannabis*. Indoor areas are commonly contaminated by bringing in *Cannabis* clones from infested mother plants. Outdoor crops may also become infested in warm climates; Cherian (1932) reported 50% losses in field crops near Madras, India. Spider mites bite into leaves and suck up exuded sap. They usually congregate on the undersides of leaves, but in heavy infestations may be found on both sides of leaves.

SIGNS & SYMPTOMS

Damage is not initially evident. Each mite puncture produces a tiny, light-coloured leaf spot ("stipple"), which appears on both sides of the leaf. Stipples are grey-white to yellow. They begin the size of pinpricks, then enlarge (see Plate 1). Many stipples arise in lines parallel to leaf veins. Ultimately, whole leaves turn a parched yellow colour, droop as if wilting, then turn brown and die (Kirchner 1906). Inspecting the underside of leaf surfaces, particularly along main veins, reveals silvery webbing, eggs ("nits"), faecal deposits ("frass"), and the mites themselves. Leaves near the bottom of the plant are usually infested first.

Spider mites tend to infest crops in a patchy distribution, so early infestations may be missed. Symptoms are the worst during flowering, when whole plants dry up and become webbed together. Short days may induce diapause in spider mites, causing them to migrate and cluster together at tips of leaves and flowering tops. Clusters of diapausing mites may contain many thousands of individuals and grow quite large (Plate 2).

TAXONOMY

This pest may consist of one species (Ravensberg, pers. commun. 1998) or two species (Gill & Sanderson 1998). We see two species, separated by differences in morphology and ecology. The species are known by at least six common names—red spider mites, carmine spider mites, two-spotted spider mites, glasshouse spider mites, simple spider mites, and common spinning mite. To compound the confusion, about 60 *Latin* names exist for these two species, described from different hosts around the world. At least four of these taxonomic synonyms appear in the *Cannabis* literature.

1. TWO-SPOTTED SPIDER MITE

Tetranychus urticae Koch 1886, Acari; Tetranychidae.

=*Tetranychus bimaculatus* Harvey 1898, =*Tetranychus telarius* of various authors, =*Epitetranycus athaea* von Hanstein 1901

Description: Eggs are spherical and 0.14 mm in diametre. Initially translucent to white, eggs turn a straw colour just before hatching. Hatching larvae have six legs and two tiny red eye spots. Protonymphs become eight-legged and moult into deutonymphs. Deutonymphs moult into yellow-green coloured adults. As adults feed on chlorophyll-rich plants, two brown-black spots enlarge across their dorsum (Fig 4.1; Plate 3). Females mites average 0.4 to 0.5 mm in length. Males are slightly smaller, with a less-rounded posterior. Their dorsal spots may not be as evident. The male's knobbed aedeagi are at right angles to the neck and symmetrical. As winter approaches, two-spotted spider mites turn bright orange-red (Plate 2), making them difficult to distinguish from carmine spider mites or even the predatory mite *Phytoseiulus persimilis*.

Life History & Host Range

T. urticae overwinters as an adult and emerges in the spring. Females lay eggs on undersides of leaves or in small webs, one at a time. They lay as many as 200 eggs. Eggs hatch into larvae, which moult three times before they are capable of reproduction. Females arise in a 3:1 ratio to males. Under optimum conditions for development (30°C with low humidity), the life cycle of two-spotted spider mites repeats every eight days. Shortened photoperiods in autumn usually induce a reproductive diapause, where females stop feeding, turn orange-red, migrate into clusters, then hibernate under ground litter. The photoperiod that induces diapause will differ in mite populations from different latitudes. Warm temperatures inhibit diapause (Fig 4.2), and food availability may play a factor. Cool temperatures also induce hibernation in outdoor populations.

T. urticae is common in North American and European glasshouses. It attacks outdoor crops in temperate climates, and infests fruit trees as far north as Canada. Frank & Rosenthal (1978) incorrectly stated this species will not infest female flowers.

2. CARMINE SPIDER MITE

Tetranychus cinnabarinus (Boisdural) 1867, Acari; Tetranychidae.
=*Tetranychus telarius* (Linnaeus) 1758

Description: Eggs and immature stages of *T. cinnabarinus* closely resemble those of *T. urticae.* Adults, however, become plum red to brick red, with dark internal markings. But in cooler climates, adults turn green, and become difficult to distinguish from two-spotted spider mites. Adult male's aedeagi are not always symmetrical, having a rounded anterior side and a sharp posterior side.

Life History & Host Range

The life history of *T. cinnabarinus* is similar to that of *T. urticae,* especially in its fecundity. Its geographic range differs, because *T. cinnabarinus* prefers higher temperatures—35°C and above—so it thrives in semitropical areas. In cool climates the pest is limited to hot glasshouses. High humidity causes all stages (larvae, nymphs, adults) to stop feeding and enter a quiescent period. Hussey & Scopes (1985) called *T. cinnabarinus* "the hypertoxic mite" and considered it more dangerous than *T. urticae.* Cherian (1932) claimed the carmine spider mite is attracted to female flowering tops. He noted plants yielding the largest flowers were the most heavily infested.

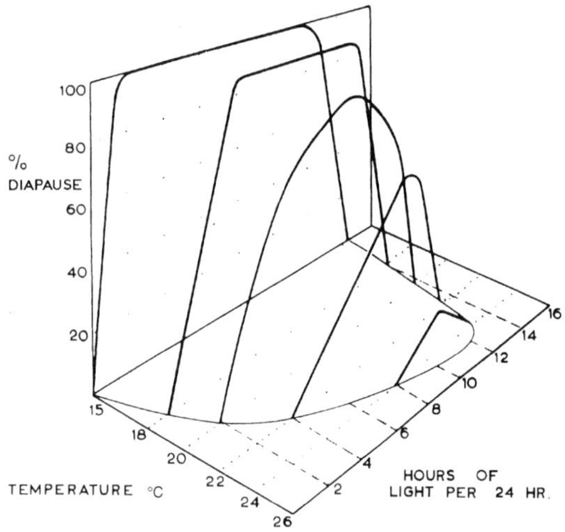

Figure 4.2: Diapause response curves in *Tetranychus urticae* from Holland—the effect of temperature on photoperiod (McPartland modified from Helle 1962).

DIFFERENTIAL DIAGNOSIS

Damage by other mites can be confused with spider mite injury. See the next section. Early aphid damage, sudden fungal wilts, and nutrient deficiencies may be confused with symptoms from spider mites. Late-season hemp borers and assorted budworms hide in webbing that is mite-like. Find the mites for a positive diagnosis.

CULTURAL & MECHANICAL CONTROL

(method numbers refer to Chapter 9)

Method 1 (sanitation) must be observed every growing season. Spider mites can be carried into growrooms on plants, people, and pets; they can even float on air currents. Glasshouses should be surrounded by a weed-free zone at least 3 m wide. Chickweed (*Stellaria* species) is an important weed host (Howard *et al.* 1994). For growers working with vegetative clones, infested mother plants are the most common source of new mite infestations. Once mites have infested your growroom or glasshouse, you will never get rid of them without removing everything from the space and disinfesting the place with steam heat or pesticides. Method 5 (genetic resistance) is a future goal—select plants that survive heavy mite infestations. Cherian (1932) compared six varieties of Indian ganja for resistance to *T. cinnabarinus;* the most resistant varieties produced the smallest female flowers. Regev & Cone (1975) compared five varieties of hops for resistance to *T. urticae;* the most resistant varieties produced the least amount of farnesol, a sesquiterpene alcohol that is also produced by *Cannabis* (Turner *et al.* 1980).

Table 4.1: Infestation Severity Index (ISI) for the two-spotted spider mite

Light	any mites seen often no symptoms
Moderate	<5 mites/leaf (not leaflet) feeding patches present
Heavy	>5 mites/leaf feeding patches coalescing
Critical	>25 mites/leaf shrivelled leaves and webbing

BIOLOGICAL CONTROL (see Chapter 10)

Biocontrol should be established before spider mite populations explode. If mite populations balloon, biocontrols never catch up. A mixture of three biocontrols, *Phytoseiulus persimilis, Neoseiulus californicus,* and *Mesoseiulus longipes,* provides excellent biocontrol for most glasshouses. These predatory mites are described below. For unique situations and outdoor crops, *Galendromus occidentalis, Galendromus pyri,* and *Neoseiulus fallacis* are also described below. Other predatory mites eat spider mites in the absence of their primary hosts (see *Neoseiulus cucumeris* and *Iphiseius degenerans,* described under thrips).

Spider mite destroyers (*Stethorus picipes,* described below) and midge maggots (*Feltiella acarisuga,* described below) are predatory *insects* that prefer eating spider mites. General predators include lacewings (*Chrysoperla carnea,* described under aphids), mirid bugs (*Macrolophus caliginosus,* under whiteflies; *Deraeocoris brevis,* under thrips), pirate bugs (*Orius* species, under thrips), lygaeid bugs (*Geocoris punctipes,* under whiteflies), and predatory thrips (*Aeolothrips intermedius,* under thrips).

Two fungi have been used to control spider mites. *Neozygites floridana* is being developed for commercial use, and *Hirsutella thompsonii* (Mycar®) was previously registered in the USA. These biocontrols require high humidity. They work well against mites in humid vegetative propagation chambers.

Biocontrol of mites must be achieved before flowering has begun. Gaining biological control after the photoperiod drops below 12 hours per day is nearly impossible (Watson, pers. commun.

1999). Diapausing spider mites cluster by the thousands; at that point, many biocontrols stop working (especially predatory mites).

Phytoseiulus persimilis "P-squared"

BIOLOGY: A predatory mite that feeds on *Tetranychus* spider mites, and is native to subtropical regions. Its eggs survive best in high humidity (70–95% RH). The adults work best at moderate temperatures (20–30°C) and >70% RH. Reproduction stops <60% RH (eggs stop hatching), and feeding stops <30% RH. A related species from Florida, ***Phytoseiulus macropilis***, is also becoming commercially available.

APPEARANCE: Adult mites are orange-red in colour, pear-shaped or droplet-shaped, 0.5–0.7 mm long, with long legs and no spots (Plate 4). Compared to the spider mite, *P. persimilis* is slightly larger, more elongate, and moves quicker. *P. persimilis* eggs are oblong and twice the size of spider mite eggs.

DEVELOPMENT: There are five stages—eggs, six-legged larvae, eight-legged protonymphs, deutonymphs, and adults. After the larval stage *P. persimilis* feeds continuously. The life cycle takes seven days in optimal conditions (Fig 4.3). Adults live another 30–40 days in the lab, but less in the field. Adults consume up to 24 immature spider mites or 30 eggs per day. Within a week of hatching, females start laying four or five eggs per day, for a total of 60 eggs per lifetime—they convert 70% of ingested food into eggs, and produce a daily egg biomass equal to their own body weight (Sabelis & Janssen 1994). *P. persimilis*, unlike some phytoseiid mites, does not enter diapause in the presence of cool temperatures and short day lengths.

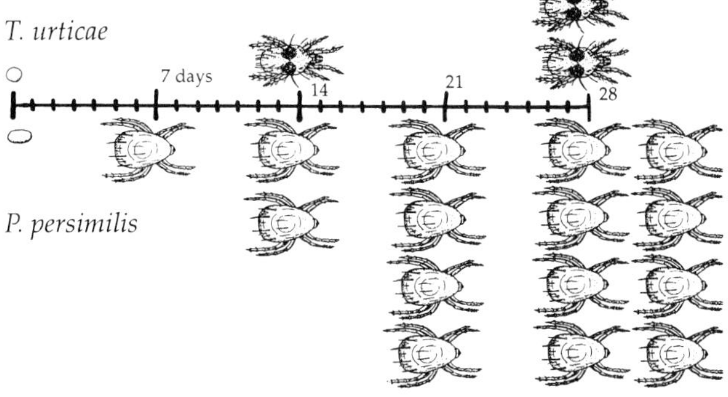

Figure 4.3: A Month'o'Mites. At 21°C, predatory mites (*Phytoseiulus persimilis*) reproduce every 7 days, twice as fast as spider mites (*Tetranychus urticae*), leading to a large buildup of predators (McPartland redrawn from Olkowski *et al.* 1991).

APPLICATION: Supplied as adults and nymphs mixed with inert materials (vermiculite, bran, corn cob grit, etc.) in pour-top bottles or tubes. Some are shipped on bean leaves in plastic tubs. As with all predatory mites, *P. persimilis* should be applied immediately; store only if necessary. Store for a maximum of four days in a cool (8–10°C), dark place. *P. persimilis* can be sprinkled into distribution boxes hanging on plants (Plate 85), or sprinkled directly on plants, the closer to spider mites the better. Recommended release rates are presented in Table 4.2. Increase rates when plants are taller than 1 m, planted more densely than six per m², or when humidity is low (<45% RH). Encourage *P. persimilis* by misting plants with water during periods of low humidity. Misting also discourages spider mites. Of course, flowering plants cannot be misted, because they may mould.

NOTES: *P. persimilis* is the most popular mail-order mite. It does best on low-growing bushy plants, where plants touch each other in a closed canopy (Helle & Sabelis 1985, vol 1B). *P. persimilis* remains active in high humidity, works in a wide range of temperatures, and reproduces fast. But the predators avoid areas of high temperature, such as flowering tops basking in bright light. The effectiveness of *P. persimilis* decreases when *Cannabis* begins flowering, because the predators find it difficult to move across sticky flower resins. Furthermore, their need for high humidity is not compatible with mould-susceptible *Cannabis* flowers.

P. persimilis can annihilate a pest population. But the voracious predators subsequently die out themselves, because they have no alternative food source. Thus, *P. persimilis* is often combined with *Neoseiulus californicus* and *Mesoseiulus longipes*, because these mites live longer without food. *P. persimilis* is also compatible with Bt, *Feltiella acarisuga*, and most parasitic wasps. *P. persimilis* can coexist with *Neoseiulus cucumeris* but the two biocontrols prey upon each other in the absence of their primary hosts (Hussey & Scopes 1985). *P. persimilis* has been mixed with lacewings (*Chrysoperla* species), but lacewings may eat predatory mites. Avoid most insecticides, miticides and even fungicides while utilizing *P. persimilis* (Table 10.1). According to van Lenteren & Woets (1988), some strains of *P. persimilis* tolerate some miticides (sulphur and fenbutatin oxide) and some insecticides (diazinon and malathion). Benomyl sterilizes predatory mites so they stop reproducing. Allow previously applied pesticides to break down for two or three weeks before introducing predators.

Neoseiulus (Amblyseius) californicus

BIOLOGY: A predatory mite that eats spider mites, broad mites, and pollen, is native to southern California and Florida, and does best in moderate humidity (≥60% RH) and moderate to high temperatures (18–35°C).

APPEARANCE & DEVELOPMENT: Adult mites are droplet-shaped and translucent to beige in colour, depending on what they eat. The life cycle may take only six days in

Table 4.2: *P. persimilis* release rate for control of two-spotted spider mites.

ISI*	NUMBER OF PREDATORS RELEASED PER M² OF GLASSHOUSE CROP**
Preventative	5 m⁻² every 3 weeks
Light	10 m⁻² initial release, then 5 m⁻² every 3 weeks
Moderate	25 m⁻² on trouble spots, 10 m⁻² elsewhere, then 5 m⁻² every 3 weeks
Heavy	200 m⁻² on trouble spots, 25 m⁻² elsewhere, then 10 m⁻² every 3 weeks
Critical	mechanical & chemical control of trouble spots, 25 m⁻² elsewhere, then 10 m⁻² every 3 weeks

* ISI = Infestation Severity Index of two-spotted spider mite, see Table 4.1.

**effectiveness of *P. persimilis* decreases during flowering.

optimal conditions. Females lay three eggs per day for up to 14 days. Adults eat five to 15 adult spider mites per day, along with some eggs and larvae. New strains of *N. californicus* do *not* diapause in short days (Ravensberg, pers. commun. 1999).

APPLICATION: Supplied as adults and nymphs mixed with vermiculite or sawdust in pour-top bottles. Store one to four days in a cool (8–10°C), dark place. *N. californicus* migrates fairly well and has been used outdoors. *N. californicus* tolerates lower humidity and lives longer without prey than *P. persimilis*, but *N. californicus* feeds slower than *P. persimilis* and its populations build slower. These attributes make *N. californicus* an excellent preventative, and fine for treating light infestations (same release rates as *P. persimilis*). But for heavy or critical infestations, switch to *P. persimilis*.

NOTES: *N. californicus* is compatible with *P. persimilis* and all biocontrols compatible with *P. persimilis*. *N. californicus* tolerates more insecticides than *P. persimilis*, but is still susceptible to most pesticides (even benomyl, a fungicide).

Mesoseiulus (Phytoseiulus) longipes

BIOLOGY: A predatory mite that feeds on many mites, is native to South Africa, and tolerates conditions that are warm (up to 38°C) and dry (as low as 40% RH at 21°C).

APPEARANCE & DEVELOPMENT: Adults resemble *P. persimilis*, as do their eggs. The life cycle can turn in seven days at optimal temperatures. Female mites lay three or four eggs per day (Sabelis & Janssen 1994), and live up to 34 days. Short photoperiods may send *M. longipes* into diapause.

APPLICATION: Supplied as adults in pour-top bottles. Adults should be used immediately upon delivery but can be stored in a cool (6–10°C), dark place for a couple of days. Release at the same rate as *P. persimilis*.

NOTES: *M. longipes* thrives in temperatures that wilt *P. persimilis* or even *N. californicus*, and it tolerates lower humidities. It migrates better than many mites, making it suitable for taller plants. *M. longipes* reproduces slower than *P. persimilis*. *M. longipes* is compatible with the same biocontrols as *P. persimilis*. Avoid insecticides, though some strains tolerate malathion.

Neoseiulus (Amblyseius) fallacis

BIOLOGY: A predatory mite that feeds on spider mites, red mites, and russet mites, and survives on pollen in the absence of prey. It is native to the Northwestern USA, and does best in moderate to high humidity (60–90% RH) and moderate temperatures (optimally 10–27°C, but up to 38°C in high humidity).

APPEARANCE: Adults are pear-shaped and white until they feed, when they take on the colour of their prey, usually pale red or brown. They are slightly smaller than other predators, such as *P. persimilis*.

DEVELOPMENT: Adults overwinter in crevices of tree bark, but not as successfully as *Galendromus pyri* (described below). The life cycle takes ten days in optimal conditions. Adults live another 20–60 days (four to six generations arise per year). Adult females eat about 15 mites per day, and lay a total of 40–60 eggs. Short days and cold temperatures interact to send *N. fallacis* into diapause. Diapause is delayed by warmer temperatures, and completely averted above 27°C (Helle & Sabelis 1985, vol 1B).

APPLICATION: Supplied as adults in bottles, or a mix of adults and nymphs on bean leaves in plastic tubs. Store a maximum of two or three days in a cool (8–10°C), dark place. Used preventively, release ten mites per m^2, every two weeks, two or three times. For light to moderate infestations, release 20–40 mites per m^2 weekly until controlled.

NOTES: Strong & Croft (1996) used *N. fallacis* to control *T. urticae* infesting hops in Oregon, released at rates of 125,000 per ha (50,000 per acre). *N. fallacis* migrates vigorously across plants of all sizes, and uses aerial dispersal to blow from plant to plant. It is probably compatible with other *Neoseiulus* species. *N. fallacis* tolerates more pesticides than any other predatory mite. Some strains are compatible with neem, sulphur, abamectin, and pyrethrin (and the synthetic pesticides malathion, dicofol, and propargite).

Galendromus (Metaseiulus) occidentalis

BIOLOGY: A predatory mite that eats spider mites and some russet mites, and is native to western North America. It tolerates a range of humidities (40–80% RH) and temperatures (26–35°C). *G. occidentalis* has been used in Washington State apple orchards since 1962.

APPEARANCE & DEVELOPMENT: *G. occidentalis* resembles *N. fallacis*. The life cycle takes seven to 14 days depending on temperature. Adults eat one to three spider mites per day or six mite eggs per day. Many strains of *G. occidentalis* go into diapause with short days and cold temperatures (Helle & Sabelis 1985, vol 1B), so their effectiveness ends when *Cannabis* begins flowering.

APPLICATION: Supplied as adults mixed in pour-top bottles. In the Pacific Northwest, *G. occidentalis* can also be obtained from nearby orchards by collecting apple leaves. Store a maximum of five days in a cool (8–10°C), dark place. On field crops, release 12,500 per ha (5000 per acre) at the first sign of pests, repeat every two weeks. Release 50 per m^2 every two weeks during indoor infestations.

NOTES: *G. occidentalis* eats less than *P. persimilis* and reproduces slower, but lives longer without food, disperses more rapidly (including aerial dispersal), and tolerates semiarid conditions (it does *not* tolerate high humidity). *G. occidentalis* does well on both short and tall plants, and has been used outdoors. It is compatible with Bt and *Galendromus pyri* in apple orchards and hops yards. A new strain of *G. occidentalis* is photoperiod neutral (nondiapausing), so it can be used while plants are flowering. *G. occidentalis* tolerates pesticides better than most predator mites; some strains tolerate sulphur, pyrethroids, abamectin, and carbaryl. Release mites at least a week after spraying. *G. occidentalis* is a small predator, so some *T. urticae* adults may be too big for *G. occidentalis* to handle.

Galendromus (Typhlodromus) pyri

BIOLOGY: A generalist predatory mite that feeds on spider mites. It occurs naturally in temperate apple orchards, and does best in cooler, humid climates.

APPEARANCE & DEVELOPMENT: Adults are indistinguishable from the aforementioned species. *G. pyri* reproduces slowly, taking nearly two weeks to mature, and lays only one egg per day (Sabelis & Janssen 1994). Short days send it into diapause.

APPLICATION: Supplied as adults in bottles or tubes. Store up to five days in a cool (8–10°C), dark place. Release the same rate as *G. occidentalis*.

NOTES: *G. pyri* overwinters better than *G. occidentalis* and other predatory mites (Helle & Sabelis 1985, vol 1B). Its evasive behaviours also enhance outdoor survival. *G. pyri* does best on medium-sized to tall plants. It is compatible with Bt and *Galendromus occidentalis* in apple orchards. A New Zealand strain is resistant to pyrethroids.

***Stethorus* species** "spider mite destroyers"

BIOLOGY: Three species of tiny ladybeetles (*S. punctillum, S. punctum, S. picipes*) prey on spider mites and their eggs, as well as red mites, aphids, whiteflies, and other soft-bodied insects. All three species are temperate Americans and do best in moderate humidity (60% RH) and temperatures (26–35°C). Other *Stethorus* species are found in tropical regions.

APPEARANCE: Adults are oval, black, shiny, pubescent, 1.5–3 mm long, with brown or yellow legs (Fig 4.4). Larvae are cylindrical, dark grey to black (turning reddish before pupation), covered with fine pale-yellow spines, 1–2 mm long. Eggs are oval, <0.5 mm long, initially white but turning grey before hatching, usually laid singly on undersides of leaves near veins.

DEVELOPMENT: Adults overwinter in leaf litter on the ground, and emerge when apple trees are blooming. They lay eggs from May to mid-August. Females lay up to 750 eggs in the presence of adequate prey. Larvae feed on eggs and young spider mites before pupating in foliage. The life cycle takes 20–40 days. Adults live another 30–40 days. Adults consume all stages of mites, up to 75–100 per day. One to three generations arise annually in temperate regions. Short days may send *S. picipes* into diapause (Helle & Sabelis 1985, vol. 1B).

APPLICATION: Supplied as adults in pour-top bottles. Store up to a week in a cool (8–10°C), dark place. Used preventively, release one or two beetles per m^2 every month. For light to moderate infestations, release three or four beetles per m^2 every two weeks. For heavy infestations, double the release rate.

NOTES: *Stethorus* beetles are good fliers and find infestations easily. They feed greedily but migrate before their job is finished. To reduce migration, use strategies described under *Hippodamia convergens* (in the aphid section). Dense mite webbing impedes newly-hatched *Stethorus* larvae (Helle & Sabelis 1985, vol. 1B). *Stethorus* species may eat predatory mites (Helle & Sabelis 1985, vol. 1B), or they may not (Cherim 1998). Adults tolerate organophosphates and juvenile growth hormones (e.g., fenoxycarb, teflubenzuron), but larvae are seriously impacted by insecticides (Biddinger & Hull 1995).

Feltiella acarisuga (Therodiplosis persicae)

BIOLOGY: A cecidomyiid gall midge similar to *Aphidoletes aphidimyza*, the aphid predator. *F. acarisuga* feeds on all spider mites, including *T. urticae* and *T. cinnabarinus*. Adult midges do best in >60% RH and 20–27°C, but the predatory larvae tolerate a wider range of conditions. The species is called *T. persicae* by some suppliers, but the correct name is *F. acarisuga* (Gagné 1995). A related species, ***Feltiella occidentalis***, is also being developed.

APPEARANCE & DEVELOPMENT: Adults look like tiny pink-brown mosquitos, with a wingspan of 1.6–3.2 mm. Adults only live three or four days, long enough for females to lay 30 eggs on plants near spider mite colonies. Eggs are tiny, oblong, and shiny-yellow. Hatching larvae are slug-like and grow to 1.7–2.0 mm long. They are yellow, cream-coloured, orange, or red; their colour depends on body contents. Larvae sink their mandibles into all stages of spider mites, from eggs to adults. They feed on 30 nymphs and adults per day or as many as 80 mite eggs per day (Gagné 1995). In one week larvae stop eating, they spin white cocoons, and pupate on undersides of leaves. The entire life cycle takes two to four weeks.

APPLICATION: Supplied as pupae on leaves or on paper. Store a maximum of one or two days in a cool (10–15°C), dark place. Sprinkle pupae on soil or in open containers attached to plants; protect against direct sunlight. Release rates for preventative control range from one pupa per m^2 to 1000 pupae per ha. For moderate infestations apply five pupae per m^2 per week or 7500 pupae per ha per week.

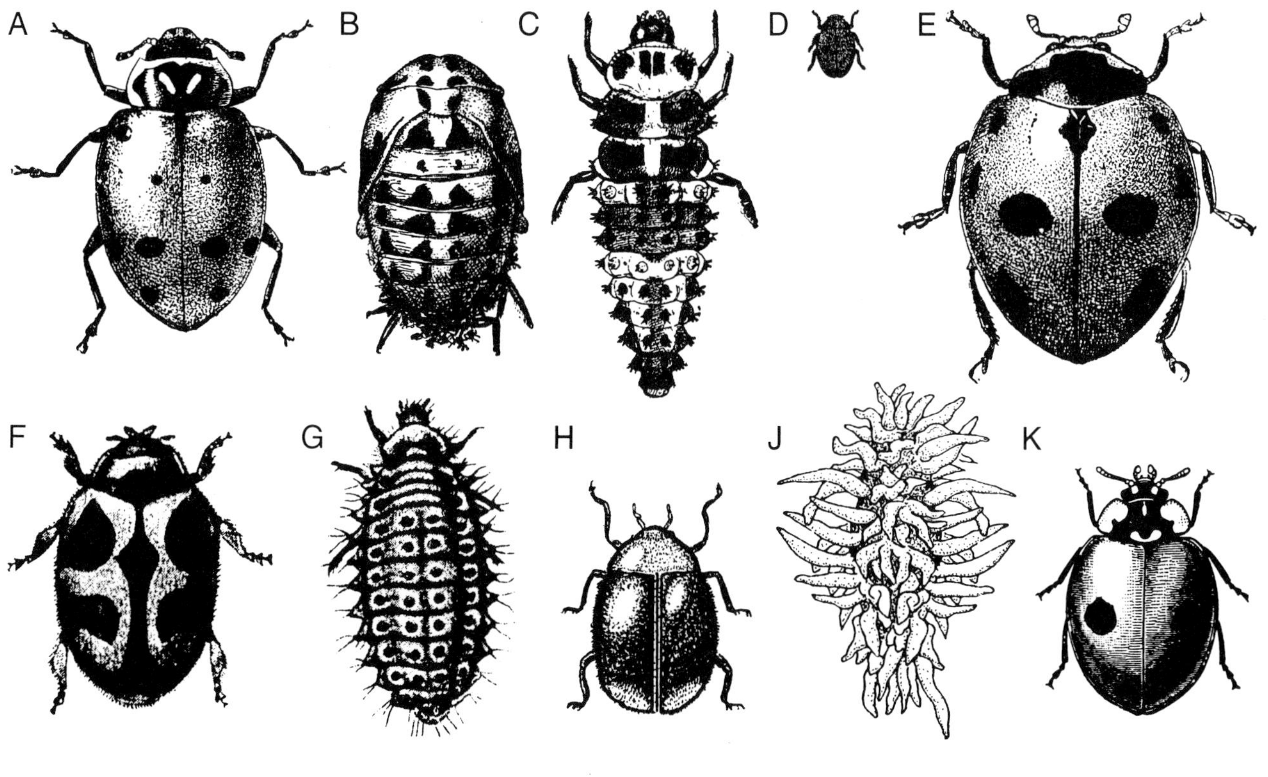

Figure 4.4: Assorted beneficial ladybug beetles (all x6). A,B,C. Adult, pupa, larva of convergent lady beetle (*Hippodamia convergens*); D. Spider mite destroyer (*Stethorus picipes*); E. Seven-spotted ladybeetle (*Coccinella septempunctata*); F & G. Adult & larva of vedalia (*Rodolia cardinalis*); H & J. Adult & larva of mealybug destroyer (*Cryptolaemus montrouzieri*); K. Two-spotted lady beetle (*Adalia bipunctata*) (Courtesy USDA except H & J, McPartland).

NOTES: Some strains of this "flying predator" do not diapause, a welcome advantage in flowering crops. The larvae move well across heavily-trichomed leaf surfaces. *F. acarisuga* larvae eat more spider mites than *P. persimilis*, but reproduce slower. The two biocontrols are compatible. *F. acarisuga* tolerates sulphur pesticides. In Europe *F. acarisuga* can be parasitized by *Aphanogmus* wasps, which hamper its effectiveness (Gagné 1995).

BIORATIONAL CHEMICAL CONTROL (see Chapter 11)

Use chemicals for "spot treatment" of heavy mite infestations. Heavy infestations arise on stressed plants, on plants located near glasshouse openings, and on plants growing along the windward edge of fields. Direct all sprays at the undersides of leaves. Since mite eggs are not harmed by many chemicals, repeat the treatment a week later. After every harvest in glasshouses or growrooms, spray all walls, floors, ceilings, and equipment (pots, tools, etc.). Of course, strong pesticides must subsequently be washed off, to prevent residues from harming biocontrols in the next crop. For growers working with clones, chemicals are recommended thrice: on mother plants before clones are cut, on clones several days after transplanting, and again on clones the day before flowering is induced.

Horticulture oil, Safer's Soap, clay microparticles, and even tap water are mildly effective, and described in Chapter 11. Cherian (1932) killed spider mites with either a lime-sulphur spray or fish oil soap. "Ganja" sprayed with lime-sulphur "...was tested by veteran smokers who gave their verdict against it," whereas ganja sprayed with soap passed the smoker's test. Pure sulphur (not lime-sulphur) may work, but mite populations started developing resistance to sulphur 90 years ago (Parker 1913b). Rosenthal (1998) sprayed mites with sodium hypochlorite (household bleach diluted to 5%). Bush Doctor (1986b) sprayed mites with a flour slurry, strained through cheesecloth. Parker (1913b) used a flour slurry against *T. urticae* in hops, mixing 3.6 kg flour in 378 l water (8 lbs/100 gallons). Flour solutions require frequent agitation to keep sprayers from clogging.

Turning to botanicals, Parker (1913b) killed 99% of two-spotted spider mites with nicotine sulphate, mixing 190 ml of 40% concentrate nicotine sulphate in 378 l water (6.5 oz/100 gallons). *T. cinnabarinus* is highly susceptible to neem seed extracts, up to 58 times more susceptible than its predator, *P. persimilis* (Mordue & Blackwell 1993).

Cinnamaldehyde, extracted from cinnamon (*Cinnamonum zeylanicum*), kills all stages of spider mites, including eggs. But it also kills beneficial mites and insects. Ironically, European researchers have controlled spider mites with leaf extracts of *Cannabis sativa* (Fenili & Pegazzano 1974). Pyrethrum works (Frank 1988), but synthetic *pyrethroids* (such as permethrin) rarely kill mites and actually induce egg laying. Imidacloprid (a nicotine derivative) and abamectin (a fermentation product) kill spider mites.

Synthetic insect growth hormones, such as flucyclozuron and methoprene, kill immature spider mites. Hexythiazox is a growth hormone that selectively kills mites. A synthetic pheromone, trimethyl docecatriene (StirrupM®), attracts spider mites. It markedly enhances the effectiveness of miticides. Kac (1976) tested 25 synthetic and systemic pesticides for controlling mites in Slovenian hemp (see the appendix).

HEMP RUSSET MITE

Hemp russet mites have infested hemp in central Europe (Farkas 1965) and feral hemp in Kansas (Hartowicz *et al.* 1971). At Indiana University, the pest thrives in glasshouses and feeds on all kinds of *Cannabis*, including European fibre cultivars, southeast Asian drug landraces, Afghan landraces, and ruderals (Hillig, unpublished data 1994). The mite population at I.U. was possibly imported on seeds from Nepal or northern India.

SIGNS & SYMPTOMS

A. cannabicola feeds primarily on petioles and leaflets. Leaflets curl at the edges, followed by chlorosis and necrosis. Petioles become brittle and leaflets break off easily. In bad infestations the mites crowd plants by the thousands, giving leaflets a beige appearance. The mites may also infest flowering tops; they selectively feed on pistils, rendering female flowers sterile.

TAXONOMY & DESCRIPTION

Aculops (Vasates) cannabicola (Farkas) 1965, Acari; Eriophyidae.

Description: Eriophyid mites are soft-bodied, sausage-shaped, and exceptionally tiny, with only two pairs of legs (Fig 4.5 & Plate 5). Their pale beige bodies are composed of two sections: the **gnathosoma** (mouthparts) and **idiosoma** (rest of the body). The legs project from around the gnathosoma, near the front. The idiosoma is covered with many minute transverse ridges and tiny spines. *A. cannabicola* females reach 200 μm (0.2 mm) long and 45 μm wide—less than half the size of *Tetranychus urticae.*

Life History

We know little about *A. cannabicola's* life cycle. Outdoor populations probably overwinter in contaminated seed. Indoor populations remain on plants year round. The mites move towards the top of dying plants, where they spread to other plants by wind or splashing water. A turn of the life cycle takes about 30 days under optimum conditions of 27°C and 70% RH. Related eriophyid mites lay ten to 50 eggs during a life-span of 20–40 days.

A related pest, *Aculops lycopersici* (Tyron), the tomato russet mite, vectors viruses. *A. lycopersici* goes into a feeding frenzy called "solanum stimulation," where it feeds until it kills its own host (Lindquist *et al.* 1996).

DIFFERENTIAL DIAGNOSIS

The hemp russet mite's morphology, colouring, and lack of webbing make it easy to discern from spider mites.

CULTURAL & MECHANICAL CONTROL (see Chapter 9)

Use clean seed to keep mites out—see method 11. Once mites infest an area, a grower's best efforts only decrease populations, not eliminate them. Method 5 (genetic resistance) is a future goal.

BIOCONTROL (see Chapter 10)

No effective biocontrol of *A. cannabicola* is known. *Phytoseiulus persimilis* does not feed on *A. cannabicola* (Hillig, unpublished data 1994). Other russet mites are controlled by ***Zetzellia mali*** (a tiny predator mite not commercially available). Some *Aculops* mites are susceptible to *Verticillium lecanii* (Olkowski *et al.* 1991), a biocontrol fungus described under whiteflies. Two other possibilities are described below:

Homeopronematus anconai

BIOLOGY: Royalty & Perring (1987) controlled *Aculops lycopersici* with this predatory tydeid mite. Royalty & Perring found *H. anconai* resistant to abamectin, so the predator and pesticide could be used together. *H. anconai* may prey on *Phytoseiulus persimilis*, so their compatibility is questionable.

Hirsutella thompsonii

BIOLOGY: *H. thompsonii* infests several eriophyid and

tetranychid mites. The fungus was released in Cuba to control an *Aculops* species, and sprayed in Florida citrus groves against other eriophyid mites (Lindquist *et al.* 1996). *H. thompsonii* does best in hot, humid conditions.

APPEARANCE & DEVELOPMENT: Conidia quickly germinate, directly infect mites, and kill them in one or two weeks. Fungal hyphae emerge from dead mites. The hyphae bear solitary phialidic conidiophores and slimy, round-to-lemon-shaped conidia. Under optimal conditions, mite cadavers sprout long, white, hairlike synnemata, which bear conidiophores and conidia.

APPLICATION: *H. thompsonii* is mass-produced in liquid media fermenters, and supplied as conidia (10^9 conidia per g). Release rate for citrus and turf is 2.2–4.5 kg ha^{-1} (2–4 lbs/acre) in optimal weather conditions (Lindquist *et al.* 1996). This fungus was previously sold in the USA (as Mycar®), but Abbott Labs cancelled its registration in 1988.

CHEMICAL CONTROL (see Chapter 11)

Spraying oil or sulphur works against some russet mites, but not *A. cannabicola* (Hillig, unpublished data 1994). Cinnamaldehyde, a new product extracted from cinnamon (*Cinnamonum zeylanicum*), may be effective against russet mites. Hillig used abamectin (avermectin B1). Lindquist *et al.* (1996) used abamectin against *Aculops lycopersici;* abamectin killed more russet mites than dicofol, yet spared the biocontrol mite *Homeopronematus anconai.* Hexythiazox is a growth hormone that selectively kills immature mites. Hillig (unpublished data 1994) killed *A. cannabicola* by enclosing potted plants in large plastic bags and filling the bags with CO_2 for two hours. Spider mites and thrips can also be killed with 100% CO_2, but require at least 12 hours of exposure. Take care while using CO_2 in growrooms and other sealed spaces.

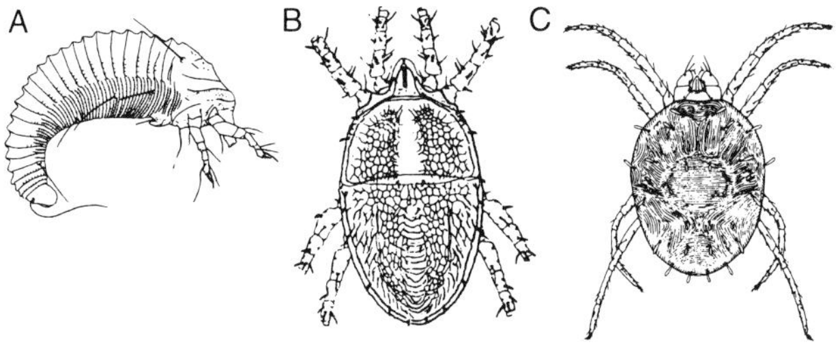

Figure 4.5: Other mites. A. *Aculops cannabicola* (from Farkas 1965); B. *Eutetranychus orientalis* (from Hill 1994), *C. Brevipalpus obovatus* (drawings not to scale).

OTHER MITES

Researchers cite other mites attacking *Cannabis:* the oriental mite, two species of privet mites, and the ta ma mite. Additionally, many species of mites parasitize hemp seeds in storage (see Chapter 8).

1. ORIENTAL MITE

Eutetranychus orientalis (McGregor) 1914, Acari; Tetranychidae.
=*Eutetranychus orientalis* (Klein) 1936

Description: Eggs are subspherical, pale brown, 0.14 mm diametre. Larvae are light brown and six-legged. Nymphs emerge from moults with eight legs, like adults. Female *E. orientalis* adults are round-bodied, greenish-brown, up to 0.5 mm long, and covered with short spines (Fig 4.5). Males are smaller and more red than females.

Life History & Host Range

Females lay eggs along main veins on *upper* leaf surfaces. Upper leaf surfaces of infested plants turn yellow, then red-brown, then die. Feeding damage peaks in April–June and September–November. The life cycle can be as short as two weeks. *E. orientalis* normally attacks *Citrus* species but infests a wide variety of hosts, including *Cannabis* in India (Dhooria 1983, Gupta SK 1985). Oriental mites live in northern Africa, the middle East, India, and southeast China.

2. PRIVET MITES

a. *Brevipalpus obovatus* Donnadieu 1875, Acari; Tenuipalpidae.
b. *Brevipalpus rugulosus* Chaudri, Akbar & Rasool 1974

Description: Eggs are bright red, 0.1 mm long. Larvae and nymphs are bright red. Adults are oval in outline, average 0.3 mm long and vary from light orange to dark red in colour (Fig 4.5).

Life History & Host Range

B. obovatus is a polyphagous pest found in temperate zones around the world. *B. rugulosus* is restricted to the Indian subcontinent. Both species attacked *Cannabis* in India (Gupta SK 1985). Their toxic saliva causes severe leaf spotting, on upper and lower leaf surfaces. Female privet mites lay eggs on undersides of leaves, one at a time. They overwinter as adults and their life cycle takes about 60 days.

3. TA MA MITE

Typhlodromus cannabis Ke & Xin 1983, Acari; Phytoseiidae.

Ke & Xin (1983) found this new species on Chinese hemp. The mite also infested horsetail rush (*Pteridium aquilinum*), clematis (*Clematis* species) and *Populus* species. No plant symptoms or control measures were described.

DIFFERENTIAL DIAGNOSIS

These other mites can be confused with spider mites, described earlier.

BIOLOGICAL & CHEMICAL CONTROL

Damage is greatest in drought-stressed plants (Hill 1983). For oriental mites, Dhooria (1983) cited two biocontrol organisms—a predator mite (*Amblyseius alstoniae* Gupta) and a thrips (*Scolothrips indicus* Priesn.). *Neoseiulus californicus* feeds on *E. orientalis* but does better against mites that produce heavier webbing, such as spider mites (described there).

According to Ravensberg (pers. commun. 1998), privet mites can be controlled by *Neoseiulus californicus* and *Neoseiulus fallacis* (described under spider mites), and *Neoseiulus cucumeris* and *Neoseiulus barkeri* (described under thrips). Privet mites are susceptible to sulphur but resistant to organophosphate and carbamate pesticides (Hill 1983). Try soap or horticultural oil.

APHIDS

Aphids, or "plant lice," are tiny, soft-bodied, and pear-shaped. They have relatively long legs and antennae. Some species have **antennal tubercles**, important landmarks in pest identification. Antennal tubercles are paired "bumps on the head," located between antennae (see Fig 4.8). Protrusions from the rear of aphids also serve in pest identification: A pair of tubelike **cornicles** (or **siphons**) project backwards, they resemble dual tailpipes. Between the cornicles, the rear end tapers to a pointed, tail-like **caudum**.

Most adult aphids do not have wings (these adults are called **apterae**), but some do (**alatae**). The wings of alatae are much longer than their bodies.

Aphids suck sap from a plant's vascular system, using long narrow **stylets**. Most aphids are phloem-feeders, but some also suck on xylem (Hill 1994). Besides sucking sap, aphids damage plants by vectoring fungi, bacteria, and especially viruses. Viruses and aphids have a symbiotic relationship (Kennedy *et al.* 1959).

At least six aphid species attack *Cannabis.* Aphid damage increases in warm, moist weather, with gentle rain and little wind. Damage decreases in hot, dry weather and in the presence of strong, dry winds (Parker 1913a).

SIGNS & SYMPTOMS

Aphids congregate on the undersides of leaflets and cause yellowing and wilting. Some aphid species prefer older, lower leaves (e.g., *Myzus persicae*), and some prefer younger, upper leaves (e.g., *Aphis fabae*). Some species even infest flowering tops (e.g., *Phorodon humuli, Phorodon cannabis*). Early damage is hard to detect—the undersides of leaflets develop light-coloured spots, especially near leaflet veins (look closely at Plate 8). Eventually, leaves and flowers become puckered and distorted (Kirchner 1906). Heavily infested plants may completely wilt and die. Surviving plants remain stunted.

Honeydew exudes from the anus of feeding aphids. On warm afternoons honeydew may be seen falling as a mist from severely infested plants (Parker 1913a). Honeydew causes secondary problems—ants eat it, and sooty mould grows on it. For the love of honeydew, ants become "pugnacious bodyguards" of aphids, and attack aphid predators. Ants must be eliminated for biocontrol to be effective. Sooty mould reduces plant photosynthesis and leaf transpiration, and hinders the movement of aphid predators and parasites.

1. GREEN PEACH APHID

Myzus persicae (Sulzer) 1776, Homoptera; Aphididae.
=*Phorodon persicae* Sulzer

Description: *Apterae* are green (sometimes yellow-green or pink), oval in outline, averaging 2.0–3.4 mm in length (Plate 8). Antennal tubercles are mammary-like bumps (less than half as long as the first antennal segment), and point inward (converge towards each other). Cornicles are the same colour as the abdomen, except for darkened tips. Cornicles are long, thin, slightly swollen near the midpoint, and grow twice as long as cauda. Cauda are lightly bristled and constricted slightly at the midpoint. *Alatae* nymphs are often pink or red, and develop wing-pads. Winged adults have black-brown heads, a black spot in the middle of their abdomens (Fig 4.6 & Plate 6), and hold their wings in a vertical plane when at rest.

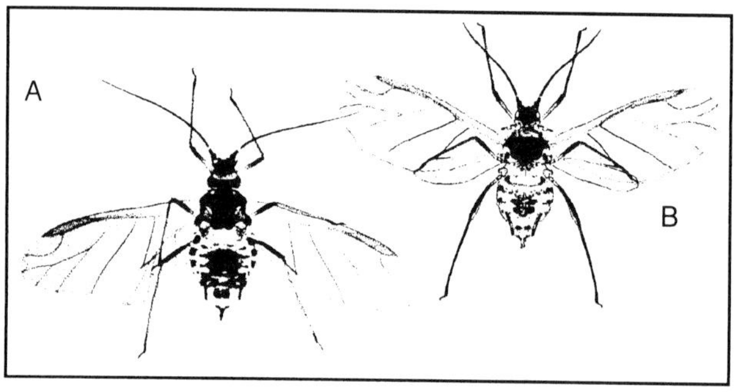

Figure 4.6: Adult aphids, alatae (winged forms), mounted on a microscope slide and cleared with potassium hydroxide. A. *Myzus persicae;* B. *Phorodon cannabis,* LM x 10.

Life History & Host Range

The complicated life cycle of aphids requires special terminology. Most outdoor aphids are **heteroecious** (or **holocyclic**) —they migrate between two hosts. The overwintering host is called the **primary** host. *M. persicae* overwinters on *Prunus,* as shiny black eggs laid on tree limbs and under the axils of tree buds. Eggs hatch into rotund "**stem mothers**" or **fundatrices.** Hatch time is temperature dependent. Fundatrices multiply parthenogenically—no sex needed. They are born fertile and within days begin giving birth. They give birth **viviparously**—eggs hatch within their reproductive tract—so larvae, not eggs, are "born alive" (another discovery by Leeuwenhoek). Fundatrices bear 60–100 **fundatrigeniae.** Soon the fundatrigeniae—which are **apterous** (wingless) females—begin giving birth to more live females (**apterous viviparae** or **apterae**). Thus stem mothers may live to see their great-great-great-granddaughters. In late spring the first **alatae** (winged aphids) develop. They are called **spring migrants** and fly off to **secondary hosts,** such as*Cannabis.*

Aphids are weak fliers, with a flight speed of only 1.6–3.2 km h^{-1} in still air. To migrate, alatae fly straight up into a moving air mass. The wind governs their direction of flight. In this way they can migrate 15–20 km. Aphids terminate migration by actively flying downward and settling on plants. Once settled on *Cannabis* the alatae give birth to apterae; apterae undergo four moults in about ten days to reach sexual maturity. Each aptera gives birth to 30–70 young. Crowded conditions or a lack of food induce new alatae, which fly off seeking unexploited *Cannabis.* They are called **summer migrants.** At the end of the summer, special alatae called **sexuparae** (or **autumn migrants** or **return migrants**) fly back to the primary host and give birth to ten **sexuales.** Sexuales are either females or males, alatous or apterous. They mate and the females become **oviparae** and lay five to ten eggs, which overwinter.

Hill (1994) claims that a single springtime fundatrix can give rise to up to 12 generations of aphids in one year—and theoretically 600,000 million offspring. *In tropical regions and in warm glasshouses, aphids do not migrate between hosts, nor do they lay overwintering eggs. They reproduce parthenogenically all year long, and remain on their secondary host.* Thus, some aphids that are normally heteroecious may become autoecious in warmer climates. *M. persicae,* for instance, overwinters as adult females on secondary hosts in the south or in warm glasshouses (Howard *et al.* 1994).

M. persicae attacks dozens of plant species, and now lives worldwide (Spaar *et al.* 1990). The pest is exceptionally restless; alatae repeatedly land on plants, probe briefly, then take off for other plants. This behaviour makes *M. persicae* "the most notorious vector of plant viruses" (Kennedy *et al.* 1959). It infests feral hemp in Illinois (Bush Doctor, unpublished data 1981) and marijuana in India (Sekhon *et al.* 1979). *M. persicae* is the most common aphid in Dutch glasshouses (Clarke & Watson, pers. commun. 1995). This species resists a broader range of pesticides than any other insect pest.

2. BLACK BEAN APHID

Aphis fabae Scopoli 1763, Homoptera; Aphididae.

Description: *Apterae* are oval to pear-shaped, olive green to dull black, averaging 1.5–3.1 mm in length (Fig 4.7 & Plate 7); legs are light green to white; antennae tubercles not prominent; cornicles relatively short (0.3–0.6 mm) and cylindrical; cauda are heavily bristled. Alatae nymphs have prominent white markings on their abdomen. *Alatae* are slightly smaller than apterae (1.3-2.6 mm long), their bodies are dark green-black with variable white stripes, wings are spotted with white wax. *Alatae* hold their wings vertically over their abdomens when at rest. *A. fabae* colonies are regularly ant-attended and they may congregate in great numbers.

Life History & Host Range

A. fabae is heteroecious—it migrates between two hosts. The overwintering host is *Euonymus* and sometimes *Virburnum* species. In the spring, eggs hatch into fundatrices, then fundatrigeniae, then apterae (see the life history of *M. persicae* for these terms). Apterae give birth to spring mi-

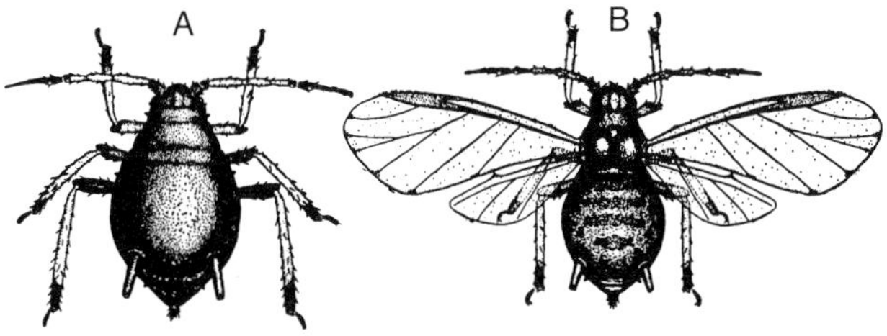

Figure 4.7: *Aphis fabae.* A. Adult aptera (wingless); B. Adult alata (winged); both x 10 (from Imms 1948, labelled *"Aphis rumicis").*

grants, which migrate to *Cannabis* and other secondary hosts. At the end of the summer, autumn migrants fly back to the overwintering host.

Blackman & Eastop (1984) listed *Cannabis* as a host for *A. fabae.* It is the second most common aphid in Dutch glasshouses, and attacks outdoor crops in China (Clarke & Watson, unpublished data 1992) and South Africa (Dippenaar *et al.* 1996). *A. fabae* may have been the "black aphid" infesting a seed-oil cultivar ('FIN-315') in Finland (Callaway & Laakkonen 1996). The taxon *A. fabae* represents a complex of at least four subspecies; exactly which subspecies attacks *Cannabis* is unknown. Black bean aphids are found in all temperate regions except Australia. They infest many crops and vector over 30 viruses.

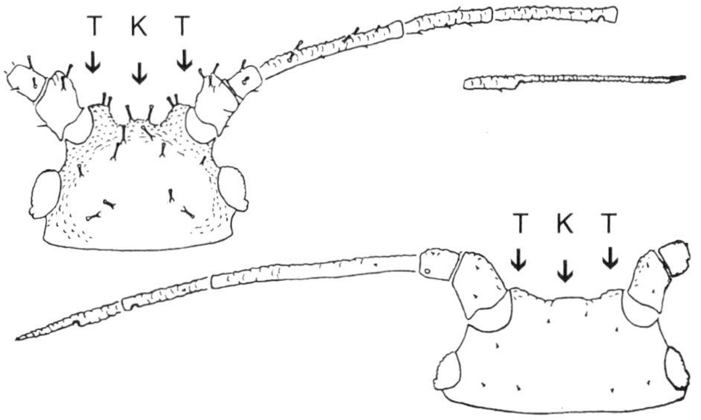

Figure 4.8: Prominent head bristles on *Phorodon cannabis* (left) compared to *Phorodon humuli* (right), from Müller & Karl (1976)."T" arrows point to tubercles, "K" points to midline knob.

3. BHANG APHID or HEMP LOUSE

Phorodon cannabis (Passerini) 1860, Homoptera; Aphididae.

=*Aphis cannabis* Passerini, =*Myzus cannabis* (Passerini), =*Paraphorodon cannabis* (Passerini), =*Diphorodon cannabis* (Passerini), =*Aphis sativae* Williams 1911, =*Phorodon asacola* Matsumura 1917, =*Capitophorus cannabifoliae* Shinji 1924, ? =*Semiaphoides cannabiarum* Rusanova 1943

Description: *Apterae* have flattened, elongate-oval bodies. They closely resemble *M. persicae* but are about 25% smaller, averaging 1.9–2.7 mm in length. They are described as nearly colourless (Ceapoiu 1958) to bright green with darker green longitudinal stripes (Kirchner 1906). Their heads are covered with tiny bristles; the bristles have *knobbed* apices which differentiate bhang aphids from hops aphids (Müller & Karl 1976, Fig 4.8). Antennae are 1.1–2.2 mm long; antennal tubercles are prominent (at least half as long as the first antennal segment), converge slightly, and several bristles sprout from each tubercle. Between the tubercles arises a smaller midline knob, also bristled. Cornicles are white, up to 0.8 mm long (nearly a third of the body length), cylindrical, and taper towards their tips. Cauda taper evenly to their tips. *Alatae* are slightly smaller than apterae, and develop black-brown patches on heads and abdomens (Fig 4.6). They hold their wings vertically over their abdomens when at rest. Males are smaller and darker than females, averaging 1.6–1.8 mm long. Overwintering eggs are ovate, shiny green-black, 0.7 mm long. Fundatrices are more oval than other apterae but slightly shorter (1.8–2.4 mm), with long antennae.

Life History & Host Range

The life history of *P. cannabis* is **autoecious**, sometimes called **monoecious**—the pest never alternates hosts (Balachowsky & Mesnil 1935, Müller & Karl 1976). Sexuparae of *P. cannabis* never fly away, so oviparous females lay eggs in the flowering tops of *Cannabis.* Most eggs are destroyed when the hemp crop is harvested.

The bhang aphid is particularly damaging to female buds; "...it sits between female flowers and seeds, sucking plant sap" (Kirchner 1906). Bhang aphids vector hemp streak virus (Goidanich 1955), hemp mosaic virus, hemp leaf chlorosis virus (Ceapoiu 1958), cucumber mosaic, hemp mottle virus, and alfalfa mosaic virus (Schmidt & Karl 1970). The species is native to Eurasia (ranging from Britain to Japan), and now lives in North America and northern Africa. Bantra (1976) suggested using the pest as a biocontrol agent against illicit marijuana crops. According to Müller & Karl (1976), *P. cannabis* infests *C. sativa, C. indica* and *C. ruderalis.* The pest also attacks hops (Blackman & Eastop 1984).

According to Martelli (1940), bhang aphids are identical to hops aphids (the next species described below). We disagree—hops aphids have different-shaped heads, with few bristles (Müller & Karl 1976, Fig 4.8). The two species also have different life histories—*P. cannabis* has reduced to an autoecious life cycle (only one host), first noted by Balachowsky & Mesnil (1935). Autoeciousness suggests *P. cannabis* evolved from *P. humuli,* not vice-versa.

4. HOPS APHID

Phorodon humuli (Schrank) 1801, Homoptera; Aphididae.

Description: Hops aphids closely resemble bhang aphids, except the head of *P. humuli* sports few bristles, if any (Müller & Karl 1976, see Fig 4.8); the few bristles on antennal tubercles have blunt or pointed apices, never knobbed; the midline knob is scarcely apparent. Apterae are larger on *Prunus* in the spring (2.0–2.6 mm) than apterae on *Humulus* in the summer (1.1–1.8 mm); alatae average 1.4–2.1 mm long.

Life History & Host Range

P. humuli is heteroecious; the overwintering host is *Prunus.* Eggs of *P. cannabis* hatch when the temperature reaches 22–26°C. Eggs hatch into fundatrices, then fundatrigeniae, then apterae, then spring migrants (see the life history of *M. persicae* for these terms). In the California Bay area, spring migration of *P. humuli* peaks on June 1st (Parker 1913a); in southern England the flight peaks in late June (Dixon 1985). *P. humuli* migrants fly straight up into a moving air mass, and can travel 150 km or more (Dixon 1985). They spend the summer on secondary hosts, then fly back to overwintering hosts. For *P. humuli* this autumn migration peaks in late September (Dixon 1985).

Blunck (1920), Flachs (1936), and Eppler (1986) reported *P. humuli* infesting *Cannabis.* These pests normally attack hops. *P. humuli* vectors many plant viruses and *Pseudoperonospora humuli* (a fungus causing downy mildew of hops and hemp, Sorauer 1958). *P. humuli* lives in Europe, central Asia, north Africa, and North America, but not Australia. In California *P. humuli* is aided by a large black ant, *Formica subsericea* Say (Parker 1913a).

5. OTHER APHIDS

Cherian (1932) and Raychaudhuri (1985) cited the cotton aphid (also called the melon aphid, ***Aphis gossypii*** Glover 1877) on marijuana in India. *A. gossypii* adults are 1–2 mm

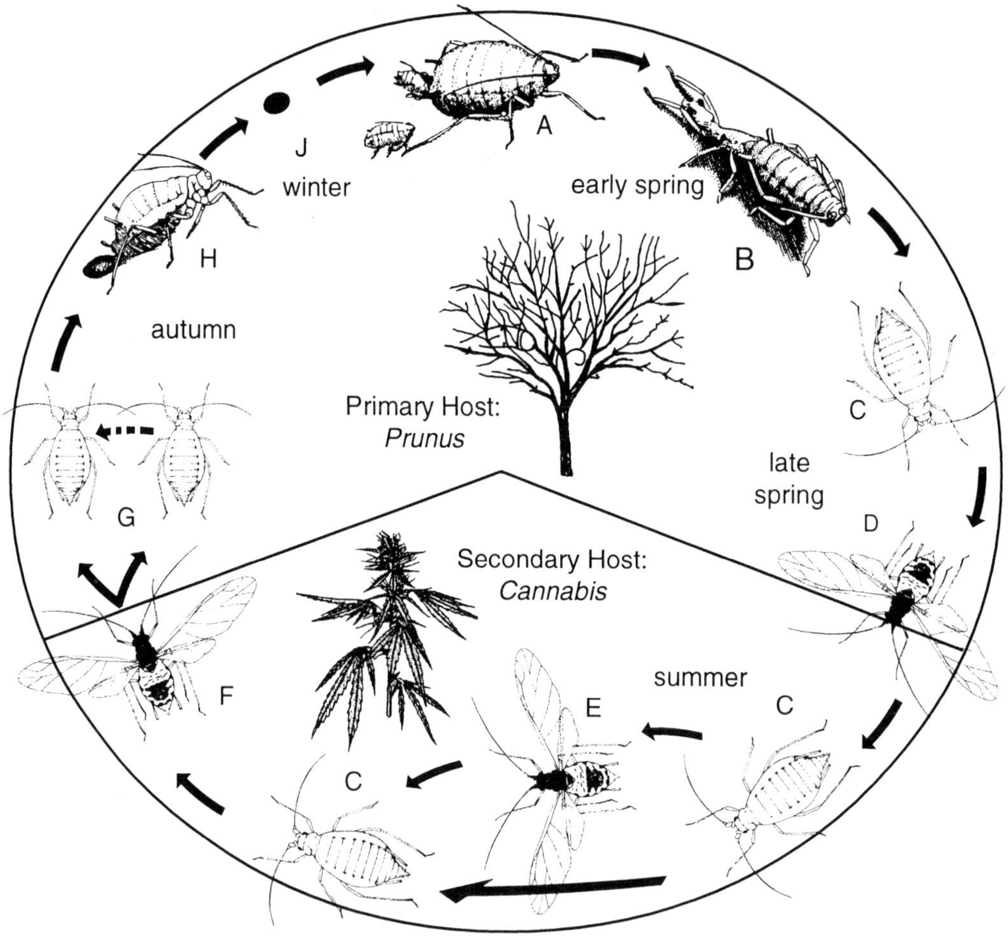

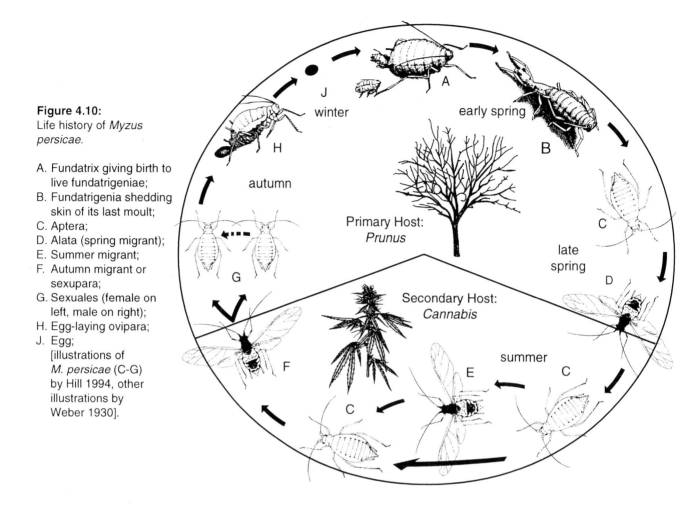

Figure 4.10: Life history of *Myzus persicae.*

A. Fundatrix giving birth to live fundatrigeniae;
B. Fundatrigenia shedding skin of its last moult;
C. Aptera;
D. Alata (spring migrant);
E. Summer migrant;
F. Autumn migrant or sexupara;
G. Sexuales (female on left, male on right);
H. Egg-laying ovipara;
J. Egg;
[illustrations of *M. persicae* (C-G) by Hill 1994, other illustrations by Weber 1930].

long, have short, bristled cauda, and lack antennal tubercles. Body colour ranges from light yellow to very dark green. Cornicles are always black, regardless of body colour (Fig 4.9). *A. gossypii* lives around the world and seriously damages cotton and cucurbit crops, but feeds on almost anything. The species prefers high temperatures; at 27°C the aphids mature in seven days. *A. gossypii* vectors over 50 plant viruses and is often ant-attended.

Raychaudhuri (1985) cited ***Uroleucon jaceae*** (Linnaeus) 1758 in India. *U. jaceae* is a large (2.5–3.5 mm long) reddish brown to brown-black aphid that infests Compositae in Europe and Asia.

DIFFERENTIAL DIAGNOSIS

Symptoms from aphids (wilting, yellowing) can be confused with damage by spider mites and whiteflies. Turn over a leaf to see what lurks beneath. Young aphids can be confused with young whiteflies, young scales, even young tarnished plant bugs. A hand lens helps distinguish these pests. The wings of aphids extend at least twice the length their bodies, which differentiates winged aphids from fungus gnats, whose wings are shorter (Gill & Sanderson 1998).

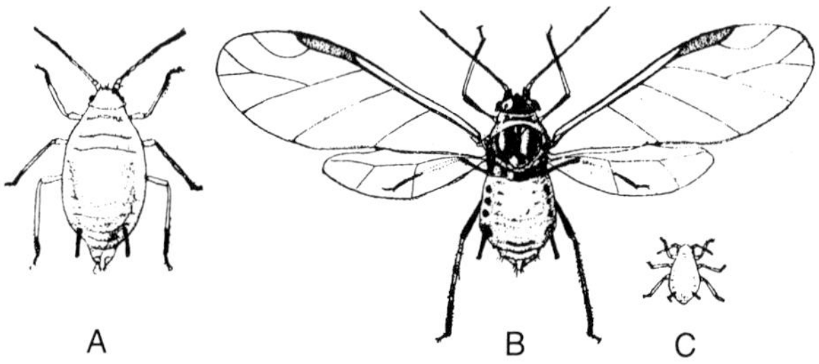

Figure 4.9: *Aphis gossypii.* A. Adult aptera (wingless); B. Adult alata (winged); C. Young nymph; all x 10 (courtesy USDA).

Differentiating aphids from each other can be difficult. If apterae are light green, consider *M. persicae, P. humuli,* or *P. cannabis. M. persicae* is about 25% larger than the other two species, its antennal tubercles are short and rounded, its midline knob between the tubercles is small, its cornicles develop a slight swelling, and its cauda is slightly constricted. *P. cannabis* tubercles are long (sometimes finger-like projections), heavily bristled, and *P. cannabis* sports the largest midline knob. *P. humuli* has short tubercles, few bristles, and lacks a midline knob. Blackman & Eastop (1984) provided an illustrated key for these species as they appear on hops.

If apterae are dark green-black and lack attennae tubercles, suspect *A. fabae.* But don't forget *A. gossypii,* which can also be quite dark and lacks tubercles. The jet-black cornicles of *A. gossypii* set it apart from *A. fabae.*

CULTURAL & MECHANICAL CONTROL

(method numbers refer to Chapter 9)

Method 1 (sanitation) is the workhorse. Mechanical repelling (method 12a) works well against winged aphids. Catching winged aphids with yellow sticky traps (method 12b) provides an early warning of aphids migrating into your area. Screen all entrances to glasshouses (method 13). Method 5 (genetic resistance) glitters with future potential—but thus far, only aphid-susceptible plants have been identified. BB

(1990) reported aphids "just love" *afghanica* cultivars developed in Holland. Exclude or kill all ants in the area.

BIOCONTROL (see Chapter 10)

Aphids can be nearly eradicated with biocontrol organisms. To control *established* aphid infestations, predators work better than parasitoids. Parker (1913a) and Campbell & Cone (1994) listed many natural aphid predators in California and Washington hops yards. Most are now commercially available—green lacewings (*Chrysoperla carnea*), convergent ladybeetles (*Hippodamia convergens*), two-spotted ladybeetles (*Adalia bipunctata*), and the tiny slug-like predator *Aphidoletes aphidimyza*. These are described below. Predators that feed on aphids in the absence of their primary hosts include *Cryptolaemus montrouzieri*, *Harmonia axyridis*, and *Rodolia cardinalis* (discussed in the section on mealybugs), *Stethorus picipes* (described under spider mites), *Orius* species (described under thrips), and *Deraeocoris brevis* (described under thrips).

The best *preventatives* are parasitoid wasps—*Aphidius matricariae*, *A. colemani*, and *Aphelinus abdominalis* (described below). For some reason, parasitoids tend to prey on rose-coloured aphid colour morphs, and predators tend to attack green morphs (Losey *et al.* 1997).

Fungi are the best microbial biocontrols of aphids. Viruses and bacteria must be *eaten* to be infective—and since aphids suck sap, there is little chance of their injesting viruses and bacteria sprayed on plant surfaces. Fungi need not be injested. They work on *contact*, and directly penetrate aphid skin. *Verticillium lecanii*, *Metarhizium anisopliae*, *Entomophthora exitialis*, and *Erynia neoaphidis* are discussed below; *Beauveria bassiana* and *Beauveria globulifera* are discussed under whiteflies.

Yepsen (1976) suggested companion-planting with coriander, anise, wormwood, or mint to drive away aphids. Aphids are also repelled by nasturtiums, marigolds, chives, onions, and garlic (Israel 1981).

Table 4.3: Infestation Severity Index for aphids.

Light	any aphids seen often no symptoms present
Moderate	< 10 aphids/leaf (not leaflet) feeding patches present
Heavy	11–50 aphids/leaf feeding patches coalescing
Critical	> 50 aphids/leaf shrivelled, discoloured leaves and honeydew

Chrysoperla (Chrysopa) carnea "Green lacewing"

BIOLOGY: The lacewing larva (also known as the "Aphid lion") is a nocturnal predator. It eats aphids, and to a lesser extent, feeds on whitefly nymphs, budworm eggs, thrips, and spider mites. Unfortunately it may also eat ladybeetle eggs (Gautam 1994). The *C. carnea* species complex lives in temperate regions worldwide. *C. carnea* does best in moderate temperatures (24–27°C) and moderate humidity (55% RH), but adapts to a wide range of humidity (35–75% RH).

APPEARANCE: Adults (called "lacewings") have slender pale green bodies 12–20 mm long, with delicate, lacy, light green wings, and bright eyes (Fig 4.11). They flutter like moths when disturbed. Larvae (called "aphid lions") have yellowish-grey bodies with brown marks, tufts of hair, and large jaws (Plate 8). They look like tiny alligators, 2–10 mm long. Eggs are oval, white, and suspended from undersides of leaves by slender threads 1 cm long (Plate 9).

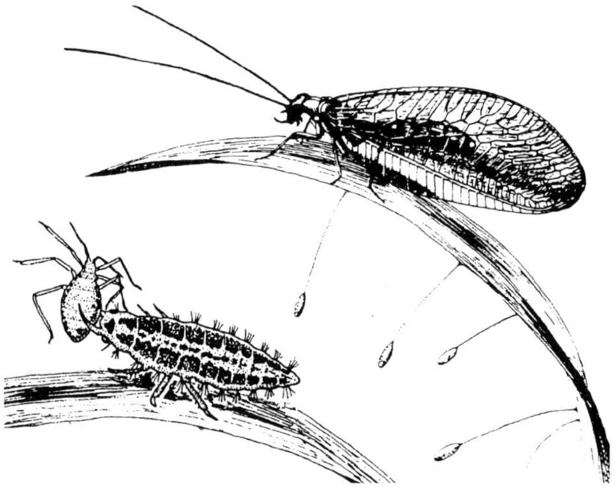

Figure 4.11: Green lacewing, *Chrysoperla carnea*, adult, "aphid lion" larva, and eggs (courtesy USDA).

DEVELOPMENT: Lacewings overwinter as pupae or adults. Adults have strong migratory instincts and may fly for three or four hours before settling down on plants. They are attracted by the scents of tryptophan and caryophyllene—two components of aphid honeydew. Caryophyllene is also part of the aroma of *Cannabis* (Brenneisen & ElSohly 1988). Adults do not eat aphids, they feed on nectar or honeydew. Females lay up to 600 eggs. Larvae consume 300–400 aphids during their two-week larval life (Hill 1994). In optimal conditions the life cycle takes 25 days. Adults live another 40 days; three to eight generations cycle per year. Short days induce diapause in some populations of *C. carnea* (Tauber & Tauber 1993), rendering them ineffective in short-day flowering crops.

APPLICATION: Eggs are the best, supplied in paper tubes with intact egg-threads, or glued onto cards, or loosely mixed with vermiculite in shaker bottles. Adults are supplied in sealed tubes, larvae are supplied in divider boxes. Adults should not be stored. Larvae can be stored up to two days, in a cool (7–14°C), dark place; eggs can be stored up to four days. Recommendations for release rates are presented in Table 4.4. Place eggs or larvae at the base of plants, in the

Table 4.4: *C. carnea* release rate for control of aphids.

ISI*	NUMBER OF PREDATORS RELEASED PER PLANT OR M^{-2} OF GLASSHOUSE CROP**
Preventative	5 larvae m^{-2} every 3 weeks
Light	5 larvae/plant **OR** 10 m^{-2} every 2 weeks
Moderate	25 larvae m^{-2} on trouble spots, 10 m^{-2} elsewhere every week
Heavy	50 larvae m^{-2} on trouble spots, 25 m^{-2} elsewhere every week
Critical	100 larvae m^{-2} on trouble spots, 50 m^{-2} elsewhere every week, use ladybeetles

* ISI = Infestation Severity Index of aphids, see Table 4.3.
**control of aphids with *C. carnea* is difficult during flowering (light cycles below 12 hours/day)

crotch of branches, or in distribution boxes hanging from plants. Spread them out to avoid cannibalism. Researchers have coated eggs with acrylamide (sticky gel), so they can be sprayed and adhere to plant surfaces.

Do not irrigate plants for a couple days after release, to avoid washing lacewings away (misting is okay and may help young larvae survive). Lacewings do not establish themselves with ease. Entice adults to stay and lay eggs by providing artificial honeydew—mix honey and brewers yeast on sticks and distribute among plants (replace when mouldy); commercial products include Wheast®, Biodiet®, or Formula 57® (see Chapter 9, method 14).

NOTES: *Green lacewing eggs are the most cost-effective biocontrol of aphids currently available.* Lacewings can even provide preventative control. For spot control of *heavy* infestations, lacewings are not as effective as ladybeetles. *C. carnea* is compatible with other *Chrysoperla* species, ladybeetles, *Trichogramma* wasps, and predatory mites, although some of these biocontrols may be eaten in the absence of aphids. Normally, foliage sprayed with Bt does not affect *C. carnea;* but the genetically modified, truncated form of Bt in transgenic plants may be harmful (Hilbeck *et al.* 1998). Some *C. carnea* strains tolerate malathion.

Chrysoperla (Chrysopa) rufilabris

BIOLOGY: Another lacewing resembling the aforementioned species. *C. rufilabris* is native to the Southeastern USA, where it commonly inhabits trees. In optimal temperatures (24–27°C) the life cycle takes 23 days. *C. rufilabris* reproduces better than *C. carnae* in humid environments (≥75% RH), it is thought to be a more aggressive predator than *C. carnae*, produces more eggs per female, may be found in higher numbers towards the end of the season. *C. rufilabris* lacks the strong migratory wanderlust of *C. carnea*, so it stays in release areas longer (Hoffmann & Frodsham 1993). Nevertheless, establishing a season-long, self-replicating population is difficult; multiple releases may be required. *C. rufilabris* is supplied like *C. carnea*, released at the same rate.

Other *Chrysoperla* and *Chrysopa* species

BIOLOGY: *Chrysoperla comanche* is new. It stays active later in the season, after *C. carnea* enters diapause. *C. comanche* is more fecund than *C. carnae* (females lay up to 1100 eggs). *Chrysopa oculata* and *Chrysopa nigricornis* are native biocontrols in the USA, not yet commercially available. Adults of these species are predatory, an added advantage.

Hippodamia convergens "Convergent ladybeetle"

BIOLOGY: A beetle that preys on aphids as well as mealybugs, scales, and other small insects. It is native to North America, and does best in moderate humidity and temperatures (>40% RH, 19–31°C).

APPEARANCE: Larvae are slim, flat, alligator-like, dark with orange spots, and up to 12 mm long (Fig 4.3 & Plate 10). Adults are tortoise-shaped, 4–7 mm long, with a black and white-lined thorax, and two white lines converging behind the head (Fig 4.3 & Plate 11). Wing covers are orange, red, or yellow, typically with six black spots (the number varies from none to 13). Eggs are orange, cigar-shaped, and laid in clusters of ten to 20 near aphid colonies.

DEVELOPMENT: Females lay up to 1000 eggs in spring and early summer. Larvae eat voraciously for a couple of weeks, pupate, then the adults also feed on aphids. Development from egg hatching to adulthood takes 28 days in optimal conditions. Adults live another 11 months. At the end of the season adults migrate to hillsides and creek beds, where they aggregate and overwinter.

APPLICATION: Supplied as adults in bottles, bags, or burlap sacks. Store in a cool (4–8°C), dark place. Beetles purchased before the month of May must be released by the end of May; beetles purchased after early June can be stored for up to three months. Every couple of weeks remove them from the cold, mist with water, allow to dry, and return to storage (Cherim 1998). Used preventatively, release 4 l (1 gallon) of adults per 4000 m^2 (=20 beetles per m^2), or refer to Table 4.5. Watson (pers. commun. 1992) applied up to 5000 adults to an extremely infested, extremely valuable plant, and reapplied weekly. Extra adults flew off to the rest of the crop. In a few weeks the aphids were 99% gone.

Commercial supplies of *H. convergens* have been harvested from winter aggregation sites in California. They harbour a strong migrational instinct which requires them to disperse before feeding and laying eggs. Thus, field releases of *H. convergens* are ineffective, because the adults quickly disperse. They work in enclosed structures such as glasshouses and growrooms. Screen vents to keep them from escaping. Delay migration releasing at night, and by "gluing" their wings shut with a spray of sugared soda (Coke, Pepsi, etc.) mixed 1:1 with water. After a week the sugar-water solution wears off.

Entice ladybeetles to stay and lay eggs by serving them "artificial honeydew"—see recipes in the *Chrysoperla carnae* section, above. Provide moisture by misting plants with water before releasing (be careful misting flowering plants—they may develop grey mould). Some commercial suppliers decrease the ladybeetles' migratory instinct by feeding them a special diet and allowing them to "fly off" in a controlled environment.

NOTES: *H. convergens* is the best-known mail-order predator of aphids. *For heavy infestations, ladybeetles are the best—they can be applied in high numbers as a "living insecticide."* But ladybeetles are a poor choice for preventative control—without food they migrate away or die. Ladybeetles may eat other predators in the absence of pests, but they are compatible with parasitoids. Ladybeetles are not fully compatible with some biocontrol fungi (e.g., *Beauveria bassiana* and *Paecilomyces fumosoroseus*). Commercial supplies of *H. convergens* have been harvested from the wild, so they have little tolerance for pesticides.

Table 4.5: *Hippodamia convergens* release rate for control of aphids.

ISI*	Number of predators released per m^2 of glasshouse crop**
Preventative	10 ladybeetles m^{-2} every 2 weeks
Light	50 ladybeetles m^{-2} weekly
Moderate	100 ladybeetles m^{-2} weekly
Heavy	200-500 ladybeetles m^{-2} twice weekly
Critical	apply chemical controls, wait 2 days, then release 200 ladybeetles m^{-2} twice weekly

* ISI = Infestation Severity Index of aphids, see Table 4.3.

Adalia bipunctata "Two-spotted ladybeetle"

BIOLOGY: A beetle that preys on aphids, is native to North America, and does best in moderate humidity and temperatures. Adults have oval, convex bodies, with red to yellowish-orange wings marked by two black dots (Fig 4.3). They overwinter as adults; several generations arise per year.

Coccinella undecimpunctata "Eleven-spotted ladybeetle"

BIOLOGY: Another beetle that preys on aphids and other small insects, introduced to North America from Europe, and does best in moderate humidity and temperatures. Adults have elongate-oval, convex bodies, with red or orange wings marked by 11 black dots, 5–7 mm long. A related European species, ***Coccinella septempunctata***, known as the seven-spotted ladybeetle, has become established in eastern North America. It can accumulate in large numbers, displacing native ladybeetles and becoming a nuisance to humans (even biting people, according to Schaefer *et al.* 1987b).

Aphidoletes aphidimyza

BIOLOGY: A midge (fly) whose larvae prey on aphids, is native to northern Europe and North America, and does best in moderate humidity and temperatures (70% RH, 18–27°C). *A. aphidimyza* does poorly in hot, dry, windy locations.

APPEARANCE: Adults look like tiny mosquitos, 2.5 mm long. Eggs are oblong, shiny orange, 0.3 mm long. Larvae are white upon hatching but turn orange as they mature, reaching 3 mm in length (Plate 12).

DEVELOPMENT: Larvae inject a paralysing poison into aphids (Fig 4.12), then suck out all body contents. Westcott (1964) claimed the maggots kill an aphid a minute. Usually the larvae kill ten to 100 aphids in a week or less, then pupate in soil. Adults are night fliers and feed on aphid honeydew. Females lay 100–250 eggs near aphid colonies. The life cycle takes about four weeks, adults live another two weeks. Unfortunately, short photoperiods send maggots into diapause, ending their effectiveness when *Cannabis* flowers. *A. aphidimyza* overwinters in soil and may become established in glasshouses with soil floors.

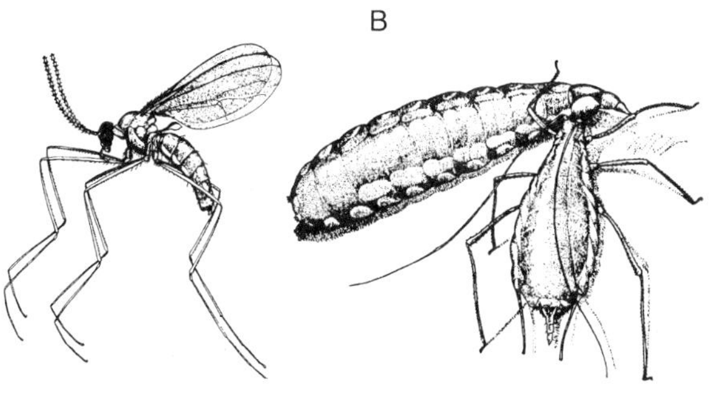

Figure 4.12: *Aphidoletes*, a midge that preys on aphids. A. Adult female; B. Maggot biting an aphid on the knee. (from Davis 1916)

APPLICATION: Supplied as pupae mixed with vermiculite or peat in shaker bottles or tubs. Store up to four days in a cool (8–15°C), dark place. Release pupae in humid, shaded areas—not on surfaces receiving direct sunlight. Hussey & Scopes (1985) released one predator per 25 aphids at five-day intervals. Clarke & Watson (pers. commun. 1995) released two to five pupae per m^2 at the first sign of aphids, or ten pupae per m^2 in heavy infestations, repeating every two weeks until control is achieved. Koppert (1998) suggested releasing *A. aphidimyza* pupae in piles of at least 10 cc (approximately 20 pupae).

NOTES: *A. aphidimyza* may be used preventatively or released against established infestations. The maggots eat less than lacewing larvae, but the maggots do not migrate away. Adult females require an established aphid population before they lay eggs. *A. aphidimyza* maggots need soil to pupate and become established. In soilless hydroponic glasshouses with concrete or plastic-covered floors, sprinkle peat moss between rows of plants for pupation sites. *A. aphidimyza* adults are *not* attracted to bright light (unlike lacewings and ladybeetles), so they won't fly into bulbs and windows. But they may be attracted to yellow sticky traps. *A. aphidimyza* is compatible with *Aphidius matricariae*, but *not* with beneficial nematodes. The maggots tolerate some pesticides, but adults are very sensitive. Adults can be released a week after spraying pyrethrum (Thomson 1992). Olkowski *et al.* (1991) suggest collecting wild maggots from Compositae weeds, but field-collected maggots may be infested by *Aphanogmus fulmeki* and other hyperparasites, which occasionally infiltrate into commercial mass-rearing units (Gilkeson 1997).

Aphidius matricariae

BIOLOGY: A braconid wasp, native to England; it does best in moderate humidity and temperatures (optimal 80% RH, 22–24°C). *A. matricariae* parasitizes over 40 species of aphids. It provides good control of *M. persicae* and *A. fabae*, but poor control of *A. gossypii* (Steenis 1995) and *Phorodon* species (Neve 1991).

APPEARANCE & DEVELOPMENT: Adults are tiny black wasps, 2 mm long (Fig 4.13). Female wasps lay 50–150 eggs, each egg is individually deposited within an aphid nymph. Eggs hatch into maggots which grow and pupate within aphids. Aphids swell and stiffen into shiny, papery, light-brown mummies. Emerging wasps leave small, round exit holes. The life cycle takes two or three weeks in optimal temperatures, and adults live another two weeks.

APPLICATION: Supplied in vials or shaker bottles as adults or pupae in mummies. Store for a maximum of one or two days in a cool (8–10°C), dark place. Release in shaded areas, from open boxes hanging in plants or on rockwool slabs. To prevent aphids, Reuveni (1995) recommended a weekly release of one mummy per 10 m^2 glasshouse. For light infestations, release three to six mummies per m^2 weekly (Thomson 1992).

NOTES: *A. matricariae* works best as a preventative in glasshouse situations. The species reproduces slowly and cannot handle heavy infestations. Some researchers report *A. matricariae* wasps travel widely in search of prey, others report they fly less than 3 m (Steenis 1995). Heavy honeydew hinders them. They do *not* diapause during flowering. Unfortunately, *A. matricariae* may be killed by its own hyperparasites, especially in late summer. Hyperparasites infest *A. matricariae* within aphids, and leave aphids via exit holes that may be confused with exit holes of *A. matricariae*. Exit holes by *A. matricariae* are smooth and round, whereas exit holes by hyperparasites have ragged, uneven edges (Cherim 1998). Purchasing adults instead of pupae will reduce or exclude hyperparasites.

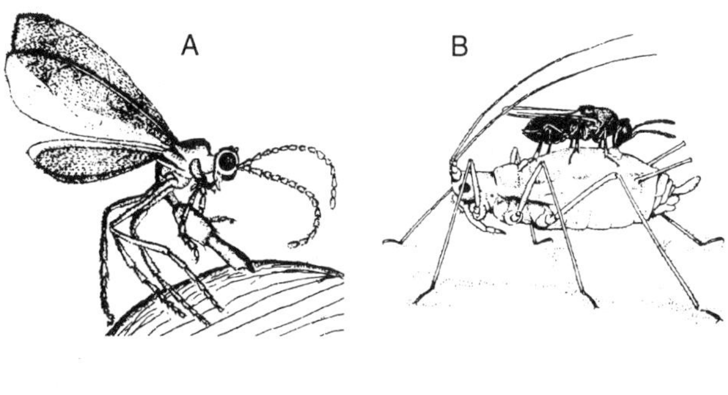

Figure 4.13: Two wasps parasitizing aphids. A. *Aphidius*, a braconid (from van Leeuwenhoek 1700); B. *Aphidencyrtus*, a chalcid (from Griswold 1926).

A. matricariae is compatible with other parasitoid wasps, *Aphidoletes aphidimyza*, ladybeetles, and lacewings. *A. matricariae* is not compatible with the biocontrol fungus *Beauveria bassiana*, which kills larvae developing inside aphids. Adults wasps are attracted to the colour yellow, so remove yellow sticky cards before releases are made. Avoid insecticides, although some strains of *A. matricariae* tolerate neem and malathion.

Aphidius colemani

BIOLOGY: This species has been confused with the aforementioned *A. matricariae*. *A. colemani* wasps lay twice as many eggs as *A. matricariae*, and they reproduce much better on *A. gossypii* (Steenis 1995). Several companies have dropped *A. matricariae* in favour of selling *A. colemani*.

APPLICATION: Supplied and applied like *A. matricariae*.

NOTES: A related species, ***Aphidius ervi***, works well against potato aphids such as *Macrosiphum euphorbiae* and *Aulacorthum solani*.

Aphelinus abdominalis

BIOLOGY: A chalcid wasp that parasitizes aphids, is native to temperate Europe, and does best in moderate humidity and temperatures.

APPEARANCE & DEVELOPMENT: Adults are 2.5–3 mm long, with a black thorax, yellow abdomen, and short antennae. Female wasps deposit eggs in adult aphids. Wasps also feed on ovipositor wounds. In about two weeks, aphids turn into black, leathery mummies. Larvae pupate within dead mummies and emerge as wasps, leaving behind a ragged exit hole at the rear of the mummy.

APPLICATION: Supplied as adults in shaker bottles. Store a maximum of one or two days at 8–10°C. The wasps migrate poorly and should be released close to aphids. Release 50 wasps per infested plant (Thomson 1992).

NOTES: *A. abdominalis* works best as a preventative. It prefers potato aphids (*Macrosiphum euphorbiae* and *Aulacorthrum solani*). This wasp is compatible with other parasitoids.

Verticillium (Cephalosporium) lecanii

BIOLOGY: A fungus that parasitizes aphids (Vertalec®) and whiteflies (Mycotal®). *V. lecanii* works best against *Myzus persicae* and *Aphis gossypii*, and worst against *Aphis fabae*. The fungus provides excellent short-term control; it knocks down heavy aphid populations to a level managed by slower-working biocontrols. For more information see the section on whiteflies.

Metarhizium anisopliae

BIOLOGY: A soil fungus first tested in 1879 against beetles in Russia (Samson *et al.* 1988). Different strains are used against aphids and whiteflies (BackOff®), termites (BioBlast®), spittlebugs (Metaquino®), cockroaches (BioPath®), thrips, beetles (chafers, weevils), ants, termites, and other insects. *M. anisopliae* does best in high humidity and moderate temperatures (24–28°C).

APPEARANCE & DEVELOPMENT: *M. anisopliae* is a hyphomycete with branching, phialidic conidiophores and simple oval spores (conidia) which aggregate into green prismatic columns. Conidia in contact with insects quickly germinate and grow into their hosts. *M. anisopliae* does not have to be eaten; it can penetrate insect skin, although it usually enters through spiracles. Infected insects stop feeding, then die in four to ten days, depending on the temperature. In humid conditions, *M. anisopliae* reemerges from dead hosts to sprout more conidia and repeat the life cycle.

APPLICATION: Supplied as conidia formulated in powder, granules, or solid cultures in foil packets, plastic bags, or bottles. Store for weeks in a cool (2–3°C), dark place.

NOTES: This fungus is compatible with biocontrol parasitoids such as *Aphidius matricariae* and *Aphelinus abdominalis* (Roberts & Hajek 1992), but may infect predators, such as *Orius insidiosis*. Disadvantages include a slow onset of action, and *M. anisopliae* does not grow through the soil (Leslie 1994). There are many strains of *M. anisopliae*, and the successful control of specific insects may require specific strains.

Entomophthora exitialis

BIOLOGY: A cosmopolitan fungus that parasitizes aphids. It does best in moderate to high humidity and moderate temperatures.

APPEARANCE & DEVELOPMENT: This fungus produces several types of spores. Ballistospores are forcibly discharged from dead insects, surrounding the cadavers with a white halo of sticky, mucus-covered spores. Resting spores arise within cadavers, and are mass-produced in liquid media fermenters. Spores that contact susceptible insects rapidly invade their hosts (*E. exitialis* does not have to be eaten by insects; it can penetrate insect skin). *E. exitialis* spores have been sprayed in California to control aphids (Yepsen 1976). Unfortunately, *E. exitialis* also infects aphid predators (but not aphid parasitoids, according to Hajek 1993).

Erynia neoaphidis (= *Entomophthora aphidis*)

BIOLOGY: This widespread fungus is related to the previous pathogen. It causes natural epidemics in *P. humuli* and other aphids (Byford & Ward 1968), and is being investigated as a biocontrol agent. In France, *E. neoaphidis* predominates in cool, humid conditions (Samson *et al.* 1988).

BIORATIONAL CHEMICAL CONTROL (see Chapter 11)

Frank & Rosenthal (1978) repelled aphids with a vegetable-based spray: grind up four hot peppers with one onion and several cloves of garlic. Let the mash sit in two quarts of water for several days. Strain the liquid, and add a half teaspoon of detergent as a spreader. Direct all sprays and dusts at undersides of leaves. Kill young aphids and repel adults with insecticidal soap, diatomaceous earth, and clay microparticles. Soaps have been sprayed up to a week before harvest without any distasteful residues discerned on finished dry flowers (Bush Doctor 1985).

Nicotine works *par excellence* against aphids, especially bhang aphids (Ceapoiu 1958) and hops aphids (Parker 1913a). Imidacloprid, a synthetic nicotine, works well against *M. persicae*, *A. gossypii*, and especially *P. humuli* (LD_{95} = 0.32 *ppm*), but not against *A. fabae* (Elbert *et al.* 1998). Cinnamaldehyde, extracted from cinnamon (*Cinnamonum zeylanicum*), kills all aphids but also kills beneficial insects. Parker (1913a) and Yepsen (1976) controlled aphids with quassia, which kills *P. humuli* but not *M. persicae*.

Two synthetic insect growth hormones, buprofezin and kinoprene, kill aphids. Deltamethrin, a synthetic pyrethroid, kills aphids, especially when sprayed in *small* droplets rather than large droplets (Thacker *et al.* 1995). Enhancing toxicity by spraying small droplets may hold true for most insecticides. Neem works poorly against aphids, and some aphids are resistant to pyrethrum. Rotenone kills insects with chewing mouthparts, not aphids. Smother overwintering eggs of *M. persicae* and *P. humuli* by spraying neighbouring *Prunus* species with dormant oil.

Winged aphids can be lured into traps baited with food attractants. *P. humuli*, for instance, is lured by a mix of volatile oils distilled from *Prunus* and *Humulus* plants (Lösel *et al.* 1996). Adding food attractants to sticky traps works well.

Winged sexuales (sexual aphids, the autumn migrants) can be baited with sex pheromones, such as nepetalactol (Lösel *et al.* 1996). Sexuales can be mass-trapped to break the reproductive cycle. *P. humuli* males can sense pheromones 6 m (20 ft) away from traps (Quarles 1999). Sex pheromones do *not* attract nonsexual aphids, so pheromones have limited value against aphids infesting *Cannabis* during the spring and summer (Howse *et al.* 1998). Sexual pheromones may attract beneficial insects that attack aphids, such as *Aphidius matricariae,* from surrounding areas (Quarles 1999).

Neve (1991) mixed an aphid alarm pheromone, (E)-β-farnesene, with pesticides or biocontrols. Alarm pheromones cause aphids to stop feeding and disperse across plants. Dispersal increases aphid contact with pesticides and biocontrols, resulting in better aphid control. The alarm pheromone is best applied with an electrostatic sprayer. (E)-β-farnesene has not been tested on *Cannabis* crops, but it may not work—*Cannabis* produces β-caryophyllene, another terpenoid which inhibits the pheromone activity of (E)-β-farnesene (Pickett *et al.* 1992).

WHITEFLIES

Whiteflies primarily cause problems in warm glasshouses. They resemble tiny moths but are neither flies nor moths. Whiteflies are related to aphids and leafhoppers. They damage plants by sucking sap and vectoring plant viruses. Three species of whiteflies infest *Cannabis.*

SIGNS & SYMPTOMS

Whiteflies produce few initial symptoms. This allows their populations to build until you are suddenly engulfed by a massive infestation. Whitefly symptoms resemble aphid damage—plants lose vigour, leaves droop, turn yellow, wilt, and sometimes die. Leaves become glazed with sticky honeydew, followed by a sooty-coloured fungus which grows on the honeydew. The adults congregate on undersides of leaves, out of sight to the casual observer. But if you look closely, the adults look like tiny specks of ash (Plate 13). When infested plants are shaken, a billowing cloud of whiteflies fills the air for several seconds before resettling.

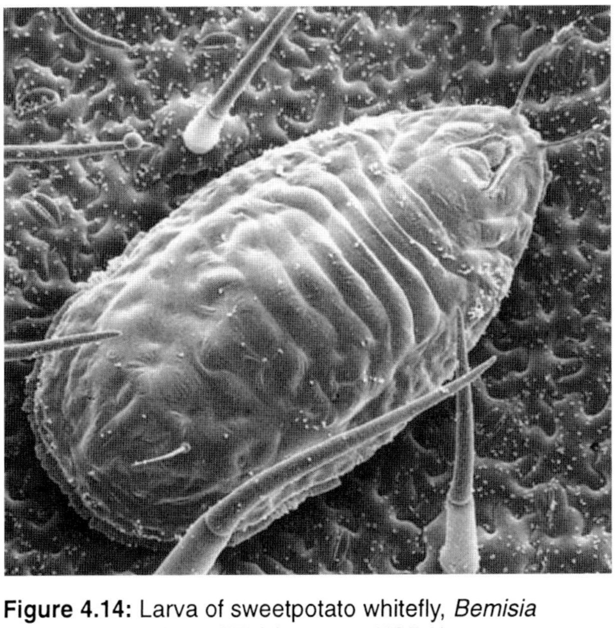

Figure 4.14: Larva of sweetpotato whitefly, *Bemisia tabaci*, seen with an SEM (courtesy USDA).

1. GREENHOUSE WHITEFLY

Trialeurodes vaporariorum (Westwood) 1856, Homoptera; Aleyrodidae.

Description: Eggs are 0.2 mm long, pale yellow, oval or football-shaped, with short stalks anchoring them to the leaf (Fig 4.16). Sometimes they are laid in circles or semicircles, covered with dust from the female's wings. Eggs turn purple-grey to brown-black before hatching into tiny 0.3 mm long "crawlers." These first-instar larvae are almost transparent, oval in outline, nearly flat, and radiate a halo of short waxy threads from their bodies. Subsequent instars lose their legs and resemble immature scale insects, reaching 0.7 mm in length. Late in the fourth stage, larvae change from transparent to an off-white colour by secreting an extra layer of wax to pupate within. Pupae project a halo of short wax threads from their palisade-like perimetre (Fig 4.15). Several pairs of longer wax threads may also arise from the top surface of pupae (Plate 15). Adult whiteflies rarely measure over 1 mm in length. Their four wings are off-white, have rounded contours, and are held flat over their abdomens almost parallel to the leaf surface (Fig 4.15 & Plate 13). The wings may become covered with a white dust or waxy powder.

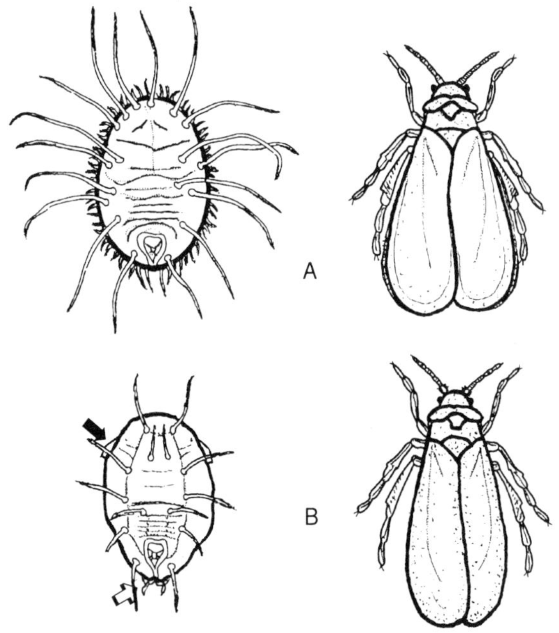

Figure 4.15: Pupae and adults of 2 whitefly species. A. Greenhouse whitefly, *Trialeurodes vaporariorum;* B. Sweetpotato whitefly, *Bemisia tabaci* (McPartland modified from Malais & Ravensberg 1992).

Life History & Host Range

Greenhouse whiteflies reproduce year-round. Females lay ≥100 eggs, on undersides of leaves near the tops of plants, often clustered in a circular pattern, anchored by short stalks inserted into leaf stomates (Fig 4.16). Eggs hatch after seven to ten days. First instar larvae crawl around plants searching for suitable feeding sites. Once feeding begins, larvae settle in one spot to suck sap, and eventually pupate. Larvae take two to four weeks to reach adulthood, depending on temperature. Adults live four to six weeks. In glasshouses the generations often overlap, so all stages occur together. The reproductive rate is dependent on temperature and host plant. Optimal conditions for *T. vaporariorum* are 27°C and 75–80% RH. The pest infests growrooms around the world (Frank & Rosenthal 1978, Frank 1988). It attacks a wide range of glasshouse crops. *T. vaporariorum* vectors many plant viruses, including the hemp streak virus (Ceapoiu 1958).

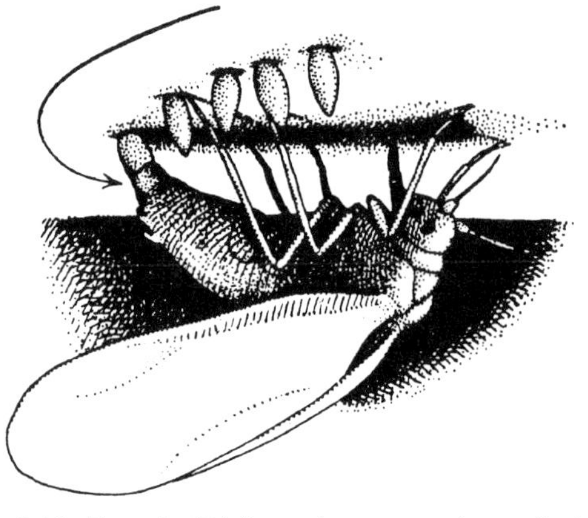

Figure 4.16: Female *Trialeurodes vaporariorum* laying eggs in a circle (from Weber 1930).

2. SWEETPOTATO (TOBACCO) WHITEFLY

Bemisia tabaci (Gennadius) 1889, Homoptera; Aleyrodidae.

= *Bemisia gossypiperda* Misra & Lamba 1929

Description: Eggs are 0.2 mm long, white, oval, with short stalks anchoring them to leaves. As eggs mature they turn yellow or amber. Larvae are nearly colourless. Pupae have reddish-coloured eyespots, their bodies are pale, marked by a caudal groove, and slightly pointed in the rear (Figs 4.14 and 4.15). Margins of pupae taper to the leaf surface, the margins lack a halo of short wax threads, and only a few longer hairs project from top surfaces (Plate 16). Pupae grow an anterior wedge of wax at tracheal folds (Fig 4.15, solid arrow), and a posterior wax fringe which extends lateral to the caudal setae (Fig 4.15, hollow arrow). Adults are light beige to yellow, with longitudinal striations and angled (not rounded) wing tips. Adults hold wings close to their bodies, in a tent-like position over their abdomens (Fig 4.15).

Life History & Host Range

Sweet potato whiteflies occur on *Cannabis* outdoors, in southern Europe (Sorauer 1958), and Brazil (Flores 1958). Outdoor populations overwinter as eggs. The pest also infests glasshouses; indoor whiteflies reproduce year-round, and have a life history similiar to that of *T. vaporariorum*, described above. Females lay ≥100 eggs. Optimal temperatures for *B. tabaci* are 30–33°C. The pest attacks many crops around the world.

3. SILVERLEAF WHITEFLY

Bemisia argentifolii (Bellows & Perring) 1994, Homoptera; Aleyrodidae.

= *Bemisia tabaci* poinsettia strain or strain B

Description: *B. tabaci* and *B. argentifolii* show subtle morphological differences—*B. argentifolii* pupae have narrower wedges of wax at tracheal folds (dark arrow, Fig. 4.15), the posterior wax fringe does not extend lateral to caudal setae (light arrow, Fig. 4.15), and they have one less pair of dorsal hairs. The adults are slightly larger than those of *B. tabaci* (Bellows *et al.* 1994).

Table 4.6: Infestation Severity Index for whiteflies.

Light	any adults seen when plant is shaken
Moderate	5–10 adults/plant, seen on more than one plant
Heavy	11–20 adults/plant, seen on many plants
Critical	> 20 adults/plant OR sooty mould and leaf discoloration present OR a few adults on all plants

Life History & Host Range

Silverleaf whiteflies have devastated growers in the southern USA, outdoors and indoors. *B. argentifolii* appeared in the USA around 1986. It was originally called the poinsettia strain of *B. tabaci* (so named because it first appeared in Florida on poinsettia plants). Summer rains in the southeastern USA dampen the pest's damage there, but it recently arrived in southern California. It can feed on anything the Imperial Valley has to offer, and is resistant to almost all pesticides. Its life history is nearly identical to that of *B. tabaci* (except *B. argentifolii* females lay 10% more eggs), and it prefers the same temperatures, conditions, and host plants.

DIFFERENTIAL DIAGNOSIS

All Homopteran leaf damage looks similar, so symptoms caused by whiteflies can be confused with those of aphids, scales, and hoppers. Immobile whitefly larvae may be confused with immature scale insects.

Whiteflies are difficult to tell apart (Fig 4.15). Eggs of *T. vaporariorum* are white and turn purple or brown before hatching, while *Bemisia* eggs turn yellow. Pupae differ in colour, hairiness, and the angle of their edges (edges of *Bemisia* pupae taper at a 45° angle to the leaf surface, whereas edges of *T. vaporariorum* pupae drop at 90° to the leaf surface—like the edges of cookie dough cut with a cookie cutter). *T. vaporariorum* adults are slightly larger and lighter coloured than *Bemisia* adults, and hold their wings differently. *B. tabaci* and *B. argentifolii* are nearly identical, their morphological differences are described above, under *B. argentifolii*. The two species are reproductively isolated (they cannot mate), genetically different (DNA sequences of the two species are as different from each other as they are from *Trialeurodes* species), and biologically distinct (*B. argentifolii* is more fecund and produces more honeydew).

CULTURAL & MECHANICAL CONTROL

(method numbers refer to Chapter 9)

Whiteflies are attracted to yellow objects, so use method 12b to remove many adults. Method 1 (sanitation) is always important, especially indoors. Severely infested plants should be rouged (method 10). Anecdotal reports claim that air ionizers alleviate moderate whitefly infestations. Frank (1988) sucked adults from undersides of leaves with a low-powered vacuum cleaner, early in the morning when whiteflies are cold and slow-moving. You can also shake plants and suck the whiteflies out of the air. Vacuuming is useful before rouging plants, otherwise you may leave behind many flying adults. Outdoors, do not plant *Cannabis* near eggplant, sweetpotato, tobacco or cotton crops. These plants are whitefly magnets.

BIOLOGICAL CONTROL (see Chapter 10)

Early whitefly infestations are hard to detect, so growers with a history of problems should release biocontrols before pests are seen. The best prophylactics are parasitic wasps—*Encarsia formosa*, *E. luteola*, and *Eretmocerus eremicus*, described below. Predators include *Delphastus pusillus*, *Geocoris punctipes*, and *Macrolophus caliginosus* (described below), pirate bugs (*Orius* species, described under thrips), *Deraeocoris brevis* (described under thrips), and green lacewings and ladybeetles (described under aphids). Researchers are evaluating two predaceous phytoseiid mites, *Euseius hibisci* (from California, described under thrips) and ***Euseius scutalis*** (from Morocco, not currently available).

Use combinations of parasitoids and predators. Heinz & Nelson (1996) achieved better control of *B. argentifolii* by releasing *Encarsia formosa* together with *Delphastus pusillus*

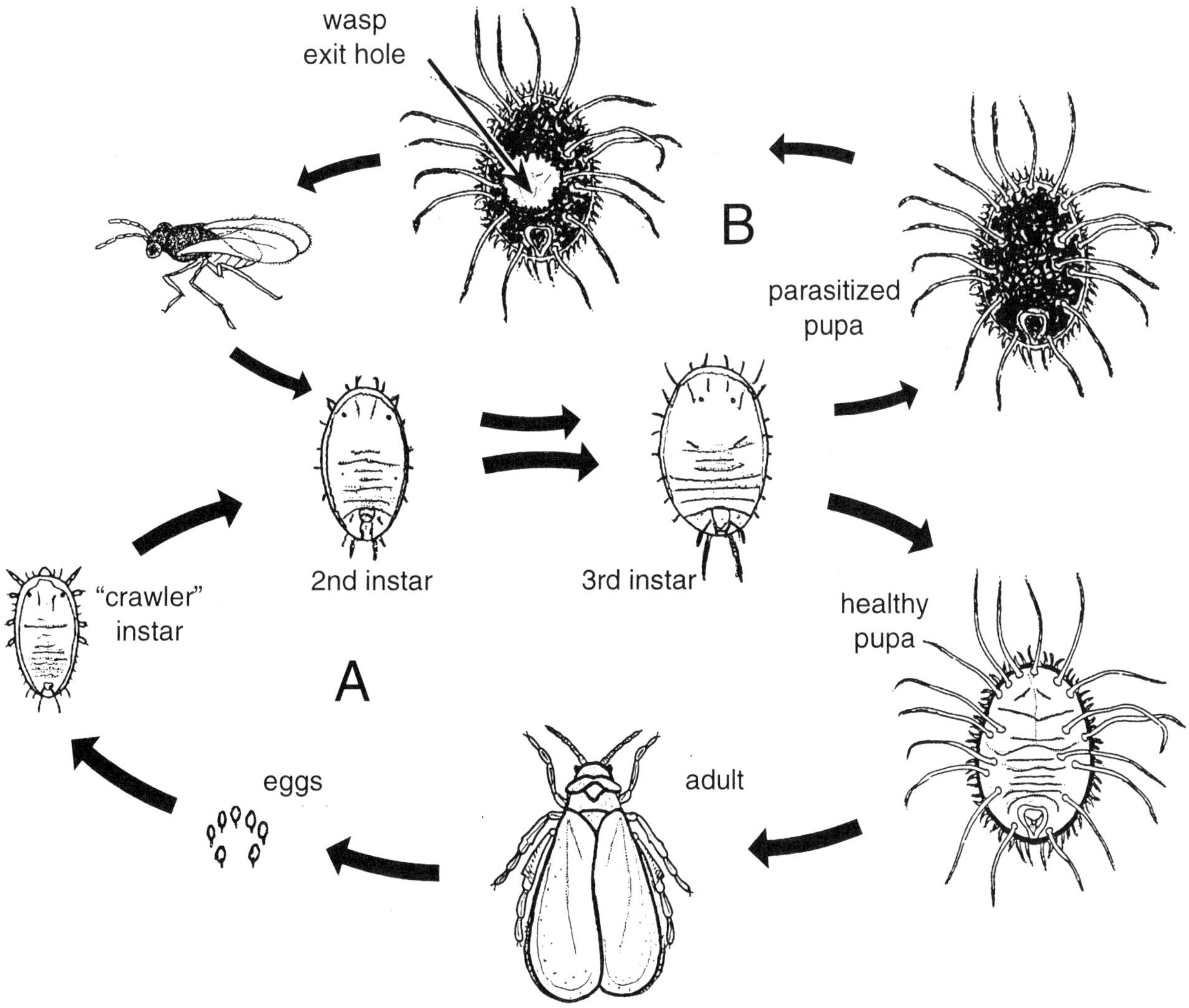

Figure 4.17: Intertwining life cycles of the parasitic wasp *Encarsia formosa* and its host, *Trialeurodes vaporariorum.* A. Normal life cycle of *T. vaporariorum;* B. Life cycle of *E. formosa,* beginning with egg insertion into 2nd instar nymph of *T. vaporariorum* (McPartland modified from Malais & Ravensberg 1992).

than by releasing either biocontrol alone. *D. pusillus* feeds on a few whiteflies containing young *E. formosa* larvae, but as *E. formosa* larvae mature within their hosts, *D. pusillus* avoids them.

Fungi are the microbial biocontrols of choice, because they infect whiteflies through their skin (other microbials such as viruses and bacteria must be eaten, which is unlikely with sap-sucking insects). Choices include *Verticillium lecanii, Beauveria bassiana, Aschersonia aleyrodis, Paecilomyces fumosoroseus* (see below), and *Metarhizium anisopliae* (discussed under aphids). Whiteflies are repelled by *Nasturtium* species (Israel 1981) and shoo-fly, *Nicandra physalodes* (Frank 1988). Cucumbers (*Cucumis sativus* L.) serve as trap crops for *B. tabaci* (Hokkanen 1991).

Encarsia formosa

BIOLOGY: This wasp parasitoid does best in bright light, 50–80% RH, and temperatures between 18–28°C (optimally 24°C). *E. formosa* occurs throughout temperate regions of the northern hemisphere, but was almost wiped out by DDT. It has been raised for biocontrol since 1926 in England (van Lenteren & Woets 1988).

APPEARANCE & DEVELOPMENT: Adults are tiny wasps, 0.5–0.7 mm long, with a black thorax and yellow abdomen (Plate 14). Wasps lay eggs in whitefly larvae (second through fourth instars) and sometimes pupae. Eggs hatch into maggots which slowly devour their host. Parasitized *T. vaporariorum* pupae turn black, making them easy to spot (parasitized *B. tabaci* pupae turn amber brown). *Encarsia* maggots moult into adults after 15–20 days and emerge from carcasses, leaving behind an exit hole (Fig 4.17). They reproduce without mating; adult females live another 30 days and lay up to 120 eggs. Females also directly kill larvae by feeding on all instars (usually two or three per day), so this species serves as both a parasitoid *and* a predator.

APPLICATION: Supplied as pupae in parasitized whitefly larvae, attached to cards, usually 80–100 *E. formosa* per card (Plate 86). Store a maximum of four days in a cool (8–10°C), dark place. Hang cards horizontally, in shade near the bottoms of plants. The card surface with pupae should face downward, to simulate the bottom of a leaf (although this is not necessary). Recommendations for release rates are presented in Table 4.7.

NOTES: *E. formosa* works best against *T. vaporariorum,* less so against *B. tabaci/argentifolii. E. formosa* works poorly in heavy infestations, because the wasps spend more time cleaning honeydew from themselves than hunting whiteflies. Thus *E. formosa* works best as a preventative. The percentage of parasitized larvae (black larvae vs. uninfested white larvae) can be monitored easily and should be ≥90% for effective control. Dense leaf hairs covering *Cannabis* may interfere with the wasp's ability to locate prey. This problem also arises on

Table 4.7: *E. formosa* release rate for control of glasshouse whiteflies.

ISI*	Number of predators released per plant or m^2 of glasshouse crop**
Preventative	1–5 pupae per plant or 10 m^{-2} every 2 weeks
Light	20 m^{-2} every 2 weeks until 80% of whitefly larvae have turned black
Moderate	30 m^{-2} on trouble spots, 20 m^{-2} elsewhere, then 10 m^{-2} every 2 weeks
Heavy	50 m^{-2} on trouble spots, 30 m^{-2} elsewhere, then 20 m^{-2} every 2 weeks
Critical	100 m^{-2} on trouble spots, 50 m^{-2} elsewhere, then 20 m^{-2} every 2 weeks

* ISI = Infestation Severity Index of greenhouse whitefly, see Table 4.6.

cucumbers. Cucumber breeders have reduced trichome density in new varieties, improving *E. formosa* effectiveness (van Lenteren & Woets 1988). Bredemann *et al.* (1956) tried breeding mutant *Cannabis* plants without trichomes, but their work was abandoned.

E. formosa adults have a bad habit of flying into HID lights and frying, which limits their use in growrooms. Yellow sticky traps *may* attract them. *E. formosa* is compatible with green lacewings (*Chrysoperla carnea*), *Delphastus pusillus*, and *Eretmocerus eremicus*. It is also compatible with Bt (used against other pests). Most *E. formosa* strains tolerate some pesticides (Table 10.1), including short-acting insecticides (soap, horticultural oil, neem, kinoprene, fenoxycarb), miticides (sulphur, abamectin), and fungicides (chlorothalonil, iprodione, and metalaxyl). Pyriproxyfen, an insect growth hormone useful against whiteflies, rarely harms adult wasps, but it kills immature *E. formosa* inside whitefly pupae.

Encarsia luteola (=*Encarsia deserti*)

This related species is being investigated for control of *B. argentifolii*.

Eretmocerus eremicus

BIOLOGY: A wasp, sometimes called *Eretmocerus* nr *californicus*, that parasitizes *T. vaporariorum*, *B. tabaci*, and *B. argentifolii*. It is native to California and Arizona, and does best in moderate humidity (optimally 40–60% RH), and temperatures (optimally 20–35°C, but tolerates up to 45°C).

APPEARANCE: Adults are tiny, 0.8 mm long, with yellow-brown heads and bodies, green eyes, and short, clubbed antennae. Eggs are translucent and brown before hatching.

DEVELOPMENT: Wasps lay eggs under sedentary whitefly larvae. Eggs hatch and larvae chew into their hosts. The hosts live long enough to pupate; infested pupae turn an abnormal beige or yellow colour. The whitefly pupae die and *E. eremicus* larvae pupate into adults within the cadavers. Adults emerge via small, round exit holes. The life cycle takes 17–20 days in optimal conditions. Adults mate and female wasps live an additional ten to 20 days, laying up to eight eggs per day. Females also host feed.

APPLICATION: Supplied as pupae within whitefly pupae on paper cards, or pupae mixed with sawdust in bottles. Store up to two or three days in a cool (8–10°C), dark place. Scatter loose pupae on dry surfaces in direct sunlight. Alternatively, place them in paper coffee cups with screened bottoms and tape the cups to stalks or stakes. Disperse pupae on cards the same way as *E. formosa* pupae. Do not allow them to get wet. Used preventively, release three females per plant per week in moderately humid areas like the Northeast (Headrick *et al.* 1995); double the rate in hot, dry climates. Koppert (1998) suggested releasing nine wasps per m^2 per week for moderate infestations, or switching to *Delphastus pusillus*. Continue weekly releases until at least 75% of whitefly pupae are parasitized.

NOTES: *E. eremicus* works at higher temperatures than *E. formosa*, tolerates pesticides better, and lays more eggs. But *E. eremicus* harbours a greater attraction to yellow sticky traps than *E. formosa* (Gill & Sanderson 1998). *E. eremicus* is compatible with the same biocontrols as *E. formosa*.

Delphastus pusillus

BIOLOGY: A ladybeetle that preys on *T. vaporariorum* and both *Bemisia* species. In the absence of whiteflies, it eats spider mites and baby aphids. *D. pusillus* is native to Florida, and does best in moderate humidity (70% RH) and moderate to warm temperatures (19–32°C).

APPEARANCE & DEVELOPMENT: Adults are shiny black and small, 1.3–2.0 mm long (Plate 15). Males have brown heads, females are all black. Larvae are elongated, pale yellow, with a fuzzy fringe of hairs, 3 mm long. They pupate on lower leaves or in leaf litter. Eggs are clear, elongate, 0.2 mm long, and laid on leaves alongside whitefly eggs. The life cycle takes 21 days at an optimal 25-30°C. Adults live another 30 to 60 days. Females lay three to four eggs a day, for a total of about 75 eggs. The adults are voracious—they devour as many as 10,000 whitefly eggs or 700 larvae. Larvae also consume whitefly eggs (up to 1000 during development) and larvae.

APPLICATION: Supplied as adults in bottles or tubes. Store up to four days in a cool (8–10°C), dark place. At the first sign of whiteflies, release one or two beetles per m^2, repeating every two weeks. For moderate infestations, double the release rate and repeat weekly.

NOTES: *D. pusillus* works best in moderate outbreaks. At low whitefly densities the beetles stop reproducing and disperse (to thwart dispersal, see strategies outlined under *Hippodamia convergens* in the aphid section). In heavy infestations *D. pusillus* bogs down in heavy honeydew. It moves poorly across hairy tomato leaves—and may find hairy *Cannabis* leaves equally disagreeable. *D. pusillus* avoids eating whitefly larvae obviously parasitized by *Encarsia* or *Eretmocerus*, making these biocontrols compatible (Heinz & Nelson 1996). *D. pusillus* can also be released with green lacewings (*Chrysoperla carnea*).

Geocoris punctipes "Big-eyed bug"

BIOLOGY: A lygaeid bug that preys on whitefly larvae and adults, aphids, spider mites and their eggs, and *Heliothis* eggs and young larvae (Hill 1994). *G. punctipes* is native to the southern USA and Mexico. It does best in moderate humidity and temperatures. Nymphs and adults prey on pests.

APPEARANCE & DEVELOPMENT: Adults have large, bulging eyes and somewhat flattened brown bodies covered with black specks, 3–4 mm long (Fig 4.39 & Plate 16). Nymphs look similar but without wings. Eggs are grey with a tiny red spot appearing shortly after being laid. Under optimal conditions, the life cycle takes about 30 days.

APPLICATION: In cases of mixed infestations, *G. punctipes* will eat the largest pests first. So even if the predators are surrounded by millions of tiny whitefly larvae, they would rather eat a plump, young budworm. *G. punctipes* is compatible with Bt but not broad-spectrum insecticides.

Macrolophus caliginosus

BIOLOGY: A mirid bug that preys on *T. vaporariorum*, *B. tabaci*, and *B. argentifolii*. In the absence of whiteflies it eats aphids, spider mites, moth eggs, and thrips. *M. caliginosus* works best on solanaceous crops; it survives on plant sap in the absence of prey.

APPEARANCE: Adults are light green bugs, 6 mm long, with long antennae and legs.

DEVELOPMENT: Nymphs and adults attack all whitefly stages but prefer eggs and larvae, which they pierce to suck out body fluids. *M. caliginosus* may feed on 30–40 whitefly eggs a day. The entire life cycle takes about a month. Each female lays about 250 eggs, inserted into plant tissue.

APPLICATION: Supplied as adults or adults and nymphs in bottles. Store one or two days in a cool (8–10°C), dark place. Gently sprinkle onto infested leaves. For light infestations, release five *M. caliginosus* per m^2; for heavy infestations, release 10 per m^2 and repeat the release two weeks later.

NOTES: *M. caliginosus* is compatible with *E. formosa* and *D. pusillus*. Avoid insecticides while utilizing this control (especially pirimicarb). In the absence of prey, large populations of *M. caliginosus* may suck enough plant sap to cause crop damage.

Verticillium (Cephalosporium) lecanii

BIOLOGY: A cosmopolitan fungus that parasitizes aphids (Vertalec®) and whiteflies (Mycotal®). Other strains infect scales, mealybugs, thrips, beetles, flies, and eriophyid mites. *V. lecanii* does best in moderate temperatures (18–28°C), and prefers high humidity (≥80% RH) for at least ten to 12 hours per day.

APPEARANCE & DEVELOPMENT: Spores germinate and directly penetrate the insect cuticle. After infection the fungus takes four to 14 days to kill the pest. Under ideal conditions, dead insects sprout a white fluff of conidiophores bearing slimy, single-celled conidia. Slime facilitates the adhesion of conidia to passing insects. Blastospores are yeast-like spores produced within cadavers and mass-produced in liquid media fermenters.

APPLICATION: Supplied as a wettable powder containing 10^{10} blastospores per g. Store in original, unopened package for up to six months at 2–6°C. Spores germinate best if soaked in water (15–20°C) for two to four hours prior to spraying. Spray around sunset—dew and darkness facilitate spore germination. Be certain to spray undersides of leaves. Keep the spray tank agitated. The manufacturer recommends spraying a 500 g container of spores over 2000 m^2 of crop area (density and height of plants not specified).

NOTES: *V. lecanii* works best as a short-lived agent to knock down heavy populations of pests, allowing *E. formosa* or other biocontrols to take over. Hussey & Scopes (1985) claim a single spray of Mycotal® will control greenhouse whiteflies for at least two or three months. But optimal spore germination requires high humidity for 24 hours. Thus *V. lecanii* cannot be sprayed when plants are flowering and susceptible to grey mould. *Encarsia formosa* can be used with *V. lecanii*, but is slightly susceptible to the fungus at high humidity levels. Most other biocontrol organisms are unharmed.

Beauveria bassiana

BIOLOGY: A fungus with a very wide host range, including sap-sucking insects (whiteflies, aphids, thrips, planthoppers, bugs, mites) and chewing insects (grasshoppers, beetles, termites, ants, European corn borers). *B. bassiana* lives worldwide. Several strains are available. The GHA strain (BotaniGard ES®, Mycotrol®, ESC 170®) is registered for use against whiteflies, thrips, aphids, and mealybugs. The ATCC 74040 strain, also known as the JW-1 strain (Naturalis®) is registered against the aforementioned insects, beetles, and other soft-bodied insects. The Bb-147 strain (Ostrinil®) is registered against European corn borers. Uncharacterized products include Ago Bio Bassiana® and Boverin®. *B. bassiana* does best in high humidity (≥ 92% RH) and a range of temperatures, 8–35°C, optimally 20–30°C.

APPEARANCE & DEVELOPMENT: Spores germinate and directly infect insects on contact, and kill them in two to ten days. Under optimal conditions, insect cadavers sprout white whorls of conidiophores, which bear globose conidia in zigzag formations (Plate 17). Both conidia and yeast-like blastospores are mass-produced in liquid media fermenters.

APPLICATION: *B. bassiana* is supplied as blastospores or conidia in emulsified vegetable oil or water-disbursable granules. Store up to two years in a cool (8–10°C), dark place. Spores are mixed with water and a wetting agent, then sprayed on all surfaces of pest-infested plants. Some strains have three to seven days of residual activity before reapplication becomes necessary. Cherim (1998) described *B. bassiana* causing some phytotoxicity in vegetable crops, but Rosenthal (1999) reported excellent success with *Cannabis* crops.

NOTES: *Beauveria* spores are sticky and they infect pests that brush against them; the spores do not have to be eaten to be infective. *Beauveria* species are not as deadly as *V. lecanii*. Nor are they as selective—*Beauveria* species kill ladybeetles, green lacewings, and other soft bodied predators. Effectiveness of *B. bassiana* depends on what pests are eating; according to Leslie (1994), pests eating plants with antifungal compounds (such as *Cannabis*) become somewhat resistant to *Beauveria* species. The fungus persists in soil, but organic material hinders survival, and nitrogen fertilizer kills it (Leslie 1994). The fungus may persist as an endophyte in some plants (see the section on endophytes in Chapter 5). Some people develop allergic reactions to *Beauveria* after repeated exposure (Hajek 1993).

Aschersonia aleyrodis

BIOLOGY: A subtropical fungus that parasitizes whitefly larvae (*T. vaporariorum*) and some scales. *A. aleyrodies* rapidly infects young larvae under conditions of high humidity. The fungus does best in high humidity and warm temperatures—conditions found in vegetative propagation (cloning) chambers and warm glasshouses. In the USA, *A. aleyrodis* was commercially available in the early 1900s until it was replaced by insecticides; now it is back (Gillespie & Moorhouse 1989).

APPEARANCE & DEVELOPMENT: *A. aleyrodis* covers *T. vaporariorum* cadavers with a slimy orange stroma, which contains pycnidia with phialidic conidiophores and hyaline, fusiform conidia. The teleomorph is an *Hypocrella* species.

APPLICATION: Supplied as conidia on solid media. Hussey & Scopes (1985) report a dose of 2×10^8 *A. aleyrodis* conidia per plant killed 75% of whiteflies, and did not infect the whitefly parasitoid *Encarsia formosa*.

Paecilomyces fumosoroseus (=*P. farinosus*)

BIOLOGY: A fungus that parasitizes whiteflies, including *B. argentifolii*. Some strains also kill spider mites, aphids, thrips, and mealybugs. *P. fumosoroseus* is native to semitropical areas and is used in the Philippines (PreFeRal®), Florida (PFR-97®), and in warm glasshouses in Europe (optimal conditions ≥ 68% RH, 22–33°C).

APPEARANCE & DEVELOPMENT: Sprayed spores infect whiteflies and begin killing them in four or five days. In humid microclimates, cadavers sprout hundreds of synnemata (bundles of conidiophores), which bear single-celled conidia.

NOTES: *P. fumosoroseus* produces conidia in solid culture, and yeast-like blastospores when cultivated in liquid fermentation tanks. Blastospores germinate faster than conidia. They have been added to mist irrigation systems in greenhouses. *P. fumosoroseus* is compatible with other fungi and with predatory mites; the fungus sometimes kills ladybeetles.

CHEMICAL CONTROL (see Chapter 11)

Heavy whitefly infestations may require chemical control before biocontrol organisms can take over. *Direct all sprays at undersides of leaves.* Small plants in pots can be dipped in spray solutions. Three or four applications (at weekly intervals) kill whiteflies emerging from eggs that escaped initial applications.

Bentz & Neal (1995) tested a sucrose ester extracted from *Nicotiana gossei*, a species of tobacco. A 0.1% spray of the ester killed nymphs of *T. vaporariorum* while sparing many *E. formosa* wasps, so the spray and the biocontrol were compatible together. Liu & Stansly (1995) compared five sprays against *B. argentifolii* in greenhouses: 96% of *adults* were killed by 2% Sunspray® Ultra-Fine horticultural oil, 68% were killed by bifenthrin (a pyrethroid), 26% by *N. gossei* sucrose ester, and 12% by M-Pede® insecticidal soap. After sprays dried on leaves, mortality from horticultural oil and bifenthrin actually increased, whereas *N. gossei* ester stayed the same, and soap lost all activity upon drying. Mortality against nymphs was not tested. In terms of repellency, bifenthrin was best (egg-laying adults were repelled for seven days), followed by horticultural oil (five days), soap and *N. gossei* (one day), and garlic spray (four hours).

Neem mimics whitefly growth hormones and effectively kills whitefly nymphs; it does not affect *D. pusillus* but may harm *Encarsia* species (Mordue & Blackwell 1993). Two synthetic insect growth hormones, kinoprene and fenoxycarb, also kill whitefly nymphs. The Rodale crowd recommends ryania (Yepsen 1976). For severe infestations, nicotine and pyrethrum kill more whiteflies, but also kill more biocontrols. A new chlorinated derivative of nicotine, imidacloprid, works very well against whiteflies. Biocontrol mortality can be reduced by shaking plants and then spraying whiteflies while they hover in the air.

EUROPEAN CORN BORERS

These pests have "a-maize-ing" appetites. European corn borers (hereafter "ECBs") have been recorded on 230 different hosts, basically any herbaceous plant with a stalk large enough to bore into. ECBs are native to eastern Europe, where hemp and hops served as original host plants. ECBs switched to corn after the introduction of maize into Europe (Nagy 1976, 1986). Nagy (1976) described ECB strains in Hungary which prefer *Cannabis* over maize, with hemp losses as high as 80–100% (Camprag *et al.* 1996). A Japanese strain also prefers *Cannabis* to corn (Koo 1940). ECBs were accidentally introduced into Boston about 1908. They have migrated across much of the USA and Canada. Another population was introduced into California via a cargo of European hemp (Mackie 1918). ECBs ruined wartime hemp in Illinois (Hackleman & Domingo 1943). ECBs attack both fe-

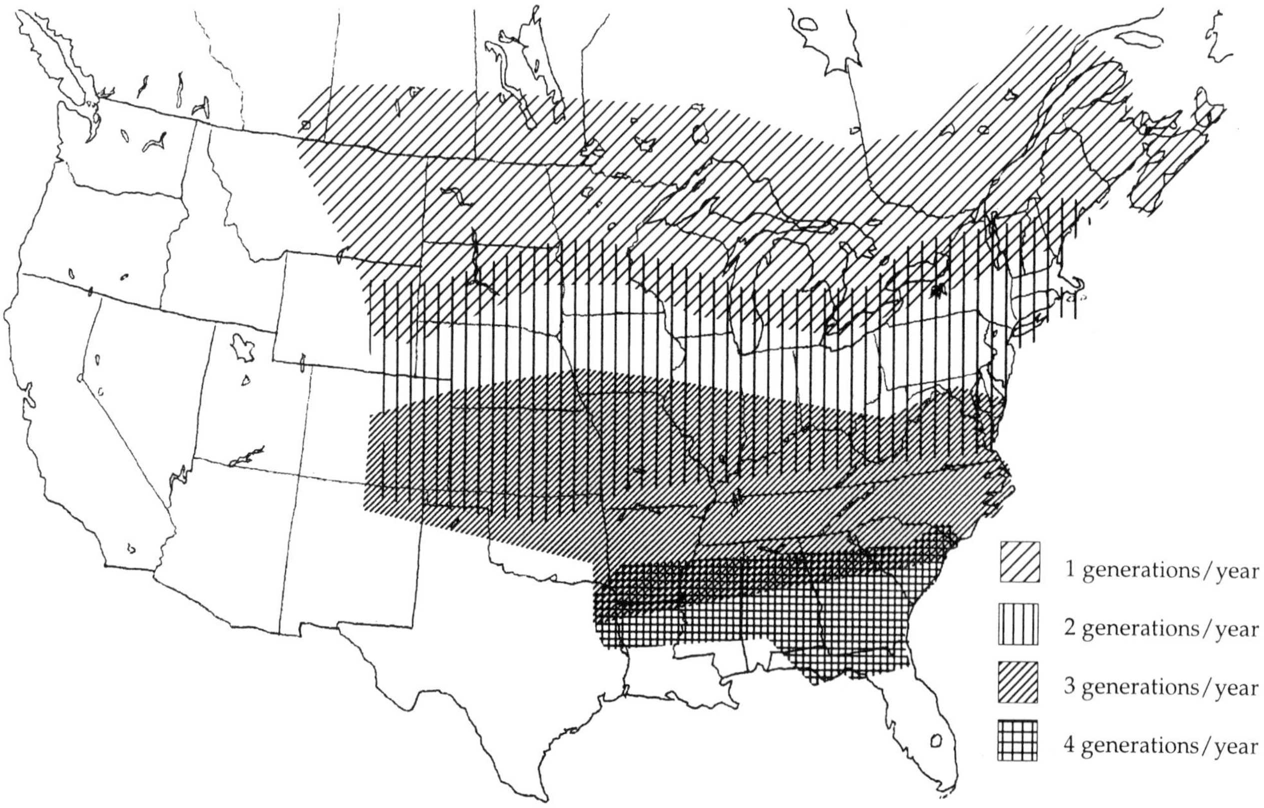

Figure 4.18: Approximate distribution of the European corn borer in the USA and Canada, circa 1997. Overlapping areas, such as Kansas, containing overlapping populations; some Kansas ECBs may produce 2 generations per year, some may produce 3 generations per year (McPartland update of Showers *et al.* 1983).

ral hemp and cultivated marijuana all over the midwestern corn belt (Bush Doctor 1987b). Today, only southern Florida, northern Canada, and sections of the western USA have yet to be invaded (Fig 4.18).

SIGNS & SYMPTOMS

Young larvae (caterpillars) eat leaves until half-grown (through the third instar). ECB larvae then bore into small branches. Their bore holes extrude a slimy mix of sawdust and frass. Bore holes predispose plants to fungal infection by *Macrophomina phaseolina* (McPartland 1996) and *Fusarium* species (Grigoryev 1998). Within one or two weeks of boring into small branches, ECBs tunnel into main branches and stalks (Nagy 1959). Their tunnels may cut xylem and cause wilting. Stalks at tunnel sites may swell into galls, which are structurally weak, causing stalks to snap (Plate 18). Snapped galls spoil the fibre and prevent seed formation. Ceapoiu (1958) and Nagy (1959) photographed badly infested hemp fields, showing half the plants toppled at odd angles. Toppled plants become tangled in harvest machinery, with yield losses up to 50% (Grigoryev 1998). According to Emchuck (1937), five to 12 ECBs can destroy a hemp plant.

ECB larvae born in late summer or autumn will change tactics—instead of boring into stems, they infest flowering tops, wherein they spin webs and scatter faeces. They selectively feed on female flowers and immature seeds. Camprag *et al.* (1996) reported seed losses of 40%.

TAXONOMY & DESCRIPTION

Ostrinia nubilalis (Hübner) 1796, Lepidoptera; Pyralidae.

= *Pyraustra nubilalis* Hübner, = *Botys silacealis* Hübner 1796

Description: Eggs are less than 1 mm long. Just before hatching, the brown heads of larvae become visible within the creamy white eggs. Caterpillars are light brown with dark brown heads (Plate 18). Along the length of their bodies are found several rows of brown spot-like plates, each sprouting setae. Mature caterpillars may grow to 15–25 mm long. They spin flimsy cocoons and transform into reddish-brown torpedo-shaped pupae, 10–20 mm long. Female moths are beige to dusky yellow, with irregular olive-brown bands running in wavy lines across their 25 mm wingspan (Plate 19). Males are slightly smaller and darker, also with olive-brown markings on their wings. Eggs are laid on undersides of leaves, stems, or crop stubble. They are laid in groups of 15–50, with the masses measuring 4–5 mm in diametre. Young eggs are translucent and overlap like fish scales (Plate 20); with age they fill out (Fig 4.19).

Prior to 1850, scientific authors preferred the taxon *silacealis* to *nubilalis*, different names for the same species. Another ECB species, ***Ostrinia scapulalis*** Walker 1859, reportedly infests Ukrainian hemp and does not attack maize (Forolov 1981). *O. scapulalis* larvae and female moths look identical to *O. nubilalis*. Male *O. scapulalis* moths exhibit large, wide tibiae; this morphological feature separates the species from *O. nubilalis* (Forolov 1981).

Life History & Host Range

Mature larvae overwinter in crop stubble near the soil line. Springtime feeding begins when temperatures exceed 10°C. Larvae pupate for two weeks and then emerge as moths in late May (or June, or even August in Canada and northern Europe). Females are strong night flyers, seeking out host plants to lay eggs. They lay up to 500 eggs in 25 days (Deay 1950). Eggs are deposited on lower leaves of the most mature (i.e., earliest planted) hosts. *Artemisia vulgaris* is a common weed host (Grigoryev 1998). Eggs of first-generation ECBs hatch in a week or less.

Larvae feed for about three weeks, then spin cocoons and pupate. Moths emerge, mate, and repeat the life cycle. A hard freeze late in the year kills all but the most mature larvae (those in their fifth instar). In Russia, ECBs live as far north as 56° latitude (Grigoryev 1998). One to four generations of *O. nubilalis* arise each year, depending on latitude and local weather (Fig 4.18). In western Europe only one generation arises north of about 45° latitude. Young (1997) predicts global warming will expand the range of ECBs, so two generations may arise as far north as 52° latitude. Summers with high humidity and little wind favour egg-laying, egg survival, and larval survival.

DIFFERENTIAL DIAGNOSIS

ECBs must be differentiated from hemp borers (*Grapholita delineana*), see Fig 4.22. Nagy (1959) reported that 91% of ECB galls are located in the lower three-quarters of *Cannabis* plants, while hemp borers tend to infest the upper thirds of plants (Nagy 1967). ECBs make longer tunnels than hemp borers (Miller 1982). The grubs of weevils, curculios, and other assorted beetles bore into stems and may be mistaken for ECB larvae. Gall midges cause small galls in small branches and male flowers.

Figure 4.19: Egg mass of *O. nubilasis* (from Senchenko & Timonina 1978).

CULTURAL & MECHANICAL CONTROL

(method numbers refer to Chapter 9)

Method 1 (sanitation) was written with ECBs in mind. Also apply methods 2a (deep ploughing), 3 (weeding, especially *Artemisia*), 4 (planting late in the spring), 6 (crop rotation with red clover, *Trifolium pratense*), 7a (proper moisture levels), 9 (hand removal), and 12d (mechanical light traps). Virovets & Lepskaya (1983) cited Ukrainian cultivars with resistance to *O. nubilalis*. Grigoryev (1998) found highest resistance to *O. nubilalis* in some Ukraine cultivars ('USO-27,' 'USO-25'), and landraces from Italy ('Carmagnola'), France ('Chenevis'), and Yugoslavia ('Domaca Province'). Cultivars with less resistance included 'USO-14,' 'USO-12,' 'Yellow Stem,' and 'Uniko B.'

Grigoryev (1998) reported less infestation in dense hemp fields (4-5 million plants per ha) than in sparse plantings (500,000 plants per ha). Crops cultivated in organically-managed soils suffer less ECB problems than crops cultivated with conventional fertilizers (Phelan *et al.* 1996). Avoid planting *Cannabis* near maize fields. Cover glasshouse vents with screens and extinguish lights at night to exclude light-attracted moths.

BIOCONTROL (see Chapter 10)

The most successful parasitoids are *Trichogramma* species, described below. Research on ECB parasitoids began around 1886 in French hemp fields (Lesne 1920). The USDA has introduced over 20 species of ECB parasitoids into the USA. Only a few became established, such as the tachnid fly *Lydella thompsonii* (described below), and the braconid wasps *Meteorus nigricollis* and *Macrocentrus grandii*.

Several predators feed on ECB eggs, such as green lacewings (described under aphids) and pirate bugs (*Orius* species, described under thrips). Yepsen (1976) claimed that ladybeetles eat up to 60 ECB eggs a day. *Podisus maculiventris* preys on young ECB larvae (described in the section on leaf-eating caterpillars).

Bacillus thuringiensis (Bt) is the best microbial pesticide, described below. The fungus *Beauveria bassiana* (described under whiteflies) is lethal to first-instar ECB larvae (Feng *et al.* 1985). *B. bassiana* is widely employed in China for control of ECBs (Samson *et al.* 1988). *B. bassiana* strain 147 (Ostrinil®) is registered for control of ECBs in France. The fungus also survives as an endophyte in some plants; maize plants inoculated with *B. bassiana* were protected against ECBs tunnelling within stalks. Kolotlina (1987) killed ECB larvae in hemp fields with a mix of *B. bassiana* and *B. globulifera*.

Nosema pyrausta, a naturally-occurring control agent, is described below. Don Jackson of Nature's Control (pers. commun. 1997) suggested killing ECBs with beneficial nematodes such as *Steinernema carpocapsae* (described under cutworms). Inject the nematodes directly into stalks.

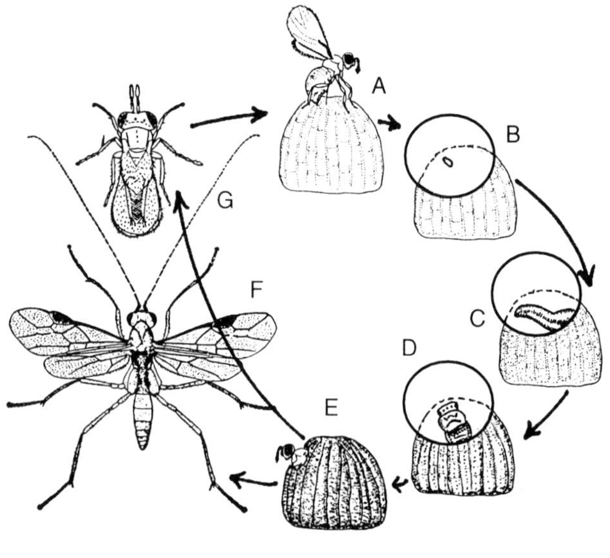

Figure 4.20: Life cycle of *Trichogramma minutum*. A. Female wasp laying egg within caterpillar egg, B. View of egg within egg, C. Larva feeding within egg, D. Pupal stage, E. emerging adult, F. Adult male with open wings, G. Adult female with closed wings (McPartland redrawn from Davidson & Peairs 1966).

***Trichogramma* species**

BIOLOGY: At least 20 *Trichogramma* species have been mass-reared for field use. They efficiently kill ECB eggs, *before* the pests can damage crops. The eight most popular species are described below. Most *Trichogramma* wasps work best at 20–29°C (range 9–36°C) and 40–60% RH (range 25–70% RH). Species of a related genus, ***Trichogramatoidea***, also provide biocontrol against caterpillars.

APPEARANCE & DEVELOPMENT: Adults are very tiny, 0.3–1.0 mm long, with a black thorax, yellow abdomen, red eyes, and short antennae (Plate 20). Females lay eggs in up to 200 pest eggs, which turn black when parasitized. Larvae pupate within pest eggs and emerge as adults in eight days (Fig 4.20), adults live another ten days.

APPLICATION: There are two application approaches, inoculation or inundation (described in Chapter 10). The inoculation approach often fails, because the wasps do not become established. The inundation approach requires the release of wasps when ECB moths lay eggs. Because wasps can only parasitize young eggs (one to three days old), and ECB moths lay eggs for several weeks, repeated inundations become necessary.

Trichogramma species are supplied as pupae within parasitized eggs, attached to cards made of cardboard, paper, bamboo, or within gelatine capsules. Store for up to a week in a cool (6–12°C), dark place. Cold-hardy species may tolerate longer storage. Avoid releasing in cold, rainy, or windy conditions. If inclement weather cannot be avoided, the rate and frequency of releases must be adjusted upward (Smith 1996). The wasps cannot fly against winds stronger than 7 km h^{-1}, which blow them out of release fields (Bigler *et al.* 1997). *Trichogramma* pupae can be manually distributed by hanging cards from plants in warm, humid places out of direct sunlight. This approach takes about 30 minutes per ha (Smith 1996). Watch out for ants, which eat the pupae. In the presence of ants, place pupae in small cartons fitted with protective screening, then attach cartons to plants. For a large-scale approach, attach *Trichogramma* pupae to carriers such as bran, and broadcast them from tractors or airplanes. Pupae and carriers can be coated with acrylamide sticky gel so they adhere to plant surfaces.

According to Orr & Suh (2000), *Trichogramma* product support in Europe is superior to that in the USA. European suppliers maintain rigid quality control procedures, and ship their products by overnight delivery in refrigerated trucks or in containers with ice packs. Furthermore, European suppliers provide their customers with extensive technical information, and even monitor temperature data and ECB populations in areas where *Trichogramma* releases take place. Customers are informed in advance of product delivery and release dates. In contrast, nearly 50% of USA companies shipped dead biocontrols or *Trichogramma* species other than that which was claimed, shipped products in padded envelopes by standard "snail mail," and provided little or no product information or instructions (Orr & Suh 2000).

NOTES: *Trichogramma* species are compatible with Bt and NPV. Avoid insecticides while utilizing the wasps. *Trichogramma* wasps search for eggs by walking across leaf surfaces, so leaf trichomes slow them down. On tomato leaves, the wasps get entangled in trichome exudates and die, especially if exudates contain methyl ketones (Kashyap *et al.* 1991). It is worth mentioning that *Cannabis* trichomes exude methyl ketones (Turner *et al.* 1980), and probably hinder *Trichogramma* wasps. Bredemann *et al.* (1956) attempted to breed *Cannabis* plants without trichomes, but their work was discontinued. *Trichogramma* wasps live longer and lay more eggs if provided with food (e.g., wild flowers with nectaries, see Chapter 9, method 14).

Different *Trichogramma* species vary in their effectiveness against different pests. Some species work best against ECBs, other work best against budworms. The relative effectiveness of different *Trichogramma* species against ECBs is described below.

T. evanescens This is a cold-hardy European species, with popular strains from Moldavia and Germany. Adults are weak fliers and usually move less than 3 m from release sites. Li (1994) tested 20 *Trichogramma* species, and judged *T. evanescens* the second-most effective control of ECBs (behind only *T. dendrolini*). At the first sight of moths in the glasshouse, release ten pupae per m^2. *T. evanescens* can be released at wider intervals than other *Trichogramma* species—every one to three weeks, using 200,000–300,000 wasps per ha (Smith 1996).

T. pretiosum* & *T. minutum These polyphagous species parasitize the eggs of many Lepidoptera. Mixtures of both species are available. Clarke (unpublished research, 1996) used a combination of *T. pretiosum*, *T. minutum*, and *T. evanescens* for best control in Dutch glasshouses. *T. pretiosum* works best on plants under 1.5 m tall, while *T. minutum* protects taller plants. Neither species is very cold-hardy. According to Losey & Calvin (1995), *T. pretiosum* parasitized ECB eggs better than *T. minutum*, but both were inferior to *T. nubilalis*. Nevertheless, Marin (1979) used *T. minutum* to protect 12,500 ha of Soviet hemp. Marin released *T. minutum* from an airplane; release rates were not given. Camprag *et al.* (1996) released 75,000–100,000 wasps per ha, and repeated the release a week later. Wasps were released from 50–60 locations per ha. Tkalich (1967) required 120,000–150,000 wasps per ha, released when oviposition began, and released again when oviposition peaked.

T. brassicae (=maidis) This cold-hardy species is susceptible to mechanical injury, so wasp pupae shipped on paper cards often die before release. French researchers improved survival rates by gluing pupae to the inner wall of capsules (Trichocaps®). This formulation makes handling easier, with less wasp mortality. It has made *T. brassicae* the favorite biocontrol of ECBs in Europe. In maize, release 200,000 wasps (400 capsules) per ha per week, whenever significant numbers of ECB moths are caught in traps or seen laying eggs. Depending on the number of ECB generations per season, the number of required releases ranges from one (Orr & Suh 2000) to four to nine (Smith 1996). Bigler *et al.* (1997) found that *T. brassicae* heavily parasitized ECB eggs within 8 m of each release site.

T. ostriniae & T. dendrolimi These species have been introduced from China. *T. dendrolini* is cold-hardy, *T. ostriniae* is not (*T. ostriniae* cannot survive winters in Vermont). Release rates and intervals are similar to those of *T. brassicae*. In maize, *T. ostriniae* prefers to parasitize eggs found in the lower and middle parts of plants—not flowering tops (Wang *et al.* 1997). Li (1994) judged *T. dendrolimi* the most effective *Trichogramma* control of ECBs, out of 20 tested species. Additionally, *T. dendrolimi* is polyphagous, so in the absence of ECBs it will parasitize other caterpillars.

T. nubilale This species is native to North America and readily parasitizes *O. nubilalis*. Losey & Calvin (1995) reported best results with *T. nubilale* against ECB in maize. According to others, however, *T. nubilale* may not reduce ECB populations consistently (Andow *et al.* 1995); *T. nubilale* was 15% less effective than *T. ostrinia*. Simultaneous release of the two species produced worse control—so *T. nubilale* and *T. ostrinia* should not be used together.

T. platneri is native to the USA west of the Rockies. It has been used in orchards and vineyards. According to Thomson (1992), it is not compatible with *T. minutum*, an East Coast species. Losey & Calvin (1995) reported poor results against ECB in maize.

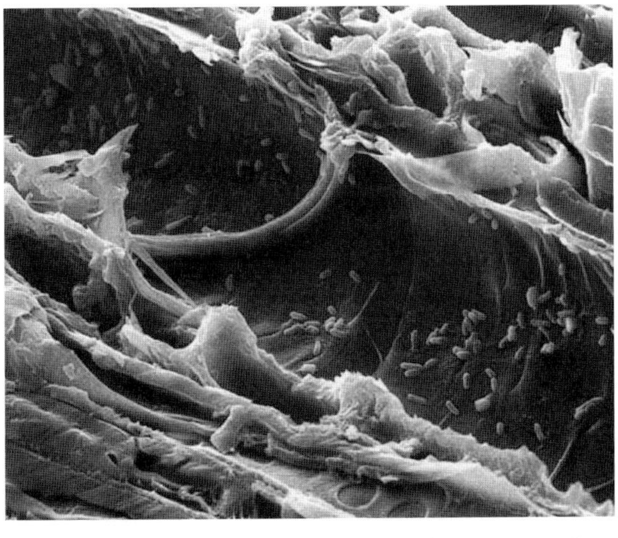

Figure 4.21: Xylem vessel colonized by *Clavibacter xyli*, a bioengineered Bt carrier (SEM x1000, McPartland).

Bacillus thuringiensis "Bt"

BIOLOGY: The Bt bacterium was described by Louis Pasteur in the 1860s (Flexner & Belnavis 2000), but Bt was not field-tested against pests until the 1920s (the first trials were against ECBs). Bt is actually a *family* of bacteria—at least 35 varieties of Bt produce at least 140 types of toxins.

The active agent in Bt is a spore toxin, which exists in the bacterium as a nontoxic crystalline protein (δ-endotoxin). The endotoxin does not become toxic until the protein is dissolved by proteinases in the gut of certain insects. The activated toxin binds to cell membranes in the gut. Part of the toxin forms an ion channel in the cell membrane, causing selective cell leakage of Na^+ and K^+, resulting in cell lysis.

Bacterial plasmids that encode toxin production (*cry* genes) have been classified by amino acid homology. Generally, Cry1, Cry2, and Cry9 toxins are active against lepidopteran caterpillars (formerly grouped as CryI), Cry3, Cry7, and Cry8 toxins kill beetle grubs (formerly grouped as CryIII), and Cry4, Cry10, and Cry11 toxins kill fly maggots (formerly grouped as CryIV).

APPEARANCE & DEVELOPMENT: After eating Bt, insects stop feeding within the hour. They become flaccid, and a foul-smelling fluid (liquefied digestive organs) trickles from their mouths and anuses. They shrivel, blacken, and die in several days (Plate 21). Dead insects can be found hanging from leaves. Unfortunately, Bt bacteria rarely multiply in pest populations; they must regularly be reapplied.

APPLICATION: Bt is formulated as a dust, granule, sand granule, wettable powder, emulsifiable concentrate, flowable concentrate, and liquid concentrate. Commercial products do not contain viable bacteria. Most Bt formulations contain the toxin and spore, but some contain only the toxin. Store powders in a cool, dry place. Store liquid formulations in a refrigerator, for a maximum of one year. Immediately before spraying, mix with slightly acidic water (pH 4–7). For better results add a spreader-sticker and UV inhibitor. For best results add a feeding stimulant (Entice®, Konsume®, Pheast®). Spray all foliage uniformly and completely.

Bt works best against young, small larvae—so spray at the first sign of caterpillar hatchout or crop damage. Bt on foliage is degraded by UV light within one to three days, so spray outdoors in late afternoon or on cloudy (not rainy) days. Repeat application *at least weekly* while pests are out and about. Spraying Bt every ten to 14 days prevents ECB infestations, *although surface sprays will not kill ECBs already inside stalks.* Frank (1988) used a large-bore syringe to inject Bt into stalk galls.

Bt* variety *kurstaki is the most popular Bt, on the market since 1961. Over 100 products contain Bt-k, largely derived from the HD-1 strain. Popular trade names include Agrobac®, Biobit®, BMP 123®, Condor®, Cutlass®, DiPel®, Full-Bac®, and Javelin®. Bt-k kills caterpillars, including ECBs, hemp borers, and budworms. Genetic engineers have transferred the Bt-k toxin gene Cry1A(b) to *Pseudomonas florscens.* This bacterium has a thick wall, so the *P. florscens* product is marketed as "microencapsulated" (M-Trak®, MVP II®, M-Peril®, Mattch®). Microencapsulation protects the toxin from UV light, so the toxin remains effective on plant surfaces for up to eight days (or two to four times longer than regular Bt-k). Bt-k genes have also been inserted into a bacterium, *Clavibacter xyli*, which lives in plant xylem (Fig 4.21). Inoculate seeds with *C. xyli*, or hand-vaccinate plants to treat ECBs already inside stalks.

Bt* variety *aizawai (Bt-a, XenTari®, Agree®, Design®) is a new Bt that kills armyworms, budworms, and some borers. Bt-a contains a different endotoxin, so it is more effective against *Spodoptera* species and insects that have developed resistance to Bt-k. Some products contain a transconjugated combination of Bt-a and Bt-k (Mattch®).

Bt* variety *morrisoni (Bt-m), the latest Bt, being marketed for use against lepidopterae with high gut pHs, such as armyworms and cabbage loopers.

NOTES: Bt toxin is normally harmless to plants and beneficial insects, though some people may develop allergic reactions. The Bt toxin is safely pyrolysed. Unfortunately, heavy reliance on Bt has created Bt-resistant caterpillars, which first appeared around 1985 (Gould 1991).

The work of genetic engineers may accelerate resistance. Bioengineers have transferred Bt genes to crop plants. Every cell in these transgenic plants produces its own cache of Bt—the equivalent of a permanent systemic insecticide. This input trait creates constant, global pressure for the selection of resistant pests. Organic farmers, who have relied on Bt for decades, may lose their best weapon against caterpillars. To delay the evolution of Bt-resistant pests, farmers who plant Bt crops are required to plant a percentage of their acreage in non-Bt crops. The idea behind this "refuge" strategy is that Bt-resistant pests arising in transgenic Bt-corn will mate with Bt-susceptible pests from neighbouring nontransgenic crops. Assuming Bt resistance is recessive, and assuming susceptible adults are homozygous, the offspring of these matings should continue to be susceptible to Bt. The problems with this strategy, however, are multifarious. The correct size of "refuges" has not been determined. The spatial proximity of transgenic plants to nontoxic plants is critical—close enough to allow mating between resistant and nonresistant moths, but far enough apart to prevent the migration of larvae from transgenic plants to nontoxic plants. Convincing farmers to create pest havens (aka, refuges) may be difficult, and there is no enforcement mechanism. Furthermore, the basic premise may be flawed: Bt resistance in ECBs may not be recessive (Huang *et al.* 1999). Pests have alread appeared with resistance to Bt-transgenic plants (Gould 1998).

To complicate the situation, transgenic Bt is expressed in a modified, truncated form. Hilbeck *et al.* (1998) demonstrated that truncated Bt may harm predators, such as lacewings (*Chrysoperla carnea*) that eat Bt-consuming ECBs. The effects of transgenic Bt upon nontarget organisms cannot be predicted; pollen from Bt-corn, for instance, contains enough Bt to kill larvae of the monarch butterfly, *Danaus plexippus*. Bt-toxic corn pollen blows onto milkweed plants (*Asclepias curassavica*) within 60 m of corn fields; caterpillars that eat pollen-dusted milkweed plants stop feeding and die (Losey *et al.* 1999).

Lydella thompsonii

BIOLOGY: This tachinid fly is native to Europe. The USDA released *L. thompsonii* in the USA between 1920 and 1938. It became firmly established and now can be found from New York to South Carolina and west to the corn belt. *L. thompsonii* parasitizes up to 75% of ECBs in the midwest (Hoffmann & Frodsham 1993).

APPEARANCE & DEVELOPMENT: *L. thompsonii* adults resemble large, bristly houseflies. Female flies run up and down hemp stalks, looking for ECB entrance holes. Females deposit living larvae, which wriggle into entrance holes. The maggots bite ECBs and bore into them (Mahr 1997). Mature maggots leave dead ECB larvae to pupate within stalks, then emerge as adults. The life cycle can be as short as 20 days. *L. thompsonii* overwinters as maggots in hibernating ECB larvae. Two or three generations arise per year. The first generation often emerges too early in the spring, and must survive on alternate hosts, such as *Papaipema nebris*, the common stalk borer (Mahr 1997).

NOTES: *L. thompsonii* is susceptible to microscopic biocontrols, such as *Beauveria bassiana* and *Nosema pyrausta* (Mahr 1997).

Nosema pyrausta

BIOLOGY: A single-celled microsporidial protozoan that infects ECBs. It is presumably native to Europe but now lives in North America.

DEVELOPMENT: *N. pyrausta* rarely kills caterpillars, but causes debilitating disease in both larvae and adults, which reduces ECB longevity and fecundity. Hence *N. pyrausta* serves as a long-term population suppressor. It produces tiny ovoid spores (4.2 x 2 μm) which spread in larval faeces, and spreads by transovarial transmission in eggs.

APPLICATION: Researchers have mass-produced *N. pyrausta* as spores, sprayed on foliage while young ECBs are still feeding on leaves. Spores persist on foliage for one to two weeks. ECB predators, such as green lacewings, can eat *N. pyrausta*-infected ECB eggs and pass the spores harmlessly in their faeces. These spores remain infective. Unfortunately, *parasitoids* that develop within ECBs, such as *Macrocentrus grandii*, may acquire the infection. *N. pyrausta* can be used with Bt—the treatments have an additive effect.

NOTES: A second species, ***Nosema furnacalis***, is currently under investigation. A related microsporidian, ***Vairimorpha necatrix***, is also being investigated. *V. necatrix* kills ECBs and various armyworms much quicker than *Nosema* species.

CHEMICAL CONTROL (see Chapter 11)

Young ECBs (those still feeding on leaves) are repelled or killed with neem, nicotine, rotenone, and ryania. Once ECBs bore into stalks, nothing sprayed on plant surfaces will affect them. Frank (1988) used a large-bore syringe to inject stalk galls with plain mineral oil. Clarke (pers. commun. 1995) retrofitted the spray nozzle and tube from a can of WD-40® onto an aerosol can of pyrethrin, and sprayed the pyrethrin into borer holes. After treatment, wipe away all frass from stems. New excrement indicates a need for repeated treatment.

Chemicals and microbial biocontrols work better when combined with a feeding stimulant. One commercially-available stimulant, Coax®, has been used against ECBs. For more information, see the section on Budworms.

Baiting with synthetic pheromones can lure male moths into traps. When many moths appear in pheromone traps, a new generation of young larvae is only a week away—get your *Trichogramma* and Bt ready. In North America, two strains of ECBs respond to different pheromone blends of tetradecenyl acetate (Howard *et al.* 1994), so growers will need two traps for the two strains. Alternatively, small rubber septums can be impregnated with pheromones and spread around hemp fields to confuse male moths and inhibit mating (Nagy 1979).

HEMP BORERS

Hemp borers are also called hemp leaf rollers and hemp seed eaters. Besides boring into stems, Kryachko *et al.* (1965) described hemp borers destroying 80% of flowering tops in Russia. Bes (1978) reported 41% seed losses in unprotected Yugoslavian hemp. Each larva consumes an average of 16

Cannabis seeds (Smith & Haney 1973). In addition to cultivated hemp, hemp borers also infest marijuana, feral hemp (*Cannabis ruderalis*), and hops. Anti-marijuana researchers considered the hemp borer "an excellent biocontrol weapon" (Mushtaque *et al.* 1973, Baloch *et al.* 1974, Scheibelreiter 1976). Forty larvae can kill a seedling that is 15–25 cm tall in ten days (Baloch *et al.* 1974). Ten larvae per plant cripple growth and seed production. A Pakistani strain is host-specific on *Cannabis;* the larvae do not feed on hops like European hemp borers (Mushtaque *et al.* 1973). Hemp borers arrived in North America around 1943 (Miller 1982).

SIGNS & SYMPTOMS

Hemp borers feed on leaves in the spring and early summer. Subsequently they bore into petioles, branches, and stalks. Feeding within stalks causes fusiform-shaped galls (Plate 22). Stalks may break at galls, although the length of tunnels within galls averages only 1 cm (Miller 1982).

Borers that hatch in late summer and autumn spin loose webs around terminal buds and feed on flowers and seeds. Senchenko & Timonina (1978) provided an illustration of seed damage. Late season larvae pupate in curled leaves within buds, bound together by strands of silk.

TAXONOMY & DESCRIPTION

Grapholita delineana (Walker) 1863, Lepidoptera; Olethreutidae.

= *Cydia delineana* (Walker), = *Laspeyresia delineana* (Walker),
= *Grapholita sinana* Felder 1874, = *Cydia sinana* (Felder)

G. delineana and *G. sinana* were considered different species until 1968 (Miller 1982). Smith & Haney (1973) and Haney & Kutscheid (1975) report ***Grapholita tristrigana*** (Clemens) infesting hemp. Miller (1982) reexamined *G. tristrigana* specimens from Minnesota, Iowa, Missouri, Wisconsin, Illinois, Kentucky and New York—all were *G. delineana,* not *G. tristrigana.* A report of ***Grapholita interstictana*** on hemp (Dempsey 1975) probably represents another misidentification. These species are differentiated by their genitalia, summarized by Miller (1982).

Description: Eggs are white to pale yellow, oval, 0.4 mm wide, and laid singly on stems and undersides of leaves. Larvae are pinkish-white to pale brown, up to 9–10 mm long (Fig 4.22). Several pale bristles per segment are barely visible. Their heads are dark yellow-brown, with black ocelli, averaging 0.91 mm wide. Larvae pupate in silken cocoons covered with bits of hemp leaf. Adults are tiny moths, with greyish- to rusty-brown bodies and brown, fringed wings. Body length and wingspan average 5 mm and 9–13 mm respectively in males, and 6–7 and 10–15 mm respectively in females. Forewings exhibit white stripes along the anterior edge with four chevron-like stripes near the centre (Fig 4.22, Plate 23).

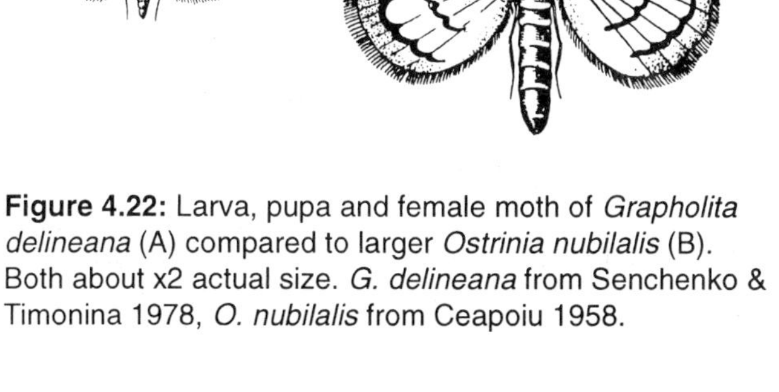

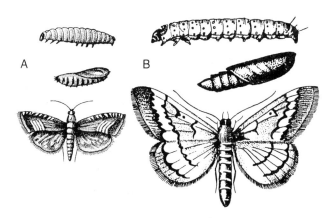

Figure 4.22: Larva, pupa and female moth of *Grapholita delineana* (A) compared to larger *Ostrinia nubilalis* (B). Both about x2 actual size. *G. delineana* from Senchenko & Timonina 1978, *O. nubilalis* from Ceapoiu 1958.

Life History & Host Range

Hemp borers overwinter as last-instar larvae in crop stubble, weeds, and sometimes stored seed (Shutova & Strygina 1969). They pupate in April, in soil under plant debris. Moths emerge in early May and migrate at night to new hemp fields. Moths are not strong fliers; Nagy (1979) calculated flight speeds of 3.2–4.7 km $hour^{-1}$ in a wind chamber. Upon finding a hemp field, females land quickly, usually within 3 m of the field's edge. After mating, females lay between 350–500 eggs. Adults live less than two weeks.

Eggs hatch in five to six days at 22–25°C, or three to four days at 26–28°C. Out of 350–500 eggs, Smith & Haney (1973) estimated only 17 larvae survived to first instar. Larvae skeletonize leaves for several days before boring into stems. Borers pupate within stems. The Primorsk region of Russia is evidently the northern limit of *G. delineana* in Eurasia. Further south, two generations occur in Hungary (Nagy 1979) and the Ukraine (Kryachko *et al.* 1965). A partial third generation arises in Armenia (Shutova & Strygina 1969) and four generations overlap in Pakistan (Mushtaque *et al.* 1973).

Larvae go into hibernation in September and October. Day length under 14 hours induces diapause (Sáringer & Nagy 1971). Temperature also influences diapause—warm weather slows photoperiodic effects.

DIFFERENTIAL DIAGNOSIS

Hemp borer damage often arises in the top 1/3 of plants (Nagy 1967), while European corn borers usually form galls in the lower 3/4ths of plants (Nagy 1959). European corn borers and other boring caterpillars drill longer tunnels than *G. delineana* larvae. Weevils, curculios, and gall midges also bore into stems and form galls. Late-season hemp borers that infest buds may be confused with late-season budworms.

CULTURAL & MECHANICAL CONTROL

(method numbers refer to Chapter 9)

Methods 1 (sanitation) and 2a (deep autumn ploughing) greatly reduce overwintering borer populations. Eliminate "sanctuary" stands of feral hemp, hops, and other weeds (method 3). Early crop harvest (method 4) is helpful. Since full-grown larvae survive in harvested stems, stems should not be transported into uninfested areas. Larvae may overwinter in seed, so method 11 is important. Methods 12d (nocturnal light traps) and 13 (screening) help. Breeding dwarf *Cannabis* may decrease hemp borer infestation—Smith & Haney (1973) rarely found larvae attacking plants less than 30 cm tall, even when plants were flowering.

BIOCONTROL (see Chapter 10)

Native organisms heavily parasitize *G. delineana* larvae in the USA, which may explain why hemp borers rarely harm feral hemp. Smith & Haney (1973) found 75% of larvae infested by *Lixophaga variablis* (Coquillett), a tachinid fly, and *Macrocentrus delicatus* Cresson (a braconid wasp, described below). *Goniozus* species attack *G. delineana* larvae in Pakistan (Mushtaque *et al.* 1973); *Scambus* species parasitize 30% of hemp borers in Hungary (Scheibelreiter 1976).

Camprag *et al.* (1996) used *Trichogramma* wasps to control the first generation infestation with "51–68% efficiency." He released 75,000–100,000 wasps per ha, and repeated the release one week later. He did not report which species he used. *Trichogramma* species are described in the previous section on European corn borers.

Peteanu (1980) experimented with *Trichogramma evanescens.* He released 80,000, 100,000, or 120,000 wasps per ha, four times per season (presumably two releases against the first generation and two against the second generation).

T. evanescens worked best at the highest release rate, and worked better against second generation larvae than first generation larvae. Peteanu combined *T. evanescens* with Bt and pesticides, with interesting results (see Table 4.8).

Smith (1996) controlled a related *Grapholita* pest, *G. molesta,* with *Trichogramma dendrolimi* released at a rate of 600,000 wasps per ha, repeated every five days while moths were laying eggs.

Bako & Nitre (1977) successfully controlled young hemp borers with aerial applications of Bt (described in the previous section). *Podisus maculiventris* preys on young hemp borers feeding in the leaf canopy (described in the section on leaf-eating caterpillars). Consider injecting stem galls with beneficial nematodes such as *Steinernema carpocapsae* (described under cutworms).

Macrocentrus ancylivorus

BIOLOGY: Several *Macrocentrus* species have been mass-reared for field use. *M. ancylivorus* is a braconid wasp native to New Jersey. It attacks many fruitworms, leafrollers, and stem borers. In the 1930s *M. ancylivorus* was reared in large numbers to control *Grapholita molesta.* The parasitoid does best between Massachusetts and Georgia, west to the Mississippi river.

APPEARANCE & DEVELOPMENT: Adults are slender wasps, 3–5 mm long, amber-yellow to reddish brown in colour, with antennae and ovipositors longer than their bodies. Female wasps are nocturnal, most active at 18–27°C and >40% RH, and lay up to 50 eggs, one egg per borer (Mahr 1998). The wasps go after borers already within branches (second and third instars preferred). *M. ancylivorus* larvae initially feed within caterpillars, then emerge to feed externally, and pupate in silken cocoons next to the body of their hosts. One generation arises per generation of the host. They overwinter as larvae in hibernating hosts.

NOTES: In orchards, *Grapholita molesta* is controlled by releasing three to six *M. ancylivorous* females per tree (Mahr 1998). A related species, *Macrocentrus delicatus,* attacks *G. delineana* in the midwest. *Macrocentrus grandii* and *Macrocentrus gifuensis* are widespread biocontrols of European corn borers.

CHEMICAL CONTROL (see Chapter 11)

Many botanicals that control European corn borers also work against hemp borers—neem, nicotine, rotenone, and ryania. Peteanu (1980) killed borers with sumithrin, a synthetic pyrethroid (see Table 4.8). Spraying pesticides only works *before* hemp borers burrow into stalks. Once inside stems, no surface sprays will affect borers. Nagy (1979) described an "edge effect" in fields infested by *G. delineana.* Weakly-flying female moths land quickly to lay eggs after discovering a hemp field; most infestations occur in the first 3 m around a hemp field. Spray this edge zone with pesticides as moths arrive. In severe infestations, the edge zone should be cut down and buried or burned.

Nagy (1979) used female sex hormones to attract and trap male moths, preventing reproduction. Fenoxycarb, a synthetic juvenoid, kills eggs of the related pest *Grapholita funebrana* (Godfrey 1995).

Table 4.8: Effectiveness of biological and chemical control against *Grapholita delineana.*[1]

Treatment	Dose per hectare[2]	Treatments per season	Fatality rate of larvae (%)
1. *Trichogramma*	120,000	4	89.5
2.. *Trichogramma*	100,000	4	82.5
3. *Trichogramma*	80,000	4	77.9
4. Bt (Dipel)	2 kg	4	74.4
5. sumithrin	750 g a.i.	4	86.1
6. diazinon 5%G	25 kg	1	55.8
7. #2 + #3	100,000 + 2 kg	2 + 2	84.9
8. #2 + #4	100,000 + 750 g	2 + 2	89.2
9. #2 + #5	100,000 + 25 kg	3 + 1	76.7

[1]Data from Peteanu (1980)
[2]Dose per individual treatment

OTHER BORING CATERPILLARS

At least five other moth larvae damage *Cannabis* stalks. These borers generally do not form galls. Infested plants appear stunted and unhealthy. Inspection of stalks reveals entrance holes, often exuding sawdust-like frass. Hollowed stalks may collapse and fall over.

1. GOAT MOTH

Cossus cossus (Linnaeus) 1758, Lepidoptera; Cossidae.

Description: *C. cossus* caterpillars are said to smell like goats (as will stalks, even after larvae exit to pupate in soil). Caterpillars are red-violet and grow 90 mm long. Pupae are reddish-brown, 38–50 mm long, covered with pieces of wood and debris. Moths are robust, brown-bodied, with olive-grey variegated wings reaching a span of 90 mm.

Life History & Host Range

Ferri (1959) described goat moth larvae attacking Italian hemp. Moths emerge from pupae in June and July. *C. coccus* takes three or four years to complete its life cycle. The pest normally infests trees in Europe, central Asia, and northern Africa. A related American species, *Cossus redtenbachi,* ("the worm in the bottle") infests *Agave* plants, the source of tequila.

2. MANANDHAR MOTH

Zeuzera multistrigata Moore 1881, Lepidoptera; Cossidae.

= *Zeuzera indica* Walker 1856, nec *Z. indica* Herrich-Schäffer 1854

Description: Adults of this species resemble Leopard moths (*Zeuzera pyrina* L.), with predominantly white bodies, three pairs of steel-blue spots on the thorax, and seven black bands across the abdomen. Wings are white with steel-blue spots and streaks, veins have an ochreous tinge, wings span 85 mm in females, 65 mm in males. Heads and antennae are black, lower legs blue, femora white. Larvae are predominantly white, with black heads and dark spots.

Life History & Host Range

Baloch & Ghani (1972) described this species causing serious damage in Pakistan. *Z. multistrigata* takes two or three years to complete its life cycle. According to Sorauer (1958), this pest normally feeds within coffee tree trunks, as does its difficult-to-distinguish cousin, *Zeuzera coffeæ* Nietn. *Z. multistrigata* occurs across the Himalayan foothills from Dharmsala to Darjeeling.

3. COMMON STALK BORER

Papaipema nebris (Guenée) 1852, Lepidoptera; Noctuidae.

=*Papaipema nitela* (Guenée) 1852

Description: Eggs are white, globular, ridged. Young larvae are reddish-brown with a pair of white dorsal stripes. Side stripes are interrupted near the middle of the body, lending a "bruised" appearance (Metcalf *et al.* 1962). When larvae approach maturity they lose their stripes and turn a dirty-grey colour. Heads and anal shields are dark brown, mandibles and true legs are black. Mature larvae reach 45 mm in length and bore large exit holes (7 mm diametre). Moths have robust brown bodies and greyish brown wings with small white dots, wingspan 30 mm.

Life History & Host Range

Tietz (1972) listed *Cannabis* among this borer's 68 herbaceous hosts. Young larvae bore in monocots (e.g., quackgrass, giant foxtail) until they outgrow these hosts; by July they migrate to larger-stemmed plants. Maize and giant ragweed are two common targets. Larvae usually enter stems near the ground, bore for 25–50 cm, then exit in search of a new host. This restless habit multiplies crop losses. Larvae enter soil to pupate in August; moths emerge in September. Females deposit eggs on monocots between curled leaves. Eggs hatch in spring. One generation arises per year. *P. nebris* lives in North America east of the Rocky Mountains.

4. BURDOCK BORER

Papaipema cataphracta (Grote) 1864, Lepidoptera; Noctuidae.

Description: Metcalf *et al.* (1962) said burdock borers resemble common stalk borers, except *P. cataphracta's* side stripes are not interrupted and they are smaller. Adult moths resemble those of common stalk borers, except for different genitalia.

Life History & Host Range

Tietz (1972) listed 31 herbaceous plants attacked by this borer, including *Cannabis*. Frequent hosts include rhubarb, burdock, and thistle. Since *P. cataphracta* larvae are smaller than those of *P. nebris*, they inhabit smaller stems and branches. Larvae pupate in stems, not soil. One generation arises per year. The species is sympatric (same geographic range) with *P. nebris*.

5. HEMP GHOST MOTH

Endocylyta excrescens (Butler) 1877, Lepidoptera; Hepialidae.

Takahashi (1919), Clausen (1931), and Shiraki (1952) cited larvae of this moth attacking Japanese hemp. Larvae feed on roots and possibly stems. Adults are large (81–90 mm wingspan) and fly swiftly. Coloration of moth forewings varies from greenish grey to brown, with black flecks.

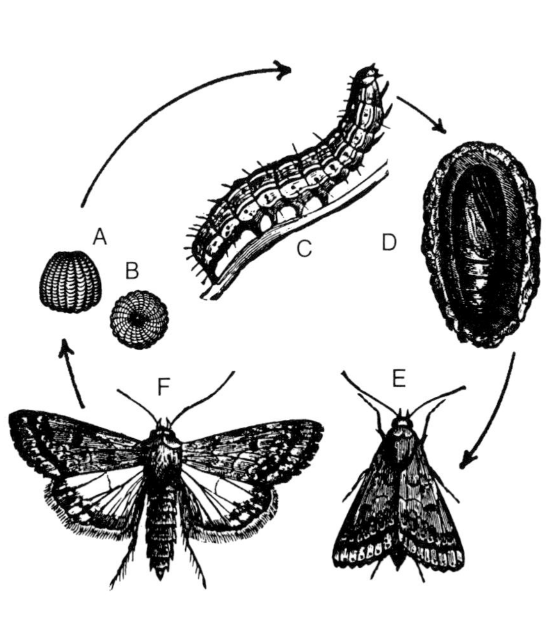

Figure 4.23: Life cycle of *Helicoverpa armigera*. A. Egg viewed from side; B. Egg viewed from above; C. Larva; D. Pupa; E & F. Adult (from Holland 1937).

DIFFERENTIAL DIAGNOSIS

European corn borers and hemp borers usually cause swellings at their feeding sites. Gall midges and boring beetles must also be differentiated. Split the stem and see what you find.

CONTROL

Follow recommendations for European corn borers and hemp borers. To reduce *P. nebris* and other omnivorous pests, mow down grass and weeds around crops. Grass and weeds need to be mowed again in August before moths emerge to lay overwintering eggs. Biocontrol researchers are testing *Scambus pterophori*, a wasp (Ichneumonidae) that parasitizes caterpillar borers and beetle grubs. *Lydella thompsoni* may attack *P. nebris* larvae (this biocontrol is described under European corn borers). All lepidoptera are killed by Bt (described under European corn borers). The two Noctuids (*P. nebris* and *P. ataphracta*) might also be sensitive to NPV (described under cutworms). To control borers with Bt and NPV requires individual injection of each plant stem. Consider injecting stems with beneficial nematodes (*Steinernema carpocapsae*, described under cutworms). Synthetic pheromone lures are available for more common pests, such as *C. cossus*.

BUDWORMS

Four budworms appear in the *Cannabis* literature. Budworms specialize in destroying plant parts high in nitrogen, namely flowers, fruits, and seeds. Some species also skeletonize leaves. Budworms spin loose webs around flowering tops and feed (and frass) therein. Sometimes they feed inside floral clusters where damage is not visible until flowers are ruined. Wounded buds and frass provide a starting point for grey mould infection.

1. COTTON BOLLWORM

Helicoverpa armigera (Hübner) 1809, Lepidoptera; Noctuidae.

=*Heliothis armigera* Hübner, =*Heliothis obsoleta* (Fabricius), =*Chloridea obsoleta* (Fabricius), = *Bombyx obsoleta* Fabricius 1775, nec *B. obsoleta* Fabricius 1793

Description: Eggs are hemispherical, shiny, with ridges that radiate from the apex like wheel spokes, white when newly laid but darkening to tan with a reddish-brown ring, 0.5 mm in diameter (Fig 4.23). Newly-hatched bollworms are pale yellow with dark longitudinal stripes, 1.5 mm in length. They grow into stout caterpillars up to 45 mm long. Mature caterpillars vary in colour from green to brown to almost black, with alternating light and dark longitudinal stripes along their bodies, pale-coloured undersides, with yellow-green heads and black legs (Fig 4.23 & Plate 24). Tiny spines cover most of their body surface (visible through a 10x hand lens), in addition to the dozen or so longer bristles found on each segment. Shiny brown pupae are found 5 cm or more below the soil surface. Moths are stout-bodied and brown; wings are yellow-brown with irregular lines and dark brown markings near the margins, wingspan up to 40 mm (Plate 25).

Life History & Host Range

H. armigera lives in Eurasia and Australia (Hill 1983), so we use the name preferred by Eurasian entomologists—*Helicoverpa*, rather than *Heliothis*. This pest does its worst damage in the tropics, where its life cycle can be as short as 28 days (Hill 1983). *H. armigera* commonly infests cotton, maize, tobacco, and chickpeas. It sometimes turns up in canned tomatoes. The pest has been reported on hemp by Goureau (1866), Riley (1885), Vinokurov (1927), Shiraki (1952), Rataj (1957), Ceapoiu (1958), Dempsey (1975), Khamukov & Kolotilina (1987), and Dippenaar *et al.* (1996). It is cited on marijuana by Rao (1928) and Nair & Ponnappa (1974). In southern India, Cherian (1932) considered *H. armigera* second only to spider mites in destructive capacity. He found that 100 bollworms could eat a pound of *Cannabis* (0.45 kg) per day!

H. armigera and the next species (*H. zea*) share similar life cycles. The pests produce one to six generations per year, depending on the latitude (six generations in the tropics, less away from the equator). Populations in temperate regions overwinter as pupae in soil. Tropical populations do not hibernate. Moths emerge from pupae as late as June in northern Russia (Vinokurov 1927). Female moths, which are nocturnal, lay over 1000 eggs, one at a time, on upper leaves of crops and weeds. According to Young (1997), *H. zea* moths lay more eggs on yellowed leaves than green leaves. Eggs hatch in three to five days. Larvae eat leaves, flowers, or seeds. The larval period lasts 14–51 days, depending on latitude (Hill 1983). Caterpillars in the tropics may seek shelter in soil during the heat of the day (Cherian 1932). Budworms are cannibalistic and feed on their brethren if suitable vegetation is unavailable.

2. BOLLWORM

Helicoverpa zea (Boddie) 1850, Lepidoptera; Noctuidae.
=*Heliothis zea* (Boddie), =*Bombyx obsoleta* Fabricius 1793, nec *B. obsoleta* Fabricius 1775

Description: *H. zea* closely resembles *H. armigera;* (compare Fig 4.24 to Fig 4.23), only a close look at genitalia tells them apart. Gill & Sanderson (1998) describe the adults as having distinct green eyes.

Life History & Host Range

A plethora of common names reflects the wide host range of *H. zea*—cotton bollworm (confused with *H. armigera*), tobacco budworm, tomato fruitworm, corn earworm, and vetchworm. Tietz (1972) cited *H. zea* on North American *Cannabis*. It may be the "bollworm" that Comstock (1879) reported "devouring heads of hemp." *H. zea* is not a permanent resident above 39°N (it cannot overwinter), but every summer the adults migrate as far north as Ontario (Howard *et al.* 1994). *H. zea* is native to the Americas, while *H. armigera* is a Eurasian species (Hill 1983). *H. zea* is easily mistaken for ***Heliothis virescens*** (F.), also confusingly known as the tobacco budworm—a horrific pest of tobacco, cotton, and other solanaceous plants. *H. zea* shares the same life cycle as *H. armigera,* described above.

3. FLAX NOCTUID

Heliothis viriplaca Hufanagel 1766, Lepidoptera; Noctuidae.
=*Heliothis dipsacea* Linnaeus 1767

Description: Larvae have greyish green to dark brown bodies up to 14–22 mm long, with green to yellow-green heads. Dorsal stripes are yellow, narrow, with dark borders described as a rust colour (Kirchner 1906). Moths and eggs resemble those of the aforementioned Noctuids.

Life History & Host Range

The Flax noctuid attacks flax, lucerne, soyabean, and at least 20 other plant hosts (Vinokurov 1927). *H. viriplaca* been found on flowering hemp by Kirchner (1906), Blunck (1920), Vinokurov (1927), Clausen (1931), Shiraki (1952), and Ceapoiu (1958). *H. viriplaca* lives in cooler Eurasian climates than *H. armigera,* and produces only one to two generations per year. Other than that, its life cycle is the same as that of *H. armigera* (see above).

4. HEMP BAGWORM

Psyche cannabinella Doumère 1860, Lepidoptera; Psychidae.

Description: Larvae slender and pale, pinkish white, with light brown heads and legs. Shortly after hatching, larvae weave a cocoon and cover it with flower fragments. Bags of mature caterpillars are spindle-shaped or conical, 20–23 mm long and 6–7 mm wide (Kozhanchikov 1956). Therein bagworms feed and eventually pupate. Doumère (1860) described moth forewings as grey with brown spots, bordered posteriorly with grey fringe; hindwings are greyish white with silver-white fringe. Head, thorax and abdomen are grey-brown. Doumère said female moths resembled males, except for smaller antennae. But according to Kozhanchikov (describing *Sterrhopteryx fusca*, see next paragraph), only males develop wings, while females are wingless, milky-white, and wormlike, with only rudimentary legs.

The identity of *P. cannabinella* is questionable. Kozhanchikov (1956) placed the species under synonymy with *Sterrhopteryx fusca* Haworth 1829, family Psychidae. Paclt (1976) considered *P. cannabinella* identical to *Aglaope infausta* (Linnaeus 1758), family Pyromórphidae (=Zygaènidae). Marshall (1917) suggested the insect was a Hymenopteron!

Life History & Host Range

Doumère (1860) originally found *P. cannabinella* in Luxembourg on hemp growing from birdseed. Larvae hatch in July and usually infest female flowers, not male flowers. According to Kozhanchikov (describing *Sterrhopteryx fusca*), mating takes place without females leaving their bags. Females lay eggs in place, then die and fall out of old bags.

DIFFERENTIAL DIAGNOSIS

Other caterpillars may infest flowering tops during part of their life cycles, such as European corn borers, hemp borers, cutworms, and armyworms.

Table 4.9: Infestation Severity Index for budworms.

Light	any budworms or webs or bud damage seen
Moderate	signs or symptoms at more than one location, moths present
Heavy	many worms and webbings seen, most buds in crop infested
Critical	every plant with bud damage, and 10 or more budworms/plant, moths common

CULTURAL & MECHANICAL CONTROL

(method numbers refer to Chapter 9)

Methods 2a (deep ploughing) and 3 (weeding) are very important. Cherian (1932) used method 9—he shook infested plants every ten days and collected all caterpillars that fell to the ground. Method 12d (nocturnal light traps) works against moths. Tomato breeders have conferred some resistance against *H. zea* by selecting plants that produce methyl ketones (Kashyap *et al.* 1991). Methyl ketones are produced by *Cannabis* (Turner *et al.* 1980).

BIOCONTROL (see Chapter 10)

Khamukov & Kolotilina (1987) controlled hemp budworms with *Trichogramma pretiosum* and *T. minutum* (described under European corn borers). They did not cite release rates. Smith (1996) said corn growers release a million *T. pretiosum* wasps per ha every two or three days while moths are laying eggs, for a total of 11–18 releases per season. *T. brassicae* and *T. dendrolimi* have also controlled budworms, at release rates of 200,000–600,000 wasps per ha

(Li 1994). Less commonly used parasitoids include *Chelonus insularis, Eucelatoria bryani, Microplitis croceipes,* and *Archytas marmoratus* (discussed below), and *Cotesia marginiventris* (described under cutworms and armyworms).

Predators work well indoors. The larvae of big-eyed bugs (*Geocoris* species) consume an average of 77 first-instar budworms or 151 budworm eggs before pupating into adults (which continue to feed!). Yet these statistics pale compared to those of adult ladybeetles (62 budworm eggs per day), larval ladybeetles (34 eggs per day), and larvae of green lacewings (28 eggs per day). Pirate bugs (*Orius* species) also eat eggs and young larvae. *Podisus maculiventris* preys on young budworm larvae (described in the section on leaf-eating caterpillars).

Effective microbial agents include Bt (see European corn borers) and NPV (discussed below). Researchers have killed budworms with the fungus *Nomuraea rileyi.* For budworms pupating in soil, consider the soil nematodes *Steinernema carpocapsae* (see cutworms) and *Steinernema riobravis* (see flea beetles).

Microplitis croceipes

BIOLOGY: A parasitic braconid wasp that lays eggs in second through fourth instar larvae of *H. zea* and *H. virescens* (Fig 4.24). The wasp is attracted to caryophyllene (Lewis & Sheehan 1997), a volatile oil produced by *Cannabis* plants (Mediavilla & Steinemann 1997). Economic mass-rearing methods are currently being developed.

Figure 4.24: A parasitic wasp, *Microplitis croceipes,* laying eggs in a budworm, *Helicoverpa zea* (courtesy USDA).

Chelonus insularis and ***C. texanus***

BIOLOGY: Two closely-related parasitic braconid wasps (some entomologists say they are the same species) that attack many Noctuids (budworms, armyworms, cutworms, etc.). Females lay eggs in budworm eggs or young larvae.

Eucelatoria bryani

BIOLOGY: A parasitic tachinid fly that lays eggs in fourth and fifth instar larvae of *H. zea.* Females deposit several eggs in each host. The species is native to the southwestern USA.

Archytas marmoratus

BIOLOGY: Another tachinid that preys on bollworms, fall armyworms, and other Noctuids. *A. marmoratus* is native to the southern USA. Each fly lays up to 3000 eggs during her 50–70 day life span. Eggs are deposited on leaves of plants, not caterpillars, so the hatching maggots must connect with their hosts.

Nuclear Polyhedrosis Virus "NPV"

BIOLOGY: Several strains of NPV infest budworms. Three are described below.

***Heliothis zea* NPV (HzNPV)** specifically kills *Heliothis* and *Helicoverpa* species. Sandoz, via IMCC, registered HzNPV for use on tobacco in 1975 (as Elcar®), but discontinued the product in 1983. A new strain (Gemstar®) is available. According to Hunter-Fujiata *et al.* (1998), the recommended dose of HzNPV is 6 x 10^{11} OBs per ha, which can be obtained by grinding up the cadavers of 25–50 infected budworms, or using 700 g of the 0.64% liquid formulation. In cotton, three applications are made at three- to seven-day intervals, beginning when pheromone traps start catching egg-laying moths.

***Autographa californica* NPV (AcNPV,** AcMNPV, Gusano®) produces mixed results against budworms—*H. virescens* is more susceptible to AcNPV than HzNPV, but *H. zea* is less susceptible to AcNPV (Hunter-Fujita *et al.* 1998).

***Mamestra brassicae* NPV (MbNPV,** Mamestrin®, Virin®) kills some *Heliothis-Helicoverpa* species. For more information on AcNPV and MbNPV, including development and application, see the next section on cutworms.

***Helicoverpa armigera* Stunt Virus** "HaSV"

BIOLOGY: An RNA virus that infests *H. armigera.* Young caterpillars are killed within three days of eating the virus. The viral genome of HaSV has been genetically engineered into tobacco plants to kill *H. armigera* (Anonymous 1996).

Nomuraea rileyi

BIOLOGY: A fungus that infects and kills many Noctuids, including *H. zea.* The fungus does best in moderate humidity and warm temperatures.

APPEARANCE & DEVELOPMENT: Spores germinate and directly infect budworms through their skins. Death occurs in about a week. Under optimal conditions, "mummified" cadavers become covered with phialidic conidiophores and oval, single-celled, greenish conidia. The cycle is repeated when conidia contact new hosts. Blastospores are yeast-like and produced in liquid media, but commercial formulations require further research.

CHEMICAL CONTROL (see Chapter 11)

Synthetic pheromone traps are available for *H. armigera* and *H. zea.* The active ingredients are (Z)-11-hexadecenal and several related aldehydes and alcohols (Howse *et al.* 1998). Pheromone traps are particularly important for monitoring the annual immigration of *H. zea* into regions north of 39°N.

In tobacco, ten budworms per plant justifies breaking out the spray gun. Rotenone kills budworms, albeit slowly. Neem also kills slowly, but it acts as an antifeedant, repelling budworms from sprayed surfaces (Mordue & Blackwell 1993). Some budworms are highly resistant to pyrethrum (e.g., *H. armigera*). Spinosad, a metabolite produced by an actinomycete, controls budworms in cotton and vegetable crops. Polygodial may be available soon. Back in the old days, Cherian (1932) sprayed plants with lead arsenate; he estimated 100 g of sprayed marijuana contained 2.0 mg of arsenic, "allowed as a medical dose."

Chemicals and NPV work better when combined with Coax®, a feeding stimulant. It was developed for budworms on cotton, but also stimulates European corn borers on corn

(Hunter-Fujita *et al.* 1998). Coax consists of cotton seed flour, cotton seed oil, sucrose, and Tween 80 (a surfactant). Coax has been used at 3.4 kg ha^{-1}, mixed with NPV to kill budworms in cotton fields (Hunter-Fujita *et al.* 1998). Feeding stimulants work even better when they contain extracts of the crop to which they are applied (i.e., try hemp seed extracts). Monitor the effectiveness of controls by looking for budworm frass—fresh frass is light green, old frass becomes dark green.

CUTWORMS & ARMYWORMS

Allen (1908) called cutworms "hemp's only enemy." Ten species reportedly attack *Cannabis.* Some of these reports are probably wrong, because cutworms and armyworms are notorious migrants and may have been collected on *Cannabis* while *en route* to a true host plant. For instance, Alexander (1984b) cited the eastern tent caterpillar, *Malacosoma americanum* (Fabricius). Other questionable observations include the clover cutworm, *Discestra (Scotogramma) trifolii* (Hufnagel) in North Caucasus hemp (Shchegolev 1929), and the oriental armyworm, *Mythimna (Pseudaletia) separata* (Walker) infesting Indian marijuana (Sinha *et al.* 1979).

On the other hand, several armyworms we would expect on *Cannabis* do not appear in the literature. These polyphagous pests include the armyworm, *Pseudaletia unipuncta* (Haworth), the fall armyworm, *Spodoptera frugiperda* (Smith), and the cotton leafworm, *Spodoptera littoralis* Boisduval.

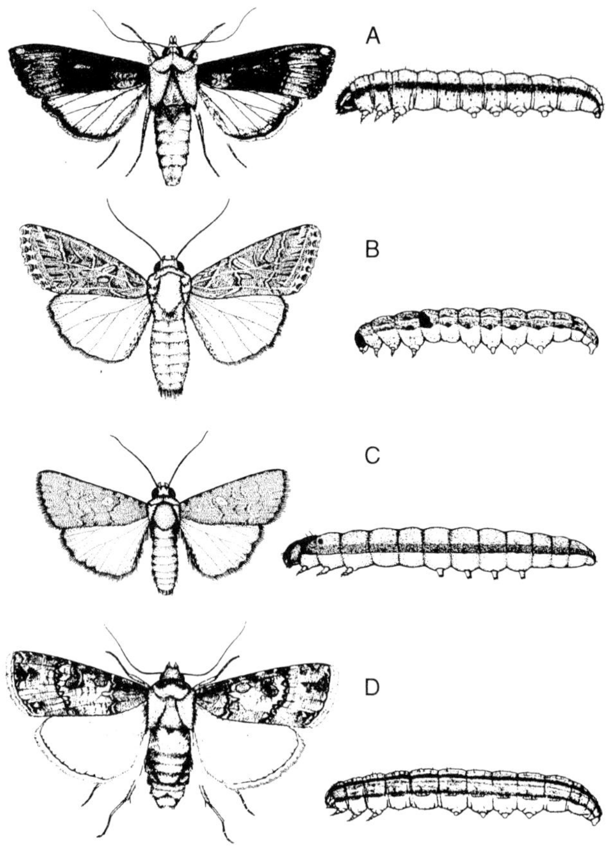

Figure 4.25: Some cutworms. A. Adult and larva of *Agrotis ipsilon;* B. *Spodoptera litura;* C. *Spodoptera exigua;* D. *Agrotis segetum* (not to scale). All illustrations from Hill (1994).

SYMPTOMS

Most farmers are familiar with cutworm damage, but few witness the culprits in action. Cutworms emerge from soil at night to feed on stems of seedlings. The next morning, dead plants are found lying on the ground, severed at the soil line. Older plants may not be completely severed—instead, they tilt, wilt, and die. Since cutworms only eat a little from each plant, they destroy a surfeit of seedlings each evening. Cutworms burrow back into the ground shortly before dawn, usually within 25 cm of damaged plants. Sometimes they drag seedlings into their burrows. Raking soil to a depth of 5 cm will uncover them. Cutworms roll into tight spirals (Plate 26) and "play 'possum" when exposed to light (Metcalf *et al.* 1962).

Larvae hatching later in the season, lacking seedlings to cut, will climb up plants to feed on leaves and flowers. Some species accumulate in large numbers and crawl *en masse* across fields, devouring everything in their path, earning their new name, armyworms. Armyworms are also called "climbing cutworms."

1. BLACK CUTWORM

Agrotis ipsilon (Hufnagel) 1765, Lepidoptera; Noctuidae.
= *Euxoa ypsilon* Rottemburg 1761

Description: Larvae are plump, greasy, grey-brown in colour, with dark lateral stripes and a pale grey line down their back, growing to a length of 45 mm (Fig 4.25). They have a dark head with two white spots. A magnifying glass reveals many convex granules on their skin, resembling rough sandpaper. Pupae are brown, spindle-shaped, with flexible abdominal segments. They are found 5 cm down in soil near host plants. Moths are small, brown, with indistinct Y-shaped markings on their dark forewing (hindwings nearly white), and difficult to distinguish from other Noctuids. Eggs are round to conical, ribbed, yellow, laid either singularly or in clusters of two to 30 on plants close to the ground.

Life History & Host Range

Black cutworms attack marijuana in India (Sinha *et al.* 1979) and hemp in Europe (Ragazzi 1954). Dewey (1914) described cutworms causing heavy losses in late-sown hemp, but noted little damage in seedlings sown at the proper time. Robinson (1952) considered cutworms a problem only in hemp rotated after sod. *A. ipsilon* frequents weedy, poorly cultivated crops.

A. ipsilon overwinters in soil as pupae or mature larvae. Moths emerge in April or early May and lay eggs on plants in low spots or poorly drained land (Metcalf *et al.* 1962). Each moth lays up to 1800 eggs. Eggs hatch in five to ten days. Larvae feed for ten to 30 days before pupating, repeating their life cycle until a heavy frost. Two generations per year arise in Canada, three in most of the USA, and four generations in southern California and Florida. *A. ipsilon* is found worldwide and despised for its pernicious habit of cutting down many seedlings to satisfy its appetite. It feeds on almost any herbaceous plant.

2. PADDY CUTWORM

Spodoptera litura (Fabricius) 1775, Lepidoptera; Noctuidae.
= *Prodenia litura* Fabricius

Description: Larvae reach 50 mm in length. They are stout, smooth-skinned, with black heads and dull grey to grey-green bodies partially covered with short setae. Mature cutworms exhibit yellow stripes down their back and sides. The dorsal stripes are bordered by a series of semilunar black spots (Fig 4.25). Paddy cutworms closely resemble cotton leafworms (*Spodoptera littoralis* Boisduval). Pupae are brown and spindle-shaped. Moths have robust brown bodies and wings coloured buff to brown (forewings are darker, with buff markings), reaching a span of 38 mm. Eggs are laid in clusters of 200–500 and covered with moth hairs.

Life History & Host Range

S. litura lives in warm climates and rarely overwinters.

In the tropics its life cycle takes 30 days; as many as eight generations arise per year (Hill 1983). Female moths lay egg clusters under leaves of host plants. Eggs hatch in three or four days. Caterpillars destroy crops for 20 days before pupating in the soil. Pupation takes six or seven days, then adult moths emerge once again.

Also called the tobacco caterpillar and the fall armyworm, *S. litura* should not be confused with *Spodoptera frugiperda* (Smith) 1797, the North American fall armyworm. *S. litura* attacks crops from Pakistan to China, and south to eastern Australia. Main hosts are tobacco, tomato, cotton, rice, and maize. Cherian (1932) considered *S. litura* a "minor pest." Nevertheless, the caterpillar attracted biocontrol researchers, who described it destroying flowering tops (Nair & Ponnappa 1974).

3. BEET ARMYWORM

Spodoptera exigua (Hübner) 1827, Lepidoptera; Noctuidae.
=*Laphygma exigua* Hübner

Description: Newly-hatched larvae are light green with black heads. Full grown larvae vary in colour, grey-green to brown, with wavy lines down the back and pale lateral stripes along each side, skin colour between the stripes often dotted and mottled, total length 25–35 mm (Fig 4.25). Larvae have dark heads and often have a dark spot on each side of the body above the second true leg. Pupae are dark brown, and found in the upper 10 mm of soil in a cell formed of soil and plant debris glued together with a sticky secretion. Moth bodies are brown, similar to those of *S. litura* but smaller (wingspan 25–30 mm). Forewings are grey-brown with a yellow kidney-shaped spot in the mid-front margin, reaching a span of 30 mm; hindwings are white with a dark fringe. Eggs are pearly-green (changing to grey), spherical, with radiating ribs, 0.5 mm in diameter, laid in clusters of 50–300, in several layers, covered by moth hairs and scales.

Life History & Host Range

Moths emerge in spring and deposit 500–900 eggs in four to ten days, which hatch in two to five days. Larvae feed for about three weeks and pupate in half that time. In warm areas four to six generations arise per year. Beet armyworms can march in enormous numbers, defoliating hundreds of hectares of sugarbeets, rice, cotton, maize, and vegetable crops.

A native of India and China, where it feeds on *Cannabis* (Cherian 1932), *S. exigua* was accidentally introduced into California around 1876 (Metcalf *et al.* 1962). It moved east to Florida and north to Kansas and Nebraska. From Florida the pest was introduced into Europe, where it attacks hemp (Ceapoiu 1958).

4. CLAYBACKED CUTWORM

Agrotis gladiaria (Morrison) 1874, Lepidoptera; Noctuidae.

Description: Mature larvae are grey with a distinctly paler dorsum, set with numerous small, flat, shining granules arranged down the back like pavement stones. The body tapers both anteriorly and posteriorly, reaching 37 mm in length. Moths resemble most Noctuids: robust brown bodies, forewings darker than hindwings, with variegated dark brown markings.

Life History & Host Range

Overwintering larvae cause their greatest destruction from May to late June. After that they pupate underground. Moths emerge in September and October, mate, lay eggs, and repeat their life cycle. Only one generation arises each year. Tietz (1972) cited this species feeding on hemp, along with 20 other hosts. *A. gladiaria* lives east of the Rocky Mountains, and becomes scarce south of Tennessee and Virginia.

5. COMMON CUTWORM

Agrotis segetum (Denis & Schiffermüller) 1775, Lepidoptera; Noctuidae. =*Euxoa segetum* Denis & Schiffermüller

Description: Mature larvae are plump, greasy, cinnamon-grey to grey-brown with dark heads and faint dark dorsal stripes, reaching 45 mm in length (Fig 4.25). Pupae are smooth, shiny brown, 12–22 mm long and found in the soil. Moths are distinguished by black kidney-shaped markings on their dark forewings (hindwings are white). They are smaller than black cutworm moths, with a wingspan of 32–42 mm. Females lay light yellow eggs haphazardly, on plants or soil, singly or in groups of two to 20.

Life History & Host Range

Also called the Turnip moth, Durnovo (1933) described *A. segetum* destroying hemp near Daghestan, Georgia. This polyphagous pests cuts down seedlings at soil level and gnaws at root crops underground. In equatorial Africa and southeast Asia up to five generations arise per year; whereas in northern Europe *A. segetum* is **univoltine** (one generation per year), and overwinters as larvae in the soil.

6. BERTHA ARMYWORM

Mamestra configurata Walker, Lepidoptera, Noctuidae.

Description: Young larvae are pale green with a yellowish stripe on each side. Mature larvae darken to a brown-black colour, with broad yellow-orange stripes on each side, and three narrow, broken white lines down their backs. Some larvae, however, remain green. Mature larvae have light brown heads and measure 40 mm in length. Pupae are reddish brown and tapered. Adults have grey forewings flecked with black, brown, olive , and white patches. A prominent white kidney-shaped spot arises near the wing margin. Forewings are edged with white and olive-coloured irregular transverse lines, wingspan 40 mm. Eggs are white (turning darker with age), ribbed, and laid in batches of 50–500, in geometric, single-layer rows.

Life History & Host Range

M. configurata overwinters as pupae, 5–16 cm deep in soil. Adults emerge and lay eggs from May to August. Each female lays an average of 2150 eggs, which hatch within a week. Young larvae congregate on the underside of leaves. Crop damage peaks between July and August. One or two generations arise per year. *M. configurata* is a serious pest of canola (rape seed), but also feeds on alfalfa, flax, peas, and potatoes. Weed hosts include lamb's quarters and wild mustard. The species is native to North America and attacks hemp in Manitoba, chewing off all leaf blade tissues, leaving only a stalk with leaf skeletons (Moes, pers. commun. 1995). The pest population is cyclical—*M. configurata* damage becomes significant for two to four years, then becomes insignificant for eight to ten years, then cycles into significance again.

DIFFERENTIAL DIAGNOSIS

Cutworm damage can be confused with post-emergent damping off, caused by fungi. Bird damage can be confused with cutworm damage, because early birds damage seedlings before growers get out of bed and see them in action. Armyworms must be differentiated from other assorted leaf-eating caterpillars (discussed in the next section).

CULTURAL & MECHANICAL CONTROL

(method numbers refer to Chapter 9)

Cold, wet spring weather causes cutworms to proliferate, so follow method 4 (plant late). A combination of methods 3 and 9 works well—hoeing, a good eye, and quick hands. Destroy egg clusters and eradicate weeds (especially grasses) from the vicinity. Observe method 6 (avoid rotating *Cannabis* after sod). Finish the season with method 2a (ploughing). Recultivate fields in early spring, and wait two weeks before planting. Cultivation destroys the cutworm's food source (weed seedlings), and a two week fallow starves them.

Light traps (method 12d) work against moths. Method 13 (mechanical barriers) is labour-intensive but effective. Place cardboard or tin collars around seedlings. Be sure to push 10-cm-tall collars at least 3 cm into the soil. Rings of sand, wood ashes or eggs shells also repel larvae. Placing toothpicks alongside stems, despite old wives' tales, will not stop cutworms. For armyworms, apply Tanglefoot® as described in method 13. Armyworms may bridge the sticky barrier by marching over their stuck brethren.

BIOCONTROL (refer to Chapter 10)

Cutworms and armyworms have been controlled by *Trichogramma* species (*T. dendolimi, T. brassicae, T. evanescens,* described under European corn borers). In sugar beets, Smith (1996) released 100,000–200,000 wasps per ha every one to two months. *Chelonus* species have killed cutworms and armyworms (see budworms). A new parasitic wasp, *Cotesia marginiventris,* is described below. Cool, rainy weather suppresses all parasitic wasps. In these conditions, switch to *Steinernema carpocapsae,* a parasitic nematode (described below). *Steinernema riobravis* kills caterpillars in warmer, drier soil (see flea beetles).

Turning to microbials—NPV kills cutworms and armyworms, albeit slowly (described below). Bt (described under European corn borers) works well against armyworms (*except S. exigua*). Bt is less successful against cutworms. *Nomuraea rileyi* is a fungus that infests Noctuids (described under budworms).

Some predatory insects (e.g., pirate bugs) eat the eggs of cutworms and armyworms. *Podisus maculiventris* preys on cutworms and armyworms (described in the section on leaf-eating caterpillars). Many toads, birds, small mammals, and other insects relish cutworms. Make their work easier by frequently tilling the soil. Yepsen (1976) repelled cutworms with tansy, *Tanacetum vulgare* (described in the next section on leaf eating caterpillars).

Steinernema (Neoapectana) carpocapsae

BIOLOGY: Of the >20 known species of *Steinernema,* this species is the best known and most widely marketed parasitic nematode (Biosafe®, Exhibit®, EcoMask®, Exhibit SC®, Sanoplant®). *S. carpocapsae* kills cutworms, armyworms, and other caterpillars, as well as beetle grubs, root maggots, and some thrips and bugs in soil. *Steinernema* nematodes are cosmopolitan, and do best in soil temperatures between 15–32°C (but the Scanmask® strain of *S. carpocapsae* is native to northern Sweden and remains active down to 10°C).

DEVELOPMENT: *S. carpocapsae* hunts for hosts with a "sit-and-wait" or "**ambush**" strategy. It stands upright on its tail, near the soil surface, and waits for mobile, surface-feeding insects to pass by. *Steinernema* species must enter hosts via natural openings; they cannot directly penetrate insect skin. Within hosts, *Steinernema* species release bacteria (*Xenorhabdus nematophilus* and subspecies *X. bovienii* and *X. poinarii*). The bacteria ooze insect-destroying enzymes, and do the actual killing. Bacteria also produce antibiotics which prevent putrefaction of dead insects, allowing nematodes to reproduce in the cadavers (Plate 31). Cutworms begin dying within 24 hours of infection, and the next generations emerges within ten days. *Steinernema*-infected hosts become limp and turn a cream, yellow, or brown colour.

APPLICATION: Supplied as third-stage larvae in polyethylene sponge packs or bottles. Sealed containers of *Steinernema* species remain viable for six to 12 months at 2–6°C. To apply to soil, follow instructions for *Heterorhabditis bacteriophora* (see the section on white root grubs). Leslie (1994) reduced black cutworms by 94% with *S. carpocapsae,* applying about 50,000 nematodes to soil around each plant. *S. carpocapsae* has also been injected into borer holes and sprayed onto foliage to control above-ground insects, but nematode survival is short. Survival can be lengthened by coating nematodes with antidesiccants (e.g., wax, Folicote, Biosys 627). Repeat applications every three to six weeks for the duration of infestations.

NOTES: *S. carpocapsae* lives near the soil surface. Its "ambush" strategy works best against active, surface-feeding insects, like cutworms, armyworms, and fungus gnat maggots. *Steinernema* species are hardier than *Heterorhabditis* nematodes, easier to mass produce, and last longer in storage. But they do not penetrate hosts as well as *Heterorhabditis* species. Nematodes living near the soil surface suffer higher predation rates from nematophagous collembolans such as *Folsomia candida.* Beneficial nematodes are compatible with Bt and NPV, but biocontrol insects that live in the soil are incompatible (e.g., pupating *Aphidoletes aphidimyza*). According to some experts, wasp parasitoids that pupate within silken cocoons are not killed.

Nuclear polyhedrosis virus "NPV"

BIOLOGY: A rod-shaped baculovirus (Fig 6.4) which contains dsDNA. Individual NPV rods are only 300 nm long, but they may clump by the hundreds into occlusion bodies (OBs) that grow to 15 μm in diameter (easily seen under a light microscope). NPV kills larvae of Noctuids (e.g., cutworms, armyworms, budworms, borers, and leaf-eating caterpillars). Of course not all caterpillars are from the Noctuid family so you must properly identify your pests. Many strains of NPV are available.

***Mamestra brassicae* NPV** (**MbNPV**, Mamestrin®, Virin-EKS®) kills at least 25 species of Noctuids, including *Mamestra, Spodoptera,* and *Heliothis-Helicoverpa* species. Mamestrin Plus® combines MbNPV with a synthetic pyrethroid, cypermethrin.

Figure 4.26: Dead armyworm killed by a nuclear polyhedrosis virus, MbNPV (courtesy USDA).

***Autographa californica* NPV (AcNPV**, VPN-80®, Gusano®), has the widest host range of any NPV. AcNPV kills at least 45 species of caterpillars, including most cutworms. A genetically engineered version of AcNPV expresses a neurotoxin from the venom of *Androctonus australis*, a scorpion (currently under investigation).

***Spodoptera exigua* NPV (SeNPV**, Spod-X®, Instar®, Ness-A®) has a restricted host range—the beet armyworm. The Yoder Brothers developed SeNPV in Florida.

***Heliothis zea* NPV (HzNPV)** was formally marketed as Elcar® and now sells as Gem-Star® (described under Budworms).

***Spodoptera littoralis* NPV (SlNPV**, Spodopterin®) kills cotton leafworms. It is not available at the time.

DEVELOPMENT: Caterpillars that eat plant material sprayed with NPV stop feeding, become flaccid, darken, then die. Unfortunately, death may take a week; slow lethality has been a limiting factor in the popularity of NPV. NPV-killed caterpillars, unlike Bt victims, generally do not smell (unless infected by secondary saprophytic bacteria). The carcasses hang from plants and spread more viruses (Fig 4.26).

APPLICATION: NPV is supplied as a liquid concentrate or wettable powder. Since NPV must be eaten to kill caterpillars, spray all foliage uniformly and completely. NPV on foliage is degraded by UV light, so spray outdoors in late afternoon or on cloudy (not rainy) days. To slow degradation, combine NPV with Pheast® or Coax®, which are lepidopteran feeding stimulants.

Shapiro & Dougherty (1993) increased the potency of NPV by adding *laundry brighteners*. Many brands worked, especially stilbene brighteners (e.g., Leucophor BS, Leucophor BSB, Phorwite AR, Tinopal LPW). They found that 0.1% of Tinopal LPW added to NPV worked the best. Brighteners protect NPV from UV light; brighteners also decrease caterpillar gut pH, which enhances NPV activity.

NOTES: Caterpillars gain resistance to NPV as they age. For instance, the LD_{50} for *M. brassicae* jumps from seven OBs in the first instar to 240,000 OBs in the fifth instar (Hunter-Fujita *et al.* 1998). Get those cutworms while they're young! NPV is compatible with Bt and all other biocontrols. NPV works well against two cutworms that are Bt-resistant, *M. brassicae* and *S. exigua*.

***Agrotis segetum* granulosis virus**

BIOLOGY: AsGV is a new OB-forming DNA virus used by European workers (Agrovir®). It is very host-specific, killing only *A. segetum* and *A. ipsilon*. AgGV has been applied as a spray or a bran bait, at a rate of 4 x 10^{13} OBs per ha (Hunter-Fujita *et al.* 1998). Young larvae are the most susceptible; use light traps or pheromone traps to determine when *Agrotis* moths are flying, and apply AsGV a week later.

***Agrotis segetum* cytoplasmic polyhedrosis virus**

BIOLOGY: AsCPV is an OB-forming RNA virus, discovered in the UK. Infected caterpillars lighten to a milky-white colour before dying.

Saccharopolyspora spinosa

BIOLOGY: A bacterium that kills *S. exigua* (a pest somewhat impervious to Bt), cabbage loopers, and other Noctuids.

Cotesia marginiventris

BIOLOGY: A tiny braconid parasitic wasp that lays eggs in young Noctuid caterpillars. *C. marginiventris* lives in temperate and semitropical regions around the world. It controls cabbage loopers, cutworms, armyworms, and budworms in cotton, corn, and vegetable crops.

APPEARANCE & DEVELOPMENT: Adults are sturdy wasps, 7–8 mm long, dark coloured, with long, curved antennae. The egg-laying wasps track down volatiles emitted by caterpillar-damaged plants. Amazingly, leaves damaged by wind or other mechanisms do *not* attract parasitoids, unless oral secretions from leaf-eating insects are added (Turlings *et al.* 1990). Females lay 150–200 eggs, inserting 20–60 eggs per caterpillar. *C. marginiventris* larvae kill their host and then pupate externally, in a yellow, silken cocoon attached to host larvae or the undersides of leaves. The lifecycle takes 22–30 days in optimal conditions.

APPLICATION: *C. marginiventris* is supplied as adult wasps. Release 800–2000 wasps per ha, weekly, while moths of pests are flying and laying eggs (at least three releases).

CHEMICAL CONTROL (see Chapter 11)

Yepsen (1976) described a cutworm bait: mix hardwood sawdust, wheat bran, molasses, and water. Scatter the sticky bait at dusk. It immobilizes larva and exposes them to their enemies. Lace the bait with an insecticide to insure results (Metcalf *et al.* 1962). For spray-gun enthusiasts, Yepsen suggested a repellent spray of garlic, hot peppers, and horseradish. Neem acts as an repellent, antifeedant, and it kills *Spodoptera* species (Mordue & Blackwell 1993). Spinosad is particularly useful against *Spodoptera* species. Rotenone controls armyworms in cotton and vegetable crops. Diflubenzuron is the synthetic form of a natural chitin inhibitor that kills armyworms. Synthetic pheromone lures are commercially available for *A. ipsilon, A. segetum, S. litura, S. exigua*, and *M. configurata*.

LEAF-EATING CATERPILLARS

The literature reports *dozens* of leaf-eating caterpillars on *Cannabis*, including many improbable pests. Some improbable pests are incidental migrants, such as the American dagger (*Acronicta americana* Harris) found on feral hemp in Michigan (McPartland, unpublished data 1995). *A. americana* is arboreal and probably fell from trees overhanging the hemp. Hanson (1963) cited three improbable pests on leaves and "pods" of Vietnamese marijuana—the long-tailed blue swallowtail, *Lampides boeticus* (Linnaeus), *Jamides bochus* (Stoll), and a *Papilionides* species.

Surprisingly, some improbable pests are cited more than once: *Amyna octo* Guen. was reported on Indian marijuana (Rao 1928, Cherian 1932) and European hemp (Sorauer 1958). Vice versa, several leaf-feeding caterpillars we'd expect to find on *Cannabis* have not been reported. Examples include the pale tussock (*Calliteara pudibunda* Linnaeus), the red admiral (*Vanessa atalanta* Linnaeus), the comma or hops merchant (*Polygonia comma* Harris), and the hops looper (*Hypena humuli* Harris).

Baloch & Ghani (1972) collected loopers from Pakistani plants, the polyphagous pests *Plusia nigrisigna* Walker, *P. (Diachrisia) orichalcea* Fabricius and *P. horticola*. Loopers have been observed on feral hemp in Illinois but not identified (Bush Doctor, unpublished data 1983).

Cherian (1932) found larvae of two tussock moths (family Liparidae) causing minor damage in Indian marijuana fields, *Euproctis fraterna* Moore and *E. scintillans* Walker. These species occupy sympatric ranges from India to Malaysia. Goidànich (1928) and Schaefer *et al.* (1987a) cited another (and infamous) member of the tussock family feeding on *Cannabis*—the gypsy moth, *Lymantria dispar* (Linnaeus).

Shiraki (1952) reported several odd species on Japanese hemp, including some improbable pests: *Eumeta japonica*

Heylaerts (Psychidae), *Plataplecta consanguis* Butler (Noctuidae) and five Nymphalids, including the Painted Lady (*Cynthia cardui* Linnaeus), *Vanessa urticae connexa* Butler, *Araschnia burejana* Bremer, *Araschnia levana* Linnaeus and *Pyrameis indica* Herbst. Anonymous (1974) also cited *Pyrameis indica* in China. Clausen (1931) added another odd east Asian nymphalid, *Polygonia c-aureum* (Linnaeus).

SIGNS & SYMPTOMS

Leaf-eating caterpillars act as leaf rollers (weaving leaves together for protection while they feed), as leaf skeletonizers (selectively eating leaf tissue between leaf veins, leaving a skeleton of veins), or they simply chew big holes in leaves. Shaking a plant, especially in the early morning, will dislodge all leaf-eating caterpillars except leaf rollers. For estimating the level of infestation, see the budworm chart (Table 4.9). For estimating the degree of damage, see Fig 1.3.

Only seven leaf-eating caterpillars are cited often enough to be considered serious pests:

1. SILVER Y-MOTH or GAMMA MOTH

Autographa gamma (Linnaeus) 1758, Lepidoptera; Noctuidae.
=*Plusia gamma* (Linnaeus), =*Phytometra gamma* Linnaeus

Description: Larvae are 12-footed "semiloopers" with only three pairs of prolegs. They are dark green, pinched towards the head, marked by a white dorsal stripe and two lateral yellow lines, and reach 20–30 mm fully grown. Head and legs have black markings. They pupate on plants inside a silken cocoon. Moths are brown bodied with forewings sporting a silvery "Y" mark (or Greek *gamma*) against a grey-brown background, wingspan 30–45 mm (Fig 4.27). Eggs are white, round, and ribbed.

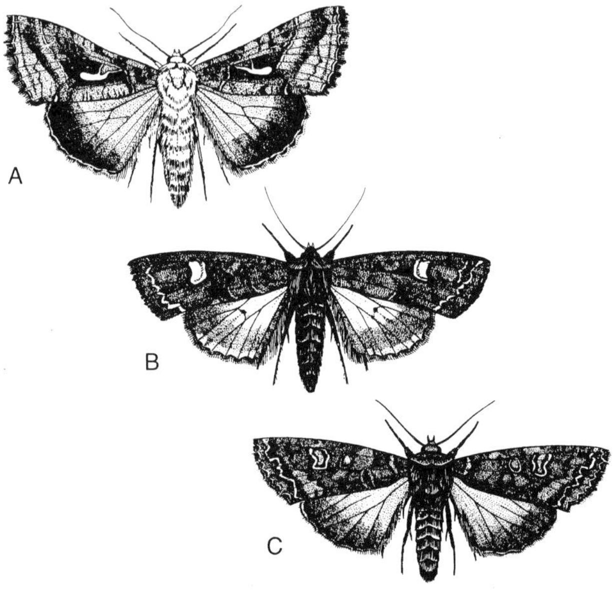

Figure 4.27: Adult moths of some leaf-eating caterpillars. A. *Autographa gamma;* B. *Melanchra persicariae;* C. *Mamestra brassicae* (not to scale). All illustrations by Hill (1994).

Life History & Host Range

Females lay 500–1000 eggs on undersides of leaves, singly or in small groups. The life cycle takes 45–60 days. Two cycles arise per year in Europe, but up to four cycles occur in Israel (Hill 1983). Caterpillars skeletonize leaves when young, but later eat the entire leaf lamina. Larvae attack seedlings (Kaltenbach 1874, Kirchner 1906), and leaves or flowers of mature plants (Blunk 1920, Goidànich 1928, Martelli 1940, Ceapoiu 1958, Spaar *et al.* 1990). Dempsey (1975) considered *A. gamma* a major pest in temperate Eurasia, and now it lives North America. The moths can migrate hundreds of miles. Hosts include sugarbeets, assorted vegetables, *Prunus, Rubus,* and *Sambucus.*

2. DOT MOTH

Melanchra persicariae (Linnaeus) 1761, Lepidoptera; Noctuidae.
=*Polia persicariae* (Linnaeus), =*Mamestra persicariae* (Linnaeus)

Description: Larvae colours vary from pale grey-green to dark green or purple-brown. They have a light green head, a pale dorsal line down their backs, a series of V-shaped markings, and a dorsal hump on the eighth abdominal segment. Setae (1–1.3 mm) cover the body giving them a velvety texture. Fully-grown larvae reach 40 mm in length. Moths are grey to bluish-black, with darker forewings marked by two kidney-shaped white dots, wingspan 38–50 mm (Fig 4.27). Eggs are hemispherical, ribbed, whitish green becoming pinkish brown, laid on undersides of leaves, singly or in small masses.

Life History & Host Range

Pupae overwinter in soil. Moths emerge in June or July and promptly lay eggs. Eggs hatch in a week and larvae feed for up to 90 days before moving underground in autumn. Only one generation arises per year. Blunk (1920) rarely found *M. persicariae* in hemp, while Dempsey (1975) considered the species a major pest. Kaltenbach (1874), Kirchner (1906), Goidànich (1928), and Rataj (1957) also report it in central Europe. *M. persicariae* prefers feeding on *Urtica, Plantago, Salix,* and *Sambucus* species, but is polyphagous.

3. CABBAGE MOTH

Mamestra brassicae (Linnaeus) 1758, Lepidoptera, Noctuidae.
=*Barathra brassicae* Linnaeus

Description: Young larvae are yellow-green, their dorsum darkening to a green-brown colour by the fourth instar. They have yellow heads and side stripes resembling a chain of dark semilunar spots; rusty brown setae cover the body. Full-grown larvae measure 40–50 mm in length and have a small hump on abdominal segment eight. Adults are brown Noctuid moths, nearly indistinguishable from other species except for a curved dorsal spur on the tibia of the foreleg (Fig 4.27). Eggs are transparent (turning light yellow to brown), ribbed, and laid in batches of 50 or more, in single-layered geometric rows.

Life History & Host Range

M. brassicae overwinters as pupae in soil. Adults emerge and lay eggs in May-June. Young larvae act as cutworms; later they eat leaves. Two generations arise per year. The species is highly polyphagous and a serious pest of vegetable crops. It attacks hemp in Europe (Martelli 1940, Rataj 1957, Ferri 1959) and Japan (Takahashi 1919, Harada 1930, Clausen 1931, Shiraki 1952). The Japanese report that it also feeds like a budworm. Yepsen (1976) suggested *Cannabis* could repel this pest from other crops. *M. brassicae* ranges across the Old World, and is somewhat tolerant to Bt.

4. GARDEN TIGER MOTH

Arctia caja (L.) 1758, Lepidoptera; Arctiidae.

Description: Larvae are black woollybears with white spots on their flanks. The woolly hairs contain high concentrations of histamine (Rothschild *et al.* 1979). Moths have variegated forewings of beige and brown, hindwings of bright orange with iridescent blue spots, wingspan 50 mm. Moth bodies are brown dorsally, orange underneath. The stiff spines growing from moths can inflict a severe sting (Rothschild *et al.* 1979).

Life History & Host Range

This species is distributed worldwide north of the equator. Harada (1930) and Shiraki (1952) reported this pest in Japan; Sorauer (1958) reported it in Europe. But when Rothschild *et al.* (1977) fed *A. caja* a pure *Cannabis* diet, the larvae became stunted and died before maturity. Given a choice, *A. caja* larvae preferred eating high-THC plants to low-THC, high-CBD varieties. Yet their survival was higher when they fed upon the latter.

A. caja is **aposematic** — the moths are conspicuously coloured, and the colours serve as a warning to predatory birds and rodents. Moths store plant poisons in their bodies which repel predators, just like their famous cousin—the monarch butterfly. Birds will eat *A. caja* without hesitation on the first occasion, but subsequently refuse to eat the distasteful moths (Rothschild *et al.* 1979). *A. caja* larvae are polyphagous, but show a predilection for poisonous plants, including *Aconitum, Senecio, Convalaria, Digitalis, Solanum*, and *Urtica* (Rothschild *et al.* 1979). One *A. caja* larva feeding on *Cannabis* can store enough THC in its cuticle to be pharmacologically active to a mouse (Rothschild *et al.* 1977).

Berenji (pers. commun. 1997) reported another Arctiid, the Fall webworm (***Hyphantria cunea*** Drury), infesting hemp in Yugoslavia. *H. cunea* larvae web leaves and flowers into an unsightly nest. Greatest damage arises from August to October. Larvae are black and white with long, whitish hairs. Two generations arise per year. *H. cunea* normally infests trees and shrubs. It is a gift from America to the Old World.

5. COMMON HAIRY CATERPILLAR

Spilosoma obliqua (Walker) 1862, Lepidoptera; Arctiidae.
=*Diacrisia obliqua* (Walker)

Description: *S. obliqua* woollybears have yellow bodies, orange stripes, dark heads, and long, light-coloured hairs. Moths are aposematic, with orange abdomens and yellow wings that are lightly spattered with black spots, wingspan 40 mm.

Life History & Host Range

This species overwinters as pupae just beneath the soil surface. Young larvae skeletonize leaves; later they completely consume leaves. *S. obliqua* sometimes infests flowering tops. Females lay eggs in large conspicuous masses, up to 1000 at a time. In warm climates the life cycle takes 30 days; three to eight generations arise per year.

S. obliqua ranges from the Himalaya foothills of India to Japan. The larvae are polyphagous, but like many aposematic insects, they prefer eating poisonous plants (Rothschild *et al.* 1979). *S. obliqua* caught the attention of biocontrol researchers in India (Nair & Ponnappa 1974) and Pakistan (Baloch & Ghani 1972). Deshmukh *et al.* (1979) offered *S. obliqua* larvae 154 different plant species, and found *Cannabis* among the six most preferred hosts. Yet, when fed a diet of pure *Cannabis*, 50% of the larvae died after 24 days.

Lago & Stanford (1989) reported another yellow woollybear, ***Spilosoma (Diacrisia) virginica*** (F.), feeding on Mississippi marijuana (Fig 4.28).

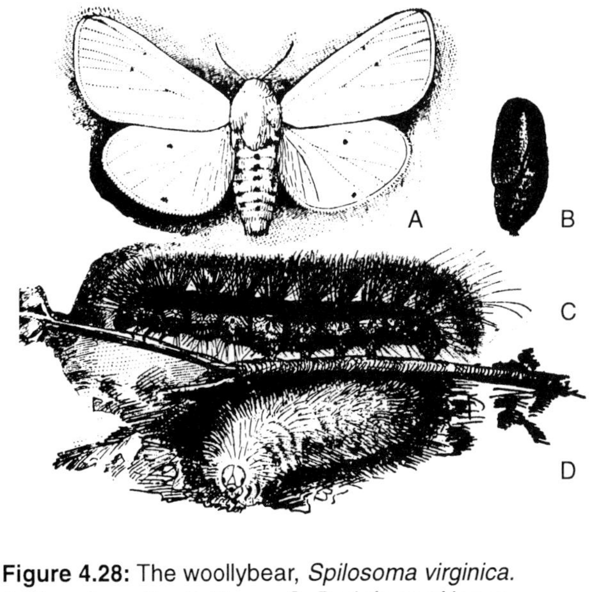

Figure 4.28: The woollybear, *Spilosoma virginica*. A. Female moth; B. Pupa; C. Dark form of larva; D. Light form of larva (courtesy of USDA).

6. BEET WEBWORM

Loxostege sticticalis (Linnaeus) 1761, Lepidoptera; Pyraustinae.
=*Phlyctaenodes sticticalis* Linnaeus

Description: Larvae are yellowish-green, darkening with age to nearly black, with a black stripe down their backs, reaching 30 mm in length. Moths are mottled brown, with straw-coloured markings and a dark margin on the underside of their hindwings.

Life History & Host Range

Larvae overwinter in silk-lined cells in the soil, pupating there in the spring. Moths emerge from March to June. Moths are strong night fliers, migrating hundreds of miles along river valleys with prevailing winds. Females lay 150 eggs, in groups of two to 20, on undersides of leaves. The entire life cycle can be as short as 35 days, with up to three generations arising each year.

This polyphagous pest is also known as the meadow moth or steppe caterpillar. It infests leaves and flowering tops throughout eastern Europe (Kulagin 1915, Ceapoiu 1958, Camprag *et al.* 1996). *L. sticticalis* ranges across temperate Asia and has been introduced into the USA and Canada (Metcalf *et al.* 1962). Young larvae skeletonize hemp leaves, often webbing several leaves together, while older larvae eat entire leaves. They occasionally march *en masse* as armyworms.

7. HEMP DAGGER MOTH

Plataplecta consanguis (Butler) 1879, Lepidoptera; Noctuidae.
= *Acronycla consanguis* Butler

Description: Butler originally placed this species in the Bombycoidea. He only described the adult moths, which are grey with lavender and brown wing markings, with 35 mm wingspans.

Life History & Host Range

Shiraki (1952) called this species the hemp dagger moth, found only in Japan. We know little about its life history and host range.

8. CHRYSANTHEMUM WEB WORM

Cnephasia interjectana (Haworth) 1829, Lepidoptera; Tortricidae. =*Cnephasia virgaureana* Pierce & Metcalfe 1922

Description: Larvae vary from brownish-cream to greyish-green in colour. They are darker dorsally, with yellow-brown heads and legs. Pupae are brown, found in webbed leaves and at the bases of stems. Moth forewings are whitish-grey with brown markings.

Life History & Host Range

Larvae overwinter in soil or crop debris. Moths emerge in late summer. They deposit eggs on plants, singly or in batches of two to three eggs. Fritzsche (1959) observed *C. interjectana* feeding on hemp in the Halle-Magdeburg region of Germany. First-instar larvae feed as leafminers. Older larvae roll and web leaves (around mid-May), and move to flowering tops in autumn. The species occurs across Europe into Siberia. It has established a North American foothold in Newfoundland. *C. interjectana* feeds polyphagously upon over 130 herbaceous species, notably *Chrysanthemum*, buttercups (*Ranunculus* species), sorrel (*Rumex* species), and plantain (*Plantago* species). Goidànich (1928) and Spaar *et al.* (1990) cited the related pest ***Cnephasia wahlbomiana*** L. on European hemp.

9. DEATH'S HEAD MOTH

Acherontia atropos (Linnaeus) 1758, Lepidoptera; Sphingidae.

= *Sphinx atropos* Linnaeus

Description: Larvae are "hornworms," initially light green, darkening with age to dark green or brown, with yellow lateral stripes and dorsal spots in the final instar, up to 130 mm in length; the dorsal "horn" grows from the back of the eight abdominal segment in an s-shape. Pupae are smooth and glossy, dark brown, 75–80 mm long, with transverse ridges and two short apical spines. Moths are heavy-bodied ("the weight of a mouse" according to Young 1997), with striped, spindle-shaped abdomens tapering to a pointed posterior; dorsal surface of the thorax with a black-and-yellow skull-like marking. Forewings are long and narrow, dark brown with mottled white and tan markings, wingspan 90–130 mm; hind wings yellow with brown-black bands. When disturbed, the moths can forcibly expel air from their pharynx, to produce a hissing or shrieking sound. Larvae click their mandibles if molested, and large ones can inflict a painful bite.

Life History & Host Range

A. atropos overwinters as larvae or pupae in Mediterranean Africa and the Middle East. Pupation occurs 15–40 cm deep in the ground. Multiple overlapping generations arise every year. Moths migrate north to Europe, beginning in June, and produce one European generation which rarely survives past October. Larvae are polyphagous, feeding on many crop plants, especially potato, tomato, and tobacco. Larvae cause minor damage on hemp (Kaltenbach 1874, Kirchner 1906, Goidànich 1928, Dempsey 1975, Gladis & Alemayehu 1995). Adult moths are night fliers and dive into hives to rob honey.

DIFFERENTIAL DIAGNOSIS

Stem borers feed on leaves before they begin boring. Budworms occasionally act as leaf eaters. Early leafmining damage may be confused with skeletonized leaves. Armyworms eat leaves, but are discussed separately due to their marching habits. Some beetles feed on leaves, but generally make smaller holes. Slugs make ragged holes. Holes from brown leaf spot (a fungus disease) and holes from bacterial blight can be confused with holes from leaf-eating insects.

CULTURAL & MECHANICAL CONTROL

(method numbers refer to Chapter 9)

Follow recommendations for cutworms and armyworms, especially methods 2a (deep ploughing), 3 (weeding), 6 (crop rotation), 8 (optimizing soil structure and nutrients), 12d (nocturnal moth traps—especially against Nocturids) and 13 (barriers coated with Tanglefoot).

BIOCONTROL (see Chapter 10)

Follow suggestions in the previous section—spray Bt when caterpillars are less than 10–15 mm long (unless you have *M. brassicae,* which is resistant). Spray with NPV (especially if your pest is a Nocturid), a NPV strain specific to *H. cunea* is described below. Release *Trichogramma* wasps (described under European corn borers), and *Steinernema* nematodes (described under cutworms). A general predator, *Podisus maculiventris,* is described below. Consider planting tansy, described below.

Podisus maculiventris Podibug®

BIOLOGY: A predatory stink bug with a wide host range; it is especially fond of caterpillars and beetle grubs in foliage of many vegetable crops. *P. maculiventris* does best in temperate climates.

APPEARANCE: Adults are called "spined soldier bugs"—they are shield-shaped, khaki coloured, 8–12 mm long, with prominent spurs behind their heads, and a swordlike snout with which they stab their prey. Young nymphs are red with black markings, older nymphs develop red, black, yellow-orange, and cream-white markings on their abdomen. Eggs vary from yellow to copper in colour, oval, spined, 0.8 mm tall, and deposited in clusters of 20–25 eggs.

DEVELOPMENT: Adults and nymphs search for prey, stab them, then suck them dry. They may kill 100 prey during their lifespan. Recorded hosts include leaf-eating caterpillars, cutworms, armyworms, budworms, European corn borers, grubs of bean beetles, and grubs of flea beetles. Adults overwinter, each female lays hundreds of eggs. Adults may live two or three months, and two or three generations arise per year.

APPLICATION: Supplied as older nymphs mixed with loose filler. Store a maximum of one to two days at 8–10°C. For light infestations, release one predator per m^2; for heavy infestations, release five predators per m^2. Repeat releases at two-week intervals.

NOTES: Adults have wings and may fly from the introduction site. *P. maculiventris* is compatible with Bt and NPV. It is very susceptible to organophosphate and carbamate pesticides, and less susceptible to synthetic pyrethroids and synthetic growth hormones.

Tanacetum vulgare Common tansy

BIOLOGY: An aromatic perennial cousin of chrysanthemums, also known as bitter buttons or golden buttons.

APPLICATION: Yepsen (1976) repelled caterpillars from crops by planting this herb nearby. *T. vulgare* grows over 1.5 m tall (the ornamental cultivar is shorter, *T. vulgare* var. *crispum*). The thujone in tansy drives away many insects.

***Hyphantria cunea* NPV** HcNPV

BIOLOGY: Different strains of this DNA virus have been used in Japan and the eastern Adriatic region (Hunter-Fujita *et al.* 1998). For more information on NPV, see the section on cutworms.

CHEMICAL CONTROL (see Chapter 11)

Spraying plants with neem repels caterpillars. Or try a repellent solution of garlic, hot peppers, and horseradish. Aqueous extracts of *Cannabis* repelled moths of *Pieris brassicae* L. trying to lay eggs on cabbage plants (Rothschild & Fairbairn 1980). To kill caterpillars, combine pyrethrum and rotenone. Sabadilla dust also works. Spinosad kills caterpillars in cotton crops. Synthetic pheromones are available for *A. gamma, C. interjectana, L. sticticalis,* and *M. brassicae.*

THRIPS

Thrips (the word is singular and plural) are tiny, slender insects. Adults have wings but fly poorly. They prefer to jump, and they spring quickly to safety when confronted. Reports indicate at least five genera of thrips attack *Cannabis.* They pose a problem in modern glasshouses that use rockwool and hydroponics. In old soil-floored glasshouses, watering with a hose kept floors damp, which encouraged the fungus *Entomophthora thripidum. E. thripidum* infects thrips when they drop to the ground to pupate. With no damp soil, there is no fungus, and no natural biocontrol.

SYMPTOMS

Immature and adult thrips puncture or rasp plant surfaces, then suck up the exuded sap. This gives rise to symptoms of white specks or streaks (sometimes coloured silver or yellow, Plate 27), appearing first on the undersides of leaves or hidden within flowers. Infested plants become covered with tiny black specks of thrips faeces. Leaves may curl up. In extreme infestations, plants wither, brown, and die.

1. ONION (TOBACCO) THRIPS

Thrips tabaci Lindeman 1888, Thysanoptera; Thripidae.

Description: Adults vary in colour from pale yellow to dark brown, are pointed at both ends, and grow to 1.2 mm in length. Female's wings are slender and fringed with delicate hairs (Fig 4.29). Eggs are white and kidney-shaped, oviposited in stems or leaves, and all but invisible to the naked eye. Larvae are small, pale, wingless versions of the adults, barely visible without a hand lens. This species pupates in the soil; pupae look like larvae with wing buds.

Life History & Host Range

The life histories of many thrips are nearly identical. Outdoor thrips overwinter in soil and plant debris. Glasshouse thrips do not hibernate. Most thrips stir into activity when the temperature reaches 16°C. Thereafter, the warmer the temperature, the worse their damage. Females lacking ovipositors usually lay eggs in cracks and crevices. Those with ovipositors insert their eggs into leaf and stem tissues, well protected from biocontrols. Eggs hatch in three to ten days under favourable conditions. First and second instars feed voraciously, and become full grown in less than a month. Third and fourth instars (called pre-pupae and pupae) generally become inactive, do not feed, and most go underground. During pupation, thrips acquire wings (in some species males do not have wings). Males are rare and reproduction is usually parthenogenic. Both mated and unmated females produce eggs; virgin females generally produce only females. In the field, four generations arise per year, and up to eight generations arise indoors.

Onion thrips attack a wide variety of vegetable and field crops. They are very active insects, especially when disturbed. Indoors, *T. tabaci* has pestered *afghanica* cultivars in glasshouses (Potter, pers. commun. 1999). Outdoors, the pest attacks fibre hemp, where it transmits hemp streak virus (Ceapoiu 1958) and the *Cannabis* pathogen Argentine sunflower virus (Traversi 1949).

Figure 4.29: Egg, larvae, and adult of the onion thrips, *Thrips tabaci* (about x25, courtesy USDA).

2. GREENHOUSE THRIPS

Heliothrips haemorrhoidalis (Bouché) 1833, Thysanoptera; Thripidae.

Description: Larvae are wingless, slender, and light yellow-brown to almost white in colour. They waddle around with a globule of black faecal fluid protruding from their anus; the globule grows until it falls off and a new one begins to form. Adults have dark brown to black bodies ornamented by a reticular pattern, light-coloured legs, and glassy-appearing wings which look white against the black body. Adults are up to 1.3 mm long.

Life History & Host Range

Greenhouse thrips are tropical, and found in warm glasshouses everywhere. They attack almost all plants growing in glasshouses. Each female lays about 45 eggs. The *H. haemorrhoidalis* life cycle turns in as little as 30–33 days at 26–28°C (Loomans & van Lenteren 1995). This species pupates on foliage of host plants (Hill 1994). The rectal fluid secreted by larvae gums up parasitoids and repels predators. Larvae and adults are rather sluggish, compared to *T. tabaci*.

3. MARIJUANA THRIPS

Oxythrips cannabensis Knechtel 1923, Thysanoptera; Thripidae.

Description: Adults are yellow, turning brown near their tails. Females reach 1.6 mm in length, and males are 1.3 mm. Larvae are smaller, lighter in colour, and wingless (Fig 4.30).

Life History & Host Range

This species appears to be host-specific on *Cannabis*. It has been isolated from hemp in Romania, the Czech Republic, Siberia, and the USA. *O. cannabensis* infests leaves and female flowers of feral hemp growing in Illinois (Stannard *et al.* 1970) and Kansas (Hartowicz *et al.* 1971). Hartowicz *et al.* considered it a possible biocontrol agent against marijuana. They noted that *O. cannabensis* is a potential vector of plant pathogens. Little is known of its life history.

Figure 4.30: Adult female marijuana thrips, *Oxythrips cannabensis* (McPartland, LM x48).

4. INDIAN BEAN THRIPS

Caliothrips indicus (Bagnall) 1913, Thysanoptera; Thripidae.

= *Heliothrips indicus* Bagnall

Description: Eggs are oval, white, and oviposited in the upper surfaces of leaves near main veins. Larvae are small, pale, wingless versions of adults; they feed along veins. Adults are blackish brown with brown and white banded forewings. Females reach 1.2 mm in length. Males are shorter (0.9 mm) and lighter brown.

Life History & Host Range

Indian bean thrips attack flax and various legumes. The species is limited to the Indian subcontinent. *C. indicus* closely resembles the North American bean thrips, ***Caliothrips fasciatus*** (Pergande). Cherian (1932) considered *C. indicus* a minor pest in Madras. Mated females produce an average of 65 eggs, unmated females average 51 eggs. In tropical zones *C. indicus* completes its life cycle in 11–14 days. Indian bean thrips are very active insects. Even the pupae can jump when disturbed.

5. WESTERN FLOWER THRIPS

Frankliniella occidentalis (Pergande) 1895, Thysanoptera; Thripidae.

Description: Eggs are kidney-shaped, 0.25 x 0.5 mm, and oviposited in leaves near main veins. Larvae are cigar-shaped, white to pale yellow, red-eyed, and rather slow moving. They feed along veins of leaflets. Adult females vary in colour from pale yellow to dark brown, with darker bands across the abdomen, light-coloured heads, and dark eyes. Forewings are light-coloured, with two longitudinal veins bearing two rows of delicate hairs. Females reach 1.3–1.7 mm in length. Males are shorter (0.9–1.1 mm) and uniformly yellow.

Life History & Host Range

F. occidentalis is California's gift to glasshouses around the world. It invaded Europe around 1983 (in Holland) and has spread throughout the continent. Surprisingly it has not yet been cited on *Cannabis*, probably for lack of recognition. Lago & Stanford (1989) described a relative, ***Frankliniella fusca*** (Hinds), on Mississippi marijuana.

F. occidentalis infests at least 250 species of plants, and vectors many viruses. Females of the species preferentially infest flowers and buds. At optimal temperatures (26–29°C), the life cycle turns in seven to 13 days. Adult females live another 30–45 days and lay 150–300 eggs. Mated females are more fecund than unmated ones. Females feeding on leaves and pollen are more fecund than females feeding solely on leaves. This species pupates in flowers or in soil (usually 1.5–2.0 cm deep).

DIFFERENTIAL DIAGNOSIS

Immature thrips nymphs are practically impossible to differentiate by species. Even the adults can be hard to tell apart. Thrips damage resembles that of other sap-sucking insects, especially spider mites. But mites cause tiny, round lesions on the upper leaf surface, whereas thrips leave irregular-shaped (often elongate) lesions on undersides of leaves, lesions often fill the spaces between leaf veins. Thrips nymphs can be confused with leafhopper nymphs.

Table 4.10: Infestation Severity Index for thrips.

Light	any thrips damage seen
Moderate	thrips damage and occasional thrips seen
Heavy	thrips damage on many plants but confined to lower leaves OR 2–10 thrips per leaf
Critical	thrips damage on growing shoots OR >10 thrips per leaf

CULTURAL & MECHANICAL CONTROL

(method numbers refer to Chapter 9)

Sanitation is important, so observe methods 1a and 1b. In thrips-infested glasshouses, remove all plant residues after harvest and heat the house for several days, to starve any remaining thrips. Water-stressed plants are particularly susceptible to thrips damage, so follow method 7a. Thrips dislike getting wet, so frequent misting may slow them down. Method 12a (repellent mulch) works for young, low-growing plants, and method 12c (sticky traps, blue or pink) works for all plants. Glasshouses can be barricaded with thrips microscreens (method 13).

BIOCONTROL (see Chapter 10)

Predators have been released indoors (*Neoseiulus cucumeris, Iphiseius degenerans, Neoseiulus barkeri, Euseius hibisci, Deraeocoris brevis,* and several *Orius* species) and outdoors (*Aeolothrips intermedius, Franklinothrips vespiformis*), all discussed below. *Hypoaspis miles* (discussed under fungus gnats) cannot provide thrips control by itself, but can reduce populations of *soil-pupating* thrips by 50%. Less effective thrips predators include lacewings (*Chryosopa carnea*) and ladybeetles (*Hippodamia convergens, Coccinella undecimpunctata*). Most thrips predators cannot be mixed together, except for a combination of *Neoseiulus, Orius,* and *Hypoaspis.*

Parasitoid wasps include *Thripobius semiluteus, Ceranisus menes,* and *Goetheana shakespearei,* described below. The life cycles of these parasitoids are longer than their hosts, so they work best as preventatives.

Beneficial soil nematodes control *soil-pupating* thrips. *Heterorhabditis bacteriophora* (described under white root grubs) works better against thrips than *Steinernema feltiae* or *Steinernema riobravis* (Cloyd *et al.* 1998). Nematodes will not reproduce in thrips cadavers, so they must be reintroduced several times during the season.

Turning to biocontrol with fungi, Cloyd *et al.* (1998) reported better control of *F. occidentalis* with *Metarhizium anisopliae* than with *Verticillium lecanii.* Both fungi are described under aphids. Hussey & Scopes (1985) killed *T. tabaci* and other thrips with the Mycotal® strain of *Verticillium lecanii.* Hussey & Scopes also controlled *T. tabaci* with *Beauveria bassiana,* spraying twice at ten-day intervals with a 0.25% water suspension of Boverin®. *B. bassiana* is described under whiteflies. *Entomophthora thripidum* and *E. parvispora* are described below. Fungi kill more adults than larvae, possibly because many adults prefer feeding in flowers (where humidity is higher and more favourable for fungi), and because larvae moult periodically, shedding spores attached to their cuticle (Cloyd *et al.* 1998).

Onions (*Allium cepa*) and garlic (*Allium sativum*) serve as trap crops for *T. tabaci.* Infested trap crops can be sprayed with pesticides or bagged and removed from the growing area.

Neoseiulus (Amblyseius) cucumeris

BIOLOGY: *N. cucumeris* was the first predatory mite observed to attack thrips, in 1939. It feeds on young (first instar) larvae and eggs; in the absence of thrips it survives on alternative food sources—pollen and mites. *N. cucumeris* does best in moderate temperatures (19–27°C) and moderately high humidity (70–80% RH). It was originally sold for control of *T. tabaci,* but works well against *F. occidentalis* and other thrips.

APPEARANCE & DEVELOPMENT: *N. cucumeris* adults are elongate-oval in outline, 0.5 mm long, white to beige to pink, with legs shorter than those of *Phytoseiulus persimilis* (Plate 28). The life cycle takes six to nine days at 25°C. Adults live another month. Adults consume up to six thrips per day and lay two or three eggs per day (Houten *et al.* 1995). Most strains of *N. cucumeris* enter diapause under short photoperiods (Gill & Sanderson 1998), but some new commercial strains do not (Ravensberg, pers. commun. 1998).

APPLICATION: Supplied as adults in shaker bottles or supplied as all stages in **sachets**. Sachets are small paper bags filled with 300–1000 mites (all stages), a food source, and a tiny exit hole. Hang sachets on plants to deliver a "slow release" of mites. When using sachets, avoid very high humidity (>90% RH), which causes sachets to mould (Cherim 1998). Store shaker bottles up to one to two days in a cool (8–15°C), dark place. Mites in sachets can be stored a little longer. *N. cucumeris* can be sprinkled directly on plants, sprinkled into distribution boxes (Plate 85), or heaped on rockwool blocks. Used preventatively, release 100 mites per m² per month (Reuveni 1995). For light to moderate infestations, release 100–500 mites per plant every two to three weeks

(Thomson 1992), or 200–300 mites per m^2 per week until the infestation subsides.

NOTES: *N. cucumeris* survives well without prey, which makes it a good preventative biocontrol. *N. cucumeris* prefers living on the undersides of leaves, rather than flowers (Cloyd *et al.* 1998). This trait, along with its need for high humidity and its propensity to diapause, make *N. cucumeris* a suboptimal choice for flowering plants. Results have been poor in glasshouse *Cannabis* (Watson, pers. commun. 1998).

N. cucumeris is compatible with *Orius tristicolor* and *O. insidiosus.* When released with other predatory mites, such as *P. persimilis* and *I. degenerans,* the predatory mites prey upon each other (in the absence of pests), but the predators can coexist (Hussey & Scopes 1985). Some *N. cucumeris* strains are susceptible to the biocontrol fungi *Verticillium lecanii* and *Beauveria bassiana,* so they are not compatible in conditions of high humidity (Gilkeson 1997). *N. cucumeris* tolerates most insecticides better than *I. degenerans* and other predatory mites, but it is very sensitive to pyrethroids and organophosphates. Some companies mass-rearing *N. cucumeris* were plagued by protozoan infections in the recent past. Protozoans cause sublethal infections of *N. cucumeris,* so the infected biocontrols are sold and exported to release sites (Gilkeson 1997). Infected biocontrols are less effective, and they may spread infection to other biocontrol mites.

Iphiseius (Amblyseius) degenerans

BIOLOGY: A predatory mite that attacks the first larval stage of *T. tabaci* and *F. occidentalis.* In the absence of thrips, *I. degenerans* preys on spider mites. In the absence of any prey, *I. degenerans* survives and reproduces on pollen. The predator is native to Eurasian and African subtropical regions, and prefers moderate humidity (55–80% RH), and warm temperatures (21–32°C).

APPEARANCE: Adults are dark-brown, oval, 0.7 mm long, and mobile. Larvae have a brown X-mark on their back. Eggs are transparent and turn brown before hatching, laid along veins on undersides of leaves.

DEVELOPMENT: *I. degenerans* does not diapause under short daylengths. The life cycle takes ten days. Adults live another month. Adults consume four or five thrips per day and lay one to two eggs per day (Houten *et al.* 1995).

APPLICATION: Supplied as adults in shaker bottles, or a mixture of adults, nymphs, and eggs on bean leaves in plastic tubs. They are stored the same as *N. cucumeris.* Used preventatively, release two to four mites per m^2 per month. For light infestations, release ten mites per m^2 every two weeks. For moderate infestations, release 20 mites per m^2 weekly (Cherim 1998).

NOTES: *I. degenerans* works better in flowering crops than *N. cucumeris*—it tolerates lower humidity, does not diapause, migrates well in tall plants, and prefers living in flowers rather than leaves (Cloyd *et al.* 1998). Nevertheless, results have been poor in sinsemilla *Cannabis,* because of the lack of pollen as an alternative food source (Watson, pers. commun. 1998). *I. degenerans* is compatible with *Orius* but may feed on *N. cucumeris.* It tolerates neem and many fungicides, but is more sensitive to insecticides and miticides than *N. cucumeris,* especially soap, sulphur, and pirimicarb.

Neoseiulus (Amblyseius) barkeri (=*Amblyseius mackenziei*)

BIOLOGY: A mite that preys on *T. tabaci, F. occidentalis,* and spider mites (*Tetranychus urticae*). It occurs naturally around the world, and does best at 65–72% RH and moderate temperatures (19–27°C).

APPEARANCE & DEVELOPMENT: Adults resemble *N. cucumeris,* perhaps a deeper tan colour. *N. barkeri* can reproduce every 6.2 days at 25°C. Adults consume two or three thrips per day and lay one to two eggs per day (Houten *et al.* 1995). Each mite eats about 85 thrips larvae during its lifetime (Riudavets 1995). Unfortunately *N. barkeri* enters diapause with short photoperiods.

APPLICATION: Supplied as adults in pour-top bottles. Store a maximum of up to two or three days in a cool (8–10°C), dark place. To control *T. tabaci,* researchers released 50,000 predators per ha every 14 days, or 15–20 per m^2 glasshouse, or 1200 per plant, or three per leaf (Riudavets 1995).

NOTES: *N. barkeri* does not work as well as *N. cucumeris* (lower predation and oviposition rates), and commercial strains diapause under short days (Houten *et al.* 1995). Nevertheless, mass rearing of *N. barkeri* is relatively easy, and the species is somewhat resistant to pesticides. *N. barkeri* may be compatible with other predatory mites. This predator is therefore best suited for preventive control on vegetative plants. *N. barkeri,* like *N. cucumeris,* has been plagued by protozoan infections that depress its effectiveness (Gilkeson 1997).

Euseius (Amblyseius) hibisci

BIOLOGY: A mite that preys on young larvae of *F. occidentalis* and other thrips, spider mites, citrus mites, and immature whiteflies. It lives in southern California avocado and citrus orchards, and loves dry, subtropical conditions.

APPEARANCE & DEVELOPMENT: Eggs are oblong and almost transparent. Larvae are also transparent. Adults are pear-shaped, very shiny, and smaller than other predatory mites. Their colour depends on what they are eating—yellow when feeding on thrips larvae, white when feeding on pollen, and red when feeding on citrus mites. The life cycle takes nine to 12 days at 26°C. Adults live another 25–30 days. Adults consume three or four thrips per day and lay one to two eggs per day (Houten *et al.* 1995). *E. hibisci* does not diapause (Houten *et al.* 1995).

NOTES: *E. hibisci* eats and reproduces slower than *N. cucumeris* and *I. degenerans.* In the absence of pests, *E. hibisci* actually reproduces faster on pollen. It tolerates drought better than any other predatory mites (Houten *et al.* 1995). Drought tolerance and lack of diapause make this mite a good preventative biocontrol in flowering plants. Natural populations of *E. hibisci* have developed resistance to many organophosphate insecticides, but are still sensitive to carbamates and pyrethroids. Unfortunately, this biocontrol must be reared on leaves, which hampers its commercial availability.

***Orius* species** "Pirate bugs"

BIOLOGY: Six different species of *Orius* are available (described below). These predators are polyphagous; after thrips are gone, pirate bugs eat other pests (aphids, mites, scales, whiteflies), then other biocontrols, then each other. They do best in temperate climates, optimally 60–85% RH and 20–30°C.

APPEARANCE: Adults are yellow-brown to black with white wing patches, 1.5–3 mm long (Plate 29). Nymphs are teardrop shaped, bright yellow to brown, with red eyes.

DEVELOPMENT: Adults overwinter; females lay 30–130 eggs in plant tissue. The life cycle takes as little as two weeks. Adults live another month. Usually two to four generations arise every year. *Orius* species move quickly and catch prey with their forelegs. Young nymphs feed primarily on pollen or young thrips. Older nymphs and adults feed on all stages of thrips; adults eat five to 20 thrips per day. Unfortunately, many *Orius* species diapause under short day lengths.

APPLICATION: Supplied as adults and nymphs in shaker bottles. Store for up to three days in a cool (10–15°C), dark place. Used preventatively, release one predator per plant (Ellis & Bradley 1992) or one predator per m^2 (Reuveni 1995), make two releases two weeks apart. Moderate infestations may require five to ten predators per m^2 per week (Riudavets 1995). Heavy pest infestations cannot be controlled by *Orius* species, the biocontrols reproduce too slowly. Koppert (1998) recommended releasing *Orius* in groups of 15–20 insects to encourage mating.

NOTES: Adults can fly and find remote infestations, and they like to inhabit flowers. In mixed pest infestations, pirate bugs will eat the largest pests first. So even if the predators are swimming in baby aphids, they would rather eat a large thrips larva. In the absence of pests, *Orius* species eat predatory mites (but *not* subterranean *Hypoaspis miles*). Young *Orius* nymphs have a hard time with hairy-trichomed plants, adults do better. All *Orius* species are sensitive to pesticides.

O. tristicolor is native to western North America (Fig 4.39) and widely available. It feeds on *F. occidentalis, T. tabaci, Caliothrips fasciatus*, spider mites (*Tetranychus* species), aphids (*Myzus persicae*), and eggs of budworms (*Helicoverpa zea*). *O. tristicolor* populations parallel thrips populations; i.e., adult longevity and fecundity increases as the thrips population increases (Riudavets 1995). The predator is reportedly compatible with *Neoseiulus cucumeris*, although *O. tristicolor* eats other beneficial mites.

O. insidiosus is native to the midwestern USA (Fig 4.39). The predator successfully controlled *F. occidentalis* in European glasshouses (Riudavets 1995). It also eats spider mites (*Tetranychus urticae*), aphids (*Aphis* species), and the eggs of budworms (*Helicoverpa zea, H. virescens*). It is compatible with *Neoseiulus cucumeris* but eats the aphid predator *Aphidoletes aphidimyza* (Riudavets 1995). *O. insidiosus* diapauses in photoperiods under 11 hours, but warm temperatures may postpone diapause.

O. albidipennis ranges across southern Europe east to Iran. It eats *F. occidentalis, T. tabaci*, spider mites (*Tetranychus* species), aphids (*Aphis gossypii*), whiteflies (*Bemisia tabaci*), budworms (*Heliothis armigera*), and cutworms (*Spodoptera litura*). It coexists with ladybeetles (*Coccinella unidecimpunctata*), lacewings (*Chrysopa carnea*), and *O. laevigatus*. Riudavets (1995) considered it a promising biocontrol agent. *O. albidipennis* prefers warmer temperatures (up to 35°C) and does *not* diapause—which makes it perfect for flowering crops in hot greenhouses (Cloyd *et al.* 1998).

O. laevigatus hails from the Mediterranean region and western Europe. It feeds primarily on *F. occidentalis* and *T. tabaci*, but also eats spider mites (*Tetranychus* species), European corn borers (*Ostrinia nubilalis*), aphids (e.g., *Aphis gossypii*), and cutworms (e.g., *Spodoptera litura*). It coexists with ladybeetles (*Coccinella unidecimpunctata*), lacewings (*Chrysopa carnea*) and *O. albidipennis*. Riudavets (1995) noted *O. laevigatus* adapts well to glasshouse conditions on many crops, and survives without thrips for a long time. *O. laevigatus* diapauses in photoperiods under 11 hours, but warm temperatures may postpone diapause (Ravensberg, pers. commun. 1999).

O. majusculus is common in central Europe (Fig 4.39). Good control of *F. occidentalis* was obtained in cucumber glasshouses using a ratio of one predator per 100 thrips. *O. majusculus* produces more eggs than any other *Orius* species. But *O. majusculus* is more sensitive to photoperiod, and diapauses in days shorter than 16 hours, although warm temperatures may postpone diapause (Ravensberg, pers. commun. 1999).

O. minutus ranges across western Europe to Siberia and China (Fig 4.39). The species associates with many plants, including *Urtica* species. *O. minutus* prefers *T. tabaci* but will eat spider mites (*Tetranychus* species), eggs of European corn borers (*Ostrinia nubilalis*), and aphids (e.g., *Aphis gossypii*).

Anthocoris nemorum is a minute pirate bug (Fig 4.39) used against thrips in England (van Lenteren 1995). It is related to *Orius* species.

Deraeocoris brevis

BIOLOGY: This predatory mirid bug eats all stages of thrips. It also eats aphids, mites, whiteflies, lygus bugs, and psyllids. Cherim (1998) said some *D. brevis* populations actually prefer aphids over thrips. *D. brevis* occurs naturally in temperate climates (optimally 30–70% RH, 18–30°C). A related species, *Deraeocoris nebulosus*, may also be commercially available.

APPEARANCE & DEVELOPMENT: Adults are mottled, brown bugs 5 mm long. Nymphs are lighter-coloured versions of adults. Adults overwinter; females lay up to 200 eggs in plant tissue. Nymphs begin preying on pests as second instars. The life cycle takes 30 days. Adults live another 21 days.

APPLICATION: Supplied as a mix of adults and nymphs in vials and bottles. Used preventatively, release one predator per 10 m^2 every two weeks. For light infestations, release one to two predators per m^2 every week. For heavy infestations, release five predators per m^2 every week. Cherim (1998) recommended releasing a "critical number" of individuals (≥300) to establish a population. In the absence of prey, adults survive on pollen. Young nymphs move with difficulty across surfaces with glandular trichomes; adults do better. Unfortunately, adults enter diapause with short days.

Aeolothrips intermedius

BIOLOGY: Seczkowska (1969) found this banded thrips preying on *T. tabaci* in Polish *Cannabis* fields. It also feeds on *Frankliniella* species, *H. hemorroidales*, and spider mites (*Tetranychus cinnabarinus, T. urticae*). The predator is native to Europe, the Middle East, and India.

APPEARANCE: Adults are yellowish to dark brown, about 1.6 mm long, with relatively broad, banded wings. Larvae are lighter coloured than adults.

DEVELOPMENT: Females lay 30–70 eggs, with three or four generations arising each year. Each *A. intermedius* larva consumes about 25 *T. tabaci* larvae. Adults feed on plant pollen and may pierce leaves and stems for water (Riudavets 1995). *A. intermedius* is eaten by other predators, so it is not compatible with pirate bugs (*Orius* species), lacewings (*Chrysopa* species), and ladybeetles (*Coccinella septempunctata*).

Franklinothrips vespiformis

BIOLOGY: Another predatory thrips, related to the previous predator, this species is found in California avocado orchards. *F. vespiformis* also feeds on immature aphids, whiteflies, scales, mites, and moth eggs.

APPEARANCE & DEVELOPMENT: Adults are black with white-banded legs; their round heads and thin waists make them look like wasps or ants. Larvae have a wide, red band covering their abdomen. These predators are less effective than predatory mites and *Thripobious semiluteus*, below.

Thripobius semiluteus

BIOLOGY: A parasitic eulophid wasp that parasitizes *H. haemorrhoidalis*, is native to Africa, Australia, Brazil, and South Africa, and does best at 23°C and 50% RH.

DEVELOPMENT: Adults crabwalk across leaf surfaces in search of first and second stage larvae. Females lay at least 40 eggs. Parasitized hosts turn into black, slightly swollen mummies. Adults emerge from thrips cadavers to repeat the lifecycle. *T. semiluteus* takes 22 days to develop (Loomans & van Lenteren 1995).

APPLICATION: Supplied as pupae in vials; attach vials to plants horizontally, as high in the plant canopy as possible (or tape to stakes if plants are small). In orchards, the standard release rate is 5000 pupae per ha or 250 pupae per tree (Loomans & van Lenteren 1995). *T. semiluteus* has been introduced into California and Hawai'i to control *H. haemorrhoidalis*. Do not use botanical insecticides within one week of release, or other pesticides within 30 days of release.

Ceranisus menes

BIOLOGY: A tiny Eulophid wasp that parasitizes *F. occidentalis* and *T. tabaci* larvae, and lives around the world.

DEVELOPMENT: Female wasps insert ovipositors into young thrips larvae and lift struggling larvae into the air; the larvae soon become paralysed and a single egg is deposited (Loomans & van Lenteren 1995). *C. menes* larvae eventually kill their hosts, then exit to pupate in soil. *C. menes* wasps also kill thrips by poking them to feed off body fluids.

APPLICATION: *C. menes* was used in a large scale effort to eliminate *T. tabaci* from Hawai'i, but has not been commercially available since then. Dense leaf hairs interfere with the parasitoid's ability to locate prey.

Goetheana shakespearei

BIOLOGY: Another Eulophid wasp (sometimes called *G. parvipennis*), smaller than the aforementioned *C. menes*. This species is native to south Asia and Australia. It was released in California to control young larvae of *H. haemorrhoidalis*. It also controls *C. indicus*. The wasp can complete its life cycle on *T. tabaci*, but attempts to control *T. tabaci* have failed (Loomans & van Lenteren 1995).

APPLICATION: *G. shakespearei* is more successful as a biocontrol in tropical areas and less effective in temperate climates (Loomans & van Lenteren 1995). Its minute size makes the parasitoid difficult to rear.

Entomophthora thripidum and *Neozygites parvispora*

BIOLOGY: Two closely-related soil fungi that infect thrips pupating underground. Both species occur naturally in damp soil in Europe and North America. *N. parvispora* is an obligate pathogen of thrips, known from *T. tabaci, F. occidentalis*, and *Limothrips* species (Carris & Humber 1998). *E. thripidum* is not so specialized—it also infects thrips predators.

CHEMICAL CONTROL (see Chapter 11)

Thwart thrips with tobacco tea or nicotine. Cherian (1932) repelled thrips with a mix of fish emulsion and soap sprays. Yepsen (1976) concocted a thrips-repellent spray from field larkspur (*Consolida regalis* or *Delphinium consolida*). Metcalf *et al.* (1962) eliminated thrips with sabadilla dust, or a mix of diatomaceous earth and pyrethrum. Spraying a slurry of clay microparticles may work. Sprays and dusts generally require two or three applications at weekly intervals. Apply at mid-day, when thrips are most active. Be sure to treat the undersides of leaves. Imidacloprid (a synthetic nicotine) and spinosad (a fermentation product) kill thrips. Meier & Madiavilla (1998) eliminated thrips with Cymbush (permethrin). Thripsticks® (polybutenes mixed with deltamethrin, a synthetic pythrethrum) can be applied to plastic around the bases of plants, so when thrips drop to the ground to pupate, they die.

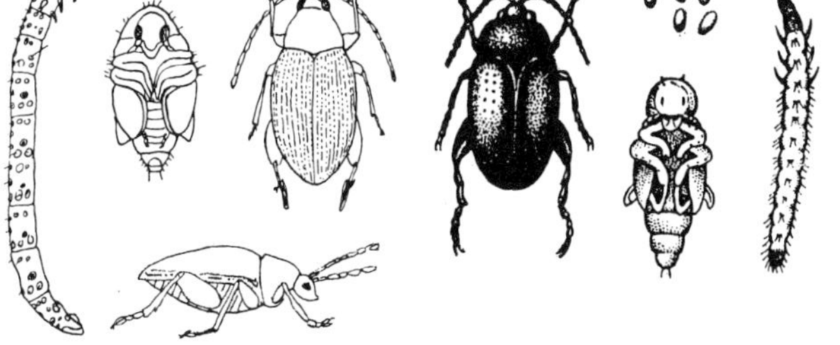

Figure 4.31: Larvae, pupae, and adults of the hemp flea beetle, *Psylliodes attenuata* (line drawing on left from Flachs 1936 except side view by McPartland, shaded version on right from Senchenko & Timonina 1978).

FLEA BEETLES

Beetles are the most common animals on earth, and the most common beetles on *Cannabis* are flea beetles (family Chrysomelidae). Flea beetles damage plants as adults and larvae (grubs). Adults feed on foliage, flowers, and unripe seeds. Grubs feed on roots or occasionally act as leafminers. Many flea beetles are polyphagous and attack a wide variety of plants. Others specialize, infesting only one or two plant species. One such specialist is the hemp flea beetle. Rataj (1957) considered the hemp flea beetle insignificant, but Ragazzi (1954), Dempsey (1975), and Bósca & Karus (1997) cited it as a major pest. Durnovo (1933) considered it the number one problem in the North Caucasus. Biocontrol researchers thought the species would be an excellent candidate for eradicating illegal marijuana (Mohyuddin & Scheibelreiter 1973).

SIGNS & SYMPTOMS

The hind legs of adult beetles are large and they leap like fleas when disturbed. Since flea beetles are small and leap frequently, they do not eat much in one place. Damage consists of many small, round to irregular holes, formed between leaf veins (Plate 30). Leaves of heavily infested plants may be completely skeletonized. Young plants are killed. Grubs of most species feed on roots, usually at depths of 4–8 cm underground. They feed on cambium, making longitudinal or tortuous galleries in roots. Grubs of one species act as leafminers, mining beneath the upper surface of leaves, tracking in a tight spiral that ends in a brown blotch.

1. HEMP FLEA BEETLE

Psylliodes attenuata (Koch) 1803, Coleoptera; Chrysomelidae.
= *Haltica attenuata* Koch, = *Psylliodes japonica* Jacoby 1885,
= *Psylliodes apicalis* (Steph.) *apud* Spaar *et al.* 1990

Description: Eggs are pale yellow, deposited singly around plant roots near the soil line. Grubs are cylindrical, white with tiny spots and bristles, with six minute legs, a brownish head, and reach 4.5 mm in length. Pupae are initially pearly white but slowly darken, beginning with the eyes, followed by other parts of the head, legs, and finally the elytra. Adult are oval in outline, black with minute bronze-grey hairs, and average 1.3–2.6 mm long (Plate 30). Wing covers are striated and punctuated. Heads exhibit sharp outgrowths crossed in an "x" at the eyes, and sprout antennae reaching half their body length (Flachs 1936). See Fig 4.31.

Life History & Host Range

Adults overwinter in soil and emerge at the end of March to feed on young seedlings. Adults mate in April and begin laying eggs ten days later. Females lay an average of 2.6 eggs per day for a total of 55 eggs (Angelova 1968). Egg laying lasts from late April to the end of July. Larvae hatch in five to 16 days, and grubs feed within roots when young. By June, grubs exit roots and live in soil. Pupation occurs from mid June to August, 4–15 cm underground. Adults emerge from pupae from late June to September. Adults feed on plant tops until autumn, then burrow into soil to depths of 20 cm (Angelova 1968). Only one generation arises per year.

Adults cause more damage than grubs. Damage peaks twice—on seedlings when adults emerge in early spring, and again in late summer when adults emerge from pupae. Late-season beetles may selectively infest female tops and feed on seeds (Silantyev 1897). *P. attenuata* crop damage increases in warm, dry weather (Flachs 1936). Adults are most active on warm sunny days; in overcast or wet weather they retreat to soil cracks, beneath clods of earth, or on the lower surface of leaves. Hemp flea beetles also feed on hops and nettles, especially in the spring (Silantyev 1897). *P. attenuata* ranges from Great Britain and France to eastern Siberia, northern China, and Japan (where it was called *P. japonica*).

2. OTHER FLEA BEETLES

The hops flea beetle, ***Psylliodes punctulata*** Melsheimer 1847, attacks hemp in Canada (Glendenning 1927). *P. punctulata* is a North American version of *P. attenuata*. Adults are small (2–2.5 mm long), oval, and greenish-black. The beetles develop bronze highlights after several days in sunshine (Fig 4.32). Adults feed on leaves of seedlings, larvae feed on roots. *P. punctulata* feeds on hops, nettles, canola, sugar beets, rhubarb, and many cruciferous crops. For more details on morphology and life cycle, see Chittenden (1909) and Parker (1910).

Phyllotreta nemorum (Linnaeus) 1758 attacks hemp in Europe (Kirchner 1906, Borodin 1915, Goidànich 1928, Kovacevic 1929). The grubs are yellow-bodied, brown-headed, and small enough (5–6 mm long) to bore into stems *or* feed as leafminers (Fig 4.39). Adults are metallic black, sport a broad yellow stripe down each wing cover, and average 3.5 mm in length (Fig 4.32). Borodin (1915) cited a related species, ***Phyllotreta atra*** (Fabricius) 1775. This small black flea beetle attacks many vegetable crops in Europe.

Dempsey (1975) considered ***Podagrica aerata*** Marsh 1802 a serious pest of hemp; Kaltenbach (1874) mentioned it in Germany. *P. aerata* usually feeds on brambles (*Rubus* species) in central Europe. The closely-related species ***Podagrica malvae*** Illiger 1807 also attacks hemp in central Europe (Rataj 1957). *P. malvae* normally infests kenaf or marsh mallows. Hill (1983) reported a *Podagrica* species chewing holes in *Cannabis* leaves in Thailand, which he called the "cotton flea beetle."

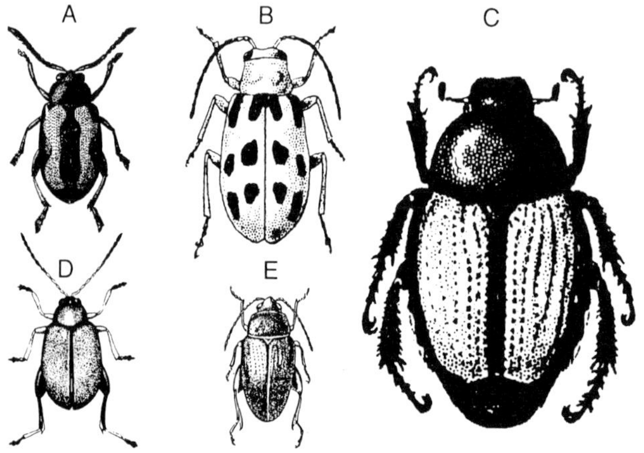

Figure 4.32: Assorted leaf beetles (all x5). A. *Phyllotreta nemorum* (from USDA); B. *Diabrotica undecimpunctata howardi* (from Westcott 1964); C. *Popillia japonica* (from Westcott 1964); D. *Chaetocnema concinna* (from Hill 1983); E. *Psylliodes punctulata* (from Senchenko & Timonina 1978).

Hartowicz *et al.* (1971) found an unspeciated *Chaetocnema* species infesting feral hemp in Kansas. Kulagin (1915) described ***Chaetocnema hortensis*** Geoffr. 1785 attacking plants near Moscow; the species normally infests grasses. Borodin (1915) and Kovacevic (1929) cited ***Chaetocnema concinna*** Marsh 1802 on hemp. But as Sorauer (1900) pointed out, this beet-eating beetle is easily confused with the hemp flea beetle. Lago & Stanford (1989) collected ***Chaetocnema denticulata*** (Illiger) and ***Chaetocnema pulicaria*** Melsheimer from Mississippi marijuana. *C. denticulata* is a tiny (2.5 mm long), oval, black beetle with bronze highlights. *C. pulicaria* looks similar but is even smaller.

Bantra (1976) described an "***Altica***" species on Indian marijuana. Takahashi (1919) cited "***Haltica flavicornis*** Baly" on Japanese hemp. Heikertinger & Csiki (*Coleopterorum Catologus* Part 116) question whether this species is a member of the subfamily Alticinae (the genus *Haltica* is no longer recognized). Goidànich (1928) cited a number of additional Chrysomelidae on *Cannabis*—***Chrysomela rossia*** Ill. 1802, ***Crepidodera ferruginea*** Scop. 1763, ***Gastroidea polygoni*** L. 1758, ***Gynandrophthlma cyanea*** F. 1775, and ***Luperus flavipes*** (L.) 1776.

DIFFERENTIAL DIAGNOSIS

Leaf damage from flea beetles can be confused with damage caused by leaf-eating caterpillars, weevils, and other beetles. No other beetles, however, leap like fleas. Grub damage on roots resembles that of maggots or other root-eating beetles. Leafmines by *P. nemorum* must be differentiated from leafmines by fly larvae. Some planthopper nymphs resemble adult flea beetles, as do short, black bugs. Grubs of weevils are indistinguishable from grubs of flea beetles, especially when young.

CULTURAL & MECHANICAL CONTROL

(method numbers refer to Chapter 9)

Follow methods 1 (sanitation), 2a (deep ploughing), and 3 (especially eliminate nettles and wild hops). Chinese farmers use method 4 by harvesting crops early to escape late-season flea beetles (Clarke 1995). European farmers use method 4 by sowing seeds early, resulting in seedlings that are older, hardier, and able to withstand flea beetles emerging in the spring (Camprag *et al.* 1996). Avoid susceptible cultivars, such as 'USO-9' ('YUSO-9') from the Ukraine. *P. attenuata*-resistant lines include southern Russian landraces and Chinese landraces (Bòcsa 1999). For crop rotations, Camprag *et al.* (1996) suggested planting hemp crops at least 0.5–1 km away from the previous year's hemp crops.

For horticultural situations, use method 12c (white sticky traps) and method 13 (protecting seedlings with fine mesh or screens). Since flea beetles prefer bright sunlight, consider planting in partial shade. Parker (1913b) stopped *P. punctulata* from crawling up stems by applying 5-cm bands of Tanglefoot®.

To catch adults on plants, fashion two "flea beetle cones" out of cardboard or metal. Make them 60 cm across and 20 cm deep. Coat the inner surface with Tanglefoot® or other adhesives. Hold the cones on either side of an infested plant and bring them together quickly, like cymbals, to surround the plant. Darkness disturbs the flea beetles and they leap onto the sticky surface. This treatment should be repeated at least once a week. Larger plants can be treated one limb at a time.

BIOCONTROL (see Chapter 10)

Adults may be susceptible to Bt-t, described below. Underground larvae are killed by beneficial nematodes, such as *Steinernema riobravis* (described below), *Heterorhabditis bacteriophora* (described under white root grubs), and perhaps *Steinernema carpocapsae* (described under cutworms). *Bacillus popilliae* may control root-feeding larvae, but this microbial primarily kills white root grubs (described there).

Two braconid wasps, *Microcronus psylliodis* and *M. punctulatus*, await commercial development. Nettles (*Urtica dioica*) attract *P. attenuata* in the spring and serve as a trap crop. In areas with lots of nettles, mow down all but a few plants. Flea beetles crowd the remaining nettles, ready to be bagged and burned. Chittenden (1909) said rhubarb (*Rheum rhabarbarum*) serves as a trap crop for *P. punctulata*. Yepsen (1976) repelled flea beetles by planting wormwood (*Artemisia absinthium*) and catnip (*Nepeta cataria*). Israel (1981) confirmed the efficacy of catnip. Howard *et al.* (1994) repelled flea beetles with marigold (*Tagetes* species).

Steinernema riobravis

BIOLOGY: This nematode was discovered in 1994, by the same researcher who unearthed *Heterorhabditis megidis* (another beneficial nematode). *S. riobravis* controls soil-dwelling chrysomelid beetles (*Diabrotica* species), root weevils, Noctuid caterpillars, and root maggots (Devour®, Bio Vector 355®).

DEVELOPMENT: *S. riobravis* exhibits the same lifecycle as other *Steinernema* species, described in detail under *S. carpocapsae* (see cutworms). *S. riobravis* utilizes "ambusher" *and* "cruiser" strategies.

APPLICATION: Supplied as third-stage larvae in polyethylene sponge packs or bottles. Sealed containers of *Steinernema* species remain viable for six to 12 months at 2–6°C. To apply to soil, follow instructions for *Heterorhabditis bacteriophora* (see white root grubs).

NOTES: *S. riobravis* is native to the Rio Grande Valley of Texas, and tolerates hot (36°C), arid soil, compared to other *Steinernema* and *Heterorhabditis* nematodes. It has been dispersed through irrigation lines, so the nematode is relatively small and tough.

***Bacillus thuringiensis* var. *tenebrionis* "Bt-t"**

BIOLOGY: Leslie (1994) claimed Bt-t (Novodor®) kills adult flea beetles. Bt-t is sold for controlling the Colorado potato beetle (*Leptinotarsa decemlineata*), a member of the Chrysomèlidae. Leslie considered the original Bt-t strain (M-One®) more effective than the bioengineered, *Pseudomonas*-encapsulated strain (M-Trak®). *B. thuringiensis* var. *san diego* is closely related and perhaps identical to Bt-t. Susceptibility to Bt-t varies greatly in beetles, presumably because of variation in their gut wall, where the Bt-t toxin must attach. For a description of Bt, see European corn borers.

CHEMICAL CONTROL (see Chapter 11)

Rotenone, nicotine, and cryolite kill flea beetles. Pyrethrins, ryania, and sabadilla are less effective. Spinosad is a new fermentation-derived insecticide that has been used against beetles. Applications must be repeated for beetles arriving from surrounding areas. Kill underground larvae with pesticide soil drenches. Bordeaux mixture repels adults and deters feeding (Chittenden 1909), as do sprays of garlic extract, insecticidal soap, and neem. Yepsen (1976) claimed that wood ashes also repel flea beetles. Sprinkle ashes on vegetative portions of each plant, two or three times a week. Methoprene, a juvenile growth hormone, works well against most beetles. Be sure to treat infested trap crops (nettles) in early spring, before pests move to *Cannabis*.

OTHER LEAF BEETLES

Besides flea beetles, other beetles eat holes in *Cannabis*. Yuasa (1927) and Shiraki (1952) cited ***Monolepta dichroa*** Harold 1877 attacking hemp leaves in Japan. Kulagin (1915) observed ***Oulema melanopa*** (Linnaeus) 1758 feeding on hemp near Moscow. Bantra (1976) collected ***Diapromorpha pallens*** Olivier 1808 (=*D. melanopus* Lacepede 1848) in India. Hartowicz *et al.* (1971) found a ***Diabrotica*** species on feral hemp in Kansas. Alexander (1980) and Frank & Rosenthal (1978) described two cucumber beetles attacking marijuana: ***Acalymma vittata*** (Fabricius) 1775—the striped cucumber beetle, and ***Diabrotica undecimpunctata howardi*** Barber 1947—the spotted cucumber beetle or southern corn rootworm. Lago & Stanford (1989) identified yet more leaf beetles on Mississippi marijuana, including ***Nodonota*** species, ***Systena*** species, ***Disonycha glabrata*** (F.), and ***Epitrix fuscula*** Crotch. In addition to these Chrysomelids, three beetles from other beetle families are frequent offenders:

1. JAPANESE BEETLE

Popillia japonica Newman 1838, Coleoptera; Scarabaeidae.

Description: Adults are attractive, with robust metallic green bodies and copper-brown wings, growing to 15 mm in length. For a description of grubs, see the next section on white grubs.

Life History & Host Range

Overwintering grubs pupate in soil in April and May. Adults emerge from pupae between mid-May and early July. Adults only fly in the daytime. They cause greatest damage on warm sunny days. Females return to the soil and lay eggs under turf. Eggs hatch in July-August and grubs feed on the fine roots of grasses and other plants. In late autumn they burrow deeper into the soil and hibernate in a "winter cell" 8–30 cm deep. One generation arises per year.

Adults skeletonize the leaves of *Cannabis* and hundreds of other hosts. Alexander (1984b) cited Japanese beetles attacking marijuana in eastern Tennessee. These pests snuck into the USA around 1916. They were first discovered in New Jersey, accidentally imported from Japan in a batch of iris bulbs. Since then they have colonized the eastern USA and southeastern Canada, east of the Mississippi River. Populations have recently appeared west of the Mississippi River to the Rocky Mountains. Isolated infestations in California and Oregon have been eradicated.

2. INDIAN BEAN BEETLE

Epilanchna dodecostigma Mulsant 1853, Coleoptera; Coccinellidae.

Description: Adults resemble ladybeetles with small (6–8 mm long) reddish-brown bodies and black spots on their elytrae. Females lay yellow eggs in clusters on undersides of leaves. Larvae are yellow, covered with spines and reach 7 mm in length.

Life History & Host Range

This species is related to the Mexican bean beetle and squash beetle, outlaws of the normally benevolent ladybeetle family. The *E. dodecostigma* life cycle takes only 33 days near the equator, where up to five generations arise per year (Hill 1983). Cherian (1932) reported this beetle feeding on marijuana leaves in Madras. Bantra (1976) collected an *Epilanchna* species on north Indian *Cannabis*. *E. dodecostigma* lives in east Asia and usually feeds on cucurbits.

3. STAUBKÄFER

Opatrum sabulosum Linnaeus 1761, Coleoptera; Tenebrionidae.

Sorauer (1958) called this polyphagous species the "large staff beetle." Grubs and adults devour seedlings of cereal plants in Europe. Durnovo (1933) said damage peaks in May, especially in hemp fields adjacent to weedy wastelands. *O. sabulosum* is related to *Tribolium confusum*, a beetle occasionally eating seeds in marijuana (Smith & Olson 1982).

DIFFERENTIAL DIAGNOSIS

Leaf beetles can be confused with weevils and curculios (see two sections below). Flea beetles are small leaf beetles with their own idiosyncracies and control methods (see the section above).

CULTURAL & MECHANICAL CONTROL
(method numbers refer to Chapter 9)

Use methods 2a (deep ploughing), 3 (weeding), 9 (shake plants early in the morning while beetles are stiff with cold), 12c (white sticky traps), and 13 (mechanical barriers). Heavy mulching around seedlings offers some protection. Several Japanese beetle traps are commercially available. The best traps use "dual lures"—food *and* sex attractants. The sex attractant is a pheromone, Japonilure®. Food attractants include eugenol and geraniol. According to Turner *et al.* (1980), *Cannabis* produces eugenol and geraniol, so it is no surprise that the plant is attractive to Japanese beetles. Baited traps work best in sunny spots during the early summer, set away from desirable plants.

BIOCONTROL (see Chapter 10)

In humid conditions, adult beetles are susceptible to the mycoinsecticide *Metarhizium anisopliae* (described under aphids). The *tenebrionis* strain of Bt kills Japanese beetles and other leaf beetles (Leslie 1994). Beetle larvae in soil can be controlled with *Bacillus popilliae* (described in the next section). Larvae in soil can also be killed with entomophagous nematodes, especially "cruiser" species, such as *Heterorhabditis bacteriophora* and *Steinernema glaseri* (described in the next section). "Ambusher" species, such as *Steinernema carpocapsae*, work less well.

Yepsen (1976) repelled Japanese beetles by planting garlic (*Allium sativum*), tansy (*Tanacetum vulgare*), and geranium (*Pelargonium hortorum*). Or reverse your strategy by planting a trap crop—white zinnias (*Zinnia elegans*) attract Japanese beetles; lure the pests onto zinnias and eliminate them. Yepsen also claimed Japanese beetles have a fatal attraction for larkspur (*Delphinium* species); they eat the foliage and die.

CHEMICAL CONTROL (Chapter 11)

Rotenone, nicotine, and spinosad kill leaf-eating beetles. Neem repels some beetles. A *Cannabis* extract weakly repelled Japanese beetles (Metzger & Grant 1932). Yepsen (1976) killed cucumber beetles with wild buffalo gourd, *Cucurbita foetidissima*. He ground gourd roots into powder, suspended the powder in soapy water, and sprayed plants.

WHITE ROOT GRUBS

Of all underground insects, white grubs are the most destructive. Larvae of four Scarab beetles cause *Cannabis* losses. All four pests are polyphagous, and commonly attack grasses. Dewey (1914) noted that white grubs cause the greatest damage in hemp rotated after sod. The grubs often attract moles which become a secondary problem.

SIGNS & SYMPTOMS

Seedlings grow 30–60 cm tall before showing symptoms. Then they wilt, yellow, and die. Damage arises in patches across infested fields. Roots are slightly gnawed or entirely eaten away by grubs. Activity by moles, sod-digging skunks, and raccoons indicates grubs are in the soil.

1. JAPANESE BEETLE

Popillia japonica Newman 1838, Coleoptera; Scarabaeidae.

Description: Adults of Japanese beetles feed on *Cannabis* leaves and are described in the previous section. Grubs are plump and white with a brown head and six dark legs, 15–25 mm long. They lie curled in a C-shaped position in the soil. Their life cycle is described in the previous section on leaf beetles.

2. EUROPEAN CHAFERS

a. *Melolontha hippocastani* Fabricius 1801, Coleoptera; Scarabaeidae.
b. *Melolontha melolontha* Linnaeus 1775, Coleoptera; Scarabaeidae.
c. *Melolontha vulgaris* Linnaeus 1775, Coleoptera; Scarabaeidae.

Description: These species are related to "June bugs," the familiar nocturnal beetles attracted to lights in the early summer. Adults have robust brown bodies with reddish-brown ridged elytra and long slender spiny legs, averaging 20 mm in length (Fig 4.33). Larvae resemble those of Japanese beetles but are larger (reaching 44 mm in length) and their posterior abdomens have smooth, thin skins with dark body contents showing through.

Life History & Host Range

Grubs hibernate in soil deep beneath the frost line (Metcalf *et al.* 1962 found grubs 1.5 m underground). In the spring, grubs return from the deep to feed on shallow roots, spend the summer close to the surface, and then return to their deep winter cells. Grubs may run this cycle a third year, depending on the species. Eventually they pupate. Adults emerge from the soil in spring, and feed at night on tree foliage. At dawn they return to soil where females lay pearly white globular eggs in batches of 12–30, under sod or weedy patches of grass. Eggs hatch in three weeks and young grubs feed on roots for the summer. *Melolontha* species range across Eurasia from Ireland to Japan, but have only been reported on *Cannabis* in Germany and Italy (Kirchner 1906, Goidanich 1928, Gutberlet & Karus 1995). These species sometimes develop synchronous life cycles—in Switzerland *M. melolontha* adults arise every three years.

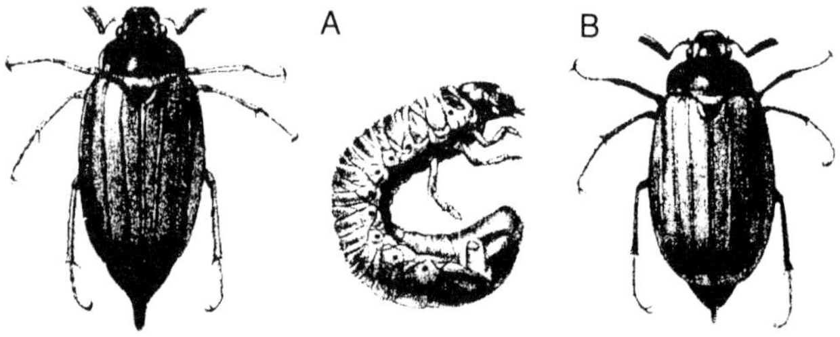

Figure 4.33: Chafers. A. Adult and grub of *Melolontha melolontha;* B. Adult of *Melolontha hippocastani* (from Sorauer 1958).

3. **BLACK SCARAB**

Maladera holosericea Scopoli 1772, Coleoptera; Scarabaeidae.

Description & Life History Adults are dark brown, 7–9 mm in length, and their elytra are marked by rows of shallow puctations. Both adults and larvae overwinter in soil. Adults emerge in spring and are active at night. Sorauer (1958) said larvae normally feed on beets and hops. The pest has been observed in hemp fields in Yugoslavia (Camprag 1961) and Korea (Anonymous 1919). Shiraki (1952) cited a related species, ***Maladera orientalis***, on hemp in Japan.

4. AFRICAN BLACK BEETLE

Heteronychus arator (Fabricius) 1794, Coleoptera; Scarabaeidae.

Description: Adults are black, rounded scarabs between 15–20 mm long. Larvae and soft-bodied and fleshy, curved in a c-shape, with well-developed thoracic legs, 35 mm long at maturity.

Life History & Host Range

Adults overwinter, emerging in the spring to chew on young seedlings near the soil line. Females lay eggs in moist soil at the base of plants. Larvae feed on roots, but their damage is rarely significant. Usually one generation arises per year. *H. arator* occurs throughout tropical Africa and southern Africa (Hill 1983). It is also found in Australia, where it is called the African Black Beetle. *H. arator* frequently infests sugarcane, maize, and wheat, and occasionally infests hemp in Western Australia (Ditchfield, pers. commun. 1997).

DIFFERENTIAL DIAGNOSIS

Above-ground symptoms caused by white root grubs (i.e., wilting) can be confused with symptoms caused by root maggots, nematodes, root-rot fungi, or drought. Dig up roots for inspection. Antennae of adult scarab beetles end in flattened, palmate segments which unfurl like a hand-fan; this feature distinguishes them from all other beetles. The hard part is differentiating root grubs from each other—they look quite similar (compare Fig 4.33 with Fig 4.34). Grubs can be differentiated by their rasters (the ends of their abdomens), see Olkowski *et al.* (1991).

CULTURAL & MECHANICAL CONTROL

(method numbers refer to Chapter 9)

Use methods 2a (deep ploughing), 3 (especially grassy weeds), and 6 (do not rotate after grass or cereal crops). Expect greater damage during damp summers. Dry soil destroys many eggs and newly-hatched grubs. Ploughing fields midsummer kills many pupae. In horticultural situations, walk around plants wearing Lawn Aerator Sandals®. The 6 cm-long spikes puncture and kill larvae.

BIOCONTROL (see Chapter 10)

Nematodes kill white root grubs, especially "cruisers" like *Heterorhabditis* species and *Steinernema glaseri* (described below). Microbial biocontrol of white root grubs began 50 years ago with *Bacillus popilliae,* described below. Gutberlet & Karus (1995) promoted a European variety, *Bacillus popilliae* var. *melolonthae,* for use against chafers. *Bacillus thuringiensis japonica* is a new Bt variety released against Japanese beetles (for more on Bt, see European corn borers). Some people have success with the microinsecticides *Beauveria brongniartii, Nosema melolonthae* (described below), and *Metarhizium anisopliae* (described under aphids).

The USDA has imported several *Tiphia* species from Japan, described below. Consider pasturing *big* biocontrols (i.e., hogs) in infested fields during summer and early autumn. Hogs happily dig up grubs and eat them. Birds also eat grubs. Flocks follow tractors, eating grubs turned up by the plough.

Heterorhabditis bacteriophora (=*Heterorhabditis heliothidis*)

BIOLOGY: These tiny nematodes, 1-1.5 mm long, parasitize many insects in soil, such as caterpillars, beetle grubs, root maggots, and thrips. Of the eight known *Heterorhabditis* species, *H. bacteriophora* is probably the favourite (Cruiser®, Heteromask®, Nema-BIT®, Nema-green®, Lawn Patrol®, Otinem®). The species was discovered in Australia but also lives in North America, South America, and Europe. *Heterorhabditis* nematodes contain gram-negative anaerobic bacteria (*Photorhabdus luminescens*). The bacteria ooze protein-destroying enzymes, and do the actual killing. Bacteria also produce antibiotics which prevent putrefaction of dead insects, allowing nematodes to reproduce in the cadavers. *H. bacteriophora* does best in soil temperatures between 15–32°C.

DEVELOPMENT: *H. bacteriophora* uses a classic "**cruiser**" strategy to find insects—third-stage larvae actively track down hosts, following trails of CO_2 and excretory products. Larvae penetrate hosts through the cuticle (body wall), as well as through natural openings (the mouth, spiracles, or anus). Cuticular penetration is important regarding white root grubs, because scarabs have sieve plates protecting their spiracles, and they defaecate so often that most nematodes entering the anus are pushed back out. Once nematodes penetrate their hosts, they release toxins which inhibit the insect immune system, then release bacteria. The bacteria kill insects and liquefy host organs within 48 hours. Nematodes then use the cadaver to reproduce and multiply (Plate 31). Insects killed by *H. bacteriophora* turn brick red, and fresh cadavers may glow dimly in the dark. A new generation of nematodes emerges from the carcass in two weeks.

APPLICATION: Supplied as third-stage larvae in polyethylene sponge packs or bottles. Sealed containers of *H. bacteriophora* remain viable for up to three months at 2–6°C, but do not store as well as *Steinernema* species. Gently stir contents of package into 5 l water (15–20°C), let stand for five minutes, then transfer to watering-can or spray tank. Some gel preparations must be mixed with warm water (<32°C) and allowed to dissolve for 30 minutes before they are released. Release within three hours, otherwise the nematodes drown. Apply in early evening, to avoid direct sunlight. To apply with a sprayer, remove all filters, use a spray nozzle at least 0.5 mm wide, spray with a pressure under 5 Bars (<73 pounds/in^2), and agitate the spray tank frequently so nematodes remain suspended. The usual recommended dose is 50,000 nematodes per plant or 500,000–1,000,000 per m^2. Water the soil before *and* after application (at least 0.64 cm of irrigation), or apply during a rainstorm, and keep the soil moist for at least two weeks after application.

NOTES: Nematodes kills insects quicker than most biological control agents. *Heterorhabditis* species live deeper in the soil than *Steinernema* species (down to 15 cm), and have superior host-penetrating abilities. *H. bacteriophora* is compatible with Bt, *Bacillus popilliae,* and predatory mites (including *Hypoaspis miles*). But biocontrol insects that live in the soil are incompatible (e.g., pupating *Aphidoletes aphidimyza*). According to some experts, wasp parasitoids that pupate within silken coccoons are not killed.

H. bacteriophora is more sensitive to physical stress than other beneficial nematodes, so avoid most insecticides while utilizing this species. Cherim (1998) noted that many nematodes packaged in gels and granules are raised on artificial nema chow. These products contain a high percentage of dead nematodes, compared to nematodes raised on insect hosts.

Heterorhabditis megidis

This species was discovered in Japanese beetle grubs from Ohio. Nevertheless it is not currently registered in the USA. Third-stage larvae are sold Europe, in a dispersible clay formulation (Larvanem®, Nemasys-H®), and used against vine weevil larvae. *H. megidis* is a large species (which makes it expensive to raise), and it has a short shelf life, like *H. bacteriophora.*

Steinernema glaseri

BIOLOGY: This *Steinernema* nematode is a cruiser (Vector®), unlike its better-known cousin, *Steinernema carpocap-*

sae (described under cutworms). It is highly mobile in sandy soil (less so in clay soil), and highly responsive to host chemoattractants. *S. glaseri* works best against sedentary root-feeders, such as white root grubs. Glaser and Fox (1930) discovered *S. glaseri* killing Japanese beetle grubs in New Jersey. Since then, *S. glaseri* has been found across the southern USA and also in Brazil. Besides scarab grubs, *S. glaseri* parasitizes soil-dwelling leaf beetles, weevils, and even some caterpillars and grasshoppers.

NOTES: *S. glaseri* is a big nematode (twice the length and eight times the volume of *S. carpocapsae*, see Plate 31). This makes *S. glaseri* expensive to produce.

Bacillus popilliae

BIOLOGY & DEVELOPMENT: A soil bacterium that infests scarab grubs and possibly other beetles. It is native to New Jersey, and does best in moderate temperatures (>16°C). Grubs eat spores, which germinate in the larvae's guts; the bacteria proliferate and eventually penetrate the haemolymph. Grubs gradually turn milky-white and die of "milky spore disease." As grub cadavers decompose, they release millions of spores into the soil.

APPLICATION: *B. popilliae* was the first bacterial insecticide available in the USA, registered in 1948, and marketed as Doom® and Japidemic®. Supplied as spores in a talc powder (100 million spores per g). It can be stored for months in a dry, cool 8–10°C) place. Broadcast soil with spores, about 5 g m^{-2}, and follow with 20 minutes of watering. Unlike Bt products, *B. popilliae* products are viable, and released bacteria can replicate and become established in stable soils, providing long-term control. Temperature is a critical factor—*B. popilliae* works only in soils warmer than 16°C. This limits its effectiveness above 40° latitude in the USA. Unfortunately, *B. popilliae* is slow-acting and requires a month or more to kill its host.

NOTE: Much of the USA supply (produced by Ringer Corporation as Grub Attack®) was tainted with other bacteria in 1992 and temporarily withdrawn from the market. Currently no product is registered in the USA. A related species, ***Bacillus lentimorbus***, is considered a strain of *B. popilliae* by some experts.

Beauveria brongniartii (=*B. tenella*)

BIOLOGY: This fungus is marketed for control of sugarcane root grubs (Betel®) and *Melolontha* species and other scarabs (Engerlingspilz®). The Japanese use this species to control longhorn beetles and other species (Thomson 1998). *B. brongniartii* is native to central Europe, and does best in moderate temperatures. Its morphology and life cycle resembles that of *Beauveria bassiana* (described under whiteflies).

APPLICATION: Formulated as liquid blastospores or in clay granules. Blastospores are sprayed on trees, so night-feeding adults become infected and then return to the soil to die. *B. brongniartii* then sporulates on cadavers underground, spreading spores throughout the field (Roberts & Hajek 1992). In years when *Melolontha* adults are not flying, the clay granules are drilled into the ground or applied as a soil drench (Hajek 1993). *Beauveria* species persist in soil, but organic material *hinders* survival and nitrogen fertilizer kills them (Leslie 1994).

Nosema melolonthae

BIOLOGY: A microscopic protozoan lethal to *Melolontha* grubs. This biocontrol is not yet commercially available. See the section on grasshoppers for discussion of a related species, *Nosema locustae.*

Tiphia **species**

BIOLOGY: White root grubs are parasitized by small wasps from the superfamily Scoliidae. The USDA has imported several wasps to control Japanese beetles. ***Tiphia popilliavora*** was first, imported from Japan in 1921. It became established in the Mid-Atlantic states. ***Tiphia vernalis*** was imported from China and Korea in 1925. It spread from New Jersey south to North Carolina and west to Indiana.

DEVELOPMENT: Adults are shiny black wasps, 5 mm long. Female *Tiphia* wasps burrow into soil to locate grubs, paralyze them with a sting, then insert a single egg in a specific location on the grubs' bodies (Fig 4.34). *T. vernalis* adults emerge in the spring (May and early June) and sting third-instar grubs in the soil. *T. popilliavora* adults emerge in autumn (August and early September), and sting third-instar larvae. Females of both *Tiphia* species lay up to 25–50 eggs in 25–30 days. Wasp larvae eventually consume their entire host (except the head and legs), and pupate in the soil cell built by their victim. One generation arises per year.

NOTES: The wasps are poor fliers. *Tiphia* adults must have a food source nearby. Adult *T. vernalis* feeds on aphid honeydew, *T. popilliavora* feeds on the nectar of wild carrot (*Daucus carota*) and other umbelliferous flowers. *Tiphia* species are compatible with *Bacillus* species.

CHEMICAL CONTROL

Drenching soil around roots with rotenone kills white root grubs.

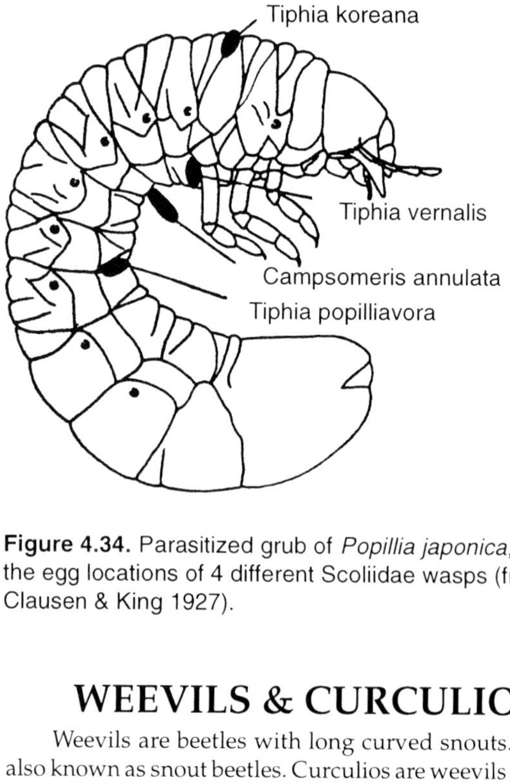

Figure 4.34. Parasitized grub of *Popillia japonica*, showing the egg locations of 4 different Scoliidae wasps (from Clausen & King 1927).

WEEVILS & CURCULIOS

Weevils are beetles with long curved snouts. They are also known as snout beetles. Curculios are weevils with even longer snouts. These snouts sprout antennae and chewing mouthparts. About ten members of the family Curculionidae reportedly attack *Cannabis;* the cabbage curculio causes the most damage.

SIGNS & SYMPTOMS

Adults chew small holes in leaves or notch leaf margins. The holes and notches often become surrounded by chlorotic halos. Most larvae (grubs) feed on pith within stems, causing a slight swelling at feeding sites. Some grubs feed on roots.

1. CABBAGE CURCULIO

Ceutorhynchus rapae Gyllenhal 1837, Coleoptera; Curculionidae.

Description: Young larvae have cylindrical white bodies and brown heads. Older larvae slightly darken in colour, become somewhat plump, and reach 4–6 mm in length. They assume a C-shape when exposed. Adults are oval to oblong, grey to black, and covered with yellow to grey hairlike scales. The curved snout is slightly longer than the head and thorax; the snout is slender, cylindrical, finely punctuated with striae on the basal half, and antennae arise near its middle. Wing covers are marked by fine longitudinal ridges and rows of shallow punctuations. Length 3.5–5.0 mm (Fig 4.35). When disturbed, adult cabbage curculios, like most weevils, draw in their legs and antennae and drop to ground and play dead.

Life History & Host Range

Adults overwinter and emerge in April-May to mate. Females insert eggs into hemp stems when seedlings are 3–8 cm tall. As a result, only the lower 50 cm of mature plants contain grubs (Nagy *et al.* 1982). Grubs leave exit holes in June to pupate in small coccoons just beneath the soil surface. Adults emerge in late June–July and feed on leaves. One generation arises per year.

C. rapae infests hemp in the Czech Republic (Rataj 1957), Hungary (Nagy *et al.* 1982), Yugoslavia (Camprag *et al.* 1996), and Italy (Goidànich 1928, Martelli 1940, Ferri 1959, 1961c). Nagy *et al.* (1982) reported up to 40% crop losses; they called *C. rapae* the "hemp curculio." The pest ranges across temperate Europe and was accidentally introduced into North America about 1855. In America, *C. rapae* primarily infests cruciferous crops. It has attacked drug cultivars in Vermont (Bush Doctor, pers. commun. 1994).

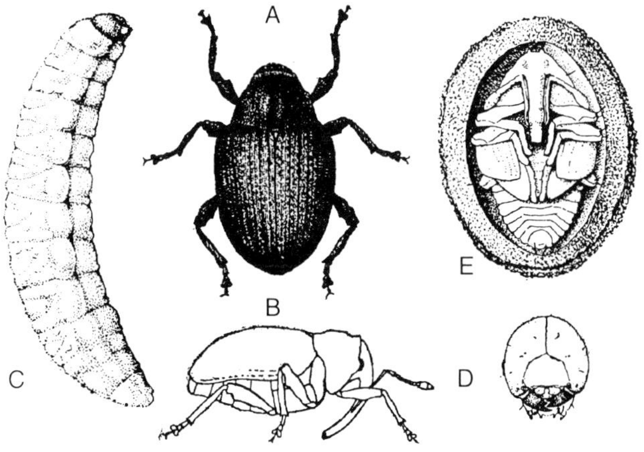

Figure 4.35: Cabbage curculio, *Ceutorhynchus rapae*. A. Adult from above; B. Adult from side; C. Grub; D. Head of grub from front; E. Pupa (from Blatchley 1916).

2. HEMP WEEVIL

Rhinoncus pericarpius Linnaeus 1758, Coleoptera; Curculionidae.

Description: Larvae are white with dark brown-black heads; they soon grow plump, reaching 4–6 mm in length (Fig 4.35A). Adults are broadly oval, dark reddish-brown to black, thinly covered with greyish-yellow hairs, length 3.5–4 mm. The scutellum sports a conspicuous triangular spot, the beak is slightly longer than the head and slightly ridged (Fig 4.35A). Eggs are oval, white, less than 1 mm long.

Life History & Host Range

Harada (1930) called *R. pericarpius* the most injurious pest of hemp in Japan. Wang *et al.* (1995) labelled it a major pest in Anhui, China. Clausen (1931) considered it less consequential. Shiraki (1952) and Wang *et al.* (1995) called *R. pericarpius* "the hemp weevil," although *R. pericarpius* also attacks bouncing bet (*Saponaria officinalis* L.), *Euphorbia*, and

Figure 4.35A: Adult and grub of *Rhinocus pericarpius* (from Harada 1930).

Polygonum species. *R. pericarpius* lives throughout temperate North America, Europe and Asia. Grubs feed within stems and cause small galls. The galls are weak and may snap. Grubs pupate in stems. Adult beetles feed on leaves and overwinter in the soil. One generation arises annually.

3. OTHER WEEVILS AND CURCULIOS

Tremblay & Bianco (1978) found the cauliflower weevil, ***Ceutorhynchus pallidactylus*** (Marsham) 1802, feeding on hemp in Italy. The life cycle of *C. pallidactylus* mimics that of *C. rapae* except *C. pallidactylus* eggs are laid on leaves. *C. pallidactylus* beetles are slightly smaller than *C. rapae* but otherwise identical; the grubs cannot be differentiated. Tremblay (1968) also reported ***Ceutorhynchus pleurostigma*** (Marsham) 1802, ***Ceutorhynchus quadridens*** (Panzer) 1795, and ***Ceutorhynchus roberti*** Gyllenhal 1837 on Italian hemp. He illustrated the morphological differences between *C. rapae, C. roberti, C. quadridens* and *C. pleurostigma*.

Mohyuddin & Scheibelreiter (1973) cited ***Ceutorhynchus macula-alba*** Herbst 1795 on Romanian hemp. They tested the species as a biocontrol agent against marijuana (the pest usually infests opium pods). Goidànich (1928) cited seven other Curculionidae on Italian hemp—***Ceutorhynchus suleicollis*** (Payk.) 1800, ***Gymnetron labile*** (Herbst) 1795, ***Gymnetron pascuorum*** (Gyllenhal) 1813, ***Polydrosus sericeus*** (Schall. 1783), ***Sitona humeralis*** (Steph.) 1831, ***Sitona lineatus*** (L.) 1758, and ***Sitona sulcifrons*** (Thumb.) 1798.

The gold dust weevil, ***Hypomeces squamous*** (Fabricius) 1794, is a pest in Thailand (Hill 1983). Adults notch leaf margins, and grubs live in the soil and feed on roots. These broad-nosed weevils attack many crops in southeast Asia. Adults are grey, 10–15 mm long, and covered with a fine golden-green "dust" of hairs. Their wings fuse together in midline, rendering the adults flightless.

Other reports include the Asiatic oak weevil, ***Cyretepistonmus castaneus*** (Rolofs) on Mississippi marijuana (Lago & Stanford 1989), the Ber weevil, ***Xanthoprochilis faunus*** Olivier 1807 (=*Xanthochelus superciliosus* Gyllenhal 1834) in India (Cherian 1932), ***Corigetus mandarinus*** Fairmaire 1888 in Vietnam (Hanson 1963), and ***Apion*** species in India (Bantra 1976) and Italy (Goidànich 1928). *Apion* species are often called weevils but belong to the family Apiónidae, not the Curculionidae.

DIFFERENTIAL DIAGNOSIS

Adult weevils and curculios may be confused with other beetles until you spot their unmistakable snouts. Grubs are harder to separate from other stem- and root-inhabiting beetles, especially the grubs of flea beetles. Stem gall symptoms must be differentiated from those caused by caterpillars and gall midges.

CULTURAL & MECHANICAL CONTROL
(method numbers refer to Chapter 9)

Observe methods 1 (sanitation), 2a (deep ploughing), 3 (eliminate cruciferous weeds), 6 (avoid vegetable crops), and 9 (hand removal of adults). Consider using crucifers as trap crops for *C. rapae.* Destroy trap crops after eggs are deposited on them. Collect adults by laying a tarp under infested plants and shaking the plants. Weevils and curculios fall and remain motionless, long enough to be squashed or gathered for soup. Or let chickens and geese at them.

BIO & CHEMICAL CONTROL

Biocontrol researchers are testing ***Scambus pterophori,*** a parasitoid of stem-boring beetle grubs (especially weevils). For grubs or adults in the ground, utilize bacteria and parasitic nematodes described in the sections on flea beetles and white root grubs. Rotenone and cryolite kill weevils and curculios. Try injecting stem galls with a large-bore syringe to kill pith-feeding larvae.

ASSORTED BORING BEETLES

In this section we grouped together a few tumbling flower beetles, longhorn beetles, and a marsh beetle. Grubs of all these species bore into *Cannabis* stems and roots.

1. TUMBLING FLOWER BEETLES

Only adult *Mordellistena* beetles tumble in flowers to elude predators. The grubs hide in stems. Flachs (1936) called the pests "barbed beetles," Lindeman (1882) and Shiraki (1952) called them "hemp beetles," and Anonymous (1974) called them "hemp flour beetles." Some *Mordellistena* species have attracted attention as potential biocontrol of illicit *Cannabis* crops (Baloch & Ghani 1972, Mushtaque *et al.* 1973).

Signs & Symptoms

Grubs of tumbling flower beetles feed on stem pith. They bore into the central stem and replace pith with fine borings and frass (Plate 32). Feeding sites often swell. The sites are structurally weak and may snap, causing a wilt of distal plant parts. Damage peaks in late July and early August. One species, *M. micans,* feeds in the lower ends of stems (Kirchner 1906) or roots (Anonymous 1974) where its tunnel can be up to 6 cm long (Lindeman 1882). Another species, *M. parvula,* feeds in upper parts of plants, within narrow branches, petioles, and even central leaf veins (Sorauer 1958).

a. *Mordellistena micans* Germar 1817, Coleoptera; Mordellidae.
=*Mordellistena cannabisi* Matsumura 1919

Description: Adults are wedge-shaped, humpbacked, black on the upper surface and covered with a dense pubescence of grey-brown hairs, brown on the underside, 2.5–3.5 mm long (Fig 4.36). They have long strong hind legs like flea beetles. The abdomen tapers to a barb which projects beyond the wings. Larvae are legless, yellow, covered with bristles, reach 3–4 mm in length, and sport a dark red barb on their posterior end (Fig 4.36). Pupae have a red-brown dorsum and a yellow abdomen.

Life History & Host Range

Grubs of *M. micans* overwinter in stems and crop stubble. They pupate there in spring. Adults emerge by May and lay eggs within a month (Krustev 1957). *M. micans* has been cited on hemp in Europe (Lindeman 1882, Kirchner 1906, Flachs 1936, Krustev 1957, Sorauer 1958, Camprag *et al.* 1996), Pakistan (Baloch & Ghani 1972, Mushtaque *et al.* 1973), and North Africa and Syria (Sorauer 1958). Sorauer said the Japanese species *M. cannabisi* is identical to *M. micans.* Takahashi (1919) called *M. cannabisi* a serious pest in Japan, but Clausen (1931) considered it less consequential. Besides hemp, the adults tumble in umbelliferous tops, such as parsley and carrot (Anonymous 1974).

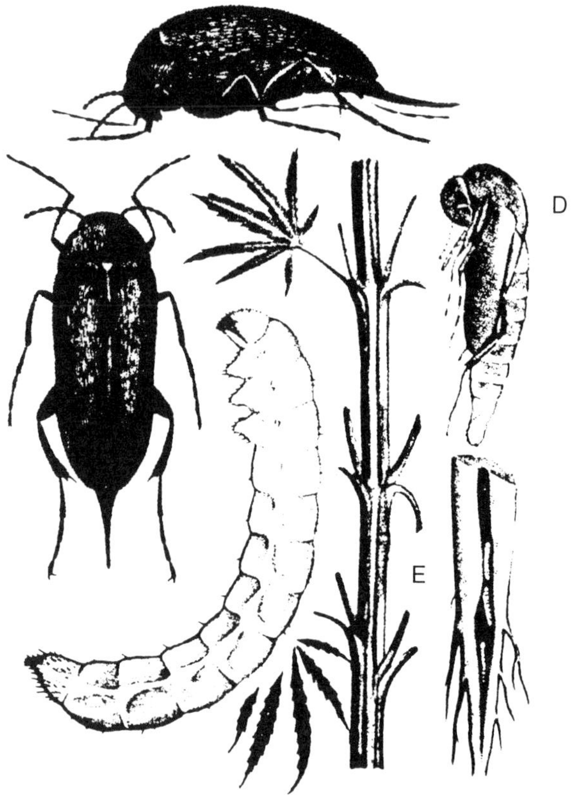

Figure 4.36: *Mordellistena micans* (=*Mordellistena cannabisi*). A. Adult, side view; B. Adult, dorsal view; C. Larva; D. Pupa; E. Damage in stems and roots (from Anonymous 1974).

b. *Mordellistena parvula* Gyllenhal 1827, Coleoptera; Mordellidae.

Description: Adults resemble those of *M. micans,* except they are shorter (2 mm long) and more slender. The larvae also look alike, except *M. parvula* grubs lack barbs.

Life History & Host Range

In the north Caucasus and Kirgizia, *M. parvula* has infested 95% of hemp sown in April or May, and 66–78% of crops sown later (Durnovo 1933). Krustev (1957) counted ten grubs per plant on severely damaged crops. In Pakistan *M. parvula* attacked marijuana and sunflowers (Baloch & Ghani 1972).

In addition to *M. micans* and *M. parvula,* Baloch & Ghani (1972) identified ***Mordellistena gurdneri*** in Pakistan, which normally infests alder (*Alnus* species) and absinthe (*Artemisia* species). Goidànich (1928) reported ***Mordellistena reichei*** Em. 1876 in Italian hemp, and Kyokai (1965) cited ***Mordellistena comes*** Marseul in Japanese hemp.

2. LONGHORN BEETLES

Grubs of most longhorn beetles (family Cerambycidae) feed within tree trunks. They are also called roundheaded wood borers. The antennae of adult males may grow several times their body length, hence their longhorn name. Some longhorn beetles live for several years as larvae.

Clarke (unpublished research, 1996) found ***Anoplohora glabripennis*** (Motschulsky) on hemp stalks in China. Adult beetles have black bodies speckled with white spots. The Chinese call them "starry night beetles." The grubs normally bore into hardwood trees. Vermont entomologists call this species the Asian longhorned beetle, and fear its accidental introduction into the USA will destroy the sugar maple industry.

Three other longhorns have been cited on *Cannabis*. An unspeciated ***Agapanthia*** was collected in Turkey (Mohyuddin & Scheibelreiter 1973); Rataj (1957) cited ***Agapanthia cynarae*** Germar 1817 on hemp in the Czech Republic. However, the most common longhorn pest is ***Thyestes gebleri*** (see below).

3. HEMP LONGHORN BEETLE

Thyestes gebleri Faldermann 1835, Coleoptera; Cerambycidae.

Description: Beetles are black with white stripes down the prothorax, elongate, cylindrical, up to 15 mm long, with striped antennae nearly as long as their bodies. Grubs are robust, cylindrical, creamy white, with prominent heads capped by a black dot, up to 20 mm long. Pupae are intermediate in size between grubs and adults, light brown.

Signs & Symptoms

Grubs feed within larger stems and stalks, ejecting excrement at intervals through frass holes. Their tunnels are much longer than those of tumbling flower beetles. Infested stems may snap at tunnel sites.

Life History & Host Range

T. gebleri grubs overwinter in plant debris. They pupate in roots at the ends of tunnels, which they plug with fibrous hemp fragments. Adults emerge in May or June. Females deposit one to two eggs at a time into hemp stems, "usually 5 inches below the first joint," according to Takahashi (1919). One generation arises per year. Shiraki (1952) called *T. gebleri* the hemp longhorn beetle. Anonymous (1974) called it the hemp beetle. Takahashi (1919) and Clausen (1931) considered *T. gebleri* the most destructive pest of hemp in Japan. *T. gebleri* lives in eastern Siberia, northern China, Korea, and Japan.

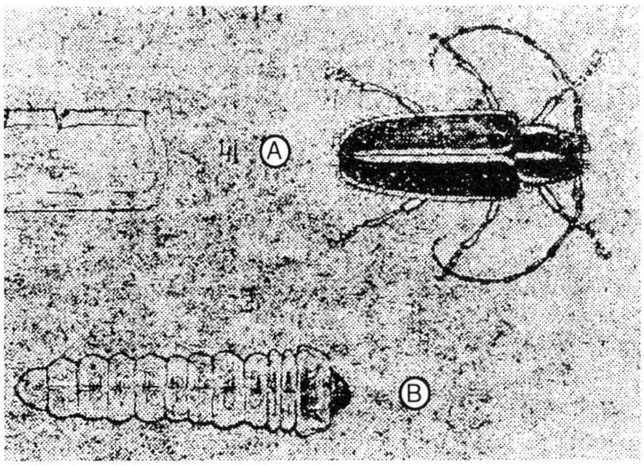

Figure 4.37: The hemp longhorn beetle, *Thyestes gebleri*. A. Adult; B. Grub (from Takahashi 1919).

4. MARSH BEETLE

Scirtes japonicus Kirsenwetter, Coleoptera; Helodidae.

Shiraki (1952) cited this marsh beetle infesting Japanese hemp. Most helodid beetles live in marshes and other swampy places. Adults are black-bodied, oval, and 2–4 mm long. Helodid grubs are aquatic. They are characterized by long slender antennae. Not much is known about *S. japonicus*.

DIFFERENTIAL DIAGNOSIS

Grubs of *M. micans* and *T. gebleri* can be confused. For further differentials, see the previous section concerning weevils and curculios.

CULTURAL & MECHANICAL CONTROL

(method numbers refer to Chapter 9)

Stem-boring beetles can be reduced by methods 1 (Clausen recommended burning crop debris to kill *T. gebleri*), 2a (deep ploughing), and 4 (plant late to avoid *M. parvula*). Avoid swampy ground and you avoid marsh beetles.

BIOCONTROL (see Chapter 10)

In the future, mordellids may be controlled by parasitoids. Baloch & Ghani (1972) found several *Tetrastichus* species, *Rhaconotus* species, and a *Buresium* species parasitizing grubs of tumbling flower beetles. The Japanese use *Beauveria brongniartii* to control longhorn beetles (described in the section on white root grubs). Try injecting stems with *Bacillus popilliae* or beneficial nematodes, described under white root grubs.

CHEMICAL CONTROL (see Chapter 11)

Treating plants with rotenone kills boring beetles. Stems must be individually injected to kill the grubs. After treatment, remove all frass from around plants. Any subsequent sign of excrement indicates a need to repeat the treatment.

PLANT BUGS

People call a variety of animals "bugs," including beetles (e.g., June bugs) and non-insect arthropods (e.g., sowbugs). True bugs are only found in the insect order Hemiptera. They are distinguished by their wings, which are half (hemi-) membranous, and half thickened and leathery. Many bugs are polyphagous and move from host to host. Pest populations build up in weeds and wild plants before the pests disperse to cultivated hosts, and vice versa (Panizzi 1997). Pest populations often follow a regular sequence of host plants as the seasons change (Fig 4.38).

About a dozen Hemipterans are regularly found on *Cannabis*, some are serious pests. One species acts symbiotically: Vavilov (1926) described a red bug (***Pyrrhocoris apterus*** Linnaeus 1758) associating with *Cannabis ruderalis* in the former USSR. *P. apterus* is attracted to a fat pad at the base of the seed, not the seed itself. In the process of fat pad feeding, the bug carries seeds "far distances" and facilitates the spread of *Cannabis ruderalis*.

SIGNS & SYMPTOMS

True bugs, like their Homopteran cousins (aphids, leafhoppers, whiteflies), have piercing-sucking mouthparts and feed on plant sap. They feed on succulent parts of plants—unripe seeds, flowering tops, leaves, and even young branches and stems. Bugs feeding on flowers cause bud abortion, which delays maturity and reduces yields. Some bugs inject a toxic saliva as they feed, turning plant tissues brown and lumpy.

1. SOUTHERN GREEN STINK BUG

Nezara viridula (Linnaeus) 1758, Hemiptera; Pentatomidae.

Description: Adults are easily recognized by their flattened shield-like shapes, 15 mm long and 8 mm wide (Fig 4.39). Green stink bugs are not always green; they may turn a russet colour before diapause. McPartland (unpublished data 1982) found *white* colourmorphs on feral hemp in Illinois (Plate 33). Adults make a stink when handled. Females lay barrel-shaped, pale yellow eggs in dense batches of 50–60 on undersides of leaves. The egg masses resemble tiny honeycombs. Nymphs are oval, bluish-green with red markings.

Life History & Host Range

Adults overwinter in secluded places in deciduous

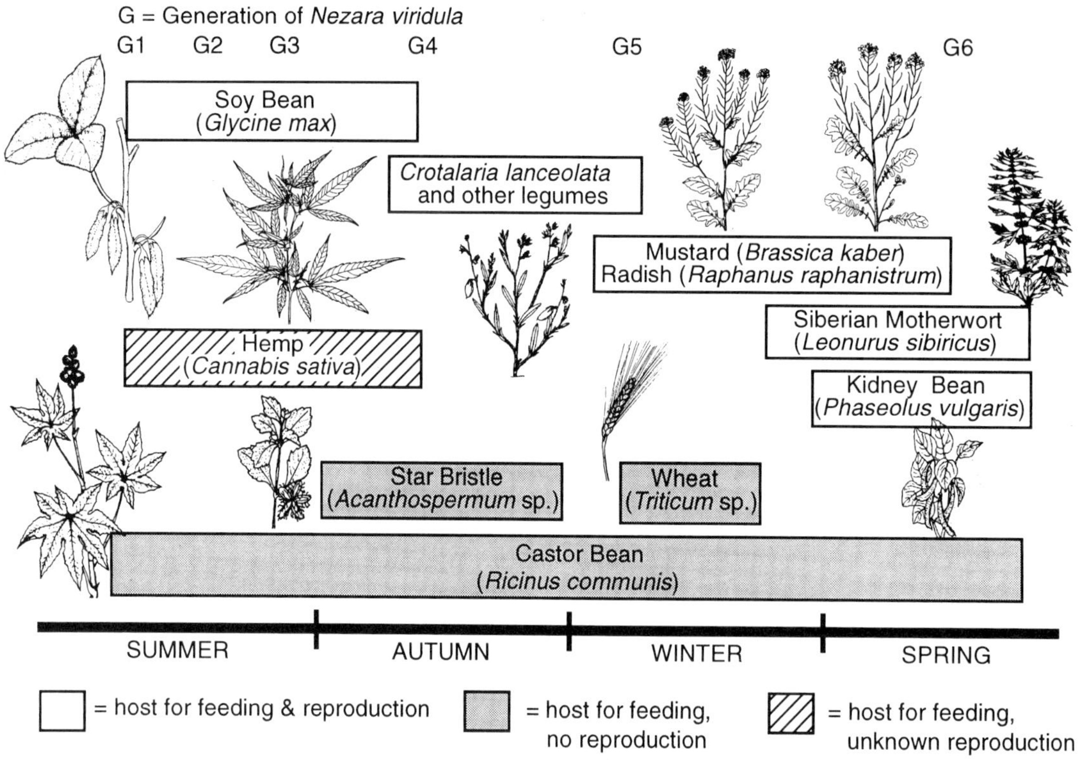

Figure 4.38: Sequence of wild and cultivated plant hosts used by successive generations of the southern green stink bug, *Nezara viridula*, in a semitropical climate where 6 generations arise per year (McPartland adapted from Panizzi 1997).

woods. They overwinter in above-ground debris, so their range is limited by winter temperatures. Females lay up to 250 eggs. Total development time averages 22–32 days in warm weather (depending on host plant) and adults live for 20–60 days (Panizzi 1977). Three generations may arise per year in temperate zones, but in tropical regions they breed continuously (Fig 4.38). *N. viridula* probably originated in Ethiopia but now lives worldwide. It is very polyphagous, with a preference for brassicas and legumes (especially soybeans). *N. viridula* feeds on hemp seeds (Ferri 1959a, Hartowicz *et al.* 1971), hemp leaves (Rataj 1957, Sorauer 1958, Dippenaar *et al.* 1996), and flowering tops of marijuana (Rao 1928, Cherian 1932, Nair & Ponnappa 1974). The bug also vectors diseases. Paulsen (1971) cultured the brown blight fungus, *Alternaria alternata,* from *N. viridula* feeding on feral hemp in Kansas. *N. viridula* can be confused with a slightly smaller green stink bug, ***Acrosternum hilare*** (Say).

2. TARNISHED PLANT BUG

Lygus lineolaris (Palisot de Beauvois) 1818, Hemiptera; Miridae.

Description: Adults are flattened and oval in outline, 4–7 mm in length, mostly greenish-brown but irregularly mottled by reddish-brown colouring, with a distinct yellowish triangle or "V" located on their backs. Their triangular wingtips are characteristically yellow, tipped by a black dot (Fig 4.39 & Plate 34). Young nymphs look like yellow-green aphids, but lack abdominal cornicles are more active than aphids. Final-instar nymphs have green bodies marked by four black dots on the thorax, one dot on the abdomen, long antennae, and red-brown legs with indistinct stripes. Eggs are elongate and curved.

Life History & Host Range

Adults overwinter in soil, weeds, and crop stubble. Females emerge in spring and insert eggs into stems of crop plants and weeds. Nymphs moult five times, gradually taking on an adult appearance. The life cycle takes only three or four weeks under optimal conditions, permitting three to five generations each year. But in northern extremes only one generation of *L. lineolaris* arises per year.

L. lineolaris is abundant on feral hemp in Illinois (McPartland, unpublished data 1981), where it feeds on new soft tissue near the apical meristem. Feeding sites become brown lesions and young plants grow malformed. *L. lineolaris* also attacks cultivated hemp in Manitoba (Moes, pers. commun. 1995), and Mississippi marijuana (Lago & Stanford 1989). Protective colouring and retiring habits make *L. lineolaris* an underestimated threat. The pest infests hundreds of wild and cultivated plants, primarily alfalfa, canola, cotton, vegetable crops, and peach trees. *L. lineolaris* damage worsens in hot, dry weather. The bugs are very mobile and move quickly between hosts. Watch for sudden infestations after neighbouring alfalfa is cut for hay.

L. lineolaris is native to North America *east* of the Rockies. Its nearly-identical homologue, ***Lygus hesperus*** Knight, is found *west* of the Rockies. *L. lineolaris* has also been reported on *Cannabis* in Europe (Goidànich 1928, Rataj 1957, Sorauer 1958); its native European counterpart is ***Lygus rugulipennis*** Poppius.

Other *Lygus* species cited on *Cannabis* include ***Lygus apicalis*** Horv. and ***Lygus pubescens*** Reut. in Italy (Martelli 1940), and ***Lygus arboreus*** Taylor and ***Lygus nairobiensis*** Poppius in Uganda (Sorauer 1958). Clarke (pers. commun. 1990) photographed unidentified *Lygus* bugs in Hungarian and Chinese hemp (Plate 34).

3. FALSE CHINCH BUG

Nysius ericae Schilling 1895, Hemiptera; Rhopalidae.

Description: Adults are small, 3–4 mm long, with light to dark grey bodies and lighter coloured wings (Fig 4.39). Nymphs are tiny reddish-brown versions of adults, darkening as they grow older.

Life History & Host Range

Adults fly south to winter quarters for hibernation. They migrate north in the spring. Females lay several hundred eggs on sheaths of grasses. Nymphs limit their damage to grasses and weedy hosts. Adults feed on a wide variety of crops. *N. ericae* infests hemp in Europe (Gilyarov 1945, Sorauer 1958). Two broods arise per year. Species of the genus *Nysius* are not easily differentiated. *N. ericae* is limited to Europe and citations of it elsewhere are probably misidentifications.

4. POTATO BUG

Calocoris norvegicus (Gmelin) 1788, Hemiptera; Miridae.
=*Calocoris bipunctatus* F.

Description: Adults are elongate, dull greenish-yellow in colour, thinly covered by fine yellow and black hairs, with two small black dots marking their pronotums, and reach 6-7 mm in length (Fig 4.39).

Life History & Host Range

C. norvegicus overwinters as an egg in stem tissues of herbaceous plants. Sorauer (1958) called *C. norvegicus* a "potato bug," but it normally feeds on legumes, mustards, and grasses. It infests leaves and flowers of hemp in Italy (Goidànich 1928, Ragazzi 1954, Ferri 1959a) and Germany (Flachs 1936, Gutberlet & Karus 1995). The species originated near the Mediterranean, but now lives in North America.

5. OTHER PLANT BUGS

Cherian (1932) considered the Indian stink bug ***Dolycoris indicus*** Stål (family Pentatomidae) as destructive

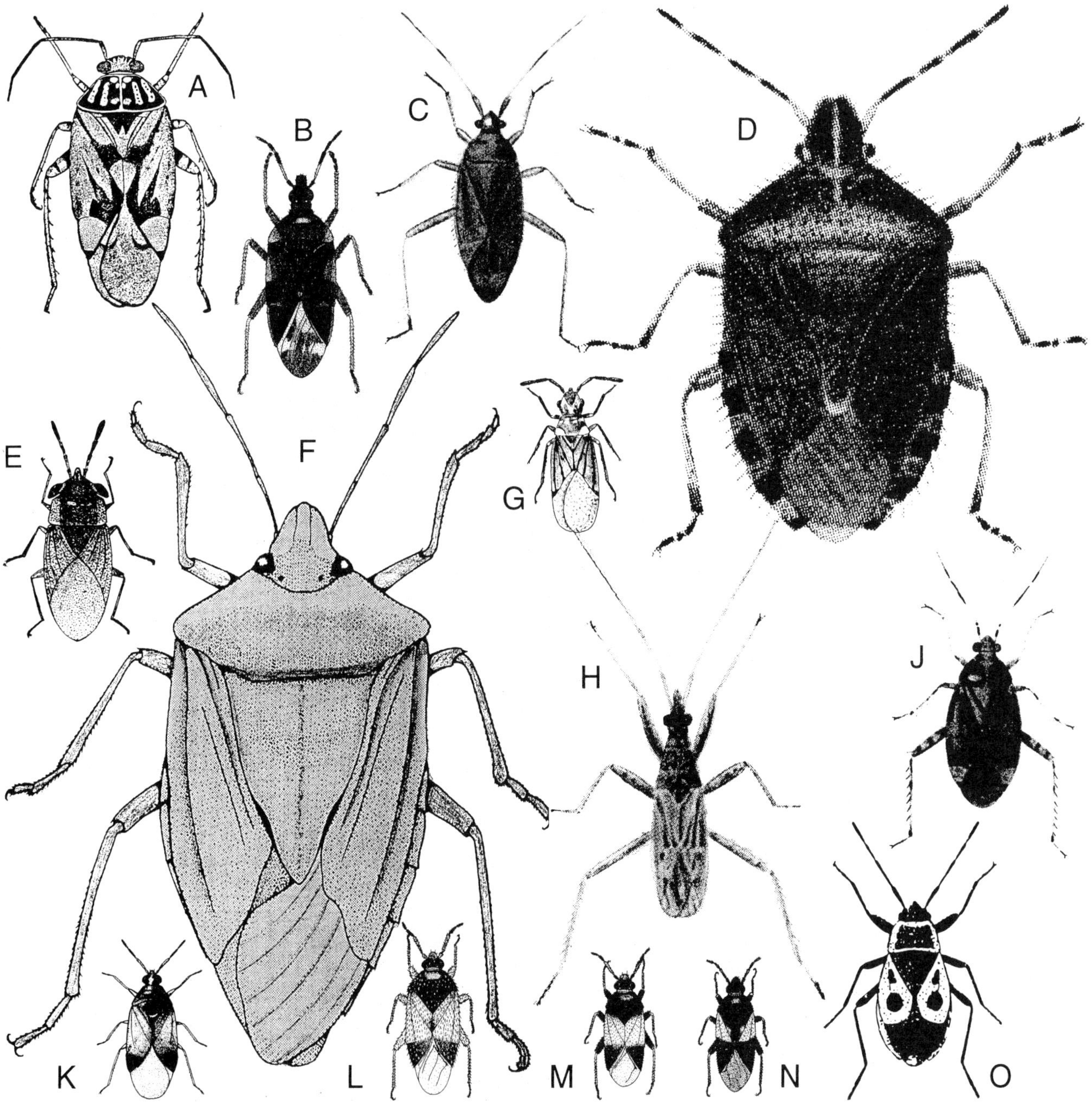

Figure 4.39: Bugs—the good, the bad, and the ugly (all x6): *A. Lygus lineolaris* (from Illinois Natural History Survey); B. *Anthocoris species* (from Kelton 1978); C. *Calocoris norvegicus* (from Southwood 1959); D. *Dolycoris* species (from Southwood 1959); E. *Geocoris* species (from I.N.H.S.); F. *Nezara viridula* (redrawn from Hill 1994); G. *Nysius ericae* (from I.N.H.S.); H. *Nabis rugosus* (from Southwood 1959); J. *Liocoris tripustulatus* (from Southwood 1959); K. *Orius majusculus;* L. *O. minutus;* M. *O. insidiosus;* N. *O. tristicolor;* D. (K–N from Kelton 1978); O. *Pyrrhocoris apterus* (from Sorauer 1958).

as *N. viridula* on Indian ganja (Fig 4.39). It also infests European hemp (Rataj 1957, Sorauer 1958). *D. indicus* usually infests jute, corn, and alfalfa.

Bantra (1976) recovered six bugs from *C. indica* in northern India: two members of the Miridae (***Paracalocoris*** species and ***Rhopalus*** species), a cotton stainer, ***Dysdercus cingulatus*** (Fabricius) (family Pyrrhocoridae), and a pentatomid (***Canthecoides*** species). Two bugs were beneficial insect predators—damsel bugs, ***Nabis*** species (family Nabidae, Fig 4.39) and assassin bugs, ***Sycanus collaris*** Fabricius (family Reduviidae). Clarke photographed an unidentified pentatomid infesting hemp flowers in Hungary, and he collected an unknown shield bug in China (subfamily Scutellerinae, possibly an ***Elasmostethus*** species).

Watson & Clarke (pers. commun. 1994) identified the mirid ***Liocoris tripustulatus*** (Fabricius) 1781 feeding on *Cannabis* pollen in Dutch greenhouses. *L. tripustulatus* adults are light yellow-brown and less than 5 mm long (Fig 4.39). They look like tarnished plant bugs, and frequently infest nettles in Europe. *L. tripustulatus* adults can be confused with *Orius* biocontrols.

DIFFERENTIAL DIAGNOSIS

Bugs (Hemipterans) can be confused with Homopterans such as leafhoppers, planthoppers, and treehoppers. Very young bugs look like aphids but lack cornicles and move faster than aphids. Short, black bugs can be confused with beetles. Bug damage is nearly identical to that of other insects with sucking mouth parts. Bud abortion caused by tarnished plant bugs can be confused with late flowering associated with late-maturing plants (i.e., Thai landraces). Observation with a hand lens will detect damaged flowers.

CULTURAL & MECHANICAL CONTROL

(method numbers refer to Chapter 9)

Control bugs with methods 3 (weeding), 1 (sanitation), 2a (deep ploughing), and 9 (some heavier bugs can be shaken off plants like beetles, best done early in the morning before they "limber up"). Tarnished plant bugs are attracted to white or yellow sticky traps (method 12c). They can be vacuumed off plants, leaving behind beneficial insects. The bugs see the nozzle coming and try to fly off. Many bugs migrate from host to host. So learn what sequence of hosts are infested, and monitor these hosts before bugs spread to crops (see Panizzi 1997).

BIOCONTROL (see Chapter 10)

Five parasitoids infest bugs. *Anaphes iole, Trissolcus basalis, Trichopoda pennipes,* and *Peristenus digoneutis* are described below. ***Ooencyrtus submetallicus***, an egg parasitoid of stink bugs, works well in laboratory studies but zero in the field.

Bugs are eaten by their predatory brethren: the mirid *Deraeocoris brevis* (described under thrips), and the big-eyed bug, *Geocoris punctipes* (described under whiteflies). Bantra (1976) discovered two predatory bugs on Indian *Cannabis*—*Sycanus collaris* and a *Nabis* species.

Biocontrol of bugs with fungi like *Beauveria bassiana* is inconsistent at best. Hill (1983) estimated that 15 chickens will keep 0.1 ha free of cotton stainers. Soyabean (*Glycine max*) serves as a trap crop for *N. viridula*. Strips of alfalfa (*Medicago sativa*) are interplanted in cotton fields at 100-150 m intervals to trap *Lygus* bugs (Hokkanen 1991). Alfalfa is also excellent habitat for many biocontrol insects.

Anaphes iole

BIOLOGY: A "fairyfly" mymarid wasp that parasitizes eggs of tarnished plant bugs (*L. hesperus* and *L. lineolaris*). It is native to Arizona and works in a range of 13–35°C.

APPEARANCE: Adults are very tiny (wingspan 1–2 mm), black or very dark brown, with brown antennae, and light-coloured bands along their legs.

DEVELOPMENT: Female wasps lays eggs in *Lygus* eggs. Development time from egg to adult can be as short as ten days. Adults emerge from *Lygus* eggs and live up to 65 days; females lay 150 eggs under optimum conditions. Nearly two generations of *A. iole* develop for each *Lygus* generation.

APPLICATION: Supplied as adults in vials. Releases of 37,500 wasps per ha (15,000 wasps per acre) per week worked well in California strawberry fields, especially when used preventatively.

Trissolcus basalis

BIOLOGY: A tiny scelionid wasp that parasitizes eggs of *N. viridula*. It is found around the world in temperate regions, and works best at 22°C.

APPEARANCE & DEVELOPMENT: Adults are minute, black wasps (wingspan 1–2 mm), with short antennae that bend downward. Adults mate immediately after emerging from host eggs. Female wasps lay up to 300 eggs; they insert one egg into each *Nezara* egg. Larvae develop and pupate in *Nezara* eggs and emerge as adult wasps. The life cycle takes about 23 days in optimal conditions.

APPLICATION: Strains from southern Europe were introduced into California in 1987, becoming established in cotton, soyabean, and vegetable-growing areas. Unfortunately, when *T. basalis* was released in Hawai'i, it moved to several native nontarget hosts and nearly caused their extinction (Howarth 1991). *T. basalis* tolerates permethrin, but is susceptible to organophosphate insecticides.

Trichopoda pennipes

BIOLOGY: A tachinid parasitoid of *N. viridula* and squash bugs, native to North and South America. A related species from Argentina, ***Trichopoda giacomellii***, has been released against *N. viridula* in Australia.

APPEARANCE & DEVELOPMENT: Adults look like houseflies with orange highlights, and a fringe of feather-like black bristles on their hind legs. Female flies produce up to 100 eggs, and lay one egg per bug (on nymphs or adults). Maggots feed within bugs for two weeks, then pupate in soil. The life cycle takes about five weeks, three generations may arise per season, maggots overwinter in hibernating hosts.

NOTES: *T. pennipes* has been released in Hawai'i and Italy, but is not currently available.

Peristenus digoneutis

BIOLOGY: A braconid parasitoid of *Lygus rugulipennis*, discovered in northern Europe. The USDA established a population in New Jersey that controls *L. lineolaris*. It has spread north to the Canadian border, but has not moved south of New York City (41° latitude).

APPEARANCE & DEVELOPMENT: Adults are small, brown-black wasps, 3 mm long. Female wasps lay a single egg into each *L. lineolaris* nymph. Eggs hatch in seven days and larvae take another ten days to kill their host and drop to the ground to pupate in soil. The second generation of *P. digoneutis* emerges in synchrony with the second generation of *Lygus* bugs. A third generation may arise, pupae overwinter.

NOTES: This biocontrol is not commercially available, but is slowly spreading to the Midwest. A related species, ***Peristenus howardi***, infests *L. hesperus* west of the Rocky Mountains in the USA.

CHEMICAL CONTROL (see Chapter 11)

Dusting plants with rotenone, ryania, or sabadilla controls green stink bugs, cinch bugs, tarnished plant bug (lygus bugs), and other hemipterans (Metcalf *et al.* 1962). Young bugs are killed by nicotine, some species are susceptible to soaps and pyrethroids (tarnished plant bugs infesting cotton have become resistant to pyrethroids). Researchers are trying to find the pheromone for *L. lineolaris,* to develop synthetic pheromone traps.

LEAFMINERS

Leafminers tunnel through tissues *within* leaves, like miniature coal miners. This way they avoid THC and other insecticidal chemicals on the leaf surface. Most leafminers are maggots—members of the fly family Agromyzidae (see exceptions listed under "differential diagnosis" below). Leafminers can become serious pests in Dutch glasshouses if allowed to get out of control.

SIGNS & SYMPTOMS

Leafminer *tunnels* can be seen through the leaf surface. They appear pale green or white, from light reflecting off air in the mines. Each species makes tunnels with the same identifiable "signature"—either linear, serpentine, or blotch-like (Fig 4.40 & Plate 35). Most mines are made on the upper sides

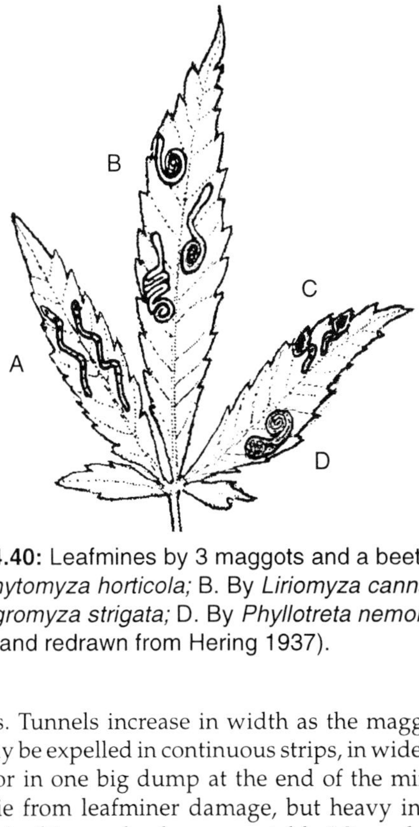

Figure 4.40: Leafmines by 3 maggots and a beetle grub. A. By *Phytomyza horticola;* B. By *Liriomyza cannabis;* C. By *Agromyza strigata;* D. By *Phyllotreta nemorum* (McPartland redrawn from Hering 1937).

of leaves. Tunnels increase in width as the maggots grow. Frass may be expelled in continuous strips, in widely-spaced pellets, or in one big dump at the end of the mine. Plants rarely die from leafminer damage, but heavy infestations cause leaf wilting and reduce crop yields. Mines also provide entrances for fungi and other pathogens. Mines in fan leaves of flowers make the flowers unattractive to consumers, especially when filled with frass.

LIFE HISTORY

Most Agromyzids have identical life cycles. They overwinter outdoors as pupae. Adults emerge in spring to mate. In warm glasshouses, leafminers do not hibernate, they breed continuously. Females drill eggs into leaves, one at a time but often clustered together. Flies feed on sap oozing from ovipositor wounds. Some species lay up to 350 eggs (Hill 1983). Eggs hatch in two to six days. Maggots of most species undergo four moults, then pupate. Some species pupate in mines, projecting two spiracular "horns" through leaf epidermis. Other species drop to the ground to pupate. Pupation takes one week to several months, depending on the species and season. Outdoors, two to six generations arise per year; in glasshouses the generations overlap so all stages can be found at any time.

Six Agromyzid species appear in the *Cannabis* literature. Surprisingly, ***Liriomyza trifolii***, the serpentine leafminer, is not one of them. *L. trifolii* is native to Florida and now lives in glasshouses worldwide, attacking a wide variety of plants. Many less-famous leafminers do not have common names.

1. *Liriomyza (Agromyza) strigata* (Meigen) 1830, Diptera; Agromyzidae.

Description: Maggots are pale, legless, seemingly headless, up to 2 mm in length. They mine slender light green serpentine tunnels that end in brown blotches. Mines begin near the leaf margin and end near the midrib. Larvae pupate externally. Adults resemble tiny houseflies. They are shiny black with yellow markings along the sides and femora, averaging 1.7–2.1 mm in length, with brown legs and white antennae. Females lay eggs near leaf edges.

This pest occurs on German hemp (Kirchner 1906, Blunk 1920, Flachs 1936, Hering 1937, Spaar *et al.* 1990), in Italy (Goidànich 1928, Martelli 1940, Ciferri & Brizi 1955, Ferri 1959a), and the Czech Republic (Rataj 1957). The species is limited to Europe. Kirchner & Boltshauser (1898) provided a fine illustration of *L. strigata* mines.

2. *Phytomyza horticola* Goureau 1851, Diptera; Agromyzidae.
=*Phytomyza atricornis* Meigen 1838, =*Phytomyza chrysanthemi* Kowarz 1892

Description: Maggots are greenish-white and 3–4 mm long. They mine long, serpentine tunnels according to Hill (1983) and Spaar *et al.* (1990), but Hering (1937) said the tunnels are short and straight. Tunnels contain small, widely-spaced pellets of frass. Larvae pupate inside leaves, visible as brown puparia at the wide end of the mine. Flies have black bodies marked by yellow lateral lines, 2–3 mm long, with pale heads and yellow "knees" (Fig 4.41). *Phytomyza* flies, unlike most other Agromyzids, lack crossveins in the posterior portions of their wings.

Known as the pea leafminer, polyphagous *P. horticola* has been cited under its different synonyms by Goidànich (1928), Hendel (1932), Hering (1937), Martelli (1940), Rataj (1957), Ferri (1959a), and Spaar *et al.* (1990). Pea leafminers live throughout Europe, Asia, and parts of Africa, but not (yet) North America. *P. horticola* should not be confused with the American species ***Phytomyza syngenesiae*** Hardy (chrysanthemum leaf miner), which was also called *P. atricornis* in earlier literature.

3. *Agromyza reptans* Fallén 1823, Diptera; Agromyzidae.
=*Agromyza haplacme* Steyskal 1972

Description: Maggots are similar to those of *P. horticola* in size and shape, but they mine linear tunnels along leaf margins, which expand into brown blotches, filled with frass. Flies are brownish black, legs black with yellowish tibiae and tarsi, and the puparia are reddish brown.

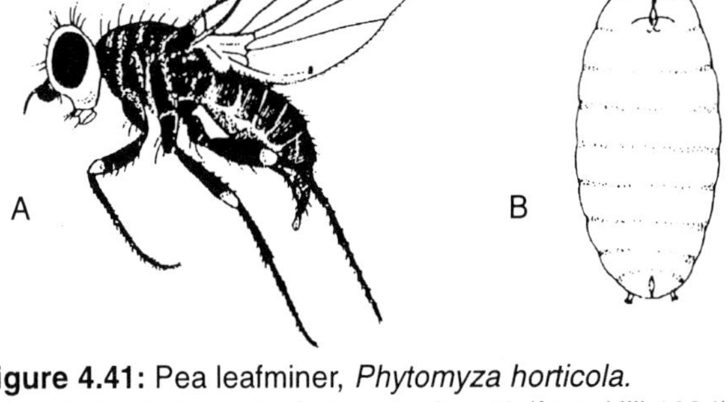

Figure 4.41: Pea leafminer, *Phytomyza horticola.* A. Adult fly; B. Larva leafminer; both x 10 (from Hill 1994).

Buhr (1937), Hering (1937, 1951) and Hendel (1936) reported *A. reptans* in northern Europe. Bush Doctor (unpublished data 1986) found this species on marijuana in Illinois (Plate 35). *A. reptans* has been reported on nettles and hops in temperate Europe, and in Michigan.

4. *Liriomyza eupatorii* (Kaltenbach) 1874, Diptera; Agromyzidae.
= *Liriomyza pusilla* Meigen 1830 subspecies *eupatorii* Kaltenbach
Description: Mines often commence in a spiral pattern, a characteristic this species shares with *L. cannabis* (the next species). The spiral, however, may be minute or even lacking. The rest of the mine is linear. Adults exhibit black bands on the lower halves of their abdomens, legs are yellow, usually with brownish-black striations; total length 1.5 mm, wing length 1.6–2.2 mm.

Hering (1921), Buhr (1937), Rataj (1957), and Spaar *et al.* (1990) cited *L. eupatorii* Kaltenbach on hemp. Hering noted high mortality in *Cannabis;* he blamed "poisonous alkaloids" present in leaves. *L. eupatorii* frequently attacks *Aster* species, goldenrod (*Solidago* species), and hemp agrimony (*Eupatorium cannabinum*). It less frequently infests hemp nettle (*Galeopsis tetrahit*) and *Cannabis.* Buhr (1937) transplanted maggots between *Cannabis* and *Galeopsis;* the pests mined either plant and pupated in soil. *L. eupatorii* lives sympatrically with *A. reptans* in North America and Europe.

5. *Liriomyza cannabis* Hendel 1932, Diptera; Agromyzidae.
= *Liriomyza eupatorii* Hering 1927 [*non* Kaltenbach 1874]
Description: Maggots form tight spiral mines as illustrated by Hering (1937). He described frass scattered along tunnels "like strings of pearls." Hendel (1932) said adults resemble *L. strigata* except for sharply defined abdominal stripes, pointed warts on the heads, and smaller size (1.5 mm long).

L. cannabis infests hemp in Germany (Hering 1937, Spaar *et al.* 1990), Finland (Hendel 1932), Italy (Ferri 1959a), and Japan (Kyokai 1965). Mushtaque *et al.* (1973) reported the pest in Afghan marijuana.

DIFFERENTIAL DIAGNOSIS Leafmining is unique and not easily confused with other symptoms. Maggot leafminers must be differentiated from caterpillar leafminers or beetle grub leafminers. One caterpillar, *Grapholita delineana,* mines within leaves when young (Mushtaque *et al.* 1973). After it outgrows this niche *G. delineana* assumes its primary role as the hemp borer. The flea beetle grub *Phyllotreta nemorum* makes tunnels in tight spirals which end in blotches. Differentiating maggots from caterpillars and grubs requires close inspection; only the maggots are legless. Leafminer eggs can be confused with thrips eggs; see what emerges.

CULTURAL & MECHANICAL CONTROL
(method numbers refer to Chapter 9)

Perform methods 1 (sanitation), 2a (deep ploughing) 2b&c (sterilizing or pasteurizing soil), 3 (for species pupating in soil), 6 (crop rotation against *Liriomyza cannabis*), 9 (pick off infested leaves and compost them or burn them), and 12a&b (repel flies with reflectant material, trap them with yellow sticky tape). Van Lenteren (1995) noted more leafminer damage in glasshouses using rockwool.

Table 4.11: Infestation Severity Index for leafminers.

Light	few mines seen
Moderate	many mines on a few plants
Heavy	leaves starting to curl on plants OR 10–99 mines per plant
Critical	plants losing vigour and wilting OR >100 mines per plant

BIOLOGICAL CONTROL (see Chapter 10)

Ferri (1959a) described several parasitoids attacking hemp leafminers. Van Lenteren (1995) controlled vegetable-mining flies (*Liriomyza bryoniae* and *L. trifolii*) with a mix of two parasitoids, *Dacnusa sibirica* and *Diglyphus isaea* (described below). Success with these parasitoids varies with the leafminer species *and* the host plant. A new parasite, *Opius pallipes,* is described below.

Hara *et al.* (1993) used a soil nematode, *Steinernema carpocapsae,* to kill leafminers in *leaves,* by spraying nematodes onto foliage at night. The nematodes entered leaf mines through oviposition holes and killed >65% of *L. trifolii* larvae (*S. carpocapsae* is described under cutworms). Leafminers pupating in soil can also be controlled with beneficial nematodes.

Dacnusa sibirica

BIOLOGY: A braconid wasp that parasitizes *Liriomyza* larvae. It is native to northern Eurasia and does best in cool to moderate temperatures (optimally 14–24°C) and moderate humidity (70% RH).

APPEARANCE: Adult wasps are dark brown-black, 2–3 mm in length, and have long antennae.

DEVELOPMENT: Wasps lay eggs in all ages of leafminer larvae, but prefer first and second instars. *D. sibirica* parasites develop and pupate within leafminer larvae. The *D. sibirica* lifecycle takes 16 days under optimal conditions; adults live another seven days, females lay 60–90 eggs.

APPLICATION: Supplied as adults in shaker bottles. Store a maximum of two days at 8–10°C. Release in the early morning or evening. Hussey & Scopes (1985) released three wasps per 1000 small plants, weekly. For light infestations, Koppert (1998) suggested releasing two wasps per m^2 crop area, with a minimum of three weekly introductions. For moderate infestations, release five wasps per m^2 weekly.

NOTES: *D. sibirica* is the most popular leafminer biocontrol, but it only works as a preventative. It reproduces too slowly to stop even a moderate infestation. *D. sibirica* is compatible with *D. isaea.* Mixtures of the two species are available. Avoid insecticides. Cherim (1998) suggested *D. sibirica* may be attracted to yellow sticky traps.

Diglyphus isaea

BIOLOGY: A predatory/parasitic chalcid wasp that feeds on leafminers. It is native to Eurasia and does best in moderate humidity and temperatures (80% RH, 24–32°C).

APPEARANCE: Adults are black with a metallic greenish tinge, 2-4 mm long, with short antennae (Plate 36). Pupae can be seen in leafminer mines as tiny black spots.

DEVELOPMENT: Wasps kill leafminer larvae, then lay eggs next to them. *D. isaea* larvae feed on carcasses and pupate in mines. The lifecycle takes 17 days under optimal conditions; adults live another three weeks. Females kill up to 360 leafminers.

APPLICATION: Supplied as adults. Store one or two days at 8–10°C. Release in the early morning or evening. Used preventively, Hussey & Scopes (1985) introduced three wasps per 1000 small plants per week. For light infestations, Thomson (1992) released one or two wasps per 10 m^2 per week for at least three introductions. Koppert (1998) suggested 1 wasp per m^2, with a minimum of three weekly introductions. Cherim (1998) suggested two to four wasps per m^2 every two weeks.

NOTES: *D. isaea* works well at higher leafminer densities than *D. sibirica*, so it serves as a preventative and a curative biocontrol. *D. isaea* prefers warmer temperatures than *Dacnusa sibirica*. It is compatible with *D. sibirica* and with beneficial nematodes. Avoid using insecticides at least a month prior to release. Cherim (1998) suggested *D. isaea* may be attracted to yellow sticky traps.

Opius pallipes

BIOLOGY: A parasitic wasp that feeds on leafminers. It is native to temperate regions and does best in moderate humidity and moderate temperatures.

APPEARANCE & DEVELOPMENT: Adults are 2–3 mm long and resemble *Dacnusa sibirica*. Wasps lay eggs in young leafminer maggots. The eggs hatch and wasp larvae slowly kill leafminer larvae.

APPLICATION: Supplied as adults. Store one or two days at 8–10°C. Release the same as *Dacnusa sibirica*.

CHEMICAL CONTROL (see Chapter 11)

Catch egg-laying flies with synthetic pheromone traps. Repel them with a foliar spray of neem or oil.

Larvae are protected within mines, so surface-sprayed insecticides may not reach them. Adding surfactants may improve the leaf-penetration of surface sprays. Leafminers in other crops have been controlled with abamectin and spinosad (two natural fermentation products), permethrin (a synthetic pyrethrin), methoprene and cyromazine (synthetic insect growth hormones) and imidacloprid (a synthetic nicotine). Try watering plants with a 0.4% neem solution. The solution is absorbed *systemically*, with activity against larvae for about three weeks.

Control species which pupate underground by covering the soil with plastic. Coat plastic with Thripstick®, which kills leafminers as well as thrips. This can be integrated with biocontrol by *Diglyphus isaea*, which does not pupate in the soil (Hussey & Scopes 1985).

LEAFHOPPERS & THEIR RELATIVES

Leafhoppers, planthoppers, treehoppers, spittlebugs, and cicadas are grouped in a Homopteran suborder, the Auchenorrhyncha. Many have been collected on *Cannabis*. Few cause serious feeding damage, but they may inflict secondary losses by spreading plant viruses. A dozen species are described below, in order of occurrence. They rarely cause serious economic damage.

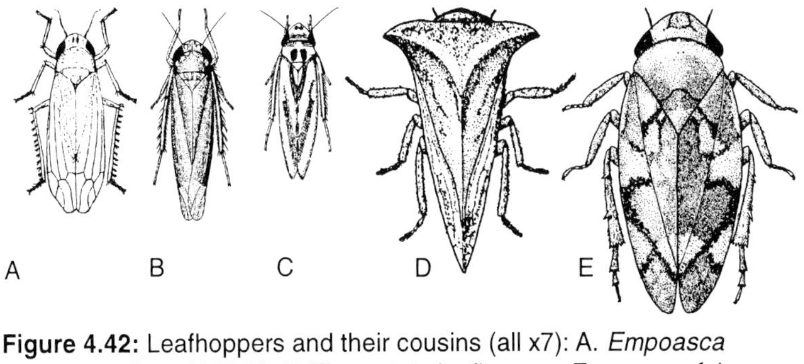

Figure 4.42: Leafhoppers and their cousins (all x7): A. *Empoasca flavescens* (from USDA); B. The potato leafhopper, *Empoasca fabae* (from I.N.H.S.); C. The glasshouse leafhopper, *Zygina pallidifrons;* D. The buffalo treehopper, *Stictocephala bubalus;* E. The spittlebug, *Philaenus spumarius* (C, D, E by McPartland).

SIGNS & SYMPTOMS

Leafhoppers and their relatives are sap suckers. Most are phloem feeders, but some suck xylem sap (e.g., cicadas, spittlebugs, many planthoppers, and some leafhoppers). Xylem sap is an extremely dilute food source (see Table 3.2), so these pests must ingest great quantities of sap. Spittlebug nymphs can suck xylem sap at a rate greater than ten times their body weight per hour. They digest what little is available, then excrete >99% of the fluid. They surround themselves with a froth of excreted spittle.

Leafhoppers and their relatives cause symptoms similar to those caused by aphids and whiteflies—wilting, a light coloured stippling of leaves, and sooty mould from honeydew. Xylem feeders also cause leaf veins to become swollen and lumpy, because their stylets are thick and stout compared to the fine flexible stylets of phloem feeders (Press & Whittaker 1993).

1. GLASSHOUSE LEAFHOPPER

Zygina (Erythroneura) pallidifrons (Edwards 1924), Homoptera; Cicadellidae.

Description: Adults are small (3–4 mm long), pale yellow-green, with iridescent wings extending beyond the rear ends of their bodies. Between the eyes are two small dark spots, and behind the eyes (near the bases of the forewings) are two larger dark spots (Fig 4.42). Legs and antennae are long. When disturbed the adults hover over plants, like whiteflies. Nymphs are small, white, and initially almost transparent; wing-pads appear during the third instar and enlarge during the fourth (final) instars. Cast skins from moults remain attached to leaves, are translucent, and are commonly called "ghost flies" (Wilson 1938). Eggs are elongated, slightly curved, white, 0.5–0.7 mm long.

Life History & Host Range

Adults overwinter on weeds outdoors. In warm glasshouses they do not hibernate, so overlapping generations arise all year long. Eggs are inserted into leaf veins, usually singly. Young nymphs feed on undersides of leaves. In optimal conditions (26°C), the life cycle takes 42 days. Adults live another four months. *Z. pallidifrons* poses a problem in English and Dutch glasshouses (Wilson 1938). It attacks many crops (especially tomato and mint) and many weeds (especially chickweed, *Stellaria media*).

2. REDBANDED LEAFHOPPER

Graphocephala coccinea (Foerster) 1771, Homoptera; Cicadellidae.

Description: Adults 8–10 mm in length, slender, with yellow, pointed heads; wings reflect alternate bands of magenta and green with yellow margins. Nymphs are yellow to green.

Life History & Host Range

Adults overwinter in leaf trash on the ground. Eggs are thrust into soft plant tissues in early spring. One or two generations arise per year. Hartowicz *et al.* (1971) and DeWitt (unpublished data) collected *G. coccinea* from feral hemp in Kansas and Illinois. Frank & Rosenthal (1978) described them in California. This leafhopper infests many hosts, but injury is rarely serious. Its size and striking colour make it a high-profile pest. Lago & Stanford (1989) described a relative, ***Graphocephala versuta*** (Say) on Mississippi marijuana.

3. POTATO LEAFHOPPER

Empoasca fabae (Harris) 1841, Homoptera; Cicadellidae.

=*Empoasca mali* LeBaron 1853

Description: Adults are green with faint white spots on the head and thorax, reaching 4 mm in length. Seen from above they appear wedge-shaped, with 1 mm wide heads tapering to pointed wingtips (Fig 4.42). Nymphs are similar in shape but even smaller, nearly impossible to see as they feed on undersides of leaves. Nymphs and adults have strong hind legs and readily jump or "crab-walk" to safety when disturbed.

Life History & Host Range

E. fabae ranges across the eastern half of North America, but cannot overwinter north of the Gulf states. Each spring, adults hitch rides on weather fronts in Louisiana. They migrate north to Manitoba, Ontario and Quebec, and east to New England. Some years they are blown as far as Newfoundland. Their numbers decrease in the west where rainfall falls below 65 cm per year.

Upon landing, females lay an average of 35 eggs in petioles and leaf veins. Up to four generations arise each year. Potato leafhoppers feed from both phloem and xylem (Press & Whittaker 1993), and frequently plug these vessels, which impairs fluid movement in leaves, causing "**hopperburn**." Hopperburn symptoms begin as a browning of leaf margins and leaf tips. Affected leaves become deformed, lumpy, and curly. In heavy infestations the leaves become badly scorched with only midveins remaining green. Dudley (1920) described hopperburned hemp in New Hampshire. *E. fabae* attacks more than 100 plant species, especially alfalfa, clover, apple, and many vegetable crops. In potato crops, the presence of ten nymphs per 100 mid-plant leaves is the threshold for control measures (Howard *et al.* 1994).

4. OTHER CICADELLIDAE LEAFHOPPERS

The "flavescent" leafhopper, ***Empoasca flavescens*** F., infests European hemp (Spaar *et al.* 1990). This species also lives in the USA; it resembles a pale (nearly white) version of *E. fabae* (Fig 4.42 & Plate 37), and like *E. fabae*, it feeds on xylem and phloem (Press & Whittaker 1993). An unidentified *Empoasca* species infests marijuana leaves in Thailand (Hill 1983). A new species, ***Empoasca uniprossicae*** Sohi, was described on *Cannabis* in India (Sohi 1977).

Frank & Rosenthal (1978) claimed Rose leafhoppers (***Edwardsiana rosae*** L.) attack marijuana in the San Francisco area. Bantra (1976) reported ***Iassus indicus*** (Lethierry) 1892 [=***Jassus chlorophanus*** (Melichar) 1905] as the most common cicadellid on Indian marijuana. DeWitt (unpublished data) collected a *Gyponana* species from feral hemp in Illinois. Lago & Stanford (1989) cited ***Gyponana octolineata*** (Say) on Mississippi marijuana, along with ***Agallia contricta*** Van Duzee and about fifteen other uncommon cicadellids. Wei & Cai (1998) described a new cicadellid, ***Macropsis cannabis*** Wei & Cai, attacking hemp in Henan Province, China.

5. SPITTLEBUG

Philaenus spumarius (Linnaeus) 1758, Homoptera; Cercopidae.

Description: Spittlebug nymphs are easily spotted—they usually occupy the crotches of small branches and surround themselves with froth (see Plate 84). Nymphs resemble tiny pale green frogs, 6 mm in length. Adult spittlebugs are the same size and shape as final-instar nymphs, but turn a mottled brown colour (ranging from straw-coloured to almost black, Fig 4.42), and no longer hide in spittle. Eggs are ovoid, 1 mm long, yellow to white, and laid in groups of two to 30 in hardened spittle, on stems and leaves near the ground.

Life History & Host Range

Eggs overwinter. Nymphs feed on xylem sap in the spring, and metamorphize into adults from late May to late June. Females lay eggs in August and September, then die. One generation arises per year. We have seen spittlebugs on feral hemp and marijuana in Oregon, Illinois, New Jersey, and Vermont. Goidànich (1928) reported the species in Italy. Spittlebugs rarely kill plants but they decrease crop yields. *P. spumarius* attacks at least 400 species of plants. The species accumulates in areas of high humidity across North America.

6. BUFFALO TREEHOPPER

Stictocephala bubalus (Fabricius) 1794, Homoptera; Membracidae.

Description: Adults are green and 6 mm long, their blunt heads support two short horns. Seen from above, adults appear triangular, pointed at the rear (Fig 4.42). Adults are shy and fly away with a loud buzzing noise. Nymphs are tiny, pale green, humpbacked and covered by spines. Eggs are yellow.

Life History & Host Range

Nymphs hatch late in the spring from overwintering eggs laid in tree branches. They drop to the ground and suck sap from herbaceous plants. Adults appear by August. Nymphs feed on *Cannabis*, adults feed on trees. The real damage is done by egg-laying females—not feeding, but slicing stems with their knifelike ovipositors to lay eggs. They perforate rings around branches, causing branches to break off. One generation arises per year.

Buffalo treehoppers infest marijuana near Buffalo, NY (Hillig pers. commun. 1993) and in southern Indiana (Clarke pers. commun. 1985). Hartowicz *et al.* (1971) reported "Membracidae treehoppers" on feral hemp in Kansas. DeWitt (unpublished) found *S. bubalus* on feral hemp in Illinois. The species lives throughout North America and causes economic damage in orchards and tree nurseries.

7. PLANTHOPPERS

Description: Planthoppers (superfamily Fulgoroidea) can be separated from leafhoppers, treehoppers, and spittlebugs by the movable spur on their hind tibia, and by the location of their antennae and ocelli. Planthoppers seldom occur as abundantly as leafhoppers or spittlebugs.

Life History & Host Range

All *Cannabis* planthoppers are reported from Asia. ***Eurybrachys tomentosa*** Fabricius (family Fulgoridae) is a minor pest of marijuana in India (Cherian 1932). ***Geisha distinctissima*** Walker 1858 (family Flatidae) attacks hemp in China, Korea and Japan (Takahashi 1919, Clausen 1931, Shiraki 1952, Sorauer 1958). The Japanese broadwing planthopper, ***Ricania japonica*** Melichar 1898 (family Ricaniidae) is also a frequent offender (Takahashi 1919, Clausen 1931, Shiraki 1952, Sorauer 1958). Kuoh (1980) collected a small planthopper from Chinese hemp, ***Stenocranus qiandainus*** Kuoh (family Delphacidae).

8. CICADAS

Bothrogonia ferruginea (Fabricius) 1787 (family Cicadidae) infests hemp in Eurasia (Clausen 1931, Shiraki 1952, Sorauer 1958). Cicadas are recognized by their large size and the males' summer mating songs. Some cicadas have life cycles as long as 17 years. Most of their damage is inflicted on trees.

DIFFERENTIAL DIAGNOSIS

Leafhopper nymphs and their damage may resemble other Homoptera such as aphids. Some planthopper nymphs (*Richania* species) look like flea beetles, and planthopper adults can be mistaken for small moths. Hopperburn symptoms can be confused with sun scald or nutrient imbalances, except for the irregular lumps on leaf veins.

CULTURAL & MECHANICAL CONTROL

(method numbers refer to Chapter 9)

All Auchenorrhyncha are controlled by methods 1 (sanitation), 2a (deep ploughing), and 3 (weeding). Leafhoppers are controlled with methods 12a&b (mechanical trapping and repelling). Potato leafhoppers are attracted to blacklights (method 12d). For spittlebugs, use methods 6 (crop rotation, and plant away from alfalfa and clover fields). Avoid planting in shaded, protected areas preferred by spittlebugs. To avoid treehoppers, avoid elm trees and orchards.

BIOCONTROL (see Chapter 10)

Fungi work best against sap-sucking insects: *Metarhizium anisopliae* (brand name Metaquino®) is a mycoinsecticide that kills spittlebugs (described under aphids). *Paecilomyces fumosoroseus* is used against planthoppers in the Philippines (described under whiteflies). *Hirsutella thompsonii* (Mycar®) was previously registered for use against planthoppers in the USA (described under hemp russet mites). ***Erynia (Zoophthora) radicans*** has been used experimentally to control potato leafhoppers (Samson *et al.* 1988). It arises naturally in the midwestern USA and kills leafhoppers in two or three days.

The mirid bug *Deraeocoris brevis* preys on young psyllids (described under thrips). A mynarid wasp, ***Anagrus atomus*** (=*Anagrus epos*?) parasitizes eggs of *Z. pallidifrons* and other leafhoppers. Commercial development is hampered by egg desiccation and the short lifespan of the adult wasp. Frank & Rosenthal (1978) repelled leafhoppers by planting geraniums around crops. Yepsen (1976) repelled leafhoppers with geraniums and petunias.

CHEMICAL CONTROL (see Chapter 11)

Damage is rarely bad enough to justify chemicals. Frank & Rosenthal (1978) used insecticidal soap, especially against leafhopper nymphs (spray on undersides of leaves). Neem and diatomaceous earth (or better, diatomaceous earth mixed with pyrethrum) kill most hoppers. Frank (1988) killed leafhoppers with pyrethrum and synthetic pesticides. Bordeaux mixture repels adults and serves as a feeding deterrent (Metcalf *et al.* 1962). Clay microparticles may also work this way.

Z. pallidifrons is controlled with nicotine sprays (two or three applications at ten-day intervals); rotenone and pyrethrum are less satisfactory (Wilson 1938). Potato leafhoppers succumb to sabadilla dust, but new waves of migrating leafhoppers may require five to eight applications per season. Spittlebugs are susceptible to rotenone, especially when nymphs are less than a month old. Alfalfa growers start spraying fields when spittlebug concentrations reach one pest per plant. Imidacloprid (a synthetic nicotine) kills all insects with sucking mouthparts.

MEALYBUGS & SCALES

Mealybugs and scales suck plant sap, as do their Homopteran cousins, the aphids and whiteflies, in the suborder Sternorrhyncha. They also gum-up plant surfaces with honeydew. Honeydew attracts ants and supports the growth of sooty mould.

1. LONG-TAILED MEALYBUG

Pseudococcus longispinus Targioni-Tozzetti 1867, Homoptera; Pseudococcidae.

=*Pseudococcus adonidum* (Linnaeus) Zimmerman 1948, nec *Coccus adonidum* Linnaeus 1767

Description: Young nymphs have tiny yellow bodies. With age, they become covered with waxy secretions and turn white; late-instar females grow four elongated wax filaments from their caudal ends. Adult females are broadly oval, segmented, 2–4 mm long and 1–2 mm wide. Tails nearly equal bodies in length (Fig 4.43).

Life History & Host Range

Long-tailed mealybugs feed on a wide range of hosts and are found worldwide. They live out-of-doors in warm climates (including Florida, Texas, and southern California). In northern latitudes *P. longispinus* becomes a serious glasshouse pest (Frank & Rosenthal 1978, Frank 1988). Unlike other mealybug species, *P. longispinus* births live larvae. New generations may emerge every month.

2. COTTONYCUSHION SCALE

Icerya purchasi Maskell 1878, Homoptera; Margarodidae.

Description: Mature scales are reddish brown, with black legs and antennae. Females are partly or wholly covered by egg sacs that are large, cottony-white, and longitudinally ridged (Fig 4.43). Total length is 8–15 mm, two-thirds of which is egg sac. Sacs contain 300–1000 oblong red eggs. The red-bodied nymphs have black legs, prominent antennae, and congregate under leaves along the midribs.

Life History & Host Range

I. purchasi overwinters in all stages. Depending on the climate, eggs hatch in a few days to two months. Larvae remain mobile, unlike most scales. Adults are gregarious and often suck sap in groups. Up to three generations arise per year in California. *I. purchasi* is native to Australia, unwittingly introduced into California around 1868. The pest posed a serious threat to the citrus industry until a natural enemy, the vedalia beetle, was imported. Bodenheimer (1944) and Sorauer (1958) reported cottonycushion scales infesting hemp. Cottonycushion scales are very polyphagous and live throughout the warmer parts of the world.

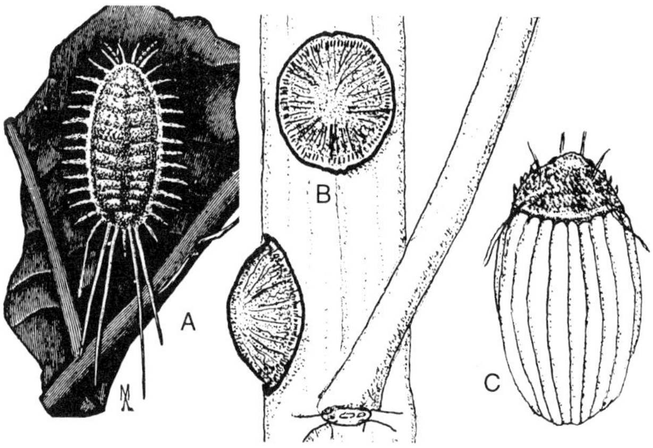

Figure 4.43: Mealybugs and scales (all x4).
A. *Pseudococcus longispinus* (from Comstock 1904);
B. *Parthenolecanium corni* (McPartland); C. *Icerya purchasi* (McPartland redrawn from Sorauer 1958).

3. EUROPEAN FRUIT LECANIUM

Parthenolecanium corni (Bouché) 1844, Homoptera; Coccidae.

=*Lecanium corni* Bouché, =*Lecanium robinarium* Douglas 1890

Description: First and second instar nymphs are tiny (<0.5 mm), green, and active. Later instars turn brown and settle on twigs and stems. Adults are oval, brown with dark brown mottling, fluted or ridged near the edges of their bodies, 3–5 mm long. Egg-bearing females become quite convex, nearly hemispherical (Fig 4.43). Eggs look like a woolly white ring appearing under the edges of the female's body.

Life History & Host Range

In warm climates, nymphs and adults overwinter on twigs and branches. In cool climates only adult females overwinter. Females lay 200–300 eggs in May-June and die (the body provides a protective covering). Eggs hatch in June–

July, and young nymphs feed on leaves near midveins. As winter approaches they move to twigs and stems. One generation arises per year.

Under different Latin names this pest has been recorded on *Cannabis* in Germany (Kovacevic 1929), Russia (Borchsenius 1957) and central Asia (Mostafa & Messenger 1972). *P. corni* usually infests fruit trees and cane crops in Europe, north Africa, south Asia, Australia, and the west coast of North America. It may be the unnamed small brown scale that Frank (1988) described on stems of Californian marijuana.

4. HEMISPHERICAL SCALE

Saissetia coffeae (Walker) 1852, Homoptera; Coccidae.
=*Saissetia hemisphaerica* (Targioni) 1867

Description: Adults look like the aforementioned *P. corni*—hemispherical in shape, like miniature army helmets, light brown becoming dark brown, 2-3 mm diameter. Eggs are tiny, oval, pink-beige, in clusters of 400–900 under the shell of dead females. First-instar nymphs (crawlers) are flat, pink-beige, 0.7 mm long, with two red eye spots. Older nymphs develop a pattern of ridges in an H-shape on their dorsal surface, and begin to dome.

Life History & Host Range

First and second instar nymphs crawl and feed on leaves. Later instars settle on branches and stems. They tend to group in clusters, and secrete a lot of honeydew. Several overlapping generation arise per year in warm glasshouses.

Bush Doctor (pers. commun. 1992) observed *S. coffeae* infesting branches of glasshouse marijuana in Becker, Michigan. *S. coffeae* lives in the tropics, subtropics, and warm enclosures. It probably came from South America. *S. coffeae* attacks a wide range of woody and herbaceous hosts, from *Ficus* to ferns.

5. WHITE PEACH SCALE

Pseudaulacaspis pentagona (Targioni-Tozzetti) 1886, Homoptera; Coccidae.

Description: Adults are broadly oval with slightly notched sides, white to reddish pink, and usually less than 10 mm long.

Life History & Host Range

White peach scales are polyphagous, but prefer the branches and trunks of woody plants. Shiraki (1952) cited *P. pentagona* on Japanese hemp. The species originated in the far East but now infests all of subtropical Asia, southern Europe, and the southern USA (from Maryland down to the Gulf of Mexico).

DIFFERENTIAL DIAGNOSIS

The long tails of *P. longispinus* and the fluted egg sacs of *I. purchasi* make these pests easy to identify, at least as adults. Nymph stages can be confused with whitefly nymphs, aphids, or mealybug destroyers. *P. corni* and *S. coffeae* can be confused with other small, sessile, brown, barnacle-like pests, such as the black scale ***Saissetia oleae*** (Bernard), and the California red scale ***Aonidiella aurantii*** (Maskell).

CULTURAL & MECHANICAL CONTROL

(method numbers refer to Chapter 9)

Methods 1 (sanitation), 2b (deep ploughing), and 9 (hand removal) are the most important cultural controls of mealybugs and scales.

BIOCONTROL (see Chapter 10)

Controlling cottonycushion scales with *Rodolia cardinalis* is an international success story. Other useful ladybeetles include *Cryptolaemus montrouzieri* and *Rhizobius ventralis* (described below). *Harmonia axyridis* and *Lindoris lophanthae* prey on many scales and mealybugs (described below). Predation of scales by *Chrysoperla carnea* seems less effective (described under aphids).

Turning to parasites, *Encarsia formosa* attacks some soft scales but much prefers whiteflies. Scales are parasitized by *Aphytis melinus* and *Metaphycus helvolus*, *Encytus lecaniorum* (described below), and ***Microterys flavus*** and ***Encytus lecaniorum*** (wasps currently under investigation). *Leptomastix dactylopii* and *Anagyrus pseudococci* are popular biocontrols of the citrus mealybug (***Planococcus citri*** Risso). Although polyphagous *P. citri* does not appear in the *Cannabis* literature, we describe the biocontrols below. The fungus *Verticillium lecanii* kills scales and mealybugs, but works better against aphids and whiteflies (described under the latter).

Rodolia cardinalis "Vedalia"

BIOLOGY: The ladybeetle that saved agricultural California from *I. purchasi*—entomologists imported only 129 vedalias from Australia, but within 18 months their offspring nearly eradicated *I. purchasi* from Californian orange groves (Metcalf *et al.* 1962).

APPEARANCE & DEVELOPMENT: Adult vedalias have round, convex bodies, wing covers red, densely pubescent, marked with irregular black spots, 3–5 mm long (Fig 4.4). Larvae are pink, wrinkled, spindle-shaped, covered with soft spines, and reach 8 mm long. Eggs are red and oval. Females lay 150–190 eggs near *I. purchasi* egg masses. Young larvae feed on eggs. Mature larvae and adults feed on all scale stages. Up to 12 generations arise per year in hot, dry, inland valleys. Pupae overwinter in cooler coastal climates.

APPLICATION: Supplied as adults. Unfortunately the pest *I. purchasi* absorbs plant poisons from certain crops, rendering it unpalatable to vedalias. *R. cardinalis* is susceptible to many insecticides (especially juvenile growth hormones); DDT almost wiped the species out during the 1950s.

Cryptolaemus montrouzieri "Mealybug destroyer"

BIOLOGY: "Crypts" eat many mealybug species, as well as young scales and aphids. Californian entomologists imported these ladybeetles from Australia in 1891, as biocontrol pioneers against citrus mealybugs. They do best at 27°C (range 17–32°C) and 70–80% RH.

APPEARANCE: Adults are 4 mm long, with red-orange heads and thoraxes, and shiny, black-brown wing covers (Plate 38). Larvae are long (up to 13 mm) and covered with woolly projections of wax—they resemble monstrous, dishevelled, mobile mealybugs. Eggs are white or yellow, oblong, and found in the cottony egg-masses of their prey.

DEVELOPMENT: Adults overwinter in moderate climates but cannot survive cold winters. Adults feed on mealybug eggs and young larvae; females lay up to 500 eggs, one per host egg-sack. Young larvae eat mealybug eggs and young larvae; older *C. montrouzieri* larvae attack all mealybug stages. Larvae eat up to 250 mealybugs and eggs, then pupate in sheltered places on plant stems or greenhouse structures. The life cycle takes 29–38 days at 27°C. Adults live another 50–60 days.

APPLICATION: Supplied as adult beetles in trays or bottles. Store up to one or two days in a cool (10–15°C), dark place. Release in the early morning or late at night to decrease dispersal. Release five beetles per plant or 20–50 per m^2 in infested areas (Thomson 1992).

NOTES: Short days *may* send this species into diapause. *C. montrouzieri* works best in heavy infestations. It tends to fly away before its job is done, like most ladybeetles. To induce it to stay, use strategies described under *Hippodamia convergens*, in the section on aphids. *C. montrouzieri* must lay its eggs in host egg masses, so it cannot reproduce in pests that give birth to live larvae, such as *P. longispinus* (Cherim 1998). This biocontrol is compatible with *Leptomastix dactylopii*. Avoid most insecticides while utilizing ladybeetles, especially diazinon. Cherim (1998) exercised caution when using white sticky traps—the beetles are attracted to white. Mass rearing of *C. montrouzieri* was hampered for years by a parasitoid of its own that proved difficult to eradicate (Gilkeson 1997).

Harmonia axyridis "Multicolored Asian ladybeetle"

BIOLOGY: This predator was imported from Japan by the USDA as a successful biocontrol against scales and aphids. *H. axyridis* also feeds on whiteflies, mealybugs, and even some mites. The species does best in moderate temperatures and humidity (optimally 21–30°C, 70% RH).

APPEARANCE & DEVELOPMENT: Adults are 8 mm long, orange with black spots (they are also called Halloween ladybeetles). Adults overwinter; females lay up to 700 eggs, larvae resemble 10 mm alligators with orange spots. Larvae take 25 days to reach adulthood; adults then live up to three years (Cherim 1998).

APPLICATION: Supplied as adults. Store for a week at 8–10°C. Koppert (1998) recommended releasing one predator per 50 pests. Other sources recommend four predators per m^2 crop area, or eight predators per m^3 foliage volume, repeating every three weeks, two or three times.

NOTES: In Japan the adults overwinter in cliffs, but in the USA, the next best thing is a house. The adults aggregate in large numbers on vertical surfaces, typically swarming on light-coloured walls with southern exposures. They enter houses through cracks and become a nuisance. Like all ladybeetles, *H. axyridis* can be "flighty." See comments regarding *Cryptolaemus montrouzieri*, above.

Lindoris (Rhyzobius) lophanthae

BIOLOGY: Yet another ladybeetle that feeds on immature soft scales, including brown and black scales (*Saissetia* and *Parasaissetia* species), and red scales (*Aonidiella* species). In the absence of soft scales, *L. lophanthae* eats immature hard scales and mealybugs. The species does best in moderate tem-peratures (15–25°C) and a wide humidity range (20–90% RH).

APPEARANCE & DEVELOPMENT: Adults are small, pubescent, black beetles with red-orange highlights, 2.5 mm long. Females lay over 100 eggs, one at a time, on plants among scales. Larvae are grey, reach 3 mm long, and prey on young scales and eggs. The life cycle takes three weeks. Adults live another six to eight weeks, eating constantly. The final generation of the season overwinters as adults.

APPLICATION: Supplied as adult beetles in trays or bottles. Store up to one or two days in a cool (10–15°C), dark place. Release in early morning or late at night. For light infestations, release three to six beetles per m^2, repeat every three weeks, two or three times. Double the rate for heavier infestations (albeit an expensive proposition). *L. lophanthae* is probably compatible with parasitic biocontrols.

Rhyzobius ventralis

BIOLOGY: A ladybeetle related to the aforementioned species; it also preys on mealybugs and scales. Adults are shiny velvety black with reddish abdomens.

Aphytis melinus

BIOLOGY: A parasitic chalcid (Eulóphidae) wasp that feeds on hard scales (*Aonidiella, Aspidiotus, Quadraspiniotus,* and *Pseudaulacaspis* species). It is native to India and Pakistan and tolerates a wide range of temperatures (13–33°C) and moderate humidity (40–60% RH).

APPEARANCE & DEVELOPMENT: Females are tiny yellow-brown wasps, 1.2 mm long. They are attracted to pheromones released by female scales when females are ready to reproduce. Ready females loosen their attachments to plants, allowing males to fertilize them. *A. melinus* uses this window of opportunity to lay eggs in about 30 scales. Eggs hatch into larvae which kill female scales, pupate therein, then emerge as adults. The life cycle takes two or three weeks in optimal conditions. Adults live another three weeks.

APPLICATION: Supplied as adults in large vials and ice-cooled boxes. *A. melinus* should be released immediately, during the mating flight of male scales. In orchards, they are released 25,000–250,000 per ha (= 2.5–25 per m^2). Indoors, Cherim (1998) controlled light infestations with 16–24 wasps per m^2 every three weeks, and treated heavy infestations with 24–48 wasps per m^2 every week. *A. melinus* tolerates Bt, sabadilla, and abamectin (Van Driesche & Bellows 1996).

Leptomastix dactylopii

BIOLOGY: A parasitic chalcid wasp that lays eggs in third-stage larvae (and sometimes adults) of *Planococcus citri*. It is native to Brazil and does best at 25°C and 60–65% RH.

APPEARANCE & DEVELOPMENT: Females are yellow-brown, 3 mm long; male wasps are smaller and have hairy antennae. When searching for mealybugs, females prefer walking to flying, so they do not disperse very efficiently. Females lay up to 200 eggs, one per mealybug. Parasitized mealybugs turn into dark, swollen mummies. Wasps emerge from small, round holes. The life cycle takes about three weeks in optimal conditions.

APPLICATION: Supplied as adults in shaker bottles or tubes. They must be dispersed the day of receipt, either in the early morning or evening. Release two wasps per m^2 as a preventative or five wasps per plant if pests are evident. The wasps work well at low pest densities. *L. dactylopii* can be combined with predators such as *C. montrouzieri*.

Anagyrus pseudococci

BIOLOGY: A parasitic Encyrtidid wasp that lays eggs in third-stage larvae (and sometimes second- and fourth-stage larvae) of *Planococcus citri*. A related species, ***Anagyrus fusciventris***, is also available.

APPEARANCE & DEVELOPMENT: Females are brown, 1.5–2 mm long, with black and white banded antennae. Male wasps are smaller, without the distinctive antenna bands. Wasps lay eggs in mealybug larvae, which turn into swollen, yellow-brown (striped) mummies. Wasps emerge from mummies via small, ragged holes. The entire lifecycle may take only three weeks.

APPLICATION: Supplied as adults in shaker bottles or tubes. Store up to one or two days in a cool (8–10°C), dark place. Koppert (1998) recommended releasing 0.25 wasp per m^2 every 14 days as a preventative. For infestations, release one wasp per m^2 per week.

Metaphycus helvolus

BIOLOGY: A parasitic *and* predatory wasp that controls many species of soft scales, especially black and brown scales (*Saissetia, Parasaissetia,* and perhaps *Coccus* species). It is native to semitropical regions and does best in warm, relatively dry glasshouses (24–32°, RH 50%). A related species from Australia, ***Metaphycus alberti***, was released in California to control *Coccus hesperidium*.

APPEARANCE & DEVELOPMENT: Wasps are tiny, yellow

and black, 1.3 mm long. Female wasps lay 100–400 eggs, one at a time, under immature scales. Larvae burrow into scales, kill them, pupate therein, and emerge as adults. The life cycle takes 24 days. Adults live another two or three months. Adults poke holes in scales and feed on body fluids.

APPLICATION: Supplied as adults, which should be released immediately. Used preventively, release one or two wasps per m^2 per month. For light to moderate infestations, release five to ten wasps per plant or ten wasps per m^2, every two weeks until controlled. *M. helvolus* is not useful in heavy infestations. According to Cherim (1998), the wasps are attracted to bright lights and yellow sticky traps. The wasps live longer if supplied with honeydew or nectar.

CHEMICAL CONTROL (see Chapter 11)

Synthetic pheromone lures are available for some mealybugs and scales (Olkowski *et al.* 1991). Puritch (1982) killed 96.9% of mealybugs with Safer's insecticidal soap spray. Neem repels mealybugs and young scales, oil kills mealybugs and young scales. Frank (1988) suggested spraying mealybugs with nicotine sulphate, or daubing mealybugs with cotton swabs dipped in rubbing alcohol. Buprofezin and kinoprene are synthetic insect growth hormones toxic to many mealybugs and scales. Imidacloprid (a synthetic nicotine) kills all insects with sucking mouthparts.

ANTS & TERMITES

Ants and white ants (termites) often curse *Cannabis* in semitropical climates. Although they look somewhat similar, they are from unrelated orders.

SIGNS & SYMPTOMS

Ants and termites tunnel into taproots and stems (Plate 39). Plants wilt and sometimes collapse. They are easily pulled from the ground. Ants also colonize aphid-infested plants. Ants protect aphids from parasitoids and predators, and the aphids supply ants with honeydew. Leafcutter ants cut pieces of leaves and carry them off to underground nests. In severe cases, leafcutter ants devour *entire* marijuana plants, leaving only the roots. Leafcutter ants are nocturnal and can destroy several plants a night.

1. ANTS

Cherian (1932) cited the red fire ant ***Solenopsis geminata*** (Fabricius) 1804 (Hymenoptera; Formicidae) chewing seedlings and tunnelling into roots of mature marijuana plants. Siegel (1989) also reported "red ants" attacking marijuana in Mississippi. *S. geminata* workers are three to six mm long, and can inflict painful bites and stings. The species is native to Central America and the Caribbean, north to Texas and Florida, and has been accidentally introduced into India, Africa, and some Pacific islands. *S. geminata* is the only creature nasty enough to slow the spread of imported fire ants, *Solenopsis invicta* and *Solenopsis richteri.*

Frank & Rosenthal (1978) described unidentified ants damaging roots of marijuana in California. Clarke (pers. commun. 1996) encountered leafcutter ants (***Atta*** species) in Mexico. Overnight, the ants cut entire plants into little pieces and carried the pieces to their underground nests.

2. TERMITES

Odontotermes obesus (Rambur) 1842 (Isoptera; Termitidae) has been a major pest of Indian marijuana (Cherian 1932). *O. obesus* occurs across the Indian subcontinent, east into Burma. The species builds spectacular mounds up to 3 m tall. Alexander (1984c, 1985) cited two instances of termites infesting marijuana in the southern USA. The first report was from a Florida grower who invited disaster by mulching his soil with wood chips. Clarke (pers. commun. 1996) found termites infesting marijuana in Mexico and equatorial Africa. Termites hollowed out the main stem and branches, up to the levels of flowers, and then plants collapsed.

DIFFERENTIAL DIAGNOSIS

Symptoms from ants and termites are similar to those caused by white root grubs, root maggots, nematodes, root rots from assorted fungi, and drought.

CULTURAL & MECHANICAL CONTROL
(method numbers refer to Chapter 9)

Observe methods 1 (remove stumps, roots and other termite-attractive debris), 2a (deep ploughing), 7a (termites attack water-stressed plants), and 8 (optimize soil structure and nutrition). Avoid mulching with woody materials except for cedar chips. Compost, humus, and manures exacerbate ant infestations. Frank & Rosenthal (1978) repelled ants and termites by flooding soil near infested plants. Clarke (pers. commun. 1996) located nests and flooded them with water. Frank (1988), a proponent of hydroponics, noted that ants and termites only cause problems in soil-grown plants, not hydroponically-grown plants.

BIOCONTROL

Fire ants have been biocontrolled experimentally with *Pyemotes tritici*, *Pseudacteon* species (described below), *Thelohania solenopsae*, and *Steinernema* nematodes. The termite *O. obesus* is parasitized by *Termitomyces striatus* (Belli) Heim, but this fungus is not commercially available. Other termites are controlled with *Metarhizium anisopliae* (see under aphids) and *Steinernema carpocapsae* (see under cutworms).

Pseudoacteon species

BIOLOGY: About 20 species of parasitic humpbacked flies (family Phoridae) are known to infest ant nests. Related flies infest termite nests. *Pseudoacteon* species from North America parasitize *S. geminata,* and *Pseudoacteon* species from South America parasitize *S. invicta* and *S. richteri.* Adult flies are tiny, half the size of their hosts. Females dive-bomb the back of ants and quickly insert a single egg, using their hypodermic-like ovipositor. Females repeat this aerial attack on dozens of ants. Eggs hatch and maggots migrate into the head of ants. Maggots do not eat vital parts, so ants remain alive until maggots are fully mature. At that point, maggots release an enzyme that causes the ant's head to fall off. Pupation occurs in the decapitated head; the adult fly emerges from the mouth. The life cycle takes four to six weeks.

Pyemotes tritici

BIOLOGY: These predatory mites are used against fire ants and other pests in stored grain. Unfortunately, *P. tritici* is known as the "straw itch mite," because it bites people who handle infested materials.

CHEMICAL CONTROL

Cherian (1932) irrigated infested soil with oil to drive away termites and ants. Drench the soil with pyrethroids (especially tralomethrin), neem oil, or abamectin. Frank & Rosenthal (1978) repelled ants from plants by dusting the surrounding soil with cream of tartar. Clarke (pers. commun. 1996) stabbed an opening into nests and drowned ants and termites by pouring water into the nests. More effectively but less ecologically, he poured a kerosene-gasoline mix into nests and burned them.

GALL MIDGES & ROOT MAGGOTS

You'd think a midge named *Chortophila cannabina* Stein would infest *Cannabis,* but no. Its host is the hemp linnet, *Carduelis cannabina* (a bird named for its love of hemp seeds). Some gall midges are carnivorous and predatory—for instance, *Aphidoletes aphidimyza* larvae devour aphids. They are used for biocontrol. Other gall midges attack plants. They are related to root maggots.

1. NETTLE MIDGE

Melanogromyza urticivora Spencer 1966, Diptera; Agromyzidae.

Description: Adults are green and black flies with 2.4 –3.0 mm wingspans. Larvae are small, white maggots without legs or distinct heads.

Life History & Host Range

This species probably overwinters as pupae in soil. Adults emerge in spring and lay eggs on host plants. Maggots hatch and form galls in stems of Pakistani marijuana (Mushtaque *et al.* 1973). Baloch *et al.* (1974) reported *M. urticivora* only attacks *Cannabis* growing near nettles (*Urtica* species, its primary host). *M. urticivora* lives in the western Himalaya.

2. OTHER GALL MIDGES

Mushtaque *et al.* (1973) described an *Asphondylia* species (family Cecidomyiidae) forming galls in male flowers in Pakistan. Baloch *et al.* (1974) reported, "It does not appear to be a promising biocontrol agent [of *Cannabis*], because pollination is almost complete by the time the infestation starts."

3. SEEDCORN MAGGOT

Delia platura Meigen 1826, Diptera; Anthomyiidae.

=*Chortophila cilicruta* Rondani 1859

Description: Female adults look like half-sized house flies, 5 mm long, with grey pointed abdomens (Fig 4.44). They lay fusiform-shaped eggs covered by net-like surface ornamations. Pupae are spindle-shaped and brown. Maggots are 5–7 mm long, white, with pointy heads and blunt posteriors, and the posteriors sport a pair of unforked tubercles.

Life History & Host Range

D. platura overwinters as pupae; adults emerge in early spring. Females lay 100 eggs, a few at a time, in disturbed soil. Hatching maggots burrow into ungerminated seeds and young seedlings. Seeds are often completely destroyed. Seedlings become honeycombed with slimy brown tunnels. Damaged

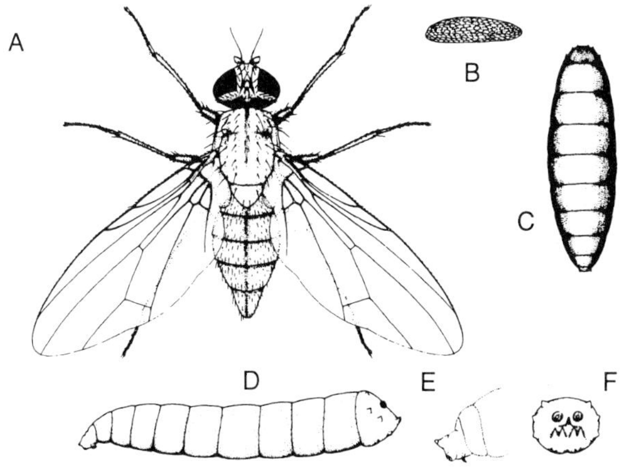

Figure 4.44: *Delia platura*, enlarged x6. A. Female adult; B. Reticulate egg; C. Puparium; D. Larva; E. Close up of larval head; F. Close up of larval posterior (from Hill 1994).

roots often rot from invading soil fungi. Plants, if they survive, become yellow and stunted. They wilt during the day but recover at night. *D. platura* pupates 2–4 cm under the soil near damaged plants; two to five generations arise per year.

D. platura infests hemp in Japan (Harukawa & Kondo 1930, Shiraki 1952), and Europe (Rataj 1957). A *Delia* species has also been reported in India (Bantra 1976). *D. platura* is cosmopolitan. It usually feeds on decaying vegetable matter, but attacks a wide range of crops (Hill 1983). Germinating seeds of corn and bean are particularly vulnerable to attack, especially in cold, wet springs when seed germination is slow (Howard *et al.* 1994).

4. CABBAGE MAGGOT

Delia radicum (Linnaeus) 1758, Diptera; Anthomyiidae

=*Chortophila brassicae* Wiedemann 1817, =*Chortophila brassicae* Bouché 1833

Description: Female flies have dark grey bristly bodies, 7 mm long, with black stripes on their thoraxes. They lay fusiform-shaped eggs covered by longitudinal striations. Cabbage maggots are spike-shaped, with pointed heads and wide flat posteriors, and the posteriors have paired tubercles that are forked at their apex.

Life History & Host Range

In early spring adults emerge from overwintering pupae 2–10 cm underground. Females lay eggs in soil abutting plant roots. Larvae hatch within a week and burrow into roots. After feeding for several weeks, larvae pupate in the plant or in nearby soil. By late June the next generation of flies emerge to lay eggs. A third generation may emerge in autumn. Because maggots thrive in cool moist weather, only the first generation causes much damage.

According to Goriainov (1914), *D. radicum* destroyed 40% of a hemp crop in the former USSR. The cabbage maggot normally attacks Crucifers; Yepsen (1976), in fact, suggested using *Cannabis* to repel *D. radicum.* This Old World pest was introduced into North America about 100 years ago. Cabbage maggots can kill 95% of unprotected cabbage crops north of the Mason-Dixon Line (39°40' latitude).

DIFFERENTIAL DIAGNOSIS

Galls can be caused by European corn borers, hemp borers, and assorted stem-boring beetles and weevils. Some fungi and bacteria also cause swellings confused with insect galls (e.g., Fusarium canker). Root maggots cause wilt symptoms similar to those caused by white grubs, nematodes, and root-rotting fungi. Their symptoms may also be confused with those caused by damping-off fungi. Digging up plants or seeds reveals the tiny maggots.

CULTURAL & MECHANICAL CONTROL

(method numbers refer to Chapter 9)

To discourage root maggots, observe methods 1 (sanitation), 2a (deep ploughing), 2b&c (sterilizing or pasteurizing soil), 3 (eliminate cabbage-family weeds), 4 (plant *late* to avoid the first wave of egg-laying *D. radicum* flies, about two or three weeks after mustard weeds flower), 7a (maintain proper soil moisture), 12c (light traps), and 13 (mechanical barriers). Protect seedlings from pregnant flies with a tent of gauze or finely-woven screen (20–30 threads to the inch). When plants outgrow tents, fit a 15 cm diametre disk (made of felt, foam-rubber, or tar paper) around the base of each stem. Plastic sheeting mulch also works if the sheets are pulled tight around the base of stems. This prevents flies from laying eggs in soil next to plants. Disks and plastic mulch also provide a humid microhabitat for predatory ground beetles, and the conserved soil moisture permits plants to better tolerate minor infestations without wilting. Organic growers pour a ring of wood ashes around seedlings, but this must be repeated after every rainstorm. Bait sticky tape and water traps with allylisothiocyanate to lure flies to their deaths. Although these traps will not provide complete control, growers may monitor traps to determine when other controls are needed.

BIOCONTROL (see Chapter 10)

Parasitic nematodes control root maggots; *D. radicum* was controlled by applying *Steinernema feltiae* (described under fungus gnats), or *Heterorhabditis bacteriophora* and *Steinernema riobravis* (see white root grubs and flea beetles). The rove beetle *Aleochara bilineata* has been used against *Delia* species (described below). Not-yet-commercially available controls include maggot predators (***Strongwellsea castrans*** and ***Trybligrapha rapae***), and the fungus ***Entomophthora muscae***, which infects many types of flies (see Plate 40). A companion crop of basil repels adult flies (Israel 1981). Marigolds, parsnips and onions also deter the maggots.

Aleochara bilineata

BIOLOGY: A predatory rove beetle whose larvae are parasitic. The species occurs in North America and Eurasia in temperate regions.

APPEARANCE & DEVELOPMENT: Adults have elongate, glossy black bodies 3–6 mm long, with small reddish-brown forewings. They look like earwigs without rear pinchers. Females lay up to 700 eggs in soil around root maggot-infested plants. Larvae hatch in five to ten days, and actively search for root maggot pupae. Larvae chew into maggot pupae and complete their life cycle therein, emerging as adults after 30–40 days. Adults eat up to five maggots per day. Two generations usually arise per year.

APPLICATION: *A. bilineata* has been mass reared in the former Soviet Union, Europe, and Canada (Hoffmann & Frodsham 1993). Some strains are tolerant of insecticides.

CHEMICAL CONTROL (see Chapter 11)

Eliminate root maggots with neem and horticultural oil, applied as soil drenches.

GRASSHOPPERS & THEIR ALLIES

The order Orthoptera includes grasshoppers, locusts, crickets, and cockroaches. Differentiating grasshoppers from locusts is like separating mushrooms from toadstools. These moderately large and elongated insects have prominent eyes, jaws, and large hind legs (Fig 3.5 & Plate 42). They may be winged or wingless. Not all Orthoptera are evil—Praying mantids (family Mantidae) are beneficial carnivorous insects. As a corollary, not all evil pests infest *Cannabis;* Rothschild *et al.* (1977) fed a swarm of locusts (***Schistocerca gregaria*** Forsk.) nothing but *Cannabis,* and the diet proved lethal.

SIGNS & SYMPTOMS

Grasshoppers, locusts, and crickets eat large, round, smooth-edged holes in leaves. In heavy infestations, Lassen (1988) reported plants being stripped to stalks in a matter of days. Swarms of locusts in western Africa do the same in a matter of minutes (Clarke, pers. commun. 1994, see Plate 41). Clarke also reported large grasshoppers in Mexico biting through stems of young seedlings to topple them so the pests could easily feed on leaves. Grasshoppers in Indiana chewed through stems and branches, causing limb tips to topple. In Canada, grasshoppers destroyed apical meristems, stunting plants and severely reducing seed production (Scheifele 1998). Crickets feed on young seedlings, causing cutworm-like damage. Mole crickets feed on roots.

1. TWO-STRIPED GRASSHOPPER

Melanoplus bivittatus (Say) 1825, Orthoptera; Acrididae

Description: These short-horned grasshoppers are members of the "spur-throated" subfamily of Acridids, they sport a short tubercle protruding from between their front legs. Females coloured olive-brown to brownish-yellow above, pale yellow below, without spots but with a distinct pale stripe on each side of the head, beginning behind the eyes and extending down the back. Hind tibiae often red or purple-brown with black spines. Females average 29–40 mm long, males average 23–29 mm.

Life History & Host Range

M. bivittatus overwinters underground in the egg stage. Each female lays one or two clusters of eggs. Southern populations reach maturity by early June, northern populations appear by late July. The species ranges across southern Canada from Newfoundland to British Columbia, south to Mexico in the west and North Carolina in the east; *M. bivittatus* predominates in the Great Plains, where it has infested hemp in Manitoba (Moes, pers. commun. 1999). Two-striped grasshoppers frequent grass pastures and prairies, as well as clover fields. The pest occasionally migrates in masses, "eating the choicest of everything." A related species, the migratory grasshopper (***Melanoplus sanguinipes*** Fabricius), also infests hemp in Manitoba.

2. SPRINKLED LOCUST

Chloealtis conspersa Harris 1841, Orthoptera; Acrididae

Description: These short-horned grasshoppers are members of the "slant-faced" subfamily of Acridids, their heads look streamlined. Females vary in colour from dull yellow to dark brown, with front wings speckled by small black spots. They average 20–28 mm in length. Males are 15–20 mm long, light brown with a black bar across their thorax, and lack sprinkled tegmina. Both sexes sport yellow to red hind tibiae, and antennae 10–12 mm long.

Life History & Host Range

C. conspersa overwinters in the egg stage, underground, in masses of 15–50 eggs encased by a hardened gummy substance. Sprinkled locusts make their home in grassy thickets alongside pastures, fields, and streams. They live east of the Mississippi, from the Canadian border down to the mountainous regions of Virginia and North Carolina. Males produce a familiar *tsikk-tssikk-tssikk* song. Bush Doctor (unpublished data 1990) found this species on marijuana in New York, and DeWitt (unpublished) collected *C. conspersa* from feral hemp in Illinois.

3. CLEARWINGED GRASSHOPPER

Camnula pellucida (Scudder) 1862, Orthoptera; Acrididae

Description: These small short-horned grasshoppers are members of the third Acridid subfamily, the "band-winged" grasshoppers—they do not have spurred throats or slanted faces. This particular species doesn't even have banded wings. Females are coloured brownish-grey with black markings until they become sexually active, and turn bright yellow. Forewings are mottled with light stripes along their angles, hindwings are transparent. Females average 21–25 mm long; males average 17–21 mm. Nymphs are strikingly coloured black with white to tan markings. Eggs are 4 mm long, laid in clusters of ten to 30.

Life History & Host Range

C. pellucida overwinters in the egg stage, underground, usually in bare ground, often along roadsides. This species emerges and matures earlier than the two aforementioned species. *C. pellucida* ranges across western North America from northern Mexico to northern Alberta; east of the Great Plains the species inhabits land along the USA–Canadian border. It has infested hemp in Manitoba (Moes, pers. commun. 1999). *C. pellucida* frequents grassy pastures on high ground. The nymphs sometimes migrate in swarms, causing trouble. This small species has a big appetite: populations greater than 25 per m^2 can completely denude rangeland forage grasses.

4. STINK GRASSHOPPER

Zonocerus elegans (Thunberg) 1773, Orthoptera; Acrididae

Description: Adults are handsome hoppers, mostly dark green with bold stripes of black, yellow, and orange, 35–55 mm long. Nymphs are black with appendages ringed with yellow or white. Eggs are sausage shaped, 6 mm long, and laid in the soil within spongelike masses of dried froth about 2.5 cm wide.

Life History & Host Range

Rothschild *et al.* (1977) fed *Z. elegans* nymphs a diet of pure *Cannabis* for six weeks, and most survived. Rothschild demonstrated that *Z. elegans* sequestered some cannabinoids in body tissues (as a predator deterrent?), and excreted the rest in frass. This colourful, aposematic grasshopper exudes an unpleasant odour when handled. Hill (1983) called it the skink grasshopper, Sorauer (1958) gave it the sinister monicker *Stinkschrecke*. The pest ranges across all of Africa below the Sahara, and feeds on many dicot crops in the seedling stage, especially cassava and finger millet (Hill 1983). One generation arises per year.

5. CITRUS LOCUST

Chondracris rosea (DeGeer) 1773, Orthoptera; Acrididae

Description: Adults are yellow to greenish brown, with rosy-violet wings. They have red hind femurs with yellow spines. Body length averages 64–85 mm (females) and 49–60 mm (males).

C. rosea attacked mature hemp crops in Taiwan (Sonan 1940). Citrus locusts damage many crops, and range from northern India to Japan.

6. OTHER GRASSHOPPERS & LOCUSTS

Lassen (1988) described unidentified grasshoppers attacking marijuana in California. Siegel (1989) cited unidentified grasshoppers attacking feral hemp in the Midwest. In South America, ***Dichroplus maculipennis*** Blanch 1851 (Orthoptera; Acrididae) caused "intense" damage on hemp (Liebermann 1944), despite predation by birds and flesh flies (Diptera; Sarcophagidae).

Atractomorpha crenulata Fabricius 1793 (Orthoptera; Acrididae) was the most common grasshopper pest in India (Bantra 1976). Sorauer (1958) called *A. crenulata* the Tobacco grasshopper, which ranges from India to Malay. Adults are spotted green with reddish wings and abdomens, 24–35 mm long. Other Asian pests include ***Atractomorpha bedeli*** Bolivar 1884 in Japan (Shiraki 1952), ***Hieroglyphus nigrorepletus*** Bolivar in India (Roonwal 1945), and ***Chrotogonus saussurei*** Bolivar 1884 in India (Cherian 1932).

In Italy, Goidànich (1928) cited ***Tettigonia cantans*** (Fuessly) 1775 and two uncommon Tettigoniidae, ***Tettigonia ferruginea*** and ***Tettigonia orientalis***. These long-horned grasshoppers are related to katydids. Clarke (pers. commun. 1994) reported minor damage by katydids in California, Indiana, and western Africa.

4. CRICKETS

Crickets (family Gryllidae) have shorter legs and longer antennae than grasshoppers and locusts. Most species are black and produce familiar nocturnal chirping sounds. Selgnij (1982) described unidentified crickets attacking marijuana seedlings. Rataj (1957) cited the house cricket ***Acheta domesticus*** (Linnaeus) 1758 infesting Czech hemp. *A. domesticus* is a European species introduced into North America by early settlers. Adults are shiny with dark crossbars and light-coloured heads, 10–20 mm long (Fig 4.45). House crickets live outdoors until the weather turns cold in the autumn.

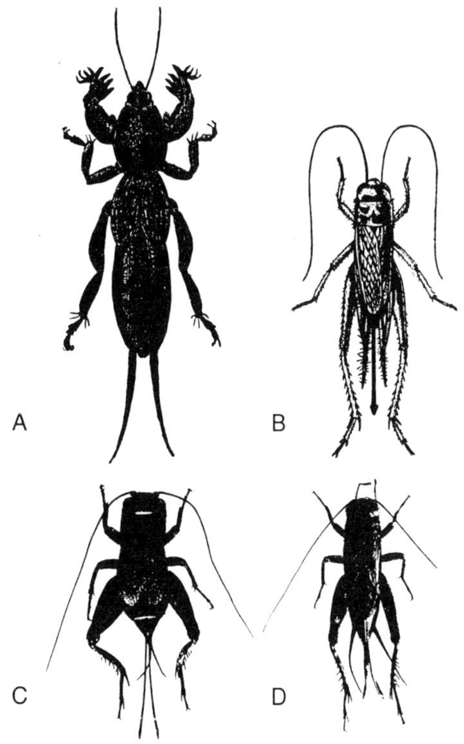

Figure 4.45: Adult crickets (all approximately life size).
A. *Gryllotalpa hexadactyla* (from Comstock 1904);
B. *Acheta domesticus* (from Ill. Nat. Hist. Surv.);
C. *Gryllus desertus;* D. *Gryllus chinensis.*
(C & D from Anonymous 1940)

Gryllus desertus Pall 1771 and ***Gryllus chinensis*** Weber 1801 are reported in Italy (Goidànich 1928, Martelli 1940, Ragazzi 1954, Ferri 1959) and the Czech Republic (Rataj 1957). These species are the same size as house crickets (14–18 mm long), but darker in colour (Fig 4.45).

The mole cricket, ***Gryllotalpa gryllotalpa*** (L.), attacks hemp in Europe (Barna *et al.* 1982, Gutberlet & Karus 1995). Clarke (pers. commun. 1994) reports mole crickets in Africa (***Gryllotalpa gryllotalpa*** or ***Gryllotalpa africana*** Pal.) and California (***Gryllotalpa hexadactyla*** Perty). Mole crickets are distinguished by their broad, shovel-like front legs (Fig 4.45). They burrow underground and eat roots and tubers.

The tree cricket ***Oecanthus indicus*** Saussure 1878 has been reported in India (Bantra 1976). Tree crickets are pale green. Females damage plants by splitting open stems with their long, strong ovipositors. *O. indicus* ranges from India to Java to Taiwan. A related species, ***Oecanthus celerinictus*** Walker, damaged Mississippi marijuana in late summer and early autumn (Lago & Stanford 1989).

5. COCKROACHES

Frank & Rosenthal (1978) described unidentified cockroaches attacking marijuana. They were probably ***Blattella germanica*** (L.) or ***Supella longipalpa*** (F.). Dass Baba (1984) described unidentified Hawaiian cockroaches eating *Cannabis* seeds in peat pots.

DIFFERENTIAL DIAGNOSIS

Grasshopper and locust damage can be confused with damage caused by caterpillars or flea beetles. Damage caused by crickets may be confused with cutworms or damping-off fungi. Root damage caused by mole crickets may be confused with root maggots or root grubs.

CULTURAL & MECHANICAL CONTROL

(method numbers refer to Chapter 9)

Observe methods 2a (deep ploughing), 3 (eliminate grassy weeds), 6 (do not rotate after pasture), 7a (avoid drought), and 9 (hand removal). Use method 13 outdoors—screened tents and row covers will keep hoppers away from tender seedlings.

BIOCONTROL

Nosema locustae is commercially available, *Metarhizium flavoviride* is registered in Africa, and *Entomophthora grylli* is being developed (all described below). *Beauveria bassiana* also kills grasshoppers (described under whiteflies). Biocontrol organisms must tolerate high temperatures, because sick grasshoppers instinctively cling to the tops of plants, seeking bright sunshine ("summit disease syndrome"), and can raise their temperatures to 40°C. To control species that live in the soil, such as mole crickets, try beneficial nematodes (*Steinernema* and *Heterorhabditis* species, described under cutworms and white root grubs, and *Steinernema scapterisci*, described below).

Lassen (1988) knocked grasshoppers to the ground by spraying *Cannabis* with water two or three times a day, then his chickens finished the job. Chickens also scratch eggs out of the ground and eat them. Almost all birds feed on hoppers (except vegetarians such as doves and pigeons). Other natural enemies include spiders, flesh flies, snakes, toads, and assorted rodents (field mice, ground squirrels).

Nosema locustae

BIOLOGY: A protozoan that parasitizes grasshoppers, Mormon crickets, black field crickets, and pygmy locusts (Locucide®, Semispore Bait®; Evans discontinued Nolo Bait® in 1993). *N. locustae* does best in temperatures above 15°C. According to Hunter-Fujita *et al.* (1998), the spores from a single grasshopper cadaver can be used to treat 4 ha.

APPLICATION: Supplied as 7.5% concentrate, 0.05% powder, and in a wheat bran bait. It can be stored for 13 weeks in a dry, cool (8–10°C), dark place. Cherim (1998) recommended *N. locustae* when grasshopper populations exceed 40 m^{-2}. He spread bait around plants at a rate of 25 bran flakes per square foot (≈270 flakes m^{-2}). Distribute bait in the morning after the temperature reaches 16°C, and when no rain is forecasted (rain or heavy dew makes the bran unpalatable to pests).

NOTES: After ingesting *N. locustae*, grasshoppers stop feeding, develop discoloured cuticles, and die a slow death. Because of this slow death, the parasite is more effective for long-term suppression of grasshopper populations than stopping sudden outbreaks. Once established in a grasshopper population, *N. locustae* spreads via cannibalism of infected individuals. *N. locustae* is compatible with all other biocontrols. A related species, ***Nosema acridophagus***, kills grasshoppers quicker and the USDA is developing it.

Metarhizium flavoviride

BIOLOGY: This soil fungus (Green Muscle®) has been used against grasshoppers and locusts in Africa and the tropics. It is closely related to *Metarhizium anisopliae* (described under aphids).

APPEARANCE & DEVELOPMENT: *M. flavoviride* is a hyphomycete with branching, phialidic conidiophores and simple oval spores (conidia). Conidia in contact with insects quickly germinate and grow into their hosts. Infected insects stop feeding, then die in four to ten days, depending on the temperature. In humid conditions, *M. flavoviride* reemerges from dead hosts to sprout more green conidia and repeat the life cycle.

Entomophthora grylii

BIOLOGY: A fungus that infests grasshoppers and locusts. It is native to North America and Africa, and does well in moderate temperatures. This fungus produces sticky, mucus-covered ballistospores, which are forcibly discharged from dead insects. Resting spores arise within cadavers and are mass-produced in liquid media fermenters. A related pathotype from Australia, ***Entomophthora praxibuli***, has a broader host range. *E. praxibuli* was released in North Dakota to suppress rangeland grasshoppers, but the fungus died out after several winters (Bidochka *et al.* 1996).

APPLICATION: *Entomophthora* spores directly penetrate grasshopper skin—spores do not have to be eaten, just touched. Hoppers infected with the fungus climb to the tops of elevated surfaces, instinctively seeking sunshine. They hang on for dear life and usually expire in the late afternoon, seven to ten days after infection (Fig 4.46).

Melanoplus sanguinipes entomopoxvirus

BIOLOGY: The MsEPV is a DNA virus being tested against grasshoppers in the western USA. The cadavers of 35 infected grasshoppers are required to produce a dose sufficient for 1 ha of crop area (Hunter-Fujita *et al.* 1998).

Figure 4.46: Grasshopper infected with the biocontrol fungus *Entomophthora praxibuli* (courtesy USDA).

Steinernema scapterisci

BIOLOGY: This species of *Steinernema* (Otinem S®, Proact®) was discovered in the 1980s. It parasitizes mole crickets (*Scapteriscus* species), house crickets (*Acheta domesticus*), and field crickets (*Gryllus* species). The gut of *S. scapterisci* contains mutualistic bacteria (*Xenorhabdus* species). *S. scapterisci* lives in the USA and temperate South America, and does best in soil temperatures between 15–33°C.

NOTES: Infective juveniles of *S. scapterisci* move through the soil (although they are also described as sedentary ambushers), and enter hosts through mouths or spiracles. The life cycle is similar to that of other *Steinernema* species. Its commercial availability varies.

CHEMICAL CONTROL

Dusting plants with diatomaceous earth or spraying plants with neem deters most grasshoppers and crickets. Yepsen (1976) used sabadilla dust against grasshoppers and pyrethrum against crickets. Hydroprene is a synthetic growth hormone targeted against cockroaches. Boric acid kills crickets and cockroaches. Sticky baits made from wheat bran, similar to cutworm baits, attract all Orthopterans. Poisoning baits with pesticides works even better.

FLOWER FLIES

Black and yellow-striped syrphid flies, also called hover flies, resemble wasps and bees. They hover over flowers to sup nectar and serve as pollinators (second in importance behind bees). They do not sting or bite. Larvae of some flower flies feed voraciously on aphids, leafhoppers, and mealybugs. Schmidt (1929) described large numbers of syrphid larvae infesting stored hemp seed in Brandenbug. He noted the preceding summer's crop was heavily infested with aphids, which led to a great syrphid boom until autumn, when the aphid population crashed. Remaining syrphid larvae, unable to pupate, hibernated in *Cannabis* flowers, ending up in seed during threshing.

A few syrphid species feed on plants. Datta & Chakraborti (1983) collected flower flies from *Cannabis* in northern India: ***Metasyrphus latifasciatus*** (Macquart) 1829, ***Episyrphus balteatus*** (DeGeer) 1776, ***Ischiodon scutellaris*** (Fabricius) 1805, ***Syritta pipiens*** (Linnaeus) 1758, ***Sphaerophoria scripta*** (Linnaeus) 1758, and unidentified adults of ***Ischyrosyphus*** species and ***Melanostoma*** species.

CRANE FLIES & FUNGUS GNATS

Larvae of crane flies and fungus gnats infest plant roots close to the soil surface. Infested plants lose vigour and colour. Close inspection reveals small surface scars on roots, and fine root hairs are eaten away. Root wounds may serve as portholes for pathogenic fungi, and the pests may transmit soilborne pathogens, notably *Pythium* and *Fusarium* species (Howard *et al.* 1994).

Crane fly and fungus gnat larvae attack plants that are stressed by nutrient imbalances or waterlogging. They prefer feeding on decaying vegetation and fungi in damp soil. Fungus gnat larvae also feed on the green algae that covers damp rockwool. The adults do not feed on plants, but may become trapped in the sticky resin of mature flowers and are aesthetically displeasing. Crane flies are common outdoors, fungus gnats dominate indoors.

1. CRANE FLIES

Adult crane flies resemble gigantic mosquitoes or "daddy-longlegs with wings." The adult's feeding habits are unknown, but they do not bite humans. Bantra (1976) caught an unidentified *Tipula* species in marijuana. Bovien (1945) reported the European crane fly, ***Tiplula paludosa*** Meigen 1830 (Diptera; Tipulidae), attacking Danish hemp crops. Watson (pers. commun. 1998) found *T. paludosa* infesting a glasshouse in the Netherlands. *T. paludosa* larvae usually feed on grasses. The maggots, known as leatherjackets, are pink to greyish-black, with black heads, 35–40 mm long. Adults have grey-brown bodies (up to 25 mm long) and delicate milky-white wings (Fig 4.47). The species was introduced into the Province of British Columbia and Washington State around 1955.

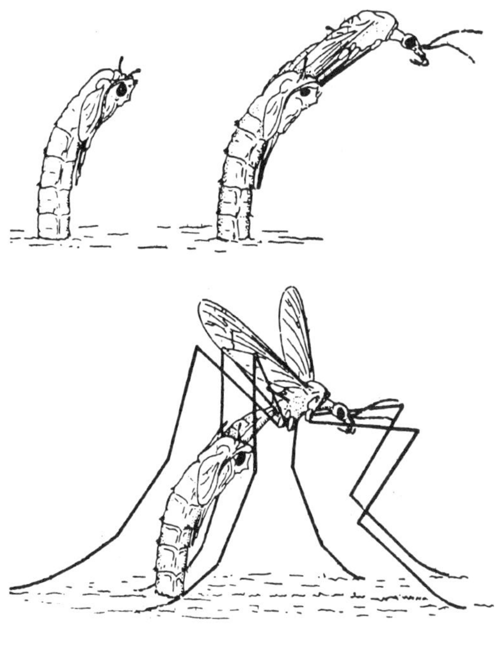

Figure 4.47: An adult crane fly (*Tiplula* species) emerging from its pupal case (courtesy USDA).

2. FUNGUS GNATS

Fungus gnats come from two fly families, the Mycetophilidae and Sciaridae. Bantra (1976) caught a ***Bradysia*** species (Diptera; Sciaridae) buzzing around Indian *Cannabis*. Arnaud (1974) found *Bradysia* species infesting confiscated stocks of midwestern marijuana. An anonymous author reported fungus gnats attacking a drug cultivar ('William's Wonder') in Ontario, Canada (*High Times* No. 233, p. 77, 1994). For some reason, plants growing in rockwool and sterile potting media are more susceptible than plants growing in soil (Howard *et al.* 1994).

Bradysia larvae are slender maggots with translucent bodies and black heads, growing to 4–5 mm in length. They pupate in soil. *Bradysia* adults are small, delicate, hump-backed flies, with long gangly legs and beaded antennae, grey to black in colour, and 2–4 mm long. They are poor fliers and usually come to one's attention when plants are moved or shaken. Fungus gnats can be confused with adult *Aphidoletes aphidimyza* (aphid biocontrols). Females lay up to 200 eggs on soil or rockwool in seven to ten days. New generations arise on a monthly basis and overlap in warm climates and warm glasshouses.

CULTURAL & MECHANICAL CONTROL

Observe methods 2b&c (sterilize or pasteurize soil), 3

(weeding), 7b (avoid overwatering), 7c (avoid excess humidity), and 8 (cover exposed soil with a layer of perlite). Method 12b works against fungus gnats. Sticky cards laid horizontally near the soil surface catch more gnats than cards hung vertically (Gill & Sanderson 1998). A grower from Ontario killed flying adults by aiming a 1500-watt hair dryer at them.

BIOCONTROL

A parasitic wasp, ***Synacra pauperi***, often arises in unsprayed greenhouses infested with fungus gnats (Gill & Sanderson 1998). It is not commercially available. Leslie (1994) reported an 87% reduction in *Bradysia* populations by inoculating soil with the predatory nematode *Steinernema glaseri* (see white root grubs), but *Steinernema feltiae* works better (see below). Other growers have reported excellent control with Bt-i and *Hypoaspis* (described below).

Bacillus thuringiensis variety ***israelensis*** Bt-i

BIOLOGY: A strain of the Bt bacterium (Vectobac®, Gnatrol®, Bactimos®) lethal to larvae of fungus gnats and other flies. Apply to the surface of soil or rockwool. Bt-i is compatible with nematodes (*Steinernema* species) and *Hypoaspis miles*. It may cause significant mortality in nontarget mayfly and dragonfly species (Howarth 1991). For a full description of Bt, see European corn borers.

Hypoaspis (Geolaelaps) miles

BIOLOGY: A nocturnal soil mite that preys on fungus gnat larvae, springtails, thrips pupae, and other small soil insects. It is native to the northern USA, and does best in moist (not soaked) soil between 15–30°C. A related species, ***Hypoaspis aculeifer***, feeds on eggs as well as larvae (Ravensberg, pers. commun. 1999).

APPEARANCE & DEVELOPMENT: Adults are brown, oblong, and slightly larger than *Phytoseiulus persimilis*, reaching 0.8 mm in length (Plate 43). They inhabit the top layer of soil (1–3 cm), and also colonize the surface of rockwool. Adults and nymphs feed on pests, consuming about five pests per day. When pests are not present, *H. miles* survives on algae and plant debris. The life cycle takes ten days at 30°C and 46 days at 15°C. *H. miles* does *not* diapause.

APPLICATION: Supplied as adults in shaker bottles or peat pots. Bottles of *H. miles* supplied with a food source (*Tyrophagus putrescentiae*) may be stored for up to two weeks in a cool, dry place. Mites without a food source should be released immediately. Used preventively against fungus gnats, release 100 adults per m^2 soil area. For light infestations, double the rate. For heavy infestations, release 500 per m^2. To control thrips, release 200–500 per m^2. *H. miles* is compatible with Bt, *Beauveria bassiana*, *Aphidoletes aphidimyza*, and beneficial nematodes (*Steinernema* species). Thanks to selective breeding and a primarily subterranean lifestyle, *H. miles* tolerates neem, imidacloprid, kinoprene, and many fungicides (Cherim 1998).

Steinernema feltiae

BIOLOGY: This nematode is marketed for control of sciarid maggots, but kills a variety of soil insects (Entonem®, Exhibit SF®, Magnet®, Nemasys-M®, Nema-plus®, Sciarid®, Traunem®) . Dutky's experiments with *S. feltiae* in the 1950s rekindled interest in nematodes as biocontrol agents.

DEVELOPMENT: *S. feltiae* exhibits the same lifecycle as other *Steinernema* species, described in the section on white root grubs. *S. feltiae* locates its prey using a strategy midway between an ambusher and a cruiser.

APPLICATION: Supplied as juvenile nematodes in polyethylene packs containing 300 million individuals, enough to treat 100 m^2 of soil area. Reapply at six week intervals. Most strains of *S. feltiae* are compatible with chlorine bleach and formaldehyde, provided they are not *tank mixed* with these pesticides.

NOTE: Some experts claim *Bradysia* maggots are too small for nematodes to reproduce within, so nematodes must be reapplied whenever fungus gnats resurge. Gill & Sanderson (1998) provided photographic evidence to the contrary. *S. feltiae* tolerates cold soil temperatures better than most nematodes (down to 10°C). But it tolerates shipping and handling less than other *Steinernema* species.

CHEMICAL CONTROL

Neem and/or insecticidal soap kill fungus gnats when applied as a soil drench. Frank (1988) recommended one or two applications of a rotenone soil drench.

SAWFLIES

Sawflies are not flies, they are wasps. Two have been reported on *Cannabis*. One eats leaves, the other bores stalks.

1. HEMP SAWFLY

Trichiocampus cannabis Xiao & Huang 1987, Hymenoptera; Tenthredinidae.

Description: Larvae resemble pale caterpillars with dark heads, with three pairs of real legs and six to eight pairs of hookless prolegs (real caterpillars have five or less pairs of prolegs, bearing tiny hooks or crochets). Larvae grow to 10 mm in length. Adult wasps are 5.5–6.8 mm long with wingspans averaging 16 mm (Fig 4.48). Eggs are white, oblong, up to 1 mm long.

Takeuchi (1949) previously cited "*Trichiocampus cannabis* Takeuchi, in litt." regarding a species infesting Japanese hemp. But we could not locate his taxon in subsequent literature.

Life History & Host Range

Wang *et al.* (1987) discovered the hemp sawfly near Liuan (Anhui Province, China). *T. cannabis* overwinters as mature larvae in soil. Pupation takes five to seven days. Adults live a week in the spring, long enough to lay eggs. Eggs hatch in four to seven days. Larvae undergo five moults before maturity, which takes 27–32 days. Two generations arise per year. Hemp sawflies selectively feed on *Cannabis*. The larvae skeletonize leaves or rip holes and rag leaf edges.

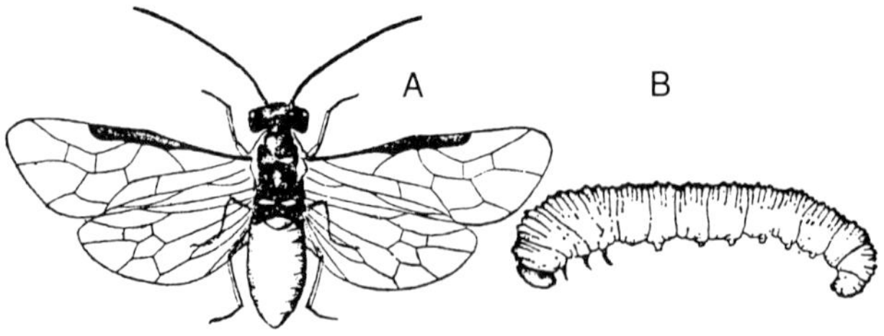

Figure 4.48: *Trichiocampus cannabis*. A. Adult; B. Larvae (from Wang *et al.* 1987).

2. PURSLANE SAWFLY

Schizocerella pilicornis Holmgren 1868, Hymenoptera; Argidae.

Description: Larvae are small and stout-bodied, body pale green with a darker green dorsal stripe, head capsules yellow-green and up to 1.3 mm wide, body length 11.3 mm. Pupae in cocoons composed of tan, silken fibres, length 6 mm, buried 2.5–5.0 cm under the soil surface. Adult wasps black with dark brown thorax, light brown legs, 5–6 mm long with a wingspan of 10 mm.

Life History & Host Range

S. pilicornis infests *Cannabis* in Mississippi, where it occurs abundantly from early May to the middle of August (Lago & Stanford 1989). The larvae act as stem borers (Sands *et al.* 1987). Previously, *S. pilicornis* was described as a host-specific pest of common purslane (*Portulaca oleracea* L.), where it acts as a leafminer (Gorske *et al.* 1976). At 24°C, the egg stage lasts 80–100 hours, larvae mature in 130 hours, pupation takes 200 hours, and the adult wasps live 25 hours (Gorske *et al.* 1977). Female wasps deposit eggs in slits at edges of leaves, generally one egg per leaf, a total of 20 eggs.

CONTROL

In China, 15–30% of *T. cannabis* are killed by ladybeetles, lacewings, and parasitoids. *Neodiprion sertifer* NPV (Virox®, Sertan®) controls *pine* sawflies, but these pests are not in the same family as *T. cannabis* (for more on NPV, see the section on budworms). Wang *et al.* (1987) killed 90–100% of young *T. cannabis* larvae with a variety of synthetic pesticides. Yepsen (1976) controlled other sawflies with rotenone and hellbore. Neem also works.

In the USA, wild populations of *S. pilicornis* are culled by a microsporidium, *Nosema pilicornis* Gorske & Maddox. *S. pilicornis* larvae are completely killed by malathion and carbaryl, and Bt causes 50% mortality (Gorske *et al.* 1976).

WIREWORMS

Wireworms are the grubs of click beetles. In German, *Saatsschnellkäfer.* Grubs feed on roots and freshly-sown seeds. Wireworms attack many plants, especially grasses (corn, lawn, etc.) in poorly-drained soil.

LINED CLICK BEETLE

Agriotes lineatus Linnaeus 1758, Coleoptera; Elateridae.

Description: Adults are light brown with lined electra, cylindrical, 7.5–11 mm long. Grubs are hard, jointed, yellow brown, covered by minute hairs, with three pairs of legs behind the head, up to 25 mm long (Fig 4.49).

Life History

This European species also lives in north Africa and the Americas. Gutberlet & Karus (1995) reported *A. lineatus* on German hemp. The species lives its life cycle underground, taking about four years. A related European species, ***Agriotes obscurus*** (L.), has become established in British Columbia (Howard *et al.* 1994). Wireworms can be confused with millipedes.

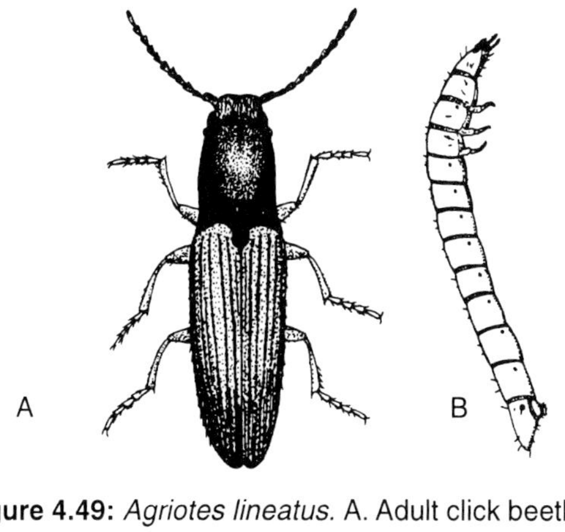

Figure 4.49: *Agriotes lineatus.* A. Adult click beetle; B. Larval wireworm (from Hill 1994).

CONTROL

Gutberlet & Karus (1995) recommended biocontrol with *Metarhizium anisopliae.* Beneficial nematodes may work (*Heterorhabditis* and *Steinernema* species, described under white root grubs). Wheat is used as a trap crop for wireworms in Switzerland (Hokkanen 1991). Yepsen (1976) suggested trapping wireworms with potatoes. Skewer potatoes on a stick, then bury them 10 cm underground, leaving the stick as a marker. Set traps 1–3 m apart. After a couple of weeks, pull up potatoes full of wireworms. Carrot slices, untreated corn seed, and mesh bags filled with 30 g of whole wheat flour also work.

SPRINGTAILS

Springtails are small, simple, and surprisingly abundant insects. They are seldom seen because of their tiny size and retiring habits in concealed soil habitats. Most springtails feed on decaying plant material, fungi, and bacteria. A few species, such as the garden springtail, ***Bourletiella hortensis*** (Fitch), damage plants in greenhouses. Another species, confusingly called a "flea," causes stippling of leaves, much like spider mites.

LUCERNE FLEA

Sminthurus viridis Linnaeus 1758, Collembola; Sminthuridae.

Description: Adults are yellow-green, wingless, globular in shape, up to 3 mm long, with a large head in relation to their body.

Life History

This European species has recently invaded Australia, where it infests hemp (Ditchfield, pers. commun. 1997). Early-sown crops suffered serious infestations, whereas later sowings were unaffected. *S. viridis* is a serious pest of alfalfa (lucerne), where up to 6000 pests may be found per square foot (Hill 1983). The pest also infests clover ("clover springtail") and various cereals.

CONTROL

Biocontrol can be achieved with *Hypoaspis miles,* a mite that preys on small soil insects (described under fungus gnats). Chemical controls have not been elucidated.

EARWIGS

Beetle-like in appearance, these insects are distinguished by a pair of prominent forceps-like cerci at their tail end. Although beneficial (they eat aphids and other pests), omnivorous earwigs can become garden pests. Large males can inflict a painful pinch with their cerci. They are nocturnal.

EUROPEAN EARWIG

Forficula auricularia Linnaeus 1758, Dermaptera; Forficulidae.

Description: Adults are dark reddish-brown, 10–15 mm long, and backed by the aforementioned cerci (Fig 4.50). Females lay round, pearly-white eggs in underground masses. Nymphs are pale brown, and their wings and cerci are much reduced or absent.

Earwigs overwinter as adults in small underground nests. Eggs are laid in late winter and females carefully guard their brood until they hatch around May. Earwigs infest all ages of *Cannabis* and all parts of *Cannabis,* including seedlings in California (Frank 1988), and flowering tops in India (Bantra 1976). Earwigs have been found in the stem galleries of European corn borers in Italy (Goidànich 1928). Earwigs can fly, but not far. To travel distances they must take off from a high place with a good tail wind. They are frequent pests of urban gardens.

CONTROL

Culpepper (1814) repelled earwigs with the juice of crushed *Cannabis* seeds. Frank & Rosenthal (1978) trapped earwigs by placing pieces of cardboard around the garden. During the day, earwigs gather in cool, moist, dark areas—find them under the cardboard. Frank (1988) eliminated earwigs with a rotenone soil drench. Puritch (1982) reported 100% mortality of earwigs with insecticidal soap sprays. Poisoned baits, the kind described for cutworms, work against earwigs if poisoned with boric acid.

Figure 4.50: Female *Forficula auricularia* in winter quarters with her eggs (from Fulton 1924).

"Achlya *species (water molds) exhibit a sexual ambivalence in which maleness and femaleness are determined in each mating by common consent of the mated.*"

—J. R. Raper

Chapter 5: Fungal Diseases

At least 8000 species of fungi attack plants (Cook & Qualset 1996). They cause more crop losses than the rest of Earth's organisms combined. How many species cause *Cannabis* diseases? The scientific literature lists 420 *Latin names* of fungi associated with *Cannabis*. Many names are taxonomic synonyms. The fungus causing grey mould, for instance, masquerades under seven different Latin names (McPartland 1995e). Other species cited in the literature are misidentifications (McPartland 1995a). Yet more names in the literature describe obligate saprophytes. Saprophytic "hemp retters and rotters" unable to attack living plants are excluded here. (See Chapter 8 concerning post-harvest problems.) After a name-by-name review, McPartland (1992) determined the 420 taxa in the literature actually represented 88 species of *Cannabis* pathogens.

The fungal diseases described here are sequenced by their economic importance. Grey mould tops the list. We use common names of diseases approved by the American Phytopathological Society (McPartland 1989, 1991). Latin names of all spore states (teleomorphs and anamorphs) are included for each species, with the most common spore state listed first. We have included synonyms for the sake of continuity and reference to earlier literature.

GREY MOULD

Grey mould has become the most common disease of *Cannabis* (Plate 44). It afflicts fibre and drug cultivars, outdoors and indoors (glasshouses and growrooms). The fungus causing grey mould attacks hundreds of other crops around the world. It is so pervasive in vineyards that some wines deliberately use grey-mouldy grapes, such as Sauterne (in France), Trokenbeerenauslesen (in Germany), and Tokay (in Hungary). The fungus thrives in high humidity and cool to moderate temperatures. Disease peaks in drizzly, maritime climates (e.g., the Netherlands, the Pacific Northwest). In these climates grey mould reaches epidemic proportions and can destroy a *Cannabis* crop in a week (Barloy & Pelhate 1962, Frank 1988).

The grey mould fungus tends to attack *Cannabis* in two places—flowering tops and stalks. Flower infestations tend to arise in drug cultivars and seed cultivars with large, moisture-retaining female buds. Scheifele (1998) reported 30-40% incidence of "head blight"in fields of early-maturing seed hemp; 'Fasamo' and 'FIN-314' also suffered sporadic infections. Stalk rot seems more common in fibre varieties (Patschke *et al.* 1998). The grey mould fungus can also infest seeds, destroy seedlings (see Damping off), and attack plants after harvest (see Chapter 8).

SYMPTOMS

Flower infestations begin within buds, so initial symptoms are not visible. Fan leaflets turn yellow and wilt, then pistils begin to brown. In high humidity, whole inflorescences become enveloped in a grey fuzz, then degrade into grey-brown slime (Plate 45). The grey fuzz is a mass of microscopic conidia (Fig 5.1). In low humidity, the grey fuzz does not emerge; infested flowers turn brown, wither, and die (Plate 46).

Stalk rot begins as a chlorotic discolouration of infected tissues. Chlorotic sections turn into soft shredded cankers. Stalks may snap at cankers. Cankers may encircle and girdle stalks, wilting everything above them. In high humidity, cankers become covered by conidia (Plate 47). Conidia are liberated in a grey cloud by the slightest breeze. Small, black sclerotia may form within stalks.

CAUSAL ORGANISM & TAXONOMY

Botrytis cinerea Persoon: Fries, *Systema mycologicum* 3:396, 1832.

=*Botrytis infestans* (Hazslinszky) Saccardo 1887, ≡*Polyactis infestans* Hazslinszky 1877; =*Botrytis felisiana* Massalongo 1899; =*Botrytis vulgaris* Link: Fries 1824.

teleomorph: *Botryotinia fuckeliana* (deBary) Whetzel, *Mycologia* 37:679 1945; ≡*Sclerotinia fuckeliana* (deBary) Fuckel 1870.

Research reveals these seven names refer to the same organism, susceptible to the same control methods. There is some dif-

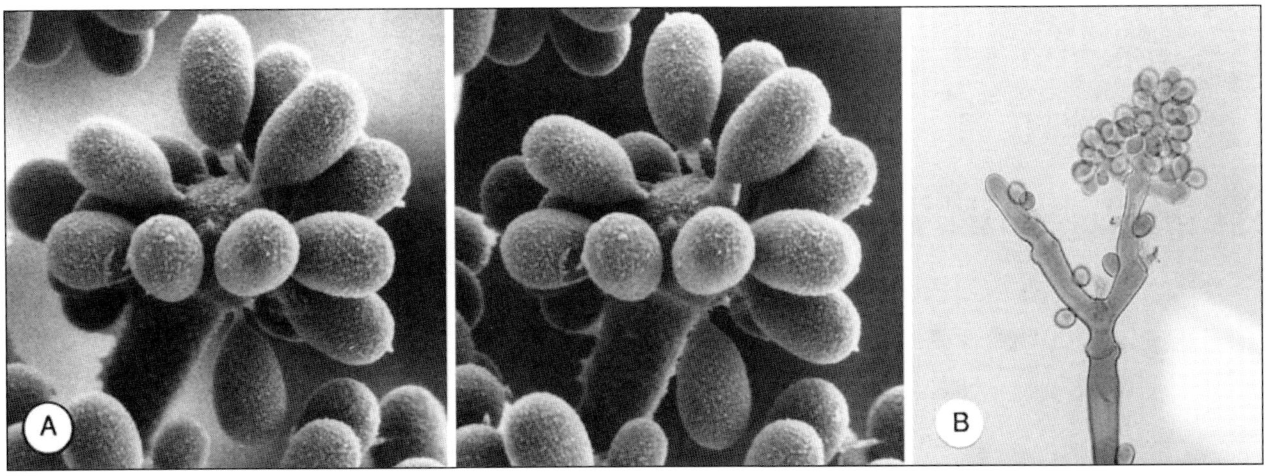

Figure 5.1: Conidiophores and conidia of *Botrytis cinerea*. A. Stereopair: view with a stereoscope or hold about 20 cm away, cross your eyes and align double vision into a single middle 3-D image, SEM x1250 (courtesy Merton Brown and Harold Brotzman, Brown & Brotzman, 1979); B. LM x260 (courtesy Bud Uecker).

ference of opinion: Ferraris (1935) considered *B. cinerea* and *B. vulgaris* different species. Gitman & Boytchenko (1934) and Barloy & Pelhate (1962) separated *B. cinerea* from *B. infestans* and *B. felisiana*. Spaar *et al.* (1990) confused the teleomorph name.

Description: Conidiophores upright, grey-brown, branching near the apex, 5–22 µm in diameter. Conidia borne on conidiophore apex in botryose clusters, hyaline to yellow-grey (grey *en masse*), aseptate, round to ovoid, 8–14 x 6–10 µm (Fig 5.1). Microconidia rare, *Myrioconium*-like, arising from phialides, hyaline, oval, aseptate, 2.0–2.5 µm in diameter. Apothecia rare, arising from sclerotia on a 3–10 mm stalk, topped by a yellow-brown disc; disc flat to slightly convex, 1.5–5mm in diameter, capped by a single layer of asci and paraphyses (Fig 5.2). Asci cylindrical, with long tapered stalks, eight-spored, 120–140 x 8 µm. Paraphyses hyaline, septate, filiform. Ascospores aseptate, uniguttulate, uniseriate, hyaline, ovoid to ellipsoid, 8.5–12 x 3.5–6 µm. Sclerotia hard, black, rough, plano-convex, irregularly round to elongate, 1–15mm long (averaging 5mm on *Cannabis*, Flachs 1936), sometimes in chains; cross-section of sclerotium reveals a thin black rind with a hyaline interior (Fig 5.4).

DIFFERENTIAL DIAGNOSIS

In flowering tops, this disease may be confused with brown blight caused by *Alternaria alternata*. A microscope easily tells them apart, the conidia are very different—see Fig 3.2. Powdery mildew looks like a white dust, not a grey mould, and powdery mildews do not cause cankers. Several fungi cause cankers, such as anthracnose fungi and *Fusarium* species, but none grow as fast as grey mould.

Sclerotia formed by *B. cinerea* can be confused with sclerotia formed by *Sclerotinia sclerotiorum*, the cause of hemp canker (see next section). Generally, *B. cinerea* sclerotia arise within stalks, whereas *S. sclerotiorum* sclerotia appear within stalks or upon the exterior surface. *B. cinerea* sclerotia are smaller than *S. sclerotiorum* sclerotia, and they have different cross sections (Fig 5.4). If sclerotia produce apothecia, the species are easily distinguished under a microscope.

DISEASE CYCLE & EPIDEMIOLOGY

B. cinerea can overwinter in infected seeds (Fig 5.2B). Infected seeds may mould in storage or give rise to seedborne infections of seedlings the following spring (Pietkiewicz 1958, Noble & Richardson 1968). *B. cinerea* also overwinters within stalk debris as sclerotia or dormant hyphae (Fig 5.2A). Sclerotia may persist for years in soil. Spring

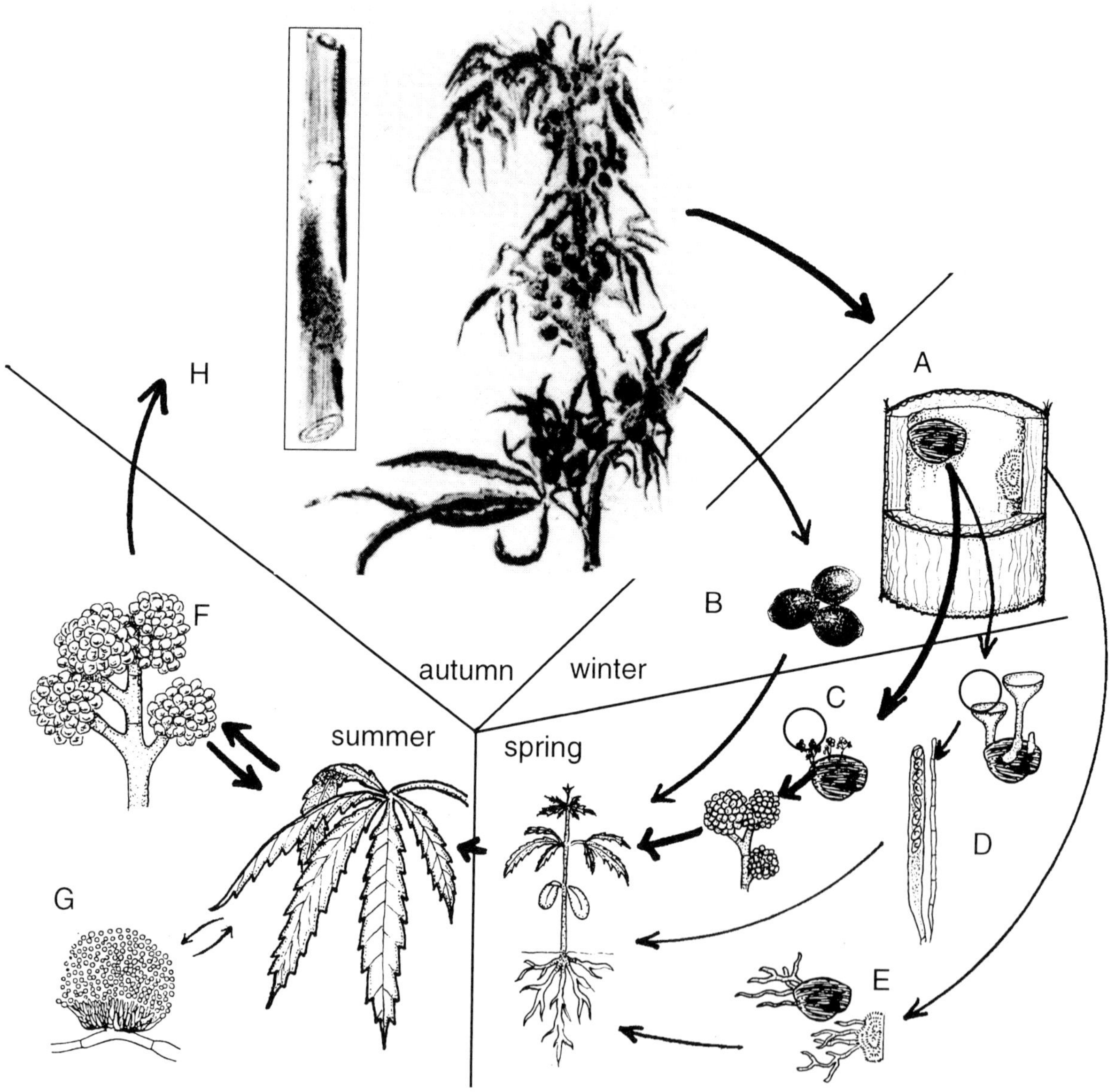

Figure 5.2: Disease cycle of gray mold; for description see text (McPartland redrawn from Agrios 1997, insert from Barloy & Pelhate 1962).

rains induce sclerotial germination, producing conidia (Fig 5.2C) or rarely producing apothecia with ascospores (Fig 5.2D). Hyphae from sclerotia may also penetrate plants directly (Fig 5.2E).

Conidia (rarely ascospores) are blown or splashed onto seedlings. High humidity or dew is needed for conidial germination. The optimal temperature for conidial germination is around 20°C. Conidia germinate and directly penetrate thin young epidermis. On older plants the fungus can only penetrate epidermis damaged by insects, rough handling, frost, or improper fertilization. After infection, the fungus grows within the host, at an optimum of 24°C.

In humid conditions, the fungus forms new conidia, which spread to sites of secondary infections (Fig 5.2F). Conidial production requires ultraviolet (UV) light, specifically at wavelengths shorter than 345 nm (Sasaki & Honda 1985). Microconidia may spread the disease (Fig 5.2G), but many experts consider microconidia noninfective. Cycles of secondary infections eventually build to epidemics in mature plants (Fig 5.2H).

Grey mould flourishes in high-density stands of hemp. De Meijer *et al.* (1995) seeded four fields at rates of 20, 40, 80, and 140 kg seed per ha. These rates initially produced 104, 186, 381 and 823 seedlings per m^2. Grey mould decimated the high-density stands. By harvest, all fields yielded about 100 plants per m^2. Crop yield (stalk biomass) was actually greatest in crops sown at the lowest seedling rate.

Fibre varieties become more susceptible after canopy closure. *Drug* varieties become most susceptible during flowering. *B. cinerea* often colonizes senescent leaves and flowers, and from these footholds it invades the rest of the plant. Watch for grey mould epidemics during periods of high humidity (>60% RH) and cooling temperatures, especially at dew point when plants cannot dry out.

CULTURAL & MECHANICAL CONTROL
(numbers refer to methods listed in Chapter 9)

The key to grey mould control is method 7c—keeping relative humidity below 50%. Also use methods 1 (sanitation), 2a (deep ploughing), 2b&c (sterilizing or pasteurizing soil), 2d (flooding soil), 3 (weeding), 4 (harvesting early), 5 (genetic resistance—see below), 8 (avoid excess nitrogen and phosphorus, add calcium, neutralize acid soils to enhance calcium absorption), 10 (careful pruning), and 11 (avoid seedborne infection).

In glasshouses, keep light intensity high and the temperature warm (>25°C) to inhibit conidial germination. Outdoors, avoid planting in shade in humid climates. Do not crowd plants—sow seed at low densities. Avoid wounding plants during susceptible periods, except to prune away injured or infected branches. Trunoff (1936) noted *B. cinerea* attacks males first (males lose vigour before females), then spreads to female plants. Rogue males if possible. Flowers should be harvested while resin glands are white or amber, not brown. Brown, wilted pistils are prime *B. cinerea* fodder. Harvested material should be cured and dried in dark rooms with good air circulation.

Sasaki & Honda (1985) slowed grey mould in glasshouse tomatoes by covering windows with sheets of UV-absorbent vinyl. Since the grey mould fungus requires UV light to produce conidia, epidemics are prevented. Sasaki & Honda (1985) filtered UV-B radiation with Hi-S vinyl by Nippon Carbide Industries; Lydon *et al.* (1987) used Mylar Type S film. Reducing UV-B also decreases production of THC (Lydon *et al.* 1987).

Breeding resistant plants is the ultimate solution for grey mould. Tall *indica* biotypes (Mexican, Colombian, and Thai plants) rarely suffer from bud rot, whereas the dense, tightly-packed buds of *Cannabis afghanica* tend to hold moisture and rot easily (Clarke 1987). Afghan biotypes evolved in very arid conditions and have no resistance to grey mould. This unfavourable trait even appears in hybrids with a small percentage of *afghanica* heritage. What marijuana breeders gained from the introduction of *afghanica* (potency, short stature, early maturity), they paid for with extreme susceptibility to bud rot.

Susceptibility to grey mould has no correlation with plant THC or CBD levels (De Meijer *et al.* 1992, Mediavilla *et al.* 1997). Cannabinoids inhibit some fungi (Elsohly *et al.* 1982, McPartland 1984), but apparently not *B. cinerea*. Perhaps *B. cinerea* evolved the ability to metabolize THC, as have other fungi (Robertson *et al.* 1975, Binder 1976).

Mediavilla *et al.* (1997) reported that 87% of 'Swihtco' plants [a drug cultivar with *afghanica* heritage] suffered grey mould—more than any fibre cultivars, including 'Felina 34' (75% diseased), 'Fedora 19' (63%), 'Uniko-B' (55%), 'Kompolti Hibrid TC' (50%), 'Futura 77' (47%), 'Secuiemi' (37%), and the most resistant, 'Livoniae' (25%). Van der Werf *et al.* (1993) said Hungarian 'Kompolti Hibrid TC' and French 'Fédrina 74' are more susceptible to grey mould than other fibre cultivars. Dempsey (1975) reported good resistance in 'JUS-1' and 'JUS-7', but these cultivars may no longer be available (see de Meijer 1995).

Several corporations produce monoclonal antibody-based assays. These detection kits can identify *B. cinerea* in plant tissues before symptoms appear.

BIOCONTROL (see Chapter 10)

Foliage, flowers, and stems can be sprayed with *Gliocladium roseum* and related *Trichoderma* species (described below). Bees have successfully delivered these fungi to flowers of other crops, by dusting bees with spores as they emerge from hives (Sutton *et al.* 1997). Unfortunately, *Cannabis* is not an attractive nectar source for this unique bud biocontrol delivery system.

Damping off caused by *B. cinerea* can be prevented by mixing *Gliocladium* and *Trichoderma* species into the soil (described under damping off). Other soil biocontrols include *Streptomyces griseoviridis* (described under damping off) and *Coniothyrium minitans* (described under hemp canker).

Post-harvest disease by *B. cinerea* has been controlled with yeasts (*Pichia guilliermondii* and *Candida oleophila*) and the bacterium *Pseudomonas syringae*, described near the end of Chapter 8.

***Gliocladium roseum* (=*Clonostachys rosea*)**

BIOLOGY: This cosmopolitan fungus lives in a wide range of soils, from tropical rainforests to subartic deserts (Sutton *et al.* 1997). *G. roseum* grows best between 20–28°C. It has been isolated from hemp stems in the Czech Republic (Ondrej 1991). The fungus protects plants in many ways: it aggressively colonizes all surfaces of plants—above and below ground—which prevents pathogens from gaining a foothold; it grows within senescent plant tissues and competes with pathogens for nutrients and space; it induces resistance in host plants against other fungi; it oozes metabolites that inhibit pathogens; and it directly parasitizes other fungi, including *B. cinerea, Sclerotinia sclerotiorum, Rhizoctonia solani, Phymatotrichum omnivorum,* and *Verticillium* species. It also parasitizes some nematodes (*Heterodera* and *Globodera* species).

APPLICATION: Spore suspensions (10^6–10^8 conidia per ml) mixed with a surfactant such as Triton X-100 have been sprayed on flowers to prevent bud rot, or sprayed on conifer stalks at canopy closure to prevent stalk rot, or daubed on clone-pruning wounds of mother plants to prevent wound infections. Protect seedlings by dressing seeds with spores mixed with a sticking agent, or by quick-dipping clones in spore slurries, or by drenching seedbed soil with spore suspensions.

Treatments should be applied shortly before nightfall, because dew and darkness facilitate spore survival and activity (Sutton *et al.* 1997). A few hours of dampness are needed for germination, which can be scary to growers familiar with grey mould. Heavy rainfall is not good, because it washes spores off plants before spores can germinate. Some researchers report a single *G. roseum* treatment can last eight to 12 weeks or more. Others reapply *G. roseum* weekly during peak grey mould seasons (Sutton *et al.* 1997).

NOTES: Spores stored in sealed containers at 3–5°C remain effective for over a year. *G. roseum* is compatible with *Trichoderma harzianum, T. viride, T. koningii,* and mycorrhizal fungi. Its compatibility with other biocontrol fungi has not been evaluated. It can probably be mixed with BT, NPV, and biocontrol insects, but these combinations remain untested. Unlike *Trichoderma* species, *G. roseum* can reproduce on the surface of plants, and the fungus bears conidia within 72 hours of initial treatment (Sutton *et al.* 1997). *G. roseum* is not harmful to bees and does not grow at human body temperatures. Because of its unique morphology, ecology, and DNA, *G. roseum* will be moved to the genus *Clonostachys* and renamed *C. rosea* (Samuels 1996).

Trichoderma species

BIOLOGY: Four cosmopolitan fungi, ***Trichoderma harzianum, Trichoderma polysporum, Trichoderma virens,*** and ***Trichoderma viride,*** control *B. cinerea* and other pathogens. These species are discussed further under Rhizoctonia sore shin disease.

APPLICATION: These fungi are marketed in a variety of biocontrol products. Most products are designed for seed and soil treatments. When applied as a *foliar spray,* the T-39 strain of *T. harzianum* (Trichodex®) controlled *B. cinerea* in grape vineyards and glasshouse vegetables, using a spray concentration of 10^8 conidia ml^{-1} (Reuveni 1995). Foliar application may need repeating, because *Trichoderma* species do not survive long on leaf surfaces (Samuels 1996). Pruning wounds can be painted with a spore suspension of the T-39 strain. Foliar application of the *T. harzianum-T. polysporum* product (Binab T®) does *not* control *B. cinerea* in flowering tops (Watson, unpublished data 1991).

CHEMICAL CONTROL (see Chapter 11)

Bordeaux mixture knocks back grey mould in early stages, and prevents infections until the next rainfall washes the mixture off plants. Doctor Indoors (1988) painted pruning wounds with alcohol or sodium hypochlorite (bleach). Mixing calcium cyanamid into the soil kills overwintering sclerotia. *B. cinerea* develops resistance to most synthetic fungicide sprays. Fungicides may still be useful as seed treatments (Patschke *et al.* 1997). Seed dressings have a low impact on the environment (see Chapters 11 and appendix).

HEMP CANKER

This disease goes by many names around the world, including cottony soft rot, watery soft rot, stem rot, white mould, and grey rot. Europeans call it hemp canker. Some consider it the #1 scourge of hemp cultivation (Rataj 1957), or #2 behind grey mould (Termorshuizen 1991). In North America, the disease hampered production of hemp in Wisconsin (Buchholtz, USDA archives 1943), in Ontario (Conners 1961, Scheifele 1998), and caused 40% losses in Nova Scotia (Hockey 1927). Hemp canker has infested Indian ganja (Bilgrami *et al.* 1981), Finnish seed-oil crops (Callaway & Laakkonen 1996), and Australian hemp (Synnott 1941, Lisson & Mendham 1995).

SYMPTOMS

Symptoms usually begin in late summer, on full-grown plants. The disease may arise near the soil line (Holliday 1980) or in the upper 2/3rds of the plant (Tichomirov 1868). Watersoaked lesions appear first, on stalks and branches. Cortical tissues beneath the lesions collapse, creating pale, light-brown cankers. In humid conditions, the surface of the stalk becomes enveloped in white mycelium (Plate 48). Black spots of sclerotic tissue emerge on the surface of cankers, usually by September. Larger sclerotia form within the hollow of stalks. Flowering, if it has begun, often ceases (Flachs 1936). Plants remain in this condition, or wilt and fall over. A secondary fungus may overgrow hemp canker, see "red boot disease."

CAUSAL ORGANISM & TAXONOMY

Sclerotinia sclerotiorum (Libert) deBary, *Vergh. Morph. Biol. Pilze, Mycet., Bact.* p.236, 1884.

≡*Whetzelinia sclerotiorum* (Libert) Korf & Dumont 1972, ≡*Peziza sclerotiorum* Libert 1837; =*Sclerotinia libertiana* Fuckel 1870; =*Sclerotinia kauffmanniana* (Tichomirov) Saccardo 1889, ≡*Peziza kauffmanniana* Tichomirov 1868 (see McPartland 1995e).

No anamorph has been named for *S. sclerotiorum*, but Brandenburger (1985) described one on *Cannabis*, albeit namelessly. Ferraris (1915) erroneously called *Botrytis vulgaris* the anamorph of *S. sclerotiorum*. His mistake followed Kirchner (1906), who called *B. fuckeliana* the hemp canker organism (*B. fuckeliana* causes grey mould).

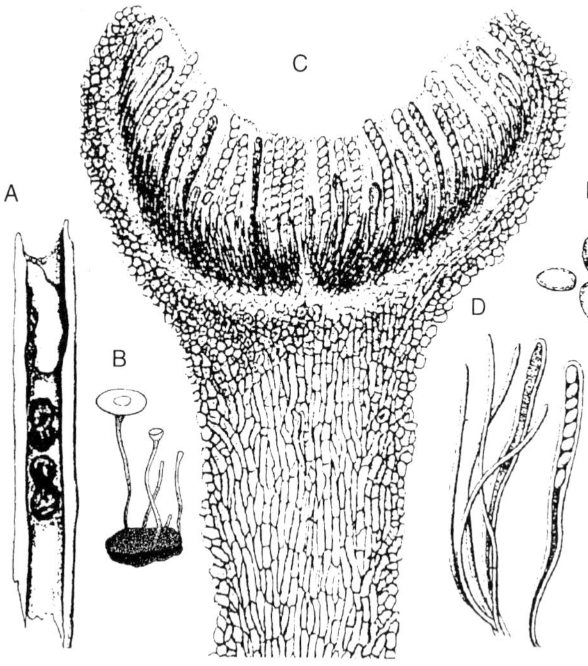

Figure 5.3: *Sclerotinia sclerotiorum*. A. Sclerotia and mycelium in pith (x1); B. Sclerotium with apothecia (x1); C. Cross section of apothecium (x150); D. Asci, paraphyses, and ascospores (x300); E. Close-up of three ascospores (x430). All from Tichomirov (1868) except B by DeBary (1887).

Description: Sclerotia within stalks are hard, yellow becoming black, smooth, oblong, 5–13 x 3–7 mm, cross section of sclerotia reveals a dark rind surrounding a soft, white centre (Fig 5.4). Apothecia consist of brown cylindrical stalks up to 25 mm long, topped with by yellow-brown cups, concave, 2–8 mm wide, filled with a single layer of asci and parapheses (Figs 3.3 & 5.3). Asci cylindrical, eight-spored, bluing with Melzer's reagent, 110–140 x 6–10 µm. Parapheses filiform, hyaline, same length as asci. Ascospores aseptate, elliptical, hyaline, usually biguttalate, 9–13 x 4–6 µm (Fig 5.3).

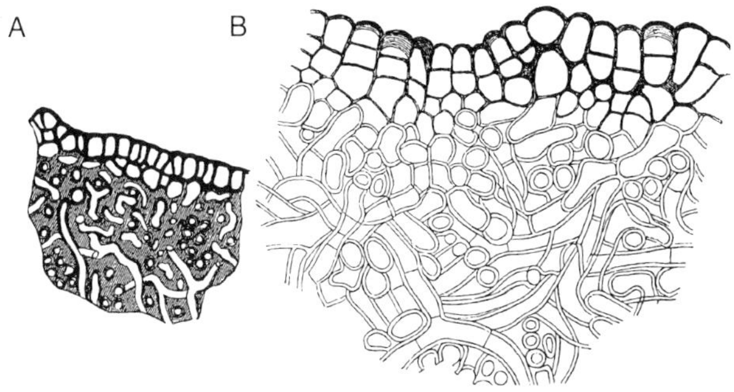

Figure 5.4: A. Thin section through a sclerotium of *Botrytis cinerea* (x310); B. Section through a sclerotium of *Sclerotinia sclerotiorum* (x300), from DeBary (1887).

DIFFERENTIAL DIAGNOSIS

Symptoms of hemp canker can be confused with stalk rot caused by the grey mould fungus, *B. cinerea*. In humid conditions, *B. cinerea* forms its characteristic blanket of grey conidia. In dry conditions, you can differentiate sclerotia. Sclerotia on stalk *exteriors* are usually caused by *S. sclerotiorum*. *S. sclerotiorum* sclerotia are larger than *B. cinerea* sclerotia. The rind covering *S. sclerotiorum* sclerotia consists of three or four layers of cells, and the cells have thick black walls. The rind of *B. cinerea* is thinner, usually one or two cells thick (Fig 5.4). The internal medulla of both species is white, but the medulla of *B. cinerea* is denser and more gelatinous.

Hemp canker can also be confused with southern blight or Rhizoctonia sore shin. Both of these diseases, however, tend to cause symptoms adjacent to soil. The fungi causing these diseases form tiny sclerotia, smaller than *S. sclerotiorum* sclerotia. Hemp canker is common in temperate climates; southern blight predominates in warmer regions.

DISEASE CYCLE & EPIDEMIOLOGY

S. sclerotiorum overwinters as sclerotia in plant debris or in soil. Springtime moisture and warm temperatures (15–20°C) initiate germination. Each sclerotium sprouts two to seven apothecia. Apothecia forcibly eject asci into the air. Ascospores germinate in the presence of moisture and directly penetrate host epidermis, or penetrate via wounds. The fungus grows inter- and intracellularly through host parenchyma and cortex. *S. sclerotiorum* invades the seeds of many plants, but this has not been documented in *Cannabis*.

S. sclerotiorum attacks over 360 species of crops and weeds, mostly herbaceous dicots. Strains isolated from sunflower, Jerusalem artichoke, potato, safflower, flax, and colza also attack hemp (Antonokolskaya 1932). *S. sclerotiorum* thrives in cool, moist conditions, just like grey mould. Both diseases accelerate after canopy closure, especially in high-density stands (De Meijer *et al.* 1995).

CULTURAL & MECHANICAL CONTROL (see Chapter 9)

Use methods 1 (especially burying), 2a (deep ploughing), 2b&c (sterilizing or pasteurizing soil), 3 (weeding), 5 (genetic resistance), 6 (rotate with grains and grasses and keep free of dicot weeds), 7c (avoid excess humidity), 8 (add calcium to soil, neutralize acid soils to enhance calcium absorption, Ferraris (1935) suggested adding phosphate), 10 (careful pruning), and 11 (avoid seedborne infection). Deep ploughing of debris is particularly effective—*S. sclerotiorum* sclerotia do not germinate if buried deeper than 6 cm underground (Lucas 1975). Rotation with nonhost crops must last three or four years to starve long-lived sclerotia. Avoid high-density stands; seed at low rates (see discussion in grey mould section). Do not use seeds from infected plants (Serzane 1962).

Soil suspected of harbouring pathogens can be tested with biotechnology-based assays. A kit detecting *S. sclerotiorum* is commercially available (though expensive, US$148 for a set of six tests). Handling plants infected with *S. sclerotiorum* may cause dermatitis. The fungus produces two compounds, xanthotoxin and bergapten, which sensitize skin to sunlight (Centers for Disease Control 1985).

BIOCONTROL (see Chapter 10)

Infected glasshouse soil can be dried for several days, and then soaked for two or three weeks. This encourages the growth of soil organisms which parasitize and kill *S. sclerotiorum* sclerotia. A parasitic fungus, ***Sporidesmium sclerotivorum***, attacks sclerotia of *S. sclerotiorum* and is being investigated (Cook *et al.* 1996). Commercially-available biocontrols of *S. sclerotiorum* include *Trichoderma harzianum* (see the section on Rhizoctonia sore shin), *Bacillus subtilis* (described under damping off), and *Coniothyrium minitans*.

Coniothyrium minitans

BIOLOGY: A fungus (ContansWG®, Koni®) that parasitizes sclerotia of *S. sclerotiorum, B. cinerea,* and a few other pathogens. *C. minitans* lives in temperate zones worldwide.

APPLICATION: *C. minitans* is formulated as water dispersable granules, applied as a soil drench to sclerotia-infested soil. It has also been sprayed directly on susceptible plants. Application rates range from 50–3000 kg ha^{-1} (Whipps & Gerlagh 1992). *C. minitans* has protected canola, sunflower, peanut, soyabean, and vegetable crops.

CHEMICAL CONTROL (see Chapter 11)

Preventative methods are more effective than curative treatments. The only effective chemicals are synthetic; they are best applied as seed dressings (see Appendix 1 and Patschke *et al.* 1997).

DAMPING OFF

At least two "water moulds," *Pythium* species, cause damping off disease of *Cannabis* seedlings. *Pythium* species occasionally attack mature plants in field soils (Frezzi 1956) and hydroponic systems (McEno 1990). Curiously, no *Phytophthora* species (closely related to *Pythium*) parasitize *Cannabis*. Hemp in fact serves as a biocontrol against *Phytophthora infestans* (Israel 1981); aqueous extracts of hemp inhibit the growth of *P. infestans* (Krebs & Jäggi 1999).

SYMPTOMS

Damping off presents itself in two scenarios: In **pre-emergent** damping off, seeds or seedlings die *before* they emerge from the soil (Plate 49). **Post-emergent** damping off hits *after* seedlings have emerged from the soil (Plate 50). Seedlings with post-emergent damping off usually develop a brown rot at the soil line, then wilt and topple over. In older seedlings (with up to eight pairs of true leaves) growth ceases, leaves turn pale yellow, then seedlings wilt and topple over (Kirchner 1906).

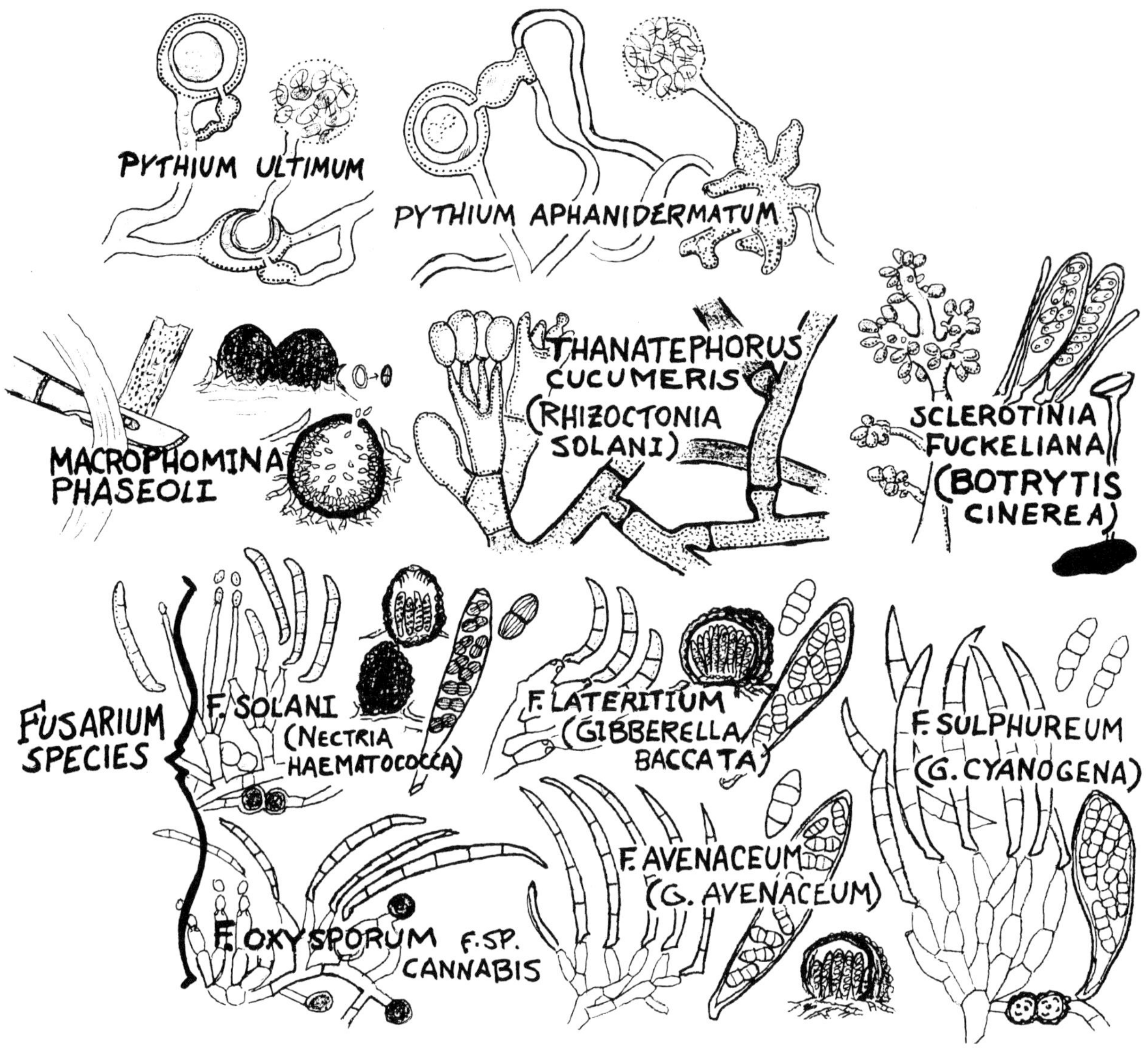

Figure 5.5: Ten organisms that cause damping off in *Cannabis* seedlings. Not drawn to scale (McPartland).

CAUSAL ORGANISMS & TAXONOMY

1. ***Pythium aphanidermatum*** (Edson) Fitzpatrick, *Mycologia* 15:168, 1923.

Description: Sporangia mature into elongated irregular swellings, branched or unbranched, 50–1000 x 2–20 μm, sometimes scarcely distinguished from the mycelium. Sporangia germinate into emission tubes of variable lengths, 2–5 μm diameter, and form vesicles at their ends. Vesicles cleave into 100 or more zoospores. Zoospores reniform, laterally biflagellate, 7.5 x 12 μm. Oogonia spherical, formed terminally (rarely intercalary), 22–27 μm diameter. Each oogonium is fertilized by one or two antheridia. Antheridia are formed on hyphae adjacent to oogonia, terminally or intercalary, doliiform to broadly clavate, 9–11 x 10–15 μm. Each fertilized oogonium forms a single oospore. Oospores spherical, containing a central vacuole, wall 2 μm thick, smooth, and do not fill oogonia, 17–19 μm diameter (Fig 5.5).

2. ***Pythium ultimum*** Trow, *Annals of Botany* 15:300, 1901.

Description: Sporangia spherical to doliiform, 20–29 x 14–28 μm, and germinate only by germ tubes. Oogonia smooth, spherical, formed terminally, 19–23 μm in diameter. Each oogonium is fertilized by one antheridium (rarely two or three antheridia). Antheridia swollen, sausage-shaped, curved, originating immediately below oogonia. Each fertilized oogonium forms a single oospore. Oospores are spherical, containing a single central vacuole and eccentric refringent body, thick walled, smooth, 14–18 μm diameter. Oospores germinate as germ tubes or (rarely) they produce an elongate emission tube, 5–135 μm long, with a vesicle at its end. Vesicles cleave into eight to 15 zoospores (Fig 5.5).

3. *Other causal organisms*

Europeans cite a third *Pythium* species attacking hemp, ***Pythium debaryanum*** Hesse (Kirchner 1906, Serzane 1962, Kirchner 1966, Vakhrusheva 1979, Barna *et al.* 1982, Gutberlet & Karus 1995, Bòsca & Karus 1997). *P. debaryanum* is not a true species; its original description was based on a mixture of fungi (McPartland 1995a). *"P. debaryanum"* citations usually indicate misidentifications of *P. ultimum* (Plaats-Niterink 1981).

Besides *Pythium*, several fungi also cause damping off—*Botrytis cinerea, Macrophomina phaseolina, Rhizoctonia solani,* and several *Fusarium* species (Fig 5.5). These pathogens usually attack mature plants. Their taxonomic descriptions are presented elsewhere.

DISEASE CYCLE & EPIDEMIOLOGY

Pythium species produce two types of spores, zoospores and oospores. Zoospores swim using flagella, and spread disease from plant to plant. Oospores are sexually produced. They do not migrate, but spread the disease over time (from season to season) as overwintering spores.

P. ultimum attacks hemp seedlings in central Europe (Marquart 1919, Schultz 1939). *P. ultimum* also causes a root rot of *mature* plants (Frezzi 1956). Root rot begins at the root-tip, eventually causing above-ground parts to wilt. *P. ultimum* lives in temperate regions worldwide. This pathogen attacks many crops, especially at cooler temperatures (12–20°C).

P. aphanidermatum is the "warm weather pythium," with optimum growth at 32°C. Although the species arises worldwide, it has only been reported on *Cannabis* in India (Galloway 1937).

Botrytis cinerea causes damping off because it can spread by seedborne infection (Pietkiewicz 1958, Noble & Richardson 1968, Patschke *et al.* 1997). Expect epidemics in seeds harvested from females infested by grey mould. *B. cinerea* conidia disperse by wind (not water like most damping off fungi).

Macrophomina phaseolina causes charcoal rot in older plants, as well as damping off in seedlings. It is common in the midwest, especially on maize. Epidemics caused by *M. phaseolina* peak in warm weather (optimal temperature 37°C).

Rhizoctonia solani causes sore shin and root rot in mature plants. It tends to damage seedlings at a later stage than *Pythium* species. Mishra (1987) reported *R. solani* causing 78% mortality in *Cannabis* seedlings (44% as post-emergence, 34% as preemergence). Dippenaar *et al.* (1996) reported heavy seedling losses by *R. solani* despite spraying with a fungicide (Rhizolex®). *R. solani* does *not* require excess moisture, unlike most damping off organisms.

Several *Fusarium* species cause damping off, including *F. solani*, *F. oxysporum* (Patschke *et al.* 1997), and less frequently *F. sulphureum*, *F. avenaceum*, and *F. graminearum*. Some of these fungi also cause cotyledon drop (Rataj 1957). In mature plants these pathogens cause Fusarium foot rot, root rot, stem canker, and wilt. *Fusarium* conidia spread via splashed rain and water runoff. They generally prefer warm climates. If introduced into sterilized soil, *Fusarium* species run wild. They cause greatest losses in damp soil, but cannot survive in *waterlogged* land.

DIFFERENTIAL DIAGNOSIS

Differentiating between *Pythium* species is difficult, and requires laboratory cultivation. Differentiating *Pythium* species from other organisms can be easier: Dig up a stricken seedling and lightly grasp the root between your thumb and forefinger. Then pull the root away from the stem. If the outer layer of the root (epidermis and cortex) slips away leaving only a thin inner cylinder (endodermis and stele), you have a *Pythium* problem. If tiny bits of soil dangle from the root of a seedling pulled from the ground, you have a *R. solani* problem—this diagnostic sign is attributed to the pathogen's coarse, clinging hyphae. Barloy & Pelhate (1962) differentiated symptoms caused by *F. solani* and *F. oxysporum*—seedlings with *F. solani* develop red discoloured roots (Fig 5.6 A), whereas seedlings with *F. oxysporum* become enveloped in a pink mass of hyphae (Fig 5.6 B).

Pre-emergent damping off can be erroneously attributed to outdated, dead seed. A lack of emerging seedlings may also be due to underground insects, such as crickets and root maggots, which kill germinating seeds. *Post-emergent* symptoms can be confused with damage from insects (cutworms) or environmental causes (frost, hail, or heavy rains). Leaf spots are *not* damping off. For instance, *Trichothecium roseum* and *Septoria cannabis* attack cotyledons. Since stems remain untouched and seedlings stay upright, this is not damping off.

CULTURAL & MECHANICAL CONTROL (see Chapter 9)

The many pathogens that cause damping off prefer

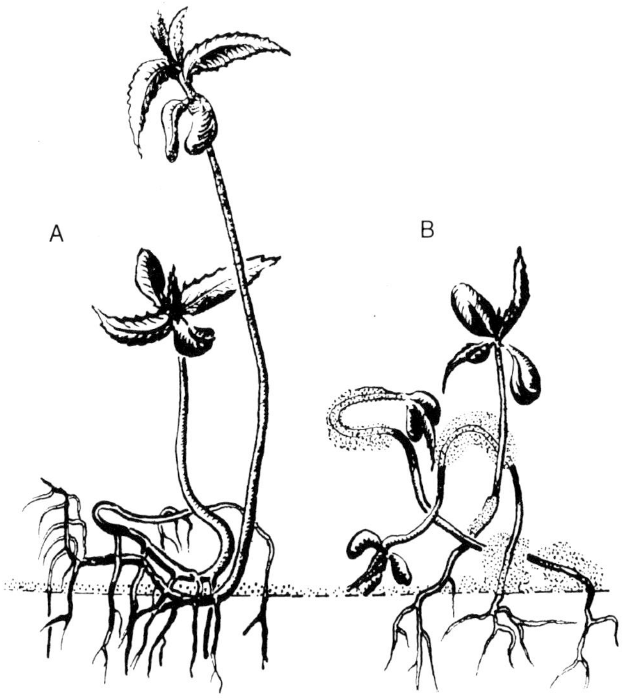

Figure 5.6: Signs and symptoms of post-emergent damping off caused by two *Fusarium* species. A. *F. solani;* B. *F. oxysporum* (from Barloy & Pelhate 1962).

many different temperatures, pH ranges, lighting conditions, and soil types. Their common denominator is excess moisture, so avoid overwatering—follow methods 7b and 7c. Irrigate soil before planting; watering afterward will pack soil around seeds. Damping off increases in seeds planted deeper than 2 cm. Some horticulturists germinate seeds between layers of lightly moistened paper towels before planting them in soil. Right out of the plastic wrap, paper towels are nearly sterile.

Do not plant in heavy, wet, poorly draining soils that tend to puddle. On questionable sites, try planting in raised beds (>10 cm high) or lighten soil with perlite. (Vermiculite is a poor second choice; it eventually decomposes into sludge, which exacerbates soil heaviness.) Some authors suggest using a sandy soil to suppress damping off. Others say a perlite and peat mix is best. The key characteristic is good drainage.

Observe method 4—do not plant too early in the spring. Most damping off pathogens thrive in cold soil. Vetter (1985) described a rule of "thumb" for *Cannabis:* "I walked into the middle of my unplanted garden plot, dropped my pants and sat... if the soil is too cold for your bare ass, it's too cold for the seeds you're getting ready to plant." Indoors, use an insulated horticulture table to provide "bottom heat" for seedlings and clone cuttings. Unfortunately, heat may encourage *M. phaseolina* and *P. aphanidermatum*. Full-spectrum lighting greater than 1000 Lumens inhibits damping off. Etiolation predisposes plants to damping off, and is prevented by high light intensity (Smith *et al.* 1981).

Follow method 11—do not plant seeds harvested from diseased females—*B. cinerea* and many *Fusarium* species can spread by seedborne infection. Robust seedling growth inhibits damping off, so follow method 8. But *excess* soil electrolytes of any type (N, P, K, Ca, Fe, Etc.) increase damping off—so avoid chemical fertilizers and concentrated organics (e.g., blood meal, chicken manure). Aim for a pH of 7.0–7.5. Hydroponic operators should take note of method 1c (disinfesting nutrient solutions). Glasshouse operators often practice method 2b (steam sterilization of soil). Tyndalization (dry heat or baking) is not effective. In the field, method 2c (solarization) controls *P. ultimum, B. cinerea, R. solani,* and many *Fusarium* species, but not heat-resistant *M. phaseolina* or *P. aphanidermatum* (Elmore *et al.* 1997).

BIOCONTROL (see Chapter 10)

Several organisms control the whole spectrum of pathogens causing damping off; they are described below. Other biocontrols work on *specific* pathogens, and these are described elsewhere (in sections concerning *Botrytis, Rhizoctonia, Fusarium,* and *Macrophomina*).

Pythium oligandrum

BIOLOGY: Unlike its phytopathogenic *Pythium* cousins, this species is an aggressive mycoparasite, attacking many pathogens. *P. oligandrum* attacks the hyphae of *P. ultimum, B. cinerea, F. oxysporum, R. solani,* even *Sclerotinia sclerotiorum* and other fungi. It also secretes metabolites that inhibit the growth of pathogens.

APPLICATION: Supplied as oospores in granules or a powder (Polygangron®), for seed treatment or soil treatment. The commercial product was developed in Slovakia, but the species lives worldwide in temperate and semitropical regions, optimal growth at 30°C and soil pH 6–7.

Streptomyces griseoviridis

BIOLOGY: An actinomycete (Mycostop®) that grows on the surface of plant roots. It is native to Finnish peat bogs and grows best in humid soil, in a wide range of soil pHs (4–9) and soil temperatures (5–45°C, optimally 10–25°C). *S. griseoviridis* controls damping off and root rot caused by *Fusarium, Botrytis, Rhizoctonia,* and *Pythium.* The organism colonizes roots of plants so pathogens cannot gain a foothold, it produces growth hormones that enhance plant growth, and it oozes metabolites that inhibit or kill pathogens.

APPLICATION: Supplied as spores and mycelial fragments mixed in a dry powder (10^8 colony forming units per g powder). Sealed packets can be stored for six months in a dry, cool (8–10°C), dark place. Apply as a seed treatment (mix 2–8 g per kg seed), root dip (in a 0.01% solution), or soil drench (mix 1 g per 1.5 l water). Soil drenches also prevent root rot in older plants if reapplied every two months. According to Cherim (1998), *S. griseoviridis* is compatible with most chemical fertilizers, pesticides, rooting hormones, beneficial nematodes, mycorrhizal fungi, and other biocontrol agents, but I'd avoid mixing it with fertilizers and pesticides.

Streptomyces lydicus

BIOLOGY: An actinomycete (Actinovate®) related to the aforementioned species. *S. lydicus* works in a similar fashion to *S. griseoviridis* under similar conditions, and suppresses similar soil pathogens.

APPLICATION: Supplied as a granular formulation, seed coat mix, or wettable powder. Follow the recommendations for *S. griseoviridis.*

Burkholderia (Pseudomonas) cepacia

BIOLOGY: A soil bacterium from Wisconsin that colonizes the roots of plants, and oozes a formidable array of antibiotics that suppress *Pythium, Fusarium,* and *Rhizoctonia* pathogens. It also works against some nematodes (lesion, sting, lance, spiral nematodes). Available as a soil mix (Bac Pack®, Intercept®) or seed coating (Deny®, Dagger®, formerly Blue Circle®, Precept®).

APPLICATION: Supplied as a powder or aqueous suspension; sealed containers can be stored for one year in a cool (8–10°C), dry place. Apply to soil as a drench or via drip irrigation, immediately prior to planting. Apply to seeds with a sticking agent. *B. cepacia* is compatible with mycorrhizal fungi and other microbial biocontrols (Linderman *et al.* 1991).

Bacillus subtilis

BIOLOGY: A bacterium (Epic®, Kodiac®, Quantum 4000®, Rhizo-Plus®, System 3®, Serenande®) that lives in soil around plant roots and suppresses *Pythium, Fusarium, Rhizoctonia,* and *Sclerotinia sclerotiorum.*

APPLICATION: Several strains are available—GBO3, FZB24, and MBI 600. They are supplied as spores in a dust, powder, or water dispersible granule. *B. subtilis* can be stored for months in a cool (8–10°C), dark place. It is applied as a seed treatment or soil drench. *B. subtilis* is compatible with mycorrhizal fungi and other microbial biocontrols (Linderman *et al.* 1991).

Trichoderma species

BIOLOGY: A bevy of fungi, ***Trichoderma harzianum, Trichoderma polysporum, Trichoderma (Gliocladium) virens,*** and ***Trichoderma viride,*** control *Pythium, Rhizoctonia, Botrytis, Fusarium,* and other pathogens. These biocontrols are discussed under Rhizoctonia sore shin disease.

Glomus intraradices

BIOLOGY: A mycorrhizal fungus that protects plant roots from pathogens. Described under Fusarium stem canker.

Bacillus cereus

BIOLOGY: Another *Bacillus* useful against damping off fungi and root-rot fungi. *B. cereus* has been called a "mycorrhizae helper bacterium" because it facilitates the growth of these friendly fungi. *B. cereus* is a facultative aerobe, so it thrives in water-saturated soils (where damping-off causes the most harm). *B. cereus* forms spores like *B. subtilis,* so it survives on seeds in soil and seeds in storage. Unfortunately, cannabidiolic acid, produced by older *Cannabis* plants, inhibits the growth of *B. cereus* (Farkas & Andrássy 1976).

Pseudomonas fluorescens

BIOLOGY: A soil bacterium (strain EG1053) related to the previous bacterium, useful against *Pythium* and *Rhizoctonia* pathogens. Other strains (A506, 1629RS) are sprayed on fruit trees to prevent frost damage and fire blight disease (BlightBan®).

APPLICATION: Supplied as a tan powder in sealed containers. Can be stored long-term in a dry, cold (0–4°C) place, and kept at 21°C for one week before use. *P. fluorescens* is compatible with mycorrhizal fungi (Linderman *et al.* 1991).

CHEMICAL CONTROL (see Chapter 11)

"Old timers" treat soil with sodium nitrate, an organic but possibly carcinogenic chemical. Apply sodium nitrate in autumn to kill pathogens; over the winter it breaks down to sodium and nitrate. Drenching soil with synthetics such as benomyl is ill advised (with the possible exceptions of fosetyl-Al and metalaxyl). Williams & Ayanaba (1975) reported *increased* damping off after applying benomyl, due to suppression of *Pythium* soil antagonists. Benomyl is also highly toxic to earthworms. An aqueous extract of hemp (50 g dried flowers soaked in 1 l water) inhibited *R. solani* growth in seed potatoes (Krebs & Jäggi 1999).

YELLOW LEAF SPOT

This disease is ubiquitous. It frequents fibre crops, but also arises on drug plants (Mushtaque *et al.* 1973, Ghani & Basit 1975, Ghani *et al.* 1978) and *Cannabis ruderalis* (Szembel 1927, Gamalitskaia 1964). Since leaf photosynthesis is the engine driving crop yields, yellow leaf spot can reduce yields of fibre, flowers, and seeds.

SYMPTOMS

Small lesions first appear on lower leaves in early June. Lesion colour is variously described as white, yellow, ochre, or grey-brown (Plate 51). Spots may remain small and round but usually enlarge to irregularly polygonal shapes, their edges partially delineated by leaf veins. Spots sometimes have reddish-brown perimeters (Peck 1884, Kirchner 1906).

Tiny pycnidia arise within leaf spots. These fruiting bodies form on the upper surfaces of leaves, not on undersides as reported by Flachs (1936). Eventually, leaf spots dry out and fragment, leaving ragged holes in leaves. In severe infections the leaves curl, wither, and fall off prematurely, defoliating the lower part of the plant (Ferraris 1935, Watanabe & Takesawa 1936, McCurry & Hicks 1925, Barloy & Pelhate 1962, Ghani & Basit 1975).

Kirchner & Boltshauser (1898) illustrated yellow leaf spot in a classic lithograph. Many subsequent illustrations are copied from this engraving (see Flachs 1936, Ceapoiu 1958, Barloy & Pelhate 1962). Although yellow leaf spot principally affects lower leaves, Bush Doctor (unpublished data 1986) saw it on fan leaves of flowering tops in Nepal (Plate 52). Spots may also arise on stems (Gitman & Boytchencko 1934, Ferri 1959b, Ondrej 1991), and seedling cotyledons (Ferri 1959b).

CAUSAL ORGANISMS & TAXONOMY

At least two species of *Septoria* cause yellow leaf spot. Rataj (1957) described a third species on *Cannabis*—***Septoria graminum*** Desmarieres. This citation is probably an error (McPartland 1995a).

1. ***Septoria cannabis*** (Lasch) Saccardo, *Sylloge Fungorum* 3:557, 1884. ≡*Ascochyta cannabis* Lasch 1846; =*Spilosphaeria cannabis* Rabenhorst 1857; =*Septoria cannabina* Westendorp 1857; =*Septoria cannabis* Saccardo, nomen nudum.

Description: Pycnidia epiphyllous, gregarious, immersed but eventually erumpent, globose to flask shaped, averaging 90 µm in diameter; peridium dark brown, thick-walled *textura angularis-globulosa;* ostiole round, 15 µm in diameter (see Figs 3.3 & 5.7). Conidiogenous cells subglobose to ampulliform, simple, hyaline, holoblastic. Conidia hyaline, filiform, pointed at both ends, straight or curved, three-septate, 45–55 x 2.0–2.5 µm. Pycnidia *in vitro* grow to 465 µm in diameter and form long necks (Watanabe & Takesawa 1936, Ferri 1959b). Conidia exude from ostioles in a slimy, ribbon-like cirrhus.

2. ***Septoria neocannabina*** McPartland, *Sydowia* 47:46, 1995. ≡*Septoria cannabina* Peck 1884 (non *Septoria cannabina* Westendorp 1857); =*Septoria cannabis* var. *microspora* Briosi & Cavara 1888.

Description: Pycnidia epiphyllous, gregarious, immersed, globose, eventually erumpent and nearly cupulate, 66 µm in diameter; peridium honey-brown near the ostiole to almost colourless at the base, thin-walled *textura angularis-globulosa;* ostiole irregular, 20 µm in diameter (Fig 5.7). Conidiogenous cells short ampulliform to lageniform, simple, hyaline, holoblastic, up to 8 µm long. Conidia hyaline, filiform, pointed at apex with a truncate base, usually curved, one to three septate, 20–30 x 1.0–2.0 µm.

DIFFERENTIAL DIAGNOSIS

Brown leaf spot causes smaller, darker spots than yellow leaf spot. White leaf spot is paler than yellow leaf spot, and dried lesions rarely flake away. Olive spot produces more symptoms on the undersides of leaves. Fungi causing these diseases look quite different under the microscope.

S. cannabis and *S. neocannabina* are distinguished by several characteristics (McPartland 1995d). Spots caused by *S. neocannabina* have a dark reddish-brown border (Peck 1884, Kirchner 1906, Kirchner 1966). Microscopically, *S. cannabis* pycnidia are larger and have thicker walls than *S. neocannabina* pycnidia (Fig 5.7). Ostioles in *S. cannabis* pycnidia are relatively small, whereas *S. neocannabina* ostioles may open to nearly half the diameter of pycnidia. Conidia of *S. cannabis* are larger and contain more septa than those of *S. neocannabina*.

DISEASE CYCLE & EPIDEMIOLOGY

Septoria species overwinter as pycnidia in crop residue near the soil surface. Seedlings and young plants become infected in early spring. Conidia of *S. cannabis* germinate rapidly and infect plants by penetrating stomates (Ferri 1959b). Optimal growth of *S. cannabis* occurs at 25°C. The incubation period—between inoculation and first symptoms—is six or seven days (Watanabe & Takesawa 1936). Pycnidia on infected plants spew copious amounts of conidia (Fig 5.8). Conidia spread by splashed rain and give rise to epidemics in the summer.

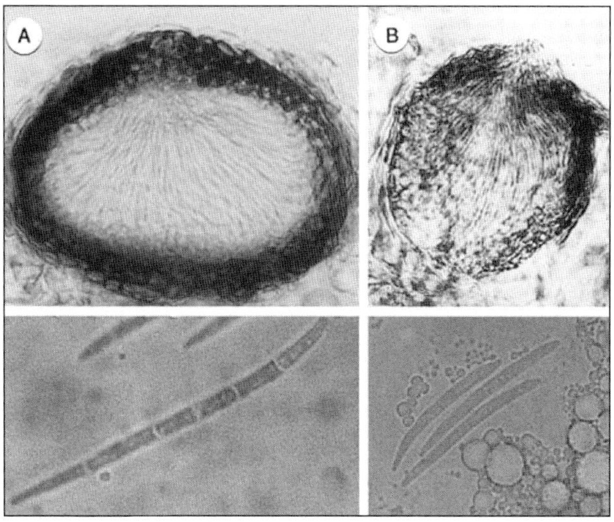

Figure 5.7: Two *Septoria* species. A. *S. cannabis* sectioned pycnidium and conidium; B. *S. neocannabina* sectioned pycnidium and conidia (pycnidia LM x500, conidia LM x980, McPartland).

Barloy & Pelhate (1962) suggested *S. cannabis* spreads via seedborne infection, but did not confirm this experimentally. Other studies of seedborne fungi (Pietkiewicz 1958, Ferri 1961b, Babu *et al.* 1977) have not implicated *Septoria* species. Ferraris & Massa (1912) cited *Leptosphaeria cannabina* as a "probable" teleomorph of *S. cannabis*. Evidence suggests they erred (see the section concerning *L. cannabina*).

CULTURAL & MECHANICAL CONTROL (see Chapter 9)

Methods 1 (sanitation) and 10 (careful pruning) are the cornerstone of control. Also use 2a (deep ploughing), 5 (genetic resistance—see below), 6 (crop rotation), 7c (especially avoid overhead irrigation), 8 (avoid excess nitrogen, add phosphorus and potassium), and 11 (unproven, but hedge your bets).

Monitor for disease during rainy seasons. Stay out of *Cannabis* fields when plants are wet, since conidia spread

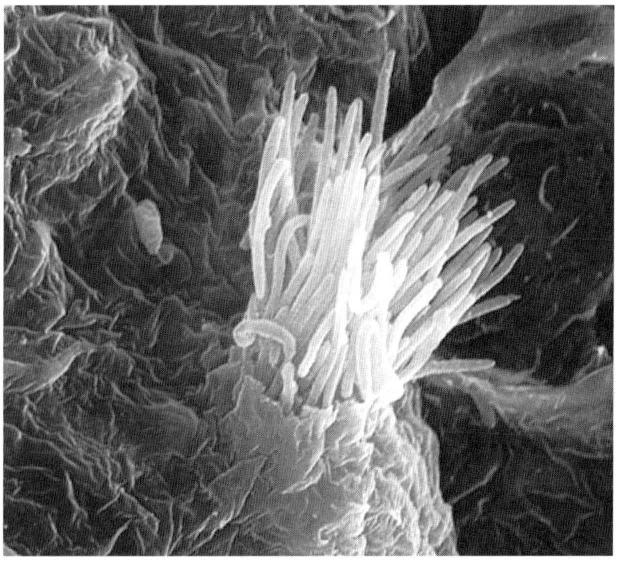

Figure 5.8: Pycnidium of *Septoria neocannabina* erupting from the leaf surface, spewing masses of conidia (SEM x500, McPartland).

easily by contact. Ferri (1959b) described resistant hemp varieties in Italy. Conversely, Bócsa (1958) said "in-bred" monoecious cultivars lose resistance to *Septoria* infection.

BIOCONTROL & CHEMICAL CONTROL

No biocontrol against yellow leaf spot fungi is available. For spray-gun enthusiasts, the disease can be slowed with Bordeaux mixture (Ferri 1959b).

RHIZOCTONIA SORE SHIN & ROOT ROT

This disease has been described on fibre and drug plants in Europe and India. Disease severity ranges from mild to severe. The two fungi causing this disease are found worldwide, and they parasitize approximately 250 plant species. They also attack fellow fungi, animals, and humans. They can live saprophytically in the soil.

SYMPTOMS

Symptoms begin as discoloured roots, usually undetected (Plate 53). Then leaf chlorosis arises, followed by wilting. A dark brown discolouration moves up from the roots to the base of the stalk. Several cm of stalk rot away, leaving a shredded "sore shin" appearance. Within six to eight weeks young plants (<three months old) topple over and die. Older plants may survive. Small black sclerotia sometimes appear in the shredded area. Occasionally a pale mat of basidiospores forms around the base of the stalk.

CAUSAL ORGANISMS & TAXONOMY

1. ***Rhizoctonia solani*** Kühn, *Krankh. Kulturgew.*, p. 224, 1858.

=*Rhizoctonia napaeae* Westendorp & Wallays 1846.

teleomorph: *Thanatephorus cucumeris* (Frank) Donk, *Reinwardtia* 3:376, 1956; ≡*Hypochnus cucumeris* Frank 1883; =*Corticium solani* (Prillieux & Delacroix) Constantineanu & Dufour 1895; ≡*Hypochnus solani* Prillieux & Delacroix 1891; =*Corticium vagnum* Matuo 1949; =*Corticium vagum* Berkeley & Curtis var. *solani* Burt *apud* Rolfs; =*Pellicularia filamentosa* (Patouillard) D.P. Rogers 1943.

Teleomorph taxonomy is unstable, see Parmeter (1970). *Rhizoctonia napaeae* is a "probable synonym" according to Parmeter (1970). Ajrekar & Shaw (1915) cited an orthographic variant, "*Rhizoctonia napi*," on *Cannabis* in India.

Description: Sclerotia deep brown to black, smooth, somewhat flattened and irregular in shape, no differentiation between rind and medulla, up to 6 mm in diameter. Hyphae wide (5–12 µm), multinucleate, with dolipore septa, no clamp connections, at first colourless but rapidly becoming brown; branches form geometrically at 45° or 90° angles from parent hyphae. Hyphae become slightly constricted at branching points; a septum always forms near the base of the branch (Fig 5.5). Basidiocarps arise on stem surface, thin, effuse, discontinuous. Basidia are barrel-shaped, 10–25 x 6–19 µm, borne on imperfectly-symmetrical racemes, with four sterigmata per basidium. Sterigmata are 6–36 µm long and bear spores. Basidiospores hyaline, ellipsoid-oblong, flattened on one side, 5–14 x 4–8 µm (Fig 5.5).

2. **Binucleate *Rhizoctonia* species**

Description: These fungi are morphologically similar to *R. solani*, but have thinner hyphae (4–7 µm) and only two nuclei per hyphal cell. Binucleate *Rhizoctonia* species, unlike *R. solani*, often produce *Ceratobasidium* teleomorphs. See Parmeter (1970) for an introductory discussion.

McPartland & Cubeta (1997) isolated a binucleate *Rhizoctonia* species from the roots of a drug cultivar ('Skunk No. 1') growing in Holland. The binucleate condition was determined by staining hyphae with Safranin O and DAPI. Amplification of hyphal rDNA with PCR, followed by digestion with four different restriction endonucleases, demonstrated that the restriction phenotype of the *Cannabis* isolate was identical to Ogoshi's anastomosis group G strain.

DIFFERENTIAL DIAGNOSIS

Symptoms of Rhizoctonia sore shin disease resemble symptoms of hemp canker and southern blight. The fungi causing hemp canker and southern blight, however, produce more prominent external hyphae and larger sclerotia than *Rhizoctonia* species. If examined under a microscope, the unique hyphae of *Rhizoctonia* species are easily separated from other fungi (Fig 5.5).

DISEASE CYCLE & EPIDEMIOLOGY

R. solani overwinters as sclerotia in soil. Sclerotia germinate in early spring and produce hyphae. Hyphae enter roots near the soil line, either by direct penetration or through wounds. After penetration, *R. solani* oozes cellulose-degrading enzymes (which disrupt xylem and cause wilting), and pectolytic enzymes (which cause cortical rot). The fungus may also kill seedlings (see damping off disease).

A virulent strain of *R. solani* destroyed 80% of *Cannabis* in an Indian epidemic (Pandotra & Sastry 1967). This strain produced neither sclerotia nor basidia—just hyphae. Most strains in Europe produce sclerotia (Rayllo 1927). A Ukrainian strain formed basidiospores (Trunoff 1936). These strains probably represent different anastomosis groups. *R. solani* encompasses at least 11 anastomosis groups, each with unique host ranges and disease cycles. Anastomosis group AG-4 usually causes sore shin in tobacco. But recently, anastomosis group AG-3 has become a problem, causing *leaf* lesions instead of sore shin, because AG-3 readily forms basidiospores which splash onto lower leaves.

Disease symptoms increase under cool, damp conditions, or paradoxically, when temperatures are elevated. Beach (in Lucas 1975) simply explained that sore shin worsens at temperatures and conditions not optimal for the host.

CULTURAL & MECHANICAL CONTROL (see Chapter 9)

Methods 2b&c (sterilize and pasteurize the soil) are very useful. Also apply methods 1 (sanitation), 4 (delay planting until soil warms up), 7b (avoid overwatering), and 8 (avoid excess nitrogen, add calcium). Acidic soils increase disease. Avoid planting in previously infested areas—sclerotia remain viable in soil for up to six years. Many weeds serve as alternate hosts. *R. solani* viability decreases in soils amended with high-carbon mulches, such as straw, corn stover, and even pine shavings (Lucas 1975, Windels 1997). Residues of cabbage and other brassicas release isothiocyanates into the soil, which are toxic to *R. solani*. Enhance the toxic effects of isothiocyanates by covering soil with a plastic tarp, as described in method 2c (Howard *et al.* 1994).

Dippenaar *et al.* (1996) tested 'Fedora-19,' 'Futura-77,' 'Felina-34,' 'Kompolti,' and 'Secuini,' and all the cultivars were equally susceptible to *R. solani*. Broglie *et al.* (1991) enhanced resistance to *R. solani* using genetic engineering. They transferred a tomato gene coding for chitinase (an antifungal enzyme) into tobacco and canola plants, which then became resistant to *R. solani*.

Root-knot nematodes (*Meloidogyne* species) act synergistically with *R. solani*. These underground "land sharks" must be eliminated before sore shin can be controlled. Disease detection kits utilizing ELISA (enzyme linked immunoabsorbent assay) biotechnology can identify *R. solani* in plants or soil. The kits are sensitive enough to detect fungi before symptoms arise.

BIOCONTROL (see Chapter 10)

Biocontrol bacteria include *Bacillus subtilis, Burkholderia cepacia, Pseudomonas fluorescens,* and the actinomycete *Streptomyces griseoviridis* (all described under damping off). Fungi are also effective, especially *Trichoderma* species (described below), and *Gliocladium roseum* (described under grey mould). The mycorrhizal fungus *Glomus intraradices* is described under Fusarium stem canker disease. The soil fungus ***Verticillium biguttatum*** has experimentally controlled *R. solani*. At least 30 other species of soil fungi parasitize *R. solani* (Jeffries & Young 1994).

Trichoderma harzianum

BIOLOGY: A soil fungus that parasitizes the sclerotial fungi *Rhizoctonia solani, Sclerotium rolfsii, Botrytis cinerea,* and, to a lesser degree, *Macrophomina phaseolina* and *Sclerotinia sclerotiorum,* as well as non-sclerotial *Fusarium, Pythium,* and *Colletotrichum* species. It is found worldwide, and grows best at 30°C (Gams & Meyer 1998). *T. harzianum* looks a lot like *T. viride* (see Fig 5.10), but with smooth-walled conidia.

APPLICATION: *T. harzianum* is formulated in a variety of biocontrol products (Binab®, Binab T®, Bio-Fungus®, Bio-Trek®, RootShield®, Supresivit®, T-22®, TopShield®, Trichodex®, Trichopel®), sold by itself or mixed with other *Trichoderma* species, such as ***Trichoderma polysporum***. Supplied as a wettable powder, granules, or planter box formulation. Unopened containers can be stored up to a year in a cool (8–10°C), dark place. Mix *wettable powder* with water and apply as a soil drench; mix *granules* directly into moistened soil or potting media. *T. harzianum* is heat-tolerant, so it can be added to soil immediately after sterilization or pasteurization. The planter box formulation is mixed with stickers and used as a seed treatment before planting. The T-22 strain is a hybrid, selected for effectiveness against a variety of pathogens on a variety of hosts. The T-39 strain (Trichodex®) can be applied as a foliar spray or painted on pruning wounds, although *T. harzianum* is not normally a phylloplane fungus (Samuels 1996).

NOTES: *T. harzianum* is compatible with *Gliocladium roseum,* mycorrhizal fungi, and other biocontrol fungi. Its compatibility with beneficial nematodes is unknown. Some *T. harzianum* isolates (e.g., T-22) tolerate fungicides such as benomyl, captan, and PCNB. T-22 has been applied as a seed coat over seeds previously treated with captan—the fungicide protects seeds as they sprout, then *T. harzianum* protects the developing root system. On the down side, some *Trichoderma* species produce trichothecenes and other highly toxic metabolites. Heat-tolerant strains may cause infections in humans (Samuels 1996).

Trichoderma (Gliocladium) virens

BIOLOGY: A soil fungus native to the USA that antagonizes and parasitizes *R. solani, Pythium ultimum, Botrytis cinerea,* and *Sclerotium rolfsii*. It is marketed for use in glasshouses as SoilGard® (formerly Gliogard®).

APPEARANCE & DEVELOPMENT: The smaller hyphae of *T. virens* coil around *R. solani* hyphae and destroy them (Fig 5.9). Many strains of *T. virens* also produce *gliotoxin,* which suppresses the growth of *P. ultimum* and *R. solani* (and mammals, the LD_{50} in rodents is 25 mg kg^{-1}).

APPLICATION: Formulated as 12–20% alginate-wheat bran prill and granules. Store up to five weeks at 10°C with no loss in efficacy. Mix granules into moistened soil at a minimum rate of 2 g l^{-1}. If granules are broadcast upon the soil surface, use a rate of 21–30 g m^{-2}. *T. virens* works best against damping off fungi if applied three or two days prior to planting. Disease control lasts at least a month. *T. virens* is compatible with mycorrhizal fungi (Linderman *et al.* 1991).

NOTES: *T. virens* produces conidia held in slime (Fig 5.10), which differs from the dry, powdery conidia produced by most *Trichoderma* species (Samuels 1996). Because of the slime, *T. virens* should disperse better in wet weather.

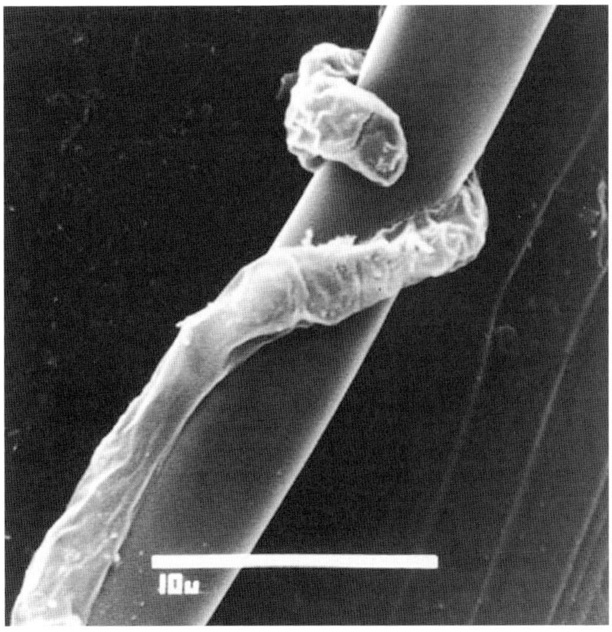

Figure 5.9: Biocontrol fungus *Trichoderma virens* coiling around hypha of *Rhizoctonia solani* (SEM x3900), courtesy USDA.

Trichoderma viride

BIOLOGY: A soil fungus sold for control of *R. solani, Pythium, Fusarium,* and other soil pathogens. It grows worldwide in temperate regions, and researchers have isolated it from hemp (see last page in this Chapter). *T. viride* prefers cooler climates, whereas *T. harzianum* grows better in warm climates (Samuels 1996).

APPLICATION: *T. viride* is formulated with *T. harzianum* as Trichopel®, Trochoseal®, and other products. Supplied as a wettable powder or granules for mixing into soil. *T. viride* is compatible with other *Trichoderma* species.

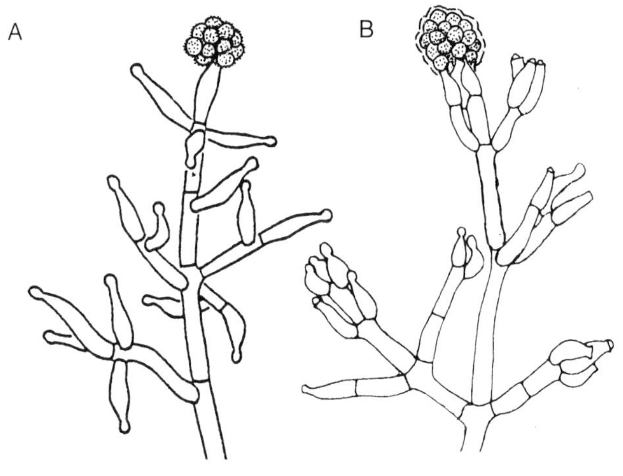

Figure 5.10: *Trichoderma* species x640: A. *T. viride* (McPartland redrawn from Rifai 1969); B. *T. virens* (McPartland redrawn from Samuels 1996).

CHEMICAL CONTROL (see Chapter 11)

Mishra (1987) controlled *R. solani* with several synthetic fungicides, by soaking seeds in fungicide solutions for 15–20 minutes before planting them in infested soil. He also root-dipped *Cannabis* seedlings in fungicide solutions before transplanting seedlings into infested soil (see Appendix 1).

BROWN LEAF SPOT & STEM CANKER

Several species of *Phoma* and *Ascochyta* cause this disease all over the world. Brown leaf spot and stem canker rivals yellow leaf spot as the most common *Cannabis* problem in Europe. Sometimes the two diseases appear on the same plant (Plate 54). Brown leaf spot, like yellow leaf spot, is a "disease of attrition"—as leaf losses increase, crop yields decrease (whether fibre, flowers, or seed).

SYMPTOMS

Leaves low in the canopy begin developing small brown spots in late May. Rarely, spots turn brick red (Shukla & Pathak 1967). Spots remain circular (Plate 57), or form a straight edge along leaf veins. They average 5 mm in diameter, rarely up to 15 mm. Sohi & Nayar (1971) described spots coalescing into large irregular lesions but this is unusual. Spots may become peppered with tiny black pycnidia (Fig 5.11). Two of the fungi causing this disease subsequently produce tiny pseudothecia (Röder 1939, Barloy & Pelhate 1962). Old spots often break apart or drop out, leaving small shot-holes in leaves. Flowering tops may also develop brown spots (Plate 55).

Stem cankers begin as chlorotic spots. They quickly turn grey or brown, and elongate along the stem axis (Plate 56). Pycnidia form on the surface of stem lesions. Two species subsequently develop pseudothecia on stem lesions (Röder 1937, 1939). Diseased plants are stunted, 30 cm shorter than healthy plants by mid-June (Röder 1939).

CAUSAL ORGANISMS & TAXONOMY

Many pathologists erroneously cite the cause of brown leaf spot as ***"Phyllosticta cannabis."*** McPartland (1995c) examined 25 herbarium specimens labelled *"Phyllosticta cannabis"* from the USA and Europe. *None* were true *Phyllosticta* species; most were misidentified *Phoma* or *Ascochyta* species. Some researchers simply list the cause as "*Phoma* sp.," from Lithuania (Brundza 1933), Iowa (Fuller & Norman 1944, 1945), Kansas (Paulsen 1971), India (Srivastara & Naithani 1979), and the Czech Republic (Ondrej 1991). Most researchers, however, attempt to label their pathogens with specific names. At least 23 names appear in the *Cannabis* literature. McPartland (1995c) sorted them out, and reduced the 23 names to nine species:

1. ***Phoma cannabis*** (Kirchner) McPartland, *Mycologia* 86:871, 1994.
≡*Depazea cannabis* Kirchner 1856, =*Phyllosticta cannabis* (Lasch) Spegazzini 1881, =*Phyllosticta cannabis* (Kirchner) Speg. apud others, =*Ascochyta cannabis* (Spegazzini) Voglino 1913 [non: *Ascochyta cannabis* Lasch 1846 (≡*Septoria cannabis* (Lasch) Sacc.)]; =*Erysiphe communis* var. *urticirum* Westendorp 1854; =*Diplodina cannabicola* Petrak 1921, ≡*Diplodina parietaria* Brun. f. *cannabina* von Höhnel 1910.
teleomorph: *Didymella cannabis* (Winter) von Arx, *Beitr. Kryptogamanflora Schwiez* 11:365, 1962; ≡*Sphaerella cannabis* Winter 1872, ≡*Mycosphaerella cannabis* (Winter) Magnus 1905, ≡*Mycosphaerella cannabis* (Winter) Röder 1937.

Conidial morphology is variable, which explains the plethora of synonyms. The fungus also produces ascospores and chlamydospores. The only known host of *P. cannabis* is *Cannabis*, both fibre and drug cultivars. Herbarium specimens are limited to the northern hemisphere (North America, Europe, the Indian subcontinent, Japan). The fungus infests flowering tops (Plate 55), leaves, and stems.

2. ***Ascochyta arcuata*** McPartland, *Mycologia* 86:873, 1994.
teleomorph: *Didymella arcuata* Röder, *Phytopath. Zeits.* 12:321–333, 1939.

Like the previous species, *A. arcuata* produces pycnidia, ascocarps, and chlamydospores. It has only been collected from European hemp, on leaves and stems. Röder (1939) erroneously connected *Didymella arcuata* with *Ascochyta cannabis*. This misconception continues to arise (Spaar *et al.*, 1990).

3. ***Phoma exigua*** Desmazieres, *Annls. Sci. Nat. (Bot.)* ser. 3, 11:282, 1849; =*Ascochyta phaseolorum* Saccardo 1878; =*Plenodomus cannabis* (Allescher) Moesz & Smarods in Moesz 1941, ≡*Plenodomus herbarum* f. *cannabis* Allescher 1899.

Sutton & Waterston (1966) listed *Cannabis* as a host of *A. phaseolorum*. Ondrej (1991) described *P. exigua* appearing "abundantly" on stalks in the Czech Republic. According to Kaushal & Paul (1989), *P. exigua* is resistant to fungitoxic extracts of *Cannabis*. Domsch *et al.* (1980) considered *P. exigua* the most common pycnidial fungus in the world. It attacks over 200 plant genera. Over 100 synonyms exist for this cosmopolitan pathogen. No teleomorph is known. *P. exigua*-infected *Cannabis* has only been found in central and eastern Europe, limited to stem infections (Plate 56).

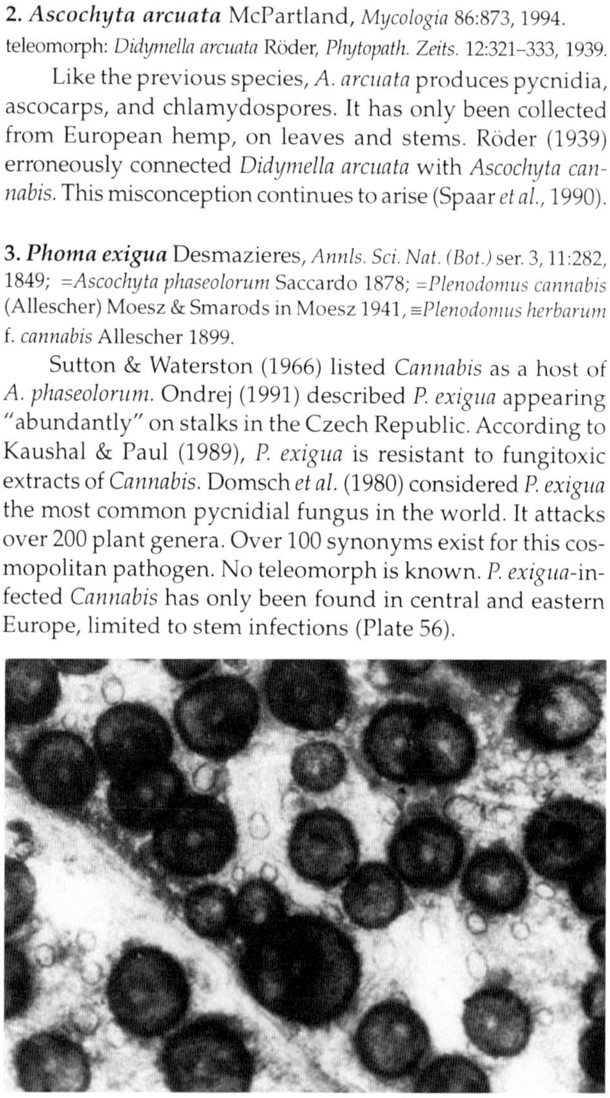

Figure 5.11: Dense growth of *Phoma cannabis* pycnidia on a leaf with brown leaf spot (LM x135, McPartland).

4. ***Phoma glomerata*** (Corda) Wollenweber & Hochapfel, *Z. Parasitkde* 8:592, 1936.

McPartland (1995c) collected this species from leaves of a *C. sativa-C. afghanica* hybrid in Illinois (Plate 57). The chlamydospores of *P. glomerata* can be confused with conidia produced by *Alternaria alternata* (Fig 5.17). *P. glomerata* parasitizes at least 94 plant genera worldwide (Sutton 1980). No teleomorph has been found.

5. ***Phoma herbarum*** Westendorp, *Bulletins Academie Royale Belgique Cl. Sci* 19:118, 1852.

This fungus reportedly causes stem cankers on hemp in Italy (Saccardo 1898), France (Brunard 1899), Denmark (Lind 1913), the Netherlands (Oudemans 1920), and Japan (Kyokai 1965, Kishi 1988). Boerema (1970) considered *P. herbarum* a very weak pathogen, but Domsch *et al.* (1980) reported it from over 35 host genera, worldwide. No teleomorph is known.

6. ***Phoma piskorzii*** (Petrak) Boerema & Loerakker, *Persoonia* 11:315, 1981; ≡*Diploplenodomus piskorzii* Petrak 1923; non: *Phoma acuta* auct., nomen ambiguum. Teleomorph: *Leptosphaeria acuta* (Hoffman:Fries) P. Karsten 1873.

Saccardo & Roumeguere (1883) reported the teleomorph on fibre plants in Belgium. They may have erred (McPartland 1995c,e). The fungus normally attacks stinging nettles, *Urtica dioica* L.

7. ***Ascochyta prasadii*** Shukla & Pathak, *Sydowia* 21:277, 1967.

Shukla & Pathak isolated this fungus from leaves of a drug plant in Udaipur, India. No teleomorph is known.

8. ***Ascochyta cannabina*** E.I. Reichardt, *Bulletin de la Station Régionale Protectrice des Plants à Leningrad* 5:46, 1925.

Reichardt described *A. cannabina* causing leaf spots on hemp near St. Petersburg. He did not report a teleomorph.

9. ***Phoma nebulosa*** (Persoon:Fries) Berkeley, *Outline of British Fungi* p. 314, 1860.

Gruyter *et al.* (1993) cited this species on *C. sativa* in the Netherlands. It is a common soilborne saprophyte, isolated from various herbaceous and woody plants in temperate climates worldwide.

MORPHOLOGICAL DESCRIPTIONS

See McPartland (1995c) for full descriptions. The nine species share many characteristics: Pycnidia dark brown, immersed then erumpent, subglobose, unilocular, ostiolated, usually 90–200 µm diametre on host substrate (up to 450 µm diameter in culture). Conidiogenous cells phialidic, discrete, determinate, simple, ampulliform to short cylindrical. Conidia hyaline, light brown *en masse*, ellipsoidal to short cylindrical, straight or curved, usually zero or one septum (rarely two septa), not constricted at the septum, sometimes finely guttulate, size ranging from small (4 x 2 µm) to very large (28 x 8 µm) on host substrate; larger in culture. Chlamydospores vary from single-celled and globose (8–12 µm diametre) to multicellular, even alternariod in appearance (Fig 5.17). Pseudothecia brown-black, immersed or erumpent, globose on leaves and flattened on stems, 130–150 µm diameter on host (up to 250 µm in culture). Ostiole simple. Asci bitunicate, clavate to pyriform, with a slight pedicle, eight-spored, 50–90 x 9–15 µm. Ascospores hyaline, subovoid to oblong, submedially one-septate with cells unequal in size, constricted at septum, ranging 16–22 x 4–6 µm. Pseudoparaphyses filamentous, zero to three septa, ranging 23–60 x 0.5–1.5 µm.

Differentiating the nine anamorphs is *not* easy on the

Table 5.1: Characteristics of nine *Phoma* or *Ascochyta* species reported from *Cannabis*.

Species	Pycnidia (µm)	Conidia (µm)	Chlamydospores (µm)	Teleomorph
P. cannabis	65–(130)–180	1–2-celled, oval-ellipsoid, guttulate, 3–8 x 2–3	globose to dictyoform, 8–17	*Didymella*
A. arcuata	90–(135)–150	2-celled, short-cylindrical, usually guttulate, 9–28 x 3.5–8.0	globose to oval, 8–12	*Didymella*
P. exigua	110–220	1–2-celled, ellipsoid-cylindrical, guttulate, 5–10 x 2–3.5	absent	none known
P. glomerata	50–300	usually 1-celled, ellipsoid-cylindrical, guttulate, 4.5–10x1.5–4	dictyoform, 30-65x15-20	none known
P. herbarum	80–260	usually 1-celled, ellipsoid, nonguttulate, 4.5–5.5 x 1.5–2.5	absent	none known
P. piskorzii	300–500	usually 1-celled, ellipsoid, nonguttulate, 6–16 x 2–3	absent	*Leptosphaeria*
A. prasadii	55–93	usually 2-celled, oblong, nonguttulate, 7.5–10 x 1.5–2.5	absent	none known
P. nebulosa	100–250	usually 1-celled, oblong, guttulate, 3.6–6.6 x 1.4–2.0	absent	none known
A. cannabina	120–130	usually 2-celled, oval-oblong, nonguttulate, 7–8.8 x 3.2–4	absent	none known

host. Teleomorphs are also difficult to tell apart. Refer to Table 5.1 for key characteristics. Most *Phoma* species must be cultured in a laboratory to confirm their identification.

In culture, the two most common species (*P. cannabis* and *A. arcuata*) are distinguished by conidial size, shape, and septation. According to Röder (1937, 1939), who cultured both fungi, the conidia of *A. arcuata* produce septa, and the conidia of *P. cannabis* do not. *A. arcuata* conidia, *in vitro* and on the host, are longer than *P. cannabis* conidia.

DIFFERENTIAL DIAGNOSIS

Yellow leaf spot and white leaf spot produce larger lesions than brown leaf spot, and they are lighter in colour. When brown leaf lesions break up and fall out, the holes may resemble insect damage. Dobrozrakova *et al.* (1956) noted "stains on stems" from this disease resembled stains caused by *Dendrophoma marconii* and *Microdiplodia abromovii.* Stem cankers may be confused with symptoms caused by *Fusarium, Coniothyrium, Leptodothiorella, Fusicoccum,* or *Phomopsis* species. These fungi are hard to distinguish without a microscope.

DISEASE CYCLE & EPIDEMIOLOGY

Pycnidia and pseudothecia overwinter in plant debris. *P. cannabis, A. arcuata,* and *P. glomerata* also survive as chlamydospores in soil. Infection occurs in early spring, but symptoms may take a month to appear. Conidia spread in splashed rain and wind-driven water. Epidemics build by late summer. The optimal temperature for mycelial growth and spore germination is 19–22°C (Röder 1939).

CULTURAL & MECHANICAL CONTROL (see Chapter 9)

Follow all measures described under yellow leaf spot. Brown spot worsens on hemp grown in peat soils (Rataj 1957); avoid *overuse* of nitrogen fertilizer. *P. exigua* can spread by seedborne infection (Sutton & Waterston 1966); do not harvest seed from infected females. Many *Phoma* diseases bloom in crops weakened by environmental stress. Be sure plants are watered and well-maintained.

BIOCONTROL & CHEMICAL CONTROL

Biocontrol against brown leaf spot and stem canker is not available. Kirchner (1966) used copper-based chemical sprays such as Bordeaux mixture.

DOWNY MILDEW

Downy mildew infests half-grown to mature plants. A world map outlining the range of downy mildew, published by the Commonwealth Mycological Institute (1989), is outdated—the disease now occurs on every continent except Antarctica (McPartland & Cubeta 1997). Two oömycetes cause downy mildew of *Cannabis.* Both pathogens attack fibre and drug cultivars. Researchers have proposed using downy mildew to destroy clandestine *Cannabis* plantations (McCain & Noviello 1985). Zabrin (1981) claimed, "A single infected plant introduced into Colombia or Jamaica during a wet season could cause complete devastation."

SYMPTOMS

Downy mildew begins as yellow leaf spots of irregular size and angular shape, limited by leaf veins. Opposite the spots, on undersides of leaves, the organisms emerge from stomates to sporulate. Mycelial growth on undersides of leaves is best seen in early morning when dew turns the mycelium a lustrous violet-grey colour. Lesions enlarge quickly, and affected leaves become contorted. Leaves soon necrose and fall off. Whole plants and entire fields may follow this course.

CAUSAL ORGANISMS & TAXONOMY

1. *Pseudoperonospora cannabina* (Otth) Curzi, *Riv. Pat. Veg. Pavia* 16:234, 1926; ≡*Peronospora cannabina* Otth 1869, ≡*Peronoplasmopara cannabina* (Otth) Peglion 1917, ≡*Pseudoperonospora cannabina* (Otth) Hoerner 1940.

Description: Sporangiophores hyaline with several arising per leaf stomate, dichotomous branching meagre, occasionally up to the third order, often with swellings on the main axis and branches, 100–350 x 4–8 µm. Sporangia ovoid to ellipsoid, grey-violet (turning brown with age), with protruding apical papillae, germinating into zoospores or hyphae, usually 26–30 x 16–19 µm (Figs 3.2 & 5.12). Zoospores reniform, laterally biflagillate, rounding up to produce hyphae. Oogonia with oospores rare, reported once within *Cannabis* cotyledon mesophyllum (by Peglion 1917, who gave no description or measurements).

2. *Pseudoperonospora humuli* (Miyabe & Takahashi) Wilson, *Mycologia* 6:194, 1914; ≡*Peronoplasmopara humuli* Miya. & Takah. 1905; =*Pseudoperonospora celtidis* (Waite) Wilson var. *humuli* Davis 1910.

Description: Sporangiophores hyaline, with two to five arising per stoma, dichotomous branching abundant, occasionally up to the sixth order, 200–460 x 6–7 µm. Sporangia ellipsoid, grey, with blunt apical papillae, germinating into zoospores, usually 22–26 x 15–18 µm. Oogonia oval, brown, 40–49 µm in diameter. Oospores spherical, brown, smooth, 25–35 µm in diameter.

Only three authors report *P. humuli* on *Cannabis*—Hoerner (1940), Ceapoiu (1958), and Glazewska (1971). The pathogen normally infests hops. Hoerner suggested *P. humuli* and *P. cannabina* represent different physiological races of one species. Berlese (1898) considered *P. cannabina* a *species dubium,* identical to *P. epilobii* Otth. Waterhouse & Brothers (1981) recognized *P. humuli* and *P. cannabina* as different species, based on sporangiophore branching patterns and sporangia papillae.

DIFFERENTIAL DIAGNOSIS

The English call this disease "false mildew," to differentiate it from powdery mildew. Downy mildew can also be confused with olive leaf spot, brown leaf spot, or gall midges. Ferraris (1935) said leaf contortions caused by downy mildew resembled symptoms by *Ditylenchys dipsaci,* the stem nematode. Careful observation with a microscope or strong magnifying lens makes these problems easy to differentiate.

DISEASE CYCLE & EPIDEMIOLOGY

Sporangia are spread by wind or water. Germination of sporangia requires a wet period (a heavy dew will do). Spo-

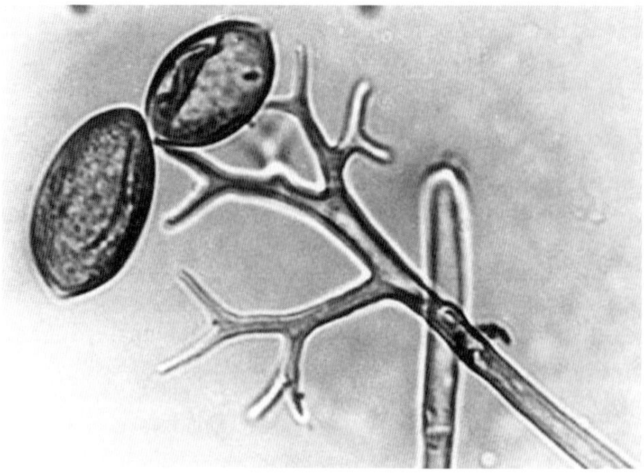

Figure 5.12: Sporangia and sporangiophores of *Pseudoperonospora cannabina* (LM x150, McPartland).

rangia germinate into hyphae or cleave into zoospores. Hyphae penetrate plant epidermis directly. Smith *et al.* (1988) noted *P. humuli* zoospores settle directly over open stomates during the day, but in darkness the zoospores settle randomly over leaf surfaces. Since zoospores can only invade via stomates, infection becomes dependent on daytime leaf wetness. Epidemics arise when warm humid days are followed by cool wet nights.

Downy mildew fungi may persist in a field, and become progressively worse over the years. Barloy & Pelhate (1962) blamed oospores as the source of these perennial infections.

CULTURAL & MECHANICAL CONTROL (see Chapter 9)

Methods 1 (sanitation), 2b&c (sterilize and pasteurize the soil), 5 (genetic resistance—see below), 6 (rotate other crops for a minimum three years), 7c (important—avoid glasshouse dew), 8 (optimize soil structure and nutrition), 10 (very important—completely rogue infected plants including roots). McCain & Noviello (1985) cited two Italian hemp cultivars, 'Superfibra' and 'Carmagnola Selezionata,' with resistance to a strain of *P. cannabina* that destroyed all drug cultivars.

BIOCONTROL & CHEMICAL CONTROL

A unique strain of *Bacillus subtilis* (Serenade®) is sold as a foliar spray for controlling downy mildew (described under damping off). Yepsen (1976) sprayed downy mildew with a solution of boiled horsetail leaves. Ferraris (1935) controlled *P. cannabina* epidemics with copper sulphate. Hewitt (1998) treated *P. humuli* with Bordeaux mixture or copper oxychloride. Be sure to spray the undersides of leaves.

FUSARIUM STEM CANKER

This section begins a subchapter on several Fusarium diseases. Fusarium stem canker is caused by six *Fusarium* species, most with *Gibberella* sexual states. Fusarium root rot, described in the next section, is caused by a *Fusarium* species with a *Nectria* sexual state. Fusarium wilt, described thereafter, is caused by two *Fusarium* species without sexual states. *Fusarium* species also cause damping off of seedlings (Figs 5.5 & 5.6).

SYMPTOMS

Fusarium stem canker usually arises on mid-season-to-mature *Cannabis*. Symptoms begin as watersoaked epidermal lesions, followed by epidermal chlorosis and cortical necrosis. Stems often swell at the site of the lesion, creating fusiform-shaped cankers that may split open (Plate 58). Leaves on affected stems wilt and necrose, without falling off plants. Cankers rarely girdle stems but, if they do, all upper leaves wilt and die. Slicing open cankers often reveals a reddish-brown discolouration.

CAUSAL ORGANISMS & TAXONOMY

Six *Fusarium* species reportedly cause canker, but only two have been isolated on a regular basis. *Fusarium* taxonomy is confusing. Even the experts disagree—Wollenweber & Reinking described 143 *Fusarium* species, varieties, and special forms; Booth cited 44 species and seven varieties, but according to Snyder & Hansen there are only nine species... see Toussoun & Nelson (1975) for an overview.

1. ***Fusarium sulphureum*** Schlechtendahl, *Flora berolinensis* p.139, 1824.

teleomorph: *Gibberella cyanogena* (Desmazières) Saccardo, Sylloge Fungorum 2:555, 1883; =*Gibberella saubinetii* (Montagne) Saccardo 1879, ≡*Botryosphaeria saubinetii* (Montangne) Niessl 1872; =*Botryosphaeria dispersa* DeNotaris 1863; =*Gibberella quinqueseptata* Sherbakoff 1928.

Description: Conidiogenous cells hyaline, cylindrical, arising palmately from metulae, phialidic, 12–20 x 3–5 µm. Macroconidia fusiform, hyaline, with a curved, pointed apical cell and a marked foot cell, usually three or four septa (range zero to six), 30–50 x 3.5–5.0 µm (Fig 5.5). Microconidia absent. Chlamydospores rare, single or in short chains within hyphae or macroconidia, smooth-walled, 8–10 µm in diameter. Perithecia superficial, scattered, blue-black, globose, peridium *textura angularis-globulosa,* with a distinct ostiole, 150–300 µm in diameter. Asci clavate, eight-spored, 65–90 x 12–20 µm. Pseudoparaphyses present but evanescent. Ascospores ellipsoid, hyaline, straight, slightly constricted at three septa, 20–25 x 5–7 µm (Fig 5.5).

F. sulphureum or its teleomorph (commonly cited as *G. saubinetti*) infest hemp in Europe (Saccardo *S.F. II* 1883, Voglino 1919, Oudemans 1920) and the USA (Scherbakoff 1928, Miller *et al.* 1960). The teleomorph has been confused with *G. zeae* (Wollenweber & Reinking 1935, Vakhrusheva 1979). McPartland (1995e) added *G. quinqueseptata* to the list of *G. cyanogena* synonyms.

2. ***Fusarium graminearum*** Schwabe, *Flora Anhaltina* 2:285, 1838. =*Fusarium roseum* Link emended Snyder, Hansen & Oswald 1957, *pro parte.*

teleomorph: *Gibberella zeae* (Schweinitz:Fries) Petch, Annales Mycologici 34:260, 1936; ≡*Sphaeria zeae* Schweinitz 1822 (non *Sphaeria zeae* Schweinitz 1832).

Description: Conidiogenous cells hyaline, short, doliiform, phialidic, simple or branched dichotomously, proliferating through previous phialides, 10–14 x 3.5–4.5 µm. Macroconidia sickle-shaped, hyaline, with a well marked foot cell, three to seven septate, 25–30 x 3–4 µm. Microconidia absent (although some are illustrated by Booth 1971). Chlamydospores rare, single or in chains, in hyphae or rarely in macroconidia, hyaline to brown, thick-walled, smooth or slightly rough surface, 10–12 µm in diameter. Perithecia superficial, clustered, blue-black, ovoid, with a rough tuberculate surfaced, 140–250 µm in diameter. Asci clavate, usually eight-spored (rarely four to six), 60–85 x 8–11 µm, but rapidly deliquesce after ascospore production. Ascospores fusoid, hyaline to light brown, curved, three-septate (20% two-septate or less), 19–30 x 3–5 µm.

F. graminearum often confuses mycologists, even so eminent a pair as Wollenweber & Reinking (they called *F. graminearum* the conidial state of *G. saubinetii,* see McPartland 1995a). *F. graminerum* or its teleomorph, *G. zeae,* infest hemp in Europe (Ceapoiu 1958, Ferri 1961b). Francis & Burgess (1977) described two groups of *G. zeae:* Group 1 is soilborne, causes crown rots, and rarely produces perithecia (recently renamed ***Fusarium pseudograminearum*** O'Donnell & Aoiki 1999). Group 2 is airborne, causes stem cankers, and readily forms perithecia. McPartland & Cubeta (1997) isolated a Group 2 homothallic strain from feral hemp in Illinois.

3. ***Fusarium lateritium*** Nees:Fries, *Syst. Pilze Schwamme* p. 31, 1817, emended Snyder & Hansen 1945, *pro parte.*

teleomorph: *Gibberella baccata* (Wallroth) Saccardo, *Sylloge Fungorum* 2:553, 1883.

Pietkiewicz (1958) cited this species infesting hemp seeds in Poland, possibly erroneously. It normally infests woody hosts. For a description, see Booth (1971).

4. ***Fusarium sambucinum*** Fuckel, *Symbolae Mycologicae* p. 167, 1869; =*Fusarium sarcochroum* (Desmazières) Saccardo 1879; =*Fusarium roseum* Link emended Snyder & Hansen 1945, *pro parte.*

teleomorph: *Gibberella pulicaris* (Fries:Fries) Saccardo, *Michelia* 1:43, 1887; ≡*Botryosphaeria pulicaris* (Fries) Cesati & DeNotaris *apud* Oudemans 1920.

G. pulicaris has been described on hemp once, by Oudemans (1920). The species causes a basal canker on hops and woody hosts. *F. sambucinum* may be identical to *F. sulphureum* (Howard *et al.* 1994).

5. ***Fusarium avenaceum*** (Corda:Fries) Saccardo, *Sylloge Fungorum* 4:713, 1886.

teleomorph: *Gibberella avenacea* R.J. Cooke, *Phytopathology* 57:732–736, 1967.

G. avenacea caused *Cannabis* disease in host-range experiments (Zelenay 1960). It attacks a wide range of hosts. For a description, see Booth (1971).

6. ***Fusarium culmorum*** (W.G. Smith) Saccardo, *Sylloge Fungorum* 11:651, 1895.

This pathogen normally infests cereal crops. Abiusso (1954) cited it on Argentinean *Cannabis*, perhaps erroneously. For a description, see Booth (1971).

DIFFERENTIAL DIAGNOSIS

Separating *Fusarium* species usually requires culturing them in lab and hoping their teleomorphs develop. *F. sulphureum* macroconidia are larger than *F. graminearum* macroconidia. The conidiogenous cells of *F. sulphureum* are cylindrical and branch in a palm-shape, whereas *F. graminearum* conidiogenous cells are stubby (doliiform), branch in pairs, and proliferate through previous conidiogenous cells.

Fusarium foot rot (see next section) may be confused with Fusarium stem canker. The cause of foot rot (*Fusarium solani*) can be distinguished from *F. sulphureum* and *F. graminearum* by its microconidia.

The reddish xylem discolouration seen in Fusarium stem canker does not appear in grey mould, hemp canker, and southern blight. Stem swelling caused by *Fusarium* species can be confused with stem swellings caused by European corn borer, hemp borer, and assorted beetles.

DISEASE CYCLE & EPIDEMIOLOGY

Perithecia overwinter in crop stubble or soil. Some *Fusarium* species also overwinter as seedborne infections. In the spring, ascospores or conidia infect seedlings. Later in the season, *Fusarium* species invade roots via wounds created by nematodes and *Orobanche* parasites. Since *Fusarium* conidia move in water droplets, they do best in damp conditions and heavy soil. Watch for *Fusarium* epidemics during seasons with above-average rainfall. Above ground, *Fusarium* conidia cause secondary infections at stem nodes and wound sites. Wounds are caused by stem-boring insects, wind, improper pruning, and from damage caused by dry, caking soil. Hydroponic cultivators should not allow rockwool blocks to dry out, because evaporated fertilizer salts accumulate around the base of the stalk and favour infections (Howard *et al.* 1994).

All six *Fusarium* species are distributed worldwide. *F. graminearum* and *F. avenaceum* predominate in cooler climates. *F. graminearum* is considered the most pathogenic of the bunch (Booth 1971). Most *Fusarium* species are facultative parasites, existing as soil saprophytes until a parasitic opportunity presents itself.

Some fusaria pose a threat to humans. *F. graminearum* produces **zearalenone**, a toxic metabolite. Consuming foods infested with *F. graminearum* may cause nausea, vomiting, diarrhea, headache, chills, and convulsions (Rippon 1988). *F. graminearum* also produces **trichothecenes**, known as "T-2 toxins." T-2 toxins cause a haemorrhagic syndrome, gaining notoriety for their reputed use in chemical warfare ("yellow rain"). Lastly, fusaria can invade human tissue and grow in dirty wounds and in the skin of burn victims (Rippon 1988).

CULTURAL & MECHANICAL CONTROL (see Chapter 9)

Booth (1971) characterized stem canker as a "disease of accumulation." Thus methods 1 (sanitation), 2a (deep ploughing), 2b&c (sterilizing and pasteurizing the soil), 2d (flooding soil), and 10 (careful pruning) are paramount for reducing inoculum. Other important methods include 7b (avoid overwatering), 11 (avoid seedborne infections), and 5 (cultivar 'Fibramulta 151' is resistant to *Fusarium* species according to Dempsey 1975). Do not plant in heavy, wet soils or low lying areas. Avoid excess nitrogen; increase potassium and phosphorus. Choose KCl as a potassium supplement instead of K_2SO_4 or KNO_3. K_2SO_4 and KNO_3 may actually increase the disease (Elmer, pers. commun. 1990). Since fusaria have a wide host range and live saprophytically, crop rotation is not effective. Diseased plants should not be harvested for human consumption.

BIOCONTROL

Biocontrols include the bacteria *Bacillus subtilis, Burkholderia cepacia, Pseudomonas fluorescens,* and the actinomycete *Streptomyces griseoviridis,* all described under damping off. *Trichoderma* and *Gliocladium* fungi are also useful (see Rhizoctonia sore shin).

Glomus intraradices

BIOLOGY: A mycorrhizal fungus (Mycori-Mix®, Nutri-Link®) that protects roots against soil pathogens such as *Fusarium* and *Pythium* species. It does best in soils with moderate moisture and moderate temperatures.

APPEARANCE & DEVELOPMENT: See the section on mycorrhizae (in Chapter 5) to understand the utility of *Glomus* species.

APPLICATION: Supplied as spores which can be stored for months in a cool (8–10°C), dark place. *G. intraradices* is used as a soil inoculant to improve plant nutrition, and it protects plants from a variety of root pathogens.

NOTES: Although mycorrhizal fungi have wide host ranges, they also show preferences for certain soils, climates, and hosts. Thus, mixtures of mycorrhizae are becoming available. Mycorrhizae work well in combination with other microbial biocontrols, such as *Trichoderma, Gliocladium, Talaromyces, Bacillus, Pseudomonas, Streptomyces, and Bacillus* species (Linderman *et al.* 1991). Other soil organisms attack mycorrhizae, including fungi (*Cephalosporium* species), nematodes (*Aphelenchoides* species), and insects (*Folsomia* species). Pesticides, synthetic fertilizers, and frequent soil tillage kill mycorrhizae.

CHEMICAL CONTROL

Seeds suspected of harbouring *Fusarium* should be disinfested. Booth (1971) soaked seeds in formalin (see Chapter 12). Spraying fungicides on plants has not proved practical against *Fusarium*. Neither has soil fumigation, except to control synergistic organisms such as nematodes. Vysots'kyi (1962) protected pine seedlings from *Fusarium* by mixing hemp leaves into soil.

FUSARIUM FOOT ROT & ROOT ROT

Fusarium foot rot and root rot has been reported in the former USSR (Gitman & Boytchenko 1934, Vakhrusheva 1979), Poland (Zelenay 1960), southern France (Barloy & Pelhate 1962), and Maryland (Miller *et al.* 1960). Data in USDA archives also described the disease affecting hemp in Illinois and Virginia. The causal organism is found worldwide. On hemp it sometimes causes stem canker instead of root rot (Wollenweber 1926).

SYMPTOMS

Plant stems turn brown at the soil line; this "foot rot" is an extension of the root rot. Common above-ground symptoms include partial or systemic wilting. Roots turn red, rotten, and necrotic (Plate 59). According to Zelenay (1960) and Barloy & Pelhate (1962), this disease knocks down plants in all stages of development, including seedlings (Fig 5.6).

CAUSAL ORGANISM & TAXONOMY

Fusarium solani (Martius) Saccardo, *Michelia* 2:296, 1881, emended Snyder & Hanson, *Am. J. Bot.* 28:740, 1941.

=*Fusarium javanicum* Koorders 1907.

teleomorph: *Nectria haematococca* Berkeley & Broome, *J. Linn. Soc. (Botany)* 14:116, 1873; ≡*Hypomyces haematococca* (Berkeley & Broome) Wollenweber 1926; =*Hypomyces cancri* (Rutgers) Wollenweber 1914, ≡*Nectria cancri* Rutgers 1913; non: *Hypomyces solani* Rein. & Bert. 1879.

Description: *Macroconidial* conidiogenous cells hyaline, short, doliiform, frequently branched, phialidic. Macroconidia fusoid, stout, broad, thick-walled, hyaline, with a pointed, somewhat beaked apical cell, one to six septa, 40–100 x 5–8 µm (Fig 5.5). *Microconidial* conidiogenous cells hyaline, long, cylindrical, phialidic, up to 400 µm in length. Microconidia hyaline, ovoid to allantoid, sometimes becoming one-septate, 8–16 x 2–4 µm. Chlamydospores usually in pairs, but sometimes single or in chains, globose to oval, smooth to rough walled, 9–12 x 8–10 µm. Perithecia superficial upon a thin pseudostroma, scattered or clustered, pale orange to brown, globose (upon drying they collapse laterally, as often illustrated in textbooks), with a warted surface, 110–250 µm in diameter. Asci cylindrical to clavate with a rounded apex, eight-spored, 60–80 x 8–12 µm. Ascospores ellipsoid to obovoate, hyaline but becoming light brown with longitudinal striations at maturity, slightly constricted at single septum, 11–18 x 4–7 µm (Figs 3.2 & 5.5).

DISEASE CYCLE & EPIDEMIOLOGY

F. solani overwinters as conidia, chlamydospores, or ascospores in crop debris or in soil. Perithecia are abundant in wet tropics and less common in temperate zones (Booth 1971). The mode of infection is similar to *Fusarium* species described in the previous section. Although *F. solani* is characterized as a weak pathogen, it acts synergistically with nematodes and parasitic plants. It easily invades root wounds created by other organisms. Barloy & Pelhate (1962) considered a combination of *F. solani* and broomrape (*Orobanche ramosa*) the greatest threat to *Cannabis* cultivation in southern France.

Like other fusaria, *F. solani* may also pose a threat to humans. Ingesting *F. solani*-moulded sweet potatoes may cause respiratory distress. *F. solani* also causes eye infections, especially in agricultural workers (Rippon 1988).

DIFFERENTIAL DIAGNOSIS

Initial symptoms (wilting) resemble wilts caused by nematodes and some soil insects (e.g., root maggots, white root grubs). *F. solani* can be distinguished from *F. oxysporum* (a related pathogen described next) by its long microconidiophores and stouter, thicker macroconidia (Fig 5.5). Grey mould, hemp canker, and southern blight produce sclerotia on stalks. No red staining is present in root rots caused by other fungi.

CONTROL

Follow recommendations in the previous section. Fusarium root rot usually appears after abundant rainfall in fields with heavy soils. Recognize disease symptoms and rogue affected plants as soon as possible. Eliminate synergistic organisms such as nematodes and broomrape.

Pseudomonas stutzeri has controlled *F. solani* root rot in kidney beans (*Phaseolus vulgaris*). This bacteria is related to *Pseudomonas fluorescens* (described under damping off).

Mixing dried *Cannabis* leaves into soil may suppress *Fusarium* species (McPartland 1998). Grewal (1989) mixed 3 kg of dried leaves into 137 kg of wheat straw compost and suppressed *F. solani* growth. Pandey (1982) reported an aqueous extract of hemp leaves inhibited *Fusarium* in petri plates. Vysots'kyi (1962) applied an aqueous extract in the field to protect pine seedlings from a *Fusarium* species. Dahiya & Jain (1977) reported that pure THC and CBD inhibited the growth of *F. solani*.

FUSARIUM WILT

The pathogen causing this disease is a xylem inhabitant, rather than a cortical pathogen like the aforementioned *Fusarium* species. It plugs plant water-conducting tissues and causes a wilt. Fusarium wilt is a serious disease in eastern Europe (Dobrozrakova *et al.*, 1956, Rataj 1957, Ceapoiu 1958, Czyzewska & Zarzycka 1961, Serzane 1962, Zhalnina 1969), Italy (Noviello & Snyder 1962), and southern France (Barloy & Pelhate 1962). It may occur in Pakistan (Ghani *et al.* 1978) and the USA (McPartland 1983a). Gitman (1968b) considered it a little-known disease in the former USSR. The causal fungus also causes damping-off in seedlings (Pietkiewicz 1958, Zelenay 1960, Barloy & Pelhate 1962).

A scientist at UC-Berkeley, Arthur McCain, suggested to President Nixon that Fusarium wilt could destroy illegal marijuana cultivation (Shay 1975). Subsequently, the Nixon administration funded research to mass-produce the wilt fungus (Hildebrand & McCain 1978). Zubrin (1981) interviewed McCain, who claimed, "Just introduce a couple of pounds [of the fungus] into an area, and while it wouldn't have much of an effect the first year, in several years it would spread throughout the country with devastating results." The project was terminated by Carter administration officials (Zubrin 1981).

Interest in eradicating "*Cannabis sativus*" (Sands 1991) with the wilt fungus resumed during the Reagan/Bush administration. A virulent isolate of the fungus was collected in Russia; Ronald Collins subsequently released the pathogen in a USDA field plot in Beltsville, Maryland. The fungus attacked *Cannabis* plants, successfully overwintered in the soil, then killed seedlings planted in the same soil the following year (Sands 1991). More recently, 25 strains of the fungus were isolated by researchers in Russia and Kazakhstan (Semenchenko *et al.* 1995). Some strains reduced *Cannabis* survival by up to 80% (Tiourebaev *et al.* 1998). The current deployment of genetic engineering to create super-wilt *Cannabis* pathogens has elicited public outrage and scientific criticism (McPartland & West 1999).

SYMPTOMS

Small, dark, irregular spots initially appear on lower leaves. Affected leaves suddenly become chlorotic. Wilt symptoms begin with an upward curling of leaf tips. Wilted leaves dry to a yellow-tan colour and hang on plants without falling off (Plate 60). Stems also turn yellow-tan. Cutting into wilted stems reveals a reddish-brown discolouration of xylem tissue. Pulled-up roots show no external symptoms. Barloy & Pelhate (1962) described the fungus wilting whole plants. Noviello & Snyder (1962) illustrated plants with only one side wilting. Dead plants may become partially enveloped in a white-pink mycelium. Surviving plants are stunted.

CAUSAL ORGANISM(S) & TAXONOMY

1. ***Fusarium oxysporum*** Schlechtendahl:Fries f.sp. ***vasinfectum*** (Atkinson) Snyder & Hansen, *Am. J. Bot.* 27:66, 1940.

≡*Fusarium vasinfectum* Atkinson 1892.

2. ***Fusarium oxysporum*** Schlechtendahl:Fries f.sp. ***cannabis*** Noviello & Snyder, *Phytopathology* 52:1315–1317, 1962.

Description: The two subspecies are morphologically identical—mycelium in culture usually floccose or felty, white to pink to purple, growing abundantly. Conidiogenous cells hyaline, short, barrel-shaped, phialidic, in tufts of one to four atop metulae, 10–12 µm long. Macroconidia hyaline, three to five septa, sickle-shaped, ends curved inward with a hooked apex and pedicellate base, as large as 45–55 x 3.5–4.5 µm (Figs 3.2, 5.5 & 5.13). Microconidia hyaline, aseptate (rarely one septum), oval to cylindrical, 5–16 x 2.2–3.4 µm. Chlamydospores hyaline, thick walled, with a rough or smooth surface, spherical, borne singularly or in pairs, formed atop conidiophores or intercalary within hyphae or macroconidia, 7–13 µm in diameter.

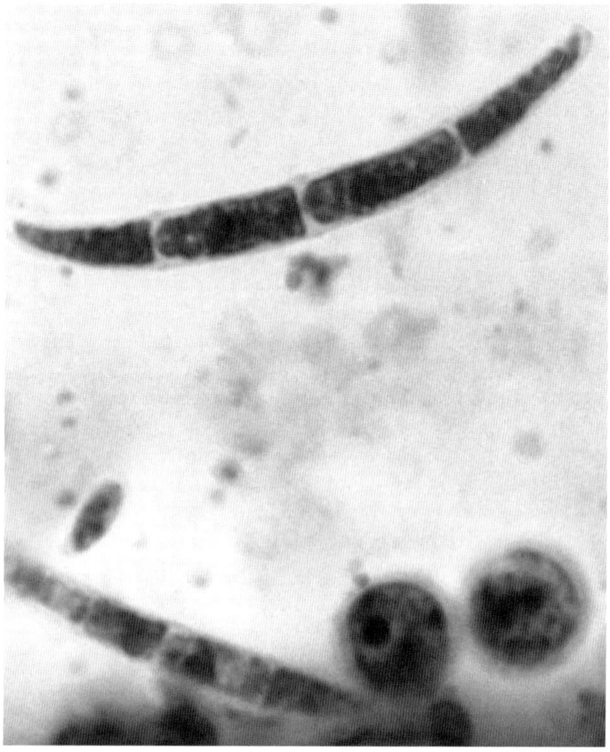

Figure 5.13: Macroconidia, microconidia, and chlamydospores of *Fusarium oxysporum* f.sp. *cannabis* (LM x950, McPartland).

In 1962 Noviello & Snyder published a new form-species, *F. oxysporum* f.sp. *cannabis*. They justified this new name saying, "...the wilt disease and its pathogen have not been described." They apparently overlooked many descriptions of *F. oxysporum* f.sp. *vasinfectum* wilting *Cannabis* (e.g., Dobrozrakova *et al.* 1956, Rataj 1957, Ceapoiu 1958, Czyzewska & Zarzycka 1961, Serzane 1962).

Armstrong & Armstrong (1975) recognized six races of *F. oxysporum* f.sp. *vasinfectum*. These races attack cotton, mung beans, pigeon peas, rubber trees, alfalfa, soyabeans, coffee, tobacco, and many other plants. The tobacco wilt fungus was originally called *F. oxysporum* f.sp. *nicotianae*, but Armstrong & Armstrong (1975) proved it to be a race of *vasinfectum* (Lucas 1975). Similar experiments may prove "f.sp. *cannabis*" to be another race of *vasinfectum*.

Mycoherbicide researchers consistently misspell the form-species as "f.sp. *cannabina*" (Tiourebaev *et al.* 1997, Tiourebaev *et al.* 1998). Ceapoiu (1958) cited ***Neocosmospora vasinfecta*** Smith 1897 on Romanian hemp. This ascomycete was *erroneously* described as the sexual stage of *F. vasinfectum*. Ceapoiu's citation of *N. vasinfecta* on *Cannabis* is probably due to this enduring misconception (McPartland 1995a).

DIFFERENTIAL DIAGNOSIS

F. solani (the cause of Fusarium foot rot and root rot) causes a reddish-brown discolouration in roots; *F. oxysporum* does not (Barloy & Pelhate 1962). Microscopically, *F. oxysporum* is distinguished from *F. solani* by its thin macroconidia and short microconidial conidiophores (Fig 5.5). *F. oxysporum* sometimes produces only microconidia (Gams 1982). Microconidia might be confused with those of *Acremonium* (=*Cephalosporium*), which also attack *Cannabis* (Fuller & Norman 1944, Babu *et al.* 1977, Gzebenyuk 1984). Fusaria differ from *Acremonium* species by faster growth, wider hyphae, and larger microconidia (Gams 1982).

Symptoms of Fusarium wilt are similar to those of Verticillium wilt, Texas root rot, southern blight, Rhizoctonia sore shin, early symptoms of charcoal rot, some nematode diseases, and injury caused by soil insects (e.g., root maggots, white root grubs). Finding a reddish discolouration of xylem is helpful, but observing the fungus with a microscope may be needed for positive identification.

DISEASE CYCLE & EPIDEMIOLOGY

F. oxysporum overwinters as chlamydospores in soil or crop debris. In the spring, chlamydospores produce hyphae which directly penetrate roots of seedlings. In older "root-hardened" plants, the hyphae must enter via wounds. Thus disease worsens in fields harbouring root-wounding broomrape or nematodes (particularly *Meloidogyne incognita*). Clarke (1993 field notes, Kompolt, Hungary) reported a wet-dry-wet summer predisposed plants to a *Fusarium* epidemic. Lack of rain caused heavy clay soil to cake and crack, which wounded plant roots, allowing *Fusarium* to invade. Clarke (1993 field notes, Kompolt, Hungary) found plants infected with Hemp streak virus exhibited resistance to *Fusarium* wilt.

After hyphae penetrate roots, *F. oxysporum* invades water-conducting xylem tissues. Microconidia arise in these vessels and flow upstream to establish a systemic infection. The fungus eventually plugs xylem vessels, interrupts water flow, and plants wilt. Wilt symptoms may appear within two weeks of fungal inoculation (Semenchenko *et al.* 1995).

Fusarium wilt is a warm-weather disease. Optimal temperature for fungal growth is 26°C (Noviello & Snyder 1962). Disease symptoms may not become evident until the advent of hot summer temperatures.

There is no windborne transmission. Conidia arising on dead plants may be rain-splashed onto neighbours. Chlamydospores also arise on dead plants. *F. oxysporum* invades *Cannabis* seeds. Seedborne infections lay dormant until seedlings sprout the following spring (Pietkiewicz 1958, Zelenay 1960). Mycoherbicide researchers discovered that spore-coated *Cannabis* seeds effectively spread *F. oxysporum* through the soil (Tiourebaev *et al.* 1997).

CULTURAL & MECHANICAL CONTROL (see Chapter 9)

Methods 1 (sanitation) and 11 (avoid seedborne infection) provide a cornerstone for wilt control. Methods 2b&c (sterilizing or pasteurizing soil) often work. Flooding soil (method 2d) kills strongly aerobic *Fusarium* species while maintaining a beneficial bacterial population. Method 5 offers control—the Romanian hemp cultivar 'Fibramulta 151' is resistant to Fusarium wilt while the Italian hemp cultivar 'Super Elite' is susceptible (Noviello *et al.* 1990). Generally, fibre plants have more resistance than drug plants (McCain & Noviello 1985). Goebel & Vaissayre (1986) devised a rapid method for screening cultivars for resistance to Fusarium wilt in the field.

Noviello *et al.* (1990) claimed different soil types affect plant resistance, but did not elaborate. Barloy & Pelhate

(1962) reported epidemics in sandy, alluvial soils. Soils deficient in calcium and potash predispose plants to wilt disease and must be corrected (see Elmer's comments regarding optimal K supplements in the section on Fusarium stem canker). Excess nitrogen increases wilt disease. Zhalnina (1969) reported that "acid fertilizers" increased wilt (he corrected this by adding lime). Keep soil pH near neutral. Mix organic material and green manure into soil to encourage the growth of natural *Fusarium* antagonists (Windels 1997).

Continuous cropping of *Cannabis* causes a buildup of inoculum and the creation of "wilt-sick soil" (Czyzewska & Zarzycka 1961). Armstrong & Armstrong (1975) found *F. oxysporum* f.sp. *vasinfectum* growing in roots of many symptomless plants. These unidentifiable hosts make crop rotation difficult. Czyzewska & Zarzycka (1961) suggested rotating hemp after wheat. Ceapoiu (1958) laid soil fallow for five years to eliminate fusaria.

Lucas (1975) cited interesting atmospheric studies: as carbon dioxide increases, *F. oxysporum* replication decreases. Supplementing a glasshouse with 20% CO_2 cut *F. oxysporum* growth rate and spore production by 50%.

BIOCONTROL

Czyzewska & Zarzycka (1961) mixed ***Trichoderma lignorum*** into soil to protect their hemp crop. Commercially available biocontrols include *Burkholderia cepacia* and *Streptomyces griseoviridis* (described under damping off), *Gliocladium* species (see Rhizoctonia sore shin), *Trichoderma harzianum* (see Rhizoctonia sore shin), and *Glomus intraradices* (see Fusarium stem canker). The latter two work well in combination against *F. oxysporum* (Datnoff *et al.* 1995). A hypovirulent strain of *F. oxysporum* is described below.

Fusarium oxysporum (nonpathogenic)

BIOLOGY: A nonvirulent strain (Biofox C®, Fusaclean®) of the wilt fungus. It prevents virulent pathogens from attacking roots. The nonpathogenic strain colonizes roots of plants and lives as a saprophyte. It has been used on tomato, basil, carnation, and cyclamen crops.

APPLICATION: Supplied as spores in a liquid or microgranule formulation. It can be stored for months in a cool (8–10°C), dark place. *F. oxysporum* can be applied as a seed treatment, mixed with potting soil, incorporated into drip irrigation, or applied as a soil drench.

CHEMICAL CONTROL

Disinfect seeds with fungicides; see the section on Fusarium canker. Spraying wilted plants with fungicides is not useful. Some farmers fumigate soil with nematicides to reduce root wounding in *Fusarium*-infested fields.

POWDERY MILDEW

Powdery mildew is known as "mould" in England and "Echter Mehltau" (true mildew) in Germany. This distinguishes it from downy mildew. Powdery mildew arises in temperate and subtropical regions. It infests outdoor hemp (Transhel *et al.* 1933, Gitman & Boytchenko 1934, Gitman 1935, Doidge *et al.* 1953) and indoor drug cultivars (Stevens 1975, McPartland 1983a, McPartland & Cubeta 1997).

Two fungi cause powdery mildew on *Cannabis*. Westendorp (1854) described a third "powdery mildew," ***Erysiphe communis* var. *urticirum***. But McPartland (1995a) examined Westendorp's original specimen. The fungus was a misidentified *Phoma* species, mixed with mite webbing (Plate 55).

SYMPTOMS

Early signs of infection include raised humps or blisters on upper leaf surfaces. From these areas the powdery mycelium arises. Mildew may remain isolated in irregular pustules or coalesce over the entire leaf. Leaves soon look like they were dusted with flour or lime (Plate 61). Infected plants remain alive, or they prematurely yellow, brown, and die. If the disease is permitted to run its course, black specks (fungal cleistothecia) arise in the powdery mycelium.

CAUSAL ORGANISMS & TAXONOMY

1. *Sphaerotheca macularis* (Wallroth:Fries) Lind, *Danish Fungi* p. 160, 1913; ≡*Erysiphe macularis* (Wallroth) Fries 1824; =*Sphaerotheca humuli* (DeCandolle) Burrill 1887.

anamorph: *Oidium* sp.; =*Acrosporium* sp.

Description: Superficial hyphae flexous, branched, with inconspicuous appressoria, cell diameters 4–7 µm, lengths averaging (37–) 64.5 (-80) µm. Conidiophores upright, simple, hyaline, 50–100 µm tall. Conidia produced in chains, hyaline (turning brown with age), containing fibrosin bodies (which disappear with age), ovate to barrel shaped, single-celled, averaging 30.2 x 14.0 µm (Figs 3.3 & 5.14). Cleistothecia globose, black, smooth, 60–125 µm in diametre, with few to many hyphal appendages (Fig 3.3). Appendages hyaline to brown, unbranched, flexous, tapering at ends, 300–500 µm in length. Asci one per cleistothecium, subglobose to broadly elliptical, indistinctly stalked, eight-spored, 50–90 x 45–75 µm. Ascospores ellipsoidal to oval, hyaline, unicellular, 18–25 x 12–18 µm.

S. macularis lives worldwide and commonly parasitizes hops (Miller *et al.* 1960). Only the anamorph *Oidium* state (subgenus *Fibroidium*) has been found on drug plants in the USA (McPartland 1983a, McPartland & Cubeta 1997), on *Cannabis* in South Africa (Doidge *et al.* 1953), and on fibre varieties in Russia and Italy (Hirata 1966).

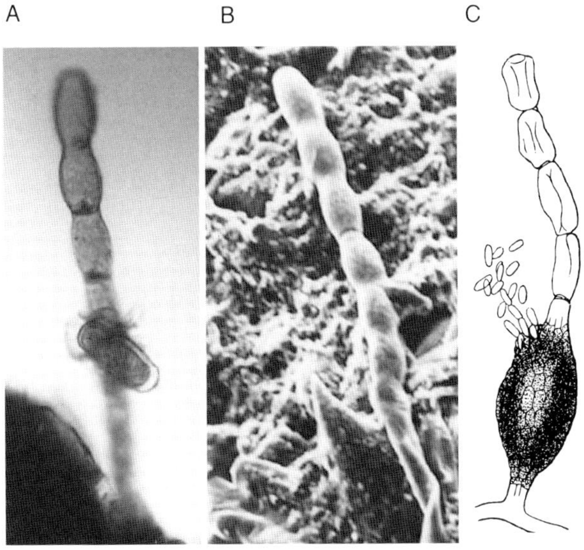

Figure 5.14: Conidia and conidiophores of *Sphaerotheca macularis*. A. LM x390; B. SEM x390; C. Drawing of conidiophore being parasitized by *Ampelomyces quisqualis* (A & B by McPartland, C by DeBary 1887).

2. *Leveillula taurica* (Léveillé) Arnaud, *Annales Épiphyties* 7:92,1919. =*L. taurica* f. sp. *cannabis* Jaczewski 1927.

anamorph: *Oidiopsis taurica* (Léveillé) Salmon 1906.

Description: Superficial hyphae often cover whole plants; pale buff to white, persistent, densely compacted, tomentose. Conidiophores often two-celled, upright, simple or occasionally branched, hyaline, emerging from host stomates, up to 250 µm long. Conidia borne singularly atop conidiophores, either cylindrical or navicular in shape, single-celled, hyaline, 50–79 x 14–20 µm. Cleistothecia rarely formed, scattered, embedded in mycelium, globose, black, smooth, 135–250 µm in diameter, with numerous hyphal appendages. Appendages hyaline to light brown, indistinctly branched, less than 100 µm in length. Asci usually 20 per cleistothecium, ovate, distinctly stalked, two-spored, 70–110 x 25–40 µm. Ascospores cylindrical to pyriform, sometimes slightly curved, 24–40 x 12–22 µm.

L. taurica is distributed worldwide and parasitizes a wide range of hosts, including *Cannabis* in southern France (Hirata 1966), eastern Europe (Jaczewski 1927, Transhel *et al.* 1933, Gitman 1935), and Turkistan (Gitman & Boytchenko 1934).

DIFFERENTIAL DIAGNOSIS

S. macularis and *L. taurica* are differentiated by microscopic examination of conidia or cleistothecia. Conidiophores of *S. macularis* grow superficially, whereas *L. taurica* conidiophores emerge through stomates.

Downy mildew and pink rot may be confused with powdery mildew. Downy mildew produces a grey mycelium on the *undersides* of leaves. Pink rot commonly affects leaves and stems, whereas powdery mildew is normally restricted to leaves. Spider mite webbings have been misidentified as powdery mildew by amateurs and experts (e.g., Westendorp). Look for mites and eggs.

DISEASE CYCLE & EPIDEMIOLOGY

The fungi overwinter as dormant mycelia or cleistothecia in plant debris. Young plants become infected in early spring but may take weeks to show symptoms. Disease generally increases as rainfall decreases. Low light intensity (indoors) or shaded areas (outdoors) increase disease severity, as does poor air circulation. *S. macularis* and *L. taurica* produce copious amounts of conidia, which spread by the slightest breeze to sites of secondary infection. Losses multiply as plants approach maturity. Succulent plants treated with excess nitrogen suffer the greatest damage.

L. taurica adapts well to xerophytic conditions; conidia can germinate in 0% RH. Films of water actually inhibit *L. taurica* germination. Its optimal growth occurs at 25°C. In contrast, *S. macularis* conidia germinate best at 100% RH (although they tolerate RH down to 10–30%), with optimal growth at 15–20°C.

CULTURAL & MECHANICAL CONTROL (see Chapter 9)

Strictly observe methods 7a (drought stress increases disease) and 7c (avoid overcrowding). Overhead irrigation may inhibit *L. taurica* but enhances *S. macularis*. Methods 8 (optimizing nutrition) and 10 (careful pruning) are also very important. Method 5 (genetic resistance) has controlled mildew in hops and other crops. Yarwood (1973) suggested blasting plants with a cold jet of water to knock back the mycelium of powdery mildew fungi, but this may predispose flowering plants to grey mould.

BIOCONTROL (see Chapter 10)

On other crops, powdery mildew has been controlled by hyperparasitic fungi such as *Ampelomyces quisqualis* (see below), a strain of *Verticillium lecanii* (see the section on whiteflies), and an experimental fungus, ***Sporothrix flocculosa*** (Kendrick 1985). A unique strain of *Bacillus subtilis* (Serenade®) is sold as a foliar spray for powdery mildew (for more information on *B. subtilis*, see damping off).

Ampelomyces quisqualis (=*Cicinnobolus cesatii*)

BIOLOGY: A fungus that parasitizes powdery mildew fungi (AQ-10®). The tiny hyperparasite has been used against *Sphaerotheca fuliginea*, *S. macularis*, *L. taurica*, and other powdery mildews, on cucurbits, grapes, strawberries, and tomatoes (Falk *et al.* 1995). *A. quisqualis* grows ubiquitously in temperate regions, especially in high humidity and temperatures between 20–30°C.

APPEARANCE & DEVELOPMENT: In optimal conditions, parasitized mildew colonies become flattened and dull grey within a week of infection. *A. quisqualis* produces pycnidia which ooze tiny cylindrical conidia (Fig 5.14).

APPLICATION: Supplied as dried, powdered conidia in extruded granules. It can be stored for months in a cool (8–10°C), dark place. Conidia are mixed with water and sprayed on plants. Conidia require free water to infect powdery mildews—this requirement is easily fulfilled in glasshouses but outdoors depends on unpredictable dew or light rain. The fungus can be applied with horticultural oil; it actually works better when mixed and sprayed with oil.

CHEMICAL CONTROL (see Chapter 11)

The superficial nature of *S. macularis* and *L. taurica* renders them susceptible to fungicide sprays. A simple solution of sodium bicarbonate or potassium bicarbonate may be sufficient. Some botanical insecticides and miticides, such as Neemguard® and Cinnamite®, also kill powdery mildews. Horticultural oil (Sunspray®) works well, especially when mixed with baking soda. Sulphur kills powdery mildew in many crops, applied at two week intervals. Copper also works, but works better against downy mildews. Physcion extract (Milsana®), a new botanical, protects plants by inducing plant resistance; technically it is not a fungicide.

CHARCOAL ROT

Charcoal rot kills plants approaching maturity. Some researchers call the disease "Premature wilt." The causal fungus also kills young seedlings (see the section on Damping Off disease). Charcoal rot has been reported on fibre varieties in Italy (Goidànich 1955, DeCorato 1997), Cyprus (Georghiou & Papadopoulos 1957), Illinois (Tehon & Boewe 1939, Boewe 1963), and Yugoslavia (Acinovic 1964). Charcoal rot also attacks drug cultivars (McPartland 1983a). The fungus is widespread and causes disease in more than 300 plant species. Maize is a particularly susceptible host.

SYMPTOMS

Plants develop a systemic chlorosis, then rapidly wilt, necrose, and die (Fig 5.30). This can happen very quickly. The pith inside the stalk becomes peppered with small black sclerotia (Figs 5.5 & 5.15).

CAUSAL ORGANISM & TAXONOMY

Macrophomina phaseolina (Tassi) Goidànich, *Annali della sperimentazione agraria Roma N.S.* 1(3): 449–461, 1947.

≡*Macrophoma phaseolina* Tassi 1901, ≡*Tiarosporella phaseolina* (Tassi) van der Aa 1981; =*Macrophomina phaseoli* (Maublanc) Ashby 1927, ≡*Macrophoma phaseoli* Maublanc 1905; =*Rhizoctonia bataticola* (Taubenhaus) Briton-Jones 1925, ≡*Sclerotium bataticola* Taubenhaus 1913.

Description: Sclerotia smooth, black, hard, oval to irregular in shape, averaging 44 x 75 µm (up to 200 µm in diameter according to DeCorato 1997). Pycnidia solitary or gregarious (when present), brown to black, subglobose, immersed in host tissue but becoming erumpent, ostiolate, averaging 180 µm in diameter. Conidiophores short, hyaline, simple (sometimes branched according to Punithalingam 1982). Conidiogenous cells at first holoblastic, becoming phialidic, with a minute collarette, lageniform to doliiform. Conidia hyaline, aseptate, obovoid to fusiform, thin walled, smooth, often guttulate, 5–10 x 14–30 µm (Fig 5.5). Punithalingam visualized a conidial appendage after treating conidia with Leifson's flagella stain. This appendage is apical, cap-like, and cone-shaped.

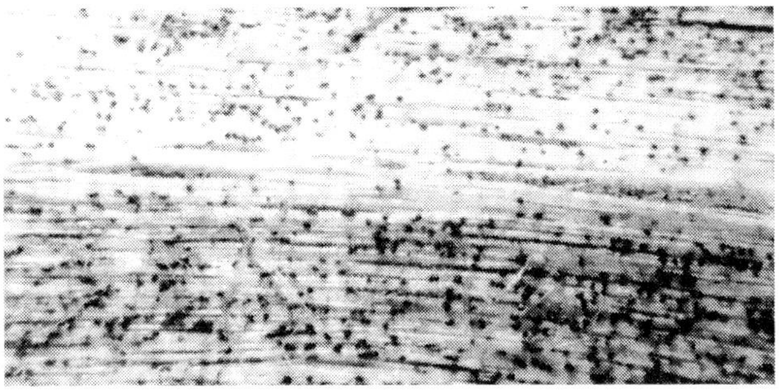

Figure 5.15: Microsclerotia of *Macrophomina phaseolina* within a *Cannabis* stalk (LM x35, McPartland).

DIFFERENTIAL DIAGNOSIS

The rapid necrosis caused by charcoal rot has been mistaken for herbicide destruction. *Fusarium* and *Verticillium* cause wilt symptoms but do not pepper the pith with sclerotia. Sclerotia are much smaller than those produced by hemp canker and southern blight organisms.

DISEASE CYCLE & EPIDEMIOLOGY

M. phaseolina overwinters as sclerotia in soil or plant debris. In dry conditions, sclerotia survive over a decade in the soil. Spring rains induce sclerotia to germinate and directly penetrate roots of young seedlings. Hyphae grow inter- and intracellularly, primarily in the plant cortex. *M. phaseolina* is an opportunist and rapidly colonizes plants stressed by drought and high temperatures. Optimal fungal growth occurs at 37°C.

CULTURAL & MECHANICAL CONTROL (see Chapter 9)

Method 7a is important—prevent drought stress by watering accessible plants during late-summer dry spells. Experiments with *Cannabis* showed that drought stress predisposed plants to charcoal rot *before* plants started wilting (McPartland & Schöneweiss, unpublished). Watering plants *after* they wilt is too late.

Use methods 1 (sanitation), 2a (deep ploughing), 2b (sterilizing soil), and 8 (optimizing soil nutrition). *M. phaseolina* is too heat-tolerant for method 2c. In the corn belt, do not plant *Cannabis* next to maize fields or in recently rotated fields. These soils often harbour a large reservoir of *M. phaseolina* sclerotia. Ghaffar *et al.* (1969) reduced *M. phaseolina* in cotton by adding organic amendments to soil. They added barley straw and alfalfa meal to increase the number (and activity) of soil organisms antagonistic to *M. phaseolina*.

BIOCONTROL & CHEMICAL CONTROL

Trichoderma harzianum infests and kills sclerotia of *M. phaseolina* (described under Rhizoctonia sore shin). The mycorrhizal fungus *Glomus intraradices* offers some protection (described under Fusarium stem canker disease). No fungicide stops charcoal rot once it has begun.

OLIVE LEAF SPOT

Two fungi cause slightly different symptoms of olive leaf spot. The disease has been reported in South America (Viégas 1961), Nepal (Bush Doctor 1987a), India, Pakistan, Bangladesh, China, Cambodia, Japan (Vasudeva 1961), Uganda, Mississippi, Missouri, and Wisconsin (Lentz *et al.* 1974). As Lentz *et al.* (1974) demonstrated, the two causal fungi are often confused in these reports. Both fungi attack fibre and drug plants.

SYMPTOMS

Spots on the upperside of the leaf are brown (Plate 62). Flip over the leaf to see the olive colour. The fuzzy olive material is the fungus itself, growing on the leaf surface. Whole leaves soon wilt, curl, and drop off. Damage escalates rapidly in August. Surviving plants are stunted with reduced yields.

Spots caused by one fungus, *Cercospora cannabis*, usually remain small and circular. Rarely, spots enlarge to margins of leaf veins, causing straight edges and somewhat rectangular shapes. The mycelium on the underside of the leaf may look yellow-brown rather than olive. The other fungus, *Pseudocercospora cannabina*, produces large irregularly-shaped spots, vein-delimited at first but eventually coalescing with other spots. Its mycelium is dark olivaceous brown and spreads effusely across the underside of the leaf. It occasionally produces conidia on upper leaf surfaces as well.

CAUSAL ORGANISMS & TAXONOMY

1. ***Pseudocercospora cannabina*** (Wakefield) Deighton, *C.M.I. Mycological Paper* No. 140, 1976; ≡*Cercospora cannabina* Wakefield 1917; =*Helicomina cannabis* Ponnappa 1977.

Description: Conidiophores simple or occasionally branched, straight, septate, olivaceous brown, narrowing to a slender apex, up to 150 x 3.5–4.5 μm. Conidia with unthickened basal scars, cylindrical, slightly or strongly curved, dilutely olivacous, usually five to seven septa, 30–90 x 3.5–5.5 μm (Vasudeva 1961 reports conidial lengths up to 120 μm). Occasionally, conidia may form microconidia. Microconidia ovoid to fusiform, unicellular, hyaline, 2–6 x 2.–3.5 μm (Fig 5.16).

2. ***Cercospora cannabis*** Hara & Fukui *in* Hara, *Zituyô sakaotubyôrigaku [Textbook in Plant Pathology]* p. 594, 1925.

≡*Cercospora cannabis* Hara & Fukui *in* Shirai & Hara 1927, ≡*Cercospora cannabis* (Hara) Chupp *apud* Green 1944, ≡*Cercosporina cannabis* Hara *Sakumotsu byorigaku [Pathology of crop plants]* p. 195, 1928, ≡*Cercosporina cannabis* Hara et Fukui *apud* Hara *J. Agricul. Soc. Shizuoka-ken* 32(364):45, 1928; =*Cercospora cannabis* Teng 1936 *apud* Korf 1996.

We have not seen Hara's 1925 text to verify priority. Many researchers cite Hara & Fukui *in* Shirai & Hara 1927. Korf (1996) incorrectly erected "*Cercospora cannabis* Teng 1936 [non Hara & Fukui]." Teng (1936) was clearly describing *C. cannabis* Hara & Fukui—he cited these authors. The synonym *Cercosporina cannabis* has similar problems with dates and names. Morphology of *C. cannabis* closely resembles that of *C. apii* Fresenius; it may be a synonym.

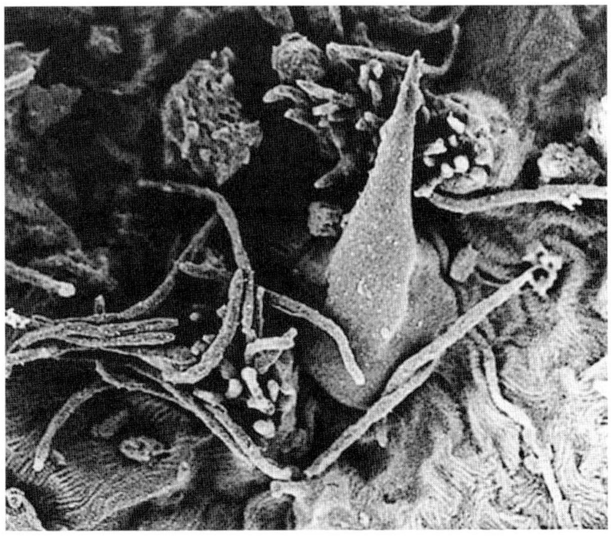

Figure 5.16: Conidia of *Pseudocercospora cannabina* erupting from leaf surface (SEM x375, McPartland).

Description: Conidiophores simple, straight or occasionally geniculate, septate, pale brown, not narrowing at the apex but broad and flat, up to 100 x 3.5–5.5 µm. Conidia bear thickened detachment scars at the base, not cylindrical but tapering toward their tips, straight or slightly curved, hyaline, indistinctly multiseptate, 25–90 x 3–5 µm.

DIFFERENTIAL DIAGNOSIS

During its early stages, olive leaf spot may be confused with yellow leaf spot or brown leaf spot. It later resembles downy mildew. The differential diagnosis is accomplished with a microscope.

DISEASE CYCLE & EPIDEMIOLOGY

The two fungi share similar life cycles. They overwinter in plant debris or in soil. Other *Cercospora* species are seed-borne, but this has not been detected in *Cannabis* (Pietkiewicz 1958, Ferri 1961b, Babu *et al.* 1977). Conidia form in the spring and spread by wind, splashing water, and farm hands. Mites can also carry conidia to sites of secondary infection. Conidia of *Cercospora* and *Pseudocercospora* can be blown great distances, unlike *Septoria* conidia, which are "sticky" and disseminate only several metres (Howard *et al.* 1994). High humidity or free water is required for conidial germination. Penetration occurs via stomates or wounds.

C. cannabis is more widespread than *P. cannabina* around the globe. *P. cannabina* seems to be more virulent than *C. cannabis*. Whereas *P. cannabina* may cause complete leaf defoliation, *C. cannabis* only causes leaf spots. McPartland (1995a) described microconidia budding from conidia of *P. cannabina* (Fig 5.16). This "microcyclic conidiogenesis" may explain why *P. cannabina* is more virulent than *C. cannabis*—because it produces copious amounts of conidia *and* microconidia.

CULTURAL & MECHANICAL CONTROL (see Chapter 9)

Follow instructions for yellow leaf spot. Calpouzos & Stallknecht (in Lucas 1975) noted that sugar beets growing in bright sunlight suffer small *Cercospora* leaf spots, but in low light intensity the spots coalesce and damage increases.

BIOCONTROL & CHEMICAL CONTROL

Dicyma may provide biocontrol (see below). A unique strain of *Bacillus subtilis* (Serenade®) is sold as a foliar spray for *Cercospora* leaf spot (described under damping off). Hawksworth *et al.* (1995) said ***Gonatobotrys simplex*** kills *Cercospora* species. This biocontrol is not yet available, although it grow naturally on hemp stems in the Czech Republic (Ondrej 1991). Yu (1973) controlled *Cercospora* with sulphur dusts and Bordeaux mixture, applied every two weeks for a total of three to five applications.

Dicyma pulvinata (=*Hansfordia pulvinata*)

BIOLOGY: A fungal mycoparasite that parasitizes hyphae and spores of *Cercospora* species, and also works against *Chaetomium* species. It has been used successfully in field experiments (Jeffries & Young 1994).

BROWN BLIGHT

Four fungi reportedly cause brown blight. The most ubiquitous fungus, *Alternaria alternata*, infests plants as an opportunistic saprophyte in necrotic tissue, *and* as a primary pathogen in female flowers (McPartland 1983a). It has destroyed up to 46% of seed (Haney & Kutsheid 1975). It also rots hemp stalks after harvest (Fuller & Norman 1944), and even destroys finished hemp products (Agostini 1927). Brown blight occurs in Europe and Asia (Vakhrusheva 1979), Tasmania (Lisson & Mendham 1995), and the USA (Hartowicz *et al.* 1971, Haney & Kutsheid 1975, McPartland 1983a).

Cigarettes made from *Alternaria*-infected tobacco generate a harsh and irritating smoke (Lucas 1975). More significantly, *A. alternata* produces toxins that mutate human oesophageal epithelial cells (Liu *et al.* 1992). Smoking *A. alternata*-infected material may cause oesophageal cancer, which is the most common cancer of marijuana smokers (McPartland 1995f, 1995g).

SYMPTOMS

Brown blight usually arises on mature plants late in the growing season. Symptoms begin as pale green or grey leaf lesions. These irregularly circular spots may or may not develop chlorotic halos, depending on the *Alternaria* species. Spots often coalesce into blight-like symptoms, as whole leaves turn brown. Undersides of spots may develop concentric zonations (especially spots caused by *Alternaria solani*). Frail necrotic tissues break up, resulting in irregular leaf perforations. Lesions may extend to leaf petioles and stems. The fungus also infests female flowers (and seeds), turning them grey-brown (Ferri 1961b).

CAUSAL ORGANISMS & TAXONOMY

1. *Alternaria alternata* (Fries:Fries) Keissler, *Beih. Bot. Cent.* 29:434, 1912; ≡*Alternaria tenuis* C. G. Nees 1816–7; ?=*Alternaria cannabis* Yu 1973.

Description: Conidiophores simple or branched, straight or curved, septate, yellow to golden brown, up to 50 µm long, 3–6 µm thick. Conidia borne in chains, straight or slightly curved, obyriform to obclavate (rarely ellipsoid), yellow to golden brown, with three to eight transverse and one or two longitudinal septa, tapering to a short beak (beak sometimes absent), length 20–63 µm, width 7–18 µm (Figs 3.2 & 5.17).

A. alternata has been isolated from hemp cultivars in Italy (Agostini 1927, Ferri 1961b), Poland (Zarzycka & Jaranowska 1977), the Czech Republic (Ondrej 1991), and the USA in Illinois (McPartland 1983a), Iowa (Fuller & Norman 1946), and Kansas (Hartowicz *et al.* 1971). Drug plants are attacked in India and the USA (McPartland 1995c

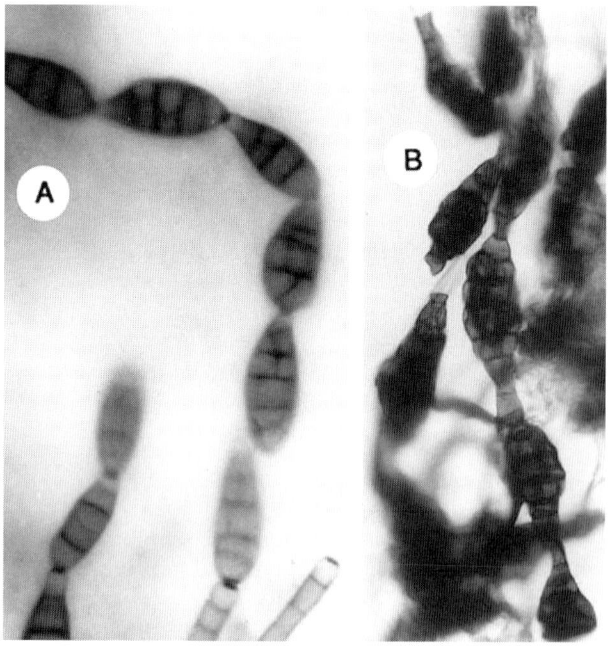

Figure 5.17: Conidia of *Alternaria alternata* (A, LM x660) compared with a chain of chlamydospores produced by *Phoma glomerata* (B, LM x490), McPartland.

and unpublished data). Paulsen (1971) cultured the fungus from green stink bugs (*Nezara viridula*) feeding on marijuana seeds. Yu (1973) described *Alternaria cannabis* from marijuana growing at the University of Mississippi. Her description of this *nomen dubium* strongly resembles *A. alternata*.

2. ***Alternaria solani*** (Ellis & Martin) Sorauer, *Zeitschrift Pflanzenkrankheiten* 6:6, 1896;

≡*Macrosporium solani* Ellis & Martin 1882, =*Alternaria porri* (Ellis) Ciferri f. sp. *solani* (Ellis & Martin) Neergaard 1945; =*Alternaria dauci* (Kühn) Groves & Skolko f. sp. *solani* (Ellis & Martin) Neergaard 1945.

Description: Conidiophores simple, straight or curved, septate, pale to dark olive-brown, up to 100 µm long, 6–10 µm thick. Conidia borne singularly, straight or slightly curved, obyriform to obclavate, light olive-brown, with eight to 11 transverse septa and zero to two longitudinal septa, tapering to a long beak which is usually equal in length to the body, overall length 140–280 µm, 15–19 µm at greatest width.

Only Barloy & Pelhate (1962) have cited *A. solani* on *Cannabis*, and the fungus they illustrated looks like *A. alternata* not *A. solani*. They erroneously synonymized *A. solani* with *Macrosporium cannabinum* (McPartland 1995a). *A. solani* attacks solanaceous crops (tobacco, potato, tomato, etc.) and *Brassica oleracea* (cabbage, cauliflower, kohlrabi, etc.).

3. ***Alternaria longipes*** (Ellis & Everhart) Mason, *Mycological Paper* 2:19, 1928;

≡*Alternaria longipes* (Ellis & Everhart) Tisdale & Wadkins 1931.

Description: Conidiophores simple or branched, straight or curved, septate, pale olivaceous brown, up to 80 µm long, 3–5 µm thick. Conidia borne in chains (rarely singularly), obclavate, rostrate, smooth to verruculose, gradually tapered to a beak (1/3rd its total length) and rounded at the base, beige to pale brown, with five or six transverse and one to three longitudinal or oblique septa, 35–(69)–110 x 11–(14)–21 µm.

A. longipes normally attacks tobacco; Lentz (1977) reported it on *Cannabis*. Joly (1964) lumped *A. longipes* with *A. solani*; Lucas (1975) lumped it with *A. alternata*.

4. ***Alternaria cheiranthi*** (Libert) Bolle [as "(Fr.) Bolle"], *Meded. phytopath. Lab. Willie Commelin Scholten.* 7:55, 1924.

≡*Helminthosporium cheiranthi* Libert 1827, ≡*Macrosporium cheiranthi* (Libert) Fries 1832.

Description: Conidiophores simple or branched, straight or curved, septate, pale olive to hyaline at the apex, up to 130 µm long, 5–8 µm thick. Conidia borne singularly (rarely in chains of two or three), ovoid to pyriform at first, later becoming irregular, generally tapered at the apex and rounded at the base, light olive-golden brown, with numerous transverse and longitudinal septa, 20–100 x 13–33 µm.

Gzebenyuk (1984) reported isolating *A. cheiranthi* from 96.4% of hemp stems sampled near Kiev. He may have misidentified *A. alternata* (McPartland 1995a).

DIFFERENTIAL DIAGNOSIS

Several microscopic characteristics separate *A. solani* from *A. alternata:* the former exhibits larger, darker conidia with longer beaks. *A. solani* produces solitary conidia whereas *A. alternata* often sporulates in chains.

Brown blight can be confused with grey mould caused by *Botrytis cinerea*. *A. alternata* conidia can be confused with chlamydospores of *Phoma glomerata*, one of the causes of brown leaf spot (Fig 5.17).

DISEASE CYCLE & EPIDEMIOLOGY

Alternaria species overwinter as mycelia in host plant debris. *A. alternata* can survive indefinitely in the soil. *A. alternata* also overwinters as a seedborne infection (Pietkiewicz 1958, Ferri 1961b, Stepanova 1975). The optimal temperature for *A. solani*, *A. alternata*, and *A. longipes* is 28–30°C. *A. alternata* infects plants during warm, wet weather but symptoms do not arise until the hot, dry summer.

Alternaria conidia disperse by wind or splashed rain. Conidia require a film of water to germinate. Conidia germ tubes penetrate host epidermis directly, or enter via stomates and wounds. Infection increases ten-fold when *Alternaria* conidia are mixed with plant pollen (Norse in Lucas 1975). Apparently the pollen serves as an energy source for conidia.

CULTURAL & MECHANICAL CONTROL (see Chapter 9)

Carefully observe methods 7c (avoid excess humidity), 10 (careful pruning), 11 (avoid seedborne infection), and 8 (fertilize with high P & K and low N; Lucas (1975) used a NPK ratio of 3–18–15). Other important measures include 1 (sanitation), 2a (deep ploughing), 2b&c (sterilizing or pasteurizing soil), and 4 (escape cropping). Sasaki & Honda (1985) decreased *A. solani* infection in glasshouse vegetables by covering glass with UV-absorbent vinyl sheets. Since *A. solani* requires UV light to produce conidia, epidemics are prevented (see the section on grey mould for more information).

BIOCONTROL & CHEMICAL CONTROL

Jeffries & Young (1994) controlled *A. solani* with *Ampelomyces quisqualis* (described under powdery mildew). A unique strain of *Bacillus subtilis* (Serenade®) is sold as a foliar spray for *Alternaria* species (described under damping off). Dipping flowers in a solution of *Pichia guilliermondii* may reduce postharvest rot (see Chapter 8). Disinfect *Alternaria*-infested seeds by soaking them in fungicides (see Chapter 12). Yu (1973) controlled brown blight caused by *Alternaria cannabis* (whatever that is) by spraying crops with Bordeaux mixture every month. Tobacco growers also use Bordeaux mixture (Lucas 1975). For farmers interested in alternative pesticides, Dahiya & Jain (1977) reported that pure THC and CBD inhibited *A. alternata*.

STEMPHYLIUM LEAF & STEM SPOT

Spaar *et al.* (1990) called this "brown fleck disease." Leaf lesions arise as light-brown spots with dark margins, often irregularly shaped and limited by leaf veins. Spots grow to 3–10 mm in diameter, become concentrically zonated, and coalesce. Disease may spread to leaf petioles and stems (Gzebenyuk 1984). Conidia (rarely perithecia and ascospores) of the causal fungi arise in necrotic plant tissue.

CAUSAL ORGANISMS & TAXONOMY

The literature is confused and confusing. Researchers cite two taxa on *Cannabis: Stemphylium botryosum* and *Pleospora herbarum*. These names are frequently misinterpreted as two stages in the life cycle of one organism. Careful research refutes this, they are different species (Simmons 1985).

1. ***Stemphylium botryosum*** Wallroth, *Flora Crypt. germ., pars post.*, p. 300, 1833; =*Stemphylium cannabinum* (Bakhtin & Gutner) Dobrozrakova *et al.* 1956, ≡*Macrosporium cannabinum* Bakhtin & Gutner 1933; =*Thyrospora cannabis* Ishiyama 1936.

teleomorph: *Pleospora tarda* Simmons, *Sydowia* 38:291, 1985.

Description: Conidiophores arise in clumps (often ten to 15 per clump), straight or flexous, simple or occasionally branched at their base, one to seven septa, cylindrical except for swelling at the site of spore formation, light olive brown darkening at swollen apex, smooth becoming echinulate at swollen apex, 4–6 µm wide, swelling to 7–10 µm at apex, usually 25 to 85 µm long, but continued proliferation through one to five previously-formed apices may extend conidiophore length to 120 µm. Conidia initially oblong to subspherical, hyaline and verruculose; at maturity oblong to subdoliiform, densely echinulate (Fig 5.18), with three (sometimes four) transverse septa and two or three (sometimes four) complete or nearly complete longitudinal septa, with a single conspicuous constriction at the median transverse septum, dilute to deep olive brown, 30–35 x 20–25 µm. Pseudothecia large (700 µm in diameter), hard, black sclerotic bodies, with formation of asci occurring slowly, often after eight months in culture. Asci subcylindrical, eight-spored, averaging 200 x 40 µm. Ascospores hyaline, oblong and constricted at three transverse septa when immature; when mature they exhibit a broadly rounded apex and almost flat base, turn yellowish-brown, develop four secondary transverse septa in addition to the three primary septa, form one or two longitudinal septa, and average 40 x 17 µm.

McPartland (1995e) synonymized *Stemphylium cannabinum*, *Macrosporium cannabinum* and *Thyrospora cannabis* under *S. botryosum* after examining leaf specimens from Russia and Japan. A misidentified "*Coniothecium* species" from Lima, Peru, also turned out to be *S. botryosum* (McPartland 1995a). Presley (unpublished 1955 data, USDA archives) described a "*Tetracoccosporium* species" on *stems* of hemp growing in Beltsville, MD. No specimen survives, but Presley's description suggests *S. botryosum.*

McCurry & Hicks (1925) reported a *Stemphylium* species causing "considerable amounts" of leaf disease in Canadian hemp on Prince Edward Island. *S. botryosum* colonizes hemp stems in Holland (Termorshuizen 1991), and the Czech Republic (Ondrej 1991). Rataj (1957) considered *S. botryosum* a dangerous *Cannabis* pathogen. Conversely, Barloy & Pelhate (1962) considered the fungus a weak pathogen, only able to invade plants previously weakened by *Alternaria* species.

2. ***Stemphylium herbarum*** Simmons, *Sydowia* 38:291, 1985.
teleomorph: *Pleospora herbarum* (Persoon:Fries) Rabenhorst, *Herb. viv. mycol. Ed. II*, No. 547, 1856.

Description: Conidiophores straight or flexous, simple or branching either at their base or subapically, one to four septa, cylindrical except for apical swelling, dilute brown, smooth, 3–5 µm wide swelling to 8–10 µm at apex, usually 15–20 µm long but occasionally much longer. Conidia initially oblong to broadly ovoid to subspherical, hyaline and minutely verruculose; at maturity becoming broadly ovoid to broadly elliptical, sometimes unequally sided, distinctly verruculose, with up to six or seven transverse septa and two or three longitudinal septa, distinctly constricted at one to three transverse septa, dilutely yellowish brown to deep reddish brown, 35–45 x 20–27 µm. Pseudothecia immersed then erumpent in host tissue (in culture, forming asci within a month), black, globose or slightly flattened due to collapse of their thin walls, 250–300 µm in diameter. Asci conspicuously bitunicate, tubular with parallel walls and a broadly rounded apex, gradually narrowing near the base, eight-spored, averaging 160 x 25 µm. Ascospores hyaline, ellipsoid to obovoid and constricted at one transverse submedian septum when immature; when mature they become medium brown in colour, obovoid, often flattened on one side, with the upper half broader than the lower half, usually exhibiting seven transverse and one or two (rarely three) longitudinal septa, averaging 32–35 x 13–15 µm.

In the Ukraine, Gzebenyuk (1984) found *P. herbarum* on 2% of hemp stems, and *S. botryosum* on 7% of stems. In the Czech Republic, Ondrej (1991) reported *P. herbarum* on some stems, whereas *S. botryosum* appeared "abundantly."

DIFFERENTIAL DIAGNOSIS

Distinguishing the *Stemphylium* stages requires careful measurements and a practiced eye. Generally, *S. botryosum* produces longer conidiophores with a series of knobby apical proliferations; *S. herbarum* produces larger conidia with a greater size range and greater complexity of septation. The teleomorphs, *P. herbarum* and *P. tarda*, can best be differentiated in culture. Stemphylium leaf symptoms can be confused with brown blight or early olive leaf spot. The stem symptoms can be confused with Cladosporium stem canker.

DISEASE CYCLE & EPIDEMIOLOGY

S. botryosum lives in temperate and subtropical regions worldwide. It attacks many hosts. The fungus overwinters in crop debris. Springtime infections begin via air-dispersed ascospores and conidia. During the growing season copious amounts of conidia form on diseased leaves. The buildup of conidia leads to epidemics of secondary infections. Ganter (1925) reported transmission of a *Pleospora* species by infected seed. Presley (unpublished data, USDA) also described "*Tetracoccosporium* species" infecting *Cannabis* seed.

CONTROL

Follow suggestions for brown blight described above. Prevention is the key to control.

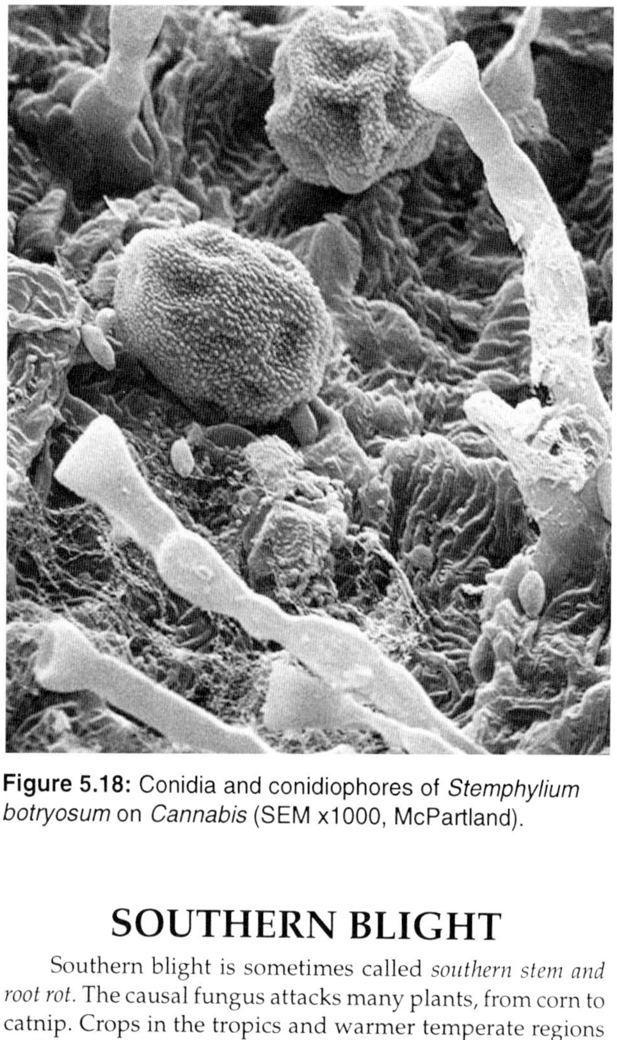

Figure 5.18: Conidia and conidiophores of *Stemphylium botryosum* on *Cannabis* (SEM x1000, McPartland).

SOUTHERN BLIGHT

Southern blight is sometimes called *southern stem and root rot.* The causal fungus attacks many plants, from corn to catnip. Crops in the tropics and warmer temperate regions suffer extensive losses from southern blight, which has been reported on *Cannabis* in India (Hector 1931, Uppal 1933, Krishna 1995), Texas and South Carolina (Miller *et al.* 1960), Italy (Ferri 1961a), Japan (Kyokai 1965, Kishi 1988), and the Ukraine (Gzebenyuk 1984). Recently the fungus was isolated in Oxford, Mississippi (Sands 1991); virulent mutants of this isolate yielded a 100% kill rate (Sands 1995).

SYMPTOMS

Disease arises during warm summer weather. Mature plants suddenly wilt and turn yellow. Systemic wilting and leaf chlorosis progress to leaf necrosis and plant death (Plate 63). Stalks decay at the soil line, turning brown and macerated. Sunken brown stalk lesions sprout sclerotia after death of the plant. A pale-brown hyphal mat sometimes radiates from the base of the stalk along the soil surface.

CAUSAL ORGANISM

Sclerotium rolfsii Saccardo, *Annales Mycologici* 9:257, 1911.

teleomorph: *Athelia rolfsii* (Curzi) Tu & Kimbrough 1978, ≡*Corticium rolfsii* Curzi 1932.

Description: Sclerotia smooth or pitted, near spherical (slightly flattened below), at first amber then turning brown to black, usually 0.7–1mm in diameter(Fig 5.19); cross-section of sclerotia reveals three cell layers—a darkly pigmented outer *rind* (three or four cells thick), a middle *cortex* composed of dense, hyaline cells (four to ten cells thick), and an inner *medulla*, comprised of loosely interwoven cells. Hyphae produce clamp connection at some septa, and branch at acute angles. Basidiocarps resupinate, effused, loosely adherent to the stem surface. Basidia clavate, usually four-spored. Basidiospores hyaline, smooth, teardrop-shaped, 5.5–6.5 x 3.5–4.5 μm (Fig 3.2).

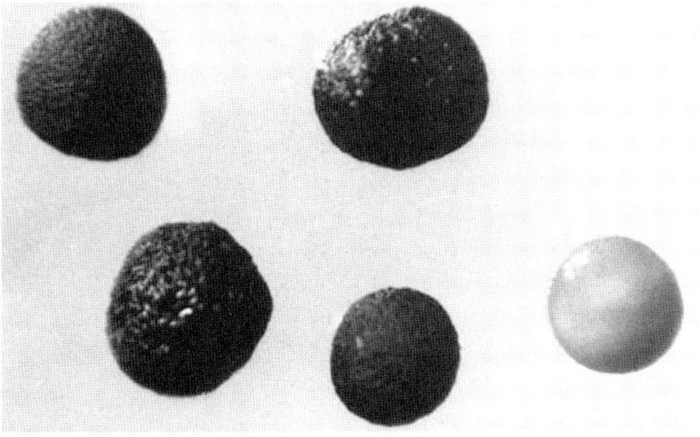

Figure 5.19: Sclerotia produced by *Sclerotium rolfsii* (LM x20, courtesy Bud Uecker).

DIFFERENTIAL DIAGNOSIS

Symptoms of southern blight may be confused with symptoms of hemp canker (caused by *Sclerotinia sclerotiorum*), charcoal rot (*Macrophomina phaseolina*), or Rhizoctonia sore shin (*Rhizoctonia solani*). Sclerotia of *S. sclerotiorum* and *R. solani* are larger and less symmetrical than those of *S. rolfsii*, and the sclerotia of *M. phaseolina* are smaller and irregular in shape. Stem cankers caused by *Fusarium* species may be confused with southern blight, but sclerotia are missing.

DISEASE CYCLE & EPIDEMIOLOGY

S. rolfsii overwinters as sclerotia in soil and plant debris. Sclerotia are spread by wind, water, contaminated farm tools, and some survive passage through sheep and cattle. Sclerotia germinate in the spring. Germination decreases in deep soil. Germ tubes directly penetrate roots or enter via wounds. The fungus grows inter- and intracellularly, primarily within the stem cortex. Optimal growth occurs near 30°C, when *Cannabis* losses may reach 60% (Krishna 1995). Basidiospores germinate best at 28°C, but are short lived and relatively unimportant. A root-boring maggot, *Tetanops luridipenis*, may vector *S. rolfsii* between *Cannabis* plants (Sands *et al.* 1987).

CULTURAL & MECHANICAL CONTROL (see Chapter 9)

Apply methods 1 (sanitation), 2a (ploughing at least 8 cm), and 2b (soil sterilization). Method 2c—soil solarization—kills *S. rolfsii* sclerotia (Elmore *et al.* 1997). Decreasing moisture at ground level inhibits disease development, so observe methods 3 (weeding), 7b (avoid overwatering), and 7c (avoid excess humidity). All weeds should be viewed as alternate hosts—method 3 again. Method 8 (balanced soil structure) is tricky, because too high a percentage of soil organic matter may *increase* the severity of southern blight. Cyclical drying and wetting of soil induces sclerotia to germinate. Germinated sclerotia die within two weeks in fallow soil (Lucas 1975). Krishna (1996) tested three ganja "cultivars" from Uttar Pradesh (India) for genetic resistance to *S. rolfsii.* Cultivar 'dwarf' [*Cannabis afghanica*?] showed the most resistance, followed by cultivar 'medium' [*Cannabis indica*?], and lastly cultivar 'tall' [*Cannabis sativa*?].

BIOCONTROL & CHEMICAL CONTROL

Kendrick (1985) reported biocontrol with the fungi *Trichoderma harzianum* and *Trichoderma (Gliocladium) virens*, described under Rhizoctonia sore shin. The mycorrhizal fungus *Glomus intraradices* offers some protection (described under Fusarium stem canker disease). Mixing urea into soil at a rate of 100 kg ha^{-1} kills many sclerotia (Krishna 1995). Drenching soil with formalin reduced hemp losses in Italy (Ferri 1961a).

BLACK MILDEW

Black mildews, like powdery mildews, are obligate parasites that grow on the surfaces of leaves. They tap host resources by immersing haustoria into host leaves. They should not be confused with sooty moulds, which are entirely superficial and saprophytic. See "Phylloplane fungi" for a discussion of sooty moulds. Black mildew is known only from Nepal.

SYMPTOMS

Black mildew forms thin grey to black colonies on otherwise healthy leaves. Colonies arise on upper sides of leaves. They remain isolated or coalesce with neighbours to form spots 5 mm or more in diameter.

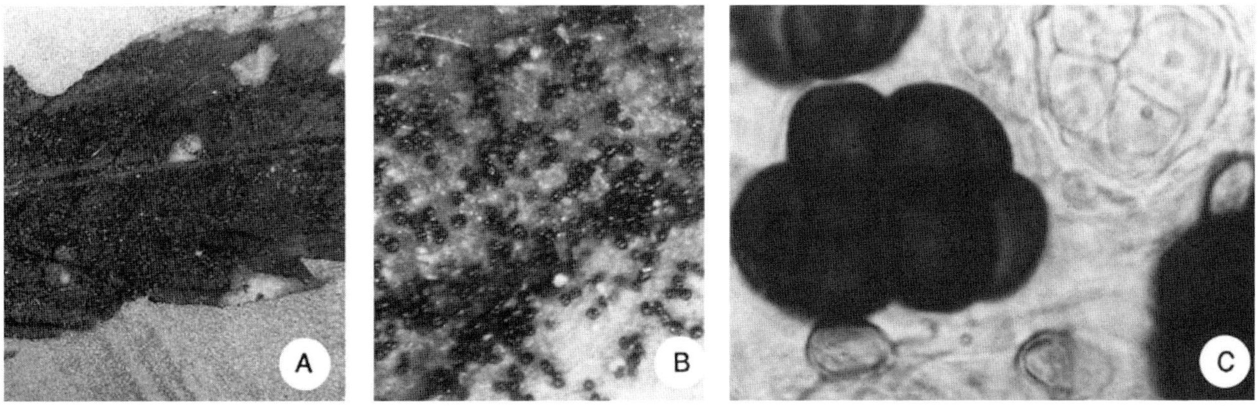

Figure 5.20: Black mildew caused by *Schiffnerula cannabis*. A. Infected leaf (LM x2); B. Leaf surface crowded with conidia (LM x70); C. Close-up of a conidium, an ascus with 2 ascospores in focus, and 2 hyphopodia (LM x1450); McPartland.

CAUSAL ORGANISM & TAXONOMY

Schiffnerula cannabis McPartland & Hughes, *Mycologia* 86:868, 1995. anamorph: *Sarcinella* sp.

Description: Mycelium composed of brown hyphae, straight or sinuate, septate (cells mostly about 16 µm long), 4–5 µm thick, branching irregular. Hyphopodia numerous, alternate, hemispherical or subglobose, brown, 9.5 x 6.5 µm. Ascomata circular in outline, 37–43 µm diameter, containing one to three asci, wall at maturity mucilaginous and breaking up in water. Asci ellipsoid to subglobose, nonparaphysate, eight-spored, 23 x 21 µm. Ascospores massed into a ball, at first hyaline, later turning light brown, oblong, smooth, one septum, scarcely constricted, the cells unequal, 17.4 x 7.3 µm. Conidiophores indistinguishable from hyphae. Conidiogenous cells intercalary, short, cylindrical, integrated, monoblastic. Conidia opaquely black-brown, ellipsoid to subglobose, dictyoseptate, composed of eight to 15 cells, bullate but smooth walled, 31.9 x 27.5 µm (Fig 5.20 & Plate 64).

DIFFERENTIAL DIAGNOSIS

Black mildew may be confused with early stages of brown blight, corky leaf spot, black dot, or sooty mould.

DISEASE CYCLE & EPIDEMIOLOGY

S. cannabis produces copious *Sarcinella* conidia but none have been observed to germinate. Their role in the life cycle is uncertain. The fungus spreads by water-transported ascospores. Schiffnerulaceous fungi occur worldwide but are more abundant in tropical and subtropical climates.

CONTROL

Measures have not been elucidated. Since most *Schiffnerula* species have limited host ranges, crop rotation should effectively control the disease.

TWIG BLIGHT

Known as *nebbia* in Italy, twig blight has been reported in Europe, Asia, North America, and South America. Dewey (1914) and Barloy & Pelhate (1962) considered the disease of no economic significance, but others consider it a severe problem (Ferraris 1935, Trunoff 1936, Ghillini 1951). Charles & Jenkins (1914) reported losses as high as 95% in Virginia. Losses multiply in drought-stressed crops—McPartland & Schoeneweiss (1984) showed how *Botryosphaeria* species opportunistically attack stressed plants. Secondary infections often follow twig blight; a *Fusarium* followed twig blight in Italy (Petri 1942), and *Alternaria alternata* followed twig blight in Illinois (McPartland, unpublished data).

SYMPTOMS

Twig blight begins with a wilting and drooping of leaves, usually late in the season (Fig 5.21). Wilted foliage turns brown and dies but remains attached to plants. Tips of young branches show symptoms first (thus "twig blight"). Within two weeks the entire plant may wilt and die. Diseased stems develop grey spots averaging 6–12 mm long and 2–6 mm wide (Flachs 1936). These spots darken and bear tiny, black pycnidia and/or pseudothecia. Spots continue to enlarge after plants are harvested (Gitman & Malikova 1933).

CAUSAL ORGANISMS & TAXONOMY

At least two species cause twig blight (McPartland 1995b, 1999b). The first species was named *Dendrophoma marconii* in 1888 by Cavara, who discovered it in Italy. The second species was discovered by Charles & Jenkins (1914), who thought it was identical to *Dendrophoma marconii*. When Charles & Jenkins's fungus produced a sexual stage, they named it after Cavara's fungus ("*Botryosphaeria marconii*").

Figure 5.21: Symptoms of twig blight (from Charles & Jenkins 1914).

But Charles & Jenkins were wrong—their species was not the same as Cavara's fungus. So we have two different species with the same *marconii* epithet.

1. *Dendrophoma marconii* Cavara, *Atti dell' Istit. Bot. di Pavia, ser. II*, 1:426; *Revue Mycologique* 10(40):205, both 1888.

Description: Pycnidia few, arising in grey spots, concealed by epidermis, flattened globose, ostiole slightly raised, 130–150 µm diameter. Conidiophores unbranched or widely dichotomous branching, septate, hyaline. Conidia pleomorphic, ovate-elliptical, teretiuscullis [tapering?], one-celled, hyaline, 4.5–6.5 x 2–2.5 µm.

The description above is from Cavara (1888). Barloy & Pelhate (1962) cited smaller *D. marconii* conidia, 3 x 2 µm. McPartland (1999b) described conidiophores as multiseptate, widely branched, with irregular monilioid swellings, growing up to 25 µm long and ramifying throughout the pycnidial locule. Conidiogenesis could not be determined. Conidia arise at restrictions along the conidiophores. Conidia are quite pleomorphic and irregular in shape, often resembling short sections of conidiophores.

Every *D. marconii* specimen we've examined has been concurrently infected by other fungi, such as *Phomopsis ganjae*, *Septoria* species, *Phoma* species, and *Botryosphaeria marconii*. No one has cultured *D. marconii* in isolation. Conceivably *D. marconii* could be a hyperparasite of other fungi, and not a plant pathogen at all. This may explain the confusion this species caused Charles & Jenkins (1914) and Petrak (1921).

2. *Botryosphaeria marconii* Charles & Jenkins, *J. Agric. Research* 3:83, 1914 [as *Botryosphaeria marconii* (Cav.) Charles & Jenkins].

anamorph 1: *Leptodothiorella marconii* McPartland, *Mycotaxon* 53:422, 1995.

anamorph 2: *Fusicoccum marconii* McPartland, *Mycotaxon* 53:421, 1995; ≡ *Macrophoma marconii* nomen nudum, various authors.

Description: Pseudothecia usually unilocular, globose, immersed then erumpent, 130–160 µm in diameter. Outer wall pale brown but darkening near the ostiole; inner wall hyaline and very thin. Asci bitunicate, eight-spored, clavate, with a short stalk, 80–90 x 13–15 µm. Paraphyses filiform. Ascospores fusoid to ellipsoid, aseptate, hyaline to pale green, 16–18 x 7–8 µm. *Leptodothiorella* conidiomata identical to pseudothecia. Conidiogenous cells simple or rarely branched, lageniform to cylindrical, 3–12 µm long, up to 3 µm wide at base tapering to 1 µm at apex, phialidic, integrated or discrete, with minute channel. Microconidia ellipsoid, single-celled, hyaline, occasionally biguttulate and swollen towards each end, 3.0–4.0 x 0.5–2.5 µm. *Fusicoccum* conidiomata slightly larger than those of *Leptodothiorella*, but otherwise indistinguishable (Plate 65). Conidiophores arise from inner wall, hyaline, smooth, simple or branched and septate near the base, 10–16 µm long, up to 4.0 µm wide at the base tapering to 2.5–3.0 µm at the tip. Conidiogenous cells holoblastic, integrated, determinate. Macroconidia fusiform, single-celled, hyaline to glaucous, smooth-walled, base truncate, 16–22 x 5–8 µm (Plate 65).

DIFFERENTIAL DIAGNOSIS

The two fungi causing twig blight are easy to differentiate microscopically. Symptoms of twig blight may be confused with symptoms of brown stem canker (Dobrozrakova *et al.* 1956) or Phomopsis stem canker. The fungi must be examined microscopically.

DISEASE CYCLE & EPIDEMIOLOGY

Both species overwinter in host tissue and spread by splashed rain in the spring. Because of confusion in the literature, the range of these two species can only be estimated: *D. marconii* infests hemp stems in Italy (Cavara 1888), France (Barloy & Pelhate (1962), and Chile (Mujica 1942, 1943). McPartland (1999b) found it on hemp stems in Austria, hemp leaves in Michigan, and leaves of drug plants in Nepal. Charles & Jenkins found *B. marconii* in Maryland and Virginia. Judging from other authors' descriptions, *B. marconii* also occurs in Russia (Gitman & Boytchenko 1934), Germany (Flachs 1936, Patschke *et al.* 1997), and Italy (Petri 1942).

CULTURAL & MECHANICAL CONTROL (see Chapter 9)

Use methods 1 (sanitation), 5 (genetic resistance—see below), 7a (avoid drought), and 7c (avoid overhead irrigation, keep plants well-spaced). Charles & Jenkins (1914) noted twig blight strikes males first, then moves to females. Male plants should be rouged. Dempsey (1975) listed Russian and Ukrainian hemp varieties with resistance to *B. marconii*—'Monoecious Central Russia,' 'Odnodomnaja 2,' 'USO-1' (variously translated as 'YUSO-1' or 'JUSO-1'), 'USO-7,' and USO-1's parents, 'USO-6' and 'Odnodomnaya Bernburga.' Unfortunately, these cultivars are gone, although some are the parents of current cultivars, such as 'USO-14' (a selection from 'USO-1,' see de Meijer 1995).

BIOCONTROL & CHEMICAL CONTROL

There is no biological control of twig blight. Bordeaux mixture controls other *Botryosphaeria* species on other hosts.

PINK ROT

The fungus causing pink rot has been isolated from hemp stems in Italy (Ghillini 1951), the Czech Republic (Ondrej 1991), and Iowa (Fuller & Norman 1944, 1945), and hemp seeds in Italy (Ciferri 1941, Ferri 1961b) and Russia (Pospelov *et al.* 1957). Pink rot is on the rise in drug cultivars. Plants of *afghanica* heritage are particularly susceptible to pink rot. The fungus has been isolated from leaves of Pakistani plants (Nair & Ponnappa 1974, Ponnappa 1977). McPartland (unpublished data, 1994) found pink rot on specimens collected in Afghanistan by Schultes, and isolated the fungus from two-month old 'Skunk No. 1' growing in a Dutch glasshouse, where it caused considerable losses.

SYMPTOMS

Pink rot is a bit of a misnomer. The fungus often presents as a white fuzz covering leaves or flowering buds (Plate 66). The pink tint arises when conidia are produced. In Holland, the fungus surrounded stems and girdled them. Girdled plants wilted and fell over. Ghillini (1951) and Fuller & Norman (1946) noted the fungus ruins hemp fibres.

CAUSAL ORGANISM

Trichothecium roseum (Persoon:Fries) Link, *Magazin, Gesellschaft Naturforschender Freunde zu Berlin* 3:18, 1809.

≡*Cephalothecium roseum* Corda 1838.

Description: Colonies quickly turn dusty pink with conidia. Conidiophores upright, unbranched, often with three septa near the base, up to 2 mm long, 4–5 µm wide. Conidiogenesis basipetal, conidia often remain in contact with each other in zig-zag chains. Conidia ellipsoidal to pyriform with truncate basal scars, two-celled with upper cell larger and rounder than lower cell, hyaline (pink *en masse*), with a thick smooth wall, 12–23 x 8–10 µm (Fig 5.22).

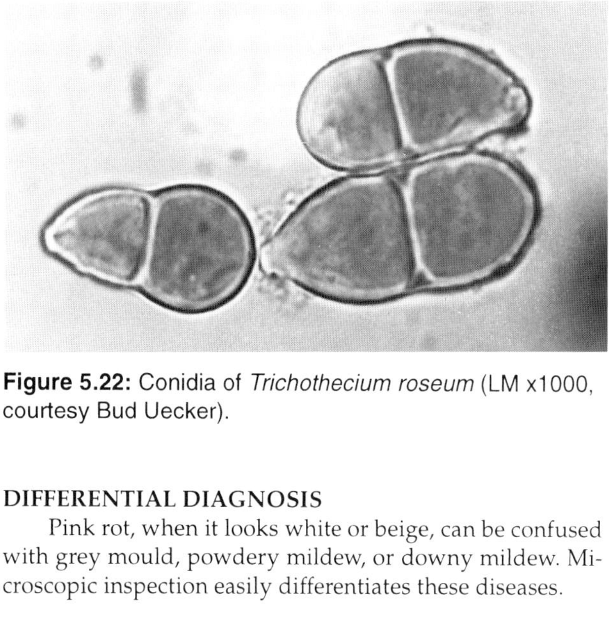

Figure 5.22: Conidia of *Trichothecium roseum* (LM x1000, courtesy Bud Uecker).

DIFFERENTIAL DIAGNOSIS

Pink rot, when it looks white or beige, can be confused with grey mould, powdery mildew, or downy mildew. Microscopic inspection easily differentiates these diseases.

DISEASE CYCLE & EPIDEMIOLOGY

T. roseum overwinters on crop debris or in the soil. It lives worldwide as a saprophyte of stored foodstuffs and a weak parasite of living plants. It also turns up in forest leaf litter, termite nests, paper mill slime, and sewage sludge. The fungus can colonize dead plant material, insect excreta, or pollen lying upon the surface of leaves, and it gains energy from these substrates to invade healthy plant tissue.

T. roseum grows and sporulates best in humid conditions. Fuller & Norman (1944, 1945) described the fungus "dominating" hemp during warm, humid weather. *T. roseum* can produce toxic metabolites called trichothecenes (see comments in the section on Fusarium stem canker).

CULTURAL & MECHANICAL CONTROL (see Chapter 9)

Remove crop debris (sanitation—method 1) to prevent *T. roseum* from establishing a saprophytic foothold. Keep plants healthy and control insect infestations. Use method 5 (genetic resistance) by avoiding susceptible *afghanica* hybrids. Control excess humidity (method 7c) and avoid seedborne infection (method 11).

BIOCONTROL & CHEMICAL CONTROL

No biocontrol is known. *T. roseum,* in fact, serves as a biocontrol against *Sclerotinia sclerotiorum* in soil. No fungicides are effective. Dahiya & Jain (1977) reported that pure THC and CBD inhibited the growth of *T. roseum.*

CLADOSPORIUM STEM CANKER

Two or more species of *Cladosporium* cause stem cankers. The most common species, *C. herbarum,* also infests leaves and seeds. *C. herbarum* produces cellulolytic enzymes which continue to ruin hemp after it has been harvested.

SYMPTOMS

On stems, this disease begins as dark green spots. The spots elongate and turn velvety green-grey in damp conditions (Fig 5.23). Cortical tissues beneath these lesions necrose, creating cankers. Affected fibres stain a dark brown-black colour and lose their tensile strength (Gitman & Malikova 1933). Leaf spots grow round to irregular in shape, covered by a green-grey mat of mycelium. The spots turn necrotic, dry out, and break up into ragged shot-holes.

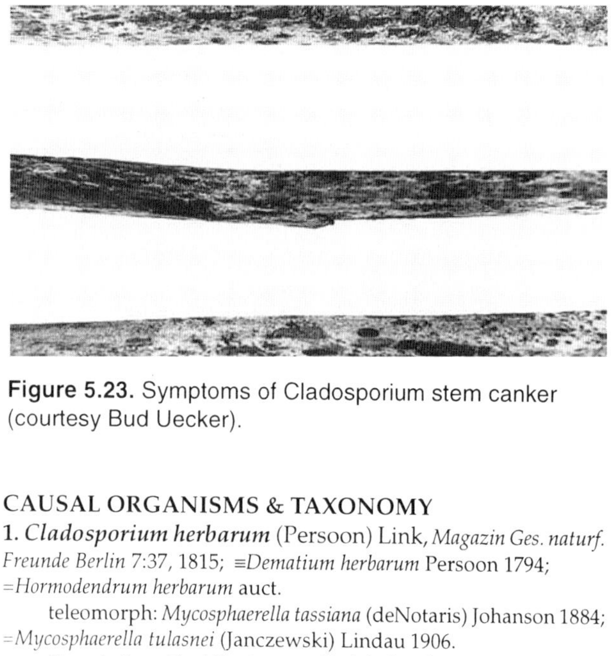

Figure 5.23. Symptoms of Cladosporium stem canker (courtesy Bud Uecker).

CAUSAL ORGANISMS & TAXONOMY

1. ***Cladosporium herbarum*** (Persoon) Link, *Magazin Ges. naturf. Freunde Berlin* 7:37, 1815; ≡*Dematium herbarum* Persoon 1794; =*Hormodendrum herbarum* auct.

teleomorph: *Mycosphaerella tassiana* (deNotaris) Johanson 1884; =*Mycosphaerella tulasnei* (Janczewski) Lindau 1906.

Description: Conidiophores cylindrical and straight (in culture) or nodose and geniculate (*in vivo*), unbranched until near the apex, smooth, pale to olivaceous brown, 3–6 µm in diameter and up to 250 µm long. Conidia blastic, produced sympodially in simple or branched chains, ellipsoidal or oblong, rounded at the ends, thick walled, distinctly verruculose, golden to olivaceous brown, with scars at one or both ends, zero or one septum, 8–15 x 4–6 µm. Teleomorph pseudothecia globose, black, scattered to aggregated, up to 160 µm in diameter. Asci bitunicate, subclavate, short stipitate, eight-spored, 35–90 x 15–30 µm. Ascospores hyaline to rarely pale brown, ellipsoid, usually one septum, slightly constricted at septum, 15–30 x 4.5–9.5 µm. Paraphyses not present.

C. herbarum attacked stalks and flowering tops of hemp in Italy (Curzi & Barbaini 1927), and stalks in the Ukraine (Gzebenycek 1984), and the Czech Republic (Ondrej 1991). It caused over-retting of hemp in Germany (Behrens 1902), and ruined harvested hemp in Russia (Gitman & Malikova 1933, Vakhrusheva 1979). *C. herbarum* has infested drug plants from Colombia (Bush Doctor, unpublished data) and India (Nair & Ponnappa 1974, Ponnappa 1977). Babu *et al.* (1977) isolated the fungus from seeds in India. Lentz (1977) reported its teleomorph on *Cannabis* stems.

C. herbarum also acts as a secondary invader of plants parasitized by *Sphaerotheca, Septoria,* or *Phomopsis* species (McPartland, unpublished data). Once established, the fungus is difficult to eradicate. It may survive after fungicides eliminate the primary pathogen. Indeed, Durrell & Shields (1960) found *C. herbarum* growing at Ground Zero in Nevada shortly after nuclear weapons testing. The fungus is one of the most common facultative parasites in the world. Farr *et al.* (1989) described *C. herbarum* from 90 host plant genera.

2. ***Cladosporium cladosporioides*** (Fresenius) deVries, *Contributions to the knowledge of the genus Cladosporium,* pg. 57, 1952.

≡*Hormodendrum cladosporioides* (Fresenius) Saccardo 1880.

Description: Conidiophores straight or flexous, simple or branched, smooth or verruculose, pale to olivaceous brown, 2–6 µm in diameter and up to 350 µm long, but generally much shorter. Ramo-conidia an extension of conidiophores, irregularly shaped, smooth or rarely minutely verruculose, 2–5 µm wide, up to 30 µm long. Conidia formed in simple or branched chains, ellipsoidal to lemon-shaped, pale olivaceous brown, mostly smooth walled, rarely minutely verruculose, one-celled, 3–7 x 2–4 µm.

Gzebenyuk (1984) isolated *C. cladosporiodes* from hemp stalks in the Ukraine. Ondrej (1991) rarely found *C. cladosporiodes* infesting stalks in the Czech Republic, whereas *C. herbarum* was abundant.

3. ***Cladosporium tenuissimum*** Cooke, *Grevillea* 5(37):140, 1877.

Description: Conidiophores flexous, branching, olive brown, swollen at the apex, 2–5 µm in diameter, up to 800 µm long. Conidia formed profusely in chains, ellipsoidal to oval, smooth walled or minutely verruculose, one-celled, 3–25 x 3–6 µm.

Nair & Ponnappa (1974) cited *C. tenuissimum* on diseased marijuana leaves in India. No one else has found *C. tenuissimum* in Asia. Nair & Ponnappa may have misidentified *C. herbarum.*

4. ***Cladosporium resinae*** (Lindau) deVries, *Antonie van Leeuwenhoek* 21:167, 1955.

≡*Hormoconis resinae* (Lindau) vonArx & deVries 1973.

teleomorph: *Amorphotheca resinae* Parberry 1969.

Description: Conidiophores straight or flexous, simple or branched, smooth or distinctly warted, pale brown to olive green, 3–6 µm wide and up to 2 mm long. Ramo-conidia, when present, are clavate or cylindrical, generally smooth, 8–20 x 3–7 µm. Conidia solitary or in chains, ellipsoid, smooth walled, pale to olivaceous brown, without prominent scars, 3–12 x 2–4 µm.

Gzebenyuk (1984) reported *C. resinae* on hemp stems. *C. resinae* normally colonizes petrochemical substrates such as jet fuel and facial creams; it has not been reported from herbaceous plants (McPartland 1995a).

DIFFERENTIAL DIAGNOSIS

Cladosporium stem canker can be confused with hemp canker, brown blight, and Stemphylium stem disease.

DISEASE CYCLE & EPIDEMIOLOGY

Cladosporium species overwinter in crop debris. They continue to cause destruction in retted hemp after harvest. *Cladosporium* species also overwinter in seeds (Pietkiewicz 1958, Ferri 1961b, Stepanova 1975). *C. herbarum* and *C. cladosporioides* grow best between 20–28°C, but they can grow at standard refrigerator temperatures. Warm humid conditions favour conidial production. A slight breeze detaches conidia and carries them for miles. Peaks of airborne conidia arise in June–July and September–October (Domsch *et al.* 1980). Airborne *C. herbarum* and *C. cladosporioides* conidia are major causes of "mould allergy." Opportunistic infections by *Cladosporium* runs the gamut from eye ulcers to pulmonary fungus balls (Rippon 1988).

CONTROL

Cornerstones of cultural control (described in Chapter 9) include method 1 (sanitation—eliminate *any* damp organic matter, especially wet paper products) and 7c (avoid excess humidity). Also follow methods 2b&c (sterilization or pasteurization of soil), 8 (optimizing soil structure and nutrition), 10 (careful pruning), and 11 (avoid seedborne infection). Pandey (1982) protected millet seeds from *C. herbarum* with a 30 minute soak in *Cannabis* leaf extract.

ANTHRACNOSE

Three species reportedly cause *Cannabis* anthracnose. All are soilborne fungi and cause disease in temperate climates worldwide. Two of the fungi parasitize a wide range of plants and also exist as saprophytes. Gzebenyuk (1986) reported a third species, ***Colletotrichum lini***, which normally attacks only flax (*Linum* species); his report is probably erroneous (McPartland 1995a).

SYMPTOMS

Leaf symptoms begin as light green, watersoaked, sunken spots. Spots enlarge to circular or irregular shapes with grey-white centres and brownish-black borders. Larger spots may become zonate. Affected leaves soon wrinkle then wilt (Hoffman 1959).

Stem lesions initially turn white. Then black, dot-like acervuli arise in the lesions, lending a salt-and-pepper appearance (Plate 67). Affected stems swell slightly and develop cankers. The periderm peels off easy. Stems sometimes snap at cankers. Distal plant parts become stunted and often wilt. Young plants die.

CAUSAL ORGANISMS & TAXONOMY

1. ***Colletotrichum coccodes*** (Wallroth) Hughes, *Can. J. Bot.* 36:754, 1958; =*Colletotrichum atramentarium* (Berkeley & Broome) Taubenhaus 1916, ≡*Vermicularia atramentaria* Berkeley & Broome 1850.

Description: Acervuli on stems round or elongated, reaching 300 µm in diametre, at first covered with epidermis; then dehiscing irregularly and exuding slimy conidia and bristling setae. Setae smooth, stiff, septate, slightly swollen and dark brown at the base and tapering to sharpened and paler apices, up to 100 µm long. Conidiophores hyaline, cylindrical, occasionally septate and branched at their base. Conidiogenous cells phialidic, smooth, hyaline, with a minute channel and periclinal thickening of the collarette. Conidia hyaline (honey coloured to salmon-orange *en masse*), aseptate, smooth, thin walled, guttulate, fusiform and straight, with rounded apices, often with a slight median constriction, averaging 14.7 x 3.5 µm on *Cannabis* stems but ranging 16–22 x 3–4 µm in culture (Fig 5.24). Appressoria club-shaped, medium brown, edge irregular to almost crenate, 11–16.5 x 6–9.5 µm, rarely becoming complex.

Over a dozen additional synonyms of *C. coccodes* are cited by von Arx (1957). The fungus is a powerful pathogen of solanaceous crops (tomatoes, potatoes, eggplants), as well as cabbage, mustard, lettuce, chrysanthemums, and many weed hosts. *C. coccodes* attacks both leaves and stems of *Cannabis*. Hoffman (1958, 1959) reported heavy hemp losses in central Europe. Conversely, Gitman (1968b) considered the pathogen of little importance in the USSR. Ghani & Basit (1975) probably collected *C. coccodes* on Pakistani drug plants (described as "*Colletotrichum* species ...cylindrical conidia with both ends rounded"). A hemp specimen collected by Vera Charles in Virginia (deposited at herb. BPI) also proved to be *C. coccodes* (McPartland, unpublished data).

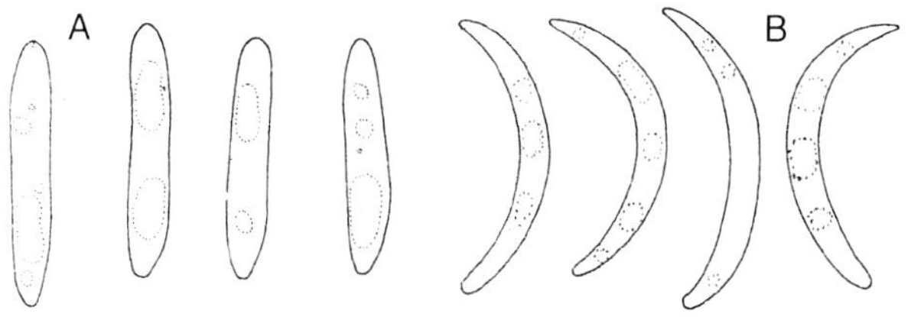

Figure 5.24: *Colletotrichum* conidia, x1500. A. *C. coccodes*; B. *C. dematium* (McPartland redrawn from Sutton 1980).

2. ***Colletotrichum dematium*** (Persoon) Grove, *J. Botany (London)* 56:341, 1918; ≡*Vermicularia dematium* (Persoon) Fries 1849; =*Vermicularia dematium* f. *cannabis* Saccardo 1880.

Description: Acervuli are round to elongated, up to 400 µm diametre, strongly erumpent through epidermis and exuding smoky grey conidial masses with divergent setae. Setae smooth, stiff, rarely curved, usually three or four septate, 4–7.5 µm wide at the base tapering to sharpened apices, 60–200 µm long (Fig 5.25). Conidiophores hyaline, cylindrical. Conidiogenous cells phialidic. Conidia hyaline, aseptate (becoming two-celled during germination), smooth, thin walled, guttulate, falcate and curved, 18–26 µm long and 3–3.5 µm wide in the middle and tapering to pointed apices (Fig 5.24). Appressoria club-shaped to circular, medium brown, edge usually entire, 8–11.5 x 6.5–8 µm, often becoming complex.

The fungus is a weak parasite on many hosts; Von Arx (1957) listed about 80 additional synonyms. *C. dematium* has been collected from hemp stems in Italy (Saccardo 1880, Cavara 1889). Saccardo named his specimen a new subspecies, but McPartland (1995e) considered it superfluous. Clarke collected a fungus infesting stalks of cultivar 'Uniko-B' growing in China, which turned out to be *C. dematium* (McPartland, unpublished data 1995).

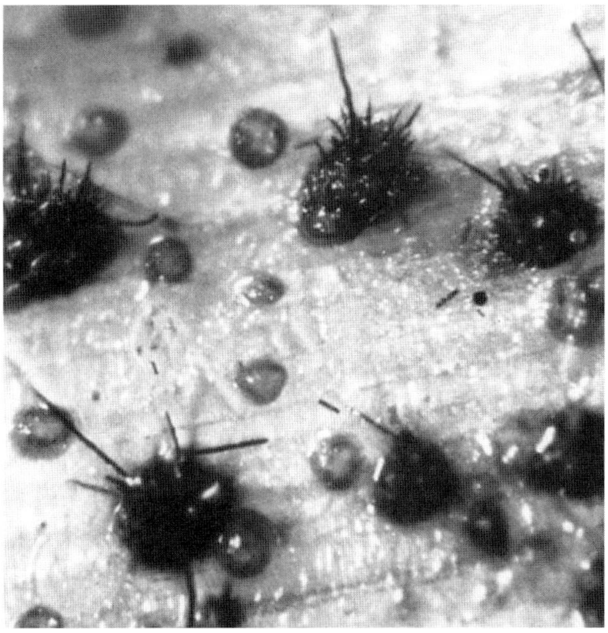

Figure 5.25: Acervuli with setae formed by *Colletotrichum dematium* (LM x40, courtesy Bud Uecker).

DIFFERENTIAL DIAGNOSIS

C. coccodes and *C. dematium* are best differentiated by their conidia (Fig 5.25). *C. coccodes* conidia are fusiform, straight, and have rounded ends. *C. dematium* conidia are falcate, curved, and have pointed ends. Both species produce abundant sclerotia and setae, and their appressoria appear similar (McPartland & Hosoya 1998).

Anthracnose can be confused with blights, leaf spots, stem cankers, wilt diseases, and plant diebacks. Anthracnose may also mimic damping-off in seedlings.

DISEASE CYCLE & EPIDEMIOLOGY

Anthracnose fungi overwinter as sclerotia in plant debris or soil. In the spring, sclerotia sporulate and conidia splash onto seedlings. Conidia form appressoria which directly penetrate epidermal tissue or enter via stomates and wounds. Seed transmission does not occur. *C. coccodes* prefers a cooler optimum temperature than *C. dematium*.

Expect anthracnose epidemics during cool damp weather, especially in heavy soils. Hoffman (1958, 1959) noted high losses in "bog" soils. Disease escalates in plants under stress from drought or frost damage. Anthracnose can rage in monocropped glasshouses, especially in hydroponic systems (Smith *et al.* 1988). Conidia spread via splashing water and wind-driven rain. Hoffman (1958, 1959) and Cook (1981) described heaviest infections after plants had flowered, with males succumbing before females. Concurrent infection by the nematode *Heterodera schactii* or the fungus *Rhizoctonia solani* increases plant susceptibility to anthracnose (Smith *et al.* 1988).

CULTURAL & MECHANICAL CONTROL (see Chapter 9)

Sanitation is the cornerstone—methods 1 (eliminating residues), 10 (careful pruning and rouging), 2a (deep ploughing), 2b&c (sterilizing or pasteurizing soil), and 3 (eliminating weeds), in that order. Also observe methods 7a&b (avoid drought or waterlogging), 8 (avoid heavy soils), and 6 (avoid solanaceous crops if *C. coccodes* prevails). Do not wet stems while irrigating. Stay out of *Cannabis* patches when plants are damp, since conidia spread by contact. Clarke (pers. commun. 1995) noted Uniko-B was susceptible to *C. dematium* in China, whereas the local Chinese landrace was resistant to the fungus.

BIOCONTROL & CHEMICAL CONTROL

According to Samuels (1996), *Trichoderma harzianum* (Trichodex®) has controlled *Colletotrichum* disease (described under Rhizoctonia sore shin disease). Yepsen (1976) suggested a prophylactic spray of lime sulphur on plants during susceptible weather. Kaushal & Paul (1989) inhibited a related species, *Colletotrichum truncatum*, with an extract of *Cannabis*.

VERTICILLIUM WILT

Two organisms probably cause this disease in *Cannabis*. Both species attack many crops, and both pathogens live worldwide.

SYMPTOMS

Leaves first turn yellow along margins and between veins, then turn grey-brown. Lower leaves show symptoms first. Slightly wilted plants often recover at night or after irrigation. These recoveries are transient as wilt becomes permanent. Dissection of diseased stems reveals a brownish discolouration of xylem tissue. If the fungus invades only a few xylem bundles, only parts of the plant may wilt.

CAUSAL ORGANISMS & TAXONOMY

1. ***Verticillium dahliae*** Klebahn, *Mycol. Centralb.*, 3:66, 1913.
= *Verticillium albo-atrum* var. *medium* Wollenweber 1929; = *Verticillium tracheiphilum* Curzi 1925.

Description: Conidiophores abundant, completely hyaline, with three or four whorled phialides arising at regular nodes along an upright branch which can reach 150 μm in height. Phialides hyaline, 19–35 μm long, 1.5 μm wide. Conidia arise singly but congregate in small droplets at tips of phialides, oval to ellipsoid in shape, hyaline, aseptate, 2.5–8 x 1.5–3.2 μm (Fig 5.26). Microsclerotia arise by lateral budding of a single hypha into long chains of cells, becoming dark brown to black, irregularly spherical to elongated, 15–100 μm in length (Fig 5.26).

Reports of *V. dahliae* come from southern Russia (Vassilieff 1933, Gitman 1968b), Italy (Noviello 1957), the Czech Republic (Ondrej 1991), the Netherlands (Kok et al. 1994), and Germany (Patschke *et al.* 1997). *V. dahliae* attacks many cultivated, weedy, and wild plants, in temperate zones and the tropics.

2. ***Verticillium albo-atrum*** Reinke & Berthier, *Die Zersetzung der Kartoffel durch Pilze*, p. 75, 1879.

Description: Conidiophores abundant, mostly hyaline but with a darkened base (especially *in vivo*), with two to four phialides arising at regular intervals along an upright branch up to 150 μm tall. Phialides hyaline, variable, 20–30 (up to 50) μm in length and 1.5–3.0 μm wide. Conidia arise singly, ellipsoid to short-cylindrical, hyaline, usually single-celled but occasionally one septum, 3.5–10.5 x 2.5–3.5 μm. No microsclerotia formed.

V. albo-atrum has attacked hemp in China (Tai 1979) and the Ukraine (Gzebenyuk 1984). *V. albo-atrum* prefers cooler climates in northern Eurasia and North America, and attacks many dicots. It causes more severe symptoms than *V. dahliae.*

DIFFERENTIAL DIAGNOSIS

V. dahliae produces slightly smaller conidia than *V. albo-atrum*, but the two species are difficult to distinguish. Only *V. dahliae* produces microsclerotia. No reports of *Verticillium* on *Cannabis* mention the presence or absence of microsclerotia. Thus either fungus could have been collected.

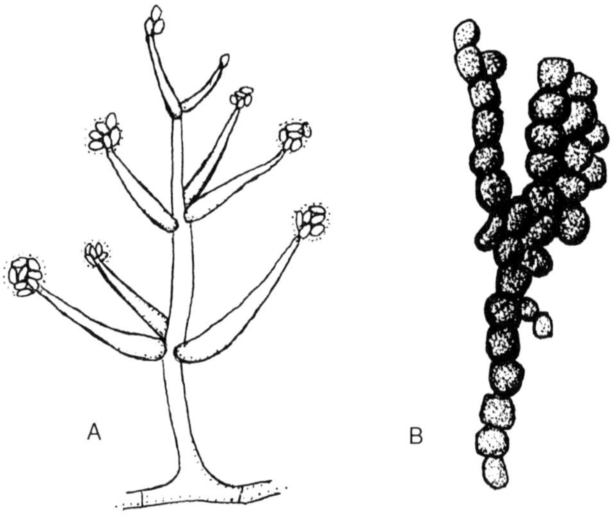

Figure 5.26: *Verticillium dahliae*, x400. A. Condiophore with conidia aggregated in droplets; B. Microsclerotium; (McPartland).

Prior to wilting, symptoms of Verticillium wilt may be confused with nutritional disorders. Verticillium wilt resembles Fusarium wilt except the xylem discolouration is different. The wilt mimics symptoms caused by nematodes and root-boring insects (root maggots, white grubs), or drought.

DISEASE CYCLE & EPIDEMIOLOGY

Both species overwinter in soil and invade roots of seedlings in the spring. Once in roots *Verticillium* species spread via xylem; they eventually clog the pipes and cause wilts. Verticillium wilt increases in moist soils rich in clay. Root-knot nematodes predispose plants to *Verticillium* infection. *V. albo-atrum* grows best at 23.5°C; the optimum for *V. dahliae* is 21°C.

CULTURAL & MECHANICAL CONTROL (see Chapter 9)

Clean the soil—use methods 1 (but no burying!), 2b (sterilize the soil), and 3 (weeding). Method 2a is not too helpful since microsclerotia can live 75 cm deep in soil. Method 2c (soil solarization) works well against *V. dahliae*, but not heat-resistant *V. albo-atrum* (Elmore *et al.* 1997). Flooding infested soil for 14 days kills *Verticillium* microsclerotia (Lucas 1975).

Avoid planting in heavy, poorly draining soils. According to Elmer & Ferrandino (1994), Verticillium wilt decreases when nitrogen is supplied as an ammonia [e.g., $(NH_4)_2SO_4$] rather than a nitrate [e.g., $Ca(NO_3)_2$].

Crop rotation with nonhost monocotyledonous plants must be long (four years for *V. albo-atrum*, twice that for *V. dahliae*). Monocot crops must be weed-free, since almost all dicot weeds serve as alternate hosts. Root-knot nematodes (*Meloidogyne* species) should be eliminated from the soil. Method 5 is a future option—hops breeders have developed varieties resistant to *V. albo-atrum*. Kok *et al.* (1994) found partial resistance to *V. dahliae* in fibre cultivar 'Kompolti Hibrid TC.'

BIOCONTROL & CHEMICAL CONTROL

Verticillium wilt constitutes a real menace to world agriculture, because no chemicals can control the disease (Smith *et al.* 1988). Many soil bacteria and fungi antagonize *Verticillium*—increase their populations by mixing green alfalfa meal into soil. Windels (1997) mixed sudangrass into soil as a green manure to control *V. dahliae* in potatoes. Grewal (1989) reduced *Verticillium* growth by mixing compost with dried *Cannabis* leaves.

A *Trichoderma* product (Bio-Fungus®) reportedly controls *Verticillium* wilt (described under Rhizoctonia sore shine disease), as does the FZB24 strain of *Bacillus subtilis* (RhizoPlus®, described under damping off). *Talaromyces flavus* is irregularly available (see below), and ***Verticillium nigrescens*** is being developed. *V. nigrescens* is a nonpathogenic species that protects plants from infection by pathogenic *Verticillium* species (Howard *et al.* 1994).

Talaromyces flavus

BIOLOGY: Kendrick (1985) reported a 76% reduction in eggplant wilt from *V. dahliae* by inoculating soil with spores of the soil-inhabiting fungus *T. flavus*. The biocontrol fungus produces hydrogen peroxide which kills *V. dahliae* microsclerotia.

APPLICATION: Supplied as spores in granules, which are mixed into soil or potting medium. *T. flavus* is tolerant of heat; heating soil (solarization) makes *V. dahliae* much more susceptible to the biocontrol.

RUST

Rust fungi cause many of our worst crop diseases. They often exhibit complicated life cycles spanning *several* plant hosts, and produce up to five distinctive spore states. A full rust life cycle includes spermagonia, aecia, uredia, telia, and basidia (Fig 3.3).

Of the four rust fungi reported from *Cannabis*, one arises from a typographical error: Rataj (1957) misspelled *Melanospora cannabis* (the cause of red boot) as ***Melampsora cannabis***—*Melanospora* is an ascomycete, *Melampsora* is a rust (McPartland 1995a). Rataj's error reappears in Barna *et al.* (1982), Gutberlet & Karus (1995), and Bósca & Karus (1997).

CAUSAL ORGANISMS & TAXONOMY

1. ***Aecidium cannabis*** Szembel, *Commentarii Instituti Astrachanensis ad defensionem plantarum*, i, 5–6, p. 59, 1927.

Description: Aecia arise in pale leaf spots on undersides of lower leaves, "globose" [cupulate?], orange coloured, gregarious, measuring 360–400 µm deep and 300–340 µm in diameter. Aecial peridium cells rectangular, often rhomboidal, 32–40 x 18–24 µm, exterior walls 6–8 µm thick. Aecidiospores round to ellipsoidal, subhyaline, verrucose, single-celled, 24–28 x 20–24 µm, with an epispore 1.5–2.0 µm thick.

Szembel collected *A. cannabis* from *Cannabis ruderalis* growing near the Caspian Sea. The *Cannabis* was concurrently infected by *Septoria cannabis*.

2. ***Uredo kriegeriana*** H. & P. Sydow, *Osterreichische botanische Zeitschrift* 52(5):185, 1902.

Description: Uredia sori borne on undersides of leaves in irregular yellow to ocher spots, either sparsely or gregariously distributed, initially subepidermal ("hidden in enclosing peridium") then erumpent and pincushion shaped, covered by yellow spores. Uredospres subglobose to ellipsoidal, finely echinulate, orange on the inside, single-celled, 21–27 x 15–22 µm, bearing many germination pores.

Sydow & Sydow (1924) amended the description, "Uredinia appear in loose aggregations within leaf spots, the covering peridium consists of roundish to obtuse-angled thin-walled cells, hyaline and membranous, 1 µm thick, 12–17 µm long by 10–13 µm wide. The uredinospores contain a hyaline epispore 1.5 µm thick, and the germination pores are difficult to see." Father and son Sydow collected *U. kriegeriana* from hemp near Schandau, Germany.

3. ***Uromyces inconspicuus*** Otth, *Mittheilungen der naturforschenden Gesellschaft in Bern.* 1:69, 1868.

Description: Leaf spots on undersides of leaves, very inconspicuous, telia growing in punctiform tufts, containing few teliospores. Teliospores borne on short hyaline pedicels; spores dark brown-black, ellipsoid, verruciform, single-celled, with a small hyaline apex, 32 x 18 µm.

Otth collected *U. inconspicuus* from hemp, potatoes, and several genera of weedy plant species. This host range is unusual for *Uromyces* species. Saccardo (*S.F.* 14:287) expressed his doubts, "Quid sit haec species pantogena non liquet, certe dubia res." Fisher (1904) thought Otth misidentified a hyphomycete.

DISEASE CYCLE & EPIDEMIOLOGY

All three fungi have been described from a single spore state. Interestingly, each represents a different stage of the rust life cycle (aecia, uredia, and telia, respectively). They could be different spore states of the same organism.

DIFFERENTIAL DIAGNOSIS

True rusts should not be confused with "white rusts," which are related to downy mildews. Manandhar (pers. commun. 1986) saw a white rust caused by an ***Albugo*** species in Nepal.

CONTROL

Cultural control method 10 (careful pruning—see Chapter 9) may be sufficient. Also follow method 7c (avoid excess humidity, avoid overhead watering and dew condensation in glasshouses). No biocontrol is known. Sulphur helps if one carefully dusts the *undersides* of affected leaves. Misra & Dixit (1979) used ethanol extracts of *Cannabis* to kill *Ustilago tritici* (Pers.) Rostr. and *Ustilago hordei* (Pers.) Lager., two smut fungi that are somewhat related to rusts.

BLACK DOT

Also called "small shot" or "black spot," this disease affects leaves and stems. Leaf disease occurs in India (Nair & Ponnappa 1974, Ponnappa 1977) and Maryland (McPartland, unpublished). Stem disease has been reported in the Ukraine (Gzebenyuk 1984) and the Czech Republic (Ondrej 1991). In Illinois, the fungus acts as a saprophyte, colonizing male flowers after release of pollen (McPartland, unpublished). The causal fungus also rots harvested marijuana in storage. It grows worldwide on a wide range of plants, animals, and processed foodstuffs.

SYMPTOMS

Black dot disease is characterized by small dark pustules of fungal growth. Pustules reach a finite size (less than 2 mm) and become covered with black conidia. Pustules may appear on healthy green tissue or in the midst of a small chlorotic ring. On leaves they concentrate near midveins or vascular tissue. On stems they scatter randomly.

CAUSAL ORGANISM & TAXONOMY

Epicoccum nigrum Link, *Magazin Ges. Naturf. Freunde, Berlin* 7:32, 1816; =*Epicoccum pururascens* Ehrenberg ex Schlechtendahl 1824.

Description: Sporodochia 100–2000 µm diametre. Conidiophores densely compacted, straight or flexuous, occasionally branched, colourless and smooth but turning pale brown and verrucose at the tip. Conidia formed singularly, monoblastically, globose to pyriform in shape, golden brown to dark brown, with a warted surface obscuring muriform septa which divide conidia into many cells (up to 15), conidia 15–25 µm diametre (Figs 3.2 & 5.27).

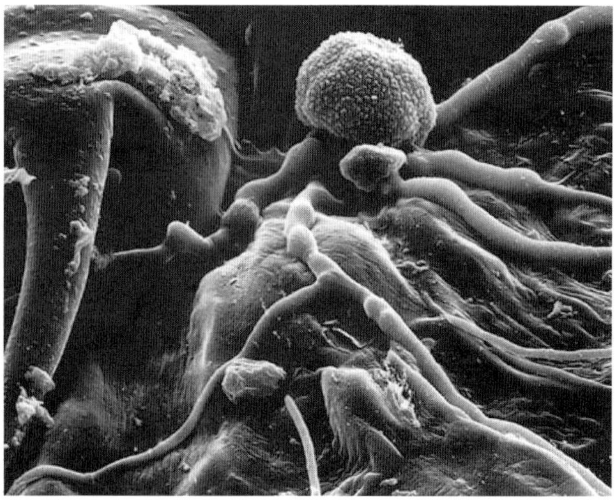

Figure 5.27: Conidium of *Epicoccum nigrum* germinating with many germ tubes, next to a deflated *Cannabis* cystolith leaf hair (SEM x1060, McPartland).

DIFFERENTIAL DIAGNOSIS

This disease can be confused with early brown blight, Curvularia leaf spot, corky leaf spot, or black mildew.

DISEASE CYCLE & EPIDEMIOLOGY

The fungus overwinters in crop debris. Conidia germinate many germ tubes (Fig 5.27) and quickly form colonies. Optimal growth occurs at 23–28°C. Conidium production increases under UV light. The fungus produces a strong reddish-brown pigment (mostly beta carotene) in host tissue. This pigment is visible in light-coloured substrates (maize kernels, apples). Metabolites of *E. nigrum* may cause hepatic and renal disorders (Rippon 1988).

CONTROL

This disease does not arise in healthy *Cannabis*—observe cultural methods 7a and 8 (described in Chapter 9). Cure harvested material in a dark (no UV light), dehumidified chamber. Biocontrol is unknown. Treat badly infected plants with copper-based fungicides.

BASIDIO ROT

Older *Cannabis* literature describes four *Corticium* species on hemp—they are all old names, now considered synonyms. Two *Corticium* species are true pathogens—*C. solani* (= *Rhizoctonia solani*) and *C. rolfsii* (= *Sclerotium rolfsii*). They are discussed under Rhizoctonia sore shin and southern blight, respectively. The remaining two species act as saprophytes or weak pathogens.

CAUSAL ORGANISMS & TAXONOMY

1. ***Athelia arachnoidea*** (Berkeley) Jülich, *Willdenowia Beih.* 7:23, 1972; ≡*Septobasidium arachnoideum* (Berkeley) Bresadola 1916, ≡*Hypochnus arachnoideus* (Berkeley) Bresadola 1903, ≡*Corticium arachnoideum* Berkeley 1844.

anamorph: *Fibularhizoctonia carotae* (Rader) Adams & Kropp 1996; ≡*Rhizoctonia carotae* Rader 1948.

Description: Basidiocarp effuse, thin, bluish white when fresh, surface arachnoid, 2–6 cm long and 1–3 cm wide, 100–200 µm thick in cross section. Subhymenial hyphae loosely interwoven, without clamp connections, 3–5 µm in diameter; basal hyphae somewhat wider, with thickened walls and occasional clamp connections. Basidia clavate, grouped like candelabra, normally producing two basidiospores borne upon pointed sterigmata, 20–30 x 5–7 µm. Basidiospores hyaline, smooth, ellipsoid to pyriform, usually 6–10 x 4–6 µm. Mycelium is similar to that of *Rhizoctonia*, but with clamp connections. Sclerotia globose, composed of dark brown parenchymatous cells, 0.2–1.0 mm diam (larger in culture, 1–5 mm diam).

Lentz (1977) cited *A. arachnoidea* on *Cannabis*. This fungus is common in North America and Scandinavia. It grows on leaf humus, parasitizes lichens, and causes a cold-storage disease of carrots (*Daucus carota*).

2. ***Athelia epiphylla*** Persoon, *Mycol. Europ.* I:84, 1822.

≡*Athelia epiphylla* Persoon:Fries 1822; =*Hypochnus cetrifugus* (Léveillé) Tulasne 1861, ≡*Corticium centrifugum* (Léveillé) Bresadola 1903; =*Butlerelfia eustacei* Weresub & Illman 1980.

anamorph: *Fibularhizoctonia centrifuga* (Léveillé) Adams & Kropp 1996; ≡*Rhizoctonia centrifuga* Léveillé 1843.

Description: Basidiocarp effuse, very thin, white to buff, surface arachnoid to byssoid, 2–6 cm long and 1–3 cm wide, 75–150 µm thick in cross section. Subhymenial hyphae loosely interwoven, without clamp connections, 2–3 µm in diameter, basal hyphae up to 6 µm in diameter rarely with clamp connections. Basidia clavate, grouped like candelabra, normally producing four basidiospores borne upon pointed sterigmata, 10–20 µm long. Basidiospores hyaline, smooth, ellipsoid to pyriform, usually 4–8 x 2.5–4 µm. Mycelium and sclerotia are identical those of *Athelia arachnoidea*.

Endo (1931) recovered *A. epiphylla* from *Cannabis* in Japan. The fungus decays forest leaves, rots wood, and causes fisheye decay of stored apples. It is common in Eurasia, less so in North America.

A. epiphylla and *A. arachnoidea* have been shuffled through many different genera. The two taxa may represent the same organism (Adams & Kropp 1996). Saccardo (*S.F.* 6:654) considered them identical 100 years ago.

DIFFERENTIAL DIAGNOSIS & CONTROL

Symptoms of basidio rot may resemble southern blight and Rhizoctonia sore shin. Follow cultural controls described for Rhizoctonia sore shin disease. No biocontrol is known and chemicals are ineffective.

RED BOOT

This disease is a secondary problem and only arises on plants previously damaged by hemp canker. Some hemp cultivators nevertheless consider red boot more dangerous than hemp canker itself. Red boot has been reported in Germany (Behrens 1891, Kirchner 1906, Flachs 1936) and the Ukraine (Gzebenyuk 1984).

SYMPTOMS

Red boot appears on plants in late July or August (Behrens 1891). The causal fungus rapidly encases stems in thick, red mats of mould. Fibre harvested from infested stems is soft, frail, stained a red colour, and valueless.

CAUSAL ORGANISM & TAXONOMY

Melanospora cannabis Behrens, *Zeitschrift Pflanzenkrankheiten* 1:213, 1891.

Description: Perithecia globose, orange red, averaging 210–230 μm diametre, ostiole atop a 60–90 μm perithecial neck. Asci described as swollen, no dimensions listed. Paraphyses not found. Ascospores black, elliptical, 22–26 x 15–17 μm. Conidiophores more or less erect, verticillately branched, multicellular, flask shaped, narrowing at tips. Conidia arise at the apices of phialides, unicellular, forming in clusters or chains, red in colour, 4.4 x 3.0 μm. Behrens noted that conidiophores bind together to form coremia in artificial culture.

The anamorph of *M. cannabis* has not been named. In a monograph of *Melanospora* species, Douguet (1955) considered *M. cannabis* similar to *M. zobelii* and *M. fimicola.* Douguet did not examine the type specimen, however, and expressed regret over Behrens's failure to measure asci.

DIFFERENTIAL DIAGNOSIS

Red boot, with its thick, red mat of fungal hyphae and red conidia, is rather unique and not easily confused with other hemp diseases.

DISEASE CYCLE & EPIDEMIOLOGY

The fungus overwinters as ascospores in crop debris or soil. Epidemics occur in low, wet fields during damp growing seasons. Weedy fields reportedly promote the disease. Conidia are splashed by rain to sites of secondary infections. *M. cannabis* cannot infect *Cannabis* in the absence of infection by *Sclerotinia sclerotiorum* (Behrens 1891). *M. cannabis* also parasitizes mycelia and sclerotia of *S. sclerotiorum.*

CONTROL

Follow cultural controls described in Chapter 9, especially methods 1 (sanitation), 2a (deep ploughing), and 10 (careful pruning). Methods 3 (destroying weeds) and 7c (avoid excess humidity) are also helpful. Eliminate hemp canker (*S. sclerotiorum*) and you eliminate red boot. See the section on hemp canker. No biocontrols or fungicides have been tested.

TEXAS ROOT ROT

Texas root rot arises in the southwestern USA and northern Mexico. The causal fungus is limited to clay soils with little organic matter but lots of calcium carbonate, a high pH, and high temperatures. In the blackland prairies of central Texas, the causal fungus attacks over 2000 species of wild and cultivated dicots. Texas root rot attacks hemp in Texas, Oklahoma, and Arizona (Chester 1941), causing 30–60% mortality rates (Taubenhaus & Killough 1923). Killough (1920) described an epidemic killing 95% of a crop during the flowering stage.

SYMPTOMS

Yellow leaves appear in late June to August, followed by wilting, leaf necrosis, and plant death. Roots develop depressed and discoloured lesions as the fungus destroys cortical tissue and invades xylem. Sifting soil around plants with a 1 mm mesh sieve reveals small brown sclerotia. A tan mycelial mat may form around the base of dead plants in warm, damp conditions.

CAUSAL ORGANISM & TAXONOMY

Phymatotrichopsis omnivora (Duggar) Hennebert, *Persoonia* 7:199, 1973; ≡*Phymatotrichum omnivorum* Duggar 1916; =*Ozonium omnivorum* Shear 1907.

Description: Mycelium with septate hyphae lacking clamp connections, forming thick-walled cruciform aerial setae and aggregating into subterranean cord-like funicles. Conidiophores borne directly on hyphae, hyaline, simple or branched, clavate to globose, hyaline, 20–28 x 15–20 μm. Conidia holoblastic, ovate to globose, smooth, thin walled, hyaline, with a broad base exhibiting a detachment scar, 6–8 x 5–6 μm. Sclerotia borne on hyphae, ovate to globose, at first yellow but turning reddish brown, 1–2 mm diameter, often aggregated in clusters reaching 10 mm diameter.

The teleomorph of *P. omnivora* is unknown. Unsubstantiated reports link *P. omnivora* to *Trechispora brinkmannii* (Bresadola) Rogers & Jackson and *Hydnum omnivorum* Shear.

DISEASE CYCLE & EPIDEMIOLOGY

The fungus is a soil organism. It grows in rope-like funicles, 20 to 60 cm underground. *P. omnivora* invades roots of seedlings in the spring. In damp conditions the fungus emerges from the ground and spreads via conidia. In dry conditions the fungus produces sclerotia in soil. These durable structures survive for years.

DIFFERENTIAL DIAGNOSIS

Initial symptoms (wilting) resemble those caused by nematodes and soil insects (e.g., root maggots, white root grubs). Texas root rot can also be confused with Fusarium wilt, Verticillium wilt, hemp canker, and southern blight. Conidia of *P. omnivora* resemble those of *Botrytis cinerea.* The cruciform setae and funicles are unique to *P. omnivora.*

CULTURAL & MECHANICAL CONTROL (see Chapter 9)

Method 2a (deep tillage) exposes funicles to desiccation. Method 6 (crop rotation with monocots) must be of sufficient duration to outlast long-lived sclerotia. Also follow methods 1 (sanitation), 3 (weeding), 7a (avoid drought), and 8 (optimize soil structure and nutrition).

BIOCONTROL & CHEMICAL CONTROL

Mixing organic matter into soil encourages natural biocontrol organisms. Use composted animal manure or a green manure of legumes or grasses. The biocontrol fungus *Gliocladium roseum* controls *P. omnivora* (described under grey mould). No fungicide is effective. Move from Texas.

OPHIOBOLUS STEM CANKER

Four species of *Ophiobolus* reportedly infest *Cannabis.* Gzebenyuk (1984) cited two species, *O. porphyrogonus* and *O. vulgaris.* These are old names of the species *Leptospora rubella,* discussed later. The remaining *Ophiobolus* species are described below.

Symptoms develop as weather warms. Lower stem surfaces turn brown-black. Plants senesce prematurely and die. After plants die, fruiting bodies of the causal fungi arise in blackened areas of stems.

CAUSAL ORGANISMS & TAXONOMY

1. ***Ophiobolus cannabinus*** G. Passerini, *Rendiconti della R. Accademia dei Lincei* 4:62, 1888.

Description: Pseudothecia scattered, globose or conical, black, immersed with conical apex and ostiole barely erumpent. Asci cylindrical, short-stalked, eight-spored, 85 x 5 µm. Paraphyses cylindrical. Ascospores filiform, aseptate, hyaline, 65–85 x 1.0–1.25 µm.

Passerini found *O. cannabinus* on hemp near Parma, Italy. His description of *aseptate* ascospores would preclude this fungus from the genus *Ophiobolus*. Many species have inconspicuous septa, however, and he may have missed them.

2. ***Ophiobolus anguillidus*** (Cooke in Cooke & Ellis) Saccardo [as *anguillides*] *Sylloge Fungorum* 2:341, 1883; ≡*Raphiodospora anguillida* (Cooke in Cooke & Ellis) Cooke & Ellis 1878–9.

Description: Pseudothecia gregarious, globose to ampulliform, black, immersed but quickly erumpent, papillate, 300–500 µm diameter. Asci cylindrical, short-stalked, eight-spored, 80–150 x 9–12 µm. Paraphyses cylindrical, slightly longer than asci. Ascospores scolecosporous, yellowish, ten to 14 septa, straight or curved, with basal cells attenuated and apical cells swollen to an ovoid shape, 80–120 x 2.5–4.0 µm (Fig 3.2).

O. anguillidus commonly parasitizes members of the Compositae, including *Bidens, Ambrosia,* and *Aster* species. Preston & Dosdall (1955) collected *O. anguillidus* from feral hemp near Judson, Minnesota.

DIFFERENTIAL DIAGNOSIS

O. anguillidus has much larger ascospores than *O. cannabinus*. Ophiobolus stem canker may be confused with Leptosphaeria blight, brown stem canker, Phomopsis stem canker, and striatura ulcerosa (a bacterial disease).

DISEASE CYCLE & EPIDEMIOLOGY

O. anguillidus and *O. cannabinus* overwinter as mycelia or teleomorphs in crop debris, infecting plants in early spring. The disease worsens in fields retaining stubble from previously infected crops. *Ophiobolus* species flourish during wet growing seasons and in poorly drained soil.

CONTROL

Little is known about this problem, but control measures can be adapted from *Ophiobolus* diseases on other crops, where cultural method 8 (see Chapter 9) is most important. Also follow methods 1 (remove or burn crop debris), and 7c (avoid excess humidity). Organic manuring and rotating with clover (method 6) encourages *Ophiobolus*-antagonistic soil organisms. No fungicides work well.

CHAETOMIUM DISEASE

Chaetomium species are common soil organisms. They aggressively decompose cellulose, including hemp and cotton fibres. During WWII the USA Army lost more tents, tarps, rope, and clothing to *Chaetomium* than to enemy forces. Most *Chaetomium* species act as saprophytes, but reports of *Chaetomium* species acting as plant pathogens appear in the literature.

SYMPTOMS

Ghani & Basit (1976) reported a root disease of Pakistani marijuana caused by *Chaetomium*, but did not identify the species nor describe symptoms. Chandra (1974) described *Chaetomium succineum* causing a leaf disease of Indian marijuana. Symptoms begin as light brown spots. Spots enlarge but remain limited by leaf veins. Spots gradually darken, dry, and drop out, leaving irregular holes in leaves. Chandra's citation is the first isolation of *C. succineum* outside of North America.

CAUSAL ORGANISM & TAXONOMY

Chaetomium succineum Ames, *Mycologia* 41:445, 1949.

Description: *Chaetomium* perithecia are covered by extravagant ornamental hairs (Fig 5.28). These hairs probably serve as deterrents against insect predation. Taxonomists use them to differentiate species. Perithecia are superficial, black, carbonaceous, ostiolated and globose to barrel-shaped. Asci of all species reported on *Cannabis* are clavate and eight-spored. Ascospores are olive brown to brown when mature.

Other *Chaetomium* species have been associated with hemp retting, such as ***Chaetomium globosum*** Kunze 1817 [cited as *Chaetomium fieberi* Corda 1837 by Oudemans (1920) but Oudemans may be wrong—see McPartland (1995a) for a discussion]. Gzebenyuk (1984) isolated five *Chaetomium* species from Soviet hemp stems: the aforementioned *C. globosum*, plus ***Chaetomium elatum*** Kunze:Fries 1817,

Table 5.2: Characteristics of six *Chaetomium* species described on *Cannabis*.

Species	Perithecium size (µm)	Ornamental hairs	Ascospore size (µm)	Ascospore shape
C. globosum	200–300 tall x 200–280 diam.	undulating, tapering to hyaline tips	9–13 x 6–10	lemon-shaped with apiculate ends
C. elatum	400–500 tall x 335–450 diam.	dichotomous branches tapering to points	11–14 x 8–10	lemon-shaped with apiculate ends
C. murorum	240–340 tall x 200–345 diam.	circinate tips with graceful arches	13–17 x 7–9	ellipsoid with a longitudinal furrow
C. piluliferum	280–560 tall x 222–480 diam.	circinate tips sinuous and smooth	13–17 x 8–10	ellipsoid with apiculate ends
C. aureum	100–140 tall x 100-130 diam.	yellowish brown with spiral ends	9–12 x 5–7	ovate and flattened on one side
C. succineum	220–340 tall x 200–230 diam.	amber when young looping in coils	12–17 x 6–9	oval and rounded at both ends

Figure 5.28: Perithecium of *Chaetomium succineum* with extravagant ornamental hairs (LM x150, McPartland).

Chaetomium murorum Corda 1837, ***Chaetomium trilaterale*** Chivers 1912 (=*Chaetomium aureum* Chivers 1912), and ***Chaetomium piluliferum*** Daniels 1961. Characteristics of all six *Chaetomium* species are compared in Table 5.2.

DIFFERENTIAL DIAGNOSIS

C. murorum, C. elatum and *C. piluliferum* resemble each other. The former is distinguished by ascospores bearing a longitudinal furrow. The latter is distinguished by the presence of its anamorph, *Botryotrichum piluliferum* Saccardo & March. *C. elatum* produces hairs with dichotomous branching. *C. succineum* is characterized by its loose cluster of amber-coloured hairs. *C. globosum* stands apart by its "permed hair" appearance.

Leaf spots described by Chandra (1974) could be confused with leaf diseases caused by *Phoma, Ascochyta, Septoria, Colletotrichum,* and *Cercospora* fungi, or *Pseudomonas* and *Xanthomonas* bacteria. Microscopy tells them apart.

DISEASE CYCLE & EPIDEMIOLOGY

Chaetomium species overwinter as perithecia in soil or crop stubble. They are excellent saprophytes and thrive on many organic materials. Ascospores are water-dispersed.

CONTROL

Control of *Chaetomium* is a simple matter of maintaining vigourous plants. Stress-free plants will not succumb to these fungi. Dead plants are another matter; anything in contact with soil is subject to attack. See Chapter 8 regarding retting and rotting of hemp fibres. The fungus ***Dicyma pulvinata*** has been used experimentally as a biocontrol against *Chaetomium* species (Jeffries & Young, 1994).

PHOMOPSIS STEM CANKER

This disease arises on senescent plants in late autumn. It has been described on hemp in Italy (Curzi 1927, DeCorato 1997), the former USSR (Gitman & Malikova 1933, Gitman & Boytchenko 1934), and Illinois (Miller *et al.* 1960). Herbarium specimens (e.g., IMI no. 128315) show the disease also strikes drug varieties in India (McPartland 1983b). In 1995, Clarke collected a specimen on hemp stems in China, coinfected with *Colletotrichum dematium*.

SYMPTOMS

Stem cankers begin light-coloured and slightly depressed, with a distinct margin. Leaves near stem cankers may wilt and become chlorotic. Stems then darken and become peppered with small black pycnidia. Symptoms may worsen after harvest (Gitman & Malikova 1933). By November, thin black "zone lines" form within stalks (Fig 5.29). Perithecia extend tiny spike-like beaks to the stem surface.

CAUSAL ORGANISM & TAXONOMY

Phomopsis arctii (Saccardo) Traverso, *Fl. Ital. Crypt.* 2:226, 1906.
≡*Phoma arctii* Saccardo 1882; ?=*Phomopsis cannabina* Curzi 1927.
teleomorph: *Diaporthe arctii* (Lasch) Nitschke, *Pyrenomycetes Germanici* 1:268, 1867; =*Diaporthe tulasnei* Nitschke f. *cannabis* Saccardo 1897.

Description: Pycnidia subglobose to lens-shaped, immersed then erumpent, ostiolated (20 µm diameter), upper peridium black and stromatic, lower peridium parenchymous and pale sooty-brown, 200–450 x 100–190 µm. Conidiophores conical to cylindrical, less than 10 µm in length. Conidiogenous cells phialidic, cylindrical to obclavate. α-conidia fusiform, straight or slightly curved, hyaline, usually one septum and biguttulate, 7–10 x 2–3 µm (up to 11.5 x 4.2 µm *apud* Gitman & Boytchenko). β-conidia unicellular, filiform-hamate, 18–25 x 1 µm. Perithecial pseudostroma widely effuse over stem surface, appearing blackened, carbonaceous; ventral margin of pseudostroma delimited by a narrow dark-celled prosenchymatous "zone line." Perithecia immersed in stroma, globose to slightly flattened, solitary or gregarious, 160–320 µm in diametre with a conical ostiolated beak 280–480 µm tall. Asci unitunicate, clavate, with an apical refractive ring, eight-spored, 40–60 x 6–10 µm. Paraphyses elongate, multiseptate, branching at base, deliquescing at maturity. Ascospores hyaline, biseriate, guttulate, one septum when mature, slightly constricted at the septum, fusoid-ellipsoidal, straight or slightly curved, 10–15 x 2.5–4.0 µm.

DIFFERENTIAL DIAGNOSIS

Phomopsis cannabina and *Diaporthe tulasnei* f. *cannabis* are synonyms of *D. arctii* (McPartland 1995e). Researchers have confused the fungus causing Phomopsis stem canker with a different *Phomopsis* causing white leaf spot (Sohi & Nayar 1971, Ghani & Basit 1976, McPartland 1983a, Gupta RC 1985). White leaf spot afflicts living leaves, and has not been isolated from stems. See the following section.

Figure 5.29: Surface of hemp stem blackened by Phomopsis stem canker, with patch of surface shaved (arrow) to expose pockets of perithecia, and stem split to reveal zone lines (McPartland).

DISEASE CYCLE & EPIDEMIOLOGY

P. arctii overwinters as pycnidia or perithecia in plant debris. The anamorph is found more frequently than the teleomorph. Both conidia and ascospores are spread by water and wind-driven rain. Although the fungus is considered a weak pathogen, it infests many Compositae, Umbelliferae, and Urticaceae in Europe and North America.

CONTROL Several cultural methods are useful (see Chapter 9)—methods 1 (sanitation), 2a (deep ploughing), 3 (weeding), 7 (avoid drought), and 5 (genetic resistance). Gitman & Malikova (1933) noted that damp storage conditions accelerate the fungal destruction of harvested stems.

Gitman & Malikova (1933) listed several resistant fibre varieties. DeCorato (1997) tested four fibre varieties: 'Yellow stem' (from Hungary) was the most susceptible, followed by 'Shan-ma,' 'Chain-chgo' (from China), and 'Foglia pinnatofida' (from Italy), but these differences were not statistically significant.

The biocontrol agent *Streptomyces griseoviridis* controls Phomopsis stem rot in other crops. It is described under damping off fungi. Bordeaux mixture inhibits *Diaporthe/Phomopsis* species on other crops.

WHITE LEAF SPOT

White leaf spot has appeared in India (Sohi & Nayar 1971, Gupta RC 1985), Pakistan (Ghani & Basit 1976), Illinois (McPartland 1983b, 1984), and Kansas (McPartland 1995a). The fungus attacks fibre and drug varieties.

SYMPTOMS

Symptoms begin as pinpoint white leaf spots on young plants in late spring. Spots enlarge, remain irregularly circular, and become slightly raised or thickened. They may darken to a beige colour. Black pycnidia arise in concentric rings within spots (Plate 68). Leaf spots coalesce together, leaf tissue between spots becomes chlorotic and necrotic, and leaves drop off (Fig 5.30). White leaf spot rarely infests flowering tops; THC and CBD inhibit the fungus (McPartland 1984). White leaf spot's ability to completely defoliate plants attracted biocontrol researchers (Ghani & Basit 1976).

CAUSAL ORGANISM & TAXONOMY

Phomopsis ganjae McPartland, *Mycotaxon* 18:527–530, 1983.

Description: Pycnidia immersed in stroma then erumpent,

Figure 5.30: Healthy seedling (C), flanked by plants inoculated with *Phomopsis ganjae* on left, and *Macrophomomina phaseolina* on right (McPartland).

globose to elliptical, ostiolate, peridium *textura angularis-globosa*, 120–220 x 120–300 µm. Conidiophores cylindrical and slightly tapering towards the apex, simple or branched, 8–15 x 1–2 µm. Conidiogenous cells phialidic, cylindrical to obclavate. α-conidia hyaline, unicellular, fusiform to elliptical, biguttulate, usually 7–8 x 2.5 µm. β-conidia hyaline, unicellular, filiform, mostly curved, 16–22 x 1.0 µm (Fig 3.2). Cultures deposited at ATCC (#52587) and CBS (#180.91).

DIFFERENTIAL DIAGNOSIS

White leaf spot can be confused with yellow leaf spot and brown leaf spot. Yellow leaf spot produces darker, more angular lesions peppered with *randomly*-arranged pycnidia. Brown spot lesions are smaller, darker, and break apart leaving holes in leaves.

Ghani & Basit (1976) collected *P. ganjae* in Pakistan and initially called it a *Phoma* species. They subsequently made a second mistake by calling the fungus *Phomopsis cannabina*. *P. cannabina* is a synonym of *P. arctii* and causes Phomopsis stem canker (described in the previous section). *P. ganjae* and *P. arctii* are distinguished by differences in morphology (smaller pycnidia and conidia), modifications in pathogenicity (attacking young vs. senescent plants; leaves vs. stems) and the inability of *P. ganjae* to form a teleomorph in culture or on the host (McPartland 1983b).

DISEASE CYCLE & EPIDEMIOLOGY

P. ganjae overwinters in crop debris. Conidia exude from pycnidia in slimy cirrhi, and spread by water dispersal. α-conidia germinate to form appressoria. Appressoria directly penetrate leaf epidermis. β-conidia of *P. ganjae* sprout short germ tubes but do not form appressoria; their function remains uncertain (McPartland 1983b). Uecker (pers. commun. 1994) studied the genetics of *P. ganjae;* he applied molecular techniques reported by Rehner & Uecker (1994). *P. ganjae* was related to *Phomopsis* isolate no. 649 reported in Rehner & Uecker (1994).

CONTROL

Follow methods utilized for control of yellow leaf spot. Male plants are more susceptible to *P. ganjae* than females (McPartland 1984)—so consider rouging males. Many *Phomopsis* species are seedborne (M. Kulik, pers. commun. 1988); do not use seeds from infected females.

PEPPER SPOT

Pepper spot appears on Russian hemp (Gutner 1933, Gitman & Boytchenko 1934) and Nepali plants (Bush Doctor 1987a). The report by Nair & Ponnappa (1974) from India is a misidentification (McPartland 1995a).

Gutner (1933) described symptoms as round yellow-brown leaf spots 2–4 mm in diameter, with pseudothecia appearing as tiny black dots within the spots. According to Gitman & Boytchenko (1934), pseudothecia form on *both* sides of leaves.

CAUSAL ORGANISM & TAXONOMY

Leptosphaerulina trifolii (Rostrup) Petrak, *Sydowia* 13:76, 1959.
≡*Sphaerulina trifolii* Rostrup 1899; =*Pleosphaerulina cannabina* Gutner 1933.

Description: Mycelium in culture becomes grey and slightly floccose, pseudothecia arise in black crusts, produced in concentric rings of satellite colonies growing from the central inoculation point. Pseudothecia globose, brown, erumpent at maturity, ostiolate, nonparaphysate, thin walled, 120–250 µm diameter. Asci bitunicate, ovate, eight-spored, 50–90 x 40–60 µm. Ascospores hyaline (becoming light brown at maturity), oval to ellipsoid, smooth, with three or four transverse septa and zero to two longitudinal septa, 25–50 x 10–20 µm (Fig 5.31).

L. trifolii displays a wide variation of spore and pseudothecium morphology, resulting in a large synonymy. To that synonymy, McPartland (1995e) added *Pleosphaerulina cannabina*.

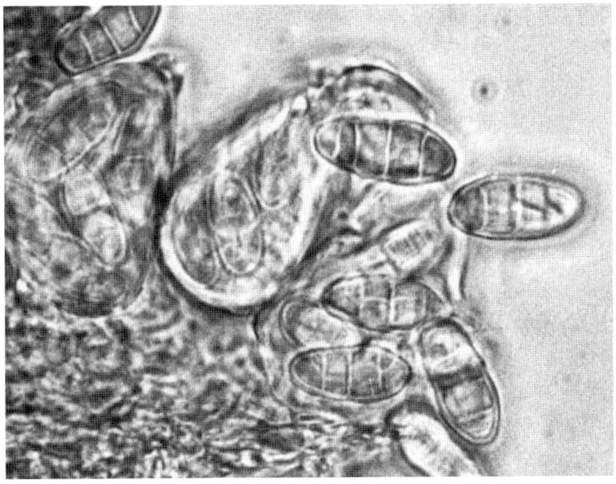

Figure 5.31: Asci and ascospores of *Leptosphaerulina trifolii* from Russian hemp (LM x500, McPartland).

DIFFERENTIAL DIAGNOSIS

Pepper spot may be confused with brown leaf spot or early stages of yellow leaf spot. Another *Leptosphaerulina* species, *L. chartarum*, rarely occurs on *Cannabis* and is described under *Pithomyces chartarum*.

DISEASE CYCLE & EPIDEMIOLOGY

L. trifolii attacks a wide range of plants. It overwinters as pseudothecia embedded in dead leaves. In the spring, ascospores germinate and directly penetrate leaf epidermis.

CONTROL

Avoid overcrowding—crop losses increase in dense stands (Smith *et al.* 1988). Sanitation is the key to control since no effective fungicides are known. Cultural methods 1, 2a, and 10 are most important.

CURVULARIA BLIGHT

Litzenberger *et al.* (1963) described a "*Curvularia* species" causing leaf spots on Cambodian *Cannabis*. Babu *et al.* (1977) recovered *Curvularia lunata* from Indian seeds. McPartland & Cubeta (1997) isolated *Curvularia cymbopogonis* from seeds in Nepal. No symptoms were described in any of these reports.

CAUSAL ORGANISMS & TAXONOMY

1. ***Curvularia cymbopogonis*** (C.W. Dodge) Groves & Skolko, *Canadian J. Research* 23:96, 1945; ≡*Helminthosporium cymbogonis* [as *cymbogoni*] C.W. Dodge 1942.

teleomorph: *Cochliobolus cymbopogonis* Hall & Sivanesan, *Trans. Br. mycol. Soc.* 59:315, 1972.

Description: Conidiophores simple, septate, brown, up to 300 µm long. Conidia acropleurogenous, smooth, straight or curved, clavate to ellipsoidal, four (sometimes three) septa, averaging 40–50 x 12–15 µm, middle cells dark brown, base cell and end cell paler, base cell obconical with a protuberant hilum. Pseudothecia scattered or aggregated in concentric zones on agar, black, globose with a long cylindrical beak, up to 575 µm diameter. Asci bitunicate, cylindrical with a short stipe, eight-spored, 210–275 x 15–23 µm. Ascospores filiform, 8–14 septate, hyaline, 195–420 x 3.5–4.5 µm.

2. ***Curvularia lunata*** (Wakker) Boedijn, *Bull. Jard. Bot. Buitenz.* III 13:127, 1933; ≡*Acrothecium lunatum* Wakker *in* Wakker & Went 1898.

teleomorph: *Cochliobolus lunatus* [as *lunata*] Nelson & Haasis 1964; ≡*Pseudocochliobolus lunatus* (Nelson & Haasis) Tsuda *et al.* 1977.

Description: Conidiophores up to 650 µm long. Conidia three-septate, third cell from the base is curved and larger and darker brown than other cells, fourth cell (end cell) nearly hyaline, surface smooth to verruculose, 13–32 x 6–15 µm (Fig 3.2). Pseudothecia solitary or gregarious, black, ellipsoidal to globose with a tall beak, ostiolate, up to 700 µm in height including a 210–560 µm long beak, up to 530 µm in diameter. Asci bitunicate, cylindrical to clavate, with a short stipe, one to eight spored, 160–300 x 10–20 µm. Ascospores filiform, 6–15 septate, hyaline, arranged either straight, or coiling in a helix within asci, 130–270 x 3.8–6.5 µm.

DIFFERENTIAL DIAGNOSIS

C. lunata conidia are smaller, less septate, and more curved than *C. cymbopogonis* conidia. They also lack the prominent hilum present in *C. cymbopogonis* conidia.

DISEASE CYCLE & EPIDEMIOLOGY

Both species overwinter in plant debris near the soil line. Infection begins in warm, wet spring conditions. *C. cymbopogonis* causes blights on dicots, monocots, and gymnosperms around the world. Its homothallic pseudothecia have only been seen in culture. *C. lunata* is cosmopolitan, and normally causes blights on monocots. On rare occasions, *C. lunata* infects humans. In one celebrated case, Schwartz (1987) wondered if the source of a patient's *C. lunata* sinusitis was contaminated marijuana. Brummund *et al.* (1987) replied, "no *C. lunata* has been cited in any *Cannabis* research we have seen."

CONTROL

Disease can be controlled with cultural methods 1 (sanitation), 4 (escape cropping), and 7c (avoiding excess humidity). Researchers have controlled *C. lunata* disease in other crops by spraying extracts of *Cannabis* leaves (Pandey 1982, Upadhyaya & Gupta 1990). Purified THC and CBD inhibit *C. lunata* (Dahiya & Jain 1977).

PHYLLOPLANE FUNGI

Phylloplane fungi live in nooks and crannies of leaves, but cause no disease. They feed on cellular leakage oozing from plant epidermis. They may also feed on aphid honeydew, pollen grains, and other airborne contaminants. Phylloplane fungi exist as **epiphytes** (living above the leaf epidermis) or **endophytes** (living in spaces below the epidermis).

Some epiphytes, such as sooty moulds, indirectly harm plants by blocking sunlight and reducing photosynthesis. Other epiphytes *protect* plants by suppressing or destroying pathogenic fungi. These helpful epiphytes thus attract attention as possible biocontrol organisms (Fokkema & Van den Heuvel 1986).

No one has systematically investigated *Cannabis* phylloplane fungi. One protective epiphyte, ***Aureobasidium pullulans*** (deBary) Arnaud, appears in the hemp literature (Lentz 1977, Ondrej 1991). Many dematiaceous "pathogens" cited by Ponnappa (1977) and Gzebenyuk (1984), such as *Alternaria alternata, Cladosporium herbarum, Epicoccum nigrum,* and *Stemphylium botryosum,* can also exist as nonpathogenic phylloplane fungi (Fokkema & Van den Heuvel 1986). As the old saying goes, "one plant's protective phylloplane fungus is another plant's latent pathogen" (paraphrased from Palm 1999).

Control of phylloplane fungi is neither necessary nor advisable. Fungicides kill epiphytes and cause a "rebound" of pathogenic organisms. As in human medicine, this is termed "iatrogenic" disease. Heavy aphid or whitefly infestations may cause an overgrowth of sooty mould, which is best eliminated by controlling the insects.

MYCORRHIZAE

About 120 years ago, scientists discovered that some fungi invaded plant roots without causing diseases. We now recognize these strange mould-plant relationships as symbiotic, not parasitic. Termed "mycorrhizae" (Latin for "fungus-roots"), these associations occur in almost all plants and are very important. Mycorrhizae nevertheless have escaped wide attention, because infected roots look normal and the fungi themselves grow with difficulty in artificial culture.

CAUSAL ORGANISMS

Mycologists describe two classes of mycorrhizae. **Ecto**mycorrhizae associate with tree species, especially gymnosperms. **Endo**mycorrhizae associate with trees and herbaceous plants. Endomycorrhizae are usually Zygomycetes. Endomycorrhizae produce swellings (**vesicles**) or minute branches (**arbuscules**) within plant cells. Botanists call them VA mycorrhizae.

In 1925 Arzberger photographed mycorrhizae in *Cannabis* roots. He died shortly thereafter, and his findings were never reported. We recently rediscovered Arzberger's notes and glass plate negatives in the USDA archives (Fig 5.32). In 1961 Mosse produced an artificial VA mycorrhizal relationship in *Cannabis* by inoculating roots with "an *Endogone* species." McPartland & Cubeta (1997) documented naturally-occurring VA mycorrhizae in feral hemp. They identified the fungus as a *Glomus* species, probably ***Glomus mosseae*** (Nic. & Gerd.) Gerd. & Trappe (Plate 69).

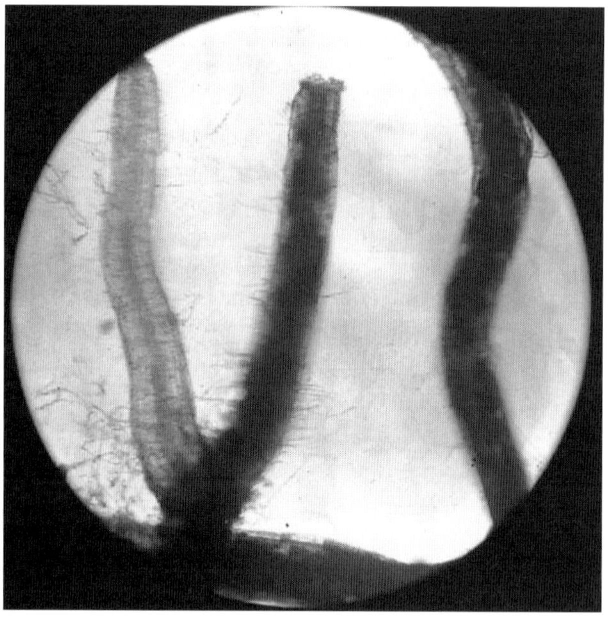

Figure 5.32. "Endotrophic mycorrhiza on roots of *Cannabis sativa*" (photo by E.G. Arzberger, ca. 1925).

EPIDEMIOLOGY

Merlin & Rama Das (1954) discovered that metabolites produced by *Cannabis* roots enhance the growth of mycorrhizal fungi. In return, mycorrhizal fungi improve *Cannabis* growth by increasing the surface area of the root network and making several soil nutrients more available. Some nutrients are immobile in soil, which causes a nutrient-depletion zone to form around roots. Mycorrhizal fungi grow beyond the root zone and draw nutrients back to the plant. These fungi form an extensive network—Tisdall & Oades measured about 150 feet of mycelium per cm^2 of soil.

Phosphate ions are the most immobile soil nutrients. Menge (1983) found that mycorrhizae-infected plants absorb *60 times* more P than uninfected plants. Uptake of zinc and copper dramatically increases, and absorption of K, Ca, Fe, Mg, Mn and S also improves. Only the absorption of nitrogen (the most mobile soil nutrient) remains unchanged in the presence of mycorrhizae. Nitrogen-fixing *Azotobacter* species, however, synergistically increase plant growth when inoculated with mycorrhizal fungi (Linderman *et al.* 1991).

The mycorrhizal mantle protects plants from some root-feeding insects, nematodes, and many fungal pathogens, including *Fusarium oxysporum, Fusarium solani, Macrophomina phaseolina, Pythium ultimum, Rhizoctonia solani,* and *Sclerotium rolfsii.* Mycorrhizae also aid plants by reducing drought stress, and even produce plant growth hormones.

Application for crops

Mycorrhizal fungi serve as "biotic fertilizers," substituting for nutrient supplements. Optimizing the growth of mycorrhizal fungi requires a balance of proper pH, moisture, light intensity, soil fertility, percentage of organic matter, and soil flora and fauna—not a project for the neophyte grower. Hayman (1982) described two pointers for beginners: adding organic fertilizers to soil improves mycorrhizal growth, whereas adding petrochemical fertilizers (e.g., ammonium nitrate) decreases mycorrhizal growth.

Pesticides raise questions of practical importance. Some growers report *stunting* of plants after soaking soil with fungicides. They attribute this to chemical toxicity. Review of the symptoms, however, suggests phosphorus deficiency—the fungicides destroyed the mycorrhizae (reducing uptake of phosphorus).

Menge (1983) tested pesticides on several *Glomus* species, including *G. mosseae.* He found six pesticides lethal to mycorrhizae: methyl bromide, metam-sodium, chloropicrin, formaldehyde, PCNB, and thiram. Benomyl was only lethal as a soil drench—if sprayed on foliage it caused little mycorrhizal destruction. Maneb was intermediate in lethality. Captan, terrazole, copper sulphate, and nematocidal fumigants caused little damage and may have improved mycorrhizal growth by eradicating mycorrhizal parasites and predators.

Why isn't crop stunting a universal phenomenon following heat sterilization of soil? Menge (1983) noted that VA mycorrhizae survive deep in organic debris, insulated from damage. In addition, earthworms, insects, and small mammals carry mycorrhizae back into sterilized areas.

Almost all soils contain natural populations of mycorrhizal fungi. Exceptions include soils laying fallow for two or more years, and soils supporting continuous crops of non-mycorrhizal plants such as Cruciferae (broccoli, cabbage, cauliflower, mustard greens, turnips, etc.) and Chenopodiaceae (quinoa, lamb's quarters, pigweed). Plastic sacks of sterile potting soil, peat moss, builder's sand, and perlite also lack mycorrhizae.

Can we inoculate sterile soil with mycorrhizal fungi? Some cultivators throw in a handful of "starter"—soil and root fragments from a previously successful crop. Wilson *et al.* (1988) compared this "sourdough starter" with pure mycorrhizae spores: after four months, test plants (*Andropogon gerardii*) grown in *sterile* soil were stunted. Average dry weight was only 0.02 g. Plants grown in sterile soil inoculated with pure spores weighed an average of 5.40 g, or *270 times* greater than plants grown in sterile soil. Plants inoculated with the sourdough method averaged 3.71 g each, 185 times greater than plants in sterile soil.

Menge (1983) harvested pure spores from roots of "mother plants." He ground up the roots and recovered the spores by wet-sieving or centrifugation. Subsequent disinfection of spores with streptomycin and sodium hypochlorite (bleach) assures a pure inoculum. The mycorrhizal fungus *Glomus intraradices* is now commercially available as a biocontrol against *Pythium* and *Fusarium* species (described in the section on Fusarium stem canker). Menge estimates the cost for mycorrhizal inoculation equals current costs for phosphorus application.

In hydroponic systems, a constant flow of soluble nutrients around roots eliminates the nutrient depletion zone. Nevertheless, some hydroponics operators inoculate their systems with mycorrhizae to optimize nutrient uptake. Ojala & Jarrell (1980) reviewed techniques for establishing mycorrhizae in hydroponic systems.

MISCELLANEOUS LEAF, STEM AND ROOT DISEASES

The 26 pathogens presented in this section rarely cause disease. They don't even have common names. They have only been mentioned once or twice in the literature. Many lack herbaria-preserved voucher specimens, and their correct identification is questionable. They are briefly described and discussed:

Arthrinium phaeospermum (Corda) M.B. Ellis, *C.M.I. Mycological Paper* No. 103, p.8, 1965; =*Papularia sphaerosperma* (Persoon:Fries) Höhnel 1916.

Description: Hyphae hyaline to pale brown, smooth or verruculose, septate, 1–6 µm thick. Conidiophore mother cells short, lageniform, smooth or verruculose, 5–10 x 3–5 µm. Conidiophores mostly long, cylindrical, flexous, simple, septate, hyaline, 5–65 x 1–1.5 µm. Conidia borne on short sterigmata along the lengths of conidiophores, lens-shaped, golden-brown with a hyaline band around the perimeter, 8–12 µm diam.

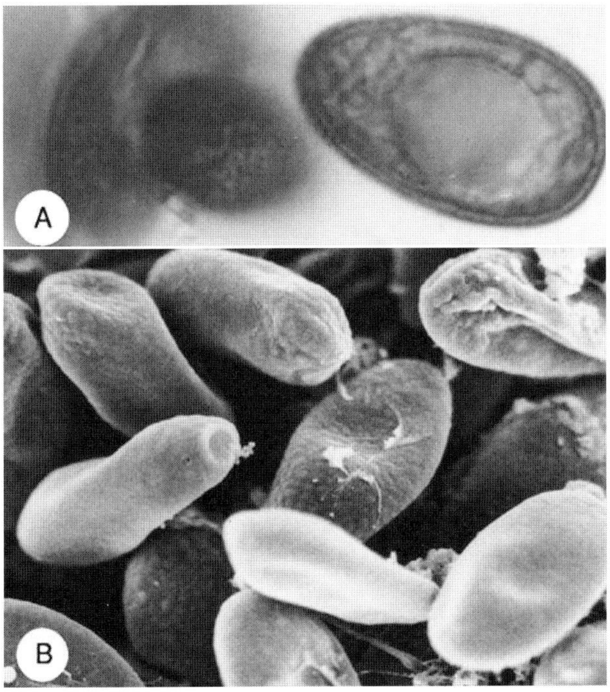

Figure 5.33: *Sphaeria cannabis* (=*Botryosphaeria obtusa*) macroconidia. A. LM optical section showing large internal guttule, (x2300); B. SEM showing surface details (x1700); McPartland.

Persoon's basionym appears to have priority. *A. phaeospermum* grows worldwide on sedges and reeds (*Carex, Glyceria, Phragmites* species). Chandra (1974) described *A. phaeospermum* erupting from *C. sativa* leaf spots in India. Spots turn light-brown, gradually necrose, and finally collapse, leaving irregular holes in the leaf.

Botryosphaeria obtusa (Schweinitz) Shoemaker, *Canadian J. Botany* 42:1298, 1964; ≡*Physalospora obtusa* (Schweinitz) Cooke 1892.
anamorph 1: *Sphaeropsis malorum* Peck, nomen nudum;
=*Sphaeria cannabis* Schweinitz 1832.
anamorph 2: unnamed.

Description: Pseudothecia embedded in stems, stromatic, dark brown-black, solitary or gregarious in botryose aggregations up to 3 mm wide; individual locules globose, 150–300 µm diameter. Asci bitunicate, clavate, eight-spored, 90–120 x 17–23 µm. Paraphyses filiform. Ascospores broadly fusoid, hyaline, unicellular or sometimes with one septum, 25–33 x 7–12 µm. Anamorph 1 pycnidia are immersed then erumpent, stromatic, black, on *Cannabis* appearing unilocular and globose, averaging 235 µm diameter. Conidiogenous cells hyaline, simple, cylindrical, holoblastic, discrete, determinate, 8–14 µm in length, 3–4 µm in width. Macroconidia borne in mucilage, elliptical or broadly clavate, base often truncate and bordered by a scar, thick-walled, verruculose to almost smooth-walled, at first hyaline and unicellular with a large central guttule; rapidly becoming brown, at length becoming two-celled and biguttulate, 19–21 x 10–11 µm (Fig 5.33). Anamorph 2 (the microconidial state) is rarely encountered; for a description see McPartland (1995b).

Historically, Schweinitz's 1832 publication of *Sphaeria cannabis* is the first description of a *Cannabis* pathogen. McPartland (1995b) examined Schweinitz's specimen and synonymized it with the anamorph of *B. obtusa.* Stevens (1933) provided an extensive synonymy for this common pathogen.

Coniothyrium cannabinum Curzi, *Atti Istit. Bot. Univer. Pavia,* ser. III(3):206, 1927.

Description: Pycnidia scattered along stems, immersed then erumpent, spherical or flattened from above, ostiole somewhat sunken, sooty brown-black, peridium consisting of small parenchymous cells finely woven together, 90–120 x 60–90 µm. Conidiophores scarcely apparent. Conidia thick-walled, olive brown, almost spherical to oval, with one large central guttule, 4–5 x 2.5–3.5 µm (Bestagno-Biga *et al.* 1958 described conidia 5–5.5 x 2.5µm).

Curzi's description of "scarcely apparent" conidiophores suggests the determinant character of *Microsphaeropsis* phialides. But the thick-walled conidia suggest a *Coniothyrium* species. Unfortunately, Curzi's type specimen is missing (Curator Dr.ssa Terzo, pers. commun. 1987).

C. cannabinum infests hemp in Italy (Curzi 1927, Bestagno-Biga *et al.* 1958) and Russia (Gitman & Boytchenko 1934). Gzebenyuk (1984) cited two other species, ***Coniothyrium tenue*** Diedicke and ***Coniothyrium olivaceum*** Bonorden, which are probably misdeterminations (McPartland 1995a). Fuller & Norman (1944) reported an unspeciated *Coniothyrium* attacking field-retted hemp stalks in Iowa. No disease symptoms are described in any of these reports. *Coniothyrium* species overwinter in crop debris. *Coniothyrium* diseases of other crops (notably *Coniothyrium fuckelii* on roses) are controlled by avoiding injury to stem surfaces.

Cylindrosporium cannabina Ibrahimov, *Akademii Nauk Azerbaidzhanskoi SSR. Izvestiya* 4:66–67, 1955.

Description: Acervuli epiphyllous, gregarious, barely conspicuous, 55–96 µm wide. Conidiophores small, hyaline, cylindricalacuminate, crowded together. Conidia unicellular, filiform, hyaline, curved, rarely straight, 18–54 x 1.2–2.0 µm.

Ibrahimov discovered this species causing leaf spots in Azerbajian. Leaf spots are yellow-brown or cinnamon coloured, 0.1–6 mm in diameter. Ibrahimov's illustration of cupulate acervuli resembles Briosi & Cavara's illustration of *Septoria neocannabina*. Ibrahimov's specimen could be a *Septoria*, *Septogloeum*, or *Phloeospora* species.

McPartland (1995a) examined the herbarium specimen of "*Cylindrosporium* species" cited by Miller *et al.* (1960) and Farr *et al.* (1989). The specimen proved to be a mix of *Septoria neocannabina* and *Pseudocercospora cannabina*. Ghani & Basit (1975) described a *Cylindrosporium* species forming a blackish fluffy mycelium on undersides of leaves in Pakistan. They described conidia as filiform, two to four septate, and bent at the apex.

Didymium clavus (Albertini & Schweinitz) Rabenhorst, *Deutschland Kryptogamen Flora* 1:280, 1844.

Description: Plasmodium grey or colourless. Sporangia stalked (sometimes appearing sessile), discoid, greyish white with darker stalk, 0.5–1 mm diameter, up to 1 mm tall. Peridium (cap) membranous, adaxial surface nearly covered with light lime crystals, under surface without lime, brown. Stalk tapers upward, longitudinally striate, dark brown to black, paler near top. Capillitium (threads) delicate, hyaline to pale purple, sparsely branched. Spores black *en masse*, pale violaceous brown under the microscope, nearly smooth, 6–8 µm diameter.

Gzebenyuk (1984) cited *D. clavus* covering 4.6% of hemp stems near Kiev. *D. clavus* occurs across Eurasia and North America. Slime moulds may crawl up plant stems in wet weather, but cause little damage.

Jahniella bohemica Petrak, *Annales Mycologici* 19:123, 1921.

Description: Pycnidia unilocular, brown to black, immersed then erumpent, flattened subglobose, ostiolated, (310–) 600 (–850) µm diam., up to 400 µm tall, thick walled (50–60 µm thick), scleroplectenchymatic, peridium *textura angularis*. Ostiolum central, papillate, up to 110 µm tall. Conidiogenous cells holoblastic, discrete, determinate, ampulliform to short cylindrical, 4–14 x 2–5 µm. Conidia hyaline, filiform, straight or curved, base truncate, smooth-walled, finely guttulate, indistinctly (2–)3–4(–5) septate, (18.5–) 45 (–55) x 1.0–2.5 µm.

Saccardo & Roumeguere (1883) misdetermined *J. bohemica* as *Leptosphaeria acuta*, on hemp stems near Leige, Belgium. McPartland (1995e) inspected their specimen and recognized the mistake. For a full account see McPartland & Common (2000). *J. bohemica* has also been collected from figwort (*Scrophularia nodosa*) and garden loosestrife (*Lysimachia vulgaris*).

Hymenoscyphus herbarum (Persoon:Fries) Dennis, *Persoonia* 3:77, 1964; ≡*Helotium herbarum* (Persoon:Fries) Fries 1849.

Description: Apothecia common, gregarious, arising on a very short stalk, pale yellow to ochraceous, minutely downy, flat to slightly convex, disc up to 3 mm in diameter. Asci cylindrical to clavate, stalked, apical pore turns blue with iodine, eight-spored, up to 90 x 8 µm. Paraphyses hyaline, filiform, 90 x 2 µm. Ascospores two-celled, hyaline, often biseriate, fusiform to cylindrical, 13–17 x 2.5–3.0 µm.

Saccardo found *H. herbarum* fruiting on *Cannabis* stems near Padova, Italy. He distributed exsiccati specimens as *Mycotheca italica* No. 119 (BPI!). *H. herbarum* infests many herbaceous hosts, especially *Urtica* species. The fungus lives worldwide in temperate climates. It sporulates on dead stems in October, spreading ascospores by wind and water. The fungus's rôle as a pathogen is questionable. Most mycologists consider *H. herbarum* a saprophyte, although some *Hymenoscyphus* species (e.g., *H. ericae*) act as mycorrhizae.

Lasiodiplodia theobromae (Pat.) Griffon & Maublanc, *Bull. trimest. Soc. mycol. France* 25:57, 1909; ≡*Botryodiplodia theobromae* Patouillard 1892.

teleomorph: *Botryosphaeria rhodina* (Berkeley & Curtis *in* Curtis *apud* Cooke) von Arx 1970; ≡*Physalospora rhodina* Berkeley & Curtis *in* Curtis *apud* Cooke 1899.

Description: Pycnidia stromatic, simple or compound (uni- or multilocular), often aggregated, carbonaceous black, ostiolate, between 300 and 5000 µm in diameter. Conidiogenous cells hyaline, simple, cylindrical, sometimes septate, rarely branched, holoblastic, 5–15 x 3 µm. Paraphyses hyaline, cylindrical, septate, up to 50 µm long. Conidia at first ellipsoid, hyaline and thin-walled, then thick walled, later developing a median septum, dark brown pigmentation, longitudinal striations, and a truncate base, 20–30 x 10–15 µm (Fig 3.2). Teleomorph perithecia stromatic, uni- or multilocular, gregarious. Asci bitunicate, clavate, eight-spored, 90–120 x 15–28 µm. Ascospores one-celled, hyaline, ellipsoidal, 24–42 x 7–18 µm.

McPartland (1995b) found *L. theobromae* infesting stems of feral hemp in Illinois. The hemp was also parasitized by *Fusarium sulphureum* (see Plate 58). *L. theobromae* prefers climates between 40° north and 40° south of the equator, but is distributed worldwide. It decays harvested mangos, bananas, avocados, melons, cocoa pods, citrus, and cotton bolls (Holliday 1980).

Leptosphaeria acuta (Hoffman:Fries) P. Karsten, *Mycologia Fennica* 2:98, 1873; ≡*Sphaeria acuta* E. F. Hoffman 1787, ≡*Cryptosphaeria acuta* (Hoffman:Fries) Greville *Fl. Edin.* p. 360, 1824; ?=*Leptosphaeria acuta* f. *cannabis* Roumeguere 1887.

anamorph: *Phoma piskorzii* (Petrak) Boerema & Loerakker, *Persoonia* 11:315, 1981; ≡*Diploplenodomus piskorzii* Petrak 1923, non: *Phoma acuta* auct., nomen ambiguum.

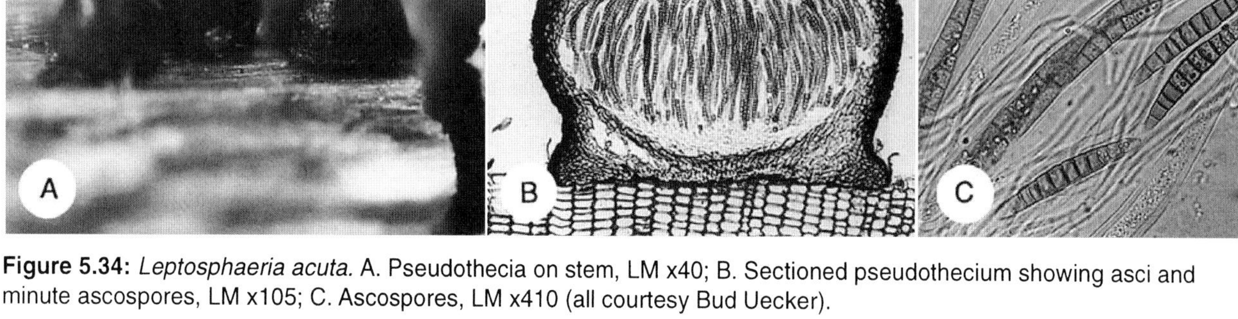

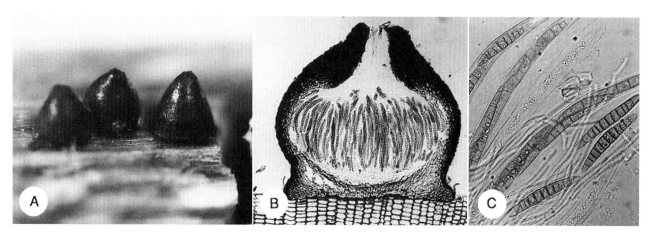

Figure 5.34: *Leptosphaeria acuta*. A. Pseudothecia on stem, LM x40; B. Sectioned pseudothecium showing asci and minute ascospores, LM x105; C. Ascospores, LM x410 (all courtesy Bud Uecker).

Description: Pseudothecia conical, surface smooth, glistening black, immersed then erumpent, ostiolate, peridium a thick-walled *textura globulosa* near ostiole, thereafter a radiating *textura prismatica*, 300–450 µm wide, beak 100–200 tall, topped by a 30–40 µm diameter ostiole (Fig 5.34). Asci cylindrical, eight-spored, base constricted to a short pedicle, 90–140 x 7–11 µm (Fig 5.34). Paraphyses filiform, hyaline, septate. Ascospores pale yellow, guttulate, fusiform, usually seven-septate and constricted at central septum, 35–45 x 4–7 µm. The anamorph is described under *P. piskorzii* in the section on Brown leaf spot and stem canker.

L. acuta is usually restricted to *Urtica* species (Shoemaker 1964, Sivanesan 1984, Crane & Shearer 1991). Saccardo & Roumeguere (1883) reported *L. acuta* on hemp stems and distributed an exsiccatus, *Reliquiae Libertianae* no. 56 (191). Four years later Roumeguere erected the taxon *Leptosphaeria acuta* f. *cannabis* as an exsiccatus, *Fungi selecti exsiccati* no. 4172. We examined several exsiccati specimens—some contain immature, unidentifiable ascomata (at BPI—see McPartland 1995e); other specimens contain misidentified *Jahniella bohemica* (at FH and BR—see McPartland & Common 1999). Taxonomically, Crane & Shearer (1991) and McPartland (1995a) recognized the taxon *L. acuta* (Hoffm.:Fr.) Karsten, whereas Boerema & Gams (1995) preferred the taxon *L. acuta* (Fuckel) Karsten.

Leptosphaeria cannabina Ferraris & Massa, *Annales Mycologici* 10:286, 1912.

Description: Pseudothecia epiphyllous, globose, black, erumpent at maturity, ostiolate, 130–140 µm diameter, peridium "membranous" [thin-walled?]. Asci clavate, straight or curved, apex rounded, base constricted to a short pedicle, 45–50 x 7–10 µm. Paraphyses not described. Ascospores honey-coloured, fusiform, three-septate, slightly constricted at septa, 19–20 x 5 µm.

Ferraris & Massa found pseudothecia of *L. cannabina* in irregularly-shaped leaf spots of wilting plants near Alba, Italy. Leaf spots were dull white with an ochre margin, 3–5 mm in diameter. The fungus also infests Russian hemp (Gitman & Boytchenko 1934, Gitman 1935, Dobrozrakova *et al.* 1956). Ferraris & Massa considered *L. cannabina* a "probable" teleomorph of *Septoria cannabis*. They did not confirm this in culture. Perhaps they found pycnidia of *S. cannabis* near *L. cannabina*. *S. cannabis* is a ubiquitous hemp pathogen; *proximity does not constitute a relationship.*

Leptosphaeria woroninii Docea & Negru *in* Negru, Docea & Szasz, *Novosti Sistematiki Nizshikh Rastenij* 9:168, 1972.

Description: Pseudothecia flask-shaped, brown-black, carbonaceous, immersed then erumpent, 150–180 µm diameter. Asci cylindrical to clavate, eight-spored, 58–100 x 18–20 µm. Paraphyses filiform, hyaline, with branching apices. Ascospores fusiform, slightly curved, four to six septa, slightly constricted at septa, guttulate, granular cytoplasm, yellow, 23–29 x 3.5–5.0 µm.

Docea & Negru (1972) described *L. woroninii* from seeds of *Cannabis sativa* near Cluj, Romania.

Leptosphaerulina chartarum Roux, *Trans. Br. Mycol. Soc.* 86:319–323, 1986.

anamorph: *Pithomyces chartarum* (Berkeley & Curtis *apud* Berkeley) M. B. Ellis 1960.

Description: Mycelium dark olive brown, septate, smooth to verruculose. Pseudothecia globose, dark olive brown, immersed or superficial, with a large ostiole, five to seven asci per pseudothecium, nonparaphysate. Asci bitunicate, short ovate but elongating with maturation, eight-spored, 100–150 x 60–100 µm. Ascospores hyaline to light brown, broadly ellipsoidal, smooth, usually three transverse septa and one longitudinal septum, slightly constricted at septa, 23–27 x 7–12 µm. Conidiophores straight or slightly curved, cylindrical, hyaline to subhyaline, holoblastic, 2.5–10 x 3–3.5 µm. Conidia borne singularly atop conidiophores, light to dark brown, broadly ellipsoidal, verruculose to echinulate, usually three transverse septa and one longitudinal septum, 18–29 x 10–17 µm.

Nair & Ponnappa (1974, Ponnappa 1977) isolated the anamorph from marijuana leaves in India. No symptoms were described. The fungus attacks many plants. It produces an oligopeptide, sporidesmin, which causes eczema in mammals (Rippon 1988).

Leptospora rubella (Persoon:Fries) Rabenhorst, *Herb. viv. mycol. Ed. II*, No. 532, 1857; =*Ophiobolus porphyrogonus* (Tode) Saccardo 1883; =*Ophiobolus vulgaris* (Saccardo) Saccardo 1881.

Description: Pseudothecia appear on blackened or purple-red stems, scattered or gregarious, immersed then more-or-less erumpent, globose to conical, often laterally compressed, black, glabrous, papillate; contain periphyses, paraphyses, and asci; 200–300 µm diameter. Asci bitunicate, borne upon short stalks, cylindrical, eight-spored, 140–160 x 4.5–6.0 µm (Holm says asci grow to 225 µm long). Ascospores slightly spiralled within the ascus, filiform, yellow, with many obscure septa and guttules, nearly as long as asci and 1 µm wide.

Holm (1957) listed ten synonyms for this common fungus. Gzebenyuk (1984) reported *L. rubella* twice on Soviet hemp stems (as two synonyms, *O. porphyrogonus* and *O. vulgaris*). The life cycle of *L. rubella* is similar to *Ophiobolus* species described under Ophiobolus stem canker.

Microdiplodia abromovii Nelen, *Nov. Sist. niz. Rast.* 14:106, 1977. ≡*Microdiplodia cannabina* Abramov 1938 or 1939.

Description: Pycnidia scattered, thin-walled, papillate, brown, cells around the ostiole become black, 130–180 µm diameter. Conidia ("stylospores") olive to pale brown, bicellular, with rounded ends, 8–11 x 4.0–4.5 µm.

Abramov found this fungus on stems of male plants in Russia. Dobrozrakova *et al.* (1956) described it again in Russia. Nelen reduced Abramov's species to a *nomen novum*. The *genus* also has taxonomic problems, see Sutton (1977).

Micropeltopsis cannabis McPartland, *Mycological Research* 101:854, 1997.

Description: Ascomata catathecioid, 45–130 µm diam., flattened ampulliform, 25–46 µm high, dark brown to black, ostiolate, margin entire; upper layer composed of radially-arranged quadrangular cells, peridium *textura prismatica;* basal layer of similar construction to upper layer but paler. Ostiole central, raised, composed of small cells and bearing a crown of divergent setae, which are arranged in an inverted cone over the ostiole. Setae straight, subulate, thick-walled, nonseptate, smooth, dark brown, 12–22 µm long. Asci bitunicate, ovoid to obclavate, four- to eight-spored, 21–40 x 4–9 µm. Ascospores hyaline, guttulate, ellipsoid, with a single median septum, nonsetulate, 11–12 x 2.5–3.0 µm.

Roumeguere's syntype of *Orbila luteola* (see two species below) contained thyriothecioid fruiting bodies which proved to be a new species.

Myrothecium roridum Tode:Fries, *Systema Mycologicum* 3:217, 1828.

Description: Sporodochia sessile or slightly stalked, cushion shaped, light coloured to dark, diameter variable, 16–750 µm. Conidiophores hyaline, cylindrical, septate, with multiple branches bearing phialides in whorls. Conidia ellipsoid to elongate, subhyaline to olivaceous, aseptate, guttulate, 5–10 x 1.5–3.0 µm.

Although *M. roridum* attacks cotton, coffee, and many other crops around the world, it has only been reported on *Cannabis* in India (Nair & Ponnappa 1974, Ponnappa 1977) and Pakistan (Ghani & Basit 1976). Gzebenyuk (1984) cited a related and easily mistaken fungus on Ukrainian hemp, ***Myrothecium verrucaria*** (Albertini & Schweinitz) Ditmar.

M. roridum causes "corky leaf spot." Symptoms begin as small brown leaf spots in July. Spots enlarge and sometimes coalesce. Masses of slimy conidia turn spots black. Upon drying, they become corky and tough. Heavy infection leads to defoliation and death. *M. roridum* overwinters in the soil. The fungus is a weak pathogen and invades plants via wounds. Optimum temperature for growth is 28–32°C; corky leaf spot flourishes in subtropical climates and warm glasshouses.

Orbila luteola (Roumeguere) McPartland, *Mycological Research* 101:854, 1997; ≡*Calloria luteola* Roumeguere 1881.

Description: Apothecia superficial, sessile, translucent yellow-orange, margin spherical to ellipsoidal, up to 0.6 mm in diameter and 100 µm in thickness. Excipulum consists of hyaline thin-walled *textura globulosa*. Asci cylindrical, eight-spored, 26.0 x 4.5 µm. Paraphyses hyaline, filiform, slightly enlarged at the apex. Ascospores hyaline, single-celled, fusiform, some indistinctly guttulate, 6.5 x 1.5 µm.

Only Dr. Roumeguere has found this fungus on hemp stems, near Villernur, France. But he found enough of it to distribute nearly 100 specimens in his *Fungi Gallici exsiccati*.

Penicillium chrysogenum Thom, Bull. *USDA Bur. Animal Ind.* 118:58, 1910; =*Penicillium notatum* Westling 1911.

Description: Conidial heads paint-brush-like, bearing conidia in well-defined columns up to 200 µm long. Conidiophores smooth, hyaline, up to 350 µm long by 3.0–3.5 µm wide. Branches (penicilli) usually biverticillate, measuring 15–25 µm by 3.0–3.5 µm. Metulae borne atop penicilli in clusters of two to five, measuring 10–12 µm long by 2–3 µm wide. Phialides borne atop metulae in compact clusters of four to six, flask-shaped, mostly 8–10 by 2.0–2.5 µm. Conidia elliptical to almost globose, smooth, pale yellow to pale green, mostly 3.0–4.0 x 2.8–3.5 µm (Figs 3.2 & 8.2).

Babu *et al.* (1977) described *P. chrysogenum* "burning" leaves of plants in India. Small branches were also affected, causing a wilt. *P. chrysogenum* also rots *Cannabis* seeds and causes a marijuana storage disease (Fig 8.2). The fungus grows worldwide but prefers an optimum temperature of 23°C. *P. chrysogenum* is a friend as well as a foe—Dr. Fleming discovered "penicillin" when this fungus floated into his lab.

Periconia byssoides Persoon, *Synopsis Methodica Fungorum* p. 686, 1801; =*Periconia pycnospora* Fresenius 1850.

Description: Conidiophores are long cylinders topped by swollen heads, erect, straight or slightly flexous, two to three septate, dark brown at base and paling to subhyaline at the apex, 12–23 µm diameter at base tapering to 9–18 µm below the head, then forming a septum with an apical head swelling to 11–28 µm diameter, entire length 200–1400 µm (up to 2 mm). Conidiogenous cells hyaline, ellipsoidal to spherical, monoblastic to polyblastic. Conidia cantenate, forming chains, spherical, brown, verrucose, 10–15 µm diameter.

Gitman & Malikova (1933) described *P. byssoides* blackening hemp stems near Moscow. Gitman & Boytchenko (1934) noted the conidiophores merged together into coremia. The fungus infests stems in the Czech Republic (Ondrej 1991). Fuller & Norman (1944) found a *Periconia* species on hemp in Iowa.

Gzebenyuk (1984) reported ***Periconia cookei*** Mason & Ellis on hemp stems near Kiev. *P. cookei* differs little from *P. byssoides*—it lacks a septum at the apical cell and produces slightly larger conidia (13–16 µm diameter).

Pestalotiopsis species

Description: Acervuli epidermal, often epiphyllous, circular to oval, up to 200 µm in diameter. Conidiophores cylindrical, septate, occasionally branched, hyaline, up to 10 µm in length. Conidiogenous cells holoblastic, annellidic, hyaline, cylindrical. Conidia fusiform, five-celled, smooth walled, averaging 24–26 x 5.6 µm; basal cell hyaline, with a simple hyaline appendage (pedicel) averaging 6.6 µm in length; three median cells umber to olivacous brown, thick-walled; apical cell hyaline, conical, with three or less commonly two simple hyaline appendages (setulae) averaging 17.1 µm in length (Figs 3.2 & 5.35).

McPartland & Cubeta (1997) reported this fungus causing leaf and stem smudge on *Cannabis* near Pokhara, Nepal. Paulsen (1971) isolated a *Pestalotia* species from hemp in Iowa.

Phyllachora cannabidis P. Hennings, *Hedwigia* 48:8, 1908.

Description: Perithecia lie tightly bunched together, developing in a blackened plano-pulvinate pseudostroma under stem epidermis, perithecial wall membranaceous, brown, 16 µm thick, ostioles erumpent, 170 µm wide x 120 µm deep. Asci clavate with rotund apices, eight-spored, 45–55 x 10–12 µm. Ascospores in one or two rows, arranged at oblique angles to ascus axis, hyaline to greyish blue, oval to almost fusiform, nonseptate, 12–17 x 5–6 µm.

Hennings discovered *P. cannabidis* on stems near Sao Paulo, Brazil. *Phyllachora* perithecia merge together atop host epidermis to form a shield-like **clypeus**. Clypei are shiny and black and called "tar spots." Theissen & Sydow (1915) noted *P. cannabidis's* peridium and excluded it from *Phyllachora*. Since then, however, peridia have been recognized in *Phyllachora* species (Parberry 1967).

Phyllachora species are obligate parasites (feeding on living cells) and cause little host damage. Tar spots rarely cause economic losses. Parberry (1967) stated these fungi have a wide host range and he proposed that many of the 1100 *Phyllachora* taxa described on different hosts are synonyms.

Rhabdospora cannabina Fautrey, *Bull. Soc. mycol. France* 15:156, 1899.

Description: Pycnidia appear on blackened stems, intensely aggregate, numerous, immersed then erumpent, surface woven and thinly reticulate, grey to black, papillate, with a round ostiole. Conidia variously curved, guttulate, 40–48 x 1.75–2.0 µm.

Fautrey isolated this fungus near Semur, France. His description of *R. cannabina* conidia resembles the conidia of *Septoria cannabis* or *Jahniella bohemica* (McPartland 1995d). Unfortunately, Fautrey's type specimen is missing.

Gzebenyuk (1984) reported two other *Rhabdospora* species on hemp stems: ***Rhabdospora origani*** (Brunand) Saccardo 1884 and ***Rhabdospora hypochoeridis*** Allescher 1897. Both species produce conidia smaller than Fautrey described for *R. cannabina*.

Rosellinia necatrix Prillieux, *Bull. Soc. mycol. France* 20:34, 1904.

anamorph: *Dematophora necatrix* Hartig 1883.

Description: Perithecia densely aggregated, globose, brown-black, ostiolate, 1000–2000 µm diameter. Asci unitunicate, cylindrical, long-stalked, eight-spored, 250–380 x 8–12 µm. Ascospores fusiform, straight or curved, single-celled, brown, 30–50 x 5–8 µm, with a longitudinal slit. Paraphyses filiform. Conidiophores bound together

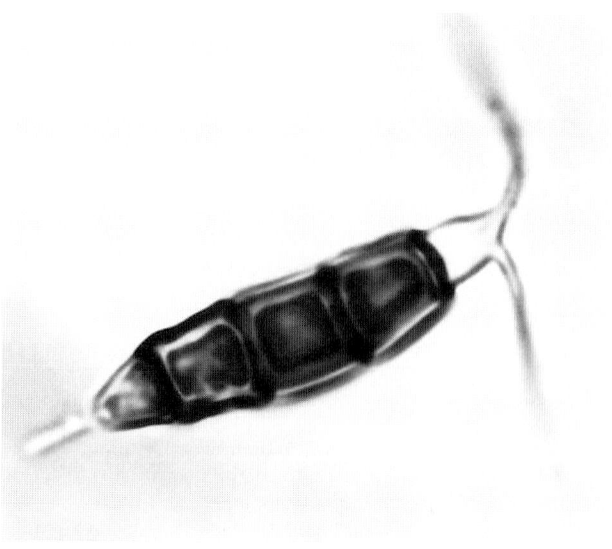

Figure 5.35: Conidium of *Pestalotiopsis* species (LM x1920, McPartland).

in upright synnemata, 40–400 µm thick and 1–5 mm tall, often dichotomously branched towards the apex. Conigiogenous cells polyblastic, integrated and terminal or discrete and sympodial. Conidia light brown, ellipsoid, one-celled, smooth, 3–4.5 x 2–2.5 µm. Hyphae brown, septate, forming pear-shaped swellings at ends of cells next to septa.

This fungus reportedly causes "white root rot" of hemp in southern China (Anonymous 1974) and Japan (Kyokai 1965, Kishi 1988). Above-ground symptoms include stunting, wilting, and early plant death. Below-ground signs include white spheres of mycelium forming on root surfaces and a feather-like mycelium growing within roots. *R. necatrix* lives in temperate and semitropical areas worldwide. It is a hardy soil fungus, difficult to eradicate once established. *Trichoderma harzianum* infests and kills *R. necatrix.*

Torula herbarum (Persoon:Fries) Link, *Magazin, Gesellschaft Naturforschender Freunde zu Berlin* 3:21, 1809; ≡*Monilia herbarum* Persoon 1801; =*Torula monilis* Persoon 1795 *apud* Hughes 1958.

Description: Colonies olivaceous when young, dark brown to black when old, velvety. Conidiophores septate, pale olive, 2–6 µm wide, enlarging at conidiogenous cells to 7–9 µm in diameter. Conidia cylindrical, resembling a chain of attached spheres, mostly four to five septa (range two to ten), constricted at septa, straight or curved, pale olive to brown, verruculose to echinulate, 20–70 x 5–9 µm.

This fungus colonizes hemp leaves as a secondary parasite and causes a post-harvest storage mould of marijuana. Gzebenyuk (1984) reported *T. herbarum* colonizing hemp stems in the Ukraine. Behrens (1902) counted the fungus among hemp retting organisms in Germany. McPartland (unpublished data 1995) identified *T. herbarum* overgrowing hemp stalks previously parasitized by *Colletotrichum dematium,* collected in China by Rob Clarke. *T. herbarum* commonly arises on dead herbaceous stems, especially on nettles (*Urtica dioica*). The fungus occurs worldwide, mostly in temperate regions.

Trichoderma viride Persoon, *Römer's Neues Mag. Bot.* 1:94, 1794. =*Pyrenium lignorum* Tode 1790.

Description: Colonies grow rapidly (covering a 9 cm petri plate in three or four days), at maturity a dark green or dark bluish green, emitting a distinctive "coconut" odour. Conidiophores septate, branching pyramidally from a central upright stalk, with short branches occurring near the apex and longer branches occurring below. Conidiogenous cells straight or bent, slender flask-shaped phialides, arranged in irregular whorls of two or three (rarely four), size variable but mostly 8–14 x 2.4–3.0 µm. Conidia green *en masse,* globose, warted surface, 3.5–4.5 µm in diameter (Fig 5.10).

Prior to Rifai's revision of the genus, all green-spored species of *Trichoderma* were cited as *T. viride.* This aggregate species flourishes anywhere from tropical forests to arctic tundra, including swamps, dunes, and deserts. Domsch *et al.* (1980) reported it from volcanic craters and children's sandpits. McPartland (unpublished data 1993) found it sporulating on soil and rockwool in Amsterdam.

T. viride colonizes the root surfaces of many plants, but disease is rarely reported. It has been isolated from hemp stems in the Czech Republic (Ondrej 1991). Fuller & Norman (1944) described a *Trichoderma* overrunning field-retted hemp.

Dried leaves of *Cannabis* suppress *T. viride* growth (Grewal 1989). The fungus produces toxic metabolites against fellow fungi; Weindling first demonstrated this antagonistic activity in 1932. *T. viride* is now used for commercial biocontrol of many pathogenic fungi (see the section on Rhizoctonia sore shin disease).

"Be very careful not to suffer weeds *of any sort to ripen their seeds on or anywhere within gunshot of your mines, or mints, for making money, which your manure-heaps and compost-beds may be styled, almost without a metaphor."* —Helen Nearing

Chapter 6: Other *Cannabis* Pests & Pathogens

In this chapter we describe nematodes, viruses, bacteria, parasitic plants, protozoa, non-insect arthropods, and vertebrates. Some of these organisms can cause significant damage, but most are uncommon problems.

NEMATODE DISEASES

About 500 species of nematodes are plant parasites (Cook & Qualset 1996), but only a few infest *Cannabis*. Most nematodes attack roots, causing subterranean damage. Nematodes indirectly damage plants by destroying beneficial mycorrhizal fungi (Hayman 1982). Not all nematodes are bad news—some are beneficial and prey on insect pests (see the chapter on biocontrol).

ROOT-KNOT NEMATODES

Root knot nematodes attack thousands of plant species around the world. They cause greatest damage in warm regions where summers are long and winters are short and mild. In the USA, root knot nematodes get nasty south of the Mason-Dixon line (39° latitude), especially in sandy soils along coastal plains and river deltas.

SIGNS & SYMPTOMS

Above-ground symptoms are nonspecific—stunting, a chlorosis that resembles nitrogen deficiency, and midday wilting with nightly recovery. Farmers may misinterpret these symptoms as mineral deficiencies or drought, mysteriously appearing despite adequate nutrients and moisture. These symptoms do not arise uniformly across fields, but in patches of scattered infestation. Greatest symptoms arise in plants near the centre of these patches, blending to healthy plants at the periphery.

Below-ground symptoms are more distinctive. Root knot nematodes embed themselves in roots and cause roots to form giant cells or *syncytia.* Syncytia swell into root galls. Galls may coalesce together to form "root knots"—conspicuous, hypertrophied roots with lumpy surfaces. Root galls also stimulate the formation of adventitious rootlets, creating a "bushy" appearance. The excessive branching of adventitious rootlets may lead to false impressions, since diseased roots look more developed than healthy roots. Nematodes can be seen by peering closely at galls with a magnifying lens, or breaking galls open (Fig 6.1 & Plate 70). Fungal root rot may arise in galls, especially late in the season.

CAUSAL ORGANISMS & TAXONOMY

Between 1855 and 1949, root knot nematodes were known by different names: *Anguillula marioni* Cornu 1879, *Heterodera radicicola* Müller 1884, *H. javanica* Treub 1885, *Anguillula arenaria* Neal 1889, and *Oxyurus incognita* Kofoid & White 1919. In 1949 B. G. Chitwood moved these nematodes to the genus *Meloidogyne* and divided them into five species—three of which attack hemp, described below. Separating Chitwood's species can be difficult; many researchers simply identify them to genus, as "*Meloidogyne* species."

1. SOUTHERN ROOT KNOT NEMATODE
Meloidogyne incognita (Kofoid & White 1919) Chitwood 1949.

Description: *Meloidogyne* species are difficult to tell apart. Only the head and anal (perineal) shields can be differentiated; everything else is nearly identical (see Table 6.1). Eggs are ellipsoidal, colourless, 40 x 80 µm in size. Larvae emerging from eggs are long and slender, averaging 15 x 400 µm. As larvae feed, a marked sexual dimorphism develops. Males enlarge into long, robust cylinders with short round tails, averaging 35 x 1300 µm (Fig 3.4). Females turn a pearly-white colour and swell into pear shapes (averaging 750 x 1300 µm), they may swell to the point of protruding from galls. Females lay eggs in a jellylike sac that is yellow-brown in colour and bulges from their genitalia.

Perineal patterns of adult females (wrinkles in the cuticle around the anus) are relatively easy to prepare for observation under light microscopy. Eisenback (1985) presented line drawing, photographs, and charts of perineal patterns. Here in Table 6.1 we describe two characters of perineal patterns—the dorsal arch and striae. Eisenback also presented drawings, photos, and charts of heads and stylets of females, males, and juveniles.

Crop susceptibility tests can identify *Meloidogyne* species. The "North Carolina Differential Host Test" uses six crop plants to differentiate *M. incognita, M. hapla,* and *M. javanica* (see Sasser & Carter 1985). Recently, polymerase chain reaction (PCR) techniques have been used to differentiate these species (Guirao *et al.* 1995).

Most "*Meloidogyne* species" in the *Cannabis* literature refer to *M. incognita,* from Brazil (Richter 1911), Tennessee (Miller *et al.* 1960), India (Johnston 1964), and the former USSR (Goody *et al.* 1965). *M. incognita* is the most widely distributed *Meloidogyne* species worldwide; it prefers a warm climate. In the USA, it remains south of the Mason-Dixon line (39° latitude), except for scattered appearances in warm glasshouses. *M. incognita* attacks hundreds of hosts. It produces larger root knots than other *Meloidogyne* species.

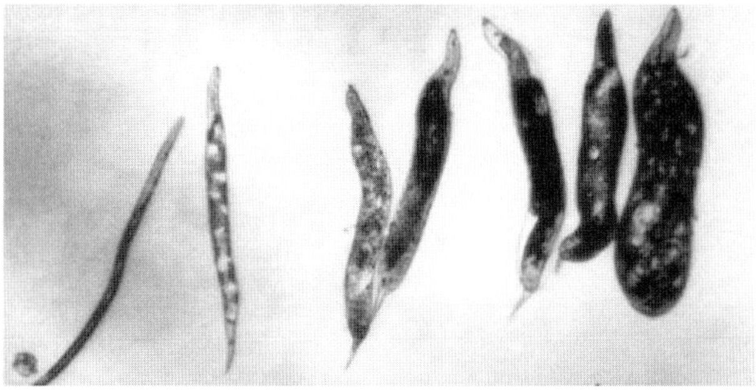

Figure 6.1: Progressive growth stages of immature *Meloidogyne incognita* larvae, from freshly hatched (on left) to nearly-mature "sausage stage" (LM courtesy G.W. Bird).

2. NORTHERN ROOT KNOT NEMATODE
Meloidogyne hapla Chitwood 1949.

Description: *M. hapla* larvae and adults are slightly smaller than those of *M. incognita,* but they are difficult to tell apart. The head and anal (perineal) shields differ, see Table 6.1.

Inoculation tests prove *Cannabis* is susceptible to *M. hapla* (Norton 1966; de Meijer 1993). *M. hapla* eggs tolerate cold

Table 6.1: Some differential characteristics of *Meloidogyne* species.

SPECIES	DORSAL ARCH OF PERINEAL PATTERN	STRIAE OF PERINEAL PATTERN	STYLET CONE OF FEMALES	HEAD CAP OF MALES
M. incognita	squarish, high	coarse, smooth to wavy to zigzaggy	large, anterior half cylindrical, curved	flat to concave
M. javanica	rounded, interrupted by lateral ridges	coarse, smooth to slightly wavy	anterior half tapered, less curved	convex, rounded, broad
M. hapla	rounded, low	fine, smooth to slightly wavy	small, narrow and delicate, slightly curved	convex, rounded, narrow

better than *M. incognita;* the northern root knot nematode lives in every state except Alaska (in Hawai'i it only occurs at higher elevations). It also lives in Canada, northern Europe, and southern Australia. *M. hapla* generally causes less damage than *M. incognita*. *M. hapla* galls remain small and spherical; compound galls rarely develop. The species is a serious pest of tomatoes and potatoes. *M. hapla* can be confused with ***Meloidogyne chitwoodi*** Golden, a species distributed in the western USA.

3. JAVA ROOT KNOT NEMATODE

Meloidogyne javanica (Treub. 1885) Chitwood 1949.

Description: *M. javanica* larvae and adults are nearly identical to those of *M. incognita* (see description above). The head and anal (perineal) shields differ, see Table 6.1.

Decker (1972) reported *M. javanica* infesting hemp in the southern USSR. *M. javanica* has no cold tolerance; it lives in semitropical regions (in the USA it has been reported in the Carolinas, the Gulf states, and southern California).

ROOT KNOT DISEASE CYCLE

Root knot nematodes overwinter as eggs. First moult occurs in eggs, so second-stage larvae hatch and penetrate roots of susceptible plants. Once in roots, the larvae undergo three more moults and develop a sexual dimorphism. Males become migratory, whereas females stay sedentary and remain in roots. Mating may occur, but *Meloidogyne* females can produce eggs parthenogenetically. Females usually produce 300–500 eggs (rarely over 2000) in egg sacs. Sacs often protrude from root surfaces (Plate 70). Females die and egg sacs decay, releasing eggs into the soil. Under optimal temperatures (25–30°C) the life cycle may turn in 20 days.

Nematode root wounds provide portholes for many root-pathogenic fungi, including *Rhizoctonia solani, Pythium ultimum, Fusarium oxysporum,* and *Verticillium* species. Studies with tobacco show that weakly-pathogenic fungi (*Curvularia, Botrytis, Penicillium,* and *Aspergillus* species) also invade roots via nematode wounds (Lucas 1975).

To estimate nematode infestation levels, two methods can be used—root examination and soil assay. *Root examination* is done at harvest. Estimate the percentage of root area galled by nematodes. Use the index outlined in Table 6.2. The second method, *soil assay,* is also done at harvest. Collect samples from soil 15–20 cm deep, mix the samples and ship one pint of soil to your agricultural extension agent. Samples must remain cool (<27°C) and moist. For soil samples, apply the index outlined in Table 6.2.

Table 6.2: Infestation severity index for root-knot nematodes, estimated by 2 methods (described in text).

	ROOT EXAMINATION (% OF ROOT AREA AFFECTED BY GALLS)	SOIL ASSAY (# OF NEMATODES PER PINT SOIL)
Very low	galls rare	0-20
Light	1-10%	21-100
Moderate	11-25%	101-300
Heavy	26-50%	301-1000
Critical	>50%	>1000

CYST NEMATODES

Two species of cyst nematodes attack *Cannabis*. They are found worldwide but cause greatest damage in temperate regions. Above-ground symptoms mimic symptoms of root-knot nematodes. Underground symptoms consist of distorted, knobbed, and bushy roots. Affected roots have cysts embedded in them (Plate 71). Cysts are the remains of female nematodes, filled with eggs. Roots damaged by cyst nematodes become predisposed to other soil pathogens, notably *Fusarium* and *Rhizoctonia* fungi.

1. SUGAR BEET CYST NEMATODE

Heterodera schachtii Schmidt 1871

Description: Cysts are spherical to lemon-shaped, coloured white-yellow-brown (depending on their age), and average 500–800 µm in length (Plate 71). Eggs within cysts are minute. First moult occurs in eggs, second-stage larvae hatch out. Once feeding begins, cyst nematodes develop a sexual dimorphism. Males reach 1600 µm in length; females enlarge to the size and shape of cysts.

H. schachtii attacks a wide range of hosts, principally plants in the Chenopodiaceae (e.g., sugar beet) and Cruciferae (e.g., cabbage). It lives in temperate zones; in North America it is found from southern California and Florida north to Ontario and Alberta. *H. schachtii* has only been reported on *Cannabis* in Europe and southwest Asia (Kirchner 1906, Goody *et al.* 1965).

2. HOPS CYST NEMATODE

Heterodera humuli Felipjev 1934

Description: *H. humuli* cysts, like those of *H. schachtii,* are lemon-shaped and yellow to brown coloured, but smaller (300–600 µm long). *H. humuli* larvae are smaller than those of *H. schachtii* (males reaching 1 mm long, females averaging 410 µm).

H. humuli normally infests hops, but Filipjev & Stekhoven (1941), Winslow (1954), Decker (1972), Spaar *et al.* (1990), and Gutberlet & Karus (1995) report the species infesting hemp. Contrarily, Kir'yanova & Krall (1971) claimed *H. humuli* did not infest hemp.

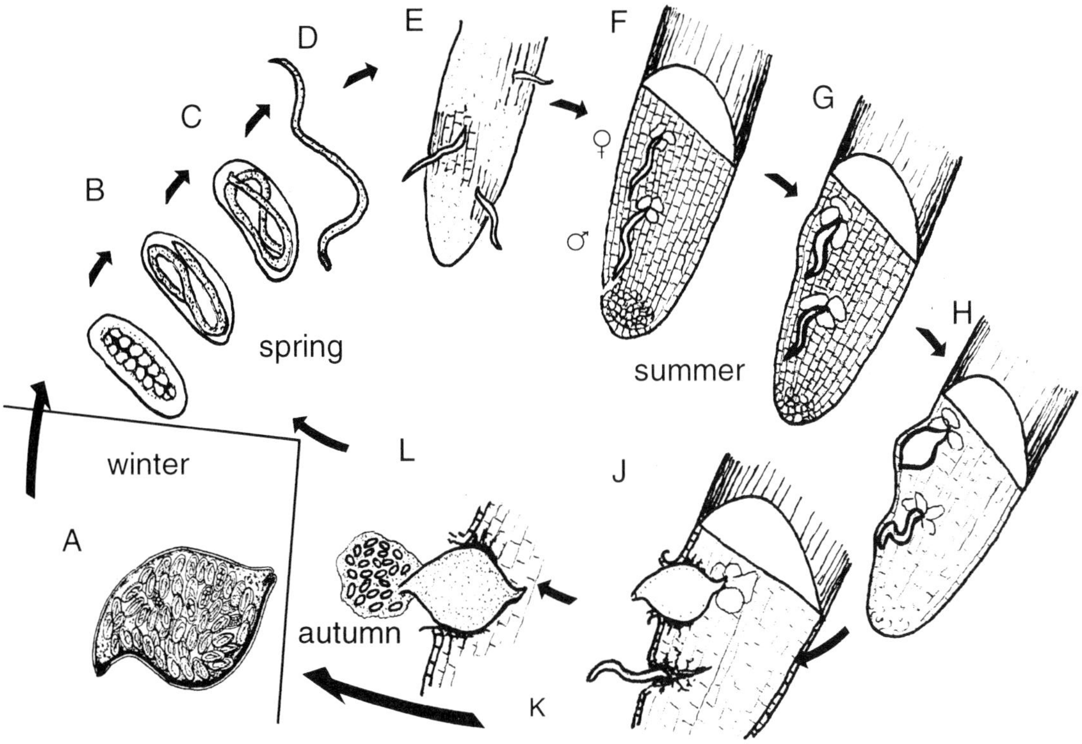

Figure 6.2: Disease cycle of *Heterodera schachtii* (McPartland redrawn from Agrios 1997). A. Eggs overwinter in cysts; B. Eggs develop into 1st-stage larvae; C. Then moult into 2nd-stage larvae while still within eggs; D. Second-stage larvae hatch in the spring; E. 2nd-stage larvae invade rootlets; F–H. After several moults, females become more rotund than males; J. Females eventually rupture roots, their posteriors exposed to the soil; K. Males exit roots to inseminate the exposed females; L. During the growing season, females extrude as many as 500 eggs in a gelatinous mass, and the cycle repeats; A. In autumn, females do not extrude eggs, they die and their cuticles harden into egg-filled cysts.

CYST DISEASE CYCLE

Eggs overwinter in a cyst (the hardened, dead shell of their mother). For the rest of the life cycle, see Fig 6.2. At an optimal temperature of 25°C, the life cycle of *H. schachtii* takes less than a month. *H. humuli* reproduces slower, only one or two generations arise per year.

STEM NEMATODE

This is one of those unusual nematodes that does *not* live in roots. It feeds on above-ground parenchymatous tissue. Symptoms begin as inconspicuous thickenings of branches and leaf petioles. Sometimes even the middle veins of leaves become swollen. Stems subsequently become twisted and distorted, with shortened internodes producing malformed, stunted plants (Fig 6.3). Infected stems feel spongy. Plants often wilt due to severed xylem vessels. Lightly-infected plants may send up new shoots from below infested areas (Mezzetti 1951).

Ditylenchus dipsaci (Kuhn) Filipjev 1936

=*Anguillulina dipsaci* Kuhn 1857; =*Tylenchus devastatrix* (Kuhn) Oerley 1880

Description: Males and females retain their filiform shape into maturity. Stylets are short and thin and difficult to see under a microscope; the stylet cone is about half of the total stylet length, stylet knobs are small and rounded. Females may reach 1.6 mm in length, with only a few eggs in the uterus at one time, and a long slender tail. Eggs are oval, 60 x 20 μm.

STEM NEMATODE DISEASE CYCLE

The life cycle of stem nematodes begins like the life cycle of other nematodes. Eggs overwinter in the soil, and first

Figure 6.3: Symptoms caused by the stem nematode, *Ditylenchus dipsaci* (from Mezzetti 1951).

moult occurs within eggs, which hatch into second-stage larvae. Thereafter, stem nematodes travel a different path—they migrate out of soil and infest above-ground plant parts. They enter stems through stomates, lenticels, or wounds. Invasion of plants is favoured by cool, moist conditions. Mating with males is necessary for reproduction. The life cycle quickens to 19 days in wet weather and optimal temperatures (15–20°C). *D. dipsaci* is one of the few nematodes that spreads by seed, but this has only been reported in bean and onion (Bridge 1996).

D. dipsaci is found in North America, southern Africa, Australia, and temperate areas of Eurasia. *Cannabis* disease has only been described in Europe (Kirchner 1906, Ferraris 1915 & 1935, Mezzetti 1951, Goidànich 1955, Rataj 1957, Ceapoiu 1958, Goody *et al.* 1965, Dempsey 1975). The nematode attacks several hundred plant species, including onion, grains, legumes, and many weeds. Taxonomists separate *D. dipsaci* into different subspecies ("races"); *Cannabis* is infected by the "flax-hemp race" (Kir'yanova & Krall 1971). The hemp flax race also infects flax (*Linum* species), rye (*Secale cereale*), and may indeed be identical to the "rye race" (Nickle 1991).

ROOT LESION NEMATODE

Symptoms arise as spots on small feeder roots, first appearing watersoaked or cloudy yellow. The spots turn dark brown and the root tissue collapses. Collapsed lesions enlarge and coalesce as secondary organisms (bacteria and fungi) invade tissues. The entire root system may become stubby and discoloured. Above-ground symptoms resemble those of drought stress or mineral deficiencies. Plants gradually lose vigour, wilt easily, turn yellow, and become stunted.

Pratylenchus penetrans (Cobb 1917) Chitwood & Oteifa 1952
=*P. penetrans* (Cobb 1917) Filipjev & Schuurmans-Stekhoven 1941

Description: *P. penetrans* adults are 400–800 µm long and 20–25 µm in diametre. They have blunt, broad heads and rounded tails, which make the nematodes look short. The stylet is short (16–19 µm) with well-developed basal knobs. Eggs are elongate, 60–70 x 20–25 µm.

LESION NEMATODE DISEASE CYCLE

Eggs overwinter in temperate soil. In warmer regions, eggs *and larvae* overwinter in diseased roots. First moult occurs in eggs. All subsequent stages (second-stage juveniles through to adults) remain migratory endoparasites, entering and leaving roots at will. As they move through roots, root lesion nematodes secrete a root-necrosing enzyme. When root necrosis becomes severe, the nematodes exit in search of new roots. Adult females lay eggs singly or in small groups in roots or soil. Because of their migratory nature, total egg production per female is unknown. The life cycle is relatively slow, taking 45–65 days.

P. penetrans predominates in temperate potato-growing regions, especially in areas with sandy soil. The species is susceptible to desiccation, and becomes dormant during droughts. *P. penetrans* predispose roots to *Verticillium* and *Rhizoctonia* fungi.

P. penetrans has been reported on hemp in the Netherlands (Kok *et al.* 1994) and South Africa (Dippenaar *et al.* 1996). *P. penetrans* attacks many crops, weeds, and wild plants. Compared to other crops, *Cannabis* is a highly susceptible host (Kok & Coenen 1996). Dippenaar *et al.* (1996) compared different cultivars, and reported more *P. penetrans* infesting 'Kompolti' roots than roots of 'Fedora-19,' 'Futura-77,' 'Felina-34,' and 'Secuini.' Potato growers test soil samples prior to planting; fields with >100 *P. penetrans* per 100 ml soil sustain significant crop damage (Howard *et al.* 1994).

NEEDLE NEMATODE

This nematode is an *ectoparasite*—most of its body remains outside the root while its head penetrates root tips. Underground symptoms include thickened and distorted root tips, with poor development of the root system as a whole. The disease is sometimes called "curly root tip." Root tips become curled or hooked, and turn a necrotic brown colour. Above ground, infected plants are stunted and develop chlorotic leaves. Heavily infested plants may wilt and die during hot dry summers.

Paralongidorus maximus (Bütschli) Siddiqi 1964
=*Longidorus maximus* (Bütschli) Thorne & Swanger 1936,
=*Dorylaimus maximus* Bütschli 1874

Description: *P. maximus* is a large nematode. Even the eggs are big—measuring 260 x 65 µm. Females average almost a centimetre (9840 µm) in length and 92 µm wide. The body tapers slightly toward both ends. The head is offset from the body by a deep constriction. Males are slightly shorter (9630 µm) and rarely encountered.

NEEDLE NEMATODE DISEASE CYCLE

Females produce few eggs. Larvae moult through four stages before reaching adulthood. Reproduction (usually parthenogenetic) occurs in late summer and autumn. *P. maximus* has been reported on *Cannabis* in western Europe (Goody *et al.* 1965). The nematode attacks a wide range of dicots, monocots, conifers, and possibly ferns. It transmits the alfalfa mosaic virus.

OTHER NEMATODES

Dippenaar *et al.* (1996) found other nematodes in roots of South African hemp, including spiral nematodes (***Heliocotylenchus*** and ***Scutellonema*** species) and reniform nematodes (***Rotylenchulus*** species). These nematodes attacked roots of the cultivar 'Secuini' more than roots of other hemp cultivars ('Fedora-19,' 'Futura-77,' 'Felina-34,' 'Kompolti').

Scheifele *et al.* (1997) assessed nematode populations in Canadian soils, before and after a hemp crop (using cultivars 'Unico B' and 'Kompolti'). The hemp crop suppressed soyabean cyst nematodes (*Heterodera glycines* Ichinohe), pin nematodes (*Paratylenchus* species), and stunt nematodes (*Tylenchorhynchus* species), but increased the soil populations of spiral nematodes (*Heliocotylenchus* species) and root knot nematodes (*Meloidogyne hapla*).

DIFFERENTIAL DIAGNOSIS OF NEMATODES

Root gall nematodes may be confused with cyst nematodes. The cuticle of female *Meloidogyne* species remains white and soft, whereas the white skin of *Heterodera* females turns yellow then brown, hard, and crusty.

Chlorosis and wilting from root nematodes mimic symptoms caused by root fungi, some soil insects (e.g., root maggots, white root grubs), nutrient deficiencies, or drought. According to Ferraris (1935), symptoms caused by *D. dipsaci* can be confused with those caused by *Pseudoperonospora cannabina*, a downy mildew.

CULTURAL & MECHANICAL CONTROL
(numbers refer to Chapter 9)

Method 1 (sanitation) controls many nematodes, especially *D. dipsaci*. Method 2a (deep ploughing) kills many nematodes (but *P. maximus* requires ploughing down to 40 cm). Method 2b (steam sterilization) can be implemented in glasshouses. Soil solarization (method 2c) controls *P. pen-*

etrans, D. dipsaci, H. schachtii, and *M. hapla,* but not *M. incognita* (Elmore *et al.* 1997). Flooding soil (method 2d) also works. Weeding (method 3) eliminates alternative hosts.

Growing vigorous plants is the best control against nematodes. Healthy plants can regenerate new roots. Avoid water stress (method 7a). Keep soil nutrients in balance. Add calcium against *D. dipsaci,* potassium against *Meloidogyne* species. Avoid planting in sandy soils that nematodes like best. Adding compost and manure encourages the growth of organisms antagonistic to nematodes.

Rotating with monocot crops will starve root knot nematodes and cyst nematodes, but may take a while (two years for root knot, seven for cyst). Finding nonhost plants for *D. dipsaci* is difficult, and nearly impossible for *P. maximus.*

Quarantine prevents the movement of nematodes into new areas. Larvae left on their own only migrate 20 cm in a lifetime. But eggs and cysts can travel many miles on tractor tyres, floating in streams, even blowin' in the wind. Quarantines must be rigorously enforced to be effective—exclude plants, soil, even dirty shoes and spades from uninfested areas. Breeding for crop resistance works has barely begun in *Cannabis.* De Meijer (1993, 1995) studied the resistance of hemp cultivars to *M. hapla.* The most resistant were Hungarian cultivars 'Kompolti Sárgaszárú' and 'Kompolti Hibrid TC.' The most susceptible were French cultivar 'Futura 77' and Ukrainian cultivar 'USO-13' ('YUSO-13').

BIOCONTROL (see Chapter 10)

Plants belonging to 57 families possess nematicidal properties (Bridge 1996). *Cannabis* plants can suppress *Meloidogyne* species (Kok *et al.* 1994, Mateeva 1995) and *Heterodera rostochiensis* (Kir'yanova & Krall 1971). Akhtar & Alam (1991) intercropped mustard greens (*Brassica juncea*) to suppress *M. incognita.* Marigold (*Tagetes* species, described below) suppresses nematodes by acting as a "decoy crop." Decoy crops cause nematode eggs to germinate, but the nematodes cannot complete their life cycle on the decoy plants, so the nematodes die out. Palti (1981) decoyed *M. incognita* and *M. javanica* with marigold (*Tagetes patula, T. minuta*), sesame (*Sesamum orientale*), castor bean (*Ricinus communis*), and *Chrysanthemum* species. *Pratylenchus penetrans* can be decoyed with marigold (*Tagetes* species) and blanket flowers (*Gaillardia* species).

Bacteria and fungi that kill nematodes are on the biocontrol horizon. According to Meister (1998), coating seeds with *Burkholderia cepacia* (described under damping off) protects seeds from lesion, sting, lance, and spiral nematodes. *Gliocladium roseum* infests cyst nematodes (described under grey mould). Seven other bacteria and fungi are described below. For more, see Carris & Glawe (1989).

***Tagetes* species**

BIOLOGY: Marigolds are annual herbs that ooze nematode-repellent metabolites into soil. Some nematodes are attracted to marigolds and bore into the roots, but are unable to feed or reproduce (Howard *et al.* 1994). The most effective species are African marigold (*Tagetes erecta*), French marigold (*Tagetes patula*), and South American marigold (*Tagetes minuta*). Crops interplanted with marigold suffer fewer nematodes, as do subsequent crops growing in the same location. But marigold is not a cure-all—some cultivars are susceptible to *Meloidogyne* infestation (Miller *et al.* 1960). Marigold works best against *Pratylenchus, Rotylenchus,* and *Tylenchorhynchus* species. It works poorly against *Ditylenchus, Heterodera,* and most ectoparasitic nematodes (Howard *et al.* 1994).

Burkholderia (Pseudomonas) cepacia

BIOLOGY: A soil bacterium that colonizes the root zone of plants and protects roots from some nematodes (lesion, sting, lance, spiral nematodes). It is described under Damping off disease.

Myrothecium verrucaria

BIOLOGY: This fungal saprophyte lives in a wide range of soils. Strains selected for biocontrol do best around 20°C.

APPLICATION: Available as an emulsifiable suspension (DiTera ES®) or granules (DiTera G®), for the control of many nematodes, including *Meloidogyne, Heterodera, Pratylenchus,* and *Xiphinema* species. The fungus may be incorporated into soil as a preplant mix or injected into the root zone of established plants.

NOTES: Gzebenyuk (1984) cited *M. verrucaria* attacking Ukrainian hemp, but he may have confused it with *Myrothecium roridum,* the cause of corky leaf spot.

***Arthrobotrys* and *Dactylaria* species**

BIOLOGY: These two genera of soil fungi strangle nematodes in hyphal nooses or trap them on sticky hyphal knobs. Commercial products containing *Arthrobotrys robusta* (Royal 300®, used against *Ditylenchus* species) and *Arthrobotrys superba* (Royal 350®, used against *Meloidogyne* species) are irregularly available. Unfortunately, these fungi grow slower than most biocontrol fungi, and they are very sensitive to desiccation.

Pasteuria (Bacillus) penetrans

BIOLOGY: A soil bacterium that infects numerous nematodes including *Meloidogyne, Herterodera, Ditylenchus, Paralongidorus,* and *Pratylenchus* species. Endospores of *P. penetrans* infect immature nematodes in soil; endoparasitic nematodes such as *Meloidogyne* species die before they can reproduce. The commercial availability of *P. penetrans* is hampered by its expensive production costs (Cook *et al.* 1996).

Verticillium chlamydosporium

BIOLOGY: A soil fungus and facultative parasite of cyst nematodes and nematode eggs. Infected cyst nematodes turn brown and shrivel. *V. chlamydosporium* takes two weeks to kill its host. Cadavers become surrounded by a halo of chlamydospores in optimal conditions. The fungus has also been isolated from plant roots, snail eggs, and other soil fungi—so it is not dependent on nematodes for survival.

Hirsutella rhossiliensis

BIOLOGY: A soil fungus whose sticky spores adhere to nematodes as they move through soil. *Heterodera schachtii* juveniles are killed within three days of infection at 20°C (Tedford *et al.* 1995). Dead nematodes sprout new spores to repeat the *H. rhossiliensis* life cycle. The fungus has a broad host range, including *Heterodera, Meloidogyne, Ditylenchus, Pratylenchus,* and *Xiphinema* species. Unfortunately, the fungus also kills biocontrol nematodes—*Steinernema* and *Heterorhabditis* species.

Nematophthora gynophila

BIOLOGY: This oömycete is an obligate parasite of nematodes in the genus *Heterodera.* It does well in the damp soil preferred by nematodes.

CHEMICAL CONTROL (see Chapter 11)

Urea is a traditional cure for infested fields, mixed into soil after the growing season. By spring, the urea has broken down and most nematodes are dead. Clandosan® combines urea with chitin from ground-up crab shells. Crab shell chitin encourages the growth of nematicidal soil organisms when mixed into soil at a rate of 2–4 kg per 10 m^2. Adding cow dung and poultry manure to soil provides some control of *M. incognita* (Bridge 1996).

Neem, normally used as an insecticide, has nematocidal properties. Akhtar & Alam (1991) mixed neem oil and castor oil into soil to repel nematodes. Olkowski *et al.* (1991) used neem oil against *M. incognita*. Neem cake also works, mixed into soil at rates as low as 100-250 kg ha^{-1} (Bridge 1996). The residual effect of soil treatments lasts for several months. Aqueous extracts of azadirachin, the active ingredient in neem, prevent *M. incognita* infection in a variety of crops when used as a seed treatment and root dip (Mordue & Blackwell 1993).

Aqueous extracts of *Cannabis* leaves caused a high mortality rate in *M. incognita* (Vijayalakshmi *et al.* 1979), *Heterodera cajani* (Mojumder *et al.* 1989), *M. javanica* (Bajpai & Sharma 1992), *Hoplolaimus indicus, Rotylenchulus reniformis,* and *Tylenchorynchus brassicae* (Haseeb *et al.* 1978). Soil mixed with 3% w/w dried *Cannabis* leaves suppressed *M. incognita* (Goswami & Vijayalakshmi 1986) and *Aphelenchoides composticola* (Grewal 1989). In human medicine, "worms" have been treated with juice extracted from crushed hemp seed (Parkinson 1640, Culpeper 1814). This practice dates to Imperial Rome (Pliny 1950 reprint).

Extracts of the pawpaw tree repel nematodes (Meister 1998). Yepsen (1976) mulched soil with latex-bearing plants (goldenrod, milkweed, dandelion, *Euphorbia* species) to repel nematodes.

Unfortunately, the egg-protecting cysts of cyst nematodes renders them impervious to nearly all nematocides. Soil fumigation (injecting soil with volatile liquids) works best against cyst nematodes but is very expensive and dangerous. Fumigants are usually injected into field soil to depths of 22–30 cm with row chisels.

VIRAL DISEASES

The virus world was discovered by Dmitri Iwanowsky in 1892. The virus he discovered infected plants, not animals. It caused "mosaic" symptoms in tobacco. Plant viruses are named by symptoms they cause and plants they infect. Iwanowsky's virus is now called the tobacco mosaic virus (TMV). For a while, viruses were named with Latin binomials (Holmes 1939). Thus TMV was named *"Marmor tabaci,"* in the Phyllum Phytophagi of the Kingdom Vira. Since biologists debate if viruses are truly living creatures (see Chapter 3), Holmes's binomial system has been abandoned.

About 1200 viruses are able to infect plants (Cook & Qualset 1996). Amazingly, only five viruses regularly infect *Cannabis*. Perhaps this is due to the presence of THC, which inactivates viruses (Blevins & Dumic 1980, Lancz *et al.* 1990, Lancz *et al.* 1991). *Cannabis* also contains terpenoids (limonene, α-pinene) and flavonoids (apigenin, luteolin, quercetin) with antiviral activity (Che 1991).

Cannabis viruses are very simple—they contain positive-sense, single strand (ss) RNA covered by a protein coat (Fig 6.4). In contrast, many human viruses contain double-stranded (ds) DNA with a protein coat *and* an outer lipid membrane. Plant viruses currently pose no threat to humans, but there is some concern of plant viruses exchanging RNA material with similar human viruses. The human viruses containing ssRNA are dreadful—Dengue, Coxsackie, Polio (all positive-sense), Rabies, Hanta, Lassa, Machupo, Marlburg, Ebola, and HIV (negative sense). Concern about mutant viruses has increased due to the engineering of genetically-altered plant viruses (Garrett 1994). RNA viruses mutate a million times faster than DNA viruses (Strauss & Strauss 1988). All RNA viruses, including plant and animal viruses, probably descended from a single ancestor virus.

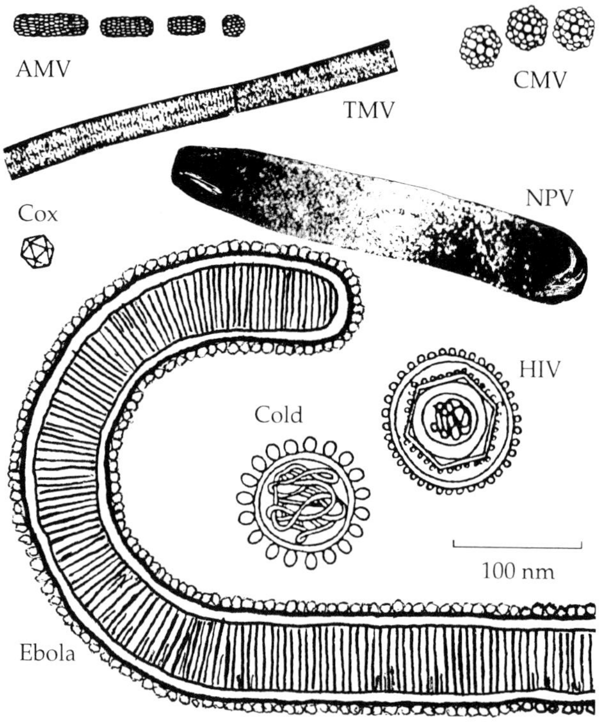

Figure 6.4: A menagerie of plant and animal viruses. AMV = Alfalfa mosaic virus; TMV = Tobacco mosaic virus; CMV = Cucumber mosaic virus; NPV = Nuclear polyhedrosis virus (insects); Cox = Coxsackie virus; HIV = Human immunodeficiency virus; Cold = human coronavirus, Ebola = Ebola (McPartland).

A separate concern is the contamination of plants by human viruses. An outbreak of Hepatitis A in Washington was linked to Mexican marijuana fertilized with faeces contaminated by the Hepatitis A virus (Alexander 1987b).

Viruses rarely kill *Cannabis*. They can, however, cause serious symptoms and reduce yields. Once acquired, viruses are nearly impossible to eradicate—they invade all plant parts, *including seeds and pollen*, enabling viruses to replicate across host generations. Insects spread many viruses as they buzz from plant to plant. According to Ceapoiu (1958), the worst vectors are the bhang aphid (*Phorodon cannabis*), green peach aphid (*Myzus persicae*), greenhouse whitefly (*Trialeudodes vaporariorum*), and onion thrips (*Thrips tabaci*).

Hartowicz *et al.* (1971) tested 22 viruses for their ability to infect *Cannabis*. Paulsen (1971) reported four of the 22 viruses caused serious mosaic symptoms and dwarfing: tobacco ringspot (TRSV), tomato ringspot (TomRSV), tobacco streak (TSV) and cucumber mosaic (CMV). Two of the 22 viruses caused mosaic symptoms without dwarfing: eunoymous ringspot (ERSV) and alfalfa mosaic (AMV). Elm mosaic (EMV) caused mild necrotic flecking. Back inoculations to indicator plants proved TMV and foxtail mosaic (FMV) cause symptomless infections in *Cannabis*. The remaining viruses did not infect *Cannabis:* barley stripe mosaic (BSMV), brome mosaic (BMV), maize dwarf mosaic strain A (MDMV-A), panicum mosaic (PMV), wheat streak mosaic (WSMV), soilborne wheat mosaic (SBWMV), bean pod mottle (BPMV), potato virus X (PVX), strawberry latent ringspot (SLRV), tobacco etch (TEV), tobacco rattle (TRV-C), soyabean mosaic (SMV) and squash mosaic (SqMV).

More recently, Kegler & Spaar (1997) tested three fibre cultivars ('USO-11,' 'YUSO-14,' 'YUSO-31') for their resistance to eight viruses: AMV, CMV, PVX, TomRSV, potato virus Y (PVY), broad bean wilt fabavirus (BBWV), arabis mosaic virus (ArMV), and raspberry ringspot nepovirus (RRV). The results of their inoculation experiments are presented in Table 6.3. Symptoms of AMV, CMV, and ArMV are described below. TomRSV caused light-green checkmarks in one cultivar ('YUSO-31') and did not infect other cultivars. PVX caused a light- and dark-green mosaic between leaf veins in one cultivar ('YUSO-31') and did not infect other cultivars. PYV caused a light- and dark-green mosaic, as well as upturning of leaf tips. BBWV caused a yellow-green mosaic at the base of young leaves. RRV caused a light-green mosaic and curling of leaf petioles.

Table 6.3: Susceptibility of 3 fiber cultivars to 8 viruses.[1]

Virus	Number of plants with symptoms (%)		
	USO-11	YUSO-14	YUSO-31
AMV	0	57.1	11.1
CMV	18.2	66.7	50.0
BBWV	37.5	0	16.7
ArMV	10.0	0	7.6
RRV	25.0	0	9.1
PVX	0	0	14.5
PVY	11.1	33.3	18.2
TomRSV	0	0	42.9

1. Results of inoculation experiments by Kegler & Spaar (1997).

A grower in Seattle reported transmitting a virus from tobacco to marijuana (Rosenthal & McPartland 1998). He used the same scissors to trim tobacco and take *Cannabis* cuttings for clones. Symptoms included dwarfing, early senescence, and reduced yields. The virus was probably TRSV, TomRSV, or TSV.

HEMP STREAK VIRUS (HSV)

Röder (1941) originally described hemp streak virus (HSV) in Germany. HSV has caused serious losses in Italy (Ferri 1963). In the Czech Republic, Rataj (1957) said the virus was limited by the range of its aphid vector. Farther north in Russia, Gitman (1968b) considered HSV a rare disease. HSV is probably the "curly leaf virus" that Clarke (pers. commun. 1993) found in Hungary. The viral genome and morphology of HSV remains unknown.

SIGNS & SYMPTOMS

Foliar symptoms begin as a pale green chlorosis. Chlorotic areas develop into a series of interveinal yellow streaks or chevron-stripes. Sometimes brown necrotic flecks appear, each fleck surrounded by a pale green halo. Flecks appear along the margins and tips of older leaves and often coalesce. Streak symptoms predominate in moist weather, whereas fleck symptoms appear during dry weather.

Eventually, leaf margins become wrinkled, leaf tips roll upward, and leaflets curl into spirals. Whole plants assume a "wavy wilt" appearance (Plate 72). Diseased plants are small and the yield of fibre and seed is greatly reduced. Röder (1941) and Ferri (1963) provided excellent black-and-white photographs of disease symptoms. Spaar *et al.* (1990) illustrated symptoms in a colour painting.

DISEASE CYCLE & EPIDEMIOLOGY

Röder (1941) and Goidànich (1955) cited hemp aphids (*Phorodon cannabis*) as the chief vectors of HSV. Seeds of diseased plants give rise to diseased progeny. Plants developed symptoms within six days of inoculation (Röder 1941). Male plants suffer higher rates of infection than females.

ALFALFA MOSAIC VIRUS (AMV)

Schmelzer (1962) infected hemp with a strain of lucerne mosaic virus from *Viburnum opulus*. Lucerne mosaic virus is a synonym for the alfalfa mosaic virus (AMV). Schmidt & Karl (1970) described naturally-occurring AMV on hemp in Germany. They identified the virus via graft and sap transmission with test plants.

SYMPTOMS & EPIDEMIOLOGY

AMV symptoms have been described as a grey leaf mosaic (Schmelzer 1962), chlorotic stripes along leaves (Schmidt & Karl 1970), or light green flecking with chlorotic veins and slight puckering of young leaves (Kegler & Spaar 1997). Spaar *et al.* (1990) illustrated symptoms of AMV in a colour painting. AMV has a wide host range and is found in nearly all temperate regions. Viral egression is primarily via aphids. Seed transmission, root grafts, and dodder (*Cuscuta* species) may also spread AMV. Virus morphology (the "viral genome") consists ssRNA encapsulated in four particles—all four particles are needed for infection—the first is spherical (18 nm diameter), the other three are bacilliform—all 18 nm wide but different lengths (58 nm, 48 nm, 36 nm).

CUCUMBER MOSAIC VIRUS (CMV)

Schmidt & Karl (1970) cited CMV causing leaf mottling in German hemp. Kegler & Spaar (1997) described symptoms as light green check-marks diffusely covering leaf surfaces, especially young leaves. Spaar *et al.* (1990) illustrated symptoms of CMV in a colour painting. Paulsen (1971) found that both CMV and CMV-X strains can infect hemp. CMV is vectored by the bhang aphid (Schmidt & Karl 1970), green peach aphid (*Myzus persicae*), and black bean aphid (*Aphis fabae*) in a nonpersistent manner. CMV also spreads in seeds and pollen of infected plants (Lucas 1975).

The virus infects at least 470 plant species and is distributed worldwide, especially in temperate regions. CMV is a multicomponent virus, consisting of ssRNA packaged in three isometric polyhedral particles, each 30 nm in diametre. CMV replication is favoured by high temperatures, long day length, high light intensity, and excess nitrogen (Lucas 1975). Breeders have developed tolerant varieties of many susceptible crops (cucurbits, tobacco, spinach).

ARABIS MOSAIC VIRUS (ArMV)

Schmidt & Karl (1969) described ArMV causing yellow-green leaf spots and stripes on "weakly growing hemp" near Potzdam, Germany. Kegler & Spaar (1997) described symptoms as yellow-green check-marks or mosaic on young leaves. Spaar *et al.* (1990) illustrated symptoms of ArMV in hemp. ArMV causes "nettlehead" disease in hops. The virus also infects many vegetables. It occurs in Europe, North America and New Zealand. ArMV is packaged in three isometric particles each 28 nm in diameter, much like CMV. ArMV is transmitted via infected seeds and soil nematodes (*Xiphinema* species).

HEMP MOSAIC VIRUS (HMV)

Not much is known about HMV. It may be a Cucumovirus such as CMV, or a Nepovirus such as ArMV,

TRSV, and TomRSV. Ceapoiu (1958) described symptoms in Romania, beginning as punctate, chlorotic lesions which turn necrotic, coalesce and, finally, the entire leaf wilts. Blattny *et al.* (1950) described the virus causing leaf enation (torsion) in Czech hemp. Ghani & Basit (1975) reported a mosaic virus in Pakistani plants causing leaf curl, bunchy top, and "smalling of leaves." Traversi (1949) inoculated hemp with the Argentine sunflower virus, and it caused HMV-like symptoms. Traversi's virus was transmitted by sap, seeds, the greenhouse whitefly, green peach aphid, and onion thrips. HMV is vectored by *Phorodon cannabis* (Ceapoiu 1958). Schmidt & Karl (1970) described a "hemp mottle virus" similar to HMV; the virus was vectored by the aphid *Phorodon cannabis*.

DIFFERENTIAL DIAGNOSIS

Viral diseases may be confused with symptoms of bacterial blight, early brown leaf spot, or yellow leaf spot. The "wavy wilt" appearance of HSV seems unique, but ArMV and HMV may cause similar symptoms. HSV, ArMV, HMV, and AMV cause mosaics, making exact identification difficult. Using an electron microscope is effective but expensive. Cross-inoculation studies with other plants are time-intensive. Serological tests for HSV and HMV are not available.

Christie *et al.* (1995) used simple stains to detect viral inclusion bodies in plants. Inclusion bodies are intracellular structures consisting of aggregated virus particles and coat proteins. Inclusions induced by a specific virus produce a characteristic appearance, even in different hosts. Christie stained inclusions with Azure A, or a combination of Calcomine Orange 2RS and Luxol Brilliant Green BL. After staining, viral inclusion bodies are easily seen with a light microscope.

CULTURAL & MECHANICAL CONTROL (see Chapter 9)

Destroy obviously infected plants and *do not* clone them or use their seeds. *Viruses migrate into developing seeds,* despite assurances to the contrary (Rosenthal *High Times* 278:95).

Virologists discovered that apical shoots of many plants grow slightly faster than viruses can migrate cell-to-cell. Thus, virus-*free* material can be obtained by cutting off a tiny piece of shoot tip and regenerating it in tissue culture. This technique has not been attempted in *Cannabis* (Mandolino & Ranalli 1999). Virus-*resistant* plants are a possibility—agronomists are testing fibre cultivars for resistance to viruses (Kegler & Spaar 1997). Other crops have been genetically engineered for resistance to AMV, CMV, and other viruses (Goodman *et al.* 1987).

If virus infestation is suspected, minimize the mechanical spread of viruses by dipping hand tools in a dilute solution of skim milk (100 g skim milk powder per l water). Milk inactivates many plant viruses (Howard *et al.* 1994). Eliminate virus vectors, especially aphids, dodder, and nematodes. Cover seedlings with mesh netting to protect them from flying insects. Many weeds harbour symptomless infections, so eradicate them (method 3). Some researchers claim viruses can be eradicated from infected seeds using thermotherapy, see method 11.

Figure 6.5: Symptoms of bacterial blight caused by *Pseudomonas syringae* pv. *cannabina* (from Sutic & Dowson 1959).

BIOLOGICAL & CHEMICAL CONTROL

No biocontrol or antiviral chemicals have been effective. Protect plants from virus vectors with insecticides. Control ArMV vectors with nematocides.

BACTERIAL DISEASES

Dozens of bacteria have been reported from *Cannabis* over the years. Most of these species are retters of harvested hemp. Others bacteria rot harvested female flowers. Retters and rotters of harvested products are discussed in Chapter 8.

Kosslak & Bohlool (1983) isolated two species of diazotrophic bacteria from the rhizospheres of marijuana plants: ***Azospirillum brasilense*** Tarrand, Kreig & Döbereiner and ***Azospirillum lipoferum*** (Beijernick) Tarrand, Kreig & Döbereiner. These mutualistic bacteria live on the surfaces of roots, where they fix nitrogen for their host.

Subtracting the aforementioned saprophytes and mutualists, we only found four species of true pathogens (with one species split into four "pathovarieties"), described below.

BACTERIAL BLIGHT

Also known as bacterial leaf spot, this disease may be limited to Europe. Bacterial blight has been described in Italy, Germany, Hungary, Bulgaria, Romania, Yugoslavia, and the former USSR. It affects plants of all ages, from seedlings to full-flowering females.

SIGNS & SYMPTOMS

Symptoms begin as small watersoaked leaf spots. Spots enlarge along leaf veins (rarely crossing them) and turn brown or grey. Dead tissue breaks apart causing leaf perforations (Fig 6.5). An Italian strain of the bacterium ("variety *italica*") produces "ulcerous striping"—small protuberances arising in rows between leaf veins, turning into dark, necrotic stripes (Sutic & Dowson 1959). Stem lesions may also form. Plants become stunted and deformed.

CAUSAL ORGANISM & TAXONOMY

Pseudomonas syringae* pv. *cannabina (Sutic & Dowson 1959) Young, Dye & Wilkie 1978, *New Zealand Journal of Agricultural Research* 21:153-177.

=*Pseudomonas cannabina* Sutic & Dowson 1959; =*Pseudomonas cannabina* Sutic & Dowson var. *italica* Dowson 1959

Description: Cultures white, convex, circular, some strains produce green fluorescent pigment on King's medium B agar, other strains exude a nonfluorescent brown pigment (Smith *et al.* 1988).

Aerobic gram (-) rods, size usually 1.5 x 0.3 µm, straight or slightly curved with rounded ends, exhibiting one to four polar flagella (Fig 3.1). Non-proteolytic, starch weakly hydrolysed, acid (but no gas) produced from xylose, dextrose, galactose, mannose, sucrose, raffinose, and glycerol. No sheaths or spores are produced. Bacteria often form in chains.

DIFFERENTIAL DIAGNOSIS

Klement & Kaszonyi (1960) and Klement & Lovrekovich (1960) noted the close relationship between *P. s.* pv. *cannabina* and *P.s.* pv. *mori*. Bacterial blight can be confused with early symptoms of Wisconsin leaf spot and Xanthomonas leaf spot, or even early wildfire. It may also be confused with some viral or fungal diseases (notably brown leaf spot), or insect damage. Stem symptoms can be confused with striatura ulcerosa (see next).

DISEASE CYCLE & EPIDEMIOLOGY

Goidànich & Ferri (1959) reported *no* seedborne infection. Subsequent researchers refute this (Noble & Richardson 1968), and consider seedborne infection a primary mode of dissemination. The bacterium overwinters in plant debris and infects seedlings in the spring. During the growing season, rain droplets pick up bacteria oozing from lesions, and splash them around.

Cross-inoculation studies with *P. s.* pv. *cannabina* by Sutic & Dowson (1959) demonstrated lethal susceptibility in kidney bean (*Phaseolus vulgaris*), and slight symptoms in common vetch (*Vicia sativa*), grass pea (*Lathyrus sativus*), pea (*Pisum sativum*) and hyacinth bean (*Dolichos lablab*). Plants related to *Cannabis*, such as hops (*Humulus lupulus*), nettle (*Urtica dioica*), and mulberry (*Morus alba*), were not susceptible.

STRIATURA ULCEROSA

Kirchner (1906) first described this disease in Italy and Germany. Gitman (1968a) called the disease "bruzone" in Russia. Ferri (1957a & b) and Goidànich & Ferri (1959) published important monographs on striatura ulcerosa.

SIGNS & SYMPTOMS

Striatura ulcerosa arises in adult plants, beginning as waxy, dark grey, oval lesions which elongate along stems (Plate73). These lesions rarely encircle more than half the stem circumference, but may run 10 cm along its length. Small pustules (1–2 x 2–3 mm) arise within lesions, filled with a yellow mucilaginous ooze of bacteria. These pustules rupture and shred the epidermis. Secondary pustules arise around the shredded remains of earlier pustules.

CAUSAL ORGANISM & TAXONOMY

Pseudomonas syringae* pv. *mori (Boyer & Lambert 1893) Young, Dye & Wilkie 1978, *New Zealand Journal of Agricultural Research* 21:153–177.

=*Pseudomonas mori* (Boyer & Lambert 1893) Stevens 1913, =*Bacillus mori* (Boyer & Lambert 1893) Holland 1920; =*Pseudomonas cuboniani* (Macchiati 1892) Krasil'nikov 1949, =*Bacillus cubonianus* Macchiati 1892 *non* Steinhaus 1942.

Description: Pathovars of *P. syringae* are morphologically indistinguishable from each other (see description of *P. s.* pv. *cannabina* above).

DIFFERENTIAL DIAGNOSIS

Klement & Lovrekovich (1960) noted a close relationship between striatura ulcerosa and bacterial blight. The two causal organisms are now considered pathovars of the same species, *P. syringae*. Striatura ulcerosa resembles early symptoms of *Cladosporium*, *Phomopsis*, or *Ophiobolus* stem cankers.

DISEASE CYCLE & EPIDEMIOLOGY

In greenhouse studies, plants with a percentage of Chinese ancestry seemed more susceptible to striatura ulcerosa than hemp of strictly Russian heritage (Hillig, pers. commun. 1998). *P. s.* pv. *mori* commonly blights mulberry, attacking leaves and stems. On trees, the disease resembles fire blight, oozing white or yellow masses of bacteria. It spreads by wind-driven rain and overwinters in crop debris. *P. s.* pv. *mori* was the first bacterium isolated and identified from diseased hemp (Peglion 1897).

XANTHOMONAS BLIGHT

Xanthomonas blight has been reported in Japan and Korea (Okabe 1949, Okabe & Goto 1965, Kyokai 1965). In Romania, the disease appears as a wilt (Sandru 1977) or as small, brown leaf spots (Severin 1978). The bacterium produces a mucilaginous, extracellular polysaccharide known as *xanthan gum*, which plugs xylem and disrupts fluid flow.

CAUSAL ORGANISM & TAXONOMY

Xanthomonas campestris* pv. *cannabis Severin 1978, *Archiv für Phytopathology und Pflanzenschutz* 14:7–15.

=*Xanthomonas cannabis* (Watanabe) Okabe & Goto 1965, =*Bacterium cannabis* (Watanabe) Okabe 1949, =*Bacillus cannabis* Watanabe 1947–8; =*Xanthomonas cannabis* Mukoo *apud* Kishi 1988.

Approximately 12 years before Severin's publication, Okabe & Goto described *Xanthomonas cannabis* (Watanabe) Okabe & Goto in Japan. Okabe's taxon was overlooked by Severin, by Bergey's 7th and 8th editions, as well as by Young, Dye & Wilkie (1978).

Description: Cultures usually yellow, smooth and viscid. Aerobic gram (-) straight rods, with single polar flagella, usually 0.7–1.8 x 0.4–0.7 µm. No denitrification or nitrate reduction. Proteolytic, nonlipolytic; acid production from arabinose, glucose, mannose, galactose, trehalose, cellobiose.

DIFFERENTIAL DIAGNOSIS

Differentiating *Pseudomonas* from *Xanthomonas* is difficult. *Xanthomonas* species produce a yellow pigment in culture and only grow one flagellum per cell. Xanthomonas leaf spot is easily confused with bacterial blight (caused by *Pseudomonas syringae* pv. *cannabina*). Bacterial leaf diseases can also be confused with some fungal problems such as brown leaf spot or yellow leaf spot.

DISEASE CYCLE & EPIDEMIOLOGY

X. campestris infests many crucifers and other plants. Okabe (1949) said the species easily infects hemp and flax. Severin (1978) found the hemp pathogen could also infect hops, mulberry, pinto bean, soyabean, cucumber, tobacco, and *Pelargonium* plants. Disease worsens in summers with high temperatures and high humidity (Okabe 1949). *X. campestris* overwinters in crop refuse and infected seed (Smith *et al.* 1988). During the growing season it spreads by splashing water, wind-driven rain, insects, and field workers.

WILDFIRE and WISCONSIN LEAF SPOT

These two diseases are difficult to distinguish from Bacterial Blight, from Xanthomonas blight, and from each other. Symptoms begin on lower leaves and spread rapidly in wet weather. Watersoaked spots turn into small necrotic lesions surrounded by chlorotic halos. Lesions may coalesce into irregular necrotic areas. Alternately, leaf lesions may grow into "angular leaf spots" without halos, limited by leaf veins. Affected leaves become twisted and distorted.

1. WILDFIRE

Pseudomonas syringae* pv. *tabaci (Wolf & Foster 1917)

Young, Dye & Wilkie 1978, *New Zealand Journal of Agricultural Research* 21:153–177.

=*Pseudomonas tabaci* (Wolf & Foster 1917) Stevens 1925; =*Pseudomonas angulata* (Fromme & Murray 1919) Stevens 1925.

Description: Cultures white, slightly raised, with translucent edges and opaque centres, producing a green fluorescent pigment. Pathovars of *P. syringae* are morphologically indistinguishable from each other (see description of *P. s.* pv. *cannabina* above).

Some authors (Lucas 1975, Smith *et al.* 1988) consider *P. angulata* a nontoxin-producing strain of *P. s.* pv. *tabaci*, so we place *P. angulata* in synonymy here. Toxin-producing strains destroy chlorophyll, causing chlorotic halos. Nontoxin strains cause angular leaf spots without halos.

P.s. pv. *tabaci* normally infects solanaceous crops (tobacco, tomato, potato), but can infest *Cannabis* (Johnson 1937). Johnson noted, however, that *Cannabis* plants require stressful water soaking for successful infection. Succulent watersoaked leaves become hypersensitive to infection. Wind-driven rain splash from infected plants creates epidemics and spreads like wildfire.

2. WISCONSIN LEAF SPOT

Pseudomonas syringae* pv. *mellea (Johnson 1923) Young *et al.* 1978, *New Zealand Journal of Agricultural Research* 21:153–177.

=*Pseudomonas mella* Johnson 1923, =*Bacterium melleum* Johnson 1923.

P. s. pv. *mellea* normally attacks tobacco in Wisconsin, Japan, and the former USSR. Gitman (1968a) reported it on hemp but did not describe symptoms. Lucas (1975) and others consider the organism a strain of *P. s.* pv. *tabaci.*

CROWN GALL

Lopatin (1936) and Gitman (1968a) described this disease in Russia. Lopatin placed *Cannabis* in the "extremely susceptible" category. The causal organism attacks dicot plants around the world; 643 species from 331 genera are recorded as hosts (Smith *et al.* 1988). Monocots and gymnosperms are rarely infected.

SIGNS & SYMPTOMS

Twelve to 15 days after infection, abnormal cancer-like growths appear on plants. These beige-coloured, granular-surfaced, spherical galls form at the soil line (crown) or grow on roots underground. Galls rarely arise on above-ground stems and rarely exceed 10 mm in diameter. Crown galls stunt plants but rarely kill them.

CAUSAL ORGANISM

Agrobacterium tumefaciens (Smith & Townsend 1907) Conn 1942, *J. Bacteriology* 44:353-360.

=*Bacterium tumefaciens* Smith & Townsend 1907.

Description: Cultures white, convex, circular, glistening and translucent. Aerobic gram (-) rod, size 1.0–3.0 x 0.4–0.8 µm, with one to five peritrichous flagella. Nonproteolytic, nonlipolytic, starch not hydrolysed, acid (but no gas) produced from glucose, fructose, arabinose, galactose, mannitol, and salicin. Some strains reduce nitrate, others do not. Optimum growth temperature is 25–30°C.

DIFFERENTIAL DIAGNOSIS

A. tumefaciens is split into three biovars; biovar 1 probably infects *Cannabis*. Symptoms of crown gall are unique.

DISEASE CYCLE & EPIDEMIOLOGY

A. tumefaciens may spread via seedborne infection, as demonstrated in hops. *A. tumefaciens* can live as a saprophyte in soil, unlike most phytopathological bacteria. Bacteria in soil are attracted to root sap oozing from wounded roots. The bacteria enter through wounds.

A. tumefaciens harbours extrachromosomal plasmids coded with "Ti" (tumour-inducing) genes. After *A. tumefaciens* infects plants, the bacterium injects Ti plasmids into plant cells, and the plasmids become incorporated into the host plant genome. The plasmid genes cause plant cells to proliferate into cancer-like galls and produce nutrients for the bacteria.

Scientists can bioengineer foreign genes into Ti plasmids, genes coding for insect and disease resistance. In the laboratory, *A. tumefaciens* then splices these desirable genes into plant protoplasts, which regenerate into resistant plants (Goodman *et al.* 1987).

BACTERIAL WILT

Gitman (1968a) cited this disease and causal organism in Russian hemp. Ghani & Basit (1975) described bacterial wilt of Pakistani *Cannabis* caused by an unidentified bacterium. Sands *et al.* (1987) described an *Erwinia* causing a vascular wilt in Oxford, Mississippi, affecting at least 1% of plants in the plantation.

Symptoms begin as dull green leaf spots, followed by sudden wilting and necrosis of leaves and stems. Some plants exhibit incipient wilting, others suffer complete wilting and die. Seedlings are especially susceptible. Bacteria ooze in a highly viscid fluid from cut stems. The bacterium normally grows in the vascular system of cucurbits (cucumber, pumpkin, squash, etc.).

CAUSAL ORGANISM

Erwinia tracheiphila (Smith 1895) Bergey, Harrison, Breed, Hammer & Huntoon 1923, *Zentralblatt für bakteriologie, parasitenkunde, infecktionskrankheiten und hygiene,* Abteilung II, 1:364–373.

=*Bacterium tracheiphilus* (Smith 1895) Chester 1897, =*Erwinia amylovora* var. *tracheiphila* (Smith) Dye 1968.

Description: Cultures grow slowly in nutrient agar. Facultatively anaerobic, gram (-) straight rods, size 1.0-3.0 x 0.5-1.0 µm, with four to eight peritrichous flagella. Nonproteolytic, nonlipolytic, acid (but no gas) produced from glucose, fructose, galactose, methylglucoside, and sucrose. Optimum growth temperature is 27–30°C.

DIFFERENTIAL DIAGNOSIS

Bacterial wilt can be confused with drought, nematodes, soil insects, some nutritional problems, or wilts caused by *Fusarium, Verticillium,* and *Sclerotium* fungi.

DISEASE CYCLE & EPIDEMIOLOGY

E. tracheiphila is transmitted via striped cucumber beetles (*Acalymma vittàta*). Bacteria overwinter in the gut of beetles and spread via insect excrement, entering through feeding wounds or leaf stomata. *E. tracheiphila* lives world wide. The *Erwinia* isolated in Mississippi grew best at temperatures above 28°C (Sands *et al.* 1987).

PHYTOPLASMAS

These organisms resemble bacteria, but lack cell walls when viewed under an electron microscope (Fig 3.1). Because of their superficial resemblance to mycoplasmas, botanists provisionally called them **mycoplasma-like organisms** (MLOs). Now they are called phytoplasmas.

SIGNS & SYMPTOMS

Typical phytoplasma symptoms include chlorosis, dwarfing, and **phyllody**, which is a distorted hypertrophy of leaves or flowers, also called a rosette or witch's broom. Phatak *et al.* (1975) described these symptoms near Delhi, India. Flowers on male plants proliferated into small branches causing a bushy appearance. Electron microscopy revealed phytoplasmas in phloem tissue of infected plants. Sastra (1973) previously described these symptoms from northeastern India but attributed the disease to viruses. Ghani & Basit (1975) reported "bunchy top" on Pakistani plants but blamed viruses. Bush Doctor (unpublished data 1983) found similar symptoms on feral hemp in Illinois.

CAUSAL ORGANISM

***Cannabis* Mycoplasma-like Organism**, Phatak *et al.*, *Phytopath. Zeit.* 83:281, 1975.

Description: Pleomorphic cells observed only in phloem elements of infected plants, membrane-bound, roughly circular and less than 0.5 µm in cross-section; cell membrane is electron dense as visualized by electron microscopy.

DIFFERENTIAL DIAGNOSIS

Symptoms can be confused with those caused by viruses, especially HSV. Some genetic disorders also produce phytoplasma-like symptoms.

DISEASE CYCLE & EPIDEMIOLOGY

Leafhoppers commonly vector phytoplasmas but Phatak *et al.* (1975) could not indict any specific insect. Phytoplasmas also spread by seed transmission.

CULTURAL & MECHANICAL CONTROL (see Chapter 9)

Exclude bacteria and phytoplasmas by employing sanitary measures (method 1). Avoid planting seeds from diseased plants; if seeds must be used, disinfest them with hot water (method 11). Long-term storage of seed (for at least 18 months) may kill *Pseudomonas* in seeds (Lucas 1975).

Losses from *Pseudomonas* increase in hemp planted on unworked soil (Ghillini 1951), so observe methods 1 (sanitation), 3 (weeding) and 2a (deep ploughing). Deep ploughing in the autumn is effective against some bacteria (e.g., *P. s.* pv. *tabaci*). Method 2c (soil solarization) works well against *A. tumefaciens*, but control of *Pseudomonas* is unpredictable (Elmore *et al.* 1997). Follow methods 7c (avoid excess humidity) and 8 (optimize soil nutrition).

Protect plants from insect vectors. To prevent crown gall, avoid wounding plants near the soil line. Bury carrot slices in soil to monitor for *A. tumefaciens*; in infected soil the carrots will develop galls. Genetic engineering has produced tobacco and tomato plants with resistance to *A. tumefaciens*. Sandru (1977) evaluated different cultivars for their susceptibility to *X. campestris*. 'Fibramulta' and 'Kompolti' were the most susceptible, the most resistant cultivars were 'USO-6' ('YUSO-6'), 'USO-13' and 'Afghan hemp.'

BIOLOGICAL CONTROL (see Chapter 10)

Agrobacterium radiobacter

BIOLOGY: A nonpathogenic soil bacterium that suppresses *A. tumefaciens* (Nogall®, Galltrol-A®, Norbac 84C®). *A. tumefaciens* and *A. radiobacter* are nearly identical except for a lack of Ti plasmids in the latter.

APPLICATION: Supplied as agar plates or aqueous suspensions, to be mixed into nonchlorinated water. Plates can be stored for 120 days in a cool (8–10°C), dark place. Apply as a pre-planting dip, or spray on seeds and seedling roots, or apply as a post-planting soil drench.

CHEMICAL CONTROL (see Chapter 11)

For chemical treatment of infested *Cannabis* seed, Kotev & Georgieva (1969) recommended seed soaks in formalin (1:300 dilution). Maude (1996) soaked infested seeds in streptomycin for two hours, using a 500 µg ml^{-1} solution against *Xanthomonas campestris* and 1000 µm ml^{-1} against *Pseudomonas syringae*. Unfortunately, streptomycin is phytotoxic and may stunt seedlings. Harman *et al.* (1987) rid *Brassica* seeds of *Xanthomonas campestris* by soaking seeds for 30 minutes in Nyolate®. This hospital disinfectant is packaged as two components. Mix ten parts of component A (2.7% sodium chlorite) and three parts of component B (15.1% lactic acid) in 90 parts water. After soaking, rinse seeds in water and air-dry.

Drenching soil with Bordeaux mixture or fixed copper will control wildfire and *Xanthomonas* disease but may cause phytotoxicity in *Cannabis*. Copper and streptomycin have been used as foliar sprays, but the copper sprays are largely ineffective. Treating phytoplasma-infected plants with tetracycline causes temporary remission of disease but does not cure plants (Lucas 1975).

Extracts of *Cannabis* suppress many bacteria (Krejcí 1950, Ferenczy 1956, Ferenczy *et al.* 1958, Kabelik *et al.* 1960, Gal *et al.* 1969, Velicky & Genest 1972, Velicky & Latta 1974, Klingeren & Ham 1976, Fournier *et al.* 1978, Braut-Boucher *et al.* 1985, Vijai *et al.* 1993, Krebs & Jäggi 1999). In the Ukraine, aqueous extracts of feral hemp were sprayed on crops to protect them from bacteria (Zelepukha 1960, Zelephukha *et al.* 1963). The extracts worked best against gram (+) *Corynebacterium* species, worked okay against gram (-) *Xanthomonas* species, and were least effective against gram (+) *Bacillus* and gram (-) *Pseudomonas, Erwinia, Bacterium,* and *Agrobacterium* species (Bel'tyukova 1962). Aqueous extracts also inhibited soft rot of potatoes, caused by gram (-) *Erwinia carotovora* and other bacteria (Vijai *et al.* 1993). An aqueous extract of hemp (50 g dried flowers soaked in 1 l water) inhibited *E. carotovora* more than the essential oils extracted from cinnamon oil (*Cinnamomum zeylanicum*), thyme (*Thymus vulgaris*), and peppermint (*Mentha piperita*) (Krebs & Jäggi 1999).

WEED PLANTS

Several plants compete with *Cannabis*, despite the oft-repeated claim by Dewey (1914) that "hemp smothers all weeds." Giant ragweed (***Ambrosia trifida***) and bamboo (***Bambusa, Dendrocalamus,*** and ***Phyllostachys*** species) can outgrow *Cannabis* and shade it from sunlight.

Cannabis subdues slower-growing plants after **canopy closure**—when leaves of adjacent *Cannabis* plants mesh together and shade the soil. Weed suppression works best when the canopy closes early, when plants are approximately 50 cm tall. This depends on proper seeding density. At lower seeding densities (<40 kg ha^{-1} or <36 lbs/acre) the canopy of fibre hemp closes late (not until plants are 100–150 cm tall), allowing weeds to proliferate (Bocsa & Karus 1997).

Short-statured seed cultivars (e.g., 'FIN-314') may never close their canopy, so they become susceptible to weed pressure. Grasses can choke the growth of young seedlings if hemp is sown late. Any weed may cause problems in thin stands, where gaps exist in the canopy.

Some weeds damage *Cannabis* by harbouring pests or spreading diseases. For instance, wild mustard (***Brassica kaber***) is frequently infected with a fungus, *Sclerotinia sclerotiorum*. The weed disease spread to hemp crops in Manitoba, Canada (J. Moes, unpublished data 1996).

Vines damage hemp by climbing stems, binding plants together, and sometimes girdling branches. Dewey (1914) and Dempsey (1975) singled out several bindweeds—***Calystegia sepium, Convolvulus arvensis,*** and ***Polygonum convolvulus,*** which are noxious cousins of morning glories. Dewey (1914) noted that bindweed seeds are the same size as hemp seeds, and difficult to separate by screening. Seed contamination commonly occurs in Chinese hemp seed (Clarke, pers. commun. 1999). Bindweeds are deep-rooted, difficult to pull up, and can reproduce from roots remaining in soil. In 1988 the USDA imported a caterpillar, ***Tyta luctuosa***, which defoliates wild morning glory and (supposedly) nothing else.

Cannabis itself can become a difficult weed, despite assurances by Hackleman & Domingo (1943): "Since the crop is an annual and is harvested before seed is formed, there is not much danger of its becoming troublesome." These researchers did not look out their laboratory windows! A field survey conducted in the same Illinois county located nearly 900 stands of feral hemp (Haney & Kutscheid 1975). Feral hemp grew in central Illinois as early as 1852 (Brendel 1887). Schweinitz (1836) found feral hemp growing near Bethlehem, Pennsylvania in the early 1800s. The feral hemp described in these early reports was probably of European ancestry, what Dewey (1902) called 'Smyrna' hemp. In the late 1800s, European landraces were replaced by Chinese hemp, from seeds obtained by American missionaries (Dewey 1902). Chinese hemp became the breeding stock for 20th century American hemp (Dewey 1927). Descendants of Dewey's Chinese hybrids now grow wild across central North America. Indeed, Chinese hemp has grown wild in China for almost 2000 years (Li 1974). Haney & Kutscheid (1975) estimated that feral hemp covered 5–10 million acres (2–4 Mha) in the USA. At that time, the acreage was expanding. Since then, law enforcement has spent millions to eradicate feral hemp (Bush Doctor 1986a).

WEED CONTROL

Do not allow weeds to flower and set seed—"One year seeds, eleven years weeds." If weeds do seed, the weed seeds must be separated from hemp seeds prior to sowing. This is accomplished by careful screening. Some weed seeds in soil can be killed using soil solarization (method 2c in Chapter 9). Solarization works best against annual weeds with low heat tolerance (Elmore *et al.* (1997).

Weeds can be controlled by timely tillage of soil. Dodge (1898) recommended tilling soil twice—when hemp plants are 5 cm tall and again when plants are 10–20 cm tall. Some weed seeds require a brief exposure to light to break dormancy. Thus, German research suggests that tilling at night reduces weed germination. USDA researchers experimented with night tillage, driving tractors while wearing military-issue night goggles. Night tillage reduced the germination of small-seeded broadleaf species (ragweed, lambsquarters, pigweed, black nightshade), but did not reduce annual grasses and large-seeded weeds.

Quackgrass, ***Agropyron repens***, is a creeping perennial problem of late-planted hemp in New England, the north central states, and the prairie provinces. It spreads by underground rhizomes. Standard tillage equipment (rototillers, rotovators) merely cut and distribute the rhizomes. Alternatively, use Danish s-tines, which pull rhizomes to the soil surface where they dry out and die.

ALLELOPATHIC PLANTS

Certain plants inhibit *Cannabis* by secreting allelochemicals, which are toxic. Allelochemicals ooze from roots and emanate from leaves as a gas or liquid. Leaf trichomes act as repositories for these "natural herbicides." Heavy rainfall collapses the trichomes, releases their contents, and washes the contents to the ground. Allelochemicals inhibit seed germination and plant growth. Good (1953) noted that *Cannabis* grows poorly near *Spinacia oleracea, Secale cereale, Cicia sativa* and *Lepidium sativum.* Peculiarly, some plants are "anti-allelopaths"—extracts of *Geranium dissectum* actually improve germination of hemp seeds (Muminovic 1990).

The ability of hemp to serve as a "smother crop" is usually attributed to physical characteristics, such as its dense canopy and root system. But *Cannabis* also produces allelochemicals—water extracts of *Cannabis* inhibit the growth of other plants (Stupnicka-Rodzynkiewicz 1970, Vimal & Shukla 1970, Srivastava & Das 1974, Pandey & Mishra 1982). Haney & Bazzaz (1970) speculated that terpenoids produced by *Cannabis* may suppress the growth of surrounding vegetation. Pulegone, 1,8-cineole, and limonene suppress plants (Asplund 1968), and these terpenoids are produced by *Cannabis* (Turner *et al.* 1980). Latta & Eaton (1975) suggested that cannabinoids may also play a role; they measured increased THC production in *Cannabis* competing with weeds.

Could allelochemicals in *Cannabis* inhibit the growth of subsequent crops grown in a crop rotation with hemp? Probably not. Barley is another smother crop, it produces an arsenal of allelochemicals, yet it serves in a variety of cropping systems and rotation schemes (Overland 1966).

Allelochemicals may overwinter better in no-till situations, where crop residues remain in place on the field. Muminovic (1991) found that a mulch of *Cannabis* straw inhibited the growth of *Agropyron repens,* a noxious weed. Allelochemicals may intensify in crops being attacked by insects; the insects feed on leaves and concentrate the allelochemicals in their frass (Silander *et al.* 1983).

PARASITIC PLANTS

Two plants, dodder and broomrape, are genuine plant parasites. They have no chlorophyll and depend on other plants for survival. Dodder and broomrape sink specialized roots (haustoria) into the host's xylem and phloem to withdraw fluids and nutrients.

DODDER

Five species of this alien-looking plant have been pulled off hemp around the world. Dodders are related to morning glories. But while morning glories spiral up fence posts, dodders have no chlorophyll and must twine around living plants. In addition to causing mechanical damage, dodders vector viruses as they grow from plant to plant.

SIGNS & SYMPTOMS

Dodders arise as conspicuous tangles of glabrous yellow filaments (Plate 74), bearing vernacular names such as "gold thread," "hair weed," "devil's ringlet," and "love vine" (Lucas 1975). Robust specimens girdle branches and pull down hosts.

1. ***Cuscuta campestris*** Yunker 1932
   This is the common field dodder in North America. It lives on marijuana in Oregon and Pennsylvania, and on feral hemp in Illinois (Bush Doctor, pers. commun. 1987). *C. campestris* has been introduced into Europe (Ceapoiu 1958). Like most dodders, *C. campestris* infests a wide range of plants, mostly herbaceous dicots such as clover (*Trifolium*) and alfalfa (*Medicago*).
2. ***Cuscuta europea*** L. 1732
   This species lives in Europe (Kirchner 1906, Transhel *et al.* 1933, Ferraris 1935, Flachs 1936, Ciferri 1941, Dobrozrakova *et al.* 1956, Barloy & Pelhate 1962, Serzane 1962, Vakhrusheva 1979, Gutberlet & Karus 1995). It frequently infests hops (*Humulus*) and nettles (*Urtica*).
3. ***Cuscuta pentagona*** Engelman
   This may be a subspecies of *C. campestris.* It is native to North America but now lives everywhere (Parker &

Riches 1993). *C. pentagona* attacked hemp in Serbia (Stojanovic 1959).

4. ***Cuscuta epilinum*** Weihe 1824
   This species has been reported on *Cannabis* in Italy (Ferraris 1935) and Kenya (Nattrass 1941). It normally attacks flax, *Linum usitatissimum* (Parker & Riches 1993).
5. ***Cuscuta suaveolens*** Seringe 1840
   Dewey (1914) reported "a Chilean dodder" (he called it *Cuscuta racemosa*) tangling hemp in California. This species normally infests *Trifolium* and *Medicago* species. It originated in Chile but now occurs across the Americas and Europe.

Description: Dodder stems are threadlike, twine up host stems, 1–2 mm wide, orange-yellow in colour, sometimes tinged with red or purple, sometimes almost white. Tendrils arise at frequent intervals along the stems, twining around host stems. Leaves reduced to minute scales, arising opposite tendrils. Separating *Cuscuta* species by their flowers is not easy. They all produce small (1–4 mm), short-lived flowers. The species cited on hemp produce two styles per flower; some stigmas have knobs, some do not. Even experts make mistakes in identification, as exampled by Rataj (1957), who cited *Cuscuta major*, a nonexistent species. Seeds are reddish-brown, 1–2 mm long. For more information see Yunker (1932) or Parker & Riches (1993).

DODDER DISEASE CYCLE

Each plant produces thousands of seeds from late summer until frost. Seeds survive in soil for up to ten years. Seeds germinate in late spring after soil is well-warmed. Seedlings grow a rudimentary root and send up a heliotropic (sun-seeking) tendril, which gropes for a host. After the tendril makes contact with a host, the dodder root atrophies and the parasite escapes from the soil (Fig 6.6). Dodders are annual plants.

DODDER CONTROL

Observe the following techniques described in Chapter 9: method 1 (burning, especially if *Cuscuta* has shed seed), 2a (deep ploughing), 2b (steam-sterilize soil), 4 (plant late after the spring flush of *Cuscuta* germination), 7c (avoid moist, shaded ground near streams or rivers), 9 (cut and pull, no fragments should remain on stems or ground). Ceapoiu (1958) said hemp seed from infected crops should not be planted. If planting such seed is necessary, carefully screen all alfalfa-sized dodder seed. Rotate with crops resistant to dodder—bean (*Phaseolus vulgaris*), squash (*Cucurbita pepo*), tomato (*Lycopersicon lycopersicum*), and cotton (*Gossypium hirsutum*) are resistant to *C. campestris*.

The Chinese released a dodder-specific biocontrol fungus, ***Colletotrichum gloeosporiodes* f.sp. *cuscutae*** (Lubao 1®). New formulations and hypervirulent strains of the fungus, Lubao 1 S22® and Lubao 2®, are now in use (Quimby & Birdsall 1995). Russian researchers have controlled dodder with an *Alternaria* species (Parker & Riches 1993).

Figure 6.6: Disease cycle of dodder, *Cuscuta campestris* (McPartland & Clarke redrawn from Agrios 1997).

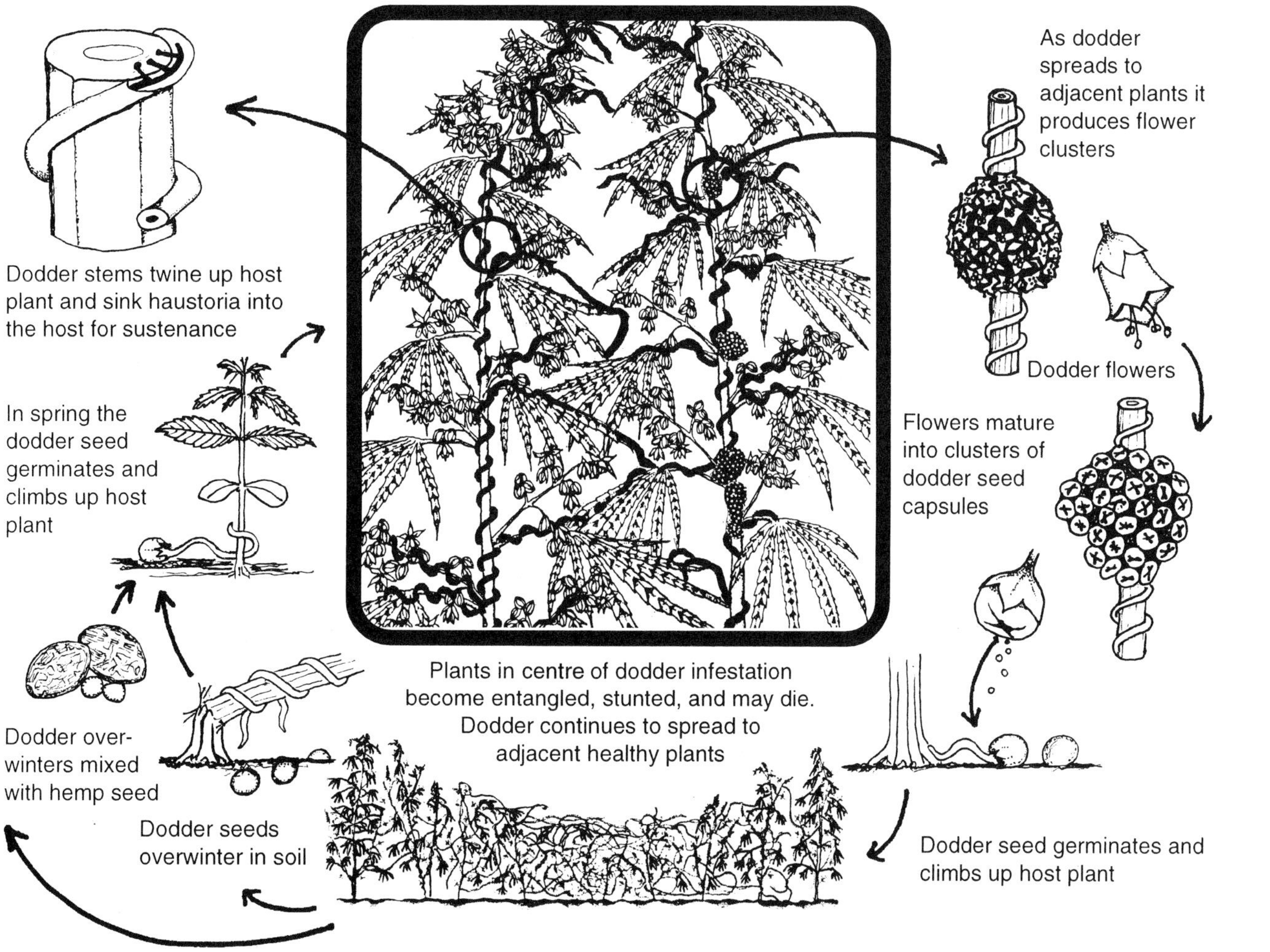

BROOMRAPE

Host plants become stunted, sickly, and die prematurely. Broomrapes starve their hosts, provide portholes for root rot fungi, and vector viruses. Dewey (1914) called broomrape "the only really serious enemy to hemp." Broomrape infestation seriously threatened the hemp industry in Kentucky (Dewey 1902, Garman 1903, Musselman 1994). According to Dodge (1898), "the Germans call this terrible weed the *Hanfmörder*."

Broomrapes do their damage underground (Fig 6.7). Only briefly do they appear above ground, sending up shoots which quickly flower and seed (Plate 75). Strangely, Fournier & Paris (1983) detected THC and CBD in flowers and seeds of broomrape parasitizing French hemp.

1. BRANCHED BROOMRAPE

Orobanche ramosa L. 1753

=*Phelipaea ramosa* (L.) C.A. Meyer

Description: Broomrape shoots arise near the base of hemp stems. They resemble short shoots of pale, branching asparagus. Shoots average 10–20 cm in height (range 8–45 cm), and bear brownish-yellow, scalelike leaves 3–8 mm long (Plate 75). They produce five to seven pale blue-purple flowers, with a tubular corolla supported by both a bract and bracteole, *less* than 20 mm long. Anther filaments not densely hairy, arising from *base* of corolla. Flowers mature into resinous capsules filled with 600–800 tiny, oblong seeds (each only 260 µm long), see Fig 6.7.

Kreutz (1995) called this species "hemp broomrape," and noted many experts place it in the genus *Phelipaea* or *Phelipanche*. Dempsey (1975) considered *O. ramosa* a major pest. Rataj (1957) described *O. ramosa* "raging" on hemp in the Czech Republic. Barloy & Pelhate (1962) considered a combination of *O. ramosa* and *Fusarium solani* the greatest hazard to hemp in southern France.

Several *O. ramosa* subspecies together attack over 60 plants (Lucas 1975). *O. ramosa* subspecies *ramosa* parasitizes hemp, tobacco (*Nicotiana tabacum*), tomato (*Lycopersicon lycopersicum*), and *Solanum* species. The *ramosa* subspecies is further divided into races; Paskovic (1941) said the Italian race virulently infested all hemp cultivars, whereas the Russian race only caused problems in Russian hemp cultivars, not Ukrainian or Italian cultivars.

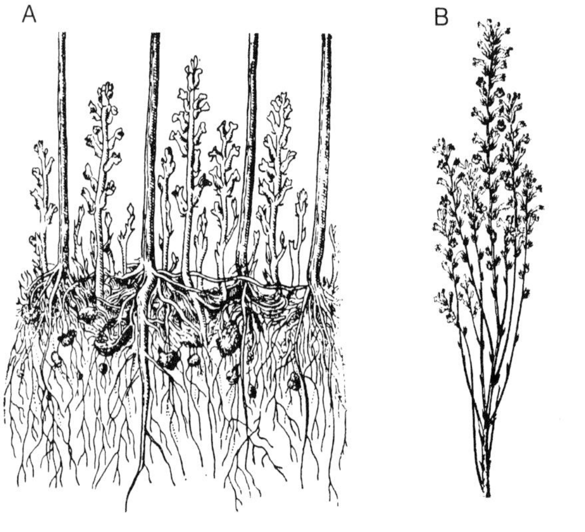

Figure 6.7: *Orobanche ramosa.*
A. Roots of hemp plants parasitized by *O. ramosa*, which is sending up shoots; B. Fully-flowering shoot (from Lesik 1958).

O. ramosa originated around the Mediterranean, but now ranges from England to South Africa, and across Asia (Kreutz 1995). *O. ramosa* was introduced into Kentucky around 1880 from hemp imported from China (Garman 1903). It still persists in Kentucky, California, and Texas (Musselman 1994).

2. PERSOON'S BROOMRAPE

Orobanche aegyptiaca Persoon 1806

Description: *O. aegyptiaca* closely resembles *O. ramosa*, but tends to be taller (20–45 cm) with larger leaves (5–12 mm). The corolla is *longer*, usually 20–35 mm; anther filaments densely hairy.

Parker (1986) reported this species on *Cannabis*. *O. aegyptiaca* lives in the eastern Mediterranean region, extending south into Africa and east to India. It attacks the same plants as *O. ramosa*. Future research may prove it to be a subspecies of *O. ramosa* with hairy anthers.

3. TOBACCO BROOMRAPE

Orobanche cernua Loefling 1758

=*Orobanche cumana* Wallroth

Description: Shoots robust, rarely branched, 20–30 cm tall, leaves 5–10 mm long, bearing numerous flowers. Deep blue-purple flowers do not have bracteoles, corolla 12–20 mm long, anthers arise well up the corolla at least 4 mm distal from base.

This species parasitizes *Cannabis* in Romania (Ceapoiu 1958) and India (Marudarajan 1950). Parker & Riches (1993) used hemp as a trap crop to induce *O. cernua* germination. The species is a serious pest of tobacco, sunflower (*Helianthus annuus*), and *Artemisia* species. It lives around the Mediterranean region, east to the Arabian Sea, India, and Australia.

BROOMRAPE DISEASE CYCLE

O. ramosa seeds can remain dormant in soil for up to 13 years (Garman 1903). Root exudates from hosts stimulate seed germination. Germinating seeds produce a "radicle" 3–4 mm long which drills haustoria into *Cannabis* roots, penetrating xylem and phloem. Ten or more parasites may be found attached to a single host. Broomrape spends most of its life cycle underground and unseen. Shoots emerge from the soil in summer and quickly flower and go to seed. Each parasite can produce up to 500,000 seeds (Barloy & Pelhate 1962). Mature seed capsules open and disperse the dust-like propagules to wind and surface water. The sticky seeds also adhere to *Cannabis* seeds; Berenji & Martinov (1997) reported an *Orobanche* species infesting seed hemp in Yugoslavia.

CULTURAL & MECHANICAL CONTROL (see Chapter 9)

Observe methods 1 (sanitation) and 2b&c (sterilize or pasteurize the soil). Method 11 (cleaning seed) requires *meticulous* attention because broomrape seeds easily adhere to hemp flowers (Dewey 1914, Dempsey 1975). Garman (1903) killed broomrape seeds by immersing contaminated hemp seeds in a 60°C water bath for ten minutes; this treatment did not harm hemp seeds.

Hand-pulling broomrape (method 10) prevents broomrape from setting seeds, but crop damage has already taken place. Broomrape shoots arise all summer and the shoots go to seed within two weeks, so hand-pulling must be repeated frequently. Be careful not to pull up the hemp with the broomrape, especially in light, sandy soil. Method 8 (nutrient supplementation) offers hope: Berger (1969) said potassium supplements promote resistance; Abu-Irmaileh (1981) suppressed broomrape with manure, ammonium nitrate (NH_4NO_3), and ammonium sulphate [$(NH_4)_2SO_4$].

Method 6 (crop rotation) takes years to be effective. Breeding for resistance (method 5) is difficult. Chinese varieties are very susceptible, according to Bòcsa & Karus (1997). Barloy & Pelhate (1962) reported a loss of resistance in French crops. Senchenko & Kolyadko (1973) tested 26 cultivars for resistance and ranked them accordingly. Resistant cultivars include the Italian cultivar 'Carmagnola' (Paskovic 1941), the Hungarian cultivar 'Kompolti,' Russian cultivars 'JUS-1' and 'JUS-87,' and the monoecious cultivar 'Juznoja Odnodomnoja' (Dempsey 1975). Most of these cultivars are no longer available (see de Meijer 1995).

BIOLOGICAL & CHEMICAL CONTROL

Garman (1903) suggested controlling broomrape with ***Phytomyza orobanchiae*** Kaltenbach. This fly lives everywhere *O. ramosa* arises (Lekic 1974). Releasing as few as 500–600 flies per ha provides significant biocontrol (Kapralov 1974). Broomrape infested with *P. orobanchiae* maggots should be pulled up and laid to the side allowing the fly to mature and reproduce (Parker & Riches 1993). Unfortunately, the maggots are hyperparasitized by chalcids. Two fungi may provide biocontrol of broomrape, ***Fusarium lateritium*** and ***Fusarium oxysporum* f.sp. *orthoceras*** (Bozoukov & Kouzmanova 1994). The latter fungus achieved commercial use in the former USSR (Quimby & Birdsall 1995).

Trap crops reduce broomrape seed banks in the soil. Traps for *O. ramosa* include bean (*Phaseolus vulgaris*) sorghum (*Sorghum bicolour*) (Parker & Riches 1993), or white mustard (*Brassica alba*) (Lucas 1975). For *O. cernua* use cowpea (*Vigna unguiculata*) or sorghum (Parker & Riches 1993). Tobacco farmers in Bulgaria allow geese to graze in infected fields. Two geese per ha eliminate broomrape (Kircev in Lucas 1975). Fumigating soil kills seeds, but this draconian measure is dangerous and expensive.

ANIMALS OTHER THAN INSECTS AND MITES

PROTOZOA

Phytomonas is a newly-discovered genus of protozoans. These species principally infect lactiferous ducts, but also invade xylem and phloem vessels. Thus far no *Phytomonas* species have been discovered in hemp, even though *Cannabis* produces vestigial lactifers (Zander 1928, Furr & Mahlberg 1981, Mesquita & Santos Dias 1984, Lawi-Berger *et al.* 1984). *Phytomonas* species attack closely-related plants in the Urticaceae (Dollet 1984). Insects vectoring these protozoans, such as *Nysius* and *Stenocephalus* species, are common *Cannabis* pests. *Phytomonas* species are found between latitudes 50° north and 50° south, girdling the globe in distribution.

Please note: until recently, several oömycetes were misidentified as fungi (e.g., *Pythium* and *Pseudoperonospora* species). Because of this historical tradition, oömycetes are still studied by mycologists and described in Chapter 5.

SOWBUGS & PILLBUGS

Members of the arthropod (but not insect) order Isopoda are also called called woodlice. Frank (1988) described sowbugs killing *Cannabis* sprouts. These crustaceans have slate-grey, segmented, robust bodies. They are omnivorous and normally feed at night on decaying vegetable matter.

Sowbugs and pillbugs are controlled with cultural methods 1 (sanitation), 2a (spring tillage before planting), 3 (weeding), and 9 (hand removal). Frank & Rosenthal (1978) trapped pillbugs and sowbugs by placing pieces of cardboard on soil. The pests gather in cool, moist areas during the heat of the day; you will find them under the cardboard, congregated for easy picking. Frank (1988) rid pillbugs and sowbugs with a rotenone soil drench.

MILLIPEDES, SYMPHYLANS & SPIDERS

According to Yepsen (1976), millipedes cause problems during dry spells in the summer. Bush Doctor (unpublished data 1986) found dark brown millipedes, probably ***Oxidus gracilis*** (Koch), chewing roots of feral hemp in Illinois.

Garden symphylans, ***Scutigerella immaculata*** (Newport), are slender, pearly-white, centipede-like arthropods, 8 mm long. Symphylans have 12 pairs of legs; centipedes have 15. Symphylans feed on plant roots and kill young seedlings. They are troublesome in glasshouses and hop yards in the Pacific Northwest. They survive for years in damp locations such as decaying logs and deep humus.

Spiders are close relatives to mites. They normally feed on other arthropods. Bantra (1976) reported finding ***Argyodes*** and ***Cyclosa*** species on Indian *Cannabis*. We do not kill spiders... most do more good than harm.

Control millipedes by applying peat compost (Yepsen 1976). Symphylans, on the other hand, are attracted to compost (move compost piles *away* from gardens). Hops growers in the Willamette valley reduce symphylans with soil solarization and repeated rototilling (Jackson, pers. commun. 1997). Eliminate millipedes with nicotine soil drenches and repel symphylans by drenching soil with garlic or tobacco tea (Yepsen 1976).

SLUGS & SNAILS

Slugs can be vexing in humid corners of the globe. Selgnij (1982) described slugs attacking young sprouts in Watsonville, California. Bush Doctor (pers. commun. 1989) found grey garden slugs destroying seedlings in Pennsylvania, ***Deroceras reticulatum*** (Miller) or ***Deroceras agreste*** (L.). Adults are grey-white or flesh-coloured with grey markings, growing up to 45 mm long. They can cause serious damage until plants are 1 m tall (Plate 76).

Most North American slug pests were introduced from Europe. The European giant garden slug (***Limax maximus*** L.) is a frequent offender in vegetable gardens. Other aggressive slugs include ***Agriolimax*** and ***Arion*** species. The brown garden snail, ***Helix aspersa***, is a problem in California.

Slugs cause two peaks of maximum damage—in April–May and September–November. One or two generations arise per year. They overwinter as eggs. Slugs prefer wet weather and crops planted adjacent to meadows, pastures, or woods. They usually feed at night. They are hermaphroditic—each slug is equipped with both male and female reproductive organs. They can mate with themselves if no other slugs are around.

MECHANICAL & BIOCONTROL OF SLUGS

Protect small gardens from slugs with a circular strip of copper (the best), zinc, or steel. The metal interacts chemically with slug slime, and repels them, if the metal is kept clean. Many commercial brands are only 5 cm broad, yet Ellis & Bradley (1992) suggested using a band 7–10 cm wide.

Chervil (*Anthriscus cerefolium*) serves as a trap crop for slugs (Ellis & Bradley 1992). Yepsen (1976) repelled slugs with a border planting of common wormwood (*Artemisia absinthium*). Many organic gardeners let ducks or chickens roam their gardens to chug slugs. Researchers are working on *Tetanocera* species, flies that parasitize various slugs.

Phasmarhabditis hermaphrodita (=*P. megidis*)

BIOLOGY: A soil nematode (Nemaslug®) that infests slugs

(mostly *Deroceras* and *Arion* species) and snails (*Helix aspersa* and many others). It does best in moist but not waterlogged soil that is loamy or sandy, not clay, and soil temperatures between 5–20°C (15°C is optimal).

APPEARANCE & DEVELOPMENT: The nematodes are approximately 1 mm long. They actively move through soil in search of slugs. Within three to five days of infection, slugs get sick and stop feeding; they die ten days later.

APPLICATION: Supplied as infective juveniles in packets. They can only be stored for one or two days in a cool (2°C), dark place. *P. hermaphrodita* must be poured or sprayed into moist soil and not exposed to UV light (apply in evening), and soil must be kept moist for at least two weeks after application. The usual recommended dose is 300,000 nematodes per m^2, every six weeks. *P. hermaphrodita* works best on young slugs and snails, so apply when juvenile slugs are present.

Rumina decollata

BIOLOGY: A decollate snail that preys on snail eggs and immature garden snails. It does best in moderate humidity and warm temperatures.

APPEARANCE & DEVELOPMENT: *R. decollata* resembles elongate versions of its prey, growing to 25–50 mm in length. It feeds at night. During the day it hides under rocks and well-drained rubble. The species reproduces slowly and may take one or two years to control large pest populations.

APPLICATION: Supplied as dormant adults. Store up to a week in a cool (8–10°C), dark place. Release one to ten snails per m^2 in moist, shaded areas with adequate organic matter. Occasionally they eat seedlings, so do not release in a freshly seeded area. To protect native molluscs, the sale of snails is restricted in some areas.

CHEMICAL CONTROL OF SLUGS

Setting a pan of stale beer in the garden, edges flush with the ground, is a time-honoured slug trap. Slugs wallow into the beer and drown. For slugs not falling for this old saw, set out toxic bait—moisten bran with beer then mix with metaldehyde, a molluscicide. Sluggo® and Escar-Go® kill slugs with ferric phosphate (the form of iron utilized in nutritional supplements). Other commercial slug traps are available. You can handpick slugs off plants at night. Kill them with a spritz of saturated salt solution or Bordeaux mixture.

BIRDS

Any ornithologist worth her weight in binoculars knows that birds devour *Cannabis* seeds. Darwin (1881) noted that hemp seeds can germinate after passing through the digestive tract of birds, if not cracked by beak or gizzard. According to old folk tales, birds sing better songs after ingesting a few hemp seeds. The American birdseed industry used 2 million tons of *Cannabis* seed per year before the Marihuana Tax Act of 1937. After lobbying by the birdseed industry, sterilized seed was exempted from restriction. Parkinson (1640) and Martyn (1792) reported, "Hemp seeds are said to occasion hens to lay a greater quantity of eggs." Lindley (1838) claimed, "Hemp seed... has the very singular property of changing the plumage of bullfinches and goldfinches from red and yellow to black if they are fed on it for too long a time or in too large a quantity."

Early reports from Kentucky described the passenger pigeon (***Ectopistes migratorius***) feeding on hemp seeds (Allen 1908). Feral hempseed is the most important food of mourning doves (***Zenaidura macroura***) in Iowa (McClure 1943). Captive doves thrive for long periods of time on hemp alone. Bobwhite quail (***Colinus virginianus***) and ringtail pheasant (***Phasiánus cólchicus***) depend on feral hemp seed in the American midwest (Hartowicz & Eaton 1971). The dependence of game birds on feral hemp has led wildlife agencies to oppose police eradication efforts (Vance 1971).

Birds caused substantial losses in Tasmania (Lisson & Mendham 1995). Sorauer (1958) cited many seed-eating European birds including the hemp linnet (***Carduelis cannabina***), magpie (***Pica pica***), starling (***Sturnus vulgaris***), common purple grackle (***Quiscalus quiscula***), tree sparrow (***Passer montanus***), English sparrow (***Passer domesticas***), nuthatch (***Sitta europaea***), lesser spotted woodpecker (***Dendrocopus minor***), and turtledove (***Streptopelia turtur***). In Nepal, Mattheissen (1978) described Himalayan goldfinches (***Carduelis spinoides***) feeding on *Cannabis* seeds.

BIRD CONTROL

French hemp breeders treat seeds with anthraquinone, a bird repellent (Karus 1997). Netting protects seedlings in small seed plots. Rataj (1957) lamented, "Birds cause the most serious damage to hemp seed. In spite of their harmfulness we lack protective measures against this danger. It is usually necessary to guard seed-plots and drive the birds away." Chinese farmers keep a constant daytime watch, hang brightly coloured cloth, or use firecrackers to scare away birds (Clarke 1995). Lisson & Mendham (1995) deployed owl decoys and used "bioacoustics"—the broadcasting of bird distress calls over loudspeakers. Stretching iridescent ribbons across fields (which vibrate in wind and make annoying sounds), floating helium balloons with large owl eyes, ringing "bird bells," firing "bird cannon," deploying scarecrows that spray water when set off by motion detectors... all work, for a while. Spraying birds with soapy water repels birds but may kill them by removing protective oils. Some birds, such as sea gulls, die after swallowing pieces of Alka Seltzer®. Synthetic bird repellents and avicides are also available.

MAMMALS

Antibacterial terpenoids and THC in *Cannabis* may harm domesticated ruminants. Cattle in India (***Bos indicus***) refused to eat cattle feed mixed with ganja (Jain & Arora 1988). Driemeier (1997) described Brazilian cattle (***Bos taurus***) dying after consuming bales of dried marijuana. Cardassis (1951) reported the deaths of horses (***Equus caballus***) and mules (***E. caballus*** crossed with ***Equus asinus***) who grazed on *Cannabis* in Greece.

Conversely, Siegel (1989) described cattle and horses happily feeding on flowering tops in Hawai'i. Patton (1998) reported cattle and horses eating hemp seed meal. Goats (***Capra*** species) devour *Cannabis* without ill effects (Clarke, pers. commun. 1996).

Dogs (***Canis familiaris***) have sickened and occasionally died from eating hashish and marijuana (Meriwheter 1969, Clarke *et al.* 1971, Goldborld *et al.* 1979). Dogs also suffer allergic reactions from breathing marijuana dust (Evans 1989). Ferrets (***Mustela furo***) become comatose and die after ingesting marijuana (Smith 1988). Cats (***Felis catus***) have no use for *Cannabis*, but may dig up transplanted seedlings, attracted to fish emulsion in soil mixes (Bush Doctor, pers. commun. 1993).

Non-domesticated mammals cause more problems. Some are protected by wildlife laws and cannot be killed except under rare circumstances. Wildlife pests include deer (***Odocoileus*** species) in the USA (Frank & Rosenthal 1978, Alexander 1987, Lassen 1988, Siegel 1989), dik-dik (***Madoqua*** species) in Kenya (Quinn, pers. commun. 1996), and unidentified monkeys in South America (Siegel 1989).

Other mammals are considered vermin and not protected by wildlife laws. Rabbits (***Sylvilagus*** species) cause problems in the USA (Hartowicz & Eaton 1971, Alexander 1987, Lassen 1988). During droughts in Hungary, rabbits strip bark from the base of stalks in search of moisture (Clarke, pers. commun. 1996, Plate 77). Siegel (1989) cited raccoons (***Procyon lotor***) and ***Rattus rattus*** attacking *Cannabis*. Rats strip bark from stalks to build nests. Field voles (***Microtus*** species) and hamsters (***Cricetus cricetus***) feed on sown hemp seeds in Europe (Sorauer 1958). Mice (***Mus musculus***) have broken into police vaults to feed on seeds in confiscated marijuana (Siegel 1989). In the USA, gophers (***Citellus*** species), moles (***Scalopus*** species), voles (***Microtus*** species), and groundhogs (i.e., woodchucks, ***Marmota monax***), attack plants (Selgnij 1982, Alexander 1987, Lassen 1988). Groundhogs can quickly destroy a young *Cannabis* stand. They seem to feed on some plants and roll around in the rest.

MAMMAL CONTROL

Contact repellents such as Hinder® and Deer Away® can be applied to stalks making them unpalatable to deer. Repel deer with scents—ring the field perimeter with small mesh bags filled with bloodmeal. For repellents to be effective, you must change scents every two or three weeks. Rotate with hair, urine, and faeces from deer predators (e.g., dogs, humans), or bars of scented soap. Hang at deer breast height (1 m above ground). Scents work best when hung before males mark their territory in early spring.

Fencing deer requires a 2.5 m tall barricade, or a three-line electric fence. Groundhogs can be fenced with wide poultry netting, 75–90 cm above ground and 30–45 cm buried. A ring of tin (i.e., cans with the tops and bottoms removed) protects seedlings from most rodents. Discourage rabbits by removing refuge sites such as brush piles, stone and trash heaps, and weed patches. Rabbits shy from bloodmeal. They can be humanely trapped in cages for removal. Set empty pop bottles in mole holes; bottles whistle in the wind and vibrate moles away (they work better than those silly plastic windmills).

Planting spurge (*Euphorbia marginata*) will stupefy moles (Yepsen 1976). Moles and rodents are repelled by castor oil (Frank & Rosenthal 1978). The biocontrol bacterium ***Salmonella enteritidis* var. *issatschenko*** kills rodents. Mice and rats can be eliminated with live traps, spring traps, or poisons. Warfarin and cholecalciferol are much less dangerous than single-dose poisons (e.g., sodium fluoroacetate, red squill, and strychnine, all of which are produced by plants). Single-dose rodenticides should be avoided for the sake of wandering dogs.

"The use of pesticides is an act of desperation in a dying agriculture. It's not the overpowering invader we must fear but the weakened condition of the victim." —William Albrecht

Chapter 7: Abiotic Diseases

Abiotic (nonliving) causes of hemp disease and injury include imbalances of soil nutrients (either deficiencies or excesses), climatic insults (drought, frost, hail), air pollution, soil toxins, pesticides, and genetic factors. These abiotic problems also predispose plants to infectious diseases caused by other organisms. Drought-stressed plants, for instance, become more susceptible to pathogenic fungi (McPartland & Schoeneweiss 1984).

Symptoms of abiotic diseases may resemble disease symptoms caused by living organisms. Conversely, symptoms caused by disease organisms have been blamed on abiotic causes. Early observers of Dutch elm disease in Europe, for example, attributed elm death to poison gas released during WWI. Some hemp diseases have unknown causes, such as "grandine" disease.

NUTRITIONAL IMBALANCES

Poor soil causes most abiotic problems. This chapter treats soil fertility in three sections: 1) symptoms caused by nutritional imbalances, 2) correction of imbalanced field soils, and 3) correction of glasshouse/growroom soils. The latter two sections come from different perspectives: the field soil section emphasizes "soil balancing" to *prevent* nutrition imbalances, whereas the glasshouse section emphasizes *treatment* of imbalances after symptoms arise in plants.

The section on field soils was contributed by Bart Hall, a soil scientist with nearly 30 years experience as an organic agronomist. His company, **Bluestem Associates**, is based in Lawrence, Kansas, USA (tel: 785-865-0195), and provides consulting for organic, biological, and conventional producers worldwide.

SYMPTOMS

Generally, **deficiency** symptoms of mobile nutrients (N, P, K, Mg, B, Mo) begin in large leaves at the bottoms of plants. Deficiency symptoms from less mobile nutrients (Mn, Zn, Ca, S, Fe, Cu) usually begin in young leaves near the tops. The most common deficiencies arise from shortages of macronutrients—N, P, K, Ca, Mg, and S.

Too much of a nutrient may cause **toxicity**. Symptoms of toxicity often mimic symptoms of deficiency. Micronutrients cause most toxicities, especially Mn, Cu, B, Mo, and Cl. Excesses of anything, however, may stress plants.

Obviously, symptoms may be misleading, because different problems give rise to identical symptoms (Bennett 1993). Thus, observing symptoms is only the first step in determining a diagnosis. Always check soil pH. Soil pH directly affects the availability of nutrients (see Chapter 2 and Fig 2.1). The next step is soil testing and plant analysis.

Soil testing and plant analysis may uncover **hidden hunger** (Jones 1998). Hidden hunger is a nutrient deficiency without symptoms. Hidden hunger affects the quantity and quality of crop yield, but no plant abnormalities are seen. Plant analysis of *Cannabis* is problematic, because analysis guidelines have not been published for this plant. Bennett (1993) proposed general guidelines for most crop plants, and these are presented in Table 7.1.

Plant analysis requires precise sampling protocols. Samples should be mature leaves, from the upper part of the plant, free of disease, insect damage, and dead tissue. Collect a composite sample by selecting one or two leaflets from at least a dozen plants (Jones 1998). Collect samples from plants with symptoms and without symptoms, for comparison in the laboratory. Samples for soil testing should be collected at the same time and location as leaf samples are taken.

Table 7.1: Proposed guidelines for nutrient levels in *Cannabis* leaf tissue.[1]

Nutrient	Deficient level	Sufficient range	Toxicity level
N, %	<2.0	2.0–5.0	
P, %	<0.2	0.2–0.5	
K, %	<1.0	1.0–5.0	
Mg, %	<0.1	0.1–1.0	
Ca, %	<0.1	0.1–1.0	
S, %	<0.1	0.1–0.3	
Fe, ppm	<50	50–250	
Zn, ppm	15–20	20–100	>400
Mn, ppm	10–20	20–300	>300
Cu, ppm	3–5	5–20	>20
B, ppm	<10	10–100	>100
Mo, ppm	<0.1	0.1–0.5	>0.5
Cl, %	<0.2	0.2–2.0	>2.0
Si, %	<0.2	0.2–2.0	
Na, %	<1.0	1.0–10	
Co, ppm	<0.2	0.2–0.5	>0.5
V, ppm	<0.2	0.2–0.5	>1

[1] Adapted from general crop guidelines proposed by Bennett (1993).

Standard plant analysis is conducted in a laboratory. It is accurate, but slow and expensive. Alternatively, leaf sap can be quickly tested in the field using leaf tissue analysis (LTA). LTA field kits contain either chemically-treated paper strips or plastic vials with indicator reagents. Battery-powered ion meters for the determination of nitrate, potassium, and calcium are also available. For information on plant analysis and LTA, see Jones (1998).

NITROGEN

Lack of nitrogen (N) is the most common deficiency in *Cannabis* (Frank 1988). Symptoms begin with chlorosis of lower leaves (Plate 78). Yellowing starts down the midrib and then expands. In extreme cases, whole plants turn pale yellow-green, and leaves grow sparse and small. Stems and leaves may accumulate a red pigment called anthocyanin, but this symptom may also indicate a phosphorus deficiency. Frank & Rosenthal (1978) claimed N deficiency shifts sex ratios towards male plants. Because of N's high mobility in soil, N deficiency is common in loose, sandy soils.

Deficiency symptoms reverse within four days of applying N fertilizer. Ammonium (NH_4^+) is best utilized by young plants, whereas nitrate (NO_3^-) is utilized during most of the plant's growth period (Bennett 1993). Manure, blood meal, fish emulsion, and guano are excellent organic sources of N. See Table 7.3. Free-living, nitrogen-fixing bacteria have been sprayed on plants to serve as "biofertilizers" (Fokkema & Van den Heuvel 1986). *Azospirillum* species are commercially available. *Azospirillum brasilense* and *Azospirillum lipoferum* have been isolated from the rhizosphere of *Cannabis* (Kosslak & Bohlool 1983).

Excess N causes lush, dark green growth, susceptible to insects and diseases. The tensile strength of fibre decreases (Berger 1969), and plants become susceptible to stalk breakage (Scheifele 1998, see Plate 79). Roots of N-overdosed plants discolour and rot. *In extremis,* N-toxic plants turn a golden-copper colour, wilt, and die. Ammonium tends to become toxic in soils with an acid pH; nitrate is a problem in neutral to alkaline soils (Agrios 1997). Toxic levels of N can be leached from soil by frequent, heavy irrigation. A common drug cultivar of *Cannabis* ('Skunk No. 1') produces lime green leaves often mistaken for N or iron deficiencies (Schoenmakers 1986, Kees 1988). Beware—'Skunk No. 1' is easily burned by N amendments (Alexander 1988).

PHOSPHORUS

Phosphorus (P) deficiency causes stunted growth, and plants produce small, dark, bluish green leaves. Branches, petioles, and main leaf veins develop reddish purple tints. But purple tints from anthocyanin are normal in some hemp varieties (Dewey 1913) and drug varieties (Frank 1988). The tips of lower leaves may turn brown and curl downward. P deficiency delays flowering, flowers become predisposed to fungal diseases, and seed yield is poor. P deficiency worsens in heavy soil, in acidic soil, and cold, wet soil (Plate 80).

Excess P reduces the availability of zinc, iron, magnesium, and calcium in new leaves. Watch for deficiency symptoms of these elements, especially zinc.

POTASSIUM

Potassium (K or "potash") deficiency symptoms change with age. Young leaves, according to Rataj (1957), exhibit a nonspecific chlorotic mottling. Older leaves develop a brown "burn" of leaf tips and margins, with curling (Plate 81). Schropp (1938) described dark-coloured foliage with brown margins. Small branches and petioles may turn red, stalks weaken and lodge easily. Flowers are small and sparse. K shortage becomes more pronounced in dry weather, cold weather, and under low light intensity. K leaches easily, so K deficiency arises in sandy soil. Recovery is slow, often weeks after applying potash.

Excessive K scorches plants and causes wilting. Excesses may lead to deficiencies of calcium or magnesium.

MAGNESIUM

Deficiency causes older leaves to develop **interveinal chlorosis**—the margins and tissues between leaf veins turn yellow, leaving only dark green veins (Frank & Rosenthal 1978). In extreme cases, leaves turn white (see Plate 61). Berger (1969) described grey-white spotting in lower leaves of stunted plants. Deficiencies worsen after heavy application of ammonia or K to soil. Sandy, acidic soils commonly lack Mg, especially soils leached by heavy irrigation or rainfall. Plants recover quickly after Mg supplementation.

Excess Mg creates small, dark green leaves with curled edges. Adding calcium acts as an antidote. Latta & Eaton (1975) considered Mg a co-enzyme required for THC synthesis, but Coffmann & Genter (1975) found a negative correlation between soil Mg and plant THC.

CALCIUM

Berger (1969) says hemp requires large quantities of calcium (Ca). Deficient leaves are distorted and withered, their margins curl, and tips hook back. Apical buds may wither and die (Clarke 1981). Plants are stunted, stalks are brittle, roots are discoloured and excessively branched. Ca deficiency predisposes plants to stem and root pathogens (*Botrytis, Fusarium,* and *Rhizoctonia* fungi; *Ditylenchus* nematodes). Deficiencies are most common in acidic, sandy soils low in organic matter. Deficiencies may arise in growing tips during periods of very high humidity, because Ca is carried by transpiration—in 100% RH, stomates close down, transpiration ceases, and tips grow deficient. Excess Ca causes wilting and symptoms of Mg, K, iron, and maganese deficiencies.

SULPHUR

Symptoms of sulphur (S) shortage resemble N deficiency—chlorosis of leaves—except S symptoms begin with small leaves at the top of the plant. Stems are thin, brittle, and etiolated. Drought worsens S deficiency, spontaneous recovery often arises after rainfall washes S from the sky. Symptoms of excess S mimic those of sulphur dioxide air pollution, described below. S overdose can be partially corrected with Ca amendments.

ZINC

Among the micronutrients, zinc (Zn) causes the most problems in *Cannabis*. Deficiency causes interveinal chlorosis of younger leaves. New leaves grow torqued and twisted, and may drop off the plant prematurely. Flowers grow small and deformed, stems are short and brittle. Zn deficiency arises in alkaline soils low in organic matter. Adding lime or excess P may precipitate Zn shortages. Excess Zn causes dark mottled leaves.

MANGANESE

Manganese (Mn) shortage leads to interveinal chlorosis of small leaves (Coffman & Genter 1977). Extreme Mn deficiency resembles Mg deficiency—leaf margins turn nearly white, while veins remain green and boldly stand out against the pale background. Weigert & Fürst (1939) and Wahlin (1944) described leaves developing necrotic spots which expanded and destroyed the plants. Weigert & Fürst (1939) and Wahlin (1944) described extremely stunted plants (20–30 cm instead of 60 cm in height), and described bad Mn shortages in hemp grown on reclaimed marshland (alkaline soils), especially during cold, wet periods.

Excess Mn arises in acidic soils and in soils that have been steam-sterilized. Excess Mn causes mottled leaves with orange-brown spots. This toxicity appeared in our plants, which grew in compost naturally high in Mn; the glasshouse air was dehumidified, causing extra transpiration, which pulled excess Mn into the leaves. Plant analysis revealed Mn levels of 1500 ppm (Potter, pers. commun. 1999). Overdosing Mn locks up Zn and iron. Mn may play a role in cannabinoid synthesis (Latta & Eaton 1975).

IRON

Young leaves deficient in iron (Fe) develop interveinal chlorosis with green veins, like symptoms of Mn deficiency. Leaves may turn completely white with necrotic margins (Storm 1987). Fe shortages arise in alkaline soils (Fig 2.1) and soils with excess P or heavy metals. Adding an acidifying agent to the soil improves Fe availability. Plant synthesis of THC requires optimal Fe levels (Kaneshima *et al.* 1973, Latta & Eaton 1975). Excess Fe causes leaf bronzing.

COPPER

Copper (Cu) deficient plants have been called "gummihanf" (Kirchner 1966). Xylem loses its rigidity, causing stems to become rubbery, and plants bend over. Berger (1969), on the other hand, described Cu-deficient stems as brittle and easily broken. Young leaves wilt and necrose at tips and margins. In extreme cases, roots become affected and whole plants may wilt. Berger described Cu shortages in peat soils. Kirchner (1966) applied 50–100 kg ha^{-1} copper sulphate to soil in autumn to correct deficiencies.

Excess Cu results in Fe deficiency, especially in acid soils. Toxicity may arise from repeated use of copper sulphate fungicides.

BORON

Boron (B) shortages cause grey specks on leaves, or chlorosis, with twisted growth. Terminal buds turn brown or grey and then die, lateral buds follow suit (Frank & Rosenthal 1978). Stalks swell at their bases, tend to crack, and then rot (Clarke 1981). B shortages arise in marshland soils (Weigert & Fürst 1939). B shortages also arise in growing tips during periods of excessive humidity. Excess B causes discolouration and death of leaf tips, then leaf margins.

MOLYBDENUM

Deficiency of molybdenum (Mo—not Mb abbreviated by Frank & Rosenthal 1978) causes "whiptail disease"—young leaves grow pale, twisted, and withered. Mo shortage tends to arise in acid soils. Excess Mo is not common, it causes Fe and Cu deficiency.

CHLORIDE

Chloride (Cl) shortage causes wilting and chlorosis or bronzing of leaves. This deficiency is rare and easily overcorrected. Excess Cl also causes bronzing! Overdosing tobacco plants with Cl has an adverse effect on cigarette combustibility; Cl fertilizers are eschewed (Lucas 1975).

SELENIUM

Selenium (Se) is worth mentioning. Plants do not require it, but we do—our RDI is about 70 μg per day. Plants absorb Se from the soil; soyabean seeds and cereals (wheat, rye, rice) are good sources, *if* grown in Se-rich soil. In general, soil in rainier areas have *less* Se (e.g., the Pacific Northwest and the east coast of the USA). Unfortunately, hemp oil is a poor source of Se (Wirstshafter 1997).

CORRECTING FIELD CROP NUTRITION: SOIL, FERTILITY, AND GROWTH

by Bart Hall
Bluestem Associates
Lawrence, Kansas, USA

Healthy soil is the basis of all successful agriculture. It is not my intention to provide the reader with a complete discussion of soils and fertility, as this information is readily available in numerous professional and introductory texts to be found in any decent university library (e.g., Donahue 1973, Berger 1978). Rather, after an introductory overview of soils and how they work, it is my intention to focus on soil management approaches oriented towards soil health and crop quality, particularly as they relate to hemp production. Those familiar with soil science and its principles may wish to skim or skip this section.

What is soil, anyway?

Soil is the highly variable result of complex interactions between parent material, climate, plant cover, and time. These interactions eventually result in the development of different layers to make a 'stack' characteristic of each soil type. The stack is called the soil profile, and families of similar profiles define soil types, groups, and orders.

Of the ten soil orders in the world, only two are generally suited to large scale commercial hemp production on a sustainable basis, the Alfisols and the Mollisols. Hemp production is certainly possible on other soil types, but is proportionally more challenging.

Table 7.2: Summary of symptoms caused by nutrient deficiencies and overfertilization.

Symptoms	Nutrient deficiency										Over-
	N	P	K	Mg	Ca	S	Zn	Mn	Fe	Cu	fertilized
Yellowing of young leaves					1	1			2		1
Yellowing of middle leaves	2				2	2					1
Yellowing of older leaves	1		1	1				1			1
Yellowing between veins				1			1	1	1		
Leaf pale green	1										
Leaf dark green or red/purple		1									
Leaf colour mottled			1								1
Leaf small, stunted	1	1					1				
Whole leaf brown				2			1	2			1
Leaf tip/edge brown		1	1								1
Leaf tip/edges curl			1	2				2			1
Leaf wrinkled							1				
Stem soft, pliable	2		2							1	
Stem brittle, stiff		2	2		1	1				1	
Root stunted		2			1					1	
Whole plant wilted					1					1	1

Chart modified from table posted by N.P. Kaye (http://npkaye.home.ml.org).
1 = primary symptoms, 2 = secondary symptoms

Alfisols are the second largest order, accounting for 13% of world soils, and primarily found in a broad band centred on 50°N in Europe and Russia, the USA maize belt, and much of the Canadian wheat belt. Significant patches are also to be found in the tropics, primarily India, sub-Saharan Africa, eastern Brazil, and southern Australia. Alfisols can be identified in soil surveys when the longer taxonomic name for a soil type includes the ending "...alf," as (for example), Typic Hapludalf.

Mollisols are the fourth largest order, accounting for 9% of world soils. These are some of the most productive soils in the world, and probably the most cultivated. Mollisols are predominant in central Asia (south of the Alfisols), southeastern Europe, the North American Heartland, and Argentina. Mollisols always end in "...oll," as in Typic Hapludoll.

Soil can be thought of as having three major components (and I depart somewhat here from classic soil science approaches)—physical, chemical, and biological. The ***physical*** component of soil is the particulate mineral matter comprising the bulk of soil mass: sand, silt, clay, gravel, and so on. Sand, silt, and clay are each precisely defined terms describing a particular size particle without reference to its composition. What soil scientists call "texture" is a shorthand description of the relative proportions of each size class in the soil unit.

Sandy soils are considered light-textured (because they're open and don't hold much water), while clay soils are considered heavy-textured because they're dense and retain a lot of water. Loamy soils have about 40% each of sand and silt, along with roughly 20% clay. Loams are medium-textured, and are generally excellent soils to farm. The most common Alfisols and Mollisols tend to be silty loams and silty-clay loams; generally easy to work (if not too wet) and highly productive soils.

The ***chemical*** component of soils is less precise, and is often included with the physical. Because particle names, however, describe only size and not chemical composition, two soils of similar texture can behave very differently, depending on their native chemistry. Sand-sized particles might be made of nearly pure silica, or calcitic limestone, or even dolomitic (magnesium-rich) limestone. Each of those soils will have a different response to the same fertilizer application because, for example, high magnesium levels may obstruct potassium uptake by the crop. The silica-based soil will be much more acid than the limestone based soil of exactly the same texture.

Additionally, the smallest (clay-sized) particles tend to be dominated by a family of minerals with surplus negative electrical charges along their rather ragged edges. Positively charged elements (cations) in the soil are understandably attracted to those negative sites and can be held there until removed by locally-altered soil chemistry. These cations can be exchanged between two negative sites or between negative sites and the water component of the soil, or 'soil solution.'

This chemical capacity of a soil to store and exchange cations depends on the amount and specific types of clay, and can be measured. The resulting measurement, called Cation Exchange Capacity (CEC), is the most important chemical feature of a soil after the composition of the parent material. Heavier soils have a higher CEC and can store and exchange more cations than lighter soils, roughly two to three times more, which is one reason the Alfisols and Mollisols are so fertile and productive.

The soil solution (watery component) contains dissolved nutrients, and although the amounts of such nutrients are limited in unfertilized systems, the solubility of those nutrients makes them readily available to soil microbes and growing plants.

As it is incorporated by a soil microbe or growing plant, a nutrient mineral transfers from the chemical to the ***biological*** component of the soil. Because it is closely interfaced with living systems, this biological component is the most dynamic of the three. The biological component consists of growing plants, their residues, stable and decaying organic matter, and a wide range of both soil microbes (fungi, bacteria, etc.) and soil macrobes (the big guys, like worms and grubs).

Soil organic matter is a delightfully complex subject, but one largely beyond the scope of this chapter. Readers interested in more information about soil organic matter should consult texts and journals in a good university agricultural library. Two of the real leaders in this field have been Maurice Schnitzer (Canada, 1960s through 1980s) and MM Kononova (Russia, 1950s and 1960s). For a recent compilation, see Jackson (1993).

Soil organic matter is stored solar energy, the remains of formerly living plants and animals, mostly plants. These remains are present in a variety of forms, ranging from fresh plant tissue, through partially decomposed plant parts, to a dark substance bearing no remnants of the original materials from which it was derived. As a general rule, the ensemble of these residues is referred to as "organic matter," and the dark, undifferentiated matter, as "humus." In addition to functioning as the substrate for soil microbial activity, organic matter has enough negative charges that it contributes significantly to overall CEC as well.

The microbial activity in healthy soil is remarkable, and 1 kg of such soil will be home to 300 million algae, 1000 million fungi, 20,000 million actinomycetes, and 4 million million bacteria. Under favourable conditions (a freshly incorporated green manure crop, for example) soil bacteria can double their population in less than half an hour.

How does it work for hemp?

With hemp, as with any crop, there is a limited amount any grower can do to alter the soil. Changing the physical component is generally impossible, or too expensive to be practically possible. The chemical fraction can be adjusted, sometimes easily, sometimes not. It is helpful to think in terms of *soil correctives* as being in a different category than *crop fertilizers*. Generally speaking, crop fertilizers can be used much more efficiently when they are not also interacting with a soil that is chemically out-of-kilter. One such example would be the case of trying to provide calcium to hemp in a very acid (low-calcium) soil. Most of the calcium will react chemically with the soil in such a way that renders it unavailable to the crop.

Adjusting the biological fraction of the soil is relatively simple, though neither easy nor quick. For the most part, more organic matter is better than less, and fewer chemicals are generally better than more. Soil organic matter levels are substantially controlled by the rotation used, and within that framework by the extent to which soil is open and/or cultivated. Soils rich in organic matter are almost always an excellent substrate for a vibrant microbial community. The one exception is poorly drained soils, rich in organic matter because there is too little oxygen to break it down.

Crops can be broadly divided into three categories — soil-builders, neutral crops, and soil-destroyers. Soil-builders are the sod crops: grass, clover, alfalfa (lucerne), and so on. Because sod crops cover the soil more or less completely, there is less exposure to heat and excess oxygen, neither of which is good for soil organic matter. Sod crops also produce prodigious quantities of roots, so on sum, they generate more organic matter than they allow to be destroyed.

Row crops like maize (corn), soyabean (soybean), and vegetables are at the other end of the spectrum. Much of the soil is exposed to the sun for substantial segments of the growing season. Row crops are often cultivated (especially on biological and organic farms), exposing the soil to degradation by excess oxygen. Furthermore, row crops tend to produce fairly minimal root mass. As a result, under row crops, more organic matter is destroyed than created, and soil health suffers as a consequence.

Drilled crops, such as grains and green manures, fall more or less halfway between the soil-destroyers and the soil-builders. On balance they will probably be generally neutral in their influence on overall organic matter levels. It is worth noting again that green manures are not particularly effective as soil builders. Their role in a healthy farm system, however, is vitally important since their incorporation furnishes soil microbes with the abundant fresh organic matter they require as their primary source of energy.

Because hemp is a row crop it could be considered a soil-destroyer, were it not for one important exception. Much of the massive above-ground production of organic matter remains in the field as a result of the retting process. If that organic matter is incorporated into the soil after the harvest (instead of being left on the surface to oxidize) hemp's soil-destroying effects can be substantially mitigated. Hemp also has a more extensive root system than other crops considered as soil-destroyers. The cautious approach is nevertheless to consider hemp as a soil-destroyer in spite of these mitigating factors, and to manage the rotation accordingly.

For a farming system to be sustainable, the crop rotation must have at least as many years of soil-building crops as of soil destroying crops. The number of years dedicated to neutral crops has no significant bearing on sustainability, provided the grower remembers that open, uncovered soil is the ultimate soil-destroying 'crop.' Some illustrative examples may serve to illustrate this dynamic.

Maize — Soyabean (the most common 'rotation' in the USA Corn Belt) is strongly soil destroying. Maize is a row crop, and the soil is left open over winter. On conventional farms the soya are most usually drilled, making them a neutral crop, but the soil is left open over winter again.

Maize — Soyabean — Oats — Clover is substantially better (and substantially more rare). Here, oats (a neutral crop) are used as a nurse for the soil-building clover. On a conventional farm with drilled beans, this is getting close to a sustainable rotation. On an organic farm, however, with its row-cropped beans and frequent mechanical cultivation of both the maize and the beans, this would not be a soil-building rotation.

Maize — Soyabean — Oats — Alfalfa — Alfalfa — Alfalfa is strongly soil-building, in comparison to both of the other crop rotations.

Substitution of hemp for maize in any aforementioned examples would be a significant improvement, although it is impossible to recommend a hemp—soyabean rotation as being any more agronomically sound than maize—soyabean.

Correcting soil "balance": the step before crop fertilization

Within that framework of general soil concerns, it becomes possible to examine some specifics as they pertain to hemp production. As stated above, hemp is most productive in the Alfisols and Mollisols so common in the agricultural heartlands of Europe and America. These are usually some variant of silty loam (best) or silty clay loam (somewhat harder to work). Local extension agents or other professional agronomists in the field can be of considerable assistance in determining the soil order, characteristics, and suitability to hemp production of the various soils found on your farm.

Fertilizing hemp for optimum production and quality will be substantially easier if the soils in which it is to be grown are reasonably well balanced. The whole topic of "**soil balancing**" has been somewhat controversial for a generation or two, and still engenders heated discussion in some circles. The concept was largely developed by William Albrecht, a professor of soils at the University of Missouri from the 1920s through the 1950s (Albrecht & Walters 1975). In general, the proponents of soil balance hold that soils are most productive when exchangeable calcium, magnesium, and potassium levels in the soil correspond to *pre-determined optimum levels of base saturation.* Opponents contend that none of it makes any difference.

Base saturation refers to "base" minerals (as in acids and bases), especially calcium, magnesium, and potassium. In virtually all cases (fruit being a notable exception) these **soil saturation optima** are 70–80% calcium, 12–15% magnesium, and 3–5% potassium. The optimum 70–80% for calcium means that calcium occupies 70–80% of the total possible cation exchange sites in the soil. While calcium saturation of 68% is nothing to worry about, calcium in the mid 50s will lead to problems with crop quality. Magnesium at 17% is similarly not a problem, but above 25% Mg the soil will indeed be much tighter and compact than at lower levels.

Particularly with Alfisols and Mollisols, which (after all) are the soils with which the theory was developed, I have worked with many field situations in which paying attention to soil balance has made a real difference. Without wishing to universalize those observations, the Albrecht principles of soil balance probably lend themselves rather well to hemp simply because hemp's optimum zone of production corresponds so closely to that in which the theory was developed.

Out-of-balance soils, according to the Albrecht model, usually have one of five problems, or some combination thereof — low calcium, low magnesium, high magnesium, low potassium, and high potassium. Many of the soil correctives that follow are not acceptable for organic certification. Growers for whom certification is important should verify acceptability of materials with their agency.

Materials generally (but not always) acceptable in organic certification are followed by a § symbol. Amounts are appropriate for silt loams and silty clay loams, and are listed in kg ha^{-1} (kilograms per hectare). Increase the amounts listed by 10% to arrive at a nearly-exact lb/ac equivalent. Reduce the amounts by 40% for sandy soils, and increase them by 50% for clays and clay loams.

CALCIUM

<u>Ca-saturation 65–75%</u> – Major corrections are unnecessary. Apply 1000 kg ha^{-1} hydrated lime (calcium hydroxide) just before planting hemp and also ahead of any legumes in the rotation. Monitor every five years.

<u>Ca-saturation 55–65%</u> – Apply 1500 kg ha^{-1} finely ground calcitic limestone§ (*not* magnesium-containing dolomitic lime) plus 1000 kg ha^{-1} hydrated lime. Monitor after two years.

Ca-saturation 45–55% – Apply 3000 kg ha^{-1} finely ground calcitic limestone§ (*not* magnesium-containing dolomitic lime) plus 1500 kg ha^{-1} hydrated lime. Monitor after two years.

MAGNESIUM

Mg-saturation >30% – This is a serious problem soil, but not an uncommon one in certain areas, such as the USA upper Midwest. The challenge is high pH, coupled with resulting low calcium. Use the same criteria as for calcium (above), except replace all limestone with fertilizer-grade gypsum§ pellets, and increase the application rate by 50%. Fertilizer-grade gypsum§ is more expensive than field grades, but because it is much more finely ground (and pelleted), the calcium is vastly more available to react with the chemical fraction of the soil. The amounts of hydrated lime are low enough that they will not significantly affect pH. Monitor after two years.

Mg-saturation 20–30% – Use the same criteria as for calcium (above), except replace all limestone with fertilizer-grade gypsum§ pellets, and increase the application rate by 20%. Monitor after two years.

Mg-saturation 10–20% – No action specific to magnesium is necessary. With soils having >15% Mg-saturation it may be desirable when adjusting calcium saturation to replace 1000 kg ha^{-1} of limestone with fertilizer-grade gypsum§. When Mg-saturation is below 15% and limestone is needed to adjust calcium saturation, dolomitic limestones§ are acceptable when their magnesium contents are below about 5%.

Mg-saturation <10% – Significant magnesium shortage. Use the same criteria as for calcium (above), except replace all limestone with dolomitic limestone§. If dolomitic limestone is unavailable in the region, or if Ca-saturation is high enough that no limestone of any sort is needed, it will be necessary to furnish potash in the form of sulphate of potash-magnesia§ in order to supply magnesium.

POTASSIUM

K-saturation >10% – This is often a serious problem soil, probably suffering from alkali difficulties. High sodium may also be a problem, and the pH is often over 8. Such situations are not at all common in Alfisols and Mollisols. The most practical solution is an attempt to flood some potash out of the soil with heavy applications of fertilizer-grade gypsum§ and modest amounts of hydrated lime. Adequate rainfall (or irrigation) is necessary for this approach to work.

K-saturation 5–10% – Good soil for fruit, but hemp fibre quality and disease resistance are likely to be compromised unless calcium levels are in the 70s or higher. Adjusting for calcium and magnesium will usually decrease potassium saturation levels, particularly under the rainfall regimes common in prime hemp production areas.

K-saturation 3–5% – Ideal levels. No soil correction for potassium is necessary, though this will not eliminate the need to furnish appropriate amounts of potash as a crop fertilizer.

K-saturation <3% – Correct simultaneously for Ca and Mg if these are out of balance. For potassium, use a 50–50 mix of potassium chloride (muriate, 0-0-60) and potassium sulphate§ (0-0-50), applying as a corrective 100 kg ha^{-1} of this mix at 3% K-saturation and increasing proportionally by 100 kg ha^{-1} for each percent K-saturation below 3%. If Mg-saturation is below 10% and Ca-saturation levels preclude use of limestone, replace each 50 kg ha^{-1} potassium sulphate§ with 100 kg ha^{-1} sulphate of potash-magnesia§ (SPM, 0-0-22-11Mg-22S). Dolomitic limestone§ is the least costly source of magnesium correction, and is to be preferred over SPM§ unless the Ca-saturation situation clearly militates in favour of SPM.

In most cases, correcting for low calcium will simultaneously correct for low pH (acid soil). A common error in conventional agriculture is to see low soil pH as the problem, rather than as a symptom of low calcium. With such an approach, calcium additions to the soil have tended to be incidental to limestone applications focused on correcting pH, rather than governed by the calcium/magnesium regime in the soil.

This is unfortunate, since calcium as a *nutrient* plays an important role in crop health and quality. Of particular interest to hemp producers is calcium's action in strengthening cell structure. This occurs both at the level of plant framework (fibre in this case) and at the level of cell chemistry. Adequate levels of available calcium significantly improve the strength of the pectin that is such an important component of plant cell walls.

Strengthening pectin with calcium is a chemical reaction independent of whether or not the pectin is part of a living plant or not. Anyone who has used soluble calcium to set the low-methoxy pectins used in making sugarless jams can attest to this fact. In living plants with abundant available calcium, pectin is much more robust. Amongst other things, it makes cell walls more resistant to the polygalacturonase enzyme used by most fungal germ tubes to break down plant defences when attempting to exploit an infection court. In plain English that means that there is a greater chance that germinating fungal spores will dry out and die before they get into the plant and start to grow. In practical terms, the plant is more resistant to fungal diseases.

If a soil has been "balanced" for calcium, soluble calcium added as a crop fertilizer will not get tied up trying to adjust a disequilibrium in the soil, but instead remains available for uptake by the plant and subsequent incorporation within pectin or other elements of plant structure.

Total soil balance requires not only a balance of "base" cations (calcium, magnesium, and potassium), but also a balance of anions, particularly phosphates and sulphates:

PHOSPHORUS

Another common way in which soils remain underproductive is a shortage of phosphorus. Phosphorus takes many forms in the soil and is highly sensitive to chemical conditions. As a result, testing for soil phosphorus can be a somewhat confusing exercise, given that there are half a dozen possible lab tests that could be used. The trick is in finding a test that will give a usable and halfway decent representation of what is actually going on in the soil.

For soils anywhere on the acid side of the ledger, the most common test is the Bray, which itself is really two tests, one for available phosphate and the other for reserve phosphorus. These tests employ two different concentrations of the same reagent and are known as Bray P_1 (weak) and Bray P_2 (strong). Other common phosphorus tests are the Olsen, Morgan, and Mehlich tests. It is important to know what test has been used, and to remember that phosphorus numbers for the same soil will vary tremendously according to which test is used. Because the Bray P_1 test is so common in USA Maize Belt soils, it will be the focus of these discussions.

Most soil scientists consider that P_1 levels above about 10 ppm are more than adequate for successful production of most crops. While a good case can be made for this position, its primary weakness is that it focuses exclusively on crop yield, ignoring not only crop quality, but the needs of a healthy soil microbial environment as well. Soil microbes need one part of usable phosphorus for every part of nitrogen, and one part of nitrogen for every 20 to 30 parts of carbon. Efficiency of carbon cycling, therefore, is ultimately dependent on phosphorus. For this reason I usually encourage growers to consider corrective action if phosphorus P_1 test levels are much below 50 ppm. If levels are below 30 ppm, I would fairly strongly urge correction of some sort.

For soils with pH below about 6.4, periodic applications of soft-rock phosphate§ at a rate of 1000 kg ha^{-1} (every three to five years, depending on the rotation) are usually adequate. Rock phosphates are not only the least expensive source of phosphate on a kilo-for-kilo basis, but supply about 24% calcium and a range of micronutrients (particularly molybdenum) as well. Above pH 6.4, however, soil availability of the phosphate in rock phosphates is limited, and above pH 6.9, virtually nil.

Now is a good time to highlight a common source of confusion in regard to phosphate fertilizers in general, and the rock phosphates in particular. In order to allow comparison of phosphate materials, many agronomists have agreed on a common basis for expressing phosphate content of any fertilizer material. The term "available phosphate" has a very specific meaning, much narrower than that used in more everyday agronomic speech, referring to the amount of phosphorus than can be extracted using a 2% solution of citric acid, expressed as P_2O_5 (= P x 2.29). Citric acid solution is used because, as a weak acid, it roughly approximates what plant scientists think might be going on as the plant root interacts with soil phosphate.

Total phosphate in a material may or may not be actually available to the plant, and ironically, the same caution applies to the standardized "available phosphate." Most highly soluble phosphate fertilizers react with the chemical fraction of the soil, up to 85% of the "available" phosphate becoming unavailable to the plant. In acid soils the phosphates react with iron and aluminium to form extremely unavailable compounds. In alkaline soils these same phosphates react instead with calcium, forming compounds quite similar to those found in rock phosphates. Suggested application rates for conventional phosphate fertilizer material account for this problem and consequently exceed actual crop needs by several fold.

Natural phosphate availability peaks in modestly acid soils (pH 6.4–6.7), with abundant organic matter and an enthusiastic microbial community. If a soil is not yet in that condition, there is an excellent corrective approach available, though it is not particularly cheap or easy. Natural rock phosphates§ can be mixed with manure and composted to render nearly all of the total phosphate available for crop growth. The organic acids in manure solubilise the calcium phosphates in the natural rock, while the phosphates stabilize and protect much of the manure nitrogen against loss during composting and application.

Chemically what's going on is that most of the nitrogen in manures is in the form of ammonium carbonate, which is unstable. Adding calcium phosphate in the presence of organic acids produces calcium carbonate (limestone) and ammonium phosphate, which is stable, yet available to plants. As a general rule it works well to add about 25 kg of rock phosphate per m^3 of manure (50 lb/yd^3), composting to a medium finish by turning two or three times over two months before field application. Ten to 15 tonnes per hectare (4–6 tons/ac) is a good application rate. It is almost certainly not worth doing this if the manure must be bought, but if manure is free for the hauling and the distances involved aren't too great, this approach can be an extraordinarily effective method for kick-starting a worn out soil.

SULPHUR

This often forgotten nutrient has two key roles in the healthy hemp production system. Most importantly for all crops in the rotation, sulphur is essential for enthusiastic microbial activity. Most microbes need one part of sulphur for each ten parts of nitrogen and phosphorus. What is commonly described as the carbon:nitrogen ratio (C:N) is really the carbon:nitrogen:phosphorus:sulphur ratio (C:N:P:S) and should be approximately 250:10:10:1 for optimal microbial health.

The second important role for sulphur is its significant influence on protein *quality*. This will affect seed yields, and in particular the feeding value of hempseed oilcake for livestock. With attentive crop fertilization, sufficient sulphur can be provided on a regular basis by adding gypsum§ or one of the potassium sulphates§ into the crop fertilizer blend. Correction of soil sulphur levels is rarely needed, and is usually ineffective in climates sufficiently humid for profitable hemp production.

MICRONUTRIENTS

Apart from rare occasions, it is not particularly wise to attempt micronutrient correction of any soil. Most micronutrients are very expensive, and several are toxic if applied to excess. Even distribution is essential to avoid localized toxicity, and this requires precisely calibrated sprayers or spreaders. Since the micronutrient requirements of each crop in the rotation are different, it makes the most sense to supply modest amounts of micronutrients with the crop fertilizer.

Fertilizing the crop

Hemp is a nutrient hog, requiring even more nutrients for productive growth than an equivalent crop of maize. While so-called nutrient extraction measurements are of limited utility in evaluating a fertility programme, they can be somewhat indicative. Whole plant nutrient uptake for a 20 tonne ha^{-1} hemp crop will be on the order of 180 kg ha^{-1} nitrogen, 50 kg ha^{-1} phosphate (P_2O_5), 185 kg ha^{-1} potash (K_2O), 120 kg ha^{-1} calcium, 20 kg ha^{-1} magnesium, and 20 kg ha^{-1} sulphur (Berger 1969).

Fortunately, the retting process is such that the nutrient-rich leaves generally remain in the field after harvest. If the leaves are quickly incorporated into the soil (and their nutrients captured by a winter cover crop such as autumn rye), most of the nutrients contained in the leaves can be retained in the biological fraction of the soil system. Nutrient removal by a 6 tonne ha^{-1} crop of retted stems is substantially lower, being approximately 50 kg ha^{-1} nitrogen, 15 kg ha^{-1} phosphate, 100 kg ha^{-1} potash, 45 kg ha^{-1} calcium, 10 kg ha^{-1} magnesium, and 10 kg ha^{-1} sulphur (Berger 1969). These nutrients must ultimately be returned to the soil, lest the system be unsustainable over the long term.

Although it is impossible (and rather foolhardy!) to provide precise recipes for fertilization of a hemp crop, it is often helpful to provide an *example* as a starting point, which can be modified as soil, rotation, climate, and crop specifics dictate. Using the above nutrient extraction data as a base, this example assumes that the hemp crop follows a healthy clover crop providing 70 kg ha^{-1} nitrogen. Furthermore, estimated nitrogen release from the breakdown of an hypothetical 3% organic matter content will provide another 25 kg ha^{-1} (Menghini, pers. commun. 1999, where the ENR is estimated at 9 kg ha^{-1} nitrogen for each percent organic matter in the soil, via the Walkley-Black test).

Allowing for the clover carryover and the natural estimated nitrogen release, a productive hemp crop will nevertheless require an additional 85 kg ha^{-1} nitrogen. This could be provided from about 10 tonnes ha^{-1} cattle manure§, or about half as much poultry manure§. Other sources of nitrogen acceptable for organic certification are not economical for hemp production. The most common dry materials supplying nitrogen (bloodmeal§, feather meal§, alfalfa meal§, and so on) are pricey enough that the cost of their nitrogen approaches £3 kg^{-1} (Euro 5 kg^{-1} or US$2 per lb). Fish solubles are more expensive, commonly costing £18 kg^{-1} nitrogen. By comparison, conventional nitrogen *and* nitrogen taken out of the air by a healthy crop of alfalfa costs the grower only around £0.40 kg^{-1} nitrogen.

In addition to 85 kg ha^{-1} nitrogen, the hemp grower should plan to provide 25 kg ha^{-1} phosphate, assuming the soil is well-balanced soil with Bray P_1 at 15 ppm (amount of P_2O_5 determined as 25[0.70–0.035 (Bray P_1 in ppm)](stem yield goal in tonnes ha^{-1}). The grower should add 65 kg ha^{-1} potash, assuming a soil with exchangeable potash at 100 ppm (amount of K_2O determined as 25[1.166–0.0073 (exchangeable K in ppm)] (stem yield goal in tonnes ha^{-1}). The grower should add at least 45 kg ha^{-1} calcium, 10 kg ha^{-1} each of magnesium and sulphur, and an array of micronutrients.

This can be provided by 250 kg ha^{-1} (banded at planting) of a 10-10-10 type fertilizer with 2 Ca-2 Mg-4S-0.5 B-0.3Cu-0.5Mn-0.4Zn. One tonne of 10-10-10 agronomically ideal for hemp production can be formulated from the following recipe.

360 kg ammonium sulphate
200 kg mono-ammonium phosphate
110 kg sulphate of potash-magnesia
90 kg fertilizer grade gypsum
80 kg potassium chloride (muriate)
55 kg potassium sulphate
50 kg sodium-calcium borate
15 kg magnesium oxide
14 kg manganese sulphate
14 kg zinc sulphate
12 kg copper sulphate

Fertilizer can be formulated more cheaply than from this recipe, but the ammonium sources of nitrogen are particularly beneficial to hemp and the three different sources of potash provide a phased release early in the season when potash demands are low.

When the crop is 30 cm tall, sidedress at cultivation with either of two blends: a) 250 kg ha^{-1} ammonium sulphate (21-0-0) plus either 90 kg ha^{-1} potassium sulphate (agronomically ideal, but a bit expensive) or 75 kg ha^{-1} muriate of potash, *or* b) 175 kg ha^{-1} ammonium nitrate (34-0-0) plus 75 kg ha^{-1} muriate of potash. Do not mix nitrate and sulphate materials, as they will usually clump.

Properly fertilized under normal field production conditions, hemp will only rarely suffer nutrient deficiencies. The most probable deficiency to occur in this example is potash deficiency, induced by an excess of magnesium in certain soils. If the grower has previously paid attention to questions of soil "balance," this is less likely to be a problem. Flushing the soil with relatively high levels of soluble calcium — for example 1500 kg ha^{-1} hydrated lime — will reduce the phenomenon in future. Shifting from one side-dressing to two (at 25 cm and 50 cm), splitting the nitrogen between the two but doubling the potash by applying the original amount in each of the two applications, will also help, especially if done in conjunction with calcium flushing.

In my experience, plant tissue analysis is probably not helpful. Although analysis guidelines can be broadly inferred from other crops, the accuracy of such tests is highly sensitive to the timing of sampling, and particularly to the specific parts of the plant sampled. We simply do not know whether it is the third leaf cluster below the top, or the seventh, or some other part of the plant entirely that will prove most helpful in assessing the nutritional state of hemp. Nutrient levels in plants also shift widely as the plant matures, and tissue analysis results clearly demonstrating deficiency in early July might indicate a very healthy crop six weeks later. Until these parameters have been defined, tissue analysis is probably more of an expensive curiosity piece than it is a useful management tool.

Before deciding that nutrient deficiency is indeed the cause of symptoms in the field, it is advisable to rule out common biotic diseases (see Chapters 4–6), as well as abiotic damage. Apart from accidental (or intentional) herbicide damage, the most common abiotic distress in hemp will probably be ozone damage, discussed later in this chapter.

Experienced field crop growers should find hemp a relatively easy crop to manage. When sufficiently supplied with nutrients, hemp provides a full, dense canopy very effective in controlling weeds during that year of the rotation. Where production is allowed, hemp is an excellent substitute for maize, while being rather less damaging to the soil. If hemp fields are planted to winter cover (such as rye) soon after retted stems are removed from the field, hemp production can strengthen the rotation while increasing sustainability of the enterprise both in terms of soil health and diversified sources of revenue.

Bart Hall is a soil scientist with nearly 30 years experience as an organic agronomist. His company, Bluestem Associates, is based in Lawrence, Kansas, USA (tel: 785-865-0195), and provides consulting for organic, biological, and conventional producers worldwide.

FERTILIZING GLASSHOUSE AND GROWROOM SOILS

As Frank & Rosenthal (1978) stated, "Most indoor growers prefer to buy their soil, while some prefer to dig it." Field soils are *too heavy* for plants grown in containers. Most bagged commercial soils are also too heavy. Field soils and bagged commercial soils benefit from the addition of soil conditioners. Soil conditioners balance water retention and drainage, and promote root growth. Popular soil conditioners (in decreasing cost) include granular rockwool, perlite, vermiculite, peat, composted manure, and sand. Three examples of soil mixtures:

- 5 parts soil/2 parts perlite/1 part composted manure
- 5 parts soil/1 part sand/1 part peat/half part blood meal/ half part wood ash
- 5 parts soil/1 part vermiculite/1 part sand/1 part peat/quarter part 12-12-12 chemical fertilizer

Manure and peat are rich sources of organic material. Organic material improves the moisture-holding capacity of sandy soils. Organic material improves the drainage of clay soils, and make root penetration and tillage easier. Particles of organic material have negatively charged surfaces, which keep K, Mg, Ca, and other positively-charged cations from leaching out of soil. Organic material brings soil to life by providing nutrients for saprophytic microorganisms. Soil should consist of 4–10% organic material. Organic material must be added to soil every year, since soil microorganisms break it down.

To decide on nutrient supplementation, first evaluate plant symptoms. Next, verify symptoms by testing with pH meters, chemical soil tests, and leaf tissue analysis. Some growers forego verification and simply supplement with equal amounts of the "big three" nutrients—N, P, and K. The results are often suboptimal.

Different crops vary in their demands upon the soil, and vary in their needs for supplementation. Fibre production requires high soil N, at least 150 kg ha^{-1}. According to Hendrischke *et al.* (1998), supplementing N over this baseline does not increase yield significantly, but restores soil N. At harvest, whole hemp plants contain about 150 kg N ha^{-1} ($\approx$ 16.6 g N m^{-2}). Field retting releases 40% of N back to the soil, but the soil is still left with a significant N deficit, and nitrate released by retting is susceptible to further loss from leaching (Hendrischke *et al.* 1998). Fibre production makes equally high demands for K, then in descending order: Ca, P, Mg, and micronutrients (especially Si).

Seed crops, compared to fibre crops, require the same amount of soil N, less K, and more P (see Table 2.2 in Chapter 2). Drug crops are probably similar to seed crops in their nutrient requirements. Coffman & Genter (1977) found positive correlations between soil N, soil P, and plant THC. Haney & Kutsheid (1973) also correlated soil N with plant THC but found a *negative* correlation between soil P and plant THC. This disagrees with most experts' advice to add P to soil as *Cannabis* begins flowering (Frank & Rosenthal 1978, Frank 1988). Coffman & Genter (1975) found a *negative* correlation between soil Mg and plant THC. This is surprising, since Mg, along with Mn and Fe, probably plays a role in cannabinoid synthesis (Latta & Eaton 1975).

In Holland, we supplement with organic fertilizers rather than chemical fertilizers. Most organic fertilizers, such as composted manures, improve soil texture and nurture soil microorganisms. Chemical fertilizers harm soil texture and kill soil organisms. Marshmann *et al.* (1976) analysed Jamaican plants cultivated with organic fertilizers, and compared them to plants cultivated inorganically. Their sample of 19 organically-grown plants contained 79% more THC than their sample of 18 plants grown inorganically. Organic crops cause less laryngitis in smokers than chemically-fertilized crops (Clarke, unpublished research 1996). Chemically-fertilized, hydroponically-grown *Cannabis* contains higher molybdenum concentrations than field crops (Watling 1998).

Not all organic fertilizers are safe. Frank & Rosenthal (1978) recommend spraying leaves with organic fish emulsion. But Farnsworth & Cordell (1976) warned against spraying marijuana plants with fish emulsion—virtually all liquid foliar fertilizers, including fish emulsion, contain nitrates. Leaves of *Cannabis* convert nitrates into carcinogenic N-nitrosamines.

Macronutrient deficiencies should be corrected with a mix of rapidly-absorbed fertilizers and slow-release fertilizers. See Table 7.3. For N deficiencies, side-dress plants with a mix of bird guano (rapid) and composted manure (slow). Supplement P with a mix of wood ash (fast) and bone meal (slow). Supplement K with wood ash (fast) and greensand (very slow). Supplement Mg with dolomitic (high-Mg) limestone. If liming is contraindicated, use magnesium sulphate (Epsom salt) or sul-po-mag (11%Mg, 22%K, 23%S). Ca can also be supplemented by liming. If you need Ca without changing pH, apply calcium sulphate (gypsum—21%Ca, 16%S). Sulphur deficiency is rare thanks to acid rain; if needed, apply the aforementioned sul-po-mag or gypsum.

Micronutrient supplementation is tricky. They are easily over-supplemented, causing toxicity. Composted manure contains adequate micro-nutrients. Maintaining organic matter and proper soil pH is usually sufficient for most micronutrients.

The pH of soil affects the availability of soil nutrients (see Fig 2.1). In many humid and semihumid areas, soil acidity increases to the point that many nutrients become unavailable (Wolf 1999). Acid soils need to be limed. Limestone (calcium carbonate, $CaCO_3$) is a cheap way to neutralize acidic soils. Wolf (1999) provides charts that estimate the amount of limestone required per acre, depending on soil pH, soil class (clay loams, silt loams, sand loams, etc.), and percentage of organic material in the soil.

Table 7.3: Some fertilizers, their N-P-K ratio, and nutrient availability.[1]

Fertilizer		Percentage (by weight) N	P_2O_5	K_2O	Availability to plant
Natural:	manure (dairy cow)	1	0.75	1	medium
	activated sludge	5	3	0.1	medium
	worm castings	3.5	1	1	medium/rapid
	blood meal	15	3	0.75	medium/rapid
	bird guano	12	8	3	rapid
	bat guano	6	9	1	rapid
	fish emulsion	8	3	3	slow/medium
	cottonseed meal	8	3	3	slow/medium
	bone meal (raw)	4	21	0.2	medium
	seaweed	4	21	0.2	medium
	wood ash	0	2	7	rapid/medium
	phosphate rock	0	30	0	slow
	greensand	0	1.5	5	slow
Chemical:	anhydrous ammonia	82	0	0	rapid
	urea	44	0	0	rapid
	ammo-phos (Grade A)	11	15	0	rapid
	potassium sulphate	13	0	39	rapid
	triple superphosphate	0	46	0	rapid
	"Rose Food"	12	12	12	rapid

[1] Data collected from various references.

POLLUTANTS, TOXINS & PESTICIDES

Cannabis does not tolerate excess **salt** (NaCl), or saline/brackish water, despite the fact that Cl is an essential nutrient and Na may be beneficial in trace amounts. Dewey (1902) applied small amounts of NaCl to soil and reported an increase in cellulose production in hemp plants. He emphasized caution, because "salt is likely to prove very injurious on light soils." Hancer (1992) claimed that "poisonous salty breezes" killed seedlings growing near the Hawaiian shoreline.

Many **air pollutants** are injurious to plants. Injury from air pollution peaks during daylight hours, and worsens in warm, humid conditions. "**Acid rain**" doubles the damage by acidifying soil and creating various nutrient deficiencies. According to Agrios (1997), the lowest rain pH reported (pH 1.7 in Los Angeles) is more acidic than vinegar (pH 3.0).

Sulphur dioxide gas (SO_2) causes interveinal leaf chlorosis in *Cannabis* (Goidànich 1959). It is toxic at concentrations as low as 0.3 ppm. SO_2 combines with water to form toxic droplets of acid rain.

Hydrogen fluoride gas (HF) causes interveinal chlorosis and necrosis of leaf tips and margins. Alternatively, leaves may develop a white speckling (Goidànich 1959). Young growth is the most susceptible, especially when wet. HF vapors are produced by various industries and factories processing ore or oil.

Nitrogen dioxide gas (NO_2) causes symptoms similar to those of SO_2 toxicity, at levels as low as 2–3 ppm. NO_2 combines with O_2 in bright sunlight to form ozone.

Ozone (O_3) is the most destructive air pollutant to plants (Agrios 1997). Ozone initially causes chlorotic spots on leaves, primarily on upper surfaces. The spots enlarge and turn brown, leaves defoliate prematurely, and plants become severely stunted.

Ozone binds with incompletely combusted gasoline to produce **peroxyacyl nitrates** (PANs, a.k.a. **smog**). PANs may combine with atmospheric water to create **nitric acid** rain. Sharma & Mann (1984) studied *Cannabis indica* growing next to a smoggy Himalayan highway. Roadside plants suffered chlorosis and necrosis. Plants produced smaller leaves with shorter petioles than plants growing in the hills. Microscopically, polluted plants produced fewer stomates but more trichomes per leaf area. Because of increased trichome density, Sharma & Mann concluded that auto pollution might increase THC concentration. Roadside plants collected in the USA were mostly female (Haney & Bazzaz 1970). This observation agrees with Heslop-Harrison (1957), who reported that **carbon monoxide** (CO) may cause a shift of sex expression in plants, from male to female.

Carbon dioxide (CO_2) is produced by automobiles, industries, and burning rainforests. Global levels of CO_2 are rising. This may promote photosynthesis, carbon accumulation, and plant growth. But it dilutes the plant's N concentration, which forces insects and other N-seeking herbivores to eat more plant tissue (Roberts 1988). Increased CO_2 also exacerbates weed problems.

Cannabis absorbs many **heavy metals** from the soil, without ill effects. These toxins deserve mention for their effects on us. Ingesting plants laced with **lead** (a contaminant found in chemical fertilizers) causes anaemia and brain damage. Ingesting **mercury**, another cumulative toxin, also causes neurologic deficits. Mercury is released from combustibles such as coal and volcanoes. Siegel *et al.* (1988) measured 440 ng mercury per gram of marijuana grown on Hawai'i's volcanic soil. They noted that mercury is absorbed ten times more efficiently by the lungs than by the gut, and calculated that smoking 100 g of volcanic marijuana per week would lead to mercury poisoning.

Jurkowska *et al.* (1990) found high levels of **lithium** in hemp (1.04 mg kg^{-1}), higher than any other crop plant tested, including barley, maize, mustard, oats, radish, rape, sorrel, spinach, sunflower, and wheat. *Cannabis* crops grown near Australian uranium mines become enriched with **uranium** and **thorium**, whereas crops grown in gold producing regions contain a significantly enhanced **gold** signature (Watling 1998).

Hemp's ability to hyperaccumulate metals may prove useful for decontaminating toxic waste sites. In Silesia, *Cannabis* is deliberately cultivated on wasteland contaminated with **cadmium** and **copper**. The crop efficiently extracts the toxic metals from soil and accumulates the metals in seeds. The crop is harvested, and the metals are recovered by leaching seeds with hydrochloric acid (Kozlowski 1995). In 1999 a hemp crop will be cultivated near Chernobyl, to extract radioactive caesium-137 and strontium-90 from the soil.

Some of the most dangerous toxins are agricultural **pesticides**. Many insecticides, miticides, and fungicides are phytotoxic. Pesticides are mixed with solvents, emulsifiers, diluents, and carriers, which may also cause plant injury. Common symptoms include petiole malformation, leaf burn, chlorosis, stunting, and bud necrosis.

Herbicides deserve special mention. Farmers apply tons of herbicides to their fields, and tons of herbicides move off target by vapour drift and water runoff. Hemp is very sensitive to herbicide drift (Bòcsa & Karus 1997). High concentrations of herbicides are sprayed on suburban lawns, fields, roadsides, railroads, and power line rights-of-way.

Herbicides are purposely sprayed on *Cannabis* for two reasons. *Hemp cultivators* occasionally apply herbicides as defoliants to make stalks easier to harvest (Goloborod'ko 1986). **Treflan**® is popular for this purpose (Nazirov & Tukhtaeva 1981). Growth-retarding chemicals are also used (Keijzer *et al.* 1990). *Law enforcement officials* spray herbicides on marijuana to render crops unusable. Horowitz (1977) tested nearly 50 herbicides against *Cannabis* and listed 20 chemicals which kill or severely injure feral hemp. Most herbicides are absorbed by roots, limiting their effectiveness to young plants. A few herbicides work directly on foliage—**glyphosate** (Roundup®), **amitrole** (Weedazol®), **phenmedipham** (Betanal®), **2,4-D**, and **paraquat**.

From 1975 to 1978 the USA government used paraquat to eradicate Mexican marijuana crops. Paraquat causes plants to completely necrose within two days of application (see photograph in *Science* 119:863). But crops harvested *immediately* after exposure look normal. As a result, normal-looking Mexican marijuana tainted with paraquat appeared in the USA. In 1983 the Drug Enforcement Agency (DEA) began spraying **glyphosate** on illicit crops in Georgia and Kentucky. Widespread outrage forced the DEA to stop spraying. Recently, the DEA has sprayed feral hemp with the herbicide **triclopyr**.

CLIMATIC HAZARDS

As summarized by Hackleman & Domingo (1943), hemp is susceptible to *moisture stress*—too little or too much. **Drought** can be corrected by irrigating accessible plants. Avoid using tap water filtered through a water softening system—it contains too much sodium. Flooding causes the same symptoms as drought—wilting. Wilting may be temporary, pending quick correction of the problem. But it often becomes permanent. **Overwatering** is common. *Cannabis* is very sensitive to water soaked, wet soil conditions (Scheifele 1998). Plants can sometimes be revived with Oxygen Plus®, a hydrogen peroxide product. Flooding is best prevented by intelligent land use and careful field drainage.

Frost is poorly tolerated by mature plants; temperatures below -2 to -3°C are harmful (Hanson 1990). Frosted flowers blacken and develop a harsh taste (Frank & Rosenthal 1978, Selgnij 1982, see Plate 82). 'Gouda's Glorie' and 'Early Bird' were two Dutch cultivars resistant to cold and early frost (Kees 1988). Seedlings are less susceptible to frost than mature plants; they tolerate temperatures down to -7°C (Urban 1979, Hanson 1990). Perhaps cold tolerance is related to the fact that hemp seeds generate more heat upon germination than other crops plants (Darsie *et al.* 1914). Seedlings that survive late frosts often have chlorotic leaves.

Cannabis tolerates **hail** and **high winds** better than other large-leaved plants. Nevertheless, high winds can be a problem during stages of rapid stalk elongation, before stalk strength is fully developed. Plants may lodge at the base, in which case they right themselves fairly readily; or the stalk may kink somewhere above the base—in this case stalks will often curve upright again, "gooseneck" fashion (Moes, pers. commun. 1999).

Hail can severely damage young hemp fields (Plate 83). Hail injury may ruin fibres for textile use, but the fibres can still be harvested for cellulose (paper). Immature seed crops suffer little damage from hail, because the injured plants branch out and produce more seed (Bòcsa & Karus 1997). Plants damaged right before harvest, however, may not have time to rebranch and recover (Moes, pers. commun. 1999). Split branches and broken stalks of valuable plants can be hand-repaired with splints and tape.

Lightning sometimes strikes hemp fields, leaving circular patches of dead plants. Do not mistake these patches for diseases or deer damage.

Poor soil aeration (low soil oxygen) causes lower leaves to turn grey-green, then wilt and fall from plants (Frank 1988). *Cannabis* grows in a wide range of relative humidity, but extremely **dry air** causes leaf tips to turn brown (Frank 1988).

Short photoperiods (daylight less than 11 hours) cause *Cannabis* to flower prematurely, before full flower and seed yields are attained. Night light as low as 0.03 Lumen keeps plants from flowering (Frank & Rosenthal 1978). *Cannabis* requires a much higher light intensity to keep growing, however. Plants become spindly and chlorotic in light intensity less than 350 Lumen (see Chapter 2). Too much light causes "**sun scald**." This may happen when indoor seedlings are transplanted outside. Undersides of fast-growing leaves are vulnerable if wind turns them over to exposure in brilliant sunlight. Scalded tissues become dry and brown, and resemble "hopperburn" caused by potato leafhoppers.

GENETICS

Cannabis is normally diploid, its $2n = 20$ chromosomes (18 + XX or XY). Polyploid strains can be artificially produced. Generally, XX plants are female and XY plants are male. But sex determination in *Cannabis* is not as rigidly genetic as it is in humans. It can be modified by the environment. Mohan Ram & Sett (1982) found the balance of ethylene and gibberellins in plants determined sex expression. They used chemicals (silver nitrate, cobalt chloride, etc.) to induce male or female flowering. Severe defoliation, cold, and photoperiods of less than six to eight hours also produce sex reversal (Clarke 1981). Laskowska (1961) reported that pollen stored for over two weeks produced more female offspring. Haney & Bazzaz (1970) suggested carbon monoxide shifts sexual expression towards females. Frank (1988) said root-bound seedlings mature into predominantly male plants.

Cannabis is normally **dioecious**—plants produce a 50:50 ratio of male (staminate) flowers and female (pistillate) flowers. Hemp breeders have developed **monoecious** plants with both female and male flowers. These are often called "hermaphrodites," incorrectly. As Borodina & Migal (1987) pointed out, monoecious plants with unisexual flowers (male and female) are **intersexual**, not **hermaphroditic**. Truly hermaphroditic (bisexual) flowers in *Cannabis* are abnormal, usually sterile, and rarely seen. Miller (1970) explained monoecism as the expression of heterozygous genes on X-chromosomes or autosomes in XX plants.

The deleterious effects of **inbreeding** were first described by Fleischmann (1934), who reported a 50% reduction in seed yield. Crescini (1956) illustrated many incestuous phenomena, such as strange pinnate phyllotaxy and ramification of hemp stems. Bócsa (1958) blamed inbreeding for losses of seed and fibre yield, short plant stature (only 68% the height of normal hemp), shortened lifespan (vegetative growth nine weeks shorter than normal plants), production of sterile seeds, and increased susceptibility to disease. Recently, Lai (1985) confirmed these mutagenic effects.

The Hungarian cultivar 'Kompolti Sárgaszárú' is a yellow-stemmed mutant (de Meijer 1995). **Yellow stem** is caused by a monogenic recessive mutation (Sitnik 1981). The gene involved has a pleiotropic effect on plant yields, decreasing biomass, fibre and seed production.

Mutations may also cause plant **fasciation**, a twisting of stems and leaves (Crescini 1956). Borodina & Migal (1987) illustrated flower fasciation and other flower teratologies (Fig 7.1). Lyster Dewey collected fasciated plants in Virginia, of the 'Kymington' variety (specimens deposited in the USDA herbarium). The cause of fasciation might be pathogenic rather than genetic—***Corynebacterium fascians*** (Tilford) Dowson, a small bacterium, causes fasciation in other crops.

For more information concerning the genetics and breeding of *Cannabis*, see Chapter 1 ("host" side of the crop damage triangle), and Chapter 9 (method 5).

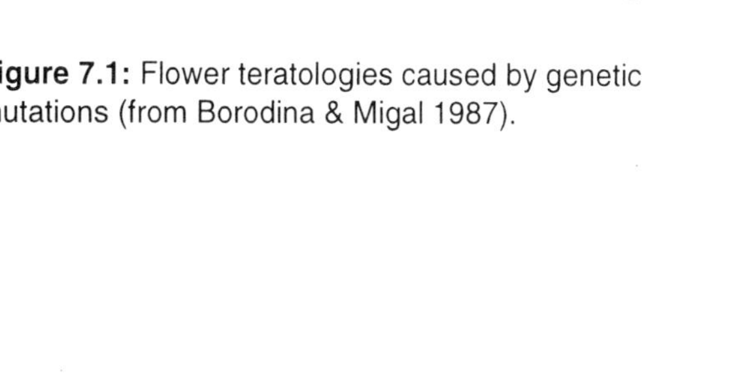

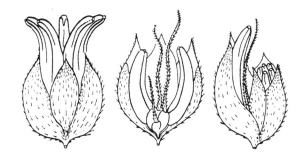

Figure 7.1: Flower teratologies caused by genetic mutations (from Borodina & Migal 1987).

"Hemp loosens the soil and makes it more mellow."
—Lyster Hoxie Dewey

Chapter 8: Post-harvest Problems

Cannabis is a plant of many uses. Nearly all above-ground parts can be utilized—*stalks* provide fibre (from phloem cells) and hurds (from xylem cells), *achenes* ("seeds") provide edible seeds and polychotomous seed oil, *female flowers* provide marijuana (with seeds) and sinsemilla (without seeds), and *resin glands* extracted from female flowers provide hashish (charas) and cannabinoids. All of these products, however, are subject to decay and destruction by fungi, insects, and other organisms.

HEMP FIBRE

The history of hemp cultivation fills many pages (Able 1980, Canapasemi 1988, Herer 1991). Agricultural fibres are currently making a comeback. Hemp fibres (bast cells) are the hemp plant's phloem or sap-conducting cells. Primary fibre cells are only 10 µm (1/150") wide, but up to 3" (7500 µm) long. Zylinski (1958) described "secondary" fibres, which are shorter, stiffer, and weaker than primary fibres. Secondary fibres are useful for paper production but undesirable for most textile production.

Primary and secondary fibres overlap in bundles containing ten to 40 cells per bundle. Fibre bundles often run the entire length of plants, from roots to tops. About 25 bundles lie around stems, embedded in a ring of phloem parenchyma. This ring is sandwiched between stem epidermis ("bark") and the cambium. The cambium layer merges with thick-walled woody xylem cells. This woody layer thins into pith, which surrounds a hollow centred stalk (Fig 8.1). Fibre bundles, cambium, and woody xylem are glued together by plant resins and pectins. Loosening fibres from the bark and wood is called **retting.**

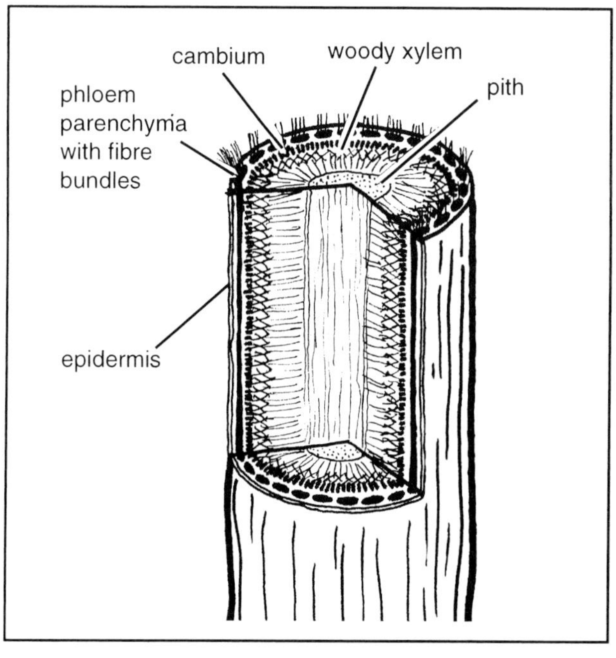

Figure 8.1: Cross section of a hemp stalk (McPartland).

Retting is accomplished by microbial, mechanical, or chemical means. Microbial water retting is the most common method, described later. Steam retting was practised in the former USSR; it is expensive, takes several hours, and produces stiff, unpliable fibre. The new "steam explosion" method is much quicker than old steam retting, but the fibre is shredded and unsuitable for spinning. Chemical retting is also called enzymatic retting or alkaline digestion. Chemical retting produces a high quality fibre, but is very expensive.

Microbial retting is a technical term for rotting. As microorganisms digest resins and pectins in hemp stems, they give off a bad odour. Microbial retting can be accomplished two ways—either dew retting or pond retting.

Dew retted stalks are cut close to the ground and spread on fields in thin, uniform layers to catch dew. Robinson (1946) noted dew retting is difficult in dry parts of the USA and Canada west of the 100th meridian (where the great plains begin). Microorganisms normally dew-ret hemp in four to six weeks (Dodge 1890). This happens faster in warmer conditions and slower in cooler climates. Farmers in northern latitudes (e.g., Wisconsin, Ontario, and Siberia) have practised "snow retting"—stalks are spread on fields and remain under snow until springtime. The disadvantage of dew retting is its dependence on ideal weather.

Pond retting has been practised in Europe and China. Pond-retted fibre is superior to dew retted fibre—the retting is more consistent and the fibre is not discoloured (dew retted fibre, in contrast, is stained grey). Henry Clay tried introducing pond retting to Kentucky but growers considered pond water poisonous and an unnecessary hazard (Dodge 1890, 1898). French peasants retted hemp in running streams and even in the Loire River, "...although public opinion is generally against river retting because it renders the waters foul" (Dodge 1890).

Hemp bundles are rafted in water 1.5 m deep, free of aquatic growth, and low in calcium and chlorides. The initial temperature should be above 15°C. Within a week the water turns acidic (from production of butyric acid) and fermentation bubbles appear around bundles. After two weeks stems lose their chlorophyll and turn white.

Retting in *warm* water speeds the process. Ponds in China are passively heated by sunlight to 23–30°C (Clarke 1995). Stalks are pond retted for one to three days and then spread on fields to continue retting a few days more. Hungarians use geothermally-heated water, contained in concrete basins. In Poland, warm water retting takes place in concrete or steel tanks for 24 hours, then fresh water is added and brought to an elevated temperature for two or three days. Approximately 50 metric tonnes of water are used per metric tonne of fibre (Kozlowski 1995). Air compressors aerate the water, and tanks are periodically inoculated with pure cultures of retting organisms.

HEMP RETTING ORGANISMS

Researchers have isolated many retting organisms. Behrens (1902, 1903) and Fuller & Norman (1944, 1945, 1946) surveyed the dew retting microflora in Europe and North America, respectively. Behrens isolated *Rhizopus stolonifer,*

Cladosporium herbarum, and several *Mucor* species from dew retted hemp. *R. stolonifer* was more active in warm weather, while *Mucor hiemalis* predominated in snow retting. Other *Mucor* species involved in retting included *Mucor plumbeus, Mucor mucedo,* and *Mucor corticola.* On the contrary, in Iowa, Fuller & Norman never isolated *Mucor* species from retted hemp. They only found *Mucor* in artificial laboratory conditions. In fields they isolated 15 genera of fungi, predominantly species of *Alternaria, Hormodendrum* (=*Cladosporium*), *Fusarium, Phoma,* and *Cephalosporium.*

Other fungi have been isolated from dew retted hemp by other researchers, including several *Acremonium* species, *Acremoniella atra, Aureobasidium pullulans, Chaetostylum fresinii* (=*Ascophora pulchra, Bulbothamnidium elegans*), *Diplodiella ramentacea, Epicoccum nigrum* (=*E. purpurascens*), *Gonatobotrys simplex, Phialophora* species, *Periconia* species, *Stachybotrys lobulata, Stachybotrys alternans, Thamnidium elegans, Torula herbarum,* and many *Mortierella* and *Ulocladium* species. Several of these fungi are cellulose-destroying organisms and therefore undesirable for retting (Gzebenyuk 1984).

Bacteria are less important than fungi in dew retting. Behrens (1903) only identified *Bacillus asterosporus* (= *B. polymyxa*). Fuller & Norman (1944) described bacteria but did not identify the organisms. Caminita *et al.* (1947) once reported *Aerobacter cloacae* (=*Enterobacter cloacae*) from retted hemp but failed to specify whether it was from dew or pond retted hemp.

In *pond* retted hemp, scientists cite about 20 different bacteria. All of these names, however, are synonyms of just five species: *Clostridium acetobutylicum, Clostridium butyricum, Clostridium pectinovorum, Clostridium pectinovorum* var. *parvum,* and *Clostridium felsineum.* The first two organisms commonly rot various substrates. In contrast, *C. felsinium* has only been isolated from the mud of Italian retting ponds.

FIBRE DECAYING ORGANISMS

Retting must be stopped the moment pectins and resins are gone. Otherwise, some microorganisms will begin rotting the fibre, ruining it. Since hemp fibre is pure cellulose, its attraction is limited to organisms with cellulolytic enzymes. Fungi are notorious destroyers of cellulose fibre (Marsh & Bollenbacher 1949). Some fungi cause double trouble—they cause disease in living plants, then rot fibre after plants die. These fungi include *Botrytis cinerea, Alternaria alternata, Trichothecium roseum, Epicoccum purpurascens,* and *Cladosporium herbarum.* Fuller & Norman (1944, 1945) considered *Trichothecium roseum* particularly destructive. Behren (1902) named *Cladosporium herbarum* as the most destructive and dangerous cellulose-degrading fungus, although Thaysen & Bunker (1927) disagreed with his opinion.

Other retters that may become rotters include *Aspergillus niger, Aspergillus glaucus, Acremoniella atra,* and *Stachybotrys atra.* Oudemans (1920) listed several rare hemp-rotting fungi: *Dendrylphium macrosporum, Diplodiella ramemtacea, Rhizophidium zylophilum, Sporotrichum sulphureum, Sporormia cannabina, Sordaria wiesneri, Perisporium dilabens, Perisporium funiculatum,* and *Perisporium kunzei.* (Some of these species require taxonomic re-evaluation.)

Some fungi rot hemp textiles, such as *Stachybotrys lobulata* and *Alternaria alternata* (Agostini 1927). Not many insects attack finished hemp fibre. Carpet beetles cause minor problems. In Italy, larvae of the Mediterranean flour moth (*Ephestia kühniella*) can infest stored fibre (Ferri 1959a).

BIOPULPING ORGANISMS

Removal of lignin from pulp is the most toxic process in paper pulping. Most paper companies use toxic chemicals. Recently a number of basidiomycete fungi have been evaluated as biopulping organisms. De Jong *et al.* (1992) degraded hemp lignin with the fungus *Phanerochaete chrysosporium.* Less effective fungi included *Trametes versicolour, Trametes villosa,* and *Bjerkandera adusta.*

CONTROL OF FIBRE DECAYING ORGANISMS

Retting must be monitored carefully to avoid rotting. Finished fibre must be dried quickly to stop the retting process. Hemp fibres should be stored in cool, dry conditions. Damp conditions permit dormant fungi to reactivate and renew their retting and rotting. Properly stored hemp does not need pesticides. Romanian growers sprayed processed fibres with **pentachlorophenol** (PCP) to prevent fungal infestations, but this was discontinued (Bòcsa & Karus 1997).

OCCUPATIONAL HAZARDS

Fields workers may develop allergies to hemp. Because of its prolific pollen production, *Cannabis* is an important cause of hayfever (Durham 1935, Wodehouse 1945, Freeman 1983). During the late summer and autumn, pollen from wild hemp constituted 17% of all pollen in the air over Nebraska (Maloney & Broadkey 1940). Significant amounts of *Cannabis* pollen appear over Ann Arbor, Michigan (Solomon 1976) and Delhi, India (Malik *et al.* 1990). The allergic agent in hemp pollen is a water-soluble agent (Añíbarro & Fontela 1996), possibly a lectin (Tumosa 1984), but pure THC and CBD also elicit allergic reactions (Liskow *et al.* 1971, Watson *et al.* 1983). Individuals allergic to hemp pollen often cross-react with hops (Lindemayr & Jäger 1980).

Factory workers handling retted hemp often suffer from **byssinosis**. This serious disease, also termed the "Monday syndrome," is common to all textile workers exposed to fibre dust. After a weekend away from work, renewed exposure causes chest tightness, shortness of breath, wheezing, and coughing. These symptoms usually dissipate as the work week progresses, only to reappear the following Monday. In Italian hemp workers, the syndrome was called "cannabosis." The British physician Schilling (1956) equated cannabosis to byssinosis, a syndrome shared by flax, cotton, jute, and sisal workers. Hemp processors, however, suffer the worst (Zuskin *et al.* 1976). Thomas *et al.* (1988) report 15% of hemp workers experienced byssinosis ("pousse" in Ireland), whereas only 2.8% of sisal workers were afflicted.

A causal agent in fibre dust has not been identified. Dimitrov *et al.* (1990) found a high concentration of fungal spores in hemp mill air (mostly *Cladosporium, Alternaria, Aspergillus, Penicillium,* and *Fusarium* species). Castellan *et al.* (1984) blamed byssinosis on an endotoxin-producing gram negative bacteria. Caminita *et al.* (1947) isolated gram (-) *Enterobacter cloacae* from hemp dust. They blamed *E. cloacae* for "hemp fever," a byssinosis-like syndrome with fever, chills, nausea, and vomiting. Nicholls *et al.* (1991) found that hemp fibres contain ten times more bacteria than cotton, flax, or jute fibres. Nicholls isolated *E. cloacae* and four other species of Enterobacteriaceae, three species of *Pseudomonas,* and assorted *Bacillus, Corynebacterium, Staphylococcus,* and *Acinetobacter* species—a very pathogenic bunch.

Many researchers fear byssinosis leads to chronic obstructive lung disease. Bouhuys & Zuskin (1976) stated, "We believe that deterioration of lung function among hemp workers begins before the age of 45 and it continues even if exposure to dust ceases." In one surveyed Spanish town, the average life span for hemp workers was only 39.6 years. Nonhemp farm workers in the same town lived an average of 67.6 years (Schilling 1956).

SEED

Plants consist mainly of carbohydrates. Compared to animals, plants contain much less protein and oil. Herbivores must therefore consume large volumes of plant material to gain sufficient protein and oil. Seeds contains higher levels of protein and oil than the rest of the plant (see Table 3.2 in Chapter 3), so seeds becomes attractive—to us and to pests.

Cannabis seed is technically an achene—a small, dry nut. *Cannabis* plants are prolific seeders. Haney & Kutsheid (1975) counted nearly 7000 seeds per plant in scrubby feral hemp. Herer (1991) claimed over half the weight of a well-pollinated female turns to seed. Field-grown crops yield an average of 400 g seeds per plant (Watson, pers. commun. 1998), or 0.5–1.0 metric t ha^{-1} (Pate 1999b). Males (staminate plants) are prolific pollen producers; an average male sheds 40 g pollen, or over 500 million pollen grains (Miller 1970).

Special landraces for seed oil have been selected in Europe, West Asia, Chile (De Meijer 1999), and China (Clarke, pers. commun. 1997). A new cultivar, 'FIN-314,' has been bred for seed production in northern climates (Callaway & Laakkonen 1996). Canadian agonomists are currently breeding new seed cultivars. The drug cultivar 'Skunk No. 1' is a prolific seed and oil producer (Latimer 1996). Many fibre varieties produce seed consisting of 25–35% oil by weight (Dempsey 1975, Deferne & Pate 1996, Pate 1999b). A Russian cultivar, 'Olerifera,' reportedly contains 40% oil (Small *et al.* 1975). Hemp yields average 455 l oil per ha (48 gallons/acre), compared to 560 l ha^{-1} from sunflower, but yields of hemp oil up to 800 l ha^{-1} have been reported.

Hemp seed oil has been used for industrial applications such as lighting, lubrication, and a base for soaps and detergents. The fatty acids in hemp oil are "quick drying," which makes hemp oil useful for paints, varnishes, and printing inks. Hemp oil can also be burned as fuel; it has combustion and viscosity ratings similar to # 2 heating oil.

Hemp seed is very nutritious. In China, whole hemp seed is commonly eaten, roasted and raw. Li (1974) reported a traditional Chinese belief regarding the consumption of hemp seeds, "...if eaten over a long term... it makes one communicate with spirits and lightens one's body."

Per 100 g, *Cannabis* seeds contain 421 calories, 21.7–27.1 g protein, 27.1–34.7 g carbohydrate (including 18.8–20.3 g fibre) and 25.6–30.4 g fat, plus many vitamins and minerals (Duke 1985). Wirtshafter (1997) offered slightly different numbers: 503 calories, 22.5% protein, 35.8% carbohydrate, and 30% fat. Wirshafter also analysed vitamins, minerals, amino acids, and fatty acids in hemp seeds.

The primary hemp seed protein, edestin, is easily digested and contains all eight essential amino acids (Pate 1999b). Hemp seed oil contains a rich array of fatty acids—80% polyunsaturated, 10% monosaturated, and 10% saturated (Pate 1999b). Some polyunsaturated fatty acids are essential nutrients, such as omega-3 fatty acids (e.g., linolenic acid) and omega-6 fatty acids (e.g., linoleic acid, gamma-linolenic acid, stearidonic acid). The percentage of omega-6 and omega-3 fatty acids in hemp oil are present in a 3:1 ratio, considered optimum for human nutrition (Pate 1999b).

SEED DETERIORATION

Many seeds are lost to other organisms. As neatly summed up in an olde English verse:

"One for the rat, one for the crow,
one to rot, and one to grow."

Fungi may ruin entire lots of poorly-stored hemp seeds. Pietkiewicz (1958) and Ferri (1961b) investigated the microflora of stored seeds. Both researchers isolated *Alternaria alternata, Rhizopus stolonifera, Cladosporium herbarum,* and several *Fusarium* species. Stepanova (1975) also investigated stored seed, but he only identified fungi to genus: *Alternaria, Cladosporium, Aspergillus, Penicillium,* and *Fusarium.* Babu *et al.* (1977) identified 16 fungi and a *Streptomyces* species growing on the *surfaces* of seeds. More importantly, they uncovered ten organisms from *internal* parts of seeds: *Penicillium chrysogenum, Penicillium frequentans, Penicillium chermesinium, Penicillium lavitum, Penicillium fellutanum, Penicillium chrlichii, Aspergillus niger, Aspergillus sulphureus, Cladosporium herbarum,* and *Cephalosporium curtipes.*

Harvey (1925) discovered that hemp seeds served as excellent "bait" to attract water moulds. He steam-sterilized seeds and suspended them in pond water. Since Harvey's discovery, many moulds new to science have been described from "baited" hemp, including a dozen *Achlya* species, *Allomyces arbuscula, Isoachlya anisospora, Pythium* species (Ogbonna & Pugh 1982), *Phytophthora, Pythium,* and other *Allomyces* species (Fatemi 1974), *Pythium perigynosum* (Sparrow 1936), *Mortierella raphani* var. *cannabis* (=*M. vantieghemi* var. *cannabis*) (Zycha *et al.* 1969), *Geolegnia inflata,* and *Geolegnia septisporangia* (Lentz 1977). *Olpidium luxurians* was isolated from similarly-treated hemp pollen (Sparrow 1960). None of these fungi cause decay in properly stored seeds.

Many insects feed on immature seeds in the field, such as hemp aphids (*Phorodon cannabis*), hemp flea beetles (*Psylliodes attenuata*), hemp borers (*Grapholita delineana*), and European corn borers (*Ostrinia nubilalis*). In Kansas, Eaton *et al.* (1972) also found *Chaetonema* flea beetles, treehoppers (Membracidae), leafhoppers (Cicadellidae), cucumber beetles (*Diabrotica* species), and stinkbugs (*Nezara* species) feeding on seeds. Seed-destroying mites include *Acarus siro* (=*Tyroglyphus farinae*), *Glycyphagus destructor* (Boczek *et al.* 1960), *Tyrophagus noxius* (Dombrovskaya 1940), *Tyrophagus longior* (Chmielewski 1984), *Tyrophagus putrescentiae* (Chmielewski & Filipek 1968), *Aculops cannabicola* (Hartowicz *et al.* 1971), and *Epitetranychus* species (Rataj 1957).

Some insects are only associated with seeds in storage, not in the field. Arnaud (1974) listed the Indian meal moth (*Plodia interpunctella*), foreign grain beetle (*Ahasverus advena*), rusty grain beetle (*Cryptolestes ferrugineus*), flat grain beetle (*Cryptolestes pusillus*), saw-toothed grain beetle (*Oryzaephilus surinamensis*), milichiid fly (*Desmometopa* species), fruit fly (*Drosophila busckii*), fungus gnat (*Bradysia* species), and scavenger flies (*Scatopse fuscipes* and *Desmometopa* species). Glover (1869) cited *Laemophloeus modestus,* a cousin of the flat grain beetle. Strong (1921) and Smith & Olson (1982) found the confused flour beetle, *Tribolium confusum,* bumbling in marijuana seeds. Schmidt (1929) reported Syrphid larvae in hemp seeds.

Rataj (1957) characterized birds as the most damaging pests of outdoor seed crops. Mice are the greatest destroyers of stored hemp seeds. See Chapter 6 for a discussion of birds and mammals in *Cannabis* crops.

CONTROL OF SEED DETERIORATION

Hemp seed, like finished fibre, should be stored in cool, dry conditions (Toole *et al.* 1960). Seed maintained near 1°C retains highest germination rates. Scheifele (1998) recommended drying seed to <12% moisture before storage, and heat used for drying seed should not exceed 30°C.

Seeds harbouring bacteria and fungi can be disinfested with thermotherapy (see method 11 in Chapter 9). Seed not destined for food use can be treated with pesticides (see "Pesticide Application" near the end of Chapter 11). Review articles by Ferri (1961), Robinson (1943a), and Koehler (1946) cite the extensive literature on this subject.

Ironically, researchers have used *Cannabis* leaves or *Cannabis* extracts to protect other crops. Zelepukha (1960) described a preparation from feral hemp that protects potato and tomato seeds from bacterial diseases. Pandey (1982) protected millet seeds from over 25 species of fungi with extracts of Indian *Cannabis*. Riley (1892) and MacIndoo & Stevers (1924) scattered *Cannabis* leaves among bags and heaps of grain to protect against grain weevils.

EUPHORIANTS & MEDICAMENTS

Several investigators believe *Homo sapiens* was originally attracted to *Cannabis* for its euphoriant qualities, not for its fibre or seed. Reininger (1946) even speculated, "Use of *Cannabis* for fibre occurred rather late." The Scythians were likely aware of its intoxicating properties. Indian use of *Cannabis* as a drug may date back 12,000 years (Morningstar 1985). Morningstar hypothesized that the meditative practices of Hindu culture helped make *Cannabis* socially acceptable. In contrast, the plant's euphoric effects were incompatible with Chinese philosophy (Li 1974). Thus, the Chinese utilized the plant for fibre, food, and medicine, but not as an intoxicant.

Use of *Cannabis* as a medicine extends into prehistory. In *Cannabinoids as Therapeutic Agents*, Mechoulam (1986) listed 20 medicinal uses of *Cannabis* employed by traditional societies. See Table 8.1, reprinted from Mechoulam's preface (with Raphael's permission).

In Western medicine, *Cannabis* enjoyed a 100-year heyday, beginning with two Irish physicians stationed in India (O'Shaughnessy 1839, Donovan 1845), and ending with the USA Marihuana Tax Act of 1937. The inaugural issue of the *Journal of the American Medical Association* praised the valuable properties of *Cannabis indica* (Brown 1883). Sir William Osler, recognized as the finest physician of his time, recommended *Cannabis indica* as the "most satisfactory remedy" for migraine (Osler 1918). By 1937, the year it was restricted, at least 28 pharmaceutical preparations in the American Pharmacopeia contained *Cannabis indica* (Sasman 1938).

Table 8.1: Medicinal use of *Cannabis* in folklore.

Analgesic*	Antiparasitic*
Anaesthetic	Antirheumatic*
Antiasthmatic*	Alleviation of memory loss
Antibiotic*	Appetite promoter*
Anticonvulsive*	Facilitation of childbirth
Antidepressive	Hypnotic*
Antidiarrhoeal*	Reduction of fatigue
Antimigraine	Sedative*

*Modern research substantiates the medicinal uses marked by an asterisk.

Alternative medical systems utilizing *Cannabis* include Ayurveda (Dash 1989a, Kapoor 1990), Tibetan medicine (Molvray 1988, Dash 1989b), Chinese medicine (Reid 1987, Bensky *et al.* 1993), and homeopathic medicine (Allen 1875, Sukul *et al.* 1986). Among 19th-century American herbalists, the Eclectics praised *Cannabis* (King 1854, Scudder 1875), but the Physiomedicalists condemned the herb as a neural toxin (Thurston 1900). To treat *Cannabis* overdoses, lemon juice has been recommended by homeopaths (Hamilton 1852) and Ayurvedic practitioners (Shanavaskhan *et al.* 1997).

The taxonomy of medicinal *Cannabis* is somewhat confused, as described in Chapter 1. True *Cannabis indica* was utilized by Ayurvedic practitioners in India; ancient healers carried this plant to southeast Asia and Africa over 1000 years ago (Clarke 1999). In the 1800s, Indian servants brought *C. indica* to British Jamaica, where it spread to Mexico and South America. *Cannabis afghanica*, in contrast, comes from Afghanistan and Pakistan. Two modern cultivars, 'Afghani #1' and 'Hindu Hush,' are stable progenies of *C. afghanica*, bred for ten years in California (De Meijer 1999). Stable, inbred varieties of *C. indica* include 'Durban,' from South Africa, and 'Haze,' a multihybrid with ancestors from Columbia, Mexico, Thailand, and southern India (De Meijer 1999). 'Skunk #1' is a cross of *C. indica* (from Mexico and Columbia) and *C. afghanica*. 'Northern Lights' is one-fourth *C. indica* (from Thailand) and three-fourths *C. afghanica* (De Meijer 1999).

Stockberger (1915), who probably grew *C. indica*, reported a yield of 400–500 lbs of dried, unseeded flowering tops per acre (448–560 kg ha^{-1}). Under intensive indoor cultivation, modern *indica-afghanica* hybrids yield 250–500 g flowering tops per m^2, or 2,500–5,000 kg ha^{-1} (D. Watson, pers. commun. 1998). Dried flowers yield 10–15% THC, or 250–750 kg THC ha^{-1}.

Field-cultivated *Cannabis* yields about 1.3 litre of essential oil per metric tonne fresh weight, or about 10 l ha^{-1} (Mediavilla & Steinemann 1997). The yield of essential oil increases at low seeding rates (optimally 5 kg $ha^{-1} \approx$ 15 plants per m^2), and increases when pollination is prevented (18 l ha^{-1} in sinsemilla crops, versus 8 l ha^{-1} in pollinated crops). In pollinated crops, yield peaks when about 50% of seeds reach maturity (Meier & Mediavilla 1998).

MEDICINAL PLANT ANATOMY

Trichomes (leaf hairs) cover nearly all above-ground parts of *Cannabis*. There are two types of trichomes, *glandular* and *nonglandular*.

Glandular trichomes are the site of cannabinoid synthesis. Glandular trichomes come in three sizes (Hammond & Mahlberg 1977). All are divisible into an upper secretory section ("head") subtended by an auxiliary section ("stalk"). The smallest and simplest glandular trichomes, termed *bulbous glands*, usually consist of two-celled heads on two-celled stalks. Their overall height is 25–30 μm, including a 20 μm diameter head. Bulbous glands arise on all aerial parts of the plant except hypocotyls and cotyledons.

The second type, *capitate-sessile glands*, produce heads two or three times larger than bulbous glands. These heads consist of eight to 16 secretory cells arranged in a flat disc, covered by a tough but distensible sheath. Accumulation of secretions beneath the sheath swell it into a spherical head up to 60 μm in diameter. Capitate-sessile glands are not truly sessile but possess a short stalk one cell in height and two to four cells thick. They arise on leaves and flowers.

The third type, *capitate-stalked glands*, produce heads similar to those of capitate-sessile glands (but larger—up to 120 μm diametre), which are subtended by large multicellular stalks 100–200 μm tall (Fig 3.2). Some researchers say capitate-stalked and capitate-sessile glands are identical, just different developmental stages of the same structure (Dayanandan & Kaufman 1978, Pate 1994). Differences exist: capitate-stalked glands do not form until plants flower, whereas capitate-sessile glands (and bulbous glands) form months earlier. Capitate-stalked glands arise only on bracts and tiny subtending leaflets of flowers (Hammond &

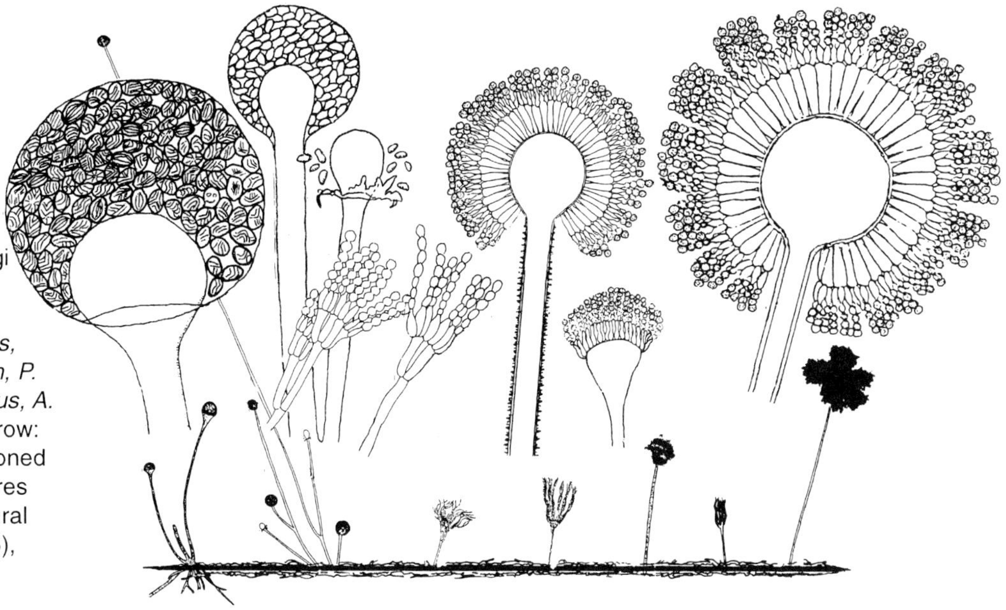

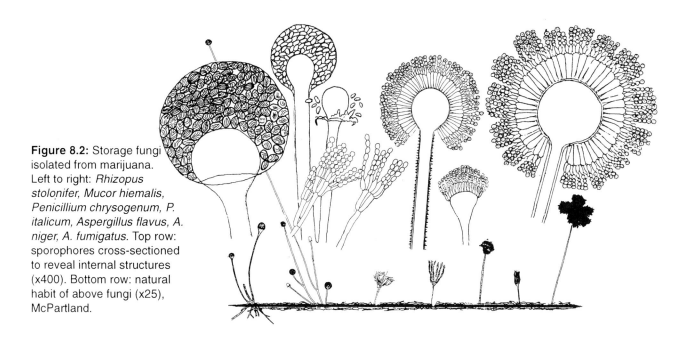

Figure 8.2: Storage fungi isolated from marijuana. Left to right: *Rhizopus stolonifer, Mucor hiemalis, Penicillium chrysogenum, P. italicum, Aspergillus flavus, A. niger, A. fumigatus.* Top row: sporophores cross-sectioned to reveal internal structures (x400). Bottom row: natural habit of above fungi (x25), McPartland.

Mahlberg counted up to 842 glands on a single bract!). Lastly, stalked glands contain higher levels of THC and CBD than sessile glands.

Nonglandular trichomes have been studied from a forensic viewpoint, since they survive combustion and serve as a marker of marijuana ashes. Nonglandular trichomes are unicellular with silicified walls and warty surfaces. Nonglandular trichomes come in two sizes. Shorter ones (70–125 µm long) contain cystoliths of calcium carbonate and are generally restricted to adaxial (upper) leaf surfaces (Fig 3.6). Long trichomes (250–370 µm in length) resemble long curved glass needles. They arise on ad- and abaxial leaf surfaces, flowers, petioles, and stems.

STORAGE DISEASES AND PESTS

Marijuana crops are unusual considering they may not reach consumers for years after harvest. This interlude provides wide berth for other "consumers" to cause decay and destruction. These "consumers" can be categorized into two groups. **Group 1** consists of **field organisms** which infest living plants and carry over as storage problems. **Group 2** consists of **storage organisms** which are saprophytes and only invade dead plants after harvest.

Group 1 fungi include *Botrytis cinerea, Sclerotinia sclerotiorum,* and *Alternaria alternata.* In living plants, these fungi cause grey mould, hemp canker, and brown blight, respectively. Less damaging Group 1 fungi include *Cladosporium herbarum, Epicoccum nigrum, Stachybotrys lobulata, Stemphylium botryosum,* and several *Fusarium* and *Mucor* species.

Group 1 bacteria contribute to the curing of harvested *Cannabis.* **Curing** represents the breakdown of chlorophyll, polyphenols, and plant starches, which eliminates the harsh, green, "homegrown" taste of fresh marijuana. Plant enzymes may also be involved; one possible enzyme may be o-diphenol oxidase (=polyphenol oxidase), the enzyme involved in the curing of tea leaves.

Group 1 insects, such as aphids and spider mites, rarely damage marijuana in storage. Like many other fastidious organisms (e.g., viruses, nematodes, phytoplasmas, etc.), these insects die with their host at harvest time. Exceptions include *Grapholita delineana* (hemp borers) and *Helicoverpa armigera* (budworms).

Group 2 insects were studied by Smith & Olson (1982). They studied confiscated Mexican marijuana, at the request of DEA agents whose offices were overrun by these insects. The predominant pest, *Tribolium confusum* (confused flour beetle), only fed on seeds (not marijuana). Two other beetles cited in the study, *Adistermia watsoni* and *Microgramme arga,* were fungus feeders (the marijuana was mouldy). It seems the researchers found no insects that actually fed on marijuana proper—no *Cannabis* equivalent to *Lasioderma serricorne,* the tobacco cigarette beetle. Arnard (1974) investigated "infested evidence" in San Francisco. He discovered scavenger flies (*Scatopse fuscipes, Drosophila busckii, Desmometopa* species, and *Bradysia* species), two fungus-eating beetles (*Microgramme arga* and *Typhaea stercorea*), and a mite, *Machrocheles muscadomesticae.* Piao (1990) examined *Cannabis* in Chinese herbal medicine stores and found three mites, *Glycyphagus destructor, Tydeus kochi,* and a *Tarsonemus* species.

Group 2 bacteria include *Clostridium* species which rot *Cannabis* in anaerobic conditions, such as damp marijuana stored in an airtight container (Bush Doctor, unpublished data 1977). Many aerobic bacteria grow on damp marijuana. Some of these bacteria harm humans—*Klebsiella pneumoniae, Enterobacter cloacae,* group D *Streptococcus* (Ungerleider *et al.* 1982) and *Salmonella muenchen* (Taylor *et al.* 1982). Kurup *et al.* (1983) isolated several actinomycetes from marijuana cigarettes—*Thermoactinomyces candidus, Thermoactinomyces vulgaris,* and *Micropolyspora faeni.* These actinomycetes cause allergenic reactions and "Farmer's lung" disease. Ramírez (1990) reported four policemen contracting pulmonary histoplasmosis (from *Histoplasma capsulatum*) after pulling up marijuana in Puerto Rico.

Group 2 **storage fungi** are the primary cause of storage losses—*Aspergillus, Penicillium, Rhizopus* and *Mucor* species (Fig 8.2). "Storage moulds" predominate in storage because they have evolved as such. They thrive under low oxygen levels, limited moisture, and intense competition for substrate. For instance, Grewal (1989) found that dried *Cannabis* leaves suppressed the growth of most mesophilic fungi, *except* the storage fungi *Aspergillus* and *Penicillium.* Dahiya & Jain (1977) tested the effects of cannabinoids on 18 species of fungi; THC and CBD inhibited the growth of all but two organisms, the storage fungi *Aspergillus niger* and *Penicillium chrysogenum.* Indeed, some storage moulds can metabolize THC (Robertson *et al.* 1975, Binder 1976).

Aspergillus

The genus *Aspergillus* includes over 500 species. These ubiquitous fungi grow on anything from rocket fuel to astronauts. The genus is millions of years old; while *Homo sapiens* may come and go, *Aspergillus* will remain. These meek moulds, like their sister genus *Penicillium*, are found worldwide. *Aspergillus* species grow better in warmer climates, *Penicillium* in cooler climates.

Westendorp (1854a) first described an *Aspergillus* species mouldering *Cannabis;* he called it *Aspergillus conoideus*. *Aspergillus niger* has been isolated from mouldy marijuana (Kagen 1983) and mouldy hemp (Vakhrusheva 1979). Margolis & Clorfene (1975) and DuToit (1980) described a mould that *increased* marijuana potency. Bush Doctor (1993a) found *A. niger* moulding marijuana and determined the mould did *not* increase potency. The mould does, however, cause human disease. Schwartz (1987) scraped *A. niger* from the sinuses of a marijuana smoker suffering severe headaches.

Kagen (1983) isolated *Aspergillus fumigatus* from mouldy marijuana. *A. fumigatus*-moulded marijuana has caused bronchopulmonary aspergillosis (Llamas *et al.* 1978). Chusid *et al.* (1975) reported *A. fumigatus* causing near-fatal pneumonitis in a 17-year-old. They noted the patient buried his marijuana in the earth for "aging." No doubt the patient hoped for a potency-enhancing mould but *A. fumigatus* found him instead.

Lastly, Kagen (1983) isolated *Aspergillus flavus* from mouldy marijuana. *A. flavus* produces nasty metabolites called aflatoxins. Aflatoxins are acutely poisonous, with an LD_{50} of 0.25–0.55 mg kg^{-1} (nearly 2000 times more toxic than parathion, the most toxic insecticide). Aflatoxins are also carcinogenic (Moss 1996). Llewellyn & O'Rear (1977) found aflatoxins in marijuana contaminated by *A. flavus* and *Aspergillus parasiticus*.

Penicillium

There are half as many *Penicillium* species as *Aspergillus* species in the world, but they are twice as difficult to tell apart. As previously mentioned, *Penicillium* species grow worldwide but prefer cooler climates. Refrigerator conditions may actually encourage *Penicillium* infestation. These fungi rot fruits and vegetables, poison livestock that eat infected grain, and cause opportunistic infections in people. On the positive side, *Penicillium* species are cultivated in huge vats to provide us with antibiotics and vitamins. And *Penicillium roqueforti* puts the "blue" in its namesake cheese.

Kagen *et al.* (1983) and Kurup *et al.* (1983) isolated *Penicillium* from marijuana cigarettes. Babu *et al.* (1977) identified *Penicillium chrysogenum* attacking marijuana. *P. chrysogenum* grows abundantly in nature. It provided Alexander Fleming with penicillin. Bush Doctor (1993a) isolated *Penicillium italicum* from marijuana that was stored with an orange peel at 3°C. Adding peels to marijuana imparts a "pleasant bouquet" (Frank & Rosenthal 1978). In this case, the peel imparted a nidus of infection. *P. italicum*, the "blue citrus mould," is notorious for its ability to spread by contact (i.e., "one bad apple spoils the whole bunch").

Mucor

About 50 *Mucor* species are recognized worldwide. These fungi resemble yellowish-grey cotton candy. *Mucor* species grow on almost any organic matter. They specialize in dung—*Mucor* spores survive passage through animal guts, which gives them first access to animal dung. A very existential life cycle.

Five *Mucor* species have been described on *Cannabis*. *Mucor hiemalis*, a common soil fungus, was originally discovered on hemp (Wehmer 1903). It has been isolated from leaves (Saccardo 1904, Oudemans 1920), stems (Behrens 1902, Wehmer 1903) and hemp fibres (Oudemans 1920). *M. hiemalis* prefers cooler climates. The Japanese use *M. hiemalis* for cleaning crud from pearls. The fungus metabolizes soyabean curd into products like tempeh. *M. hiemalis* regrettably bioconcentrates (and cannot metabolize) paraquat from tainted substrates (Domsch *et al.* 1980).

The "*Mucor* species" recovered by Kagen *et al.* (1983) and Kurup *et al.* (1983) may have been *M. hiemalis*. Other possibilities include *Mucor mucedo*, a hemp retter cited by Lentz (1977), or three species that Gzebenyuk (1984) collected from Soviet hemp stems: *Mucor corticola*, *Mucor genevensis*, and *Mucor plumbens*.

Figure 8.3: A young *Aspergillus* conidiophore with many conidia (SEM x800, McPartland).

Rhizopus

This genus contains approximately 30 species. These fungi live in soil, on ripe foodstuffs and, occasionally, on people (especially diabetics). *Rhizopus* sporangiophores arise in clusters that grow from rootlike rhizoids. These clusters are connected by stolons, like strawberry plants. The presence of rhizoids and stolons differentiates *Rhizopus* species from *Mucor* species.

Gzebenyuk (1984) isolated *Rhizopus stolonifer* from hemp stems. This species is an important hemp-retting organism (Behrens 1902, 1903). Bush Doctor (1993a) infected damp marijuana with a colony of *R. stolonifer* found growing on stale bread. The fungus is found worldwide.

SIGNS & SYMPTOMS OF CONTAMINATION

Marijuana consumers identify microbiological contaminants using several crude but seemingly effective screening techniques (Bush Doctor 1993a, McPartland 1994, Conrad 1997). The techniques mimic those used by tobacco growers (Lucas 1975).

Visual screen: A golden colour means the herb was cured in sunlight; green or purple material was cured in darkness (Conrad 1997). Contaminated marijuana often darkens in colour and becomes crumbly. Anaerobic bacteria turn marijuana into brown slime. Fungi appear in mouldy material as tufts and strands of fungal hyphae. Hyphae look white to light grey when marijuana is stored in complete darkness. But if exposed to light, the tufts turn into velvet clumps of coloured spores. A slight tap sends these spores into billowing clouds. Clumps of *Rhizopus* and *Mucor* spores generally appear grey-black, *Penicillium* spores (conidia) are light blue-green, and *Aspergillus* conidia are dark green-black.

Improperly prepared hashish may also mould (Clarke 1998). Grey veins of mould become visible when hashish is broken. Cherniak (1979) illustrated a piece of mouldy hashish—see figure 8.17 in his book.

Fluorescence screen: Plant material freshly contaminated by aflatoxin will fluoresce a bright greenish-gold hue under ultraviolet light (Bush Doctor 1993a). Wood's lamps ("black lights" emitting a spectrum of 365 nm) made specifically for aflatoxin detection are sold in agriculture catalogues.

Olfactory & temperature screen: Rotting marijuana produces a spectrum of odours, from stale to musty to mouldy. *P. italicum* produces a lavender or lilac odour. Lucas (1975) reported *A. flavus* as smelling particularly offensive. Anaerobic bacteria stink like carrion. Marijuana undergoing rapid decay may feel warm to touch. (At this stage the product is ready for composting.)

Smoke screen: Infested material produces a brown, black, or sooty smoke. Healthy marijuana produces milky white or light blue smoke (Conrad 1997).

Screen screen: To expose insects, shake samples in a No. 10 steel sieve. Not all bugs found in marijuana cause storage damage. Some are simply "innocent bystanders" caught during harvesting. Crosby *et al.* (1986) found *fig-pollinating wasps* in confiscated marijuana. These rare insects helped police determine the marijuana grew in Thailand or southern Burma. Living insects in marijuana are more suspicious. Spreading samples on a sheet of white paper often reveals live insects as they crawl away. A hand lens is often helpful for the identification of small insects.

CULTURAL & MECHANICAL CONTROL

Avoid damaging plants before they completely dry. Wounded tissues release exudates upon which fungi establish a foothold. Lucas (1975) said diseased and nutrient-deficient leaves also release exudates. Expect more storage mould problems in poorly-grown plants.

The secret to stopping bacteria and moulds is *moisture control*. Flue-cured (or oven-dried) marijuana suffers less contamination than air dried or sweat-cured crops. Sweat-cured *Cannabis* (common in Colombia) maintains a "tradition" of *Aspergillus* contamination (Bush Doctor 1993a). Hashish must be pressed carefully, with no air trapped inside. Poorly hand-rubbed charas nearly always contains air pockets with moisture, resulting in fungal and bacterial spoilage (Clarke 1998).

Oven drying inevitably leads to a harsh-smoking product. So most people air-dry marijuana. Drying rooms should be cool and dry, preferably in *uninterrupted* darkness (most storage fungi require UV light to sporulate). Electrical fans improve air circulation. Drying marijuana in a 10% CO_2-supplemented atmosphere will reduce mould and insects (Wilson & Chalutz 1991).

Living *Cannabis* plants are about 80% water. According to Bush Doctor (1993a), perfectly dried marijuana contains about 10–12% water or moisture content (MC). Material below 10% becomes too brittle and disintegrates. Fungi cannot grow below 15% MC. Unfortunately, many growers prefer to market their crop above 15%. *Cannabis,* like corn flakes, is sold by weight, not volume. Tobacco farmers also allow their products to gain weight by resorbing moisture before sale. They term this risky business "coming into order" (Lucas 1975). Freezer storage will not protect material above 15% MC. Placing lemon peels in stored marijuana is discouraged; the peels may raise the MC above 15% and inoculate the marijuana with pathogens. Properly dried marijuana also discourages insects and mites.

BIOLOGICAL CONTROL

None of the organisms listed below has been tested on harvested marijuana. They have great potential as biocontrols, but may taste bad or pose a threat to immunocompromised individuals.

Pichia (Candida) guilliermondii (?=*Debaryomyces hansenii*)

BIOLOGY: An epiphytic yeast that colonizes numerous plants and controls postharvest rot in fruit caused by *Botrytis cinerea, Alternaria alternata, Penicillium digitatum, Penicillium italicum,* and *Rhizopus stolonifer.*

APPLICATION: Supplied as a powder or aqueous suspension. Plant material is sprayed or dipped in a yeast suspension, then dried. Adding $CaCl_2$ (2% w/v) to the solution improves control (Wilson & Ehalutz 1991).

Candida oleophila

BIOLOGY: An epiphytic yeast (ASPIRE®) that protects apples and citrus fruits from postharvest decay caused by *Botrytis* and *Penicillium* species.

APPLICATION: Supplied as a wettable powder. Plant material is sprayed or dipped in a yeast suspension, then dried.

Pseudomonas syringae

BIOLOGY: A bacterium (Bio-Save 11®) that protects apples and citrus fruits from postharvest decay caused by *Botrytis cinerea, Mucor* species, and *Penicillium* species. It is a nonpathogenic variety of a common plant bacterium. Strains ESC 10 and ESC 11 protect against postharvest decay; strain 742RS has a different purpose—protecting orchards from ice-crystallizing *Pseudomonas* species.

APPLICATION: Supplied as spores in pellets or a wettable powder. Plant material is sprayed or dipped in a suspension, then dried.

CHEMICAL CONTROL

Dipping *Penicillium*-infested marijuana in a baking soda solution will inhibit the acid-loving fungi, but the marijuana must be rapidly re-dried. *Synthetic pesticides should never be sprayed on marijuana.* Spector (1985) reported youngsters spraying marijuana with formaldehyde. The treated weed caused anoxia and psychomotor retardation in smokers.

Fumigants (gases, not sprays or aerosols) may remain as gaseous residues in air pockets of marijuana flowers. Ungerleider *et al.* (1982) fumigated marijuana with ethylene oxide. They reported complete sterilization of fungi with no loss of THC, but did not comment on gaseous residues. Harvested tobacco is fumigated with high pressure CO_2, either 4000 kPa for 30 minutes or 3000 kPa for 50 minutes. This "Carvex process" leaves no dangerous residues.

Ungerleider *et al.* also sterilized marijuana with high-dose Cobalt 60 irradiation (15,000 to 20,000 Gray Units). *This method is not recommended for novices.* Seeds are killed by 20,000 Gray Units, and the presence of radiolytic substances in irradiated marijuana has not been evaluated.

CONSUMER HAZARDS

Immunosuppressed individuals and asthmatics should never be exposed to moulds, especially *Aspergillus* (Figs 8.2 & 8.3). Kurup *et al.* (1983) showed that spores of *Aspergillus* and *Mucor* species survive in smoke drawn from pyrolysed marijuana cigarettes. People using medical marijuana should take extra precautions. McPartland (1984) suggested patients switch to synthetic THC. But Doblin & Kleiman (1991) reported dissatisfaction with synthetic THC among patients and their physicians. Nearly half of all oncologists polled by Doblin & Kleiman recommended their patients switch to illegal marijuana.

Moody *et al.* (1982) tested water pipes with *Aspergillus*-contaminated marijuana and found only a 15% reduction in transmission of fungal spores. McPartland & Pruitt (1997) concluded that vaporizer devices are superior to water pipes for protecting smokers. Vaporizers heat marijuana to 180-190°C, a temperature range which vaporizes THC, but is below the ignition point of combustible plant material. Gieringer (1996) determined that vaporizers deliver a higher cannabinoid-to-tar ratio than cigarettes or water pipes. The best vaporizer delivered a 1:10 ratio, whereas cigarettes averaged 1:13, and water pipes averaged 1:27 (the worst pipe delivered 1:40).

Levitz & Diamond (1991) suggested baking marijuana in home ovens at 150°C (300°F) for five minutes before smoking. This temperature kills *Aspergillus* spores without vaporizing THC. Unfortunately, this temperature is not sufficient to degrade *Aspergillus* antigens, so sensitized patients may still develop asthma (McPartland 1994).

"*Since insects for the most part lack cunning or intelligence, insect traps are often surprisingly simple.*"
—C. L. Metcalf

Chapter 9: Cultural & Mechanical Methods of Controlling Diseases and Pests

Cultural and mechanical methods encompass a wide variety of techniques that are used by both organic farmers and conventional farmers. Cultural and mechanical methods can be applied to any aspect of the crop damage triangle described in Chapter 1. *Cultural* methods are usually preventative, and consist of ordinary farm practices that encourage healthy plant growth. These methods alter the landscape and make it less favourable for pest reproduction and survival, such as tilling the soil, crop rotation, regulating moisture levels, and balanced use of fertilizers. *Mechanical* methods can be preventative (e.g., steam sterilization of glasshouse soil) or curative (e.g., vacuuming bugs off leaves or pruning away fungus-infected branches). Mechanical methods can be as simple as handpicking pests from plants or as complicated as electronic insect traps.

Some of the most useful cultural and mechanical methods are listed below. Many of these methods are ancient. Many were abandoned with the advent of chemical pesticides, but are enjoying a resurgence in popularity. David West points out that some of these methods pertain to *agronomic* crops (e.g., fibre and oil seed crops), whereas other methods are better suited for *horticultural* situations (e.g., pharmaceutical *Cannabis*).

For sources of sticky materials, traps, electric fencing, etc., obtain the annual **Directory** published by BIRC (Bio-Integral Resource Centre), P.O. Box 7414, Berkeley, California 94707, telephone: (510) 524-2567.

1. Sanitation:

a. Destroy crop residues after harvest. Shredding of stalks and roots exposes overwintering pests and pathogens to their natural enemies and the environment. Shredding hemp stalks requires special tillage equipment, however, because standard shredders cannot adequately cope with the fibre (Moes, pers. commun. 1999).

   Covering crop residues with soil, using a mouldboard plough, will kill many pests and pathogens (see comments under 2a, below). Cart residues off site for burial, burning, or composting. Composting must be done properly—compost piles may become trash heaps and turn into pest nurseries (Jarvis 1992). Remove anything that may shelter overwintering pests, such as weeds and surface trash (dead leaves, brush heaps, boxes, equipment, etc.).

b. Wash residues from glasshouse walls, cultivation equipment, and clothing (especially boots). Heat-sterilize hand tools, previously-used rockwool, and polystyrene trays. Pests and pathogens are killed at different temperatures, see Table 9.1. Rockwool should be as dry as possible before it is steam-sterilized, since high temperatures are attained more quickly; slabs stacked 1.5 m high on pallets (with plastic covers removed) take two hours to reach 100°C in an autoclave (Runia 1986).

c. Hydroponic cultivators should disinfect recirculating nutrient solutions. Plant diseases caused by water moulds can quickly shut down an entire hydroponic farm (McEno 1990). Several methods for sterilizing large quantities of nutrient solutions are reviewed by Jarvis (1992). Precipitation of calcium is a problem common to all methods. Jarvis also reviewed ultraviolet irradiation, which is feasible for small operations but it precipitates iron. Ultrafiltration of nutrient solutions through pores of about 10 µm at 300 kPa eliminates fungi but not viruses. Ozone bubbled through nutrient solution at a concentration of 1.5 mg l^{-1} for 30 minutes kills pathogens but may be phytotoxic to *Cannabis*. Sodium hypochlorite at concentrations of 1–5 mg l^{-1} kills many pathogens but not all. Sterilization of nutrient solutions with chlorine gas kills pathogens but causes phytotoxicity if concentrations of free chlorine exceed 10 mg l^{-1} (Jarvis 1992).

Table 9.1: Thermal inactivation of selected pathogens and pests. (from Jarvis 1992).

Organism	Temperature (°C)	Exposure time (minutes)
Botrytis cinerea	55	15
Sclerotinia sclerotiorum	50	5
Pythium ultimum	46	20–40
Fusarium oxysporum	60	30
Rhizoctonia solani	53	30
most other fungi	60	30
most bacteria	60–70	10
most viruses	100	15
most nematodes	48–56	10–15
most insects and mites	60–70	30
most weed seeds	70–80	15

2. Work the soil after harvest:

a. Cultivation with tillage equipment exposes soil pests and pathogens to their enemies and to the weather. Properly done, tillage also reduces soil compaction, which reduces root rot. Improperly done, tillage leads to a loss of soil structure, organic matter, moisture, and beneficial microorganisms. Repeated tilling may accelerate soil erosion from wind and water (Cook *et al.* 1996).

   Rotary tillers fluff surface soil nicely, but they are slow and not very energy efficient. Chisel ploughs with 10 cm twisted chisels work the same depth as rotary tillers, but tend to get plugged with crop residues. Disk ploughs with standard-sized disks penetrate soil 15 cm or more; unfortunately, disks may compact soil worse than mouldboard ploughs. Mouldboard ploughs, considered *outré* by many people, suck up and turn over the top 20 cm of soil, but mouldboards capable of reaching a metre down have been used against deep-living soil pathogens (Wolf 1999). Soil damage caused by annual ploughing has prompted the use of "no-till" techniques. No-till has soil benefits (especially on land with little organic material), but pests, pathogens, and weeds also benefit.

b. Heat-sterilize the soil. Heating soil to 80°C for 30 minutes kills most soil organisms (Table 9.1). Steam heat is economical. Batches of soils, compost, and rock wool can be steamed in autoclaves. Autoclaves come in various sizes, including walk-in models. For small fields, steam can be injected into soil through hollow-tined soil injection devices, or through perforated pipes buried permanently underground.

Small batches of soils can be heat-sterilized in conventional ovens: Place moist soil in a shallow pan, cover the pan with aluminium foil, and insert a meat thermometre into the soil. Ellis & Bradley (1992) suggested preheating the oven to 93°C (200°F), and heating soil to a temperature between 60–82°C (140–180°F) for 30 minutes. If soil temperature exceeds 82°C, remove the pan from the oven and let it cool below 82°C before returning it to the oven. To use a microwave oven, heat 1 kg of damp soil for 150 seconds, using a 625 watt microwave oven at full power. Rockwool slabs can also be microwaved, but microwaves may be poorly distributed in slabs, preventing effective sterilization (Runia 1986).

Soil sterilization has drawbacks—it eliminates beneficial organisms and creates a biological vacuum. An accidental reintroduction of pathogens will cause epidemics in sterilized soil. Avoid reintroduction by washing all equipment touching soil—including shovels and (s)hoes—in a solution of sodium hypochlorite (bleach). But even with the best precautions, pathogen reintroduction may occur via seedborne infection.

Another drawback to sterilization is nutrient toxicity. Manganese in soil heated above 80°C becomes available to plants in toxic amounts, particularly in acidic, clay soils (Jarvis 1992). Unless this free manganese is leached from the soil, it may remain toxic for 60 days or more and contribute to iron deficiency. Nitrogen in steamed soil also undergoes changes. Nitrifying bacteria are killed, so ammonia may build up to phytotoxic levels (Jarvis 1992).

c. Pasteurization may be better than sterilization—it kills pathogenic fungi while maintaining a rich microflora of beneficial organisms. Pasteurization is accomplished by mixing steam with air at 60°C. Pasteurization is performed in the field by *soil solarization,* where the soil is covered by sheets of plastic and heated under the sun for four to six weeks (Elmore *et al.* 1997). Intense summer sunlight can heat the top 5 cm of soil to 55°C, and to 37°C as deep as 45 cm. Solarization works best in sunny climates, obviously. The field must be carefully disked and its surface smoothed, removing any rocks or clods that might raise the plastic or puncture it. Treated soil must be kept moist so that the heat will penetrate evenly (consider laying drip irrigation lines before laying plastic). Plastic sheets should be polyethylene, transparent, and at least 76 cm wide. Proper thickness ranges from 1 mil (0.001 inch or 0.0025 mm thick) to 2 mils, depending on the wind. Plastic sheets must be laid tightly against the soil; row edges are anchored by burying the edges in a shallow trench.

Solarization controls many soilborne fungal and bacterial pathogens, as well as some nematodes, weed seeds, and insects. Beneficial fungi, such as *Trichoderma* and mycorrhizae, are very heat resistant. Earthworms survive by escaping to lower soil depths. Carefully disking a *Brassica* crop into the soil before laying plastic provides a combination of mechanical and chemical control. Winter rape (*Brassica napus*) and certain broccoli cultivars ('Brigadier') produce high levels of glucosinolates. In the soil, glucosinolates decompose into isothiocyanates and nitriles, which kill fungal spores, weed seeds, and some insects.

d. Flooding the soil. This technique is like pasteurization—it kills plant-pathogenic fungi, nematodes and soil insects while maintaining a population of nonpathogenic microorganisms. Artificial inundation can be costly for large fields. But some land, such as fields alongside rivers and lakes, may be naturally flooded or neighbouring waters can be diverted easily (Bridge 1996).

Figure 9.1: Row cultivation of hemp (from Lesik 1958).

3. Work the soil during the early growing season, between seedling emergence and canopy formation. Soil tillage exposes soil pests and eliminates weed seedlings. Many pests and pathogens are attracted to weeds first, then move to *Cannabis.* Weeds also hold humidity which favours mould development. Keeping the soil clean eliminates insects that cannot lay eggs on bare ground. Of course, bare ground may also accelerate soil erosion from wind and water (Cook *et al.* 1996). Rarely, pests *luxuriate* in weed-free fields, because certain weed-dependent biocontrol organisms are lacking in these situations (Altieri *et al.* 1977).

For small plots, a standard hoe works well. Cultivate large fields with tractor-drawn tines, spike-toothed harrows, or rotary brush weeders (Fig 9.1). Alternatively, cover rows of soil with rolls of black woven plastic, or plastic ground cover ("plastic mulch"), and grow plants through holes punched in the plastic.

4. Escape cropping—vary the time of planting and/or harvesting. Early planting permits susceptible seedlings to harden before pests arrive. Late planting avoids cool temperatures that predispose seedlings to damping off and root rots. Late planting also eludes the egg-laying period of some pests. Early harvest avoids the endless autumn rains that predispose plants to grey mould in the Pacific Northwest and the Netherlands. Early harvest also reduces flea beetles and theft.

5. Use resistant varieties. Most hemp breeders qualify and quantify their cultivars by fibre yield and THC content, with little regard for pest resistance (Bòcsa & Karus 1997). Nevertheless, a few varieties have been selected for their resistant traits. In most cases, the mechanism of resistance is physiological and poorly understood. But sometimes it is simple—for instance, slimmer buds hold less moisture which protects them from grey mould. Another mechanism of resistance is chemical—the terpenoids, ketones, and cannabinoids synthesized in *Cannabis* trichomes. Selecting plants for trichomes poses a dilemma, however, because these features also suppress biocontrol organisms, including predaceous mites, ladybeetles, *Chrysoperla carnea, Encarsia formosa, Geocoris punctipes,* and various *Trichogramma* and *Orius* species (Campbell & Duffey 1979).

Traditionally, we have manipulated the *Cannabis* genome using sexual propagation. Sex introduces genetic variability, the cornerstone of breeding programmes. Breeders promote desired traits by methodically hand-pollinating choice females with pollen from choice males. *Cannabis* is wind-pollinated, so contamination of females with unselected pollen must be avoided. Lemeshev *et al.* (1995) recommend a distance of at least 2 km between fields to provide spatial pollen isolation. If this is not possible, Clarke (1981) detailed the use of pollen-collecting bags and protection of female flowers. Since most desirable traits (e.g., high yield) are quantitative, more than two plants are needed for a breeding programme. Breeding techniques well-suited for *Cannabis* include the Ohio technique, Bredemann method, modified Bredemann method (Bòcsa & Karus 1997), and reciprocal recurrent selection (Seven Turtles 1988). The latter strategy maintains two select populations while crossing them to produce hybrid offspring.

Once a choice hybrid is selected, it can be preserved by asexual propagation. Clones from a single plant are genetically identical. Clones are propagated by cutting-rooting or by air-layering. Clarke (1981) and Frank (1988) described both propagation methods in detail. Clones are maintained under artificially created long daylength.

To induce mutations in a search for useful traits, hemp breeders have blasted pollen and seeds with gamma (γ) radiation (Zottini *et al.* 1997). Genetic engineers have begun sequencing *Cannabis* DNA (Gillan *et al.* 1995, Faeti *et al.* 1996, Jagadish *et al.* 1996, Siniscalco Gigliano *et al.* 1997, Linacre & Thorpe 1998). In the future we will manipulate the *Cannabis* genome via recombinant DNA technology. Researchers are currently having trouble regenerating *Cannabis* callus tissue into whole plants (Mandolino & Ranalli 1999).

6. **Crop rotation keeps pests moving,** which is a lethal inconvenience. It eliminates many pests and pathogens that specifically attack *Cannabis.* Hemp crops cultivated continuously in monoculture suffer more pests than crops grown in rotation (Bòcsa & Karus 1997). Crop rotation is less effective on general feeders, migratory pests, and pathogens that survive for long periods in the soil.

7. **Maintain proper moisture levels** that are optimal for plant growth:

a. Drought kills plants outright and also predisposes plants to diseases and pests.
b. Overwatering also kills plants and predisposes them to diseases and pests. It is a common problem with novice growers. Allow soil to dry between waterings, and do not plant in poorly-draining soils. If poorly-draining soils are unavoidable, plant in raised rows.
c. Avoid excess atmospheric humidity. Excess humidity is a common problem and permits many fungi and some pests to flourish.

Outdoors, increase air circulation between plants by not overcrowding them. Humidity always increases after canopy closure, which is when leaves of adjacent plants touch each other and shade the soil.

Canopy closure is unavoidable in fibre crops, but can be avoided in seed crops and seedling beds. Properly-spaced plants keep the canopy open, increase air circulation, and reduce humidity and leaf wetness. Plant rows in the direction of prevailing winds, or in an east-west or northeast-southwest orientation to promote solar drying. Air circulates better on sloped hillsides or high points than on low or flat ground. Avoid overhead irrigation during flowering.

Indoors, proper ventilation is critical so humidity can escape. *In glasshouses, narrowing the temperature differential between night and day will lower the chance of dew formation.* Keep glasshouses cool during the day and heat them at night. Turn on heat (electric, propane, butane or natural gas, *not* kerosene or gasoline) before sundown. This slows the drop in temperature and prevents moisture condensation on plants. Heat all night and keep all vents open—although a "waste of heat," this is the only way to drive moisture out of the night air. During the day use air conditioning—cool air holds less water and lowers the atmospheric humidity. Avoid dew at all costs during flowering. Do not irrigate plants late in the day or at night.

Humidity increases in low-ceiling structures such as Dütch Venlo™ glasshouses. Never allow plants to fill more than 1/3 the volume inside a glasshouse or growroom. Have a high roof and short plants. Humidity increases in glasshouses using polythene blackouts (which restrict day length and induce flowering). Try using blackouts made of porous woven cloth, instead of plastic sheeting. Humidity also increases in glasshouses using thermal screens to reduce heat losses, and in glasshouses with microscreens installed to exclude outdoor pests. Keep all screens clean to optimize airflow.

8. **Optimize soil structure and nutrition:**

a. Balance nutrients carefully. Excess nitrogen predisposes plants to bacteria, fungi, and leaf-eating insects. Indiscriminate use of phosphorus to promote flowering is ill-advised for the same reason.
b. Improve soil structure (see Chapter 2). Adding organic amendments to soil augments naturally occurring biological control of pests and pathogens. This tactic works particularly well against soilborne fungi and nematodes. Adding well-composted material is safe and effective. Conversely, poorly composted materials may contain pests and pathogens. Adding municipal sludge to soil has *increased* the incidence of root-rot diseases (Windels 1997). Patented formulations of organic materials and minerals (e.g., S-H Mix™) have been sold for the suppression of soil pathogens. Some formulations also contain urea and other pesticidal compounds (e.g., Clandosan™) and are described under chemical control.

9. **Remove insect pests by hand**—eggs, larvae, pupae, and adults. Crush them or drop them into a bucket of soapy water. (Be sure they are pests and not beneficial insects!) Chilly mornings are a good time to shake beetles and bugs off plants and onto a ground cloth. Stem borers can also be removed by hand—carefully split galls lengthwise, remove borers,

then bind the stem back together. Some weakly-flying insects can be removed with a hand-held vacuum cleaner. Whiteflies, for instance, will conspicuously hover in the air next to a plant that is shaken. Evolution did not prepare whiteflies for vacuum cleaners. Suck them out of the air. Avoid tearing foliage in the vacuum by covering the nozzle with gauze. Vacuum cleaners are available in gas-powered backpack models.

10. Limbs with isolated fungal infections may be pruned from otherwise healthy plants. Many moulds establish a foothold in senescent plant tissues, so pruning yellow leaves and injured branches is preventative. Coat all wounds and pruning cuts with a fungicidal tree sealer or biocontrol agent (e.g., *Trichoderma harzianum* or *Gliocladium roseum*). Whole plants can be rouged from otherwise healthy fields or glasshouses. When rouging an infested plant, prevent the spread of airborne fungi or insects by quickly covering the plant with a big plastic bag, and cinching the bag tightly around the stalk before removing the plant.

11. Avoid seedborne infection. Do not save seeds from plants infested by viruses, bacteria, seedborne fungi, or mites. If seeds must be used, try disinfecting them with heat. Soaking seeds in hot water (50°C) for 30 minutes eliminates most pests and pathogens with minimal damage to seeds. But thermal damage may delay seed germination or stunt seedlings. Minimize damage by plunging seeds into cold water after treatment, then carefully dry them (Maude 1996). Maude also disinfected seeds in microwave ovens (try 625 W for three minutes then plunge into cold water). For chemical seed treatments, see Chapter 11.

12. Mechanically trap or repel insects with light and colour:

a. Cover soil with reflective material laid between plants to confuse flying insects and prevent them from depositing eggs. Aphid alatae (winged forms), female thrips, leafminer flies, and some leafhoppers are particularly susceptible. Unroll aluminium foil on the ground and perforate it with holes through which seedlings can grow. Keep the foil in place by covering its edges with soil. Foil must extend 10 cm beyond the plant dripline. Reflective material loses its effectiveness on taller plants.

b. Some winged pests have a fatal attraction for yellow objects, especially aphids, whiteflies, leafhoppers, leafminer flies, and fungus gnats (Plate 6). Catch them with "sticky traps," which are rectangular yellow cards coated with mineral oil, vaseline, Tanglefoot® or other sticky materials. The rectangles should be 25–50 cm wide and constructed of heavy cardboard, Masonite, or thin plywood. Rustoleum Yellow No. 659 reflects a wavelength of 550 nm and works the best (Olkowski *et al.* 1991). Hang rectangles vertically, with the centre of the rectangles level with the top of the crop. Support the squares on stakes or wire holders, and raise them as the crop grows. Yellow sticky ribbons (tapes) are also commercially available.

   Sticky traps work best in areas with little wind, otherwise they become covered by dust. Stir up winged insects by shaking plants periodically. For *monitoring* purposes, distribute one trap per 200 m² crop area. For *control* purposes, distribute one trap at least every 20 m², to a maximum of one trap per 2 m², and replace the traps frequently. Monitoring for pests is not easy—sticky traps may reap a heap of dead, distorted insects. Their twisted bodies look different than living insects or standard illustrations.

   Gill & Sanderson (1998) assembled over 25 photographs of insects on sticky traps. You must sort through the mix to separate pests from other insects. Unfortunately, some beneficial insects may also be attracted to yellow, including *Aphidoletes aphidimyza, Aphidius matricariae, Dacnusa siberica, Diglyphus isaea, Encarsia formosa, Eretmocerus eremicus,* and *Metaphycus helvolus.* Cherim (1998) removed sticky traps before releasing these biocontrols, and thereafter only hung traps for two days per week for monitoring purposes.

   Pimentel *et al.* (1991) recommended tilting yellow traps 45° off vertical to avoid trapping syrphids. Alexander (1984a) warned that yellow clothing also attracts pests; unwitting growers carry them into uninfested areas.

c. Attract thrips and root maggot flies with yellow sticky traps or *light blue* traps (reflecting a 440 nm wavelength). Blue traps are less attractive to beneficial insects.

   Recently, Cherim (1998) recommended using *hot pink* sticky traps to catch thrips. Prepare sticky traps as described above. Hang traps near tops of plants to catch flying thrips. Catch flea beetles and tarnished plant bugs with *white* sticky traps. Suspend traps near the top of the canopy to catch flea beetles; suspend traps beneath the canopy to catch tarnished plant bugs. Unfortunately, some predators that eat white pests may also be attracted to white sticky traps (e.g., *Cryptolaemus montrouzieri*).

d. Attract nocturnal moths with white light (a mix of wavelengths) or ultraviolet light. If coupled with an electric grid, a light trap can zap hundreds of moths a night. Each female moth dies with dozens to hundreds of eggs. Many electrocutor traps are simply light bulbs surrounded by an electric grid. People find them vaguely entertaining.

   Light traps can affect plant photoperiod so they must be used carefully during flowering. Different moth species tend to fly at different heights and speeds; they are caught preferentially by different trap designs (Young 1997). The attraction of light traps is limited to a radius of about 25 m; they are least effective at full moon (Young 1997), and most effective between the hours of 11 PM and 3 AM (Ellis & Bradley 1992).

e. Bait traps with synthetic insect pheromones, which are described in Chapter 11. Pheromones can be applied to sticky traps or placed within rubber septums, nylon or wire mesh cones, wing traps, aerial water pans, or funnel-bucket traps. Funnel-bucket traps hold the most carcasses.

13. Mechanical barriers protect plants from insects and other pests. Yepsen aptly described mechanical barriers as "traps without the power of attraction." To stop crawling insects, wrap stems with a sheath of aluminium foil and coat the foil with Tanglefoot® (do not apply Tanglefoot directly on stems—the product is a mixture of castor oil, gum resins, and wax—it stresses plants and may translocate to flowers).

Protect seedlings from flying pests and their egg-laying mothers. Outdoors, use tent netting, tightly woven cloth, polyethylene screens (Green-Tek®), or floating row covers (Harvest-Guard®, Reemay®). Indoors, install micro-screening to seal all openings, including louvres (vents), open windows, and exhaust fan openings.

To screen the smallest pests (thrips and mites), mesh holes need to be about 192 μm in diameter; finer mesh impedes glasshouse ventilation (Gill & Sanderson 1998). Immediately repair any tears in the screening. Install tight-fitting entrance doors.

A wire fence bars most mammals from crops. Mole traps guard against underground marauders, whereas "bird bells" protect against avian attack. Australians have taken mechanical barriers to the limit by stretching "vermin fences" across their entire continent to control the spread of rabbits.

14. Provide food, water, and habitat for biocontrol organisms. These cultural and mechanical techniques enhance the effectiveness of biocontrols described in the next chapter. Entice some predators (e.g., ladybeetles and lacewings) to stay in crops and lay eggs by providing artificial honeydew—mix honey and brewers yeast, apply to sticks or wax paper, and place among plants. Commercial products include Wheast®, Formula 57®, and Enviro-Feast®.

Provide vegetational diversity by intercropping or row cropping other plants with *Cannabis*. Mixed assemblages of plants disrupt pests and enhance biocontrols. Intercropping with flowers provides nectar and sanctuary for biocontrols. Biocontrols can only sup nectar from certain flowers, however, because their mouthparts differ from bees and butterflies. Suitable flowers are small, relatively open (not tubular), and have relatively long blooming periods. The most suitable flowers are found in four families: Umbelliferae choices include flowering fennel (*Foeniculum vulgare*), Queen Anne's lace (*Daucus carota*), and wild parsnip (*Pastinaca sativa*). Recommended Compositae include tansy (*Tanacetum vulgare*), sunflower (*Helianthus* species), yarrow (*Achillea* species), cosmos (*Cosmos* species), and coneflower (*Echinacea* species). Among the Leguminosae we suggest alfalfa (*Medicago sativa*), sweet clover (*Melilotus* species), fava bean (*Vicia fava*), and hairy vetch (*Vicia* species). These legumes also fix nitrogen from the atmosphere and improve soil nutrition. Beneficial Brassicaceae include sweet alyssum (*Lobularia martitima*), other alyssums (*Aurinium* and *Berteroa* species), and mustards (*Brassica* species). Other flowers attractive to biocontrol organisms include buckwheat (*Fagopyrum sagittatum*), milkweeds (*Asclepias* species), and cinquefoil (*Potentilla* species).

Of course, these general recommendations must be tailored to your specific needs—some of the aforementioned flowers are also *attractive* to pests in some situations. Also, some of the flowers have cultural and environmental requirements that may be contraindicated for *Cannabis* crops. Lastly, be sure to include plants that flower early in the season, such as *Antennaria* species, *Senecio* species, *Chrysogonum* species, coltsfoot (*Tussilago farfara*), and goldenseal (*Hydrastis canadensis*).

1

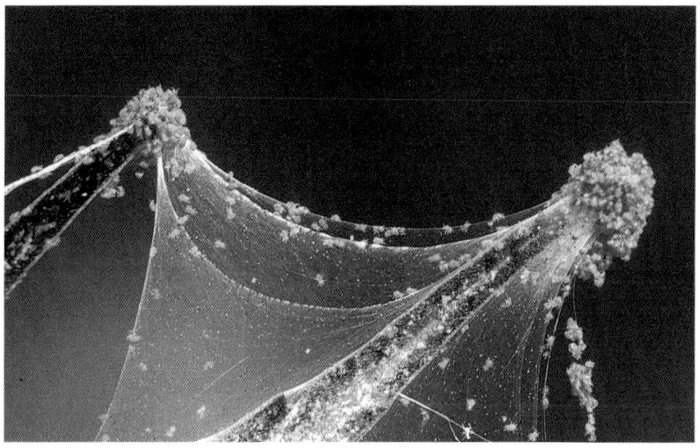

2

3

4

5

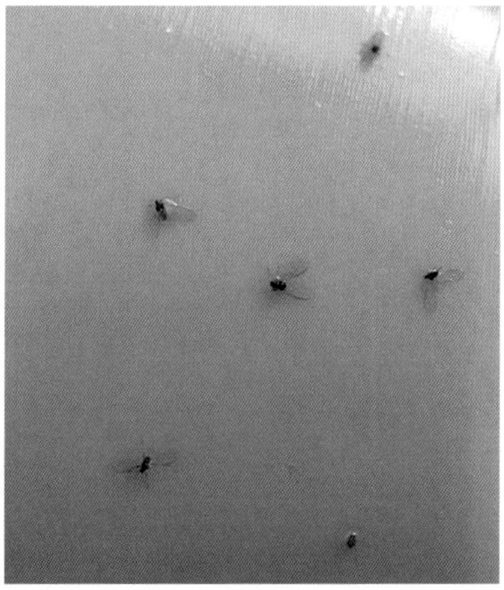

6

7

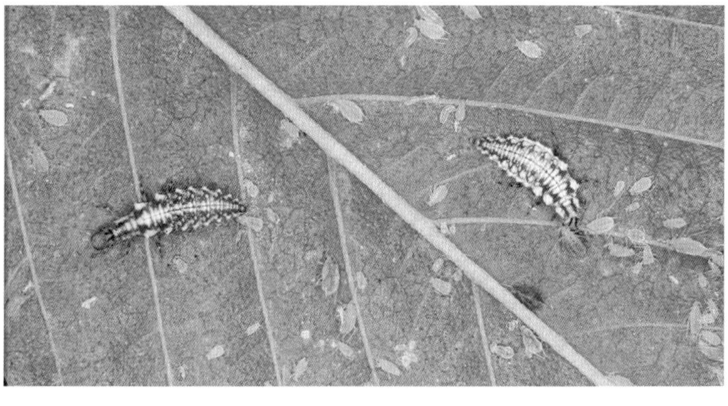

8

Plate 1. Leaf stippling caused by the two-spotted spider mite, *Tetranychus urticae* (Clarke).
Plate 2. Webbing and golf-ball-sized cluster of diapausing *T. urticae mites* (Clarke).
Plate 3. Different stages of the two-spotted spider mite, *T. urticae* (courtesy Koppert B.V.).
Plate 4. Predatory mite, *Phytoseiulus persimilis*, feeding on *T. urticae mite* (courtesy Koppert B.V.).
Plate 5. Hundreds of hemp russet mites, *Aculops cannabicola*, covering a leaf petiole (Hillig).
Plate 6. Winged migrants (alatae) of green peach aphid, *Myzus persicae*, caught on a yellow sticky trap (McPartland).
Plate 7. Wingless females (aptera) of black bean aphid, *Aphis fabae*, on axil of male flower (Clarke).
Plate 8. Aphid lions, larvae of *Chrysoperla carnae*, feeding on aphids (Clarke).

9

10

11

13

12

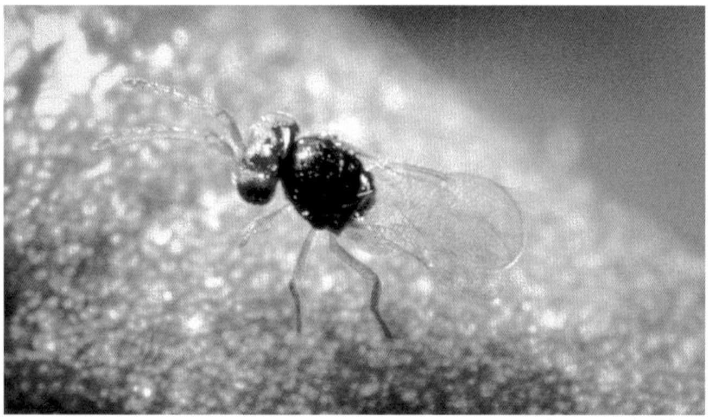

14

15

16

Plate 9. Eggs of green lacewing/aphid lion, *Chrysoperla carnae* (Clarke).
Plate 10. Ladybeetle larva, *Hippodamia convergens*, feeding on aphids (Clarke).
Plate 11. Ladybeetle adult, *Hippodamia convergens*, feeding on aphids (Clarke).
Plate 12. Slug-like larvae of *Aphidoletes aphidimyza* feeding on aphids (Clarke).
Plate 13. Whitefly adults (*Trialeuroides vaporariorum*) on underside of leaf (Clarke).
Plate 14. *Encarsia formosa*, a parasitoid of whitefly larvae (Koppert B.V.).
Plate 15. Predatory ladybeetle, *Delphastus pusillus*, next to pupa and larva of greenhouse whitefly, *Trialeuroides vaporariorum* (Koppert B.V.).
Plate 16. Big-eyed bug, *Geocoris punctipes*, feeding on larvae of sweetpotato whitefly, *Bemisia tabaci* (USDA).

17

18

19

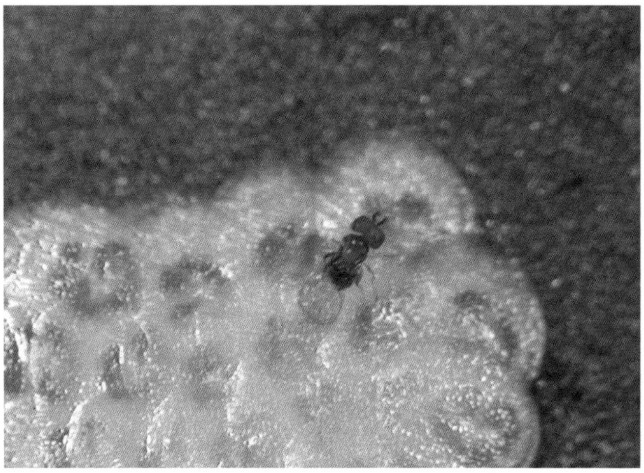

20

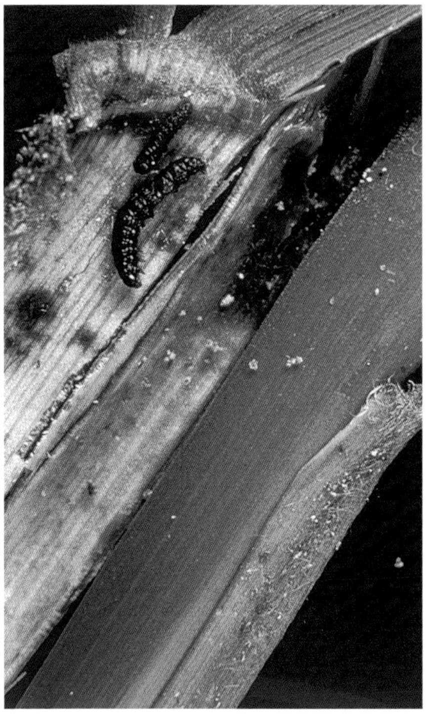

21

23

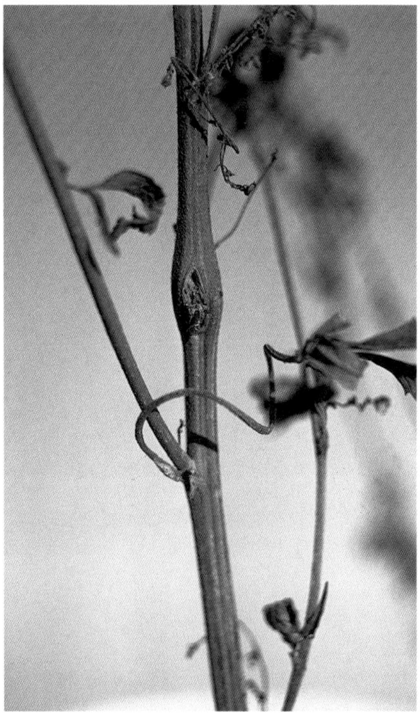

22

Plate 17. Greenhouse whiteflies, *Trialeurodes vaporariorum*, infested by biocontrol fungus *Beauveria bassiana* (USDA).
Plate 18. European corn borer (ECB), *Ostrinia nubilalis* (Vogl).
Plate 19. Female (L) and male (R) moths of ECB, *Ostrinia nubilalis* (McPartland).
Plate 20. Beneficial wasp, *Trichogramma ostrinae*, parasitizing ECB eggs (Chenus).
Plate 21. ECB larvae killed and discoloured by *Bacillus thuringiensis* (Bt)(Koppert B.V.).
Plate 22. Stem gall caused by hemp borer, *Grapholita delineana* (McPartland).
Plate 23. Moth of hemp borer, *Grapholita delineana* (McPartland).

24

25

27

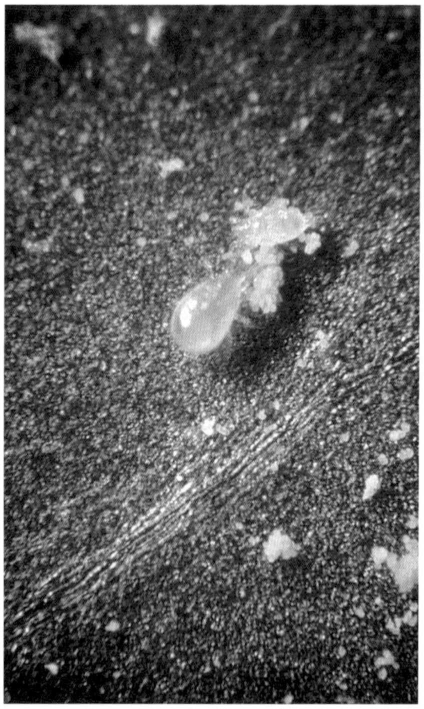

28

26

29

30

Plate 24. Budworms in bud, young and old, *Helicoverpa armigera* (Clarke).
Plate 25. Adult moth of budworm, *Helicoverpa armigera* (McPartland).
Plate 26. Cutworm, *Spodoptera litera*, assuming characteristic "C" shape when disturbed (Clarke).
Plate 27. Thrips damage, caused by *Thrips tabaci* (Clarke).
Plate 28. Predatory mite, *Neoseiulus* (*Amblyseius*) *cucumeris*, attacking a thrips (Koppert B.V.).
Plate 29. Pirate bug, *Orius insidiosus*, impaling a thrips (Koppert B.V.).
Plate 30. Hemp flea beetle, *Psylliodes attenuata* (Clarke).

31

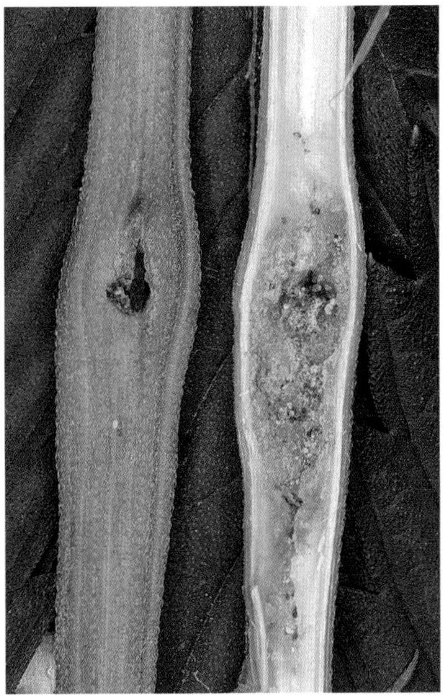
32

33

34

35

37

36

38

39

Plate 31. Cadaver of a white root grub (Japanese beetle, *Popillia japonica*) filled with parasitic nematodes, *Steinernema glaseri* (Klein, USDA).
Plate 32. Stem damage by grubs of tumbling flower beetle, *Mordellistena micans* (Clarke).
Plate 33. Southern green stink bug, *Nezara viridula*, with russet and white colourmorphs (McPartland).
Plate 34. *Lygus* bugs, related to tarnished plant bugs (Clarke).
Plate 35. Leafminer tunnels by *Agromyza reptans* (McPartland).
Plate 36. *Dighlyphus isaea*, a parasitoid of leafminers (Koppert B.V.).
Plate 37. Hungarian leafhopper, family Cicadellidae (Clarke).
Plate 38. Predatory beetle, *Cryptolaemus montrouzieri*, feeding on scale larvae (Koppert B.V.).
Plate 39. Termite damage in equatorial Africa (Clarke).

40

41

42

43

45

44

Plate 40. Halo of white spores from *Entomophthora muscae*, a fungus that infects flies (McPartland).
Plate 41. Flowering tops pruned by locusts in Africa (Clarke).
Plate 42. Damage caused by Hungarian slant-faced grasshopper, subfamily Acridinae (Clarke).
Plate 43. Predatory mite, *Hypoaspis miles*, preying on thrips larva (Koppert B.V.).
Plate 44. Hemp field infested by grey mould, caused by *Botrytis cinerea* (Vogl).
Plate 45. Grey mould of flowering tops. In high humidity, flowers and leaves become enveloped in a grey fuzz (de Meijer).

46

47

48

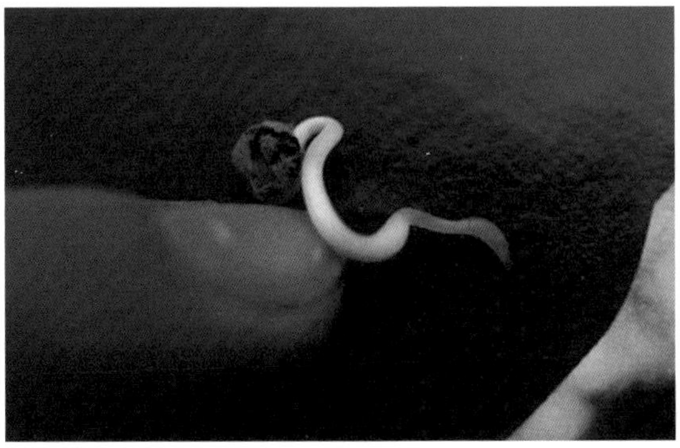

49

51

50

52

Plate 46. Grey mould of seed head. In low humidity, flowers and seeds turn brown and dry out, without characteristic grey colour (Vogl).

Plate 47. Grey mould of stalks. In high humidity, infested stalks become enveloped in a grey fuzz (de Meijer).

Plate 48. Hemp canker, caused by *Sclerotinia sclerotiorum*, stalk covered with white mycelium and black sclerotia (Scheifele).

Plate 49. Damping off of seedling (pre-emergent), note water-soaked appearance of rootlet (McPartland).

Plate 50. Damping off (post-emergent) hits seedlings after they emerge from the soil (McPartland).

Plate 51. Yellow leaf spot, caused by *Septoria neocannabina* on the left, and *Septoria cannabis* on the right (McPartland).

Plate 52. Yellow leaf spot (*Septoria cannabis*) on fan leaves of flowering tops in Nepal (McPartland).

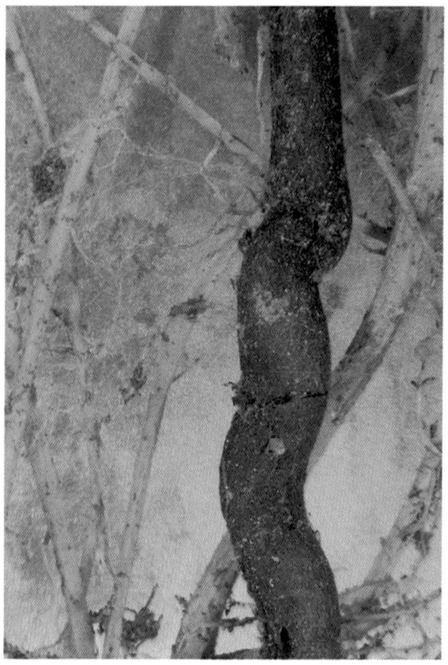

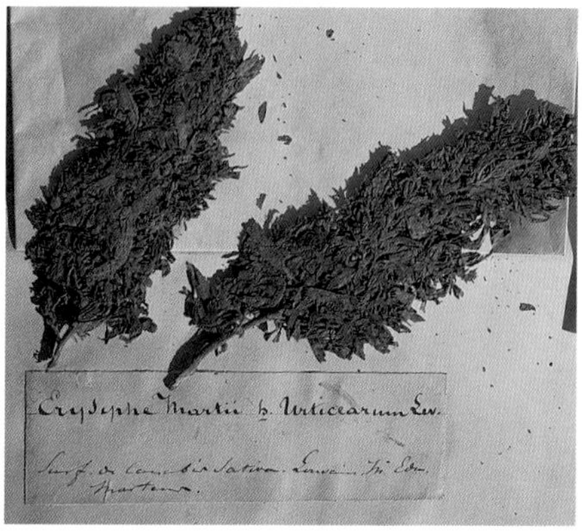

55

53

54

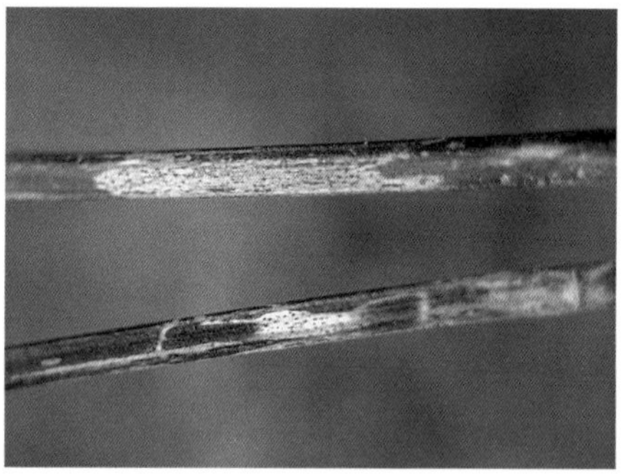

56

57

59

58

60

61

Plate 53. Root rot caused by a binucleate *Rhizoctonia* species (McPartland).
Plate 54. Yellow leaf spot (*Septoria cannabis*) and brown leaf spot (*Ascochyta arcuata*), mixed infection on the same plant (McPartland).
Plate 55. Brown leaf spot, *Phoma cannabis*, infesting flowering tops, a herbarium specimen collected by Westendrop in 1854 (McPartland).
Plate 56. Brown stem canker, *Phoma exigua* (McPartland).
Plate 57. Brown leaf spot, *Phoma glomerata* (McPartland).
Plate 58. Fusarium stem canker, with black dot-like perithecia of the sexual stage, *Gibberella cyanogena* (McPartland).
Plate 59. Fusarium root rot, *Fusarium solani* (McPartland).
Plate 60. Fusarium wilt, *Fusarium oxysporum* (McPartland).
Plate 61. Powdery mildew, *Sphaerotheca macularis* (McPartland).

62

63

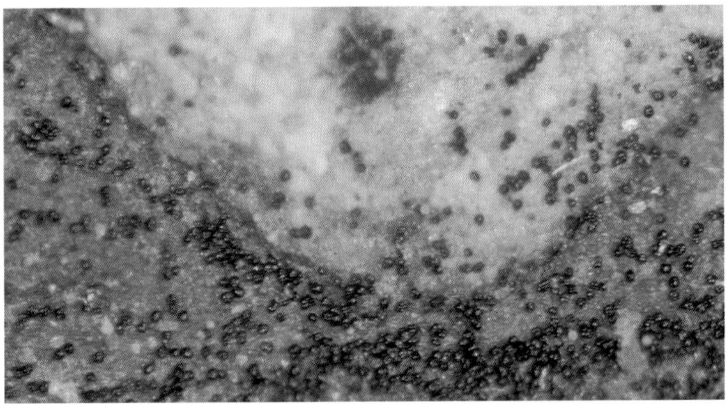

64

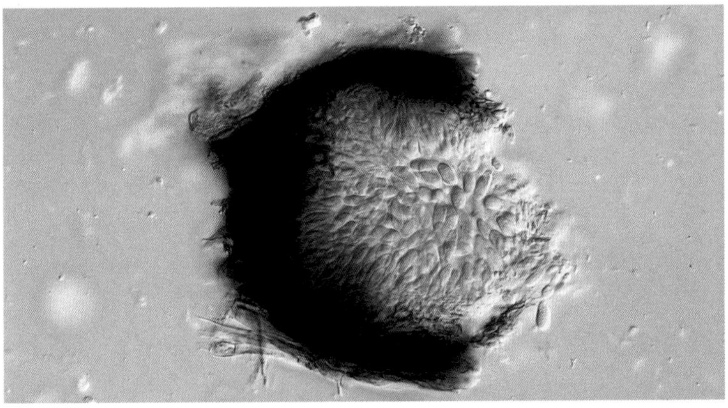

65

66

67

68

Plate 62. Olive leaf spot, *Pseudocercospora cannabina*, upper side of leaves (McPartland).
Plate 63. Southern blight, caused by *Sclerotium rolfsii* (Grassi).
Plate 64. Microscopic leaf pustules of black mildew, *Schiffnerula cannabis*, next to a white leaf spot caused by *Phomopsis ganjae* (McPartland).
Plate 65. Microscopic view of *Fusicoccum marconii*, a cause of twig blight (McPartland).
Plate 66. Pink rot, *Trichothecium roseum* (Clarke).
Plate 67. Anthracnose, *Colletotrichum dematium*, on stalk (Clarke).
Plate 68. White leaf spot, *Phomopsis ganjae* (McPartland).

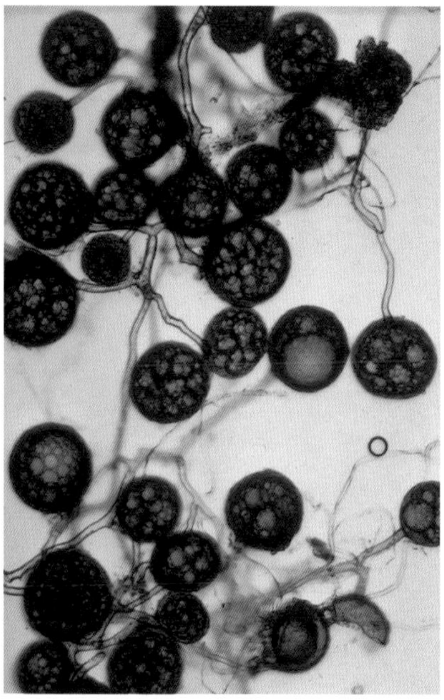

69

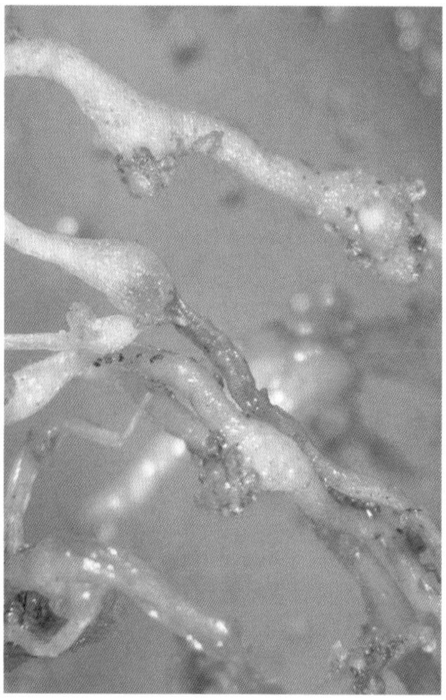

70

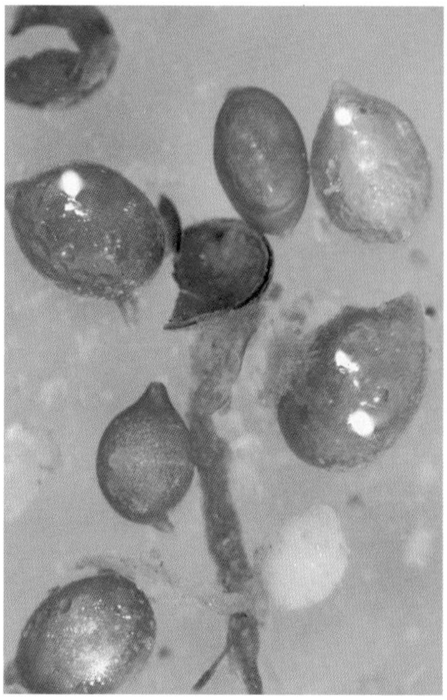

71

72

73

74

75

Plate 69. Spores of a mycorrhizal fungus, *Glomus mosseae* (McPartland).
Plate 70. Root-knot nematodes, *Meloidogyne incognita*, embedded in roots, with egg sacs protruding (McPartland).
Plate 71. Cysts of the cyst nematode, *Heterodera schachtii*, assorted ages
Plate 72. Symptoms of the hemp streak virus (Clarke).
Plate 73. Striatura ulcerosa, caused by *Pseudomonas syringae* (Hillig).
Plate 74. Dodder, *Cuscuta europea*, a parasitic plant (Berenji).
Plate 75. Broomrape, *Orobanche ramosa*, a parasitic plant (Berenji).

76

77

78

79

80

81

Plate 76. Slug, *Limax maximus* (Vogl).
Plate 77. Stalk damage caused by rabbits, *Sylvilagus* species (Clarke).
Plate 78. Nitrogen deficiency, yellowing of leaves (Clarke).
Plate 79. Nitrogen excess, causing stalk breakage (Scheife).
Plate 80. Phosphorus deficiency – small, dark leaves and petioles – caused by low temperatures (Clarke).
Plate 81. Potassium deficiency, brown edges and tips in older leaves (McPartland).

82

83

84

85

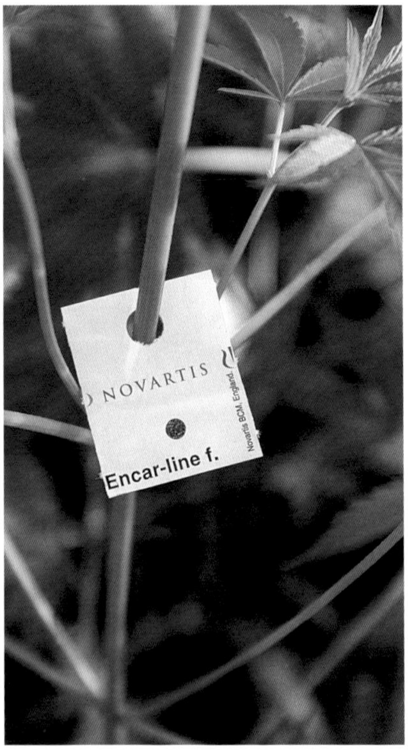

86

Plate 82. Frost damage of mature flowering tops (Clarke).
Plate 83. Hail damage (Vogl).
Plate 84. Spittle mass of the spittlebug, *Philaenus spumarius* (Clarke).
Plate 85. Koppert B.V. biocontrols for thrips and spider mites (Clarke).
Plate 86. Novartis *Encarsia formosa* biocontrol for whitefly (Clarke).

"Biocontrol uses living organisms to kill pests—turn your garden into a bug-eat-bug world."

—Bush Doctor

Chapter 10: Biological Control

Many companies sell biocontrol organisms. Unfortunately, many companies quickly go out of business. For an up-to-date *list of companies, please obtain the* **Annual Directory** *published by BIRC (Bio-Integral Resource Centre), P.O. Box 7414, Berkeley, California 94707, telephone: (510) 524-2567.*

Biocontrol organisms must be reintroduced into our unnatural, annual monocropping systems. We employ two release strategies—inoculative release and inundative release. See "Application rates for biocontrol" later in this chapter.

The best example of **inoculative release** happened over 100 years ago. Albert Koebele released a handful of Vedalias (Australian lady beetles) in a Californian orange grove, to control the marauding pest *Icerya purchasi.* The Vedalias thrived in Orange County and, in a few years, multiplied and nearly eradicated *Icerya purchasi.* This strategy is also called **permanent introduction** or **classical biological control**—a long-term solution to pests. This strategy works best on pests with steady-state populations, not those with boom-and-bust cycles (17-year cicadas being an extreme example). Permanent introduction can be applied to glasshouses if proper environmental conditions are careful maintained. But many biocontrol organisms cannot overwinter. Vedalias cannot overwinter in northern California, so the beetles must be reintroduced every spring. This is called **seasonal inoculation**, comparable to an annual "booster shot."

To achieve *rapid* pest control, a different tactic is used—**inundative release**. This tactic involves the repeated release of many, many biocontrol organisms. Inundative biocontrol uses beneficial organisms as "living pesticides." The strategy is relatively expensive, but appropriate for high-income crops, such as medical marijuana. Inundative release works best for controlling localized areas of heavy pest infestation, especially in enclosed spaces.

Inundative and inoculative biocontrol strategies benefit from "habitat management"—uniform temperatures and year-round plant cultivation—which makes glasshouses ideal for biocontrol. Biocontrol under glass began in 1926 with the discovery of *Encarsia formosa,* the whitefly parasitoid (Plate 86). But growers abandoned *E. formosa* for cheap DDT in the 1940s. The appearance of DDT-resistant spider mites around 1968 revived interest in biocontrol, using predatory mites (*Phytoseiulus persimilis*). Van Lenteren (1995) estimated that glasshouses covered 150,000 ha around the globe; the glasshouse area treated with *P. persimilis* was greater than 7500 ha. *E. formosa* was reintroduced in 1970 and has been released in over 2500 glasshouse ha. *Cannabis* cultivators in the USA began using beneficial insects, such as ladybugs and mantids, in the mid-1970s (D. Watson, pers. commun. 1995).

To work effectively, biocontrols should be introduced *before* the appearance of pests, or *early* in a pest infestation. Most biocontrols reproduce faster than pests, and soon overtake and consume a pest infestation. But if pests get a head start, crop damage occurs before the biocontrols can catch up. Hence, *preventative* biocontrol works best. Heavy populations of pests may need chemical controls before biocontrols can work effectively. To meet these emergencies, entomologists have selected biocontrols with resistance to pesticides, so the two controls can be combined (see Table 10.1).

Several categories of biocontrols are useful—**predators, parasitoids, microbial pesticides, companion plants, trap crops**, and **autocidal controls**. Some categories can be combined. Heinz & Nelson (1996) achieved much better control of whiteflies by combining a parasitoid wasp (*Encarsia formosa*) with a predatory beetle (*Delphastus pusillus*) than by releasing either biocontrol alone. *E. formosa* does not parasitize *D. pusillus,* and *D. pusillus* does not prey on *E. formosa* adults. *D. pusillus* does eat some young *E. formosa* larvae within whiteflies, but as the parasitoids mature within their hosts, *D. pusillus* avoids them.

Knipling (1992) stated that biocontrol works as a *ratio* of biocontrol organisms to pest organisms. The *density* of pests is of little consequence. This is a radical departure from traditionalists who claim a pest's density on plants determines the success of biocontrol. For inundative biocontrol to be effective, Knipling estimated that the ratio of adult *parasitoids* to adult pests must be at least 2:1. For *predators,* the optimal predator-to-pest ratio is probably reversed, 1:2 or greater.

Some biocontrols arise naturally, others are purchased, and some can be reared by growers (see Scopes & Pickford 1985). The USDA has recently encouraged the use of biocontrols over chemicals. The current approval time for new biocontrols is two to ten months, while chemical pesticides take 25–30 months for approval (Reuveni 1995).

PREDATORS

Pitting insect against insect is a most satisfying form of biocontrol. By definition, a predator must consume *more than one* pest before reaching its adult stage. Predators have either chewing mouthparts (e.g., lady beetles) or piercing-sucking mouthparts (lacewing larvae, assassin bugs). Piercing-sucking predators suck the fluids out of pests rather than eat them whole, so dead pests remain attached to plants, which may detract from the aesthetic value of ornamental plants.

Many predators arise naturally in outdoor crops: centipedes, spiders, predatory mites, lady beetles, ground beetles, rove beetles, tiger beetles, lacewings, bees, and wasps. Some predators are general consumers. Praying mantids (*Mantis religiosa*) are well-known generalists (Fig 10.1). Generalists, unfortunately, may eat other biocontrols and beneficial honeybees. Lacewings (e.g., *Chrysoperla carnea*) and ladybugs or ladybirds (more properly, lady*beetles*) are less general and more finicky. They rarely eat other beneficials. But given a

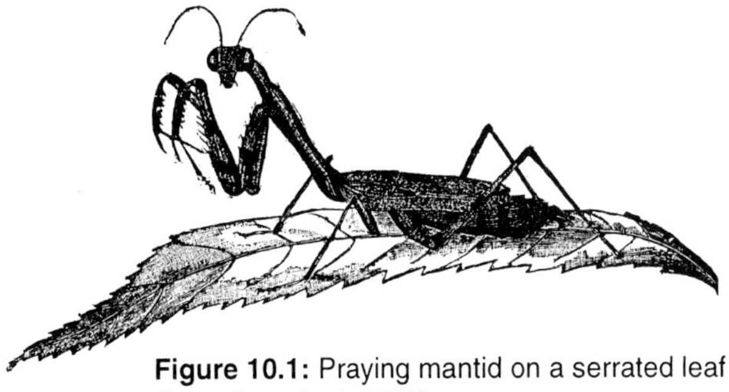

Figure 10.1: Praying mantid on a serrated leaf (from Comstock 1904).

choice between large and small prey, they'll eat the largest ones first. A lacewing larva, even if surrounded by millions of baby aphids, would rather eat a plump caterpillar. This is one reason why *selective predators* are the best, such as predatory mites and *Aphidoletes aphidimyza*.

The most popular mail-order predators are mites and ladybeetles. Popular mites include *Phytoseiulus persimilis, Mesoseiulus longipes, Neoseiulus californicus,* and *Hypoaspis miles*. Popular ladybeetles include *Hippodamia convergens, Cryptolaemus montrouzieri, Rodolia cardinalis,* and *Stethorus picipes*. See Fig 4.4 for illustrations.

Some predators go dormant (diapause) when crops begin to flower in autumn. This is triggered by a short photoperiod and affected by temperature (see Fig 4.2). Scientists have overcome this trait in some species by selective breeding. Nondiapausing breeds are becoming available.

Winged predators may ultimately fly away from crops but, hopefully, they leave behind eggs which hatch into more biocontrols. Entice them to stay and lay eggs by providing water and artificial honeydew or nectar.

PARASITOIDS

Parasitoids, in contrast to predators, kill their prey from within. Parasitoids only consume *one* individual host to reach their adult stage. Adult parasitoids usually insert individual eggs into multiple hosts. The eggs hatch into larvae which eat hosts alive, leaving vital organs for last. Parasitoid larvae usually pupate in pest cadavers and emerge as adults, off to lay more eggs. Examples of commercial parasitoids include *Encarsia formosa, Trichogramma* wasps, and braconid flies (e.g., *Chelonus texanus*). These wasps and flies do not bite, sting, or otherwise bother people or plants. Some biocontrols serve as parasitoids *and* predators—larvae only consume one individual host, but moult into adults which feed on many pests.

Parasitoids are more efficient at finding prey than predators. Parasitoids aggressively hunt until prey are nearly eradicated. Many predators, on the other hand, prefer being surrounded by many prey. When the prey population abates a bit, the predators migrate in search of happier hunting grounds, and leave many pests behind. Parasitoids stick around. Thus they are well suited for preventative control, but tend to work too slowly for large infestations.

Parasitoids are more pest-specific than most predators. Some may be pest-*crop*-specific. As parasitoids emerge from pupae, they imprint the odour of a specific crop damaged by a specific pest. Imprinted adults remain in that crop to search for prey, and will not fly off for prey in another crop. In the future we may purchase "customized biocontrols" that have been mass-reared from pests specifically raised on *Cannabis*, or raised on substrates sprayed with *Cannabis* terpenoids.

MICROBIAL PESTICIDES

If handling predators and parasitoids makes you squeamish, try *biocontrol in a can*. Containers of microbial pesticides (MPs) contain millions-to-trillions of freeze-dried bacteria, viruses, protozoans, or nematodes. Using MPs is like using chemical control—mix with water (*non-chlorinated* water) and spray onto foliage or pour into soil. In fact, many growers and governments treat microbials as pesticides rather than biocontrols. Some new MPs are genetically engineered organisms.

MPs rarely harm beneficial organisms. Nearly all MPs, such as *Bacillus thuringiensis* (Bt, a bacterium) and NPV (a virus), must be *ingested* to kill pests. Thus, they do poorly against sucking insects (e.g., aphids, whiteflies, leafhoppers). Fungal MPs are the exception. Fungi such as *Verticillium lecanii* do not have to be ingested. They work on contact, infecting insects right through their skin. Unfortunately, most fungi require high humidity, and some are infected by dsRNA viruses that decrease their effectiveness.

The use of MPs against other microorganisms is termed **antibiosis**. Bacteria such as *Bacillus subtilis, Pseudomonas fluorescens,* and *Agrobacterium radiobacter* are commercially available. They produce antibiotics that suppress the growth of other bacteria and fungi, and they ooze lytic enzymes which puncture the cell walls of pathogens. In the biocontrol future we will select specific strains of these microorganisms for their performance on *Cannabis*, so the host plant and MP work as a unit against pathogens.

MPs formulated for *soil application* survive best if applied during a heavy rain (or through an irrigation system). MPs can be applied to soil five different ways (see Table 11.7). MPs formulated for *foliage application* often require high humidity for germination and survival. This is bad, because high humidity encourages the growth of plant pathogens such as the grey mould fungus (*Botrytis cinerea*). New glasshouse research has shown that cycling two night's elevated humidity with two night's normal humidity is adequate for reliable biocontrol without increasing mould problems (van Lenteren 1995). Of course, MPs requiring high humidity work great in cloning chambers, which are always humid.

To spray MPs on foliage requires the use of compatible **spray carriers** and **spray adjuvants**. Spray carriers safe to most MPs include non-chlorinated water and spray oils. Categories of spray adjuvants include spreaders (wetting agents), stickers (adhesives), extenders (UV protectants), buffers, and feeding attractants. MPs are more chemically-sensitive than most pesticides, so they have their own special array of adjuvants. See the approved lists published in Hunter-Fujita (1998), or read your MP label.

Recently, MPs have been added to moth pheromone traps. The moths become coated with MP spores in the traps, then are released to fly to fields where they deliver the MPs to larval populations (Hajek 1993).

EPIPHYTE ANTAGONISTS

Epiphytic organisms are microscopic. They colonize the surface of plants, living off cellular leakage oozing from plant epidermis. They do no harm to plants even though they may be present in huge numbers. Wilson & Ehalutz (1991) estimated 1000 to 10,000,000 epiphytic bacteria live per cm^2 of leaf surface.

Using epiphytes to suppress pathogens is a lot like using microbial pesticides but the *modus operandi* is different. Epiphytes do not infect pathogens or produce toxins, they simply *outgrow* pathogens. Many pathogens depend on cellular leakage for spore germination. Epiphytes consume the cellular leakage, leaving the pathogens starved for nutrients and squeezed for space. Since epiphytes do not produce toxins, they may be safer than microbial pesticides. Some organisms act as both epiphytes and microbial pesticides, such as *Pichia guilliermondii* (Wilson & Ehalutz 1991).

The use of epiphyte antagonists, although attractive, is exacting. Epiphytes differ from plant to plant (Fokkema & Van den Heuvel 1986). No one has investigated the epiphytes of *Cannabis*. One plant's epiphyte is another plant's pathogen (see the section on Phylloplane Fungi in Chapter 5). *Cannabis* may not support a rich diversity of epiphytes since it produces antifungal and antibacterial compounds, including THC and CBD (McPartland 1984). Non-glandular trichomes also restrict microbial growth (Fokkema & Van den Heuvel 1986).

Table 10.1: Effects of some pesticides on some biocontrol organisms

PESTICIDE NAME AND CLASSIFICATION*	*APHIDOLETES APHIDIMYZA* LARVA / ADULT	*CHRYSOPERLA CARNEA* LARVA / ADULT	*ENCARSIA FORMOSA* PUPA / ADULT	NEMATODE SPECIES† LARVA	*PHYTOSEIULUS PERSIMILIS* EGG / ADULT	*TRICHOGAMMA* SPP. PUPA / ADULT	*VERTICILLIUM LECANII* SPORES
abamectin (I-M, B)	2/4 [-] §	1/4 [-]	1/4 [3]	1 [1]	1/4 [2]	-/4 [-]	-
acephate (I, S)	2/4 [8–12]	4/4 [6–8]	4/4 [8–12]	2 [1]	-/4 [3–4]	2/4 [>4]	3
aldicarb (I-M-N, S)	-/4 [8–12]	-	4/4 [8–12]	4 [>8]	4/4 [8–12]	-	-
Bacillus thuringiensis (I, B)	1/1 [0]	1/1 [0]	1/1 [0]	1 [0]	1/1 [0]	1/1 [0]	1
benomyl (F, S)	1/1 [0]	1/2 [-]	1/1 [0]	1 [0]	1/3 [2–3]	-	4
Bordeaux mixture (F, B)	-	1/3 [-]	-	-	-	-	-
captan (F, S)	1/1 [0]	1/1 [0]	1/1 [0]	-	1/1 [0]	-	4
carbaryl (I, S)	3/4 [-]	3/4 [4]	3/4 [4]	1 [0]	-/4 [2]	4/4 [-]	4
chlorthalonil (F, S)	1/1 [0]	1/1 [0]	1/1 [0]	1 [0]	1/1 [0]	-	4
copper compounds (F, B)	-/1 [0]	2/2 [-]	1/1 [0]	1 [0]	-/4 [0]	-	4
cypermethrin + polybutene (Thripstick + Cymbush) (I, B)	-	-	1/2 [0]	-	1/3 [2–4]	-	-
deltamethrin (I, B)	4/4 [8–12]	4/4 [8–12]	4/4 [8–12]	1 [0]	4/4 [8–12]	4/4 [8–12]	1
diazinon (I, S)	4/4 [6–8]	4/4 [4]	2/4 [4–6]	2 [-]	2/2 [1]	4/4 [2]	4
dichlorvos (I, S)-fumigant	-/4 [0.5]	4/4 [0.5]	4/4 [1]	2 [-]	1/4 [1]	-/4 [1]	1
dicofol (M, S)	1/4 [-]	1/2 [1]	1/4 [1–2]	1 [0]	3/4 [2]	3/3 [3]	1
dienochlor (M, S)	1/2 [0]	-	4/4 [6–8]	1 [0]	2/3 [2]	-	1
diflubenzuron (I, B)	1/1 [0]	4/3 [-]	1/1 [0]	1 [0]	1/1 [0]	-/1 [-]	1
fenbutatin oxide (M, S)	1/1 [0]	1/1 [0]	1/1 [0]	-	1/1 [0]	1/1 [0]	1
fosetyl-aluminium (F, S)	3/- [-]	-	-	1 [0]	1/1 [0]	-	4
horticultural (petrol) oil (I-M, B)	1/1 [0]	1/1 [0]	1/1 [0]	1 [0]	-/3 [0]	-	1
imidacloprid (I-M, B)	4/4 [0]	4/- [4]	4/4 [>2]	1 [0]	1/4 [0]	3/4 [-]	-
iprodione (F, S)	1/1 [0]	1/1 [0]	1/1 [0]	1 [0]	1/1 [0]	1/1 [0]	2
kinoprene (I, B)	-	1/- [-]	2/1 [0.5]	-	-/1 [0]	-	-
malathion (I, S)	3/2 [3–4]	4/4 [-]	4/4 [8–12]	1 [0]	2/2 [1–2]	4/4 [8–12]	3
maneb (F, S)	1/1 [0]	1/1 [0]	1/1 [0]	1 [0]	1/1 [0]	-	4
metalaxyl (F, S)	-/1 [0]	-	1/2 [-]	1 [0]	-/3 [-]	-	1
methomyl (I, S)	4/4 [8–12]	4/4 [8–12]	4/4 [6–10]	4 [-]	4/4 [4]	4/4 [8–12]	1
methoprene (I, B)	-	1/- [0]	1/2 [0]	1 [0]	1/1 [0]	-	-
neem (azadirachtin I, B)	2/2 [0]	1/- [<3]	1/3 [-]	1 [0]	1/1 [0]	-	-
nicotine sulphate (I, B)	4/4 [4]	2/3 [-]	1/3 [0.5]	-	-/4 [1]	-	-
nicotine (I, B)-fumigant	-/4 [0]	-	1/3 [0.5]	-	1/1 [0]	-	-
parathion, ethyl (I, S)	4/4 [8–12]	4/4 [-]	4/4 [8–12]	3 [-]	2/2 [0.5]	-	-
permethrin (I-N, B)	4/4 [6–8]	4/4 [6–8]	4/4 [8–12]	2 [0]	4/4 [8–12]	4/4 [8–12]	1
pirimicarb (I, S)	1/4 [1]	2/2 [-]	1/3 [0.5]	-	2/2 [0.5]	1/4 [1]	1
propargite (M, S)	2/- [0]	1/1 [0]	3/3 [1]	3 [-]	4/3 [0]	1/1 [0]	4
pyrethrum + PBO (I, B)	-/4 [6–8]	2/2 [1]	2/4 [1]	1 [0]	1/4 [1]	-/4 [-]	-
pyridaben(I-M, S)	-	1/1 [0]	4/4 [-]	-	-/3 [-]	-	-
Soap (potassium salts of fatty acids) (I, B)	-/4 [0]	4/4 [0]	2/4 [0]	4 [-]	2/4 [0]	-	-
rotenone (I, B)	-	2/4 [-]	4/4 [2]	-	3/4 [-]	-	-
sulphur (F, B)-spray	2/2 [-]	1/1 [0]	1/4 [>4]	3 [0]	1/1 [0]	1/4 [-]	4
sulphur (F, B)-dust	-	-	1/3 [3–4]	3 [-]	1/2 [1]	-	-
sulphur (F, B)-fumigant	-/1 [-]	-/1 [0]	-/3 [0.5]	-	1/2 [1]	-	-
thiram (F, S)	2/1 [-]	2/1 [-]	1/3 [0.5]	1 [0]	2/1 [0]	2/3 [-]	4
triforine (F, S)	1/2 [-]	1/1 [0]	1/1 [0]	1 [0]	1/2 [0]	-/1 [-]	3
vinclozolin (F, S)	-/1 [0]	1/1 [0]	1/1 [0]	-	-/1 [0]	-/1 [0]	1
zineb (F, S)	-/1 [-]	1/1 [0]	1/1 [0]	-	-/1 [-]	-/1 [-]	4

*Classification: I=insecticide, M=miticide, N=nematocide, F=fungicide, B=biorational, S=synthetic.

†Nematode species include *Steinernema* spp. and *Heterorhabditis* spp.

§Toxicity of pesticide to **immature / mature** biocontrol organisms. Numerical ranking: 1=harmless (<25% biocontrol organisms affected), 2=slightly harmful (25–50% affected), 3=moderately harmful (51-75% affected), 4=very harmful (>75% affected). Bracketed [numerals] estimate the number of *weeks* a pesticide remains harmful after application. A dash (-) indicates unknown data. Pesticide toxicity includes *mortality* and *decreased fertility,* when the pesticide is applied at its proper recommended rate. All applications are sprays unless otherwise noted. Compiled from information by Koppert (1998, 1999). Data presented here are only estimations and do not guarantee safety if followed.

COMPANION PLANTS & TRAP CROPS

Plants cannot run away from their enemies. Forced to stand their ground and fight, plants have evolved incredible defence mechanisms, including many repellent chemicals. The chemicals produced by **companion plants** are powerful enough to repel pests from the entire neighbourhood. Two popular companion plants are marigolds (*Tagetes* species) and tansy (*Tanacetum vulgare*). Many people intercrop companion plants with crop plants to repel pests. Counterintuitively, Bush Doctor (pers. commun. 1984) suspected pest-repellent chemicals may detrimentally affect *Cannabis*. He thought the allelopathic effects of *Tagetes* on *Cannabis* may outweighed the beneficial effects of repelling pests—a potential Master's Thesis research project.

Others plants work by their *attractiveness* to pests, such as the Japanese beetle's affinity for *Zinnia elegans*. These **trap plants** draw pests from neighbouring crops. Trap plants can then be sprayed with pesticides or removed, taking the pests with them. Many trap plants are only attractive to pests during a *specific* plant growth stage. Planting new trap crops in "waves" every two weeks presents pests with a variety of plant stages to colonize. An effective trap crop will attract 70–85% of a pest population while only covering 1–10% of the total crop area (Hokkanen 1991). Please note trap crops require careful monitoring, otherwise they turn into *pest nurseries*. Spray or remove trap crops before a new generation of pests hatch out.

Decoy crops are nonhost crops planted to control nematodes. Decoy crops cause nematode eggs to germinate, but the nematodes cannot complete their life cycle on the decoy plants, so they die out. Palti (1981) killed *Meloidogyne incognita* and *Meloidogyne javanica* with marigold (*Tagetes patula, Tagetes minuta*), sesame (*Sesamum orientale*), castor bean (*Ricinus communis*), and *Chrysanthemum* species.

AUTOCIDAL CONTROL

This technique is heavy-handed but effective. Male insects raised in captivity are sterilized with gamma radiation, then released to mate with normal females, producing infertile eggs. If the ratio of sterile males to normal males is 2:1, then 67% of females will mate with sterile insects, assuming sterile males are fully competitive with native males (Knipling 1992).

The greatest disadvantage of autocidal control over other forms of biocontrol is the lack of progeny; sterile insects must be recreated each pest generation. Another autocidal technique utilizes studly male insects which are *related* to pests. When *Heliothis subflexa* mates with the budworm *Heliothis virescens*, the offspring are sterile. These "mules" are released to mate with native insects and produce infertile eggs.

APPLICATION RATES FOR BIOCONTROL

As described earlier, there are two approaches to releasing biocontrols, inoculation and inundation. The *inoculation approach* releases biocontrols in the beginning of the season, allowing them to establish a sustainable breeding population. The inoculation approach works better in warm climates (e.g., glasshouses) that favour biocontrol reproduction over a long season. Inundation works better in cooler, short-seasoned situations.

Inoculation is frequently done *prophylactically*, before pests become a problem. Several prophylactic releases must be scheduled per season ("dribble release"), since biocontrols will *die out* if no pests are available. To prevent die out, Hussey & Scopes (1985) deliberately released small populations of pests to assure the survival of biocontrols. Many people reject this "simultaneous introduction" or "pest-in-first" concept, but it works. Most glasshouse growers already have sufficient pest populations to maintain biocontrols. But for growers who don't, Koppert sells 500 ml bottles containing 40,000 spider mites (Spidex-CPR®), "to provide a controlled infestation of the crop."

Using "banker plants" is another inoculative strategy that prevents die out (Van Lenteren 1995). Banker plants are noncrop plants infested with noncrop pests, which serve as alternative food sources for biocontrols. For instance, if you anticipate *Myzus persicae* in your *Cannabis*, release *Aphidoletes aphidimyza*. To prevent die-out, Koppert sells boxes of wheat plants infested with grain aphids (AphiBank®). The bank of barley and grain aphids will sustain *A. aphidimyza* until *M. persicae* appears.

The *inundation approach* releases biocontrols *after* pests have appeared. "Hot spots" of pests become release sites for biocontrols. The biocontrols inundate the hot spots, then diffuse into the surrounding crop. Inundation provides an immediate but nonsustainable reduction in the pest population. Because most biocontrols only parasitize pests during part of their lifecycle, repeated inundations are necessary.

Application rates for inundative biocontrol depend on many factors. First in importance is the number of pests, measured as the **Infestation Severity Index** (ISI) for each pest. Other factors include crop biomass, crop location (field, glasshouse, or under artificial lights), local environmental conditions (do temperature and humidity favour the biocontrol or the pest?), and the longevity of biocontrol organisms versus the longevity of pests.

If biocontrol organisms are released in *field* crops, much care must be taken in providing an immediate food source for them; otherwise the biocontrols will exit crops in search of shelter and food. This is especially true with winged biocontrols. Dispersion is less of a problem in glasshouses with screened vents and limited exits.

Many nonwinged biocontrols such as predatory mites have the opposite problem. They cannot disperse *enough*. If plants are touching it is much easier for biocontrols to walk in search of prey. If plants are not touching it may be necessary to hand-disperse biocontrols onto each plant. Small cups or sacks attached to each plant can hold small doses of biocontrols (assuming they are not cannibalistic, like lacewing larvae), which crawl out onto each individual plant (see Plate 85).

Biocontrols can be released by hand in small glasshouses. It is easy and inexpensive to walk around a small garden and shake beneficial insects out of their little bottles. But hand-releasing biocontrols in large glasshouses or field crops becomes expensive. Furthermore, the even distribution of biocontrols across large acreages is difficult to do by hand (Mahr 1999). Thus, mechanized, calibrated delivery systems have been devised. The mechanized delivery of some biocontrols, such as microbial bacteria and fungi, is simple—they can be sprayed on crops using conventional pesticide equipment. Mechanized delivery of insect predators and parasites is more difficult. Mahr (1999) reviewed the mechanized delivery of lacewing eggs and *Trichogramma*-parasitized caterpillar eggs. The eggs can be glued with mucilage to bran flakes or vermiculite, and scattered with hand-cranked applicators or compressed air blowers. Unfortunately, the biocontrols end up on the ground instead of remaining on the foliage. This problem can be alleviated by mixing biocontrols in sticky liquid formulations and spraying them with guns fitted with large nozzles (the Bio-Sprayer®). Liquid carriers, however, also have problems—such as keeping biocontrol organisms uniformly suspended in the liquid, and keeping the organisms from suffocating in the fluid. For very large-scale operations, no form of ground application may be practical. In the former USSR, hemp growers scattered *Trichogramma* from an airplane (Marin 1979). *Phytoseiulus persimilis* has been mixed with corncob grits, refrigerated, and released by conventional light aircraft onto spider-mite infested fields (Mahr 1999).

SOME BIOCONTROL ORGANISMS

Below is a list of biocontrol organisms mentioned in the text. They are described at length under the pests they control. Many biocontrols listed here have not yet been tested on *Cannabis*. *Please mail us feedback for the next edition of our book. Cannabis* can be tough on biocontrol organisms. The leaf hairs (trichomes) make it difficult for small predators to find and pursue prey. This also happens on leaves of tomatoes and cucumbers; entomologists have bred cucumbers with fewer trichomes for improved biocontrol activity. Bredemann *et al.* (1956) began breeding hemp plants with fewer glandular trichomes, but their work was abandoned. Furthermore, the anti-pest arsenal produced by *Cannabis*—such as terpenoids, ketones, and cannabinoids—may also repel biocontrol organisms. Some *pests* absorb plant chemicals and turn the chemicals against their *biocontrol enemies* (Campbell & Duffey 1979, Clayton & Wolfe 1993).

Certain biocontrols require permits for shipment to certain areas (especially to Hawai'i). Some biocontrols listed below are not yet commercially available, but their availability is changing rapidly. Rather than list currently available biocontrols with an asterisk (and render this book instantly outdated), please obtain the *Annual Directory* published by BIRC (Bio-Intergral Resource Centre), P.O. Box 7414, Berkeley, California 94707, telephone: (510) 524-2567. UDL addresses are ephemeral things, but four current sources of excellent information on the Internet are Koppert (www.koppert.nl/english), Cornell University (www.nysaes.cornell.edu/ent/biocontrol), Midwest Biocontrol News (entomology.wisc.edu/entomology/mbcn.html), and UC-Davis (www.ipm.ucdavis.edu/).

Adalia bipunctata, two-spotted ladybeetles control aphids
Agrobacterium radiobacter, nonpathogenic soil bacteria control crown gall disease
***Agrotis segetum* Nuclear Polyhedrosis Virus** is a NPV strain that kills *A. segetum*
Aeolothrips intermedius, predatory thrips control *Thrips tabaci*
Aleochara bilineata, staphylinid rove beetles whose larvae parasitize root maggots
Allium sativum, garlic, repels leaf beetles
***Ampelomyces quisqualis* (=*Cicinnobolus cesatii*)**, a fungus infecting powdery mildews
Anagrus atomus, mynarid wasps control leafhoppers
Anagyrus pseudococci, parasitoid wasps of mealybugs
Anaphes iole, wasps that parasitize eggs of tarnished plant bugs
Anthocorus nemorum, minute pirate bugs kill thrips
Aphelinus abdominalis, parasitic wasps control aphids
Aphidius colemani, braconid wasps control aphids
Aphidius matricariae, braconid wasps control aphids
Aphidoletes aphidimyza, maggots of syrphid flies control aphids
Aphytis melinus, wasps control hard scales
Archytas marmoratus, tachinid flies control budworms
Artemisia absinthium, common wormwood, repels flea beetles and slugs
Arthrobotrys species, predaceous soil fungi control nematodes
Aschersonia aleyrodis, fungi control whiteflies and scales
Bacillus cereus, bacteria suppress damping off fungi
Bacillus popilliae, bacteria control flea beetles
Bacillus subtilis, bacteria suppress damping off fungi
Bacillus thuringiensis (Bt-k, Bt-a, Bt-t, Bt-i, Bt-m), bacteria control many caterpillars, some beetles, and a few flies
Beauveria bassiana and ***B. globulifera***, two related fungi control many insects
Beauveria brongniartii (=*B. tenella*), fungi control beetle grubs
Brassica alba, white mustard, serves as a trap crop against broomrape
Burkholderia (Pseudomonas) cepacia, bacteria suppress *Pythium*, *Fusarium* and *Rhizoctonia* fungi
Candida oleophila, yeasts control postharvest decay
Cannabis sativa has been used as a companion plant to repel insects from neighbouring crops (Riley 1885, Berling 1932, Stratii 1976, Pakhomov & Potushanskii 1977). *Cannabis* plants also suppress nematodes in soil (Kir'yanova & Krall 1971, Kok *et al.* 1994).
Chelonus insularis, C. texanus, two related parasitic braconid wasps control budworms
Chrysoperla (Chrysopa) carnea, green lacewings control aphids, whitefly nymphs, budworm eggs, thrips, and spider mites
Chrysoperla comanche, Comanche lacewings control aphids and other small insects
Chrysoperla (Chrysopa) rufilabris, lacewings control aphids and other small insects
Coccinella undecimpunctata, eleven-spotted ladybird beetles control aphids and other small insects
Colletotrichum gloeosporiodes f.sp. ***cuscutae***, fungi control dodder (parasitic plants)
Coniothyrium minitans, fungi control *Sclerotinia sclerotiorum*
Cotesia marginiventris, braconid wasps control caterpillars
Cryptolaemus montrouzieri, a ladybeetle known as the mealybug destroyer
Cucurbita foetidissima, wild buffalo gourds control cucumber beetles
Dacnusa sibirica, braconid wasps control leafminers
Dactylaria species, predaceous soil fungi control nematodes
Delphastus pusillus, beetles control whiteflies
Delphinium exaltatum, larkspur plants attract Japanese beetles, fatally
Deraeocoris brevis, predatory mirid bugs kill aphids and thrips
Dicyma pulvinata (=*Hansfordia pulvinata*), fungi control *Cercospora* species and other fungi
Diglyphus isaea, chalcid wasps control leafminers
Encarsia formosa, mighty little wasps control whiteflies
Encarsia luteola, wasps control the whitefly *Bemisia argentifolii*
Entomophthora exitialis, fungi control aphids
Entomophaga grylli, fungi control grasshoppers
Entomophthora praxibuli, fungi control grasshoppers
Entomophthora thripidum, E. parvispora, fungi control thrips
Eretmocerus eremicus, parasitic/predatory wasps control whiteflies
Eucelatoria bryani, tachnid flies control budworms
Euphorbia marginata, spurge plants repel moles
Euseius hibisci, predatory mites control thrips
Erynia neoaphidis (= *Entomophthora aphidis*), fungi control aphids
Erynia (Zoophthora) radicans, fungi control leafhoppers and aphids
Franklinothrips vespiformis, parasitic wasps control thrips
Fusarium lateritium, fungi control broomrape
***Fusarium oxysporum* (nonpathogenic)**, nonvirulent strain controls hemp wilt fungi
Fusarium oxysporum f.sp. ***orthoceras***, fungi control broom rape (parasitic plants)
Galendromus (Metaseiulus) occidentalis, predatory mites control spider mites and some russet mites
Galendromus (Typhlodromus) pyri, predatory mites control spider mites
Geocoris punctipes, predatory lygaeid bugs feed on white flies, aphids, and spider mites; a related species from India, *G. tricolour*, may also be useful.

Gliocladium roseum (=***Clonostachys rosea***), fungi control other soil fungi
Glomus intraradices, mycorrhizal fungi suppress *Fusarium* fungi
Gonatobotrys simplex, fungi control other fungi, such as *Cercospora* species
***Heliothis zea* Nuclear Polyhedrosis Virus** is a NPV strain that kills *H. zea* and *H. armigera*
***Helioverpa armigera* Stunt Virus** (HaSV) kills *H. armigera* and other lepidopteran caterpillars
Heterorhabditis heliothidis (=*H. bacteriophora*), ***H. megidis***, two nematodes control white root grubs and many other soil insects
Hippodamia convergens, convergent ladybeetles control aphids
Hirsutella rhossiliensis, fungi control nematodes
Hirsutella thompsonii, fungi control spider mites and eryophid mites
Hypoaspis aculeifer, mites prey on thrips
Hypoaspis miles, mites control fungus gnats
Iphiseius (Amblyseius) degenerans, predatory mites control thrips
Leptomastix dactylopii, parasitic wasps control *Planococcus citri* mealybugs
Lydella thompsonii, tachnid flies control European corn borer
Macrocentrus ancylivorus, braconid wasps control hemp borers and other caterpillars
Macrolophus caliginosu, mirid bugs that prey on whiteflies
***Melanopus sanguinipes* Entomopoxvirus**, a virus that kills migratory grasshoppers
Mesoseiulus (Phytoseiulus) longipes, predatory mites control spider mites
Metarhizium anisopliae, fungi control spittlebugs, aphids, whiteflies, termites, ants, beetles, and other insects
Metaphycus alberti, parasitic wasps control scale insects
Microcentus grandii, braconid wasps control European corn borers
Microcronus psylliodis and ***M. punctulatus***, two braconid wasps control flea beetles
Microplitis croceipes, wasps control budworms
Microterys flavus, parasitoids control brown scales
Myrothecium verrucaria, fungi control nematodes
Nematophthora gynophila, fungi control nematodes
Nicandra physalodes, shoo-fly plant, repels whiteflies
***Neoplectana* species** see Steinernema
Neoseiulus (Amblyseius) barkeri(=*A. mackenziei*), predatory mites that prefer thrips but also eat spider mites
Neoseiulus (Amblyseius) californicus, predatory mites control spider mites and aphids
Neoseiulus (Amblyseius) cucumeris, predatory mites control thrips
Neoseiulus (Amblyseius) fallacis, predatory mites control spider mites
Neozygites floridana, fungi control spider mites
Nepeta cataria L., catnip plants, repel flea beetles
Nomuraea rileyi, fungi control *Spodoptera litura* and other nocturid caterpillars
Nosema locustae, microscopic protozoans control grasshoppers and crickets
Nosema acridophagus, a related species, also kills grasshoppers
Nosema melolonthae controls certain beetle grubs (*Melolontha* species)
Nosema pyrausta controls European corn borers
Nuclear polyhedrosis virus (NPV) controls cutworms, budworms, and many other Nocturids (sub-types include MbNPV, AcNPV, HzNPV, SeNPV, HcNPV, EtcNPV)
Ocypus olens, devil's coachmen, staphylinid beetles control garden snails
Ooencyrtus submetallicus, parasitoids of stink bugs
Orius insidiosus, pirate bugs, control thrips and spider mites, as have *O. albidipennis, O. laevigatus, O. majusculus, O. tristicolour,* and other related species
Paecilomyces farinosus (=*P. fumosoroseus*), fungi control whiteflies
Pasteuria (Bacillus) penetrans, bacteria control nematodes
Pelargonium hortorum, geranium, repels leaf beetles and leafhoppers
Phasmarhabditis hermaphrodita, nematodes control slugs
Phytoseiulus persimilis, popular predatory mites control spider mites
Pichia guilliermondii [=*Candida guilliermondii, Debaryomyces hansenii*], yeasts control postharvest rot caused by *Botrytis cinerea, Alternaria alternata, Penicillium digitatum, P. italicum,* and *Rhizopus stolonifer*
Podisus maculiventris, predatory bugs control caterpillars and beetle grubs
Pseudacteon species, parasitic flies control fire ants
Pseudomonas fluorescens, bacteria suppress *Pythium, Fusarium* and *Rhizoctonia* fungi
Pseudomonas syringae, bacteria control postharvest decay
Pyemotes tritici, parasitic mites control fire ants
Pythium oligandrum, a biocontrol against the damping off pathogen *Pythium ultimum*
Rhizobius ventralis, black ladybeetles control mealybugs and scales
Rodolia cardinalis, Australian ladybeetles control mealybugs and scales
Rumina decollata, predatory decollate snails control garden snails.
Saccharopolyspora spinosa, bacteria control armyworms and other Noctüidae
Salmonella enteritidis* var. *issatschenko bacteria control rodents
Scambus pterophori parasitize beetle grubs and caterpillars
Steinernema (Neoapectana) carpocapsae, S. feltiae, S. glaseri, S. riobravis, S. scapterisci, nematodes control caterpillars, beetle grubs, root maggots, cricket nymphs, and other soil insects
Stethorus picipes, S. punctum, S. punctillum, ladybeetles control spider mites
Streptomyces griseoviridis, actinomycetes control *Rhizoctonia* and *Pythium* fungi
Sycanus collaris, assassin bugs control caterpillars, beetles and many other pests. They take out an occasional honeybee and sometimes bite a gardener's finger, but do far more good than harm.
***Tagetes* species,** marigolds repel nematodes.
Talaromyces flavus, fungi control *Verticillium dahliae*
***Tanacetum vulgare* L.**, tansy plants repel many insects, especially caterpillars and leaf beetles
Therodiplosis persicae, midges control spider mites
Thripobius semileuteus, wasps control greenhouse thrips
Tiphia popilliavora, T. vernalis, wasps control Japanese beetles
Trichoderma viride, fungi control pathogenic fungi, especially *Rhizoctonia solani*.
Trichoderma harzianum controls *Rhizoctonia solani, Sclerotium rolfsii* and, to a lesser degree, *Macrophomina phaseolina* and *Sclerotinia sclerotiorum*
Trichoderma lignorum controls *Fusarium* species
Trichoderma (Gliocladium) virens, fungi control damping off fungi *Pythium ultimum* and *Rhizoctonia solani*, as well as *Botrytis cinerea*—the grey mould plague.
Trichogramma maidis, parasitic wasps control European corn borers
Trichogramma pretiosum* & *T. minutum, wasps control corn borers, budworms, and other caterpillars
Trichopoda pennipes* & *T. giacomellii, parasitic wasps control stink bugs
Trissolcus basalis. a tiny wasp that parasitizes eggs of stink bugs
Tropaeolum majus, nasturtiums are nasty to whiteflies.
Urtica dioica, nettles attract hemp flea beetles as a trap crop
Vairimorpha necatrix, a microsporidial protozoan that infects European corn borers and other caterpillars
Verticillium biguttatum, fungi control damping off fungi such as *Rhizoctonia solani*

Verticillium chlamydosporium, soil fungi that infest soil nematodes

Verticillium (Cephalosporium) lecanii, fungi control aphids and whiteflies and help control scales, mealybugs, thrips, beetles, flies and eriophyid mites

Zinnia elegans, zinnias attract Japanese beetles as a trap crop

DARK SIDE

Biocontrol is not a panacea. *All* pest-control techniques affect other organisms, and pose some degree of environmental risk. Overuse of biocontrols, especially *inundative* biocontrols, may reduce their effectiveness. Pests can develop resistance to biocontrol organisms if confronted with sufficient selection pressure (Gould 1991). Biocontrols may ignore pests and attack nontarget hosts, including other biocontrol organisms. This side effect has actually led to the extinction of nontarget organisms. Howarth (1991) documents dozens of extinctions of nontarget species caused by rogue biocontrols, far more than the number of extinctions caused by chemical controls. Off-target biocontrols may become pests themselves. This is especially true among insects and fungi released for the control of weedy plants (McPartland & West 1999).

Biocontrols, like chemical controls, may disrupt local ecologies, allowing an outbreak of secondary pests. The European ladybeetle *Coccinella septempunctata*, for instance, may disrupt established biological control systems in the USA (Schaefer *et al.* 1987b). Lastly, biocontrols may become public health problems: the predatory snail *Euglandina rosea*, released to control plant-eating snails, is now known to carry a lung worm which infects humans (Howarth 1991). Microbial biocontrols, especially fungi, may infect humans (Rippon 1988, Samuels 1996).

According to Orr & Suh (2000), biocontrol product support in Europe is superior to that in the USA. European suppliers maintain rigid quality control procedures, and ship their products by overnight delivery in refrigerated trucks or in containers with ice packs. Furthermore, European suppliers provide their customers with extensive technical information, and even monitor temperature data and pest populations in areas where biocontrol releases take place. Customers are informed in advance of product delivery and release dates. In contrast, nearly 50% of USA suppliers shipped dead biocontrols or species other than that which was claimed, shipped products in padded envelopes by standard "snail mail," and provided little or no product information or instructions (Orr & Suh 2000). The biocontrol industry in the USA needs to improve its self-regulation; perhaps government regulation is required to implement tighter quality control and product support.

BIOCONTROLS & PESTICIDES

We present the effects of some pesticides on some biocontrol organisms in Table 10.1. For a complete list of pesticides and biocontrol organisms, see the "side effects" list available on Koppert's web page (www.koppert.nl/english). You will note many dash-marks (-) in Table 10.1, indicating unknown data. The Koppert list is updated monthly with new information.

Theiling & Croft (1988) summarized 1000 publications citing the effects of pesticides on biocontrol arthropods (biocontrol insects and mites). Overall, pesticides were less toxic to predators than parasitoids. The least susceptible biocontrols were lacewings and predatory bugs (Miridae, Lygaeidae, Anthocoridae). For most biocontrols, egg and pupal stages were less susceptible than larvae and adults.

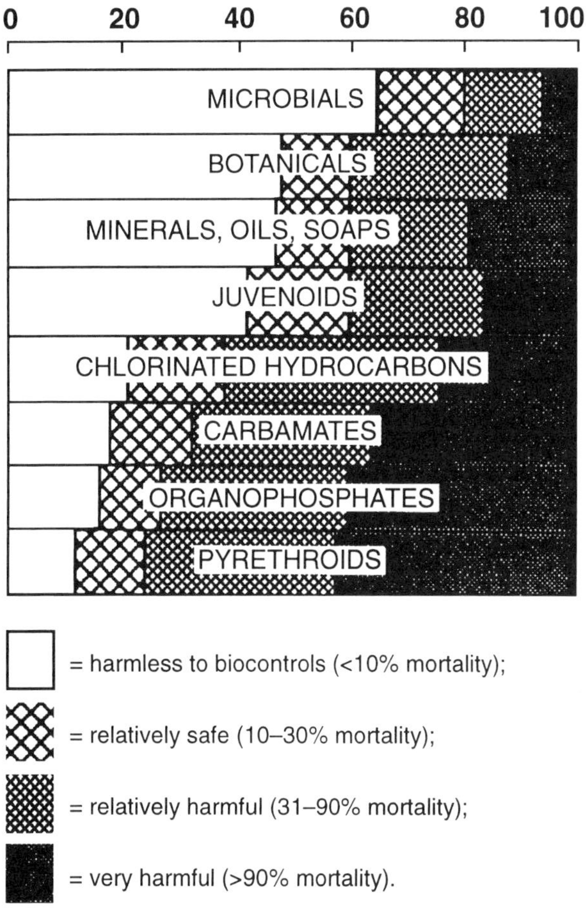

Figure 10.2: Average toxicity of pesticides on biocontrol arthropods, from data computed by Theiling & Croft (1988).

The average toxicity of different *groups* of pesticides is presented in Fig 10.2. Theiling & Croft (1988) lumped *microbial biocontrols* (viruses, bacteria, fungi, and nematodes) with *microbial fermentation products* (abamectin, streptomycin, etc.) into one group, "Microbials." This was the least-toxic group, naturally. Within "Microbials," Theiling & Croft reported that viruses, bacteria, and fungi were safer to biocontrol arthropods than were predatory nematodes and fermentation products. Within the group labelled "Juvenoids," the chitin inhibitors were safer than juvenile growth hormones. Botanicals and minerals (e.g., Bordeaux mixture, copper, sulphur) were moderately toxic to biocontrols. The most toxic pesticides were synthetic pyrethroids. Of course, within each group we find exceptions, which is why the Koppert list is so valuable.

Theiling & Croft also presented pesticide "selectivity ratios," calculated by dividing the LD_{50} of pests by the LD_{50} of their biocontrols. These selectivity ratios are valuable information for IPM. The more selective a pesticide, the better. The most selective pesticides included pirimicarb and dicofol; some of the least selective pesticides were DDT, parathion, and cypermethrin. Theiling & Croft's literature review regarding biocontrols and pesticides has been compiled on computer diskettes. The database is called NERISK (Natural Enemy Risk Assessment, formerly called SELCTV), and is sold by the National Technical Information Service, Springfield, Virginia.

"All substances are poisons; there is none which is not a poison. The right dose differentiates a poison from a remedy."
—Paracelsus

Chapter 11: Biorational Chemical Control

Chemical controls are also known as pesticides. Use them only when all else fails. All pesticides cause side effects. Some are worse than others. *See the warnings below.* Some pesticides can be used on industrial crops (i.e., fibre hemp) but should never be used on food and drug crops such as seed oil or marijuana.

We define **biorational** chemicals as naturally-occurring compounds or synthetic analogues of naturally-occurring compounds, such as synthetic pyrethroids (Djerassi *et al.* 1974). *Natural* means these chemicals occur in nature. It does not mean they are safe or belong in baby food. Natural chemicals can be quite toxic, *naturally* toxic.

Most of the biorational chemicals we discuss here are permitted in *The National List* of materials mandated by the Organic Foods Production Act of 1990. But some chemicals listed here are restricted by the *List*, such as nicotine and sodium nitrate, and eschewed by organic farmers.

Some chemicals *kill* pests, but other chemicals provide us with other options—like repelling pests, or confusing them. Baiting a field with an artificial sex pheromone, for instance, confuses males and makes reproduction impossible. **Repellents** drive pests away from plants. Some repellents do not harm pests, they are simply offensive to pests. Bordeaux mixture and copper compounds repel insects without killing them (these chemicals, however, are lethal to fungi). Garlic extracts are classic **olfactory repellents**—flying

Table 11.1: Pesticide formulations. The abbreviations for formulations are used throught the text.

Formulation	Abbrev.	Description
Aerosol	A	Pesticide dissolved in volatile solvent and pressurized in a can by a propellant gas like CO_2, a convenient but expensive formulation.
Bait	B	Pesticide impregnated into a substrate attractive to pests, such as food or sex pheromones.
Dry flowable	DF	Dry pesticide formulated as granules, to be mixed with water and sprayed; much like a WP, but the heavier DF granules reduce a person's exposure to airborne particles during handling and mixing.
Dust	D	Nearly microscopic particles of powder-dry pesticides diluted in a dry carrier; dusts adhere well to plant surfaces, useful for treating small indoor areas but tend to drift outdoors.
Emulsifiable concentrate	EC	Petroleum-based liquid plus emulsifier, mix with water and spray; like all sprays, may ineffectively roll off plant surfaces. Phytotoxicity hazard usually greater than other liquid formulations, may leave visible residue on plants.
Flowable concentrate	FC	Solid or semisolid pesticide wet-milled into a pudding-like consistency to be mixed in water and sprayed; requires frequent agitation to remain suspended, may clog spray equipment, may leave visible residue on plants.
Fumigant	F	Vapour usually stored under pressure in metal bottles as liquified gas; vaporizes when released, quickly dissipates, dangerous, expensive.
Granule	G	Prepared by applying liquid pesticides to coarse particles of porous material; granules are like dusts, but larger so they wind-drift less.
Plant-Pesticide	P^2	A new category proposed by the EPA for bioengineered *crop plants* yielding toxins normally produced by other organisms, such as plants that produce BT—a toxin ordinarily produced by *Bacillus thuringiensis.*
Slow-release	SR	Pesticide embedded in polychlorovinyl resin (e.g., No-Pest Strip®) to slow the rate of volatilization; other SRs include paint-on pesticides and microencapsulation in semi-permeable membranes.
Soluble powder	SP	Soluble powder that dissolves in water to form a true solution; packaged as powders or concentrated solutions; unlike FCs, no further agitation is needed after SPs are dissolved.
Wettable powder	WP	Insoluble pesticide mixed with mineral clay into tiny particles <25 μm in diameter (WPs look like Ds but are more concentrated); suspended in water with a surfactant; act like FCs (i.e., sprays may clog), safer to plants than ECs, may leave visible residue on plants.

Table 11.2: EPA pesticide acute toxicity classification.

Class	LD_{50} for the rat oral (mg kg^{-1})	LD_{50} for the rat dermal (mg kg^{-1})	LD_{50} for the rat inhaled (mg l^{-1})	Eye Effects	Skin Effects
I	50 or less	200 or less	0.2 or less	corrosive opacity not reversible	corrosive
II	50–500	200–2000	0.2–2.0	corneal opacity reversible within 7 days, irritation persisting for 7 days	severe irritation at 72 hours
III	500–5000	2000–20,000	2.0–20	no corneal opacity, irritation reversible within 7 days	moderate irritation at 72 hours
IV	greater than 5000	greater than 20,000	greater than 20	no irritation	mild irritation at 72 hours

aphids "smell" garlic extracts sprayed on plants and turn away before landing. **Irritant repellents** such as pyrethroids cause insects to perform exaggerated grooming behaviours; the insects become stressed and agitated and go away. **Oviposition repellents** deter females from depositing eggs on treated surfaces. THC and CBD act as oviposition repellents against *Pieris brassicae* butterflies (Rothschild & Fairbairn 1980).

Antifeedant chemicals inhibit feeding behaviour, such as polygodial and azidarachtin (from the neem tree). Some antifeedants are not repellents, so pests remain on plants until they starve. Antifeedants and repellents can be applied to flowering tops, which drive pests down to older, unharvested leaves. On unharvested leaves, the pests can then be killed with a localized application of a nasty chemical.

The nasty chemicals are **pesticides**, which are designed to *kill* pests—not confuse them—and kill them quickly. Pesticides work when pests injest them (**stomach poisons**) or touch them (**contact poisons**). Stomach poisons only kill pests that chew and swallow their food. Stomach poisons are not very effective against insects with piercing-sucking mouth parts (e.g., aphids). Aphids push their stylets right past poisons adhering to plant surfaces. Sucking insects must be controlled with contact poisons.

Pesticides are named by their target: **miticides** (**acaricides**) kill mites, **insecticides** kill insects, **fungicides** kill moulds, **antibiotics** kill bacteria, **herbicides** kill plants, **nematocides** kill nematodes, **avicides** kill birds, and **rodentocides** kill rodents. **Disinfestants** eradicate surface organisms, whereas **disinfectants** work systemically, killing pests and pathogens within plants.

Most older pesticides work on the surface. They do not penetrate plants, so they cannot control pests or pathogens already established within plant tissue. Surface pesticides are frequently called **protectant pesticides** — they repel insects off plants, or prevent fungal spores from germinating.

New pesticides may work deeper than the surface. **Systemic pesticides** can penetrate tissue and redistribute within plants. Diffusion of pesticides via the xylem is called **apoplastic mobility**, and movement cell-to-cell, such as through phloem cells, is called **symplastic mobility**. Thus, systemic pesticides provide surface protection *as well as* deep and durable eradication of established pests and pathogens. The development of systemic pesticides in the 1960s was hailed as a breakthrough. But systemics tend to remain in plant tissues without decomposing, which is not good for products destined for human consumption. Furthermore, many systemics kill pests by a "site-specific" action upon *one* metabolic pathway. As a result of this specific action, pest resistance arises by a simple mutation of a single gene. Resistant pests arise soon after systemic pesticides are deployed (Agrios 1997).

Chemicals come in an assortment of formulations, see Table 11.1. **Fumigants** are gases. They are very effective but they must be used in enclosed areas, they require expensive equipment, and most are carcinogenic. **Sprays** are liquids and much easier to apply, but may be ineffective if they bead up and roll off plant surfaces. Spray formulations include emulsifiable concentrates (ECs), flowable concentrates (FCs), soluble powders (SPs), dry flowables (DF), and wettable powders (WPs). **Dusts** adhere firmly to plant surfaces but are susceptible to *wind drift* during application. **Granules** are similar to dusts but consist of larger particles, so they wind-drift less. Granules pose a serious hazard to birds, who mistake them for food or grit and eat them. **Baits** are often formulated as granules but have the power of attraction (e.g., food attraction, colour attraction, sex attractant). **Plant-pesticides** are a new category the EPA proposed for plants genetically manipulated by recombinant DNA technology (EPA 1994). Bioengineered plants produce toxins normally produced by microbes or other organisms—a grey zone between biocontrol and chemical control (Cook & Qualset 1996). Much remains unknown about bioengineered plants, yet USA regulatory agencies have approved them, so the USA has become a large-scale experiment for transgenic field studies. Unpredicted side effects can be ecological, such as the death of monarch butterflies from pollen blown off transgenic Bt corn (Losey *et al.* 1999). Medical side effects also arise, such as changes in the gastrointestinal tract after eating potatoes engineered for insect and nematode resistance (Ewen & Pusztai 1999), and lethal allergic reactions from eating transgenic soyabeans (Nordlee *et al.* 1996).

The amount of active ingredient (a.i.) present in a pesticide formulation is listed on the product label. *Solid* formulations list the amount as a percentage of weight. For instance, a 50%WP contains 50% a.i., or 50% pesticide and 50% "inert ingredients." *Liquid* formulations list the a.i. concentration two different ways: as a percentage of weight, *or* as the number of pounds per gallon (lbs/gal or lb gal^{-1}). For instance, a 4 lb/gal EC contains 4 pounds a.i. per gallon of product. This equals a 40–50%EC, depending on specific gravity of the "inert ingredients."

Chemical controls, whether natural or synthetic, can cause serious side effects. Technically, there is no such thing as a pesticide. "Pesticide" implies that a chemical selectively kills a pest, leaving everything else alone. Pesticides are really *biocides*—they harm many living things. They may even harm plants they are supposed to protect. They often damage or kill *nontarget* organisms. Honeybees and bald eagles are famous nontarget victims. Mycorrhizae are less known nontarget victims, killed by fungicides as well as insecticides and nematocides (Menge 1983).

Damage to honeybees can be reduced by spraying crops after sundown. Protecting bald eagles is not so easy. Over 1 billion pounds of pesticides are applied annually in the USA, which represents 34% of total worldwide application (Pimentel 1991). A cynic might point out this averages 20 g of pesticide per acre—pretty thin. But persistent pesticides often become concentrated and "biomagnify" in the food chain. One ppm DDT in pond water might accumulate to 100 ppm within plankton and accumulate to 1000 ppm in plankton-feeding fish. Eagles eat an accumulation of fish and become an endangered species.

Chemicals also affect us. Farmers face acute poisoning as a job hazard. Organophosphates cause the greatest number of pesticide poisonings in the USA because of their potency and widespread use. Most are EPA Class I toxins (see Table 11.2). A lethal dose can be absorbed through the skin, eyes, lungs, skin, or stomach (don't use your mouth to clear a spray line or prime a siphon!). In one remarkable report, parathion spilled on a pair of coveralls hospitalized three workers over a two-week period despite repeated launderings (Clifford & Nies 1989).

Approximately 40,000 people, mostly children, are treated for pesticide poisoning in the USA every year. Worldwide, pesticides *kill* 250,000 people annually. To quote Cynthia Westcott (1964), "The use of chemicals by amateurs is hazardous in any event." Always read the label and never act in haste.

Many pesticides are carcinogenic (causing cancer), mutagenic (causing gene damage), teratogenic (causing birth defects), or oncogenic (causing tumours). Agricultural workers suffer an increased incidence of brain tumours, testicular cancers, leukaemias and lymphomas. The EPA ranks pesticides on a multi-tiered scheme adapted from the World Health Organization's International Agency for Research on Cancer (IARC). Many pesticides fall in IARC Group 2B: *Substances Possibly Carcinogenic To Humans*. Rockwool and saccharin also fall into this category.

Some pesticides are not acutely poisonous nor carcinogenic, but act as female hormones (they are oestrogenic), and disrupt our endocrine systems. Oestrogenic compounds are especially harmful to developing fetuses. "Hand-me-down poisons" absorbed by pregnant females may harm their offspring *post utero*, at any time of their offspring's lives, and may even harm offspring of their offspring. A male exposed before birth to oestrogenic compounds may have undescended testicles at birth, a low sperm count at puberty, testicular cancer in middle age, or prostate cancer as an old man—all from his prenatal exposure (Colborn *et al.* 1996). Certain plastics produce oestrogenic effects. Nonylphenol plastics are added to polystyrene and polyvinyl chloride (PVC) to make these plastics less brittle. Bisphenol-A is added to polycarbonate for stability. Polystyrene, PVC, and polycarbonate are frequently used in hydroponic systems, and they leach oestrogenic nonylphenol and bisphenol-A into the water (Colborn *et al.* 1996). In 1992, fourteen European countries agreed to ban these chemicals by the year 2000.

Consumers face unknown exposures; pesticide residues on foodstuffs are a major concern. The EPA accepts pesticide "tolerances" (maximum allowable residues), then tests foodstuffs in random market samplings. If residues exceed tolerance levels, the crop cannot be sold. Residues in marijuana are a different story. Few pesticides have been tested for their effects when burned and inhaled. Only tobacco researchers seem to be interested in these questions, and little research has been done (Lucas 1975).

Our final Faustian bargain with pesticides is the rebound of resistant pests. In 1950, shortly after DDT was invented, only 20 insects were resistant to the new panacea. Since then, the number of resistant pests has risen to 535 insects, 210 fungal pathogens, and 200 weed species; many of these pests are resistant to multiple pesticides (Pimentel 1991). The search for effective pesticides places farmers on a "pesticide treadmill." Pesticides eventually become counterproductive—Pimentel (1991) argues that pesticide application on USA croplands has grown *33-fold* since the 1940s, yet crops losses from pests have actually *increased* in that time, from 31% to 37%.

Despite these serious caveats, chemicals will continue to be used on crops. If chemicals are to be used, we want people to use them correctly, to minimize their toxic side effects. We dedicate this chapter (and Appendix 2) to the spirit of harm reduction.

Some pesticides are less toxic than others. Less-toxic biorational pesticides are preferable, but these require more precise timing to be effective, work more slowly, must be applied more often, and pests must be more closely monitored. Nevertheless, biorational pesticides are safer, more ecological, and morally imperative. A classification of biorational pesticides is presented in Table 11.3; a selection of biorational pesticides is presented in Table 11.4. For sources see the *Directory of Least-Toxic Pest Control Products*, published annually by BIRC (Bio-Intergral Resource Centre), P.O. Box 7414, Berkeley, California 94707, telephone (510) 524-2567. For more information on pesticides, see the series written by Thomson or the annual *Farm Chemicals Handbook* edited by Meister.

Table 11.3: Classification of Biorational Pesticides.

Natural	Non-botanicals Mineral-based Carbon-based Botanicals Fermented products
Synthetic (or semi-synthetic)	Synthetic botanicals Pest-growth regulators Synthetic pheromones

NONBOTANICALS

Nonbotanical natural pesticides are either mineral-based or carbon-based. Mineral-based pesticides usually work on contact. A few work as stomach poisons, such as once-popular but now banned arsenic insecticides (e.g., Paris green) and mercurous insecticides (e.g., calomel). Mercurous compounds also served as fungicides, such as Ceresan® - methoxyethyl mercury chloride. Ceresan was popular in European Community countries from 1929 until it was banned in 1992 due to adverse toxicology (Maude 1996). Many mineral-based pesticides are permitted on "Certified-Organic" farms if they come from a mined source, not a chemical laboratory.

Bordeaux mixture is a foliar fungicide (introduced by Millardet in 1883), but also kills bacteria and repels caterpillars, beetles, and other insects. Bordeaux is a mix of $Ca(OH)_2$ (calcium hydroxide or lime) and $CuSO_4$ (copper

Table 11.4: Some biorational pesticides.

Generic Name	US Trade Name(s) ™ or ®	Chemical Type	EPA Class
abamectin	Avid, Vertimec	fermented product	III
boric acid	Tim-Bor, Borid, Roachkil	mineral-based	III-IV
calcium cyanamide	Cyanamid	carbon-based	
Cannabis	Muggles	botanical	IV
carbon dioxide	CO_2	carbon-based	
castor	castor oil	botanical	III
cholecalciferol	Muritan, Rampage, Quintox	carbon-based	I
copper sulphate	Basicop, Tribasic, Bordo	mineral-based	III
copper sulphate & lime	Bordeaux mixture	mineral-based	III
copper oxychloride	Champ F, C.O.C., C.O.C.S	mineral-based	III
cryolite	Kyrocide, Prokil	mineral-based	III
cupric hydroxide	Kocide 101, Blue Shield	mineral-based	III
cuprous oxide	Cuprocide, Perenox	mineral-based	II
cyromazine	Aurmor, Trigard	growth regulator	III
diatomaceous earth	Celite, Celatom	mineral-based	IV
diflubenzuron	Dimilin, Vigilante	growth regulator	III
formaldehyde	Formalin	carbon-based	II
griseofulvin	Fulvicin, Grisactin	fermented product	II
hellebore	Hellebore	botanical	III
horsetail	Horsetail	botanical	IV
hydroprene	Gencor	growth regulator	IV
imidacloprid	Confidor, Admire, Provado	synthetic botanical	III
kinoprene	Enstar	growth regulator	II
methoprene	Kabat, Minex, Altosid	growth regulator	IV
neem (azadirachtin)	Bioneem, Azatin, Neemayad	botanical	IV
nicotine	Black-Leaf 40	botanical	II
oil	Sunspray, Scalecide	carbon-based	IV
piperonyl butoxide	PBO, Pybuthrin	botanical synergist	III
polygodial	Polygodial	botanical	
pyrethroid	Pokon, Safer's Bug Killer	synthetic botanical	III
pyrethrum	Buhach, Insect powder	botanical	III-IV
quassia	Bitterwood	botanical	IV
red squill	Rodine, Dethdiet	botanical	II-III
rotenone	Derris, cubé	botanical	II-III
ryania	Ryan 50, Ryanicide	botanical	III
sabadilla	Red Devil, Veratran D	botanical	IV
soap	M-Pede, Safer's, Dr. Bronner's	carbon-based	IV
sodium bicarbonate	Baking Soda	carbon-based	IV
sodium hypochlorite	Bleach	mineral-based	II-III
sodium nitrate	Chilean nitrate	carbon-based	
spinosad	Conserve, Tracer, Naturalyte	fermented product	
streptomycin	Agrimycin 17, Agri-Strep	fermented product	IV
strychnine	Strychnine	botanical	I
sulphur	Cosan, Hexasul	mineral-based	IV
trimethyl docecatriene	Stirrup M	pheromone	
urea	ClandoSan	carbon-based	

sulphate) in water. Only the copper is fungitoxic; the lime serves as a "safener" to protect plants. Preparations of Bordeaux are labelled as a ratio of copper sulphate (in pounds) and quick lime (in pounds) to water (in gallons). The most popular mixture is 4:4:50 (= 8:8:100). Bordeaux is widely available in SP or WP formulations. To make Bordeaux, suspend copper sulphate snow (not fixed copper) in half the water, suspend quick lime in the other half, then mix the two solutions. Spray immediately after mixing. Agitate the mix frequently to keep it from settling in the spray tank. Metcalf *et al.* (1962) said a gallon of Bordeaux covers 1000 square feet of leaf surface. Persistence is long. Bordeaux causes some phytotoxicity in plants, especially on seedlings when applied in cool, wet weather. To reduce phytotoxicity, decrease copper or increase lime (mix 2:6:100 or 8:24:100), and spray on a warm, dry day. Bordeaux is slightly toxic to honeybees, practically nontoxic to mammals, but highly toxic to fish. The REI is 48 hours.

Boric acid (H_3BO_3), is a stomach and contact poison useful against roaches, earwigs, crickets, and ants. Available as a 98%D, 18%G, and 1–5%B formulations. These formulations are persistent in dry conditions, but wash out rapidly after rainfall. Boric acid is phytotoxic. Borax is not dangerous to honeybees or birds. The dust irritates eyes (but boric acid *solution* is used as an eyewash). *Sodium borate* ($Na_2B_8O_{13}\cdot 4H_2O$) is a mineral borate used on wood products.

Copper is a fungicide, useful against grey mould, leaf spots, blights, anthracnose, powdery mildews, downy mildews, and also some foliar bacterial diseases. Fixed copper is less soluble than the copper used in Bordeaux mixture, so it is less phytotoxic, but also less effective. It is not toxic to birds and bees, but very toxic to fish, and acts as a eye and skin irritant to mammals. The REI for all copper compounds is 48 hours. Copper can bioaccumulate in food chains. Below we present several useful inorganic salts of fixed copper. Collectively, copper compounds are the second most widely-selling fungicides, behind only mancozeb (Hewitt 1998).

Copper sulphate ($CuSO_4\bullet 5H_2O$) is the most popular form of fixed copper. It is often called "tribasic copper." Available as a 1–82%WP, 6–8%FC, or 7%D. Residuals last five to 14 days.

Copper oxychloride ($Cu_2Cl(OH)_3$) is sold as a 45–85%WP and 10–25%D. A sulphated version (*copper oxychloride sulphate*), is available as a 50–53%WP and 3–15%D. Copper oxychloride is more popular in Europe, copper oxychloride sulphate is more popular in the USA.

Cupric hydroxide ($Cu(OH)_2$) is sold as a 21–77%WP and 3–55%D. This formulation dissolves in water better than most fixed coppers, causing less clogging of spray nozzles.

Cuprous oxide (Cu_2O) is sold as a 19–80%WP, 64%EC, 60–65%FC, 4–83%D, and G. It is less phytotoxic than other copper compounds, but more toxic to us. Residual period and reapplication interval is seven to ten days. Cuprous oxide is also used as a seed treatment.

Clay microparticles, modified from kaolin, can be sprayed onto crops in a thin, watery slurry. The material sticks to plants and dries to a white powdery film. The microscopic clay particles attach to the bodies of insects. The microparticles repel small, thin-skinned pests, such as mites, aphids, thrips, and leafhoppers. The coating does not interfere with photosynthesis; sunlight diffuses into leaves with little reduction of light. Clay microparticles are nontoxic and do not harm predatory biocontrol insects.

Cryolite, sodium aluminofluoride ($Na_3Al_1F_6$) works as a stomach poison against insects with chewing mouthparts (flea beetles, weevils, and some caterpillars). It also has some contact activity against mites and thrips. Available as a 96%D or 96%WP, applied with a duster or sprayer. Cryolite is persistent in dry weather. Do not combine cryolite with lime or other alkali. Some phytotoxicity of corn and fruit trees occurs in damp climates. Nontoxic to birds and bees, moderate toxicity to fish. Natural (organic-approved) cryolite is mined in Greenland, synthetic sources are manufactured in the USA and other countries.

Diatomaceous earth (DE) is a contact insecticide, often combined with pyrethrins. DE tears microscopic holes in the surfaces of soft-bodied insects, and slices between the exoskeleton plates of hard-bodied insects. In hot, dry weather the injured insects rapidly dehydrate and die. DE is formulated as a talc-like D. It is persistent in dry weather. DE can irritate eyes. It should not be applied to drug crops destined for inhalation, because DE is a serious respiratory hazard.

Sodium bicarbonate, $NaHCO_3$ (baking soda) works well against powdery mildew. Spray a 0.5% solution of baking soda (15 ml per 4 l or 3 tsp/gallon). Some growers double the concentration. Sodium bicarbonate works better when combined with light horticultural oil (also 15 ml per 4 l). *Potassium* bicarbonate (Kaligreen®, Armicarb 100®) can be used the same way as sodium bicarbonate. Nontoxic, low persistence, little phytotoxicity.

Sodium hypochlorite, a 5% solution of NaOCl is sold as household bleach. Bleach disinfects viruses, bacteria, fungi, and small arthropods from equipment, glasshouse walls, and the air. A solution of one part household bleach in nine parts water (0.5% NaOCl) is used as a seed soak or soil drench against fungi. Undiluted bleach is toxic to plants and caustic to our eyes and skin.

Sulphur (sulfur) is applied as a 98%D (fungicidal against rusts and powdery mildews) or a 30–92%WP (miticidal, but not against spider mites). Mix and spray at a rate of 600 g sulphur per 100 l water. Available from many manufacturers. Sulphur is the fourth most-popular fungicide, behind mancozeb, copper and chlorothalonil (Hewitt 1998). Sulphur persists on plants until washout, then it becomes a component of the soil environment. Sulphur can injure plants, especially in hot, dry weather. Do not apply if the temperature is expected to exceed 30°C. Sulphur mixed with lime kills more insects, but this mixture causes greater phytotoxicity. Honeybees, birds, and fish are safe. Sulphur dust is irritating to eyes and lungs; ingested sulphur acts as a laxative. The REI is 24 hours.

Water, full-strength, 100%, controls some pests and pathogens. Parker (1913b) washed 50–70% of spider mites off hops plants by directing a strong but fine stream of cold water at undersides of leaves. This also destroys their webbings. Gentle misting works if predatory mites (e.g., *Phytoseiulus persimilis*) are present—the mist slows spider mites yet speeds the performance of predators. Unfortunately, spraying water increases the likelihood of grey mould in flowering plants.

CARBON-BASED NON-BOTANICALS

The other category of nonbotanical chemicals is carbon-based pesticides. We define carbon-based pesticides as naturally-occurring compounds that contain carbon but are not extracted from plants. Some are carcinogenic and not permitted in "Certified Organic" operations.

Calcium cyanamide, $CaCN_2$, also known as lime nitrogen, is made from heating lime and coal. It has been used since the turn of the century as a nitrogen fertilizer, a weed killer, and a fungicide. Available as a G, mix 562 kg ha^{-1} (500 lbs/acre) in soil to kill soil pathogens. It is phytotoxic and must be applied in late autumn or very early spring, but is broken down into urea by seeding time.

Carbon dioxide is a fumigant gas that kills nearly anything not green. Seal plastic bags over plants and inflate bags with the gas for several hours. CO_2 is useful against recalcitrant insects and fungi. High concentrations are lethal to humans, so use caution in growrooms.

Cholecalciferol, a sterol hormone, is used as a rat poison, with an oral LD_{50} of 43 mg kg^{-1}. Available as a 0.075%B. It causes hypercalcemia and kills rats in two to four days. A lethal dose can be consumed in a single feeding or accumulated in smaller, multiple feedings. Unfortunately, dogs and cats may be more sensitive to cholecalciferol than rats. In humans, the RDA of cholecalciferol (vitamin D_3) is 10 µg. Vitamin D_2, ergocalciferol, is also sold as a rodenticide.

Formaldehyde is carcinogenic and not used much anymore. It works as a liquid/fumigant, lethal to damping-off oömycetes, fungi, bacteria, insects, nematodes, and viruses. Formalin is a 38% solution of formaldehyde; paraformaldehyde is a solid polymer. Using formaldehyde as a seed disinfectant began in the 1890s, to eliminate seedborne fungi. It has also been used as a soil drench, or mixed into porous materials (oat hulls, sawdust) and sprinkled in field furrows to kill soil organisms. Formaldehyde rapidly oxidizes to formic acid. It is phytotoxic, toxic to honeybees, fish, and mammals. The vapour irritates eyes and lungs.

Oil was traditionally sprayed on trees in winter as a **dormant oil** to kill aphid eggs and mites. Dormant oil is heavy (with a 50% distillation temperature or "flashpoint" of ≥460°F) and herbicidal to herbaceous plants. Improved refining techniques have produced lighter **horticultural oils** (**summer oils**) that can be sprayed on sensitive foliage. Light "hort oils" flash as low as 412°F (oils below 400°F do not kill insects). Hort oils are ranked by their *UR rating* (unsulphonated residue, the percentage of oil that is free of phytotoxic residues after flashing)—look for 92% or greater. Oil *gravity* (% unsaturated hydrocarbons) is also important—look for 32(%) or less.

Hort oil is derived from animals (fish oil), vegetables (seed oil), or minerals (petroleum). The oil plugs up spiracles (breathing tubes). It suffocates mites, whiteflies, aphids, mealybugs, scales, and other soft bodied insects. Hort oil also kills many beneficial insects, especially immatures (adults may escape the nozzle). Hort oil rarely causes phytoxicity if sprayed in a 1% solution. A 3% solution may cause damage—always test-spray. To reduce phytotoxicity, crops should be well-watered and then sprayed on a warm day (21–38°C) with low humidity (<40% RH), so oil can kill insects and then evaporate quickly. Dried oil is inactive. Hort oil can be used against fungi if mixed with 0.5% baking soda (15 ml per 4 l or 3 tsp/gallon). Do not mix with sulphur. Oil can also be used to form soap (see below).

Soap is oil combined with sodium or potassium alkali. Most soaps sold today consist of potassium oleate (the potassium salt of oleic acid, a fatty acid). Soap suffocates many insects (but not their eggs). Safer's® Insecticidal Soap causes 98.6% mortality in aphids and 91% mortality in two-spotted spider mites (Puritch 1982). Dr. Bronner's Soap® also works well (mix 2.5 ml per 4 l or 0.5 tsp/gallon). The USDA is making soap from a tobacco extract that kills sweetpotato whiteflies. A 1% solution of M-Pede® soap enhances the activity of synthetic miticides (abamectin, dicofol) and insecticides (diazinon, malathion), according to the manufacturer.

Soaps work best if mixed in soft water; soften the water if hardness exceeds 300 ppm (or 17.5 grains per gallon). Plants should be sprayed at five-day intervals. Soaps lose activity after drying, so spray early in the morning when temperatures are cool. Follow each soap spray after several hours by a water rinse, especially in warm weather. Soap is not selective and kills many beneficial insects. Soap kills predatory mites but their eggs survive, so soap and predatory mites can work together. Soap can be used on medical *Cannabis* up to a week before harvest without any distasteful residues discerned on finished dry flowers.

Sodium nitrate, $NaNO_3$, kills damping off fungi and nematodes (which act synergistically and should be eradicated together). Available from many manufacturers as a G. Mix into topsoil at a rate of 135-270 g m^{-2} (4-8 oz/$yard^2$). It is phytotoxic but not persistent; mix into soil several weeks before setting seed.

Urea, $CO(NH_2)_2$, kills some damping off fungi and nematodes like sodium nitrate, and is also available as a G. Mix 540 g m^{-2} (16 oz/$yard^2$) into topsoil several weeks before sowing seed. Clandosan® combines urea with **chitin** from ground-up crab shells. Chitin is a polymer of the amino sugar N-acetyl-glucosamine (NAG is a building block of cartilage in humans). **Chitosan** is a nonacetylated form of chitin. Crab shell chitin encourages the growth of nematicidal soil organisms when mixed into soil at a rate of 2-4 kg per 10 m^2. However, chitin may cause phytotoxicity if mixed into soil at levels exceeding 1% (Mian *et al.* 1982).

Human urine repels many verte*brats*, such as deer and other varmints. Mark your territory.

BOTANICAL POISONS

Many companion plants mentioned in Chapter 10, such as garlic, retain their repellency after they are harvested and dried. They can be crushed up, mixed with water, and sprayed on plants as bug repellents. Repellent sprays are commercially available (Guardian®, a garlic extract; Hot Pepper Wax®, a *Capsicum* extract).

"High powered" botanicals do more than repel—they kill. They act as contact or stomach poisons. Some botanicals are *extremely* lethal and *not* permitted on Certified-Organic farms, such as nicotine and strychnine.

In their simplest form, botanical pesticides are simply dried plants, ground to a fine powder. The powders can be used full-strength or diluted in a carrier such as clay or talc. Pesticidal extracts can be produced by soaking fresh or dried plants in a solvent, such as water or alcohol. After a while the solid material is filtered out, leaving the liquid extract. Extracts can be sprayed full-strength on crop plants, boiled down to liquid concentrates, distilled to pure pesticidal chemicals, or applied to clay for use as dusts.

Rosenthal (1999) described a home-made botanical brew he sprayed on marijuana plants. To 1 l (1 quart) of boiled water add 30 ml (2 tablespoons) of the following: ground cinnamon (*Cinnamonum zeylanicum*), ground chili pepper (*Capsicum annuum*), ground black pepper (*Piper nigrum*), mint or peppermint leaves (*Mentha* species), fresh crushed garlic (*Allium sativum*), fresh chopped onion (*Allium cepa*), and orange peel (*Citrus sinensis*). When the mixture cools but is still warm, add 500 ml (2 cups) isopropyl alcohol, 250 ml (1 cup) of strong coffee, and 125 ml (0.5 cup) low-fat milk). Strain through a fine sieve, add 30 ml liquid soap, and add enough water to make 2 l of pesticide.

Some botanical poisons have been used for centuries, such as neem and ganja in India, pyrethrum in the Middle East, and rotenone in South America—all described below. New commercial products are being developed from calamus (*Acorus calamus*), basil (*Ocimum basilicum*), big sagebrush (*Artemisia tridentata*), chilcuan (*Heliopsis longipes*), and mamey (*Mammea americana*).

Cannabis flowers, leaves, seeds, and their extracts have been used as repellents or pesticides against **insects** (Culpepper 1814, King 1854, Riley & Howard 1892, Indian Hemp Drugs Commission 1894, MacIndoo & Stevers 1924, Metzger & Grant 1932, Chopra *et al.* 1941, Bouquet 1950, Abrol & Chopra 1963, Reznik & Imbs 1965, Fenili & Pegazzano 1974, Khare *et al.* 1974, Stratii 1976, Pakhomov & Potushanskii 1977, Rothschild *et al.* 1977, Rothschild & Fairbairn 1980, Prakash *et al.* 1987, Kashyap *et al.* 1992, Bajpai & Sharma 1992, Jalees *et al.* 1993, Sharma *et al.* 1997), **mites** (Reznik & Imbs 1965, Fenili & Pegazzano 1974, Surina & Stolbov 1981), **bacteria** (Ferenczy 1956, Ferenczy *et al.* 1958, Schultz & Haffner 1959, Zelepukha 1960, Bel'tyukova 1962, Radosevic *et al.* 1962, Zelepukha *et al.* 1963, Gal *et al.* 1969, Veliky & Genest 1972, Veliky & Latta 1974, Farkas & Andrássy 1976, Klingeren & Ham 1976, Fournier *et al.* 1978, Braut-Boucher *et al.* 1985, Vijai *et al.* 1993, Krebs & Jäggi 1999), **fungi** (Vysots'kyi 1962, Bram & Brachet 1976, Dahiya & Jain 1977, Misra & Dixit 1979, Pandey 1982, Gupta & Singh 1983, McPartland 1984, Singh & Pathak 1984, Kaushal & Paul 1989, Grewal 1989, Upadhyaya & Gupta 1990, Krebs & Jäggi 1999), **nematodes** (Kir'yanova & Krall 1971, Haseeb *et al.* 1978, Vijayalakshmi *et al.* 1979, Goswami & Vijayalakshmi 1986, Grewal 1989, Mojumder *et al.* 1989, Kok *et al.* 1994, Mateeva 1995), **protozoans** (Nok *et al.* 1994), **"worms"** (Parkinson 1640, Culpeper 1814, Pliny 1950), and **weeds** (Stupnicka-Rodzynkiewicz 1970, Srivastava & Das 1974, Muminovic 1991).

Many terpenoids in *Cannabis* inhibit acetylcholinesterase, including limonene, limonene oxide, α-terpinene, γ-terpinene, terpinen-4-ol, carvacrol, *l*- and *d*-carvone, 1,8-cineole, *p*-cymene, fenchone, pulegone, and pulegone-1,2-epoxide (McPartland & Pruitt 1999). These cholinergic terpenoids paralyse and kill insects the same way as malathion and other synthetic organophosphate pesticides, although plant terpenoids are 4.4- to 17.1-fold less potent than malathion (Ryan & Byrne 1988). Research by Rothschild *et al.* (1977) suggests that cannabinoids are insecticidal. Cannabinoids are safe to mammals; the oral LD_{-5} of THC in mice is >21,600 mg kg^{-1} (Loewe 1946), safer than neem. Working with THC may, however, elicit an allergic contact dermatitis (Watson *et al.* 1983).

Castor, from seeds of *Ricinus communis*, suppresses nematode populations when mixed in the soil. Seed oil and leaf extracts have been used against a variety of insects (Jacobsen 1990). Persistence and phytoxicity unknown.

Creosote is distilled from wood tar or coal tar, useful as a wound treatment against fungi. It is a benzene compound, quite carcinogenic. A synthetic derivative, dinitro-o-cresol, is also used. Both are best avoided.

Hellebore, a hot water decoction of rhizomes from several *Veratrum* species, acts as a stomach poison against many insects. Available as a D. Hellebore loses potency quickly after exposure to air and sunlight; short residual. It is toxic to livestock.

Horsetail, from *Equisetum hycmale*, kills epiphytic fungi and powdery mildews. Boil 40 g of dried leaves and stems in 4 l water (1.5 oz/gallon) for 20 minutes (Yepsen 1976). Cool, strain, and spray on plants.

Limonene is a terpenoid distilled from citrus peel oil. It also occurs in *Cannabis*. Its mode of action is similar to pyrethrum (Thomson 1998). Limonene, particularly *d*-limonene, repels and paralyzes insects; this activity is synergized by PBO. Available as a 37%EC, and sold in aerosol cans (DeMize®) and shampoos for ridding house pets of fleas and ticks. Limonene is volatile and evaporates rapidly from treated surfaces. It harms honeybees and other beneficial insects if sprayed directly on them. Limonene irritates the eyes, mucous membranes, and skin.

Neem is extracted from seeds and foliage of the Indian neem tree (*Azadirachta indica*) or the chinaberry tree (*Melia azedarach*). The primary active ingredient, **azadirachtin**, is a steroid-like triterpenoid. Neem is a broad-spectrum agent that mimics insect growth hormones, so it primarily kills immature insects. Neem also serves as an antifeedant and repellent. Poisoned larvae fail to complete the moulting process and die slowly (typically three to 14 days) *but* neem deters larvae from feeding almost immediately (unlike rotenone). Neem works best against caterpillars, but also controls immature whiteflies, leafminers, fungus gnats, mealybugs, leafhoppers, and some thrips and beetles. It is less effective against aphids and some grasshoppers. Neem is available as an EC from several manufactures at different concentrations. Azatin® is the most concentrated (mix 4 ml per 4 l or 3/4 tsp/gallon).

Neem breaks down rapidly in standing water, so don't mix more than needed. Diluting water should be acidic (pH 3–7). Use a spreader/sticker. Spray when humidity is high (morning or evening). Neem only persists four to eight days on plant surfaces in bright sunlight. It persists longer, for several weeks, when plants absorb neem *systemically*, when it is applied as a soil drench (Mordue & Blackwell 1993). Some neem oil products leave a sticky residue on plants. Under laboratory conditions, normal doses of neem (<20 ppm azadirachtin) do not harm most beneficial insects (Mordue & Blackwell 1993). Neem does not harm foraging honeybees, earthworms, fish or birds; in mammals it irritates eyes.

Neem oil, a clarified hydrophobic extract of neem seeds (which does *not* contain azadirachtin), is marketed as Neemguard® and Triact®. Neem oil has fungicidal, insecticidal, and miticidal activity. Locke *et al.* (1993) found that a foliar spray of 1% Neemguard *protected* plants against powdery mildews and rust fungi better than benomyl (but benomyl worked better at *eradicating* established infections). Neem oil kills aphids and mites more effectively than regular neem. It may also harm beneficial insects (Cherim 1998).

Nicotine, derived from tobacco (*Nicotiana tabacum*), first served as an insect control in the 1690s. It is useful against aphids, mites, whiteflies, young borers, thrips, young beetles and bugs, and other soft-bodied insects. Nicotine works as a stomach poison, contact poison, and by inhalation (no surprise there). There are two forms — free nicotine alkaloid and conjugated nicotine salts.

Nicotine alkaloid is volatile and serves as a fumigant. Years ago, the New York Agricultural Society suggested placing a box over infested plants and smoking aphids to death with a cup of burning tobacco. Today this can be accomplished with a "smoke generator" formulation. Tobacco dust is also available (0.5% nicotine alkaloid). Hot water can extract the free alkaloid, but do not boil it. Parker (1913a) killed over 95% of hops aphids (*Phorodon humuli*) with two types of "tobacco tea": 1) he soaked 11.3 kg of "tobacco waste" in 378 l water (25 lbs/100 gallons); 2) he soaked 6.1 kg of commercial blackleaf tobacco in 378 l water (13.5 lbs/100 gallons). Free nicotine alkaloid is very unstable, has a very short half-life, and is not very popular.

Nicotine sulphate (NS) is the most common conjugated salt. It is sold as a 40% aqueous solution. Parker (1913a) mixed 130 ml of 40% NS in 378 l water (4.4 oz/100 gallons), and killed 99.9% of hops aphids (*Phorodon humuli*). The potency of NS increases when it is combined with an alkaline activator (e.g., soap or calcium caseinate). Parker (1913a) mixed 1.8 kg soap in 378 l water (4 lbs/100 gallons). The alkaline activator reverts NS to a free nicotine alkaloid. The half-life of NS, without an alkaline activator, is 4.5 days. Phytotoxicity is rare. Nicotine kills honeybees if sprayed directly on them. So spray late, after bees have retired for the evening. Nicotine on plants repels honeybees before they are harmed. Nicotine is moderately toxic to fish and birds, and very toxic to mammals (LD_{50} = 50-60 mg kg^{-1}), so it is prohibited from most Certified Organic farms. High dosages of nicotine cause tremors, convulsions, paralysis, then death.

Oil from peanut, safflower, citrus, corn, soyabean or sunflower, kills small insects. Vegetable oil consists primarily of fatty acids and glycerides. See more information in the description of carbon-based pesticides, above.

Pawpaw extracts, from the pawpaw blowtorch tree, *Asimina triloba,* have been applied as a soil drench to kill soil nematodes (Meister 1998).

Physcion, extracted from a European weed, giant knotweed (*Reynoutria sachalinensis*) is marketed as a foliar spray (Milsana®) for protecting fruits and vegetables against the powdery mildew fungus *Sphaerotheca fuliginia.* Physcion works by inducing plant resistance via phytoalexin production; technically it is not a fungicide. Physcion takes one or two days to induce resistance, weekly treatments may be needed.

Polygodial, derived from *Warburgia* and *Polygonum* plants, works as a feeding deterrent against aphids (e.g., *Myzus persicae*), armyworms (*Spodoptera* species), and budworms (*Heliothis* species). A 0.1% solution of polygodial sprayed on plant surfaces kills less than 20% of these pests, but it deters them from feeding on treated surfaces, so they migrate elsewhere. Polygodial is a sesquiterpene dialdehyde and unstable in sunlight; its half-life on plant surfaces is short.

Pyrethrum, a contact insecticide, kills aphids and whiteflies best, and also works against young bugs, beetles, and caterpillars. Frank (1988) killed spider mites with a pyrethrum-PBO product, Holiday Fogger®. Pyrethrum is a powder produced by grinding dried flowers of *Chrysanthemum* plants, predominantly *C. cinerariifolium.* The active ingredients in pyrethrum are collectively called *pyrethrins* (e.g., pyrethrins, cinerins, and jasmolins). Synthetic pyrethrins are collectively called *pyrethroids* (e.g., permethrin and cypermethrin). Pyrethrum originated in the Middle East and was introduced into the USA around 1870. It is now our most popular botanical pesticide.

Formulations include pyrethrum powder (powdered flowers, 1% pyrethrins), pyrethrin dust (20–60%D, mixed with clay or talc carriers), pyrethrin sprays (1–20%EC, 1–20%WP), and aerosol products. Many formulations contain the synergists PBO or MGK 264. Disadvantages include high cost and pyrethrum's rapid deterioration after exposure to air (about one day). Sunlight accelerates its deterioration, so apply in early evening or during cloudy weather. To slow degradation, combine pyrethrum with Pheast®, a lepidopteran feeding stimulant. Do not mix pyrethrum in alkaline water; it is not compatible with Bordeaux, lime, and soaps. Pyrethrum exerts a rapid paralytic action and can "knock down" flying insects, but they may recover. Some insect resistance has appeared. Pyrethrum is not phytotoxic. It kills honeybees if sprayed directly on them (foraging bees who encounter previously-sprayed plants are repelled but not killed). Pyrethrum is highly toxic to fish. Mammals safely ingest pyrethrum (it is rapidly hydrolysed by stomach acids); pyrethrum is more toxic to humans when inhaled. Some people develop skin allergies when handling it.

Quassia is extracted from the wood of bitterwood (*Quassia amara*) and the lowly tree-of-heaven (*Ailanthus altissima*). Parker (1913a) chopped up 1.4 kg quassia chips and 1.8 kg soap into 378 l water to kill 96% of *Phorodon humuli* and other aphids. For small batches, he mixed 140 g chips per 4 l (5.3 oz/gallon). Chips should be soaked for 24 hours then boiled for two hours. Warning: the quality of chips may vary. Yepsen (1976) potentized quassia by adding larkspur seeds. Mixing quassia with potassium soap also enhances its activity. Quassia supposedly spares ladybeetles and bees, and is nontoxic to mammals. It has substituted for hops in beer manufacturing (Meister 1998).

Red squill, *Urginea maritima,* is an onion-like plant from the Mediterranean region. Bulb extracts and dried powder have been used for rodent control since the 13th century (Godfrey 1995). Red squill extract has an oral LD_{50} of 500 mg kg^{-1} in rats; the purified active ingredient, scilliroside, has an oral LD_{50} of 0.43 mg kg^{-1} in female rats. It affects the cardiovascular and central nervous systems, causing convulsions and death after 24–48 hours. Some countries ban red squill as a cruel poison. The extract is very bitter; when administered to nontarget animals, red squill usually induces vomiting (thus it is considered relatively safe).

Rotenone has been used since ancient times to kill fish. The Chinese discovered rotenone's insecticidal qualities, which attracted Western attention around 1848. Rotenone is a resin extracted from Peruvian cubé root (*Lonchocarpus* species) or, less frequently these days, from Malaysian *Derris* or *Tephrosia* species. It was the most popular pesticide in the USA until DDT in the 1940s. Rotenone works best against insects with chewing mouthparts, such as budworms, cutworms and beetles, but it also works against thrips, aphids, and occasional mites. Formulations include a D, 1–5%WP, and 1.1 lb/gal EC. Unfortunately, rotenone has a slow onset of action—fast-feeding insects may consume up to 30 times their lethal dose before rotenone affects them. Consider adding pyrethrum to rotenone—fast-acting pyrethrum slows insect feeding rates, allowing rotenone time to kill. Rotenone persists longer than most botanicals. It breaks down after three to ten days on plant surfaces in bright sunlight, quicker in the presence of soaps or lime. Phytotoxicity is rare. Rotenone is highly toxic to honeybees, fish, and mammals (rotenone is more toxic to humans than malathion). Chronic use causes skin irritation, and possibly liver or kidney damage. Some organic growers refuse to use rotenone; the Netherlands banned its use in 1980. Rats given chronic, low doses of rotenone develop a selective apoptosis of nigrostriatal dopaminergic neurons, a syndrome akin to Parkinson's disease (Friedrich 1999).

Ryania is extracted from roots and stems of *Ryania speciosa* (Jacobsen 1990). It is a slow-acting stomach poison, used against European corn borers, hemp borers, other caterpillars, and thrips. Ryania was "discovered" at Rutgers, by researchers screening extracts of plants used by Amazonian natives. It works better in hot weather, whereas sabadilla works better in cool weather (Ellis & Bradley 1992). Available as a D or WP. Ryania is relatively persistent on treated surfaces. The residual remains toxic much longer than pyrethrum; it persists almost as long as rotenone. Phytotoxicity rare. Ryania is moderately toxic to mammals, fish, and birds.

Sabadilla, an alkaloid like ryania, is extracted from seeds of a New World lilly (*Schoenocaulon officinale*) and European white hellebore (*Veratrum album*). It serves as a contact and stomach insecticide against bugs, thrips, beetle grubs, caterpillars, and grasshoppers. Yepsen (1976) claimed sabadilla becomes more powerful with storage; heating it to 75–80°C for four hours also markedly activates it. Applied as a D or F. Once dusted, sabadilla breaks down quickly in sunlight (about two days). It is highly toxic to honeybees. The dust irritates mucous membranes and may cause violent sneezing.

Strychnine, derived from seeds of *Strychnos nux-vomica*, is formulated as a 0.5–1%B rodenticide. It causes violent muscular spasms in birds and mammals, killing them within 30 minutes. Strychnine is not humane, and intensely poisonous to humans.

Wild buffalo gourd, roots of *Cucurbita foetidissima*, is toxic to cucumber beetles. Grind roots into powder, suspend in soapy water, and spray on plants.

FERMENTATION PRODUCTS

Natural metabolites oozed by fungi and actinomycetes have many uses. These are not botanicals, because the organisms that produced them are not plants. The most popular products of fermentation are antibiotics, used in human medicine and agriculture. These products should be used sparingly, because widespread antibiotic use increases pressure upon bacterial populations to select for resistant mutants. Resistant genes may pass from plant bacteria to human pathogens. This passage of genes is accomplished by bacterial conjugation, plasmid exchange, or DNA scavenging.

Abamectin is a mix of *avermectin B1a* and *avermectin B1b*, produced by an actinomycete, *Streptomyces avermitilis*. Synthetic derivatives are called ivermectins, related products include *emamectin* and *milbemectin*. Abamectin is very potent against russet mites and spider mites on hops, and kills fire ants, leafminers, and nematodes. It can be diluted to concentrations as low as 0.01 ppm for spraying on plants. Abamectin is not truly systemic, but has *translaminar activity* in some plants (especially in young leaves), meaning it is absorbed from exterior surfaces to internal leaf parts. It is available as a 0.15 lb/gal EC (mix 1 ml per 4 l or 1/4 teaspoon/gallon); be sure to add a wetting agent. Abamectin degrades readily in the environment (half life eight to 12 hours when exposed to sunlight). It does not bioaccumulate. The REI is 24 hours. Repeat application may be needed in seven to ten days. Abamectin works poorly in cold weather conditions. High concentrations are toxic to mammals, fish, and honeybees (nontoxic to birds). Abamectin may harm immature biocontrol organisms. The translaminar absorption protects beneficial insects, while phytophagous insects are exposed to the pesticide. Abamectin is restricted is some localities. Several products related to abamectin are now sold—*emamectin* and *milbemectin*.

Griseofulvin, a fungicide produced by the fungus *Penicillium griseofulvum*, is used to treat human fungal infections and may cause urticaria in sensitive individuals. Brian *et al.* (1951) eliminated *Botrytis cinerea* and *Alternaria solani* by watering vegetables with a 100 mg l^{-1} solution of griseofulvin. It is systemic.

Streptomycin, an antibiotic secreted by *Streptomyces griseus*, kills bacteria (*Pseudomonas* and *Xanthomonas* species) and some oomycetous fungi (Hewitt 1998 reported activity against *Pseudoperonospora humuli* in hops). Available as streptomycin or streptomycin sulphate, as a 0.1–0.2%D, 8–62%WP. It is mixed 200 ppm and applied as a soil eradicant or foliar spray on tobacco and many other crops. It works systemically. The REI is 12 hours. Some phytotoxicity (chlorotic flecking) may occur—reduce this by spraying in slow-drying conditions, like at night. Streptomycin is nontoxic to birds and mammals, and slightly toxic to fish. Applicators may develop allergenic reactions.

Spinosad is produced under fermentation by *Saccharopolyspora spinosa*, an actinomycete. It activates nicotinic acetylcholine receptors (nAChRs), but with remarkable selectivity for insect nAChRs and not mammal nAChrs. Spinosad works as a contact and stomach insecticide against caterpillars, as well as thrips, beetles, and leafminer maggots. Spinosad is used on fruits, vegetables, and cotton. It degrades rapidly from UV light (persistence up to seven days), has no phytotoxicity, is safe to beneficial organisms (except some adult wasp parasitoids), and exhibits low toxicity to mammals, fish, and birds.

Tetracycline is an antibiotic produced by several *Streptomyces* species, useful against bacteria and phytoplasmas. A semisynthetic derivative, oxytetracycline (Terramycin®), is more popular for controlling plant diseases. The REI is 24 hours.

SYNTHETIC ANALOGUES

Here we discuss synthetic chemicals that are considered biorational, because they closely mimic naturally-occurring compounds. They include synthetic botanicals, pheromones, and growth regulators.

SYNTHETIC BOTANICALS

Cinnamaldehyde (Cinnamite®, Valero®) is a new contact insecticide and miticide, a synthetic oil of cinnamon (*Cinnamonum zeylanicum*). It is currently registered for use on fruits, vegetables, and hops, against aphids and mites. Cinnamaldehyde, however, is a broad spectrum agent capable of killing most insects, including beneficials. It also kills powdery mildew. Cinnamaldehyde provides rapid knockdown of spider mites and has a short half-life (beneficials can be reintroduced after 24 hours). Some phytotoxicity may occur at the recommended 2% concentration; test on a small number of plants for potential damage.

Imidacloprid is a chlorinated derivative of nicotine. It activates nicotinic acetylcholine receptors (nAChRs), like nicotine. But unlike nicotine, imidacloprid has greater affinity for insect nAChRs than mammal nAChRs, so imidacloprid is safer. It acts as a contact insecticide, most effectively against insects with sucking mouthparts, such as aphids, whiteflies, and leafhoppers. It is also used against thrips, leaf beetles, and stem borers. Imidacloprid does not harm spider mites. It is applied as a foliar spray, soil drench, or seed treatment; available as a 10%WP, 35%FC, and 1–5%G. Imidacloprid is less toxic to mammals than nicotine (LD_{50} = 2000 mg kg^{-1}), and serves as a flea spray for pets. It can kill predatory biocontrols, such as *Chrysoperla*, *Coccinella*, and *Orius* species, as well as bees, but does not affect beneficial mites and spiders (Elbert *et al.* 1998). Plants absorb it systemically and persistently, up to 70 days. When imidacloprid is used against *Phorodon humuli* in hops, a pre-harvest interval of 28 days is required. The REI is 24 hours. Imidacloprid is expensive and Bayer keeps increasing its price. New nicotine derivatives are coming: acetamiprid and thiamethoxam.

Piperonyl butoxide (PBO) is derived from sesame (*Sesamum indicum*). PBO inhibits the microsomal detoxification of many compounds, including insecticides and THC (Gill & Jones 1972). PBO usually serves as a synergist, mixed with insecticides to enhance their impact. PBO is present in most pyrethrum products, and is also mixed with rotenone, ryania, sabadilla, and some oils. PBO may be carcinogenic.

Synthetic pyrethroids serve as rapid-acting, broad-spectrum, contact insecticides against aphids, whiteflies, bugs, beetles, and caterpillars. Resistance has arisen in some aphids and spider mites; these pest populations explode after pyrethroids are used, because pyrethroids kill off their predators (almost all biocontrol insects are killed by pyrethroids). At last count, 33 different pyrethroids were available in many formulations and concentrations. Pyrethroids are more toxic to mammals and more persistent in the environment than their natural cousins. Synthetic pyrethroids are known to disrupt the endocrine system (Colborn *et al.* 1996). Many pyrethroids are mixed with piperonyl butoxide, a possible carcinogen.

Permethrin is commonly sold as a 0.2% aerosol, easy to apply. Also available as a 37%EC, 25%WP, and 0.5%D. Permethrin breaks down quickly on sun-exposed plants (half-life 4.6 days); in soil the half-life is longer (three to six weeks). The REI is 24 hours. To delay degradation, combine permethrin with Pheast®, a lepidopteran feeding stimulant. Permethrin is rarely phytotoxic. It is extremely lethal to honeybees if sprayed directly on them, so spray in the evening (bees find dried permethrin very irritating, so it repels them from sprayed plants before they absorb a lethal dose). Permethrin is highly toxic to fish, nontoxic to birds, low toxicity to most mammals (very toxic to cats), irritating to the eyes and skin. A tingling of nerves in the fingers may arise from prolonged occupational exposure.

Deltamethrin is formulated with polybutene as Thripsticks®. The mix is sticky—paint it on sticks and place sticks in soil between plants. Thrips hop on the Thripsticks, stick, and die. It is also available as a 2.5 lb/gal EC, but not in the USA. It is considered the most potent pyrethroid.

Less common pyrethroids include *allethrin, bifenthrin, cyfluthrin* (10–20%WP, used on hops), *cypermethrin* (10–20%EC, used on tobacco), *esfenvalerate* (10–30%EC, used on hops), *fenpropathrin, phenothrin (sumithrin), resmethrin, tefluthrin* (a systemic pyrethroid), and *tralomethrin* (used on tobacco). Some are sold unrestricted in aerosol cans (e.g., Raid® Flying Insect Killer), while others are restricted-use pesticides (e.g., bifenthrin). Many of the older pyrethroids are less toxic and degrade quickly (e.g., resmethrin), whereas the newer pyrethroids tend to be more persistent (e.g., permethrin).

SEMIOCHEMICALS

Semiochemicals are chemical messengers which can be used to disrupt pest behaviour or growth. They are also called growth and reproduction regulators (GRRs). Synthetic semiochemicals mimic natural compounds produced by pests. GRRs are called "third generation pesticides" by some researchers. Semiochemicals fall into two groups, pheromones and allelochemicals.

Pheromones are chemical signals emitted by animals, and act on others of the same species, such as reproductive pheromones, alarm pheromones, and aggregation pheromones. **Allelochemicals** are emitted by animals and plants, and act on other species, such as **allomones** (chemicals which favour the emitter, including juvenoids, antifeedants, repellents, and seed germination inhibitors), and **kairomones** (chemicals which favour the receiver, such as feeding stimulants).

Reproductive (sex) pheromones are emitted by insects to attract the opposite sex for mating. In most cases, females emit the pheromones and males follow. Night-flying insects, such as nocturnal moths, make the greatest use of these odour signals (Howse *et al.* 1998). Sex pheromones are scented aldehydes or esters, usually 14–16 carbon atoms long. Synthetic sex pheromones for dozens of insects and mites are now commercially available, and described in Chapter 4.

Sex pheromones are primarily used for pest monitoring. When pheromones lure males into traps, this indicates females are mating in the area and eggs are on their way. Traps monitor the timing *and* intensity of pest invasions. Pheromones can lure males onto Tanglefoot® (a non-drying adhesive), or into traps containing pesticides or biocontrol agents. A variety of pheromone traps have been designed (Howse *et al.* 1998), a selection is shown in Fig 11.1. Delta traps and wing traps utilize non-drying adhesives to hold trapped insects; they are sufficient for most pests. For larger pests, such as noctùrid moths and Japanese beetles, funnel traps or bucket traps are recommended. These have a one-way entrance with no exit, and may be laced with poisons (Lure N Kill® traps). Most pheromone lures should be replaced every four to six weeks, depending on temperature. Traps should be checked daily and dead pests must be removed before their bodies saturate the trap.

In addition to monitoring pests, sex pheromones can be applied for direct control of pests, by mass-trapping or "confusing" pests. Scattering about 500 sex pheromone drops per ha will saturate the field in a "pheromone fog." Male moths become confused and overwhelmed, leaving the female's eggs unfertilized. Hemp researchers have used the "confusion" tactic against European corn borers, cutworm, armyworms, budworms, and hemp moths (Nagy 1979). Small rubber septums, hollow-fibre "twistees," and other simple devices can be impregnated with pheomones and serve as "controlled release dispensers," lasting for several weeks.

As an added benefit, sex pheromones may also attract biocontrol organisms to the field. Lewis *et al.* (1982) showed that parasitization of budworm eggs by *Trichogramma* wasps *doubled* in fields treated with budworm pheromones, compared to untreated fields.

Aggregation pheromones attract pests to traps where they can be removed from the environment. Japanese beetle traps, for instance, contain Japonilure® (a pheromone bait, which attracts males) and geraniol (a food attractant, which predominantly attracts females). These chemicals lure the pests into a cul-de-sac. The cul-de-sac contains malathion which kills them. This strategy is called the *attract-annihilate* tactic (Foster & Harris 1997).

Alarm pheromones or **dispersal agents** such as β-farnesene cause *Myzus persicae* and other aphids to drop to the ground, where they may be eaten by predators (Howse *et al.* 1998). β-farnesene signals other pests to disperse across the surfaces of plants. Dispersal increases pest exposure to pesticides or biocontrol organisms. But dispersal agents don't always work as planned (Dombrowski *et al.* 1996). Interest-

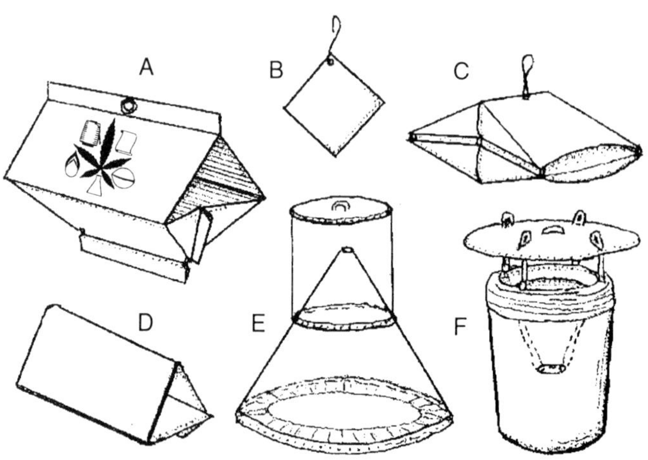

Figure 11.1: Examples of pheromone trap designs. A. Tent or Diamond trap, B. Vertical sticky trap, C. Wing trap with spacers, D. Delta or Jackson trap, E. Funnel or Heliothis trap, F. Bucket trap (McPartland).

ingly, the glandular trichomes of *Cannabis* produce β-farnesene (Nigam *et al.* 1965, Hood *et al.* 1973, Hood & Barry 1978, Ross & ElSohly 1996, Mediavilla & Steinemann 1997). Unfortunately, *Cannabis* also produces β-caryophyllene, which inhibits the pheromone activity of β-farnesene (Pickett *et al.* 1992).

Juvenoids or **juvenile growth hormones** are produced by insect larvae for purposes of moulting and growth. Plants produce allelochemicals which mimic insect hormones. The neem tree, for instance, produces azadirachtin, which arrests larval growth, prevent pests from maturing, and kills them before they can reproduce.

We now have synthetic juvenoids. Applying these semiochemicals to a crop results in a *gradual* reduction of pest populations. Juvenoids work slowly. Most juvenoids have little effect against adults, except to render adults sterile. Juvenoids work well against insects with short life cycles and high reproductive rates, such as aphids and other Homopterans. When first introduced in 1967, juvenoids were considered "resistance proof." Unfortunately, insects have developed resistance to their own hormone mimics (Gould 1991). Some *selective* juvenoids exhibit great toxicity to target species, and zero toxicity to other organisms. Other GRRs interfere with insect immune systems, rendering pests defenceless against bacteria (Reuveni 1995).

Methoprene works best against aphids, leafhoppers, caterpillars, beetles, flies (including leafminers), and dog fleas. Formulated as a 1–5 lb/gal EC and an aerosol. Methoprene has a half-life of two days in water, two weeks on tobacco leaves in sunlight, and persists longer on tobacco in storage. Nontoxic to honeybees, birds, and mammals. Moderately toxic to fish.

Kinoprene specifically kills Homopterans (aphids and whiteflies) and has little effect on most beneficial predators and parasites (Godfrey 1995). It has a very short half-life on foliage; it is currently limited to use on ornamentals.

Hydroprene is currently limited to use against cockroaches, applied indoors. It is very effective against *Myzus persicae,* but has not been registered for this application (Godfrey 1995).

Fenoxycarb is a carbamate that acts as an insect growth regulator. It is used on fruit trees and grapes against aphids and scales, and has ovicidal activity against a *Grapholita* species (Godfrey 1995). Regrettably, fenoxycarb harms the larvae of lacewings, lady beetles, and *Aphidoletes aphidimyza.* Furthermore, pet use (to control fleas) has been curtailed due to possible carcinogenic effects on dogs.

Pyriproxyfen is a flea control for dogs, used agriculturally against aphids, whiteflies, scales, and thrips on fruit and vegetable crops. Unfortunately, pyriproxyfen may kill immature *Encarsia formosa* biocontrols inside whitefly pupae; but it does not harm adult *E. formosa* wasps, or predators such as *Orius* and *Macrolophus* species.

Buprofezin is also used against whiteflies, mealybugs, and scales, on rice, vegetables, and fruit trees. *Cyromazine* controls dipterous leafminers in vegetable crops.

Diflubenzuron is technically not a juvenoid but a **chitin inhibitor**. It is used against armyworms but disrupts moulting in *many* insects. The pupal stages of parasitoids within affected insects are not killed. Formulated as a 25%WP, 2–4%FC. Half-life is one week on plant surfaces. The REI is 24 hours. Diflubenzuron is highly toxic to crustaceans, nontoxic to birds and bees, and slightly toxic to mammals. Restricted-use.

Several new chitin inhibitors have appeared: *Triflumuron* is used against caterpillars, beetles, and flies in Europe. *Chlorflurazuron* works well against caterpillars in Europe, it is not effective against Homopteran pests. *Flucycloxuron* and *diafenthiuron* work against insects as well as spider mites; they are applied to many crops outside the USA. *Teflubenzuron* is used against European corn borers, budworms, grasshoppers, whiteflies, and many others.

Hexythiazox is a highly selective growth hormone that kills mite eggs; immature mites are rarely killed and adult mites are immune (eggs laid by treated adult females, however, are not fertile). Hexythiazox only affects mites in the mite family Tetranychidae (e.g., *Cannabis* pests *Tetranychus* and *Eutetranychus*), so beneficial mites are not harmed, but *Aculops* pests are not harmed, either. Hexythiazox and a related compound, clofentezine, are used outside the USA on fruits, vegetables, cotton, and other crops. Formulated as a 10–50%WP and 10 lb/gal EC. Hexythiazox has a long residual on plants. Non-phytotoxic, honeybees and insect predators are not affected, nontoxic to birds, slightly toxic to mammals (LD_{50} > 5000 mg kg^{-1}).

Cowpea trypsin inhibitors (CpTI), interfere with the digestion of plant proteins by insects. A CpTI gene has been bioengineered into other crops to control various insects, including *Heliothis, Spodoptera, Diabrotica,* and *Tribolium* species (Hilder *et al.* 1987). CpTIs have little effect against phloem-feeding pests, such as leafhoppers, planthoppers, and some aphids. These sap sucking pests can be controlled with a group of proteins called lectins.

Lectins can bind the surface glycoproteins of two cells together; they are also called agglutinins. Lectins bind up insect gut epithelium and block nutrient uptake. The cDNA for a lectin from snowdrops, called *Galanthus nivalis* agglutinin (GNA) has been engineered into potatoes and other plants. Unfortunately, a diet of these transgenic potatoes also causes changes in rat gut mucosa (Ewen & Pusztai 1999), and GNA binds together human blood cells (Fenton *et al.* 1999).

Feeding stimulants encourage pest feeding behaviour, so they are kairomones. But kairomones can be used to trick pests, by mixing the feeding stimulants with pesticides or biocontrols (e.g., Bt or NPV), or adding them to attract-annihilate traps. Feeding stimulants range from simple carbohydrates to complex triterpenoid compounds such as cucurbitacin. Pheast®, a lepidopteran feeding stimulant, is a mix of nutritional yeast, flour, vegetable oil, and polysaccharides, formatted as a wettable powder. It also slows the photodegradation of Bt and NPV, and masks the repellency of synthetic pyrethroids.

Pests themselves produce kairomones, which attract their natural enemies, such as parasitic wasps. Pest sex pheromones, mentioned above, act in this fashion. Plants also produce volatile chemicals which attract the natural enemies of their pests, described at length in Chapter 3 under the section on insects.

Feeding repellents and oviposition repellents are naturally produced by plants, and we now have synthetic versions. These are described earlier in this chapter.

PESTICIDE REGULATION IN THE USA

Pesticides are regulated in the USA under the Federal Insecticide, Fungicide and Rodenticide Act of 1988 (FIFRA '88), as amended by the Food Quality Protection Act of 1996. About 800 pesticidal active ingredients are made into 19,000 registered pesticide products. The EPA, USDA, FDA, and OSHA enforce pesticide regulations. FIFRA '88 limits accessibility to certain pesticides. These "restricted-use" chemicals can only be purchased by persons certified by the state.

Spraying any registered pesticide in a manner inconsistent with its labelling is illegal. *No restricted-use pesticides cited in this text are registered for use on Cannabis.* Penalties for using pesticides in a manner inconsistent with labelling range up to US$1000 per offence.

One of the mandates of FIFRA '88 is the **re**registration of all pesticides. The EPA must retest pesticides that were registered years ago when requirements were less stringent than they are today. Thus, some chemicals used today will be banned tomorrow—so far the reregistration rejection rate has been 45% (Hembra 1993). The EPA has suspended sale of DDT, BHC, mercuric and arsenic compounds, mirex, vinyl chloride, and has curtailed use of other pesticides, such as dieldrin, aldrin, chlordane, endrin, and heptachlor.

Cynics claim the EPA's testing of synthetic chemicals is biased—the EPA tests chemicals in cancer-prone $B_6C_3F_1$ mice. Captan, for instance, induced carcinomas in three of 47 mice fed *more than their weight* in pesticides (Abelson 1994). When the dose was reduced to *half their weight* in captan, no females developed cancer, suggesting a threshold level is required for carcinoma development. Nevertheless, captan may be banned despite the fact that it is relatively nontoxic to humans and decomposes quickly (it is rarely detected on food or in ground water). Meanwhile, coffee remains unrestricted. Coffee contains caffeine, which is a neurotoxin (McPartland & Mitchell 1997). Coffee also contains caffeic acid, a compound *much more carcinogenic than captan,* and the 8 mg of caffeic acid in a single cup of coffee is 1000 times greater than the amount of captan an average American receives in food residues per year (Ames *et al.* 1995).

Registration of pesticides for new crops requires an average of ten years of expensive research, costing $50 million or more (Abelson 1994). This price keeps many low-cost botanicals from being registered. Rotenone, for instance, may lose its registration because no rotenone suppliers can afford the cost of EPA reregistration. Pesticide manufacturers will not get involved unless predicted sales exceed the cost of registration. For minor crops neglected by private industry, the USDA established the Interregional Research Project No. 4 ("IR-4") Program. Submitting an IR-4 Pesticide Clearance Request permits the experimental use of pesticides on unregistered minor crops. "Minor" is defined as any crop grown on less than 121,400 ha (300,000 acres). In the USA, this definition includes kenaf and hops, two beneficiaries of the IR-4 program; hopefully in the near future it will also include hemp. The National IR-4 Program is headquartered at Rutgers University, New Jersey.

Organic farmers restrict their use of pesticides to *The National List* mandated by the Organic Foods Production Act of 1990. The National List approves the use of some synthetic materials, such as synthesized versions of CO_2, bleach, insect pheromones, micronutrient fertilizers, fish emulsions (which contain synthetic pH stabilizers), petroleum oils and soaps, and plastic mulch and row covers. The *List* disapproves the use of some natural substances, such as strychnine.

Conventional pesticide regulations may change with implementation of the Food Quality Protection Act of 1996. This new law gutted the Delaney "zero cancer risk" clause with a compromise defined as "reasonable certainty of no harm." The law passed because it will consider, for the first time, children's increased vulnerability to pesticides. President Clinton presented it to the public, saying, "I call this the Peace of Mind Act, because parents will know that the foods children eat are safe." But two weeks later Congress killed funding for the Pesticide Data Program, the very program the EPA needs to set new pesticide standards for children (Colby 1996).

PESTICIDE SAFETY

Pesticide containers must be labelled. Those emblazoned with "Danger/Poison" are extremely toxic, categorized by the EPA as Class I poisons (see Table 11. 2). Simply tasting these chemicals can be lethal. A "Warning" label marks pesticides of moderate toxicity (EPA Class II); a teaspoon to a tablespoon will kill. Chemicals carrying the signal word "Caution" (EPA Class III & IV) are also lethal, but at bigger doses. For a list of labels, see www.cdms.net.

When working with pesticides, follow these common-sense precautions:

1. Read the pesticide label and follow instructions.
2. Handle all used spraying/dusting equipment with as much caution as the pesticide itself.
3. Store pesticides and equipment in lockable storage away from children and animals, and away from water, food, and feed.
4. Keep pesticides in original containers. Discard unlabelled pesticide containers; do not guess at contents. Dispose of pesticide wastes properly. Triple-rinse empty pesticide containers before throwing them away or recycling them. Do not reuse containers in the home.
5. Prepare pesticides and spray equipment in a protected area away from children and animals, where spills can be cleaned up easily. Clean up spilled chemicals with absorbent materials such as sawdust. Do not hose down the area. Decontaminate most pesticides with household bleach and hydrated lime.
6. Do not combine two pesticides unless you are certain they are compatible (e.g., mixing the insecticide pyrethrum with a lime/sulphur fungicide deactivates the former). To avoid cross-contamination of pesticides, use separate labelled sprayers for each class of pesticide (fungicide, insecticide, herbicide, etc.).
7. Do not work alone when handling dangerous pesticides, especially in enclosed areas like storage sheds, preparation areas, or glasshouses and growrooms.
8. Always wear protective clothing, including long-sleeved shirts and long-legged pants made from tightly woven cotton or hemp fabric. Protect your scalp with a waterproof, brimmed hood or hat. Rubber or neoprene footwear should fit over (or replace) leather or fabric shoes. Wash or discard protective clothing immediately following use.
9. Some pesticides require additional Personal Protective Equipment (PPE): an apron, coveralls, or chemical-resistant suit (especially when pouring and mixing chemicals or cleaning pesticide equipment). Gloves should be made of polyvinyl chloride or rubber (nitrile, neoprene, etc.), unlined (lining fabric absorbs pesticides), and extend to mid-forearm. Protect your eyes with safety glasses, goggles, or a full-face shield. Wear a well-fitting dust mask or respirator. Take steps to avoid heat illness when wearing respirators and spray suits.
10. Never eat, drink, or smoke while working with pesticides or in treated areas. Wash your hands before using the toilet. Avoid inhaling pesticide fumes; if pesticides drift nearby, get away. If eyes or skin become contaminated, quickly flush with large volumes of water. Remove contaminated clothing; handle clothing cautiously; wash in hot water with maximum detergent. Wash work clothes separate from the family laundry. Be sure that cleaners know clothes are contaminated and dangerous.
11. Know the symptoms of pesticide poisoning—irritated skin/throat/eyes, nausea, vomiting, dizziness, headache,

Table 11.5: Personal Protective Equipment (PPE) recommended for some Botanicals and Biocontrols.**

Active Ingredient	Recommended PPE
Bacillus species	Gloves (latex is sufficient), dust mask, goggles, LSS*, LLP*
Copper	Gloves (latex or rubber), dust mask, goggles for mixing powders or safety glasses for mixing liquids, coveralls or LSS, LLP
Diatomacious earth	Goggles, dust mask
Neem	Safety glasses, dust mask, LSS, LLP
Oil, dormant or horticultural	Safety glasses, mask
Pyrethrins	Gloves (latex or rubber), safety glasses, dust mask or respirator, coveralls or LSS, LLP
Rotenone	Gloves (latex or rubber), goggles for powders, safety glasses for liquids, dust mask, LSS, LLP
Ryania	Safety glasses, dust mask, LSS, LLP
Sabadilla	Gloves (latex or rubber), safety glasses, dust mask, LSS, LLP
Soaps	Gloves (latex or rubber), safety glasses, respirator, LSS, LLP
Sulphur	Gloves (latex or rubber), goggles for powders or safety glasses for liquids, dust mask (if using micro-sul formulation, use respirator with HEPA filter), coveralls or LSS, LLP

**Use these recommendations only in the absence of specific label requirements. Chart adapted from John Berino.
*LSS=long-sleeved shirt, LLP=long-legged pants with shoes and socks.

trouble breathing, muscle pains, pinpoint pupils. Post telephone numbers of the local hospital and in the USA, the national Centre for Poison Control, 1-301-443-6260.

THE WORKER PROTECTION STANDARD

Most of the common-sense precautions listed above are now law. The EPA revised its Worker Protection Standard in 1992 (WPS '92). Employers who hire agricultural workers must meet new federal guidelines for pesticide safety. (Some states have additional requirements.) The provisions in WPS '92 are complicated. Employers should contact their regional EPA office or a state agricultural extension agent for full information regarding compliance. In summary:

Table 11.6: Methods of applying pesticides to soil.

Method	Description
Broadcast	Pesticide broadcasted on the surface and mixed into soil before sowing seed
Furrow	Pesticide selectively poured into soil behind plough or drill, before sowing seed
Root zone	Pesticide mixed into soil of intended root zone, before transplanting seedlings
Irrigation	Pesticide added to trickle irrigation system
Seed coating	Seeds coated with pesticide using an adhesive

1. The WPS is label-driven. Requirements are pesticide-specific and listed on labels. Both general-use and restricted-use pesticides are covered by WPS '92. If you are using a pesticide with labelling that refers to the Worker Protection Standard, you must comply with WPS '92. Many home-use pesticides and biocontrols are exempt. But we recommend using precautions even in the absence of specific label requirements (see Table 11.5).
2. All workers potentially exposed to pesticides must be trained in pesticide safety—not just pesticide applicators. The EPA provides a checklist of required topics, including the precautions described above.
3. Before spraying, a fact sheet must be posted that describes the pesticide (its EPA number and active ingredients), the area treated, date of application, and the **REI** (restricted-entry interval or re-entry interval). In general, pesticides classified as Class I poisons (Table 11. 2) have an REI of 48 hours, the Class IIs have an REI of 24 hours, Class IIIs and IVs have an REI of 12 hours. Fact sheets and other pesticide information must be displayed at a central location. Some pesticides also require "Keep Out" signs be posted during the REI.
4. Information on pesticide safety and emergency assistance must also be displayed at a central location.
5. Establish decontamination stations within 400 m (1/4 mile) of all agricultural workers. These should be mobile, so field workers can clean up before meals and at the end of the workday. Supply water for routine and emergency washing, plenty of soap, and paper towels.
6. Workers applying pesticides require additional training, decontamination supplies, and personal protective equipment (PPE). PPE must be clean and in operating condition, worn and used correctly, and washed or replaced as needed.

REI posters and warning signs, pesticide fact sheets, worker training manuals, compliance record books, PPE equipment, and decontamination equipment can be purchased at farm supply stores. For one-stop shopping, contact Gempler's for their master catalogue (1-800-382-8473, web site http://www.gemplers.com).

PESTICIDE APPLICATION

Pesticides can be applied directly *on* plants, or applied *near* plants. Pesticides are applied *on* plants as foliar treatments, seed treatments, or onto roots as soil drenches. Off-plant applications include pest baits, pest barriers, and pre-season soil treatments. Off-plant applications are safest for *Cannabis* consumers, but are still hard on the environment.

BAITS, BARRIERS & SOIL TREATMENTS

Most baits are sold ready-to-use. Examples include pheromone baits for moths, food baits for cutworms, Japanese beetle traps, and rodent baits. Tanglefoot® is a popular barrier, useful against crawling insects. Soil treatments can be applied by any method in Table 11.6. Furrow, root zone, and seed coating methods are most commonly used.

SEED TREATMENT

Treating seeds with pesticides may be the best way to control some hemp diseases (Patschke *et al.* 1997). Planting pesticide-coated seeds is much easier on the environment than applying soil drenches. Pesticides work at different levels within seeds. Some only kill pests and pathogens on the surface of seeds. Other pesticides penetrate seeds and eliminate deep-seated pathogens. Some deeply penetrating *systemic* pesticides persist in young seedlings, protecting them from pests and pathogens lurking in soil or air. Maude (1996) noted that metalaxyl was detected for up to 20 days in the tissues of pea seedlings grown from treated seeds; benomyl lasted up to four weeks in lightly-dosed seeds and up to nine weeks in heavily-dosed seeds; carboxin only lasted a few days before oxidizing into nontoxic sulphoxide. Pesticides applied to seeds are long gone by harvest time.

Seeds may be coated with adhesive dusts, sprayed with slurries, or immersed in solutions (see Table 11.1). The machinery used for coating seeds is reviewed and illustrated by Maude (1996). Seeds are dusted in drum mixers or auger mixers. Mixers can be modified to apply liquid formulations (WP and FC). Improved machinery for spraying seeds with pesticides include *perforated drums* (drum mixers with tiny holes which draw warmed air to speed drying), *fluidized beds* (seeds are tossed in the air by vertical columns of forced hot air while being sprayed with fluids), *spinning disk techniques* (fluid is poured onto a spinning disk which breaks the fluid into fine droplets by centrifugal force—seeds fall past the disc through a peripheral curtain of pesticide spray), and *film-coating* (applying pesticides mixed with special binders and adhesives—often polysaccharides or synthetic polymers—to create a durable film coating).

Toole *et al.* (1960) reported good viability when seeds were dusted with calcium chloride and stored for 14 years. Maude (1996) mixed 1.5–5 g of thiram (a fungicide dust) per kg seeds. In comparison, old mercurial dusts like calomel and Ceresan® required 250–500 g per kg seeds (Maude 1996). Many hemp researchers recommended mercurial dusts to control damping-off fungi (Flachs 1936, Robinson 1943a, Wilsie & Reddy 1946, Andrén 1947, Ferri 1961b, Kotev & Georgieva 1969). Robinson (1943a) tested the germination rate of hemp seeds treated with eight different pesticides, at various concentrations. He presented *copious* data, but most of the compounds he tested are no longer available. Two effective compounds still in use are copper oxide, 12 g per kg seeds (9 ounces per bushel), and thiram (1.4–2.8 g kg^{-1} or 1–2 oz/bushel).

The trouble with dust treatment is poor adhesion to seeds, resulting in the environmental and health hazards of loose pesticide dust in seed bags. Spraying seeds with a fast-drying liquid pesticide results in better adherence. But the liquid may adhere irregularly, with too little on some seeds and heavy phytotoxic loads on other seeds. "Film coating" is a technology that provides good adherence and uniform coating.

A second approach to seed treatment is seed *immersion*—steeping seeds in liquids or fumigant gases for various periods of time, at ambient or elevated temperatures. Many hemp researchers have utilized this method. Soaking seeds in plain hot water eradicates viruses (see Chapter 9, method 11). To eradicate bacteria, Kotev & Georgieva (1969) soaked hemp seeds in 0.3% formalin. Booth (1971) soaked seeds in 0.1% formalin to eliminate fusaria. But for how long? Chinese growers soak seeds in formalin for one hour.

Soaking seeds in 10% household bleach for one minute kills many moulds. After immersion in bleach, rinse seeds in water at least three times before drying or planting. Mishra (1987) tested seven fungicides on *Cannabis* seeds—benomyl, carboxin, thiram, captan, zineb, and PCNB. Benomyl and carboxin were mixed at a rate of 100 mg active ingredient (a.i.) per litre water, and the other fungicides were mixed at 1000 mg a.i. per litre. Mishra soaked seeds for 15–20 minutes. In contrast, Maude (1990) soaked seeds in 0.2% thiram at 25–30°C for six to 24 hours. Obviously, more research needs to be done. Ferri (1961b) soaked hemp seeds in 0.2% sulphuric acid for up to 30 minutes, or in 90% isopropyl alcohol for ten minutes, apparently without seed damage. Kryachko *et al.* (1965) killed European corn borers overwintering in stored seeds by fumigating with methyl bromide. Another popular seed fumigant was chloropicrin, but Miège (1921) noted chloropicrin reduces hemp seed germination by at least 30%.

FOLIAR TREATMENT

Pesticides may be applied directly on plants as dusts or sprays. Be sure to treat the undersides of leaves, where many foliar pests congregate.

Dusts adhere firmly to plant surfaces, particularly if electrostatically charged or applied to dew-covered plants. Dusts are more susceptible to *wind drift* than are sprays. Do not dust when wind speed exceeds 5 mph. (Sprays tolerate breezes up to 12 mph.)

Sprays are easier to apply to undersides of leaves than dusts, but spray droplets often roll off plant surfaces, *especially* off plants covered with trichomes such as *Cannabis*. **Spray adjuvants** must be added to improve the effectiveness of spray applications. There are many types of spray adjuvants (over 150 adjuvants are registered in 17 categories for use in Washington state). Here are a few:

Spreaders (i.e., *wetting agents*) keep sprays from beading up and rolling off leaves. They reduce the surface tension of liquids, so the sprayed droplets spread out and hold onto sprayed surfaces. High-rounded drops on leaf surfaces indicate the need for a spreader. Flat drops that slide off leaves indicate too much spreader. There are *nonionic, anionic,* and *cationic* spreaders. Nonionics are the most common. They do not ionize in water, so they do not react with most pesticides (but they tend to remain as residues on plant surfaces). Two common nonionic spreaders are alkyl-aryl-poly-oxy-ethylenate (AAPOE) and alcohol-poly-oxy-ethylenate (APOE). Triton AG-44M® is a popular nonionic spreader. Mix

0.25 ml of 50% Triton solution per 50 ml pesticide solution. About 10% of spreaders fall into the *anionic* category; they are ionized with a strong negative charge. Examples include fatty acids and linear alkyl sulphonate (LAS). *Cationic* spreaders (ionized with a strong positive charge) are rarely used.

Stickers provide an adhesive effect after the spray has dried on the plant. Otherwise, the pesticide would quickly wash off from heavy dew, irrigation, or rain, as well as from wind erosion and leaf abrasion. Stickers work by three mechanisms: 1) increasing adhesion, 2) slowing evaporation, and 3) providing a waterproof coating. Some stickers are also spreaders, such as AAPOE, LAS, fatty acids, and some Triton formulations. These spreader-stickers utilize the first two mechanisms. Waterproofing stickers include latex, polyethylene, resins, and menthene polymers.

Extender adjuvants (also called *stabilizing agents*) protect spray residues against UV radiation and heat, which degrade many pesticides. *Emulsifiable oil* adjuvants enhance the penetration of sprays into waxy plant surfaces. They are also called *plant penetrants* or *translocators*. *Activator* is a general term indicating any adjuvant that increases the effectiveness of a pesticide; all of the aforementioned adjuvants are activators.

Buffers include *acidifying* adjuvants and *softening* agents. Many pesticides require mixing in water that is slightly acidic (optimal pH = 6.0) and soft. The most common acidifying adjuvant is phosphoric acid. Muriatic acid and vinegar may not be suitable. Hard water contains metal ions (Ca, Fe, Mg, etc.) which precipitate many pesticides (especially fatty acids and soaps) into a scum, curd, or "ring around the tub." Water hardness is measured in ppm or grains; if water hardness exceeds 300 ppm or 18 grains, softening may be required. Softening agents include Spray-Aide®, Blendex®, and Triton AG-44M®.

Defoamer adjuvants keep foam from forming in spray tanks. The most common defoamer is a silicone/carbon polymer called dimethylpolysiloxane. *Drift retardants* keep sprays from evaporating into ultrafine droplets that float away. *Compatibility agents* facilitate the mixing of pesticides to keep them from separating or curdling in spray tanks. *Feeding attractants* and *feeding stimulants* encourage pests to eat pesticides (e.g., Coax®, Gustol®). Their ingredients consist of sugars, starches, seed oils, and proteins. Some attractants/stimulants use extracts of the crop to which they are applied (Hunter-Fujita *et al.* 1998).

FOLIAR APPLICATION EQUIPMENT

Many pesticide application devices are sold at hardware stores and garden centres. Or contact Gempler's for their master catalogue (1-800-382-8473, web site http://www.gemplers.com). Selecting equipment adapted to your individual needs makes the choice less overwhelming.

Liquid pesticides can be atomized into different droplet sizes. The smaller the droplet the better the coverage, and smaller droplets are more toxic to insects (Thacker *et al.* 1995). But smaller droplets are more susceptible to drift. Fogs and smokes are the smallest, with a droplet size of <5 μm diameter. Fogs and smokes almost act as fumigants—they cover all plant surfaces (including the undersides of leaves). Fogs are produced by ultrasonic or thermal fogging machines, and smokes are generated by combustion. Aerosol droplets (5–50 μm) are produced by blasting pressurized air over a liquid to atomize the spray. Mists (50–100 μm) also require high pressure. Fine sprays (100–400 μm) and coarse sprays (>400 μm) can be generated at low pressures.

Here is a partial list of foliar application equipment, with comments:

Spray equipment:

1. Aerosol can: prepackaged, very convenient, but high cost.
2. Trigger-pump sprayer: inexpensive, reusable, Windex-type hand sprayers work best with water-soluble pesticides for small problems. They are too small and tiring for large gardens.
3. Slide-pump sprayer: similar to the trigger-pump sprayer, but generates pressure via a telescoping plunger. More applicable to larger gardens. Usually limited to 1 nozzle, and most sprayers only discharge during half of the pump cycle. They are leaky and often contaminate the applicator.
4. Compression sprayer: a manual pump compresses air above liquid in a tank. Air pressure forces the liquid out through openings at the bottom of the tank to a hand-held nozzle. The tank can be carried in the other hand or in a backpack. Pressure slowly discharges as pesticides are applied, requiring frequent repumping. These sprayers clog easily, and require frequent cleaning and maintenance to prevent corrosion of working parts.
5. Power compressed-air sprayer: a gas or battery-powered engine pumps air pressure into a tank, and the rest works like a compression sprayer. Usually mounted on a tractor. Higher maintenance costs.
6. Hydraulic sprayer: engine drives a hydraulic pump that draws liquid up from a tank and forces it under pressure through multiple discharge tubes (spray booms) and out any number of nozzles. Usually mounted on a tractor—expensive maintenance. Nozzles come as cones, flat fans, whirl chambers, and many other configurations.

Dusts and granules:

7. Bulb or bag duster: squeeze by hand to propel dust onto plants. Simple, easy, and many are reusable.
8. Hand-crank dust or granule applicator: dust or granules are stored in a hopper, beneath which spins a fan, powered by a hand-crank. Air currents or the fan itself discharge the granules. Hole size at the bottom of the hopper regulates application rate, and fan speed controls the range of discharge. The granules often contaminate the applicator.
9. Power duster: similar to the above, but an engine generates compressed air to blow dust. Usually tractor-mounted, generating considerable drift and collateral contamination (Fig 11.2).
10. Power granule applicator: like the power duster, but spreads granules, much less drift.

Figure 11.2: Tractor-drawn duster used in the Ukraine (from Lesik 1958).

Many soil pesticides are applied per ha or acre. To convert to square-foot gardening, refer to conversion factors provided in the front of this book. Foliar pesticides need not be measured in gallons per ha, but simply applied until "run off." For calibration and use of tractor-drawn power equipment, refer to Pimentel *et al.* (1991) or Matthews (1992) for information concerning droplet size, nozzle types, pump capacity and other aspects of application technology.

Here is an example of calibration: Using tractor-drawn power sprayers, the amount of insecticide applied per ha (or acre) depends on age of the crop. For plants under 45 cm tall (18 inches), only one hollow-cone spray nozzle is required per row. Spraying rows 90 cm apart (36 inches) at a pressure of 4.2 kg cm^{-2} (60 pounds/$inch^2$), while moving 6 km h^{-1} (4 miles/hour) will use 65 l ha^{-1} (7 gallons/acre). For tall plants, use three hollow cone nozzles per row, same pressure, same tractor speed, and apply 196 l ha^{-1} (21 gallons/acre). Always position the centre nozzle 15–30 cm (6–12 inches) above plant height. The two side nozzles should be positioned on either side of the row, low to the plants, with the nozzles angled upward—so the sprays hit pests that congregate on undersides of leaves.

"*My Golden Rule to keep out of trouble: never do two stupid things at the same time.*"
—Frank LeCase

Appendix 1: Synthetic Chemicals

According to Metcalf *et al.* (1962), synthetic pesticides were first used in 1892. They run a gamut from analogues of naturally-occurring pyrethroids to mutant DDT. Much of our information on pesticide toxicity was gleaned from ExToxNet (http://ace.ace.orst.edu/info/extoxnet/ghindex.html), a cooperative project sponsored by Oregon State University and several other agricultural universities in the USA. For a database of current pesticide labels with full text, updated on a daily basis, see http://www.cdms.net.

No pesticides have been labelled in the USA for use on *Cannabis*. Using pesticides in a manner inconsistent with their labelling brings penalties, up to US$1000 per offence. European hemp researchers have tested many synthetic chemicals on fibre crops. Synthetic chemicals, however, should not be used on crops destined for human consumption, such as seed oil or medical marijuana. We say this knowing authors of marijuana books recommended using chemicals in the past. For instance, Rosenthal (1998) recommended spraying marijuana with a tree nursery product, Wilt-Pruf®, which suffocates spider mites. Wilt-Pruf's generic name is polyvinyl chloride, the same as Saran Wrap. Not a desirable product to smoke!

Chemical toxicity depends on set and setting. Many synthetic chemicals are *systemic*. Systemics are particularly dangerous on plants destined for human consumption. The poisons remain in plant tissues without breaking down. For instance aldicarb, a systemic pesticide used to treat spider mites on hops, produces a peak concentration in hops flowers *60 days* after application (Duke 1985).

Many new pesticides are desirable because they are very *selective*. They kill pests while sparing beneficial organisms and other innocent bystanders. Many selective pesticides described in this chapter are used in the USA on *hops* and *tobacco: Hops* are related to *Cannabis* (so phytotoxicity data regarding hops should be applicable to hemp), and *tobacco* pesticides have been tested for their by-products of combustion.

Placing synthetics in *baited traps* is a relatively safe application of pesticides. Baited traps (such as Japanese beetle traps) keep the chemicals off plants, and keeps chemicals containerized for effective disposal. Using synthetics for *seed treatments* (coating seeds with pesticides before they are planted) is also relatively safe—little pesticide is introduced into the environment, and the pesticide is long gone by the time crops are harvested (see discussion in Chapter 12).

Pesticides come in a variety of formulations (see Table 11.1), including *aerosols* (abbreviated A), *baits* (B), *dusts* (D), *emulsifiable concentrates* (EC), *flowable concentrates* (FC), *fumigants* (F), *granules* (G), *slow release* (SR), *soluble powders* (SP), and *wettable powders* (WP). The amount of active ingredient (a.i.) in *solid* formulations is presented as a percentage of weight. For example, a 50%WP contains 50% a.i. by weight (the remaining 50% consists of "inactive ingredients"). Liquid formulations are expressed as *either* a percentage of weight *or* as the number of pounds a.i. per gallon of product. For example, 4 lb/gal EC contains 4 pounds of a.i. per gallon, which equals about 45%EC.

Synthetic pesticides are classified by their chemical structures. The major groups include chlorinated hydrocarbons, organophosphates, carbamates, dipyridilium herbicides, and the most popular group, "miscellaneous." Synthetic pyrethroids and synthetic pheromones are considered "biorational" and treated in Chapter 11. A selection of synthetic pesticides is provided in Appendix Table 1. For more information on pesticides, see the series by W.T. Thomson or the annual buyer's guide edited by R.T. Meister, *Farm Chemicals Handbook*.

CHLORINATED HYDROCARBONS

Chlorinated hydrocarbons are also called organochlorines. Many are relatively safe to mammals and rarely cause acute poisoning in humans (the antidote is atropine). But their long residual (*i.e.*, resistance to degradation), initially a desired feature, makes them environmentally dangerous. Some have been banned, such as DDT, chlordane, and dieldrin. Spraying hemp fibres with pentachlorophenol (PCP) is now illegal (Bòcsa & Karus 1997). In 1993 the American Public Health Association urged a phaseout of all chlorinated organic compounds.

Chloropicrin (CCl_3NO_2) was developed as tear gas but found use as an agricultural fumigant in 1919, manufactured in the USA by Great Lakes Chemical Corp. Soil fumigation with chloropicrin is expensive, technically difficult, and generally nasty—but it eliminates "thick-skinned" sclerotia-producing fungi such as *Sclerotium rolfsii*. It provides better control of soil fungi, oomycetes, and bacteria than methyl bromide, whereas methyl bromide is better against nematodes, insects, and weed seeds (Agrios 1997). Chloropicrin reduces *Cannabis* seed germination by at least 30% (Miège 1921). Restricted-use.

Dicofol is a miticide used against *Tetranychus urticae, T. cinnabarinus,* and *Brevipalpus obovatus* on hops, although many mite populations are now resistant. Dicofol is mite-specific—it spares beneficial insects (but kills most predatious mites) and is nearly nontoxic to humans. Frank (1988) considered dicofol harmful and did not recommend it; in 1986 dicofol supplies were contaminated by DDT. Made by Rohm & Hass, and is available as a D, 35%WP, and 18.5–42%EC. Its residual on plants is two weeks, with a soil half-life of 60 days. The REI is 12 hours. Dicofol is not toxic to plants and honeybees, slightly toxic to birds, and highly toxic to fish. It is carcinogenic to male (not female) mice, and may disrupt the endocrine system (Colborn *et al.* 1996). Strangely, Wu *et al.* (1978) showed that *Papaver* poppies sprayed with dicofol produced higher concentrations of opiates; all other pesticides caused decreases in opiate production.

Pentachloronitrobenzene (PCNB), a fungicide, has been used as a seed treatment or soil drench, applied to *Cannabis* at planting time to control hemp canker and southern blight (Ashok 1995). Mishra (1987) did not find PCNB helpful against *Rhizoctonia* disease. Made by Olin and Uniroyal, and is available as a 10%G, 2 lb/gal EC, and 35–75%WP. PCNB is persistent in soil (five to ten months). It is slightly phytotoxic,

Appendix Table 1: Some synthetic chemicals.

Generic name	USA trade name(s) ™ or ®	Chemical type	EPA acute toxicity[1]
acephate	Orthene, Ortho-12420	organophosphate	II
aldicarb	Temik	carbamate	I
4-aminopyridine	Avitrol 200	miscellaneous	I
benomyl	Benlate, Bavistin	carbamate	IV
carbaryl	Sevin	carbamate	III
carboxin	Vitavax	amide	III
chloropicrin	Larvacide	chlorinated hydrocarbon	II
chlorothalonil	Daconil 2787, Bravo	organophosphate	II
chlorpyrifos	Dursban, Raid Hornet	organophosphate	II
diazinon	Spectracide	organophosphate	II
dicofol	Kelthane	chlorinated hydrocarbon	IV
dichlorvos (DDVP)	No-Pest Strip, Vapona	organophosphate	II
dienochlor	Pentac	organophosphate	III
dimethoate	Dexol, Cygon, Dimate	organophosphate	II
etridiazole	TruBan, Terrazole	carbamate	II
fenbutatin oxide	Vendex, Torque	miscellaneous	III
ferbam	Carbamate	carbamate	IV
fosetyl-Al	Aliette	organophosphate	IV
iprodione	Chipco, Roval	amide	IV
malathion	Malathion	organophosphate	III
maneb	Dithane M-22	carbamate	III
mancozeb	Dithane M-45, Manzate 200	carbamate	III
metalaxyl	Ridomil, Subdue	miscellaneous	III
metaldehyde	Slugit, Deadline, Metason	miscellaneous	III
metam-sodium	Vapam	carbamate	III
methomyl	Kipsin, Lannate	carbamate	III
methyl bromide	Brom-O-Gas, Ozoneholer	miscellaneous	I
parathion, ethyl	Parathion, Thiophos	organophosphate	I
parathion, methyl	Metaphos, Penncap-M	organophosphate	I
pentachloronitro- benzene	PCNB, Brassicol, Terraclor	chlorinated hydrocarbon	III
pirimicarb	Pirimor	carbamate	II
propargite	Comite, Ornamite	chlorinated hydrocarbon	I
pyridaben	Sanmite, Dinomite	miscellaneous	III
sodium fluoroacetate	Compound-1080	miscellaneous	I
thiram	Arason, Tersan	carbamate	III
tolclofos-methyl	Rizolex, Risolex, Basilex	organophosphate	III
triadimefon	Bayleton, Fung-Away	miscellaneous	III
triforine	Funginex	miscellaneous	I
vinclozolin	Ronilan	amide	IV
warfarin	Coumadin, d-Con	miscellaneous	II
zineb	Dithane Z-78, Parzate	carbamate	III

[1] for a description of EPA acute toxicity, see Table 11.2.

slightly toxic to honeybees and birds, and highly toxic to fish. Restricted-use. Kotev & Georgieva (1969) used a related compound, Germisan® (hexachlorocyclohexane), to disinfect hemp seeds of *Pseudomonas* infections. These compounds may disrupt the endocrine system (Colborn *et al.* 1996).

Propargite is a miticide useful against the same pests as dicofol (see above), as well as mites resistant to dicofol. It is used on hops and many fruits and vegetables, and works best when the daily temperature is above 21°C. Kac (1976) sprayed it in Slovenian hemp fields. Propargite is specific; it does not affect insects (beneficial or otherwise), and spares predatory mites (Helle & Sabelis 1985). Made by Uniroyal as a 30%WP and 6 lb/gal EC, currently suspended in the USA and Canada pending results of additional research. Propargite is moderately persistent in soil. It is slightly phytotoxic (especially to young plants), very toxic to fish, slightly toxic to birds, slightly toxic to mammals by ingestion and harmful to eyes.

ORGANOPHOSPHATES

The organophosphates, like chlorinated hydrocarbons, act as contact or stomach poisons. But unlike chlorinated hydrocarbons, organophosphates degrade quickly in the environment. Lucas (1975) monitored the degradation of organophosphates on tobacco plants: Ten days after spraying parathion, diazinon, and malathion, only malathion residues exceeded 3 ppm. The malathion was gone after another ten days. On the dark side, organophosphates are far more poisonous to mammals than chlorinated hydrocarbons. Parathion, a common insecticide, is half-strength Sarin, a chemical warfare nerve gas. The antidote for both is atropine or pralidoxime chloride. Organophosphates also cause delayed neurobehavioural effects. Fryday *et al.* (1994) exposed house sparrows to chronic, low levels of organophosphates. The sparrows' feeding behaviour became clumsy, they dropped more *Cannabis* seeds than unexposed sparrows, and they lost weight.

Conversely, low levels of organosphosphates are prescribed for the treatment of Alzheimer's disease (e.g., tacrine, donepezil). Metrifonate, which is metabolized into dichlorvos (described below), is being tested for the treatment of dementia (Anonymous 1998). Organosphosphates inhibit acetylcholinesterase—a mechanism lethal to insects (Ryan & Byrne 1988) but beneficial to Alzheimer's patients. Many terpenoids in *Cannabis* also inhibit acetylcholinesterase (McPartland & Pruitt 1999).

Acephate is a frequently used *systemic* insecticide; Frank (1988) applied a foliar spray against spider mites (sprayed three times at seven-day intervals) and thrips (two applications seven days apart); he also sprayed aphids, whiteflies, mealybugs, scale insects, and caterpillars. Kac (1976) applied acephate against mites in Slovenian hemp fields. It has been used against flea beetles, leafminers, and grasshoppers in hops, tobacco, fruit, and vegetable crops. Made by Chevron, and is available as a 0.25%A, 2%D, 50%WP, and 75–90%SP. Acephate dissipates rapidly with residual activity of ten to 15 days on plants and one to ten days in soil. Highly toxic to bees, moderately toxic to birds, fish, and mammals.

Chlorothalonil is a broad-spectrum foliar nonsystemic fungicide recommended by Frank (1988) to control mould and rot (not effective against *Pythium* damping off). It is the third most-popular fungicide, behind mancozeb and copper (Hewitt 1998). Chlorothalonil works well against *Septoria* species. Made by ISK Biotech since the mid-1960s, as an 11–40%FC or 75%WP. Persistence in soil is 30–90 days, and nearly as persistent on plants. The REI is 48 hours. Nontoxic to plants and bees, somewhat toxic to birds and mammals (especially eye and skin irritation), and highly toxic to fish.

Chlorpyrifos has agricultural uses against hemp borers, cutworms, flea beetles, termites, flies, and even nematodes in field, fruit, hops, and vegetable crops. Made by DowElanco, and is available as a 15–50%WP, 1–4 lb/gal EC, and 30%FC. Short residual on foliage (ten to 14 days), but persists longer in soil (half-life of 60–120 days). The REI is 24 hours. Chlorpyrifos can be phytotoxic. It was used safely on hemp by Sandru (1976) against *Grapholita delineana*. It is extremely poisonous to honeybees, very toxic to fish and birds, and moderately toxic to humans. General and restricted uses.

Diazinon is an insecticide first marketed in 1953. Frank (1988) considered it the best control of scale insects, flea beetles, and whiteflies; he also used it against aphids, mealybugs, spider mites, leafhoppers, and caterpillars. It works very well against hemp borers (Sandru 1975, Peteanu 1980) and leafminers. Made by Novartis (formerly Ciba-Geigy), and is available as a 5%D, 5–14%G, and 25%EC, and 40–50%WP. Residual activity one or two weeks on plants and two months in soil. The REI is 24 hours. Frank & Rosenthal (1978) recommended a safety period of 35 days between application of diazinon and harvesting of treated plants. Phytotoxicity is rare. Honeybees are poisoned. Birds are quite susceptible to diazinon poisoning (especially the G formulation, which they mistake for food).

Dichlorvos (DDVP) is an insecticide with contact and vapour action (i.e., poison gas), used since the 1950s. Physicians prescribe dichlorvos for the treatment of schistosomiasis. Wang *et al.* (1987) killed hemp sawfly larvae with a dichlorvos spray. Marijuana growers eliminated spider mites and thrips by hanging No-Pest® resin strips in their growrooms (Alexander 1982, 1988b). Made by Shell, and is available as a 1–4 lb/gal EC, 10%A, 0.5%B and 20% impregnated resin strips. The EPA estimates that resin strips pose a 1-in-50 cancer risk if used over a lifetime (*Journal of the American Medical Association* 260:965).

Dienochlor, a selective miticide, was introduced in the 1960s but *Tetranychus urticae* remains uniquely susceptible to it. Treated mites do not die for one to three days (but stop feeding within hours), so dienochlor is relatively slow-acting. Made by Sandoz as a 50%FC. Dienochlor is less toxic than malathion and decomposes quickly (twice as fast as dicofol), especially in direct sunlight or temperatures above 130°C. It is not phytotoxic, does not harm beneficial insects (but is toxic to predaceous mites), is nontoxic to honeybees and birds, but highly toxic to fish. General use.

Dimethoate is a *systemic* insecticide recommended by Frank (1988) against sap-sucking insects such as aphids, whiteflies, thrips, and planthoppers in hops and vegetable crops. Made by BASF Corp. as an A, D, 10%G, 2–4 lb/gal EC, and 25–50%WP. Dimethoate degrades in 15–45 days on plants and four to 16 days in soil. The REI is 24 hours. Some phytotoxicity occurs. Dimethoate is highly toxic to honeybees, fish, and birds.

Fosetyl-Al selectively kills oömycetes such as *Pythium* (damping off disease) and *Pseudoperonospora* (downy mildew) in hops, applied as a foliar spray, root dip, or soil drench. It also treats bacterial diseases caused by *Pseudomonas syringae* and *Xanthomonas* species. Fosetyl-Al was the first product to be successfully reregistered under FIFRA'88. Made by Rhône-Poulenc, and is available as a 15%G and 80%WP. It is moderately persistent in plants (systemically absorbed), but decomposes quickly in soil. The REI is 12 hours. It is not phytotoxic, and nearly nontoxic to honeybees, fish, and birds.

Malathion, an insecticide and miticide, was first synthesized in 1950. Sandru (1976) found it effective against hemp borers. Mountain Girl (1998) used it against spider mites. Frank & Rosenthal (1978) applied it as a foliar spray against spider mites, cucumber beetles, and thrips, or as a soil drench against fungus gnats and other soil pests. Frank (1988) also recommended it against aphids, whiteflies, mealybugs, scale insects, leafhoppers, and flea beetles. Unfortunately, many pests are now resistant to malathion, especially spider mites and whiteflies. It still works well against aphids and their ant masters, hemp borers, budworms, cutworms, beetles, weevils, leafminers, leafhoppers, root maggots, grasshoppers, crickets, and fungus gnats. Made by Rhône-Poulenc, and is available as a D, G, 5–10%B, 4–10 lb/gal EC, and 25–50%WP. The average half-life for malathion is six days; residues persist longer in plants with high lipid contents (e.g., seeds). The REI is 24 hours. Malathion is highly toxic to honeybees, moderately toxic to birds, and exhibits a range of toxicities to fish—slightly toxic to goldfish, but very toxic to trout. In mammals, malathion is less acutely toxic than aspirin (LD_{50} levels between 1375–10,700 mg kg^{-1}, versus 1200 mg kg^{-1} for aspirin).

Parathion (ethyl) kills insects and careless farmers. It is much more toxic than malathion and possibly carcinogenic to rats; in 1992 its use was voluntarily restricted to field crops (cereals and soyabeans). Ethyl parathion has been used as a foliar spray and soil drench against beetles, weevils, leafminers, leafhoppers, mealybugs, scales, and grasshoppers on hops and other crops. Made by Cheminova and others, and is available as a D, 10%G, 2–8 lb/gal EC, and 15–25%WP. It is nonpersistent. It is not phytotoxic, moderately toxic to fish, and extremely toxic to birds, bees, and mammals. Restricted-use.

Parathion (methyl) is less acutely lethal to mammals than ethyl parathion, but it may cause chronic nerve damage in humans, especially children. It dissipates so rapidly its insecticidal efficacy is questioned—half-life on foliage is a couple hours. Nevertheless, the REI is 48 hours. No persistence in soil. It is used on hops against hops aphids, requiring a 15-day pre-harvest interval. Bósca & Karus (1997) controlled *Psylliodes attenuata* and *Grapholita delineana* with methyl parathion. Made by Cheminova and others, and is available as a 2–8 lb/gal EC and 20–40%WP. It is not phytotoxic but extremely toxic to bees, fish, and animals that eat fish. Restricted-use.

Tolclofos-methyl is a nonsystemic fungicide, applied as a seed treatment and soil drench to protect against soilborne diseases caused by *Rhizoctonia solani* and *Sclerotium rolfsii.* Its half-life on sunlit surfaces is two days, but it persists longer in soil. Dippenaar *et al.* (1996) used it on hemp seedlings in South Africa, without much success. Made by Sumitomo and others, available as a 5%G, 10–20%D, 20%EC, and 50–75%WP. It is not phytotoxic, practically nontoxic to birds and mammals, but very toxic to fish.

CARBAMATES

Most carbamates, like organophosphates, are very poisonous (antidote: atropine) but biodegrade rapidly in the environment. Some sulphur-containing dithiocarbamates (e.g., maneb, mancozeb, thiram) biodegrade into *ethylene thiourea,* a potent carcinogen. Lucas (1975) reports ethylene thiourea is not found in cigarette smoke generated from carbamate-treated tobacco. To confuse the issue, some dithiocarbamates may have medical applications, such as deactivating HIV viruses (Schreck *et al.* 1992). Nevertheless, sulphur-containing carbamates should not be used on plants destined for human consumption,

Aldicarb is an extremely lethal *systemic* miticide, insecticide, and nematicide (especially against cyst nematodes). Made by Rhône-Poulenc, and is only available as a 15G. Aldicarb is persistent, with residues in hops flowers peaking 60 days after application (Duke 1985). Residual in soil is ten weeks. It is not phytotoxic, repellent to honeybees, moderately toxic to fish, and extremely toxic to birds and mammals. Restricted-use.

Benomyl is a *systemic* fungicide, first available in 1968. Foliar sprays control grey mould, hemp canker, and powdery mildew. Soil drenches control southern blight (Ashok 1995). Mishra (1987) soaked seeds and root-dipped seedlings in benomyl to kill *Rhizoctonia solani.* Seed soaks also prevent *Fusarium* and *Macrophomina* diseases in other crops (Maude 1996). Benomyl distorts β-tubulin, preventing fungal mitosis. Because of this "site-specific" action, resistance arises by a simple mutation of a single gene. Benomyl-resistant fungi appeared just two years after the fungicide became available (Hewitt 1998). To inhibit resistance, combine benomyl with a nonsystemic fungicide, such as iprodione or captan (Agrios 1997). Made by DuPont, available as a 50%WP. Persistent on plants, residual six to 12 months in soil. The REI is 24 hours. Benomyl is not directly toxic to beneficial predatory mites (*Phytoseiulus* species), but may cause sterility in them (Hussey & Scopes 1985). Benomyl is not phytotoxic, nearly nontoxic to honeybees and mammals, slightly toxic to birds, and moderately toxic to fish. General use.

Carbaryl is the most commonly-used carbamate insecticide worldwide, useful against flea beetles, root grubs, grasshoppers, and many other insects. Wang *et al.* (1987) used it to kill hemp sawfly larvae. Frank & Rosenthal (1978) recommend a 35 day safety period between carbaryl application and harvest of treated marijuana plants. Half-life is three to ten days on plants, seven to 28 days in soil. The REI is 24 hours. Made by Rhône-Poulenc, and is available as a 5–10%D, 5–10%B, 4%FC, 50%WP, and 22–40%SP. It is not phytotoxic, moderately toxic to birds and fish, very toxic to honeybees, nephrotoxic to mammals (although carbaryl is used in some flea shampoos), and kills earthworms. Repeated use may stimulate spider mite infestations.

Etridiazole is a *systemic* fungicide recommended by Frank (1988). It cures damping off caused by *Pythium* in cotton, cucurbits, tomatoes, and other crops, applied as a soil drench or seed treatment. Made by Uniroyal as a 2.5%D, 5–8%G, 30–40%WP, and 4 lb/gal EC. Moderately persistent in soil. Phytotoxicity is rare. It is moderately toxic to fish, and slightly toxic to birds. General use.

Ferbam is a sulphuric fungicide containing iron, useful against seedling and foliage diseases in tobacco. Made by UCB Chemical Corp., and has been available since 1931 as a 1–25%D and 3–98%WP. Ferbam is not phytotoxic, nontoxic to honeybees, and slightly toxic to fish and birds. It leaves a black spray residue on foliage, and is irritating to the nose and throat. The REI is 24 hours.

Mancozeb is a fungicidal mix of maneb and zineb, with benefits and problems common to both compounds (see below). Since mancozeb has gone off patent, it currently outsells all other fungicides (Hewit 1998). Mancozeb has controlled yellow leaf spot in hemp, hops, and tobacco. It should not be applied within 77 days of harvest, which limits its use to seed treatment and foliar treatment of seedlings. Made by Rohm & Haas and others, and is available as a D, G, 37–75%FC, and 80%WP. Residual is moderately persistent, half-life four to eight weeks in soil. The REI is 24 hours. It is not phytotoxic, with low toxicity to bees, slightly toxic to birds, and moderately toxic to fish.

Maneb is a sulphur- and manganese-containing fungicide, used as a foliar spray or seed treatment. Bósca &

Karus (1997) recommended maneb against grey mould and rust fungi; it controls a wider spectrum of foliar diseases than any other single fungicide (Thomson 1997). Made by Rohm & Haas and others since 1950, and is available as a 1–20%D, 4%FC, and 70–80%WP. It is moderately persistent, with a half-life of four to eight weeks in soil. Maneb is not phytotoxic, with low toxicity to birds and bees, but highly toxic to fish. General use.

Metam-sodium is a very effective nematicide and fungicide, applied as a soil drench or as a fumigant injected into the soil. Made by Stauffer, and is available as a soluble concentrate. Metam-sodium is very phytotoxic, requiring an interval of two or three weeks between soil application and crop planting. It degrades to methyl isothiocyanate, with a residual 21–60 days. It is nontoxic to bees, moderately toxic to fish, and very toxic to birds. Restricted-use.

Methomyl is a broad-spectrum *systemic* insecticide used on tobacco and hops to control spider mites, budworms, and flea beetles. Made by DuPont, and is available as a 1–2 lb/gal SC and 90%WP. Methomyl takes three to six weeks to degrade in soil, with a half-life of three to five days on plants. It is not phytotoxic, highly toxic to honeybees, birds, and biocontrol insects, and moderately toxic to fish. Restricted-use.

Pebulate is a systemic herbicide used against grasses and some broadleaf weeds. Zabrodskii (1968), Tarasov (1971), and Tarchokov (1975) tested pebulate and other herbicides (trifluralin, linurom, dalapon, etc.) for use in hemp against *Ambrosia, Polygonum, Setaria, Echinochloa, Chenopodium, Amaranthus,* and *Raphanus* weed species. Made by Zeneca, in G and EC formulations. It is slightly toxic to honeybees, moderately toxic to mammals, and very toxic to fish.

Pirimicarb is a *selective* aphidicide, useful against organophosphate-resistant *Myzus persicae* on hops. It is nontoxic to most beneficial insects. Pirimicarb has fumigant properties. It is translocated by xylem (taken up by plant roots), with a relatively short residual. It was voluntarily withdrawn in the USA because of marketing problems (even in the 1970s farmers rejected selective pesticides in favour of broad-spectrum pesticides), but pirimicarb may be making a comeback. Made by Zeneca, and is available as an A, 5 lb/gal EC, and 50%WP. It is not phytotoxic, nontoxic to honeybees and fish.

Thiram, a sulphuric fungicide used since 1931, is the methyl analogue of disulfuram (Antabuse®). It is used against damping off fungi, especially *Pythium* species, *Rhizoctonia solani,* and *Botrytis cinerea,* applied as a seed treatment or soil drench. It is less effective against *Fusarium* (Maude 1996). Thiram also exhibits deer-repellent activity, and has been used in sunscreen ointments. Made by DuPont and others, and is available as a 60%D, 2–5%G, 4%FC, and 3–90%WP. Thiram's half-life in soil is one or two weeks, the REI is 24 hours. Thiram is not phytotoxic, nontoxic to honeybees, but highly toxic to fish. In mammals it is more toxic inhaled than ingested.

Zineb is a sulphur- and zinc-containing fungicide, applied to foliage or seeds. Bósca & Karus (1997) recommended it against rust fungi. Mishra (1987) did not find zineb helpful for controlling *Rhizoctonia* disease. Made by DuPont and others since 1943, and is available as a 3–15%D, 4%FC, and 1–75%WP. Residual one or two weeks on foliage. Zineb is more phytotoxic than maneb. It is nontoxic to honeybees, and more toxic to fish than to birds.

AMIDES

Amides are ammonia derivatives, fairly new agrochemicals. Some are nonsystemic or weakly systemic (e.g., iprodione and vinclozolin, two dicarboximide amides), others are absorbed systemically (e.g., carboxin, a phenylamide). They are useful for seed treatments. Some amides may be contaminated with dioxins (a group of about 75 notoriously toxic chemicals).

Carboxin is a *systemic* fungicide. Mishra (1987) soaked seeds and root-dipped seedlings in carboxin to kill *Rhizoctonia solani,* but carboxin did not work as well as benomyl. Seed treatments also control *Sclerotium rofsii, Macrophomina phaseolina,* and *Fusarium* species (Maude 1996). Made by Uniroyal, and is available as a 25–75%WP and 3%FC. Carboxin has a half-life of 24 h in soil, but longer in plants. It is not phytotoxic, nontoxic to honeybees, slightly toxic to birds, and very toxic to fish. It causes eye irritation.

Iprodione is a fungicide used as a foliar spray and seed treatment against *Botrytis, Sclerotinia, Rhizoctonia, Macrophomina, Alternaria,* and many other fungi (Maude 1996). Made by Rhône-Poulenc, in 50%WP and 41%FC formulations. Residues on plants have a half-life of seven days, with a persistent half-life of 20–120 days in soil. The REI is 12 hours. Iprodione is not phytotoxic, nontoxic to honeybees, slightly toxic to birds, but very toxic to fish. See comments under *vinclozolin,* below.

Vinclozolin is a fungicide used as a foliar spray against sclerotia-producing fungi, such as *Botrytis* and *Sclerotinia* species, and a seed treatment against *Macrophomina phaseolina* (Maude 1996). De Meijer *et al.* (1995) sprayed hemp fields with vinclozolin and iprodion, at rates of 500 g ha^{-1}. They alternated the two fungicides, spraying at 14-day intervals from June through August (after canopy closure and during Holland's high-humidity summer). The treatment reduced grey mould, but did not significantly increase crop yield—11.2 t ha^{-1} for control plots versus 12.3 t ha^{-1} for treated plots (Van der Werf 1994). De Meijer *et al.* (1995) also sprayed vinclozolin and iprodion on fibre crops suffering from hemp canker. This treatment did *not* significantly improve crop yield (Van der Werf 1994). Made by BASF as a 50%WP. Residues on plants have a half-life of seven days, and are moderately persistent in soil. The REI is 12 hours. Vinclozolin is not phytotoxic, nontoxic to honeybees, slightly toxic to birds, but very toxic to fish. It blocks testosterone receptors, which may cause feminization in males (Colborn *et al.* 1996).

BIPYRIDILIUM HERBICIDES

These herbicides act on contact, they are not translocated. Paraquat is a notorious example. Bipyridiliums cause harm if inhaled or splashed on the skin and kill if swallowed. Symptoms do not arise until *after* treatment can be effective.

HETEROCYCLIC COMPOUNDS

This group includes captan and captafol, nonsystemic fungicides introduced in 1952, and commonly found in home fruit-tree spray mixtures. Clarke (1981) used captan to protect clone cuttings from root rot (made by Zeneca as a 50–80%WP). High doses are carcinogenic; Frank (1988) discouraged the use of captan. Web scuttlebutt suggests that the seed treatment used by French hemp producers to protect seeds against damping off fungi may contain captan (captan also repels seed-eating birds). The usual dose is 0.5–6 g a.i. per kg seed. Captan's half-life in soil ranges one to ten days, residues on plants drop to 150 ppm after 27 days and cannot be detected after 40 days. The REI is 96 hours. Captan is moderately toxic to birds, bees and mammals (EPA class II due to eye irritation), but highly toxic to fish.

MISCELLANEOUS

4-aminopyridine is a bird repellent formulated as a grain bait. It causes victims to emit distress calls, scaring off the flock. The avicide Starlicide® kills starlings and blackbirds, yet spares seed-eating sparrows and pigeons. All avicides require permits, and you run the risk of killing untargeted birds. Overdoses kill mammals too. Restricted-use.

Fenbutatin oxide, a nonsystemic miticide organo-*tin* compound, is applied to hops, glasshouse crops, and fruit trees. It is rotated with pyrethrins against spider mites in Holland and Canada (*High Times* #257 p. 8, 1997). Made by DuPont and Shell, and is available as a 50%WP. Relatively persistent in soil. The REI is 48 hours. Fenbutatin oxide is nontoxic to honeybees. It is less toxic to predator mites than to spider mites, so it has been used in IPM. It can be phytotoxic, is toxic to birds, and very toxic to fish. Organotin compounds may disrupt the endocrine system (Colborn *et al.* 1996).

Fenazaquin is a broad spectrum miticide. It is a quinazoline with a novel mode of action (mitochondrial inhibition), used on fruit and grape crops. Made by DowElanco, in EC and SC formulations. Fenazaquin is not phytotoxic, does not affect beneficial insects, and is safe to mammals and birds.

Metaldehyde, an aldehyde, serves as a molluscicide. It causes slugs to secrete a heavy trail of mucus, so they dehydrate and die. Death comes sooner in warm weather. At higher doses metaldehyde also acts as a neurotoxin. Made by Chevron, and is available as a 1–20%B and 4G. Metaldehyde has a short half-life in soil. It is phytotoxic (apply bait *near* plants, not on them), nontoxic to honeybees, nontoxic to fish, and moderately toxic to birds and mammals. Restricted-use.

Metalaxyl, an alanine ester, is a widely-used systemic pesticide. It selectively kills oömycetes such as *Pythium* (cause of damping off disease) and *Pseudoperonospora* (cause of downy mildew) in hops and tobacco. Made by Novartis (formerly Ciba-Geigy), and is available as a D, 2%G, 25%WP, and 2 lb/gal EC. Metalaxyl is applied as a seed treatment or soil drench, with a residual of 70 days in soil. The REI is 12 hours. Metalaxyl may also be applied to foliage if combined with another broad-action fungicide such as mancozeb or chlorothalonil (to inhibit the rise of resistant organisms). Metalaxyl is not phytotoxic, nontoxic to honeybees and birds, and moderately toxic to fish.

Methyl bromide, a deadly fumigant, is particularly dangerous because it is odourless. The EPA recently listed it as a Class I ozone-depleting substance, 120 times more potent than chlorofluorocarbon-111. Thus, methyl bromide is being phased out to a total ban by January 1, 2005. Kryachko *et al.* (1965) used it to kill hemp borers overwintering in stored seed. Methyl bromide has also been used as a soil fumigant to kill nematodes, and at higher concentrations kills damping off fungi and seeds of weeds and broomrape (*Orobanche ramosa*). Restricted-use.

Pymetrozine is a *systemic* pyridine azomethine, Fulfill®, Relay®, and Sterling® in the USA, for use on vegetables, tobacco, and other crops. It selectively controls sucking insects (aphids, whiteflies, planthoppers) by blocking their stylets so they die from starvation about 48 h posttreatment (Fuog *et al.* 1998). Made by Novartis (formerly Ciba-Geigy), half-life in soil is two to 29 days. Nontoxic to mammals (EPA class III), birds, fish, earthworms, bees. It is very safe to biocontrols—5x to 50x safer than pirimicarb to *Orius*, *Chrysoperla*, and *Coccinella* species (Fuog *et al.* 1998).

Pyridaben is a pyridozinone compound. It controls mites and sucking insects (aphids, whiteflies, leafhoppers, and thrips) on vegetables and fruit trees. It is not effective against caterpillars, grubs, and maggots. Pyridaben has little effect on predators, which makes it useful in mite management. Made by Nissan Chemical Industries as a 20%WP and 15%EC. Moderately persistent. Slight toxicity to mammals and birds.

Sodium fluoroacetate is an odourless, tasteless, intensely poisonous rodenticide. It does not cause cancer, it just kills. Formulated as a B. It is extremely toxic to warm-blooded animals. Restricted-use.

Triadimefon is a new *systemic* triazole fungicide. It inhibits sterol production in powdery mildews (exhibiting curative properties on established infections), with less activity against other fungi. It is used on hops, tobacco, and vegetable crops, as a foliar spray and seed treatment. Made by Bayer and Miles as a 1–50%FC and 25–50%WP. Triadimefon's half-life is one to four weeks on foliage. The REI is 24 hours. It can be phytotoxic, is nontoxic to bees and birds, and toxic to fish.

Triforine is a *systemic* fungicide recommended by Frank (1988). It controls powdery mildew and other foliar diseases on hops, tobacco, and fruit trees. It has curative properties on established infections. Triforine also suppresses spider mites. It is a derivative of piperazine (a deworming drug). Made by American Cyanamid as a 6.5–18%EC and 50%WP. Half-life of triforine in soil is approximately three weeks. It causes little phytotoxicity, is nontoxic to honeybees and birds, moderately toxic to fish, but highly toxic to mammals, with moderate inhalation toxicity. Restricted-use.

Warfarin, $C_{19}H_{15}NaO_4$, is a coumarin derivative used as a rodenticide. Made by Hacco and others as a 0.13%B or in paraffin blocks; formulations over 3% are restricted. Many rodents in North America are now resistant to warfarin. This has led to the development of "superwarfarins" such as brodifacoum, which is occasionally mixed with marijuana (LaRosa *et al.* 1997). Warfarin is nontoxic to honeybees, and very toxic to warm-blooded animals. In human medicine it serves as an anticoagulant. The antidote is Vitamin K_1.

"We will either find a way or make one."
—Hannibal

Appendix 2: A Dichotomous Key of Diseases and Pests

This dichotomous key includes over 90 of the most common pests and diseases of *Cannabis*, which collectively cause approximately 99% of crop losses. To save the forest from the trees, rarely-encountered organisms are omitted from the key. These organisms can still be identified by referring to Chapters 4–6.

To begin using this key, the user must first turn to the section concerning the part of the plant showing signs and symptoms (**I.** Seedlings and seeds in soil; **II.** Roots; **III.** Lower stems and crowns; **IV.** Upper stems and branches; **V.** Leaves; **VI.** Flowers; **VII.** Entire plants). Every level of the key presents the user with two (sometimes three) choices marked by the same letter. Trace key elements down to a specific disease or pest. Check the index to find further descriptions of your tentative diagnosis. Illustrations are often provided to further evaluate the accuracy of your diagnosis.

I. Seeds and seedlings

A_1. Seed does not germinate, no seedlings appear

- B_1. Seed no longer present in soil
  - C_1. Soil site disturbed .... Birds, Rodents
  - C_2. No soil disturbance seen .... Root maggots, Cutworms
- B_2. Seed present
  - C_1. Seed gnawed or other mechanical damage .... Root maggots, Rodents
  - C_2. Seed covered with fuzz .... Damping off fungi
  - C_3. Seed appears normal. .... Drought, Poor seed

A_2. Seed germinates and produces a seedling

- B_1. Above-ground parts show signs of feeding damage or insects present
  - C_1. Stem completely severed from root
    - D_1. Severed seedling lying nearby .... Cutworms, Crickets, Slugs
    - D_2. Severed seedling absent. .... Birds, Mammals
  - C_2. Plant upright but with holes in leaves or other insect symptoms
    - D_1. Leaves or cotyledons with holes and notched edges
      - E_1. Holes small, leaf edges rarely notched .... Flea beetles
      - E_2. Large holes and notched edges
        - F_1. Leaping insect with large rear legs .... Grasshoppers
        - F_2. Caterpillars present on plant or nearby in soil
          - G_1. Pink-skinned, often spin webs .... Hemp borers
          - G_2. Grey or brown cuticle, rarely spin webs. .... Cutworms, Armyworms
    - D_2. No missing pieces but webbing or honeydew present
      - E_1. Webbing present with tiny red or green mites .... Spider mites
      - E_2. No webbing, much honeydew, pear-shaped insects .... Aphids

B_2. Seedling obviously sick but no feeding damage evident on above-ground parts

C_1. Seedling wilted and/or toppled

D_1. Brown discolouration of root or stem near soil line

E_1. Roots discoloured and gnawed .... Root maggots

E_2. Root and/or stem discolouration, no insect damage .... Damping off fungi

D_2. No root discolouration

E_1. Fungal hyphae present microscopically .... Damping off fungi

E_2. No hyphae present

F_1. Bacteria present in sectioned tissue .... Bacterial wilt

F_2. No bacteria present .... Drought, Frost

C_2. Seedling upright but cotyledons and leaves discoloured

D_1. Discoloured areas light in colour, produce tiny, black-dot fruiting bodies

E_1. Discoloured areas greenish-yellow, form pycnidia .... Yellow leaf spot

E_2. Discoloured areas grey, form acervuli. .... Anthracnose

D_2. Discoloured areas dark, no fruiting bodies present

E_1. Cotyledons and true leaves covered with brown spores. .... Brown blight

E_2. Cotyledons or leaves covered with dark green spores ....

F_1. Primarily cotyledons, spores two-celled and oval .... Trichothecium spot

F_2. Primarily true leaves, spores 4–5 celled, elongate .... Curvularia blight

II. Roots (not including crown, see III below)

A_1. Insect signs and symptoms (gnawed roots or insects present)

B_1. Root damage an extention of stalk infestations

C_1. Insect a green to brown caterpillar. .... assorted stem borers

C_2. Insect a pale beetle grub with large head .... Longhorn beetle grubs

B_2. Root damage alone, no stem feeding

C_1. Insect a 6-legged pale-bodied beetle grub

D_1. Grub fat, assuming a C-shape when disturbed .... White root grubs

D_2. Grub cylindrical, resembles a caterpillar (but no prolegs).... .... Flea beetle grubs

C_2. Insect not a beetle grub

D_1. Insect a slow-moving legless maggot

E_1. Maggot spike-shaped with pointy head, <7 mm long .... Root maggots

E_2. Maggot cylindrical or thread-like

F_1. Maggot pink-grey in colour, up to 35 mm in length .... Crane fly maggots

F_2. Maggot white, thread-like, 3 mm or smaller .... Fungus gnat maggots

D_2. Insect a quick-moving 6-legged ant or termite

E_1. Soft-bodied, white, almost always underground .... Termites

E_2. Hard-bodied, red or black, often above ground .... Ants

A_2. No signs or symptoms of insects

B_1. Abnormal swellings (galls, knots) and bushy-appearing root

C_1. Deformed sections with small round cysts (<1 mm) .... Cyst nematode

C_2. Large irregular galls contain nematodes but no cysts .... Root knot nematode

C_3. Parasitic plant attached to deformed roots .... Branched broomrape

B_2. Roots discoloured, rotten in appearance

C_1. Sclerotia present

D_1. Mycelium white, sclerotia black, oblong, 5–12 mm long .... Hemp canker

D_2. Mycelium pale brown, sclerotia black, spherical, 1 mm dia. .... Southern blight

D_3. Mycelium forms funicles, sclerotia yellow-brown .... Texas root rot

C_2. Sclerotia not present

D_1. Root stele dark brown, no spores seen in culture .... Rhizoctonia root rot

D_2. Root stele reddish, spores present in culture .... Fusarium root rot

III. Lower stalk, including crown (stalk at soil level)

A_1. Stalks swollen into galls, puncture holes often present, frass or insect visible

B_1. Insect a caterpillar with 6 true legs in addition to 8 prolegs

C_1. Caterpillar pale brown with dark head .... European corn borer

C_2. Caterpillar darker brown, white, green or red-violet .... Other stem borers

B_2. Insect a pale white beetle grub with 6 true legs, no prolegs

C_1. Grubs <7 mm long, plump, assume a C-shape when disturbed .... Weevil, curculio grubs

C_2. Grubs >7 mm, cylindrical, with enlarged heads .... Longhorn beetle grubs

A_2. No swollen stalks, puncture holes, or other signs and symptoms of insects

B_1. Epidermis shredded, sclerotia present on surface or in pith

C_1. Mycelium white, sclerotia oblong, 5–12 mm in length .... Hemp canker

C_2. Mycelium pale brown, sclerotia nearly spherical, 1 mm dia. .... Southern blight

B_2. No sclerotia

C_1. Crown turns dark brown, epidermis shredded .... Rhizoctonia sore shin

C_2. Crown dark brown, epidermis intact, root stele discoloured red .... Fusarium foot rot

C_3. Crown with beige-coloured, lumpy, cancer-like growths .... Crown gall

IV. Upper stem and branches

A_1. Stem galls, puncture holes, frass or insect visible
- B_1. Insect on surface of stem, feeding externally
  - C_1. Insect sedentary or slow moving
    - D_1. Insect covered in froth of spittle .... Spittlebugs
    - D_2. Insect flattened, often with waxy filaments
      - E_1. Filaments disappear with age, insects brown, round .... Scales
      - E_2. Filaments persist, insects white, slowly mobile .... Mealybugs
  - C_2. Insect quickly leaps away when spotted
  - D_1. Insect small, spindle-shaped, with fringed wings. .... Thrips
  - D_2. Insect flattened shield shape, wings leather-like .... Tarnished plant bugs
- B_2. Larvae feed within stem gall
  - C_1. Insect a 3-legged pale-bodied beetle larva
    - D_1. Grubs <7 mm long, plump, assume C-shape when disturbed .... Weevil, curculio grubs
    - D_2. Grubs >7 mm, cylindrical, with enlarged heads .... Longhorn beetle grubs
    - D_3. Grubs different than above .... Assorted boring beetles
  - C_2. Larvae with prolegs (caterpillars) or no legs (maggots)
    - D_1. Small (<7 mm long) legless maggots .... Nettle midges
    - D_2. Larva a caterpillar
      - E_1. Pale brown bristly body with dark head, 10–20 mm long .... European corn borers
      - E_2. Plump pinkish-white body <10 mm long with dark head .... Hemp borers
      - E_3. Caterpillar darker brown, white, green or red-violet .... Other stem borers

A_2. No galls, puncture holes, frass or insect visible
- B_1. Stem or branch epidermis sunken or discoloured
  - C_1. Canker covered with a mat of mycelium and spores
    - D_1. Spores grey, round .... Grey mould
    - D_2. Spores white or light pink, ellipsoid .... Pink rot
    - D_3. Spores green, oblong .... Cladosporium canker
  - C_2. Canker not covered with mycelium, usually light brown
    - D_1. Fruiting bodies present in older lesions
      - E_1. Initial fruiting body a simple pycnidium
        - F_1. Fungus only produces pycnidia in culture .... Yellow spot
        - F_2. Other spore stages sometimes form in culture .... Brown spot, Twig blight
      - E_2. Initial fruiting body an acervulus or perithecium
        - F_1. Grey to pink acervuli bristling with setae .... Anthracnose
        - F_2. Perithecia with long cylindrical ascospores .... Ophiobolus canker
    - D_2. No fruiting bodies
      - E_1. Fungal spores canoe-shaped .... Fusarium stem canker
      - E_2. No fungal elements, bacteria seen under microscope .... Striatura ulcerosa

B_2. Epidermis normal but stem misshapened and often spongy

C_1. Grub or caterpillar present in pith .... go to section IV A_1B_2

C_2. No insects present

D_1. Microscopic nematodes present .... Stem nematode

D_2. No nematode present

E_1. Stem tissue discoloured red .... Fusarium stem canker

E_2. No discolouration, plant stunted .... Nutrient deficiency

V. Leaves

A_1. Insect damage (leaf cutting, puncture wounds) or other signs (honeydew, webbing)

B_1. Leaf skeletonized or holes present and/or notched leaf edges

C_1. Leaf skeletonization

D_1. Skeletonization alone

E_1. Insect a conspicuous green and bronze beetle .... Japanese beetles

E_2. Insect a hyperactive, small, black beetle .... Flea beetles

E3. Insect a white larva <12 mm long, resembles a caterpillar .... Sawfly larvae

D_2. Skeletonization followed by large holes and notched edges

E_1. Only leaves attacked .... Leaf-eating caterpillars

E_2. Other plant parts also attacked, webbing often present

F_1. Caterpillars bore into stems and branches .... Stem-boring caterpillars

F_2. Caterpillars attack flowering tops .... Bollworms, Borers

C_2. No skeletonization, only leaf holes and/or notched edges

D_1. Caterpillars present

E_1. Individuals usually solitary .... Leaf-eating caterpillars

E_2. Individuals gregarious .... Armyworms

D_2. Insect not a caterpillar

E_1. Insect with large hind legs, conspicuous.......... Grasshoppers

E_2. Insect a less-conspicuous beetle

F_1. Only edges notched, beetle with long snout. .... Weevils & Curculios

F_2. Notched edges and leaf holes, no long snout .... Flea beetles, leaf beetles

B_2. Tiny puncture wounds, rasp marks, or subsurface tracks—no leaf holes or skeletonization

C_1. Wounds punctate (often appear as white specks), plants often wilted

D_1. Honeydew present, insects <3 mm long

E_1. Webbing produced by small red, green or brown mites .... Spider mites, Other mites

E_2. No webbing present

F_1. Insects usually green, pear-shaped .... Aphids

F_2. Insects resemble tiny white moths. .... Whiteflies

D_2. Little or no honeydew, insects >3 mm long

E_1. Insects flattened shield-shaped, often slow moving .... Plant bugs

E_2. Insects elongated, quickly leap or fly away .... Leafhoppers

C_2. Wounds linear, on surface or subepidermal

D_1. Short rasp marks on surface, sometimes oozing sap .... Thrips

D_2. Epidermis intact, white or brown tracks beneath surface .... Leafminers

A_2. No insects or feeding damage but leaf spots or other discolourations present

B_1. Leaf spots with distinct edges

C_1. Spots white, yellow, light-green to light-brown

D_1. Fruiting bodies present in older lesions

E_1. Fruiting bodies pycnidia or perithecia

F_1. Spots yellow, large, irregular, with scattered pycnidia .... Yellow leaf spot

F_2. Spots white, large, irreg., pycnidia in concentric circles .... White leaf spot

F_3. Spots brown, small, circular, often breaking apart

G_1. Only pseudothecia with ascospores present .... Pepper spot

G_2. Pycnidia, rarely with perithecia present .... Brown leaf spot

E_2. Fruiting bodies other than pycnidia or perithecia

F_1. Grey to pink acervuli bristling with setae .... Anthracnose

F_2. Bright yellow or orange fruiting bodies. .... Rust

F_3. Black pin cushion-like sporodochia .... Black dot disease

D_2. No fruiting bodies in older lesions

E_1. Fungal spores present

F_1. Spores borne mostly on upper leaf surface

G_1. Leaf lesions often with chlorotic ring .... Brown blight

G_2. Leaf lesions without chlorotic ring .... Stemphylium leaf spot

F_2. Spores mostly on underside of leaf

G_1. Spores emerge from stomates only on undersides. .... Down mildew

G_2. Spores borne in tufts, mostly on undersides .... Olive leaf spot

E_2. No fungal spores or mycelium present

F_1. Bacteria present in sectioned material. .... Bacterial blight

F_2. No bacteria present .... Abiotic diseases

C_2. Leaf spots dark, nearly black

D_1. Fruiting bodies present within fungal stroma .... Tar spot

D_2. No fruiting bodies present, spores borne on leaf surface .... Cladosporium spot

B_2. Leaf lesions or discolourations without distinct edges

C_1. Mycelium on leaf surface

D_1. Mycelium white to grey mycelium, sometimes light pink

E_1. Mostly on upper surfaces. .... Powdery mildew

E_2. On either surface, even growing on branches .... Pink rot

D_2. Black mycelium, either surface

E_1. Insects present, leaf surface often sticky .... Sooty mould

E_2. No insects present, not sticky .... Black mildew

C_2. Leaf lesion without mycelium, consists of chlorosis or necrosis in streaks, mosaics

D_1. Bacteria seen in sectioned tissue .... Bacterial blight

D_2. Bacteria not seen

E_1. Leaves exhibit phyllody (distorted growth) or rosettes . Mycoplasma-like-organisms

E_2. Little distorted growth, mostly leaf chlorosis

F_1. Chlorosis in streaks or chevrons .... Hemp streak virus

F_2. Chlorosis in small rings or mosaics .... Mosaic viruses

F_3. Chlorosis without a pattern .... Nutrient deficiencies

VI. Flowers and developing seeds

A_1. Male flowers

B_1. Petioles swell into galls, flowers wilt .... Gall midges

B_2. Petioles normal, flowers covered with mycelium

C_1. Mycelium produces oval, grey spores. .... Grey mould

C_2. Mycelium produces elongated brown spores. .... Brown blight

A_2. Female flowers and developing seeds

B_1. Webbing, honeydew, leaf holes or insects present

C_1. Holes or edges notched in flowers, fan leaves, and seeds

D_1. No webbing, holes small, insect a small black beetle .... Flea beetles

D_2. Webbing and caterpillars present

E_1. Larvae pink, immature seeds partially shelled .... Hemp borers

E_2. Larvae light brown, seeds partially shelled .... European corn borers

E_3. Larvae green or dark brown, whole seed sometimes eaten .... Budworms

C_2. No holes in flowers or fan leaves, no shelled seeds

D_1. Webbing present with minute red or green mites .... Spider mites

D_2. No webbing

E_1. Insects <2.5 mm long, pear-shaped, with honeydew .... Aphids

E_2. Insects conspicuous, 15 mm long, flattened shield-shape .... Green stink bugs

B_2. No signs or symptoms of insects

C_1. Wilted flowers, no other symptoms

D_1. Whole plant wilted. .... go to section VII

D_2. Only apical bud or occasional branches wilted .... go to section IV

C_2. Discoloured flowers with/without wilting

D_1. Fungal mycelium and spores present

E_1. Fan leaves wilted and grey spores present .... Grey mould

E_2. Less conspicuous mould with brown spores present .... Brown blight

D_2. No signs of mycelium or spores

E_1. Flowers and fan leaves with chlorotic markings .... Viruses, Nutritional diseases

E_2. Flowers turn black overnight .... Frost damage

VII. Whole plant

A_1. Wilt (sometimes one-sided) leaves evenly chlorotic or normal colour

- B_1. Stems discoloured internally (xylem or pith)
  - C_1. Pith peppered with microsclerotia. .... Charcoal rot
  - C_2. No microslerotia but xylem discoloured
    - D_1. Reddish-brown discolouration, spores canoe-shaped .... Fusarium wilt
    - D_2. Brown discolouration, spores round to oval .... Verticillium wilt
- B_2. No discolouration of stem tissue
  - C_1. Roots abnormal .... go to section II
  - C_2. Roots normal
    - D_1. Bacteria in sectioned plant tissue .... Bacterial wilt
    - D_2. No bacteria seen .... Drought, Herbicides

A_2. Systemic chlorosis and necrosis

- B_1. Chlorosis gradual, plants often stunted
  - C_1. Insects or nematodes found upon close inspection of plant
    - D_1. Insects on undersides of leaves .... go to section V
    - D_2. Insects or nematodes on roots .... go to section II
  - C_2. No insects. .... Air pollution, Nutrients
- B_2. Chlorosis and necrosis sudden .... Toxins, Herbicides

"All concepts, theories, and general ideas are thin and ineffectual unless they are grounded in the concrete reality of things which specifically enter into our lives and which we steadily deal with." -John Dewey

"No ideas but in things." -William Carlos Williams

Appendix 3: Conversion Factors

Both English and Metric measures are used in this text, depending on the system used by original authors. Use charts and equations below to convert:

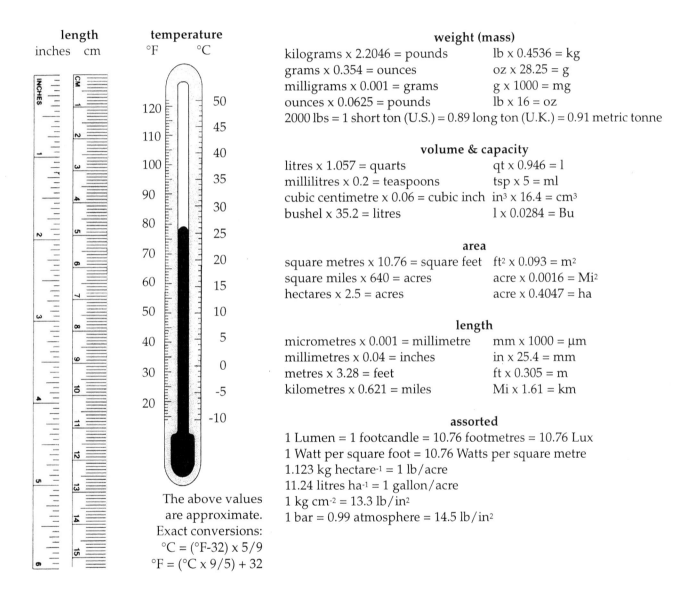

length

inches cm

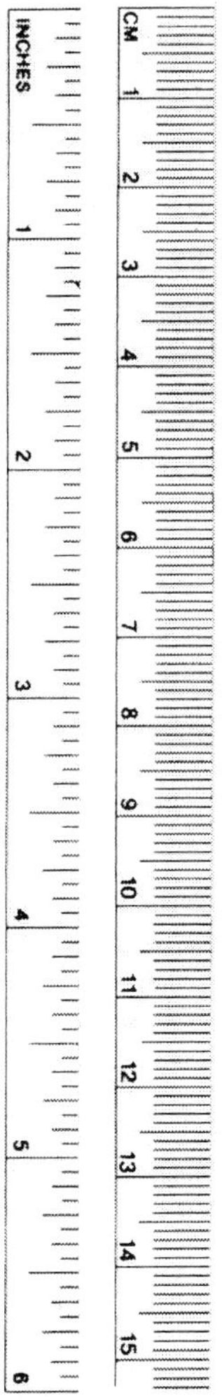

temperature

°F °C

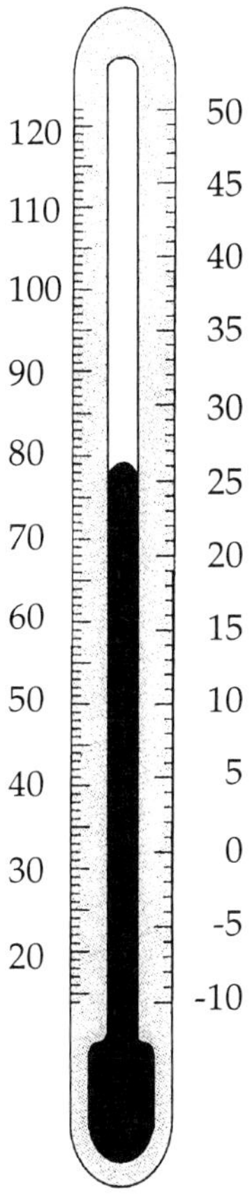

The above values are approximate. Exact conversions:

°C = (°F-32) x 5/9

°F = (°C x 9/5) + 32

weight (mass)

kilograms x 2.2046 = pounds	lb x 0.4536 = kg
grams x 0.354 = ounces	oz x 28.25 = g
milligrams x 0.001 = grams	g x 1000 = mg
ounces x 0.0625 = pounds	lb x 16 = oz

2000 lbs = 1 short ton (U.S.) = 0.89 long ton (U.K.) = 0.91 metric tonne

volume & capacity

litres x 1.057 = quarts	qt x 0.946 = l
millilitres x 0.2 = teaspoons	tsp x 5 = ml
cubic centimetre x 0.06 = cubic inch	in^3 x 16.4 = cm^3
bushel x 35.2 = litres	l x 0.0284 = Bu

area

square metres x 10.76 = square feet	ft^2 x 0.093 = m^2
square miles x 640 = acres	acre x 0.0016 = Mi^2
hectares x 2.5 = acres	acre x 0.4047 = ha

length

micrometres x 0.001 = millimetre	mm x 1000 = µm
millimetres x 0.04 = inches	in x 25.4 = mm
metres x 3.28 = feet	ft x 0.305 = m
kilometres x 0.621 = miles	Mi x 1.61 = km

assorted

1 Lumen = 1 footcandle = 10.76 footmetres = 10.76 Lux

1 Watt per square foot = 10.76 Watts per square metre

1.123 kg $hectare^{-1}$ = 1 lb/acre

11.24 litres ha^{-1} = 1 gallon/acre

1 kg cm^{-2} = 13.3 lb/in^2

1 bar = 0.99 atmosphere = 14.5 lb/in^2

"An absolutely clear and exhaustive understanding of any single thing in the world would imply a perfect comprehension of everything else."
—Arthur Schopenhaurer

References

Abbreviation of journal titles below follows a list compiled by Alkire in *Periodical Title Abbreviations*, Sixth Edition (1987), Gale Research Co., Detroit.

Abelson PH. 1994. Adequate supplies of fruits and vegetables. *Science* 266:1303.

Able E. 1980. *Marijuana: The First 12,000 years.* Plenum Press, NY. 289 pp.

Abiusso NG. 1954. Fungitoxicidad de productos quimicos en ensayos de laboratorio. *Eva Peron Univ. Nac. Facultad de Agron., Rev.* 30:149-161.

Abrol BK, Chopra IC. 1963. Development of indigenous vegetable insecticides and insect repellents. *Bulletin Regional Research Laboratory Jammu* 1:156.

Abu-Irmaileh BE. 1981. Response of hemp broomrape (*Orobanche ramosa*) infestation to some nitrogenous compounds. *Weed Science* 29:8-10.

Acinovic M. 1964. The occurrence of *Macrophomina phaseolina* on some agricultural crops and morphological and ecological properties of the parasite. *Savr. Poljopr.* 12:55-66.

Adams GC, Kropp BR. 1996. *Athelia arachnoidea*, the sexual state of *Rhizoctonia carotae*, a pathogen of carrot in cold storage. *Mycologia* 88:459-472.

Adams R. 1942. Marihuana. *Bulletin New York Academy of Medicine* 18:705-730.

Agostini A. 1927. Observazioni informa a due ifomiceti saprofiti dannosi di tessuti di canapa. *Atti della Reale Accademia dei Fisiocritici* 1(3):25-33.

Agrios G. 1997. *Plant Pathology*, 4th Ed. Academic Press, NY. 635 pp.

Ainsworth GC, Sparrow FK, Sussman AS. 1973. *The Fungi, An Advanced Treatise*, Vol. 4A. Academic Press, NY. 621 pp.

Ajrekar SL, Shaw FJF. 1915. Species of the genus *Rhizoctonia* injurious to plants cultivated in India. Mem. *Indian Dept. Agric., Botanical Series* 7(4):177-194.

Albrecht WA, ed. by Walters C. 1975. *The Albrecht Papers, Vol. I.* Acres USA, Kansas City, MO. 515 pp.

Alexander T. 1980. Spring pests. *Sinsemilla Tips* 1(1):7.

Alexander T. 1982. Sinsemilla tips. *Sinsemilla Tips* 3(1):13.

Alexander T. 1984a. Indoor sinsemilla tips. *Sinsemilla Tips* 4(3):26.

Alexander T. 1984b. Tennessee tips. *Sinsemilla Tips* 4(4):42-43.

Alexander T. 1984c. Florida reader relates. *Sinsemilla Tips* 5(1):27.

Alexander T. 1985. Southern grower. *Sinsemilla Tips* 5(4):46.

Alexander T. 1987. Rodent problems. *Sinsemilla Tips* 7(1):9-10.

Alexander T. 1987b. Hepatitis outbreak linked to imported pot. *Sinsemilla Tips* 7(3):22.

Alexander T. 1988. Rating the strains. *Sinsemilla Tips* 8(2):30-33.

Alexander T. 1988b. Spider mites. *Sinsemilla Tips* 8(2):45.

Alexander T. 1991. Publisher's notes. *Sinsemilla Tips* 9(2):2-5.

Allen JL. 1908. *The Reign of Law, a Tale of the Kentucky hemp fields.* Macmillan Co., NY. 290 pp.

Allen TF. 1875. *Cannabis indica. Encyclopedia of Pure Materia Medica*, Vol. II:448-489. Boericke & Tafel, Philadelphia.

Altieri MA, van Schoonhoven A, Doll J. 1977. The ecological role of weeds in insect pest management symptoms: a review illustrated by bean (*Phaseolus vulgaris*) cropping systems. *PANS* 23:195-205.

Ames BN, Gold LS, Willett WC. 1995. The causes and prevention of cancer. *Proceedings National Academy Science, USA* 92:5258-5265.

Añíbarro B, Fontela JL. 1996. Allergy to marihuana. *Allergy* 51:200-201.

Andow DA, Klacan GC, Bach D, Leahy TC. 1995. Limitations of *Trichogramma nubilale* (Hymenoptera: Trichogrammatidae) as an inundative biological control of *Ostrinia nubilalis* (Lepidoptera: Crambidae). *Environmental Entomology* 24:1352-1357.

Andrén F. 1947. Betningsförsök med Lin och Hampa. *Växtskyddsnotiser, Växtskyddsanst., Stockholm* 1947 6:85-87.

Angelova R. 1968. [Characteristics of the bionomics of the hemp flea beetle, *Psylliodes attenuatus* Koch.] *Rastenievudni Nauki* 5(8):105-114.

Anonymous. 1919. *Gaichu ni Kwansura Chosa* (Researches on injurious insects.) Bull. Industry Model Station, Korea. 82 pp.

Anonymous. 1974. *Chung-kuo nung tso wu ping ch'ung t'u p'u* [Diseases and pests of cotton and hemp]. Beijing Press, Beijing. 93 pp.

Anonymous. 1996. Arming plants with a virus. *Science* 271:145.

Anonymous. 1998. Organophosphorus compounds: good, bad, and difficult. *Lancet* 352:499.

Antonokolskaya MP. 1932. The races of *Sclerotinia libertiana* Fckl. on the sunflower and other plants. *Bull. Plant Protection, Leningrad* 5(1):39-62.

Armstrong G, Armstrong J. 1975. Reflections on the wilt fusaria. *Annual Review Phytopathology* 13:95-103.

Arnaud PH. 1974. Insects and mites associated with stored *Cannabis sativa* Linnaeus. *Pan-Pacific Entomologist* 50:91-92.

Arx JA von. 1957. Die artender gattung *Colletotrichum* Cda. *Phytopathologische Zeitschrift* 29:28-468.

Asplund RO. 1968. Monoterpenes: relationship between structure and inhibition of germination. *Phytochemistry* 7:1995-1997.

Babu R, Roy AN, Gupta YK, Gupta MN. 1977. Fungi associated with deteriorating seeds of *Cannabis sativa* L. *Current Science* 46(20):719-720.

Bajpai NK, Sharma VK. 1992. Possible use of hemp (*Cannabis sativa* L.) weeds in integrated control. *Indian Farmers' Digest* 25(12):32, 38.

Bako L, Nitri I. 1977. Pokusy s ochranou proti obalovaci konopnému (*Grapholitha sinana* Feld). *Len a Konopí* [Sumperk, Czech Rep] 15:13-31.

Balachowski AS, Mesnil L. 1936. *Les Insectes Nuisibles aux Plantes Cultivées.* Imprimé Busson, Paris. Vol. 2, pp 1429-1430.

Balduzzi A, Gigliano GS. 1985. Influenza dell'intensità luminosa sull'accumulo di cannabinoli in *Cannabis sativa* L. *Atti dell' Istituto Botanico e del Laboratorio Crittogamico Dell-Università di Pavia,* Series 7, 4:89-92.

Baloch GM, Ghani MA. 1972. *Natural Enemies of Papaver somniferum and Cannabis sativa.* Annual report, Commonwealth Institute of Biological Control, Pakistan station, pp. 55-56.

Baloch GM, Mushtaque M, Ghani MA. 1974. *Natural Enemies of Papaver spp. and Cannabis sativa.* Annual report, Commonwealth Institute of Biological Control, Pakistan station, pp. 56-57.

Bantra SW. 1976. Some insects associated with hemp or marijuana (*Cannabis sativa* L.) in northern India. *J. Kansas Entomological Society* 49:385-388.

Barloy J, Pelhate J. 1962. Premières observations phytopathologiques relatives aux cultures de chanvre en Anjou. *Annales des Épiphyties* 13:117-149.

Barna J, Lencsepeti JS, Rközy P, Zsombokos G. 1982. *Mezögazdasági Lexikon,* Vol 1 (A-K). Mezögazdasági Kiadó, Budapest. 897 pp.

Barnett JA, Payne RW, Yarrow D. 1983. *Yeast characteristics and identification.* Cambridge University Press, Cambridge. 512 pp.

BB. 1990. The micro-security garden. *Sinsemilla Tips* 9(1):39-41.

Behrens J. 1891. Uber das auftreten des hanfkrebses in Elsass. *Zeitschrift Pfanzenkrankheiten* 1:208-215.

Behrens J. 1902. Untersuchungen ueber die gewinnung der hanffaser durch natuerliche roestmethoden. *Centralbl. für Bakt.* 11(8):264-268, 295-299.

Behrens J. 1903. Ueber die taurotte von flachs und hanf. *Centralbl. für Bakt.* 11(10):524-530.

Beling I. 1932. Schädlingsbekämpfung im 18. Jarhhundert. *Anz. Schädlingbekämpfung* 8(6):66-69.

Bellows TS, Perring TM, Gill RG, Headrick DH. 1994. Description of a species of *Bemisia* (Homoptera: Aleyrodidae). *Annals Entomological Society America* 87:195-206.

Bel'tyukova KI. 1962. [Sensitivity of phytopathogenic bacteria to cansatine 4.] *Mikrobiolohichnyi Zhurnal (Kiev)* 24(5):62-65.

Bennett WF. 1993. *Nutrient Deficiencies and Toxicities in Crop Plants.* Ameican Phytopathological Society, St. Paul, MN.

Bensky D, Gamble A, Kaptchuk T. 1993. *Chinese Herbal Medicine Materia Medica,* Revised Ed. Eastland Press, Seattle WA. 556 pp.

Bentz J, Neal JW. 1995. Effect of a natural insecticide from *Nicotiana gossei* on the whitefly parasitoid *Encarisia formosa. J. Economic Entomology* 88:1611-1615.

Berenji J, Martinov M. 1997. Hemp in Yugoslavia: past, present and future, p. 20 in *Symposium Magazin,* 2nd Biorohstoff Hanf Technisch-wissenschaftliches Symposium. Nova Institut, Köln, Germany.

Berger J. 1969. *The World's Major Fibre Crops: Their Cultivation and Manuring.* Centre d'Etude de l'Azote. Zurich. 294 pp.

Berger KC. 1978. *Sun, Soil & Survival.* University of Oklahoma Press, Norman, OK. 371 pp.

Berlese AN. 1898. Species dubiae vel excludendae. *Icones Fungorum* 4:41.

Bes A. 1978. Prilog poznavanju izgledga ostecenja i stetnosti konopljinog savijaca — *Grapholitha delineana* Walk. *Radovi Poljoprivrednog Fakulteta Univerzita u Sarajevu* 26(29):169-189.

Bestagno-Biga ML, Ciferri K, Bestagno G. 1958. Ordina mento artificiale dolle species del genere *Coniothyrium* Corda. *Sydowia* 12:258-320.

Biddinger DJ, Hull LA. 1995. Effects of several types of insecticides on the mite predator, *Stethorus punctum* (Coleoptera: Cocinellidae), including insect growth regulators and abamectin. *J. Economic Entomology* 88:358-366.

Bidochka MJ, Walsh SRA, Ramos ME, St.Leger RJ, Silver JC, Roberts DW. 1996. Fate of biological control introductions: monitoring an Australian fungal pathogen of grasshoppers in North America. *Proceedings National Academy Science* 93:918-921.

Bigler F, Suverkropp BP, Cerutti F. 1997. "Host searching by Trichogramma and its inplications for quality control and release techniques," pp. 240-253 in *Ecological Interactions and Biological Control.* Andow DA, Ragsdale DW, Nyvall RF, eds. Westview Press, Boulder, CO. 334 pp.

Bilgrami K, Jamaluddin S, Rizwi MA. 1981. *Fungi of India: Part II, Host Index and Addenda.* Today & Tomorrow's Printers and Publishers, New Delhi. 128 pp.

Binder M. 1976. Microbial transformation of tetrahydrocannabinol by *Cunninghamella blakesleena* Lender. *Acta Helvetica Chimica* 63:1674-84.

Blackman RL, Eastop VP. 1985. *Aphids on the World's Crops.* John Wiley & Sons, NY. 470 pp.

Blatchley WS. 1916. *Rhynchophora or Weevils of North Eastern America.* Nature Publ. Co., Indianapolis. 682 pp.

Blattny C, Osvald CV, Novak J. 1950. Virosy a z viros podezrelé zjevy u konopí. *Ochrana Rostlin* 23:5-9.

Blevins RD, Dumic MP. 1980. The effect of delta-9-tetrahydrocannabinol on herpes simplex virus replication. *J. General Virology* 49:427-431.

Blunck H. 1920. Die niederen Tierischen feinde unserer Gespinstpflanzen. Ill. *Landw. Zeig.* 40:259-260.

Bòcsa I. 1958. A kender beltenyésztésémek újabb jelenségei. *Növénytermelés* 7:1-10.

Bòcsa I. 1999. "Genetic improvement: conventional approaches," pp. 153-184 in *Advances in Hemp Research.* P. Ranalli, ed. Hawthorn Press, NY. 272 pp.

Bòcsa I, Karus M. 1997. *Der Hanfanbau: Botanik, Sorten, Anbau und Ernte.* Müller Verlag, Heildelberg. 173 pp. [1998 English translation: The Cultivation of Hemp: Botany, Varieties, Cultivation and Harvesting. Hemptech, Sebastopol, CA].

Boczek J, Golebiowska Z, Krzeczkowski K. 1960. Roztocze szkodliwe w przechowalniach siemienia lnianego i konopi w Polsce. *Prace nauk. Inst. Ochr. Rosl.* 2(1):57-86.

Bodenheimer FS. 1944. Note on the Coccidea of Iran, with descriptions of new species (Hemiptera-Homoptera). *Bull. Soc. Fouad Ier Ent.* 28:85-100.

Boerema GH. 1970. Additional notes on *Phoma herbarum. Persoonia* 6(1):15-48.

Boerema GH, Gams W. 1995. What is *Sphaeria acuta* Hoffm.:Fr.? *Mycotaxon* 53:355-360.

Boewe GH. 1963. Host plants of charcoal rot disease in Illinois. *Plant Disease Reporter* 47:753-755.

Booth C. 1971. *The Genus Fusarium.* Commonwealth Mycological Institute, Kew, U.K. 237 pp.

Borchsenius NS. 1957. Fauna of USSR Coccidae. *Akad. Nauk Zool. Inst.* 66(9):365.

Borodin DN. 1915. *First Report on Work of the Entomological Bureau and a Review of the Pests of the Govt. of Poltava in 1914.* The Entomological Bureau of the govt. of the Zemstvo of Poltava. 87 pp.

Borodina EI, Migal ND. 1987. Flower teratology in intersexual hemp plants. *Soviet J. Developmental Biology* 17(4):262-269.

Bosik JJ. 1997. *Common Names of Insects.* Entomological Society of America, Washington, D.C. 232 pp.

Bouhuys A, Zuskin E. 1976. Chronic respiratory disease in hemp workers. *Annals Internal Medicine* 84:398-405.
Bouquet RJ. 1950. *Cannabis. Bulletin on Narcotics* 2(4):14-30.
Bovien P. 1945. Plantesygdomme i Danmark 1944. *Tidsskr. Planteavl.* 50:1-76.
Bozoukov H, Kouzmanova I. 1994. "Biological control of tobaco broomrape (Orobanche spp.) by means of some fungi of the genus Fusarium," pp. 534-538 in *Biology and Management of Orobanche.* AH Pieterse, JAC Verkleij, SJ ter Borg, eds. Royal Tropical Institute, Amsterdam.
Brady NC, Weil R. 1999. *The Nature and Properties of Soils,* 12th Ed. Prentice Hall, NJ. 896 pp.
Bram S, Brachet P. 1976. "Inhibition of proliferation and differentiation of *Dictyostelium discoideum* amoebae by tetrahydrocannabinol and cannabinol," pp. 207-211 in *Marihuana: Chemistry, Biochemistry, and Cellular Effects.* GG Nahas, ed. Springer-Verlag, New York.
Brandenburger W. 1985. *Parasitische Pilz an Gefäßpflanzen in Europa.* Gustav Fisher Verlag, Stuttgart. 1248 pp.
Braut-Boucher F, Cotte J, Fleury C, Quero AM, Besnier MO, Fourniat J, Courtois D, Pétiard V. 1985. Exemple de variabilité spontanée mis en évidence par l'activité biologique des extraits tissulaires de *Cannabis sativa* L. *Bull. Soc. Bot. France* 132 (3/4):149.
Bredmann G, Schwanitz F, Van Sengbusch R. 1956. Problems of modern hemp breeding, with particular reference to the breeding of varieties of hemp containing little or no hashish. *Bulletin on Narcotics* 8(3):31-35.
Brendel F. 1887. *Flora Peorina: The Vegetation in the Climate of Middle Illinois.* J.W. Franks & Sons, Printers and Binders, Peoria, Illinois. 89pp.
Brenneisen R, ElSohly MA. 1988. Chromatographic and spectroscopic profiles of *Cannabis* of different origins: Part 1. *J. Forensic Sciences* 33:1385-1404.
Brian PW, Wright JM, Stubbs J, Way AM. 1951. Uptake of antibiotic metabolites of soil microorganisms by plants. *Nature* 167:347-349.
Bridge J. 1996. Nematode management in sustainable and subsistence agriculture. *Annual Review Phytopathology* 34:201-225.
Broglie K, Chet I, Holliday M, *et al.* 1991. Transgenic plants with enhanced resistance to the fungal pathogen *Rhizoctonia solani. Science* 254:1194-1197.
Brown J. 1883. *Cannabis indica;* a valuable remedy in menorrhagia. *JAMA* 1:51-52.
Brown MF, Brotzman HG. 1979. *Phytopathogenic Fungi: a Scanning Electron Stereoscopic Survey.* University of Missouri Press, Columbia, MO. 355 pp.
Brummund W, Kurup VP, Harris GH, Duncavage JA, Arkins JA. 1987. Fungal sinusitis and marijuana: reply. *JAMA* 257:2915.
Brunard P. 1899. Sphaeropsidees recoltees jusqu'a ce jor dans la Charente-Interieure. *Academie de la Rochelle, Sociètè des Sciences Naturelles de la Charente-Interieure* 26:51-140.
Brundza K. 1933. Kai kurie parazitiniai grybeliai, surinkfi Liefuvoje 1927-1932. *Zemes Ukio Akademijos Leidinys* 1932:199-208.
Buhr H. 1937. Parasitenbefall und pflanzenverwandtschaft. *Botanische Jahrbucher* 68:142-198.
Bush Doctor, The. 1984. Premature wilt. *Sinsemilla Tips* 5(1):46.
Bush Doctor, The. 1985. Aphids or plant lice. *Sinsemilla Tips* 5(2):22-23.
Bush Doctor, The. 1985b. Gray Mold. *Sinsemilla Tips* 5(3):27-28.
Bush Doctor, The. 1985c. Damping Off. *Sinsemilla Tips* 5(4):35-39.
Bush Doctor, The. 1986a. Interview with John McEno. *Sinsemilla Tips* 6(1):33-34, 84-85.
Bush Doctor, The. 1986b. A closer look at spider mites. *Sinsemilla Tips* 6(2):31-33, 84.
Bush Doctor, The. 1987a. Roadtrip research. *Sinsemilla Tips* 6(4):46-48.
Bush Doctor, The. 1987b. European Corn Borers. *Sinsemilla Tips* 7(2):45-47.
Bush Doctor, The. 1987c. Storage diseases and pests. *Sinsemilla Tips* 6(3):41-46, 90.
Bush Doctor, The. 1987d. Roadtrip research: Nepal. *Sinsemilla Tips* 6(4):46-48.
Bush Doctor, The. 1987e. The seven wonders, Part I. *Sinsemilla Tips* 7(3):40-45.
Bush Doctor, The. 1988a. The seven wonders, Part II. *Sinsemilla Tips* 7(4):45-50.
Bush Doctor, The. 1988b. The seven wonders, Part III. *Sinsemilla Tips* 8(1):37-42.
Bush Doctor, The. 1988c. The secret life of plant mycorrhizae. *Sinsemilla Tips* 8(2):34-36.
Bush Doctor, The. 1989a. An update on common garden pests. *Sinsemilla Tips* 8(3):58-59.
Bush Doctor, The. 1989b. Whiteflies and other bad t(h)rips. *Sinsemilla Tips* 8(4):47-51.
Bush Doctor, The. 1990. Mildew Review. *Sinsemilla Tips* 9(1):32-35.
Bush Doctor, The. 1991. Anthracnose. *Sinsemilla Tips* 9(2):36-38.
Bush Doctor, The. 1992a. Post-Green Merchant growing. *High Times* No.197, pg.49.
Bush Doctor, The. 1992b. Pot on the edge. *High Times* No. 200, pg 46-47, 52.
Bush Doctor, The. 1992c. *Cannabis*-crunching caterpillars. *High Times* No. 204, p. 48, 55, 58.
Bush Doctor, The. 1993a. Stash alert: how to preserve pot potency. *High Times* No. 213, pp. 75-79.
Bush Doctor, The. 1993b. A heavy essay on lights. *High Times* No. 220, pp. 36-38.
Bush Doctor, The. 1994. War on Herbs. *High Times* No. 227, p.8.
Byford WJ, Ward LK. 1968. Effect of the situation of the aphid host at death on the type of spore produced by *Entomophthora* spp. *Trans. Br. mycol. Soc.* 51:598-600.
Callaway JC, Laakkonen TT. 1996. Cultivation of *Cannabis* oil seed varieties in Finland. *J International Hemp Association* 3:32-34.
Caminita BH., Schneiter R, Kolb RW, Neal PA. 1947. Studies on strains of *Aerobacter cloacae* responsible for acute illness among workers using low-grade stained cotton. *Public Health Reports* 58(31):1165-1183.
Campbell CA, Cone WW. 1995. Influence of predators on population development of *Phorodon humuli* (Homoptera: Aphididae) on hops. *Environmental Entomology* 23:1391-1396.
Campbell BC, Duffy SS. 1979. Tomatine and parasitic wasps: potential incompatibility of plant antibiosis with biological control. *Science* 205:700-702.
Camprag D. 1961. Observations on the occurence and injuriousness of *Maladera holosericea* Scop. with special reference to sugar-beet. *Zborn. prirod. Nauke Matica Srpska* 21:122-131.
Camprag D, Jovanic M, Sekulic R. 1996. Stetocine konoplje i integralne mere suzbijanja. *Zbornik Radova* 26/27:55-68.
Canapasemi G. 1988. Hemp fiber cultivation. Sinsemilla Tips 8(3):49-57.

Cardassis J. 1951. Intoxication des equides par *Cannabis indica*. *Rec. Med. Vet.* 127:971-3.
Carris LM, Glawe DA. 1989. Fungi colonizing cysts of *Heterodera glycines*. *Illinois Agricultural Experiment Station Bulletin* No. 786. 93 pp.
Carris LM, Humber RA. 1998. *Neozygites parvispora*, a pathogen of *Limothrips* sp. associated with *Lolium multiflorum* in Oregon. *Mycologia* 90:565-568.
Castellan RM, Olenchock SA, Hankinson JL, *et al.* 1984. Acute broncho-constriction induced by cotton dust: dose related responses to endotoxin and other dust factors. *Annals Internal Medicine* 101:157-163.
Cavara F. 1889. Materiaux de mycologie Lombarde. *Revue Mycologique* 11:173-193.
Ceapoiu N. 1958. *Cinepa, Studiu monografic*. Editura Academiei Republicii Populare Romine. Bucharest. 652 pp.
Centers for Disease Control. 1985. Phytophotodermatitis among grocery workers. *Morbidity and Mortality Weekly Report* 34(1):11-13.
Chandra S. 1974. Some new leaf-spot diseases from Allahabad. *Nova Headwigia Beih.* 47:35-101.
Charles VK, Jenkins AE. 1914. A fungus disease of hemp. *J. Agricultural Res.* 3:81-84.
Che CT. 1991. Plants as a source of potential antiviral agents. *Economic & Medicinal Plant Research* 5:167-251.
Cherian MC. 1932. Pests of ganja. *Madras Agricultural Journal* 20:259-265.
Cherim MS. 1998. *The Green Methods Manual*, 4th Ed. The Green Spot, Ltd. Publishing, Nottinham, NH. 238 pp.
Cherniah L. 1979. *The Great Books of Hash*, Vol. 1, Book 1. And/Or Press, Berkeley, CA. 140 pp.
Chester K. 1941. Cotton root rot or Texas root rot. *Oklahoma Agricultural Experiment Station Circular* No. 53. 4 pp.
Chittenden FH. 1909. Some insects injurious to truck crops: the hop flea-beetle. *USDA Entomology Bulletin* 66(4):71-92.
Chmielewski W. 1984. *Tyrophagus longior*—bioekologia, wystepwanie i szkodliwosc. *Prace Naukowe Instytutu Ochrony Roslin* 26:69-85.
Chmielewski W, Filipek P. 1968. Wplyw zaprawiania nasion lnu i konopi panogenem na roztocze. *Roczn. Nauk roln.* 93 (A) 4:701-710.
Chopra RN, Badhwar RL, Nayar SL. 1941. Insecticidal and piscicidal plants of India. *J. Bombay Nat. Hist. Soc.* 42:854-902.
Christie RG, Edwardson JR, Simone GW. 1995. "Diagnosing plant virus diseases by light microscopy." pp. 31-52 in *Molecular Methods in Plant Pathology*. RP Singh, US Singh, eds. CRC-Lewis Publishers, Boca Raton, FL. 523 pp.
Christou P. 1996. *Particle Bombardment for Generic Engineering of Plants*. Academic Press, San Diego, CA. 199 pp.
Chusid MJ, Gelfand JA, Nutter C, Fauci AS. 1975. Pulmonary aspergillosis, inhalation of contaminated marijuana smoke, and chronic granulomatous disease. *Ann. Internal Med.* 82:682-683.
Ciferri R. 1941.*Manuale di Patologia Vegetale*. Societa Editrice Dante Alighieri, Roma. 730 pp.
Ciferri R, Brizi A. 1955. *Manuale di Patologia Vegetale*, Vol. III, Table 29. Societa Editrice Dante Alighieri, Roma.
Clarke EGC, Greatorex JC, Potter R. 1971. *Cannabis* poisoning in the dog. *Veterinary Record* 88:964.
Clarke RC. 1981. *Marijuana Botany*. And/Or Press, Berkeley, CA. 197 pp.
Clarke RC. 1987. *Cannabis* evolution. MS thesis, Indiana University, Bloomington, IN. 233 pp.
Clarke RC. 1995. Hemp (*Cannabis sativa* L.) cultivation in the Tai'an district of Shandong province, Peoples Republic of China. *J. International Hemp Association* 2(2):57, 60-65.
Clarke RC. 1998. *Hashish!* Red Eye Press, Los Angeles, CA. 387 pp.
Clarke RC. 1999. "Botany of the genus *Cannabis*," pp. 1-19 in *Advances in Hemp Research*. P. Ranalli, ed. Haworth Press, NY.
Clausen CP. 1931. Insects injurious to agriculture in Japan. *USDA Circular* No. 168. 108 pp.
Clausen CP, King JL. 1927. The parasites of *Popillia japonica* in Japan and Chosen (Korea) and then introduced into the United States. *USDA Bulletin* No. 1429. 56 pp.
Clayton DH, Wolfe ND. 1993. The adaptive significance of self-medication. *Trends in Ecology & Evolution* 8(2):60-63.
Clifford NJ, Nies AS. 1989. Organophosphate poisoning from wearing a laundered uniform previously contaminated with parathion. *JAMA* 262:3035-3036.
Cloyd RA, Brownbridge, Sadof CS. 1998. Greenhouse biological control of western flower thrips. *IPM Practitioner* 20(8):1-8.
Coffman CB, Genter WA. 1975. Cannabinoid profile and elemental uptake of *Cannabis sativa* L. as influenced by soil characteristics. *Agronomy Journal* 67:491-497.
Coffman CB, Genter WA. 1977. Responses of greenhouse-grown *Cannabis sativa* L. to nitrogen, phosphorus, and potassium. *Agronomy Journal* 69:832-836.
Colborn T, Dumaroski D, Peterson-Myers J. 1996. *Our Stolen Future*. Dutton Books, NY. 306 pp.
Colby M. 1996. The great pesticide compromise: deaths vs. dollars. *Food & Water Notes*, Summer 1996, pp. 6-7. Food & Water, Walden, VT.
Commonwealth Mycological Institute. 1989. *Pseudoperonospora cannabina* on *Cannabis sativa*. *Distribution Maps of Plant Disease* No. 478, Edition 2.
Comstock JH. 1879. *Report upon Cotton Insects*. Government Printing Office, Washington D.C. 511 pp.
Comstock JH. 1904. *Manual for the Study of Insects*, 5th Ed. Comstock Publ. Co., Ithaca, NY. 701 pp.
Conners IL. 1967. *An Annotated Index of Plant Diseases in Canada*. Canadian Department of Agriculture, Research Publication No. 1251. 381 pp.
Conrad C. 1994. *Hemp: Lifeline to the Future*. Creative Xpressions Publications, Los Angeles, CA. 314 pp.
Conrad C. 1997. *Hemp for Health*. Healing Arts Press, Rochester, VT. 264 pp.
Cook A. 1981. *Diseases of tropical and subtropical field, fiber, and oil crops*. Macmillan Publ., NY. 450 pp.
Cook RJ, Qualset CO. 1996. *Appropriate Oversight for Plants with Inherited Traits for Resistance to Pests*. Institute of Food Technologists, Chicago. 35 pp.
Cook RJ, Bruchkart WL, Coulson JR, *et al.* 1996. Safety of microorganisms intended for pest and plant disease control: a framework for scientific evaluation. *Biological Control* 7:333-351.
Crane JL, Shearer CA. 1991. A nomenclator of *Leptosphaeria* Cesati & de Notaris. *Illinois Natural History Survey Bulletin* 34(3):195-355.
Crescini F. 1956. La fecondazione incestuosa processo mutageno in *Cannabis sativa* L. *Caryologia (Florentinea)* 9(1):82-92.
Crombie L, Crombie WML. 1975. Cannabinoid formation in *Cannabis sativa* grafted interracially, and with two *Humulus* species. *Phytochemistry* 14:409-412.
Crosby TK, Watt JC, Kistemaker AC, Nelson PE. 1986. Entomological identification of the origin of imported *Cannabis*. *J. Forensic Science Society* 26(1):35-44.
Culpepper N. 1814. *Complete Herbal*. Richard Evans Publisher, London, reprinted 1990 by Meyerbooks, Glenwood, IL. 398 pp.

Curzi M, Barabaini M. 1927. Fungi aternenses. *Atti dell' Istituto Botanico della Università di Pavia,* Series 3, 3:147-202.

Czyzewska S, Zarzycka H. 1961. Ergebnisse der Bodeninfektionsversuche an *Linum usitatissinum, Crambe alyssinica, Cannabis sativa* und *Cucurbita pepo* var. *oleifera* mit einigen *Fusarium*-Arten. *Instytut Ochrony Roslin, Reguly, Polen.* Report No. 41:15-36.

Dahiya MS, Jain GC. 1977. Inhibitory effects of cannabidiol and tetrahydrocannabinol against some soil inhabiting fungi. *Indian Drugs* 14(4):76-79.

Darsie ML, Elliott C, Peirce GJ. 1914. A study of the germinating power of seeds. *Botanical Gazette* 58:101-136.

Darwin CR. 1881. *The Power of Movement in Plants.* Da Capo Press, NY. 444 pp.

Dash VB. 1989a. *Fundamentals of Ayurvedic Medicine,* 7th Ed. Konark Publishers Ltd, Delhi. 228 pp.

Dash VB. 1989b. *Illustrated Materia Medica of Indo-Tibetan Medicine,* 2nd Ed. Classics India Publications, Delhi. 647 pp.

Dass Baba S. 1984. Truth, fact and hearsay from the big island. *Sinsemilla Tips* 4(3):1213, 39.

Datnoff LE, Nemec S, Pernezny K. 1995. Biological control of Fusarium crown and root rot of tomato using *Trichoderma harzianum* and *Glomus intraradices. Biological Control* 5:427-431.

Datta M, Chakraborti M. 1983. On a collection of flower flies (Diptera: Syrphidae) with new records from Jammu and Kashmir. *Records of the Zoological Survey of India* 81:237-253.

Davis JJ. 1916. *Aphidoletes meridionalis,* an important dipterous enemy of aphids. *J. Agricultural Research* 6:883-887.

Davidson RH, Peairs LM. 1966. *Insect Pests of Farm, Garden, and Orchard,* 6th Ed. John Wiley & Sons, NY. 675 pp.

Dayanandan P, Kaufman PB. 1978. Trichomes of *Cannabis sativa* L. *Amer. J. Bot.* 63:578-591.

Deay HO. 1950. Control of European corn borers with electric grids in Indiana. *Proc. N. Centr. Br. Entomol. Soc. Am.* 5:48-50.

DeBary A. 1887. *Comparative Morphology and Biology of the Fungi, Mycetozoa and Bacteria.* Clarendon Press, Oxford. 525 pp.

Decker H. 1972. *Plant Nematodes and their Control.* Kolos Publ. Co., Moscow. 539 pp.

DeCorato U. 1997. Le malattie della canapa in Basilicata. *Informatorie Fitopatologico* 47(5):57-59.

Deferne JL, Pate DW. 1996. Hemp seed oil: a source of valuable essential fatty acids. *J. International Hemp Association* 3(1):1, 4-7.

De Groot RC. 1972. Growth of wood-inhabiting fungi in saturated atmospheres of monoterpenoids. *Mycologia* 64:863-870.

de Jong E, de Vries FP, Field JA, van der Zwan R, de Bont JAM. 1992. Isolation and screening of basidiomycetes with high peroxidative activity. *Mycological Research* 96: 1098-1104.

de Meijer EPM. 1993. Evaluation and verification of resistance to *Meloidogyne hapla* Chitwood in a *Cannabis* germplasm collection. *Euphytica* 71:49-56.

de Meijer EPM. 1995. Fiber hemp cultivars: a survey of origin, ancestry, availability and brief agronomic characteristics. *J. International Hemp Association* 2(2):66-73.

de Meijer EPM. 1999. "*Cannabis* germplasm resources," pp. 133-151 in *Advances in Hemp Research.* P. Ranalli, ed. Haworth Press, NY.

de Meijer WJM, vand der Kamp HJ, van Eeuwijk FA. 1992. Characterization of *Cannabis* accessions with regard to cannabinoid content in relation to other plant characters. *Euphytica* 62:187-200.

de Meijer WJM, van der Werf HMG, Mathijssen EWJM, van den Brink PWM. 1995. Constraints to dry matter production in fibre hemp (*Cannabis sativa* L.). *European J. Agronomy* 4:109-117.

DeMoraes CM, Lewis WJ, Paré PW, Alborn HT, Tumlinson JH. 1998. Herbivore-infested plants selectively attract parasitoids. *Nature* 393:570-573.

Dempsey JM. 1975. "Hemp" pp. 46-89 in *Fiber Crops.* University of Florida Press, Gainesville, FL.

Deshmukh PD, Rathore YS, Bhattacharya AK. 1979. Larval survival of *Diacrisia obliqua* Walker on several plant species. *Indian J. Entomology* 41(1):5-12.

Dewey LH. 1902. "The hemp industry in the United States." pp. 541-554 in *U.S.D.A. Yearbook 1901,* United States Department of Agriculture, Washington D.C.

Dewey LH. 1913. A purple-leaved mutation in hemp. *U.S.D.A. Plant Industry Circular* 113:23-24.

Dewey LH. 1914. "Hemp." pp. 283-347 in *U.S.D.A. Yearbook 1913,* United States Department of Agriculture, Washington D.C.

Dewey LH. 1928. "Hemp varieties of improved type are result of selection." pp. 358-361 in *U.S.D.A. Yearbook 1927,* United States Department of Agriculture, Washington D.C.

Dewey LH. 1943. "Hemp" pp. 63-69 in Fiber production in the Western Hemisphere. *U.S.D.A. Miscellaneous Publication* No. 518. Washington, D.C. 95 pp.

Dewey LH, Merrill JL. 1916. Hemp hurds as paper-making material. *U.S.D.A. Bulletin* No. 404. Washington, D.C. 25 pp.

Dhooria MS. 1983. An outbreak of the citrus mite, *Eutetranychus orientalis* (Klein) in Delhi. *Pesticides* 17(11):36.

Dimitrov M, Ivanova-Dzhubrilova S, Nikolcheva M, Drenska E. 1990. Study of mycotoxical and dust air pollution in enterprises for preliminary processing of cotton and hemp. *Problemi na Khigienta* 15:121-127.

Dippenaar MC, du Toit CLN, Botha-Greeff MS. 1996. Response of hemp (*Cannabis sativa* L.) varieties to conditions in Northwest Province, South Africa. *J. International Hemp Association* 3(2):63-66.

Dixon AFG. 1985. *Aphid Ecology.* Blackie/Chapman & Hall, NY. 157 pp.

Djerassi C, Shih-Coleman C, Diekman J. 1974. Insect control of the future: operational and policy aspects. *Science* 186:596-607.

Doblin R, Diamond RD. 1991. Medical use of marijuana. *Annals Internal Medicine* 114:809-810.

Dobrozrakova TL, Letova MF, Stepanov KM, Khokhryakov MK. 1956. "*Cannabis sativa* L." pp. 242-248 in *Opredelitel' Bolesni Rasteniî,* Moscow. 661 pp.

Docea, Negru. 1972. "*Leptosphaeria woroninii*" in Negru, Docea, Szasz. *Novosti Sistematiki nizshikh Rasteniî* 9:168.

Doctor Indoors. 1988. Bud rot... big buds rot first. *Sinsemilla Tips* 8(1):33.

Dodge CR. 1890. "The Hemp Industry in France" pp. 27-31 and "The Hemp Industry" pp. 64-74 in *U.S.D.A. New Series Miscellaneous* (later called USDA Fiber Investigations Series) Report No. 1, Government Printing Office, Washington D.C. 104 pp.

Dodge CR. 1898. "A report on the culture of hemp in Europe" pp. 5-29 in *USDA Fiber Investigations Series,* Report No. 11, Government Printing Office, Washington D.C. 29 pp.

Doidge EM, Bottomley AM, van der Plank JE, Pauer GD. 1953. A revised list of plant diseases in South Africa. *So. African Dept. Agr., Sci Bull.* No. 345. 122 pp.

Dollet M. 1984. Plant diseases caused by flagellate protozoa. *Annual Review Phytopathology* 22:115-132.

Dombrovskaya EV. 1940. [Description of a new Cecidomyiid, *Silvestrina tyrophagi* sp.n., destroying the mite, *Tyrophagus noxious* Zachv.] *Leningrad Bull. Plant Protection* 1940 (3):87-88.

Dombrowski JA, Kolmes SA, Dennehy TJ. 1996. Behaviorally active compounds may not enhance pesticide toxicity: the case of dicofol and amitraz. *J. Econ. Entomology* 89:1130-1136.

Domsch KH, Gams W, Anderson TH. 1980. *Compendium of Soil Fungi.* 2 volumes. Academic Press, NY.

Donahue RL, Shickluna JC, Robertson LS. 1973. *Soils: An Introduction to Soils and Plant Growth.* 3rd ed. Prentice-Hall, Inc., Englewood Cliffs, NJ. 587 pp.

Donovan M. 1845. On the physical and medicinal qualities of Indian hemp (*Cannabis indica*), with observations on the best mode of administration, and cases illustrative of its powers. *Dublin J. Medical Science* 26:368-402.

Douguet G. 1955. Le genre *Melanospora. LeBotaniste* 39:1-313.

Driemeier D. Marijuana (*Cannabis sativa*) toxicosis in cattle. *Veterinary & Human Toxicology* 39:351-2.

Dudley JE. 1920. Control of the potato leafhopper (*Empoasca mali* Le B.) and prevention of "hopperburn." *J. Economic Entomology* (Concord N.H.) 13(4):408-415.

Duffey SS. 1980. Sequestration of plant natural products by insects. *Ann. Rev. Entomology* 25:447-477.

Duke JA. 1982. "Ecosystematic Data on Medicinal Plants." pp. 13-23 in *Utilization of Medicinal Plants.* CK Atal & BM Kapur, eds. United Printing Press, New Delhi. 877 pp.

Duke JA. 1985. *CRC Handbook of Medicinal Herbs.* CRC Press, Boca Raton, FL. 677 pp.

Durham OC. 1935. The pollen content of the air in North America. *J. Allergy* 6:128-149.

Durnovo ZP. 1933. [Results of work on the maize moth and other pests of newly cultivated annual fibre plants], pp. 85-106 in *Bolyezni i Vredit. nov. lubyan. Kul'tur* [Diseases and Pests of newly cultivated Fibre Plants]. Institut Novogo Lubianogo Syriia [Institute of New Bast Raw Materials], Moscow.

Durrell LW, Shields L. 1960. Fungi isolated in culture from soils of the Nevada test site. *Mycologia* 52:636-641.

DuToit BM. 1980. *Cannabis in Africa.* A.A.Balkema Press, Rotterdam. 512 pp.

Eaton BJ, Hartowicz LE, Latta RP, *et al.* 1972. Controlling wild hemp. *Kansas Ag. Exp. Station, Report of Progress* No. 188. 10 pp.

Edwards CA, Bohlen PJ. 1992. The effect of toxic chemicals on earthworms. *Reviews of Enviornomental Contamination & Toxicology* 125:23-99.

Egertová M, Cravatt BF, Elphick MR. 1998. Phylogenetic analysis of cannabinoid signalling. *Symposium on the Cannabinoids,* Burlington, VT: International Cannabinoid Research Society, 1998:101.

Eisenbach JD. 1985. "Diagnostic characters useful in the identification of the four most common species of root-knot nematodes," pp. 95-112 in *An Advanced Treatise on Meloidogyne,* Volume I. JN Sasser, CC Carter, eds. North Carolina State University Graphics, Raleigh, NC. 122 pp.

Elbert A, Nauen R, Leicht W. 1998. Imidacloprid, a novel chloronicotinyl insecticide: biological activity and agricultural importance," pp. 50-73 in *Insecticides with Novel Modes of Action.* I Ishaaya, D Degheele, eds. Springer, Berlin. 289 pp.

Ellis BW, Bradley FM. 1992. *The Organic Gardener's Handbook of Natural Insect and Disease Control.* Rodale Press, Emmaus, PA. 534 pp.

Elmer WH, Ferrandino FJ. 1994. Comparison of ammonium sulfate and calcium nitrate fertilization effects on *Verticillium* wilt of eggplant. *Plant Disease* 78:811-816.

Elmore CL, Stapleton JJ, Bell CE, DeVay JE. 1997. Soil Solarization. *University of California Division of Agricultural and Natural Resources Publication* No. 21377. 12 pp.

Elsohly HN, Turner CE, Clark AM, Elsohly MA. 1982. Synthesis and antimicrobial activities of certain cannabichromene and cannabigerol related compounds. *J. Pharmaceutical Sciences* 71:1319-1323.

Emboden WA. 1974. *Cannabis* — a polytypic genus. *Economic Botany* 28:304-310.

Emchuck EM. 1937. Some data on the injurious entomofauna of the truck farms and orchards of the Desna river region. Travaux de l'Institut de Zoologie et Biologie. *Académie des Sciences d'Ukraine* 14:279-282.

Endo S. 1931. The host plants of *Hypochnus centrifugus* (Lév.)Tul. ever recorded in Japan. *Transactions Tottori Society of Agricultural Science* 3:254-270.

EPA. 1994. Plant-pesticides subject to the Federal Insecticide, Fungicide, and Rodenticide Act and the Federal Food, Drug, and Cosmetic Act. *Federal Register* 59:60496-60518.

Eppler A. 1986. Untersuchungen zur wirtswahl von *Phorodon humuli* Schrk. I. Besiedelte Pflanzenarten. *Anzeiger für Schädlingskunde, Pflanzenschutz, Umweltschutz* 59:1-8.

Evans AG. 1989. Allergic inhalant dermatitis attributable to marijuana exposure in a dog. *J. American Veterinary Medicine Association* 195:1588-1590.

Ewen SWB, Pusztai A. 1999. Effect of diets containing genetically modified potatoes expressing *Galanthus nivalis* lectin on rat small intestine. *Lancet* 354:1353-1355.

Faeth SH. 1986. Indirect interactions between temporally separated herbivors mediated by the host plant. *Ecology* 67:479-494.

Faeti V, Mandolino G, Ranalli P. 1996. Genetic diversity of *Cannabis sativa* germplasm based on RAPD markers. *Plant Breeding* 115:367-370.

Fairbain JW. 1976. "The pharmacognosy of *Cannabis*," pp. 3-19 in *Cannabis and Health.* JDP Graham, ed. Academic Press, London and New York. 481 pp.

Falk SP, Gadoury DM, Pearson RC, Seem RC. 1995. Partial control of grape powdery mildew by the mycoparasite *Ampelomyces quisqualis. Plant Disease* 79:483-490.

Farkas H. 1965. "Family Eriophyidae, Gallmilben," p. 84 in *Die Tierwelt Mitteleuropas,* Band III, Lief 3. 155 pp.

Farkas J, Andrássy É. 1976. The sporostatic effect of cannabidiolic acid. *Acta Alimentaria* 5:57-67.

Farnsworth NR, Cordell GA. 1976. New potential hazard regarding use of marijuana—treatment of plants with liquid fertilizers. *J. Psychedelic Drugs* 8:151-155.

Farr DF, Bills GF, Chamuris GP, Rossman AY. 1989. *Fungi on Plants and Plant Products in the United States.* APS Press, St. Paul, MN. 1252 pp.

Fatemi J. 1974. A simple technique for isolation of *Phytophthora* and *Pythium* species from the soil. *Phytopathologie Mediterranea* 13:120-121.

Fenili GA, Pegazzano F. 1974. Metodi avanzati di lotta contro gli acari fitofagi. *Noti ed Appunti Sperimentali di Entomologia Agraria* 15:33-41.

Feng Z, Carruthers RI, Roberts DW, Robson DS. 1985. Age-specific dose-mortality effects of *Beauveria bassiana* (Deuteromycotina: Hyphomycetes) on the European corn borer, *Ostrinia nubilalis* (Lepidoptera: Pyralidae). *J. Invertebrate Pathology* 46:259-264.
Fenton B, Stanley K, Fenton S, Bolton-Smith C. 1999. Differential binding of the insecticidal lectin GNA to human blood cells. *Lancet* 354:1354-1355.
Ferenczy L. 1956. Antibacterial substances in seeds. *Nature* 178:639-640.
Ferenczy L, Gracza L, Jakobey I. 1958. An antibacterial preparatum from hemp (*Cannabis sativa*). *Naturwissenschaften* 45:188.
Ferraris T. 1915. *I Parassiti Vegetali*. Ulrico Hoepli Press, Milano, Italy. 1032 pp.
Ferraris T. 1935. *Parassiti Vegetali della Canapa*. Riv. Agric., Rome. 715 pp.
Ferraris T, Massa C. 1912. Micromiceti nuovi o rari per la flor micologica Italiana. *Annales Mycologici* 10:285-302.
Ferri F. 1957a. La "striatura ulcerosa" della canapa. *Informatore Fitopatologica* 7(14):235-238.
Ferri F. 1957b. La "striatura ulcerosa" flagello della canapa. *Progresso Agricolo* 3(10):1194-1194.
Ferri F. 1959a. *Atlante delle Avversità della Canapa*. Edizioni Agricole, Bologna, Italia. 51 pp.
Ferri F. 1959b. La Septoriosi della canapa. *Annali della sperimentazione agraria N.S.* 13:6, Supplement pg. CLXXXIX-CXCVII.
Ferri F. 1961a. Sensibilitá di *Sclerotium rolfsii* avari funghicidi. *Phytopathologie Mediterranea* 3:139-140.
Ferri F. 1961b. Microflora dei semi di canapa. *Progresso Agricolo* (Bologna) 7(3):349-356.
Ferri F. 1961c. Le avversità delle piante viste alla lente: Canapa. *Progresso Agricolo* (Bologna) 7:764-765.
Ferri F. 1963. Alterazioni della canapa trasmesse per seme. *Progresso Agricolo* (Bologna) 9:346-351.
Filipjev IN, Stekhoven JHS. 1941. *A Manual of Agricultural Helminthology*. Brill, Leiden. 878 pp.
Fisher E. 1904. *Die Uredineen der Schweiz*. K. J. Wyss Publishing, Bern, Switzerland. 590 pp.
Foster SP, Harris MO. 1997. Behavioral manipulation methods for insect pest-management. *Annual Review Entomology* 42:123-146.
Flachs K. 1936. Krankheiten und Schädlinge unserer Gespinstpflanzen. *Nachrichten über Schädlingsbekämpfung*. 11:6-28.
Fleischmann R. 1934. Beiträge zur Hanfzüchtung. *Faserforschung* 11:156-161.
Flores E. 1958. Relations between insect and host plant in transmission experiments with infectious chlorosis of Malvaceae. *Ann. Acad. Brasil Cienc.* 30:535-560.
Fokkema NJ, Van Den Heuvel J. 1986. *Microbiology of the phyllosphere*. Cambridge University Press, Cambridge. 392 pp.
Forolov AN. 1981. Genetic analysis of the "massive tibia"—a taxonomic character of *Ostrinia scapulalis* Walker (Lepidoptera, Pyraustidae). *Soviet Genetics* 17:1401-1405.
Fournier G, Paris M. 1983. Mise en evidence de cannabinoïdes chez *Phelipaea ramosa*, parasitant le chanvre, *Cannabis sativa*, Cannabinacees. *Planta Medica* 49:250-251.
Fournier G, Paris MR, Fourniat MC, Quero AM. 1978. Activité bactériostatique d'huiles essentielles de *Cannabis sativa* L. *Annales Pharmaceutiques Françaises* 36:603-605.
Francis RG, Burgess LW. 1977. Characteristics of two populations of *Fusarium roseum* 'Graminearum' in eastern Austrailia. *Trans. Br. Mycol. Soc.* 68(3):421-427.
Frank M. 1988. *Marijuana Grower's Insider's Guide*. Red Eye Press, Los Angeles, CA. 371 pp.
Frank M, Rosenthal E. 1978. *Marijuana Grower's Guide*. And/Or Press, Berkeley, CA. 330 pp.
Freeman GL. 1983. Allergic skin test reactivity to marijuana in the southwest. *Western J. Medicine* 138:829-31.
Frezzi MJ. 1956. Especies de *Pythium* fitopatogenas identificadas en las República Argentina. *Rev. Invest agric. Buenos Aires* 10 (2):113-241.
Friedrich MJ. 1999. Pesticide study aids Parkinson research. *JAMA* 282:2200.
Fritzsche R. 1959. Der Schattenwickler als Schädling an Lein und Hanf. *Wiss. Z. Univ. Halle (Math. et Nat.)* 8(6):1117-1119.
Fryday SL, Hart ADM, Dennis NJ. 1994. Effects of exposure to an organophosphate on the seed-handling efficiency of the house sparrow. *Bulletin Environmental Contamination & Toxicology* 53:869-876.
Fuller WH, Norman AG. 1944. The nature of the flora on field-retting of hemp. *Soil Sci. Soc. Amer., Proceedings* 9:101-105.
Fuller WH, Norman AG. 1945. Biochemical changes involved in the decomposition of hemp bark by pure cultures of fungi. *J. Bact.* 50:667-671.
Fulton BB. 1924. The European earwig. *Virginia Agricultural Experiment Station Bulletin* No. 294. 29 pp.
Furr M, Mahlberg PG. 1981. Histochemical analyses of laticifers and glandular trichomes in *Cannabis sativa*. *J. Natural Products* 44:153-159.
Fuller WH, Norman AG. 1946. Biochemical changes accompanying retting of hemp. *Iowa Agr. Exp. Sta., Research Bulletin* 344:927-944.
Fuog D, Fergusson SJ, Flückiger C. 1998. "Pymetrozine: a novel insecticide affecting aphids and whiteflies," pp. 40-49 in *Insecticides with Novel Modes of Action*. I Ishaaya, D Degheele, eds. Springer, Berlin. 289 pp.
Gagné RJ. 1995. Revision of tetranycid (Acarina) mite predators of the genus *Feltiella* (Diptera: Cicidomyiidae). *Annals Entomological Society America* 88:16-30.
Gal IE, Vajda O, Bekes I. 1969. A kannabidiolsav néhány tulajdonságának vizsgálata élelmiszertartósítási szempontból. *Elelmiszervizsgalati Közlemenyek* 4:208-216.
Galloway LD. 1937. Report of the imperial mycologist. *Annual Science Report Agricultural Research Institute, New Delhi*, 1935-1936:105-111.
Gamalitskaia NA. 1964. Micromycetes of the Southwestern part of the Central Tien Shan Mountains. *Akad. Nauk Kirghiz S.S.R.* 175 pp.
Gams W. 1982. Generic names for synanamorphs? *Mycotaxon* 15:456-464.
Gams W, Meyer W. 1998. What exactly is *Trichoderma harzianum*? *Mycologia* 90:904-915.
Ganter G. 1925. The determination of plant diseases transmitted by seed. *Rept. Fourth International Seed Testing Congress*, Moscow, USSR. English abstract pp. 110-114.
Garman H. 1903. The broom-rape of hemp and tobacco. *Kentucky Agricultural Experiment Station Bulletin* No. 105. 32 pp.
Garrett L. 1994. *The Coming Plague*. Penguin Books, NY. 750 pp.
Gautam RD. 1994. Survival of aphidophagous ladybird (*Coccinella septempunctata* L.) on non-aphid hosts together with its natural enemy complex. *Annals Agricultural Research* 15:71-75.

Georghiou GP, Papadopoulos C. 1957. A second list of Cyprus fungi. *Cyprus Dept. Agr. Tech. Bull.* TB-5. 38 pp.
Ghaffer A, Zentmeyer GA, Erwin DE. 1969. Effects of organic amendments on severity of *Macrophomina* root rot of cotton. *Phytopathology* 59:1267-1269.
Ghani M, Basit A. 1975. *Investigations on the Natural Enemies of Marijuana, Cannabis sativa L. and Opium Poppy, Papaver somniferum L.* Annual Report, Commonwealth Institute of Biological Control, Pakistan station. 8 pp.
Ghani M, Basit A. 1976. *Investigations on the Natural Enemies of Marijuana, Cannabis sativa L. and Opium Poppy, Papaver somniferum L.* Annual Report, Commonwealth Institute of Biological Control, Pakistan station. 10 pp.
Ghani M, Basit A, Anwar M. 1978. *Final Report: Investigations on the Natural Enemies of Marijuana, Cannabis sativa L. and Opium Poppy, Papaver somniferum L.* Commonwealth Institute of Biological Control, Pakistan station. 26 pp. + 12 illus.
Ghillini CA. 1951. I parassiti nemici vegetali della canapa. *Notiz. sulle Malatt. delle Piante* 15:29-36.
Gieringer D. 1996. Marijuana research: waterpipe study. *MAPS [Multidisciplinary Association for Psychedelic Studies) Bulletin* 6(3):59-66.
Gilkeson LA. 1997. "Ecology of rearing: quality, regulation, and mass rearing," pp. 139-148 in *Ecological Interactions and Biological Control.* DA Andow, DW Ragsdal, RF Nyvall, eds. Westview Press, Boulder, CO. 334 pp.
Gill EW, Jones G. 1972. Brain levels of tetrahydrocannabinol and its metabolites in mice—correlation with behaviour, and the effect of the metabolic inhibitors SKF 525A and piperonyl butoxide. *Biochemical Pharmacology* 21:2237-48.
Gill S, Sanderson J. 1998. *Ball Identification Guide to Greenhouse Pests and Beneficials.* Ball Publishing, Batavia, IL. 244 pp.
Gillan R, Cole MD, Linacre A, Thorpe JW, Watson ND. 1995. Comparison of *Cannabis sativa* by random amplification of polymorphic DNA (RAPD) and HPLC of cannabinoids: a preliminary study. *Science & Justice* 35:169-177.
Gilyarov MS. 1945. [A new and dangerous insect pest of the seeds of kok-saghyz and krym-saghyz.] *Dopov. Akad. Nauk URSR* 1945 (1-2):47-55.
Gitman LS. 1935. A list of fungi and bacteria on new bast crops in the USSR. *Za Novoe Volokno* 6:36-39.
Gitman LS. 1968a. "Bakterial'nye bolezni konopli," pp. 264-267 in *Bakter. bol. rast. i metod. bor's. nimi,* Kiev.
Gitman LS. 1968b. Maloizvestnye bolezni konopli. *Zashchita Rasteniî, Mosk.* 13(3):44-45.
Gitman L, Boytchenko E. 1934. "*Cannabis*" pp. 45-53 in *A Manual of the Diseases of the New Bast-fibre Plants.* Inst. New Bast Raw Materials, Moscow. 124 pp.
Gitman LS, Malikova TP. 1933. [Stem infection of hemp according to varieties], pp. 51-57 in *Bolyezni i Vredit. nov. lubyan. Kul'tur* [Diseases and Pests of Newly Cultivated Fibre Plants]. Institut Novogo Lubianogo Syriia [Institute of New Bast Raw Materials], Moscow.
Gladis T, Alemayehu N. 1995. Larven von *Acherontia atropos* L. (Lep., Sphingidae) neuerdings auch an Hanf (*Cannabis sativa* L.) - oder bislang übersehen? *Entomologische Nachrichten und Berichte* 39 (4):209-212.
Glazewska S. 1971. Rosliny zywicielskie grzyba *Peronospora humuli* (Miy. et Tak.) Skel. *Pamietnik Pulawski Prace Inst. Upr. Nowoz. Gleb.* 49:191-204.
Glendenning R. 1927. The cabbage flea beetle and its control in British Columbia. *Pamphlet No. 80, New Series,* Canada Department of Agriculture. 10 pp.
Glover T. 1869. "The food and habit of beetles," pp. 78-117 in *Report of the Commissioner of Agriculture,* 1868. Government Printing Office, Washington D.C.
Godfrey CRA, ed. 1995. *Agrochemicals from Natural Products.* Marcel Dekker, NY. 418 pp.
Goebel S, Vaissayer M. 1986. Mise au point d'un test rapide de résistance variétale à fusariose du cotonnier. *Coton et Fibres Tropicales* 41:63-65 [abstract in *Rev. Plant Pathol.* 65:5544].
Goidànich A. 1928. Contributi alla conoscenza dell'entomofauna della canapa. I. Prospetto generale [Contributions to the knowledge of the insect fauna of hemp. 1. General Survey]. *Bollettino del Laboratorio di Entomolgia del R. Istituto Superiore Agrario di Bologna* 1:37-64.
Goidànich G. 1955. *Malattie Crittogamiche della Canapa.* Associazione Produttore Canapa, Bologna-Naples. 21 pp.
Goidànich G. 1959. *Manual di Patologia Vegetale.* Edizioni Agricole, Bologna. 713 pp.
Goidànich G, Ferri F. 1959. La batteriosi della canapa da *Pseudomonas cannabina* Sutic & Dowson var. *italica* Dowson. *Phytopathologische Zeitschrift* 37:21-32.
Goldbold JC, Hawkins J, Woodward MG. 1979. Acute oral marijuana poisoning in the dog. *J. American Veterinary Medical Association* 175:1101-2.
Goloborod'ko PA. 1986. Defoliation and desiccation of hemp. *Zashchita Rastenii "Agropromizdat"* 1986(8):53.
Good R. 1953. *The Geography of Flowering Plants.* 2nd Ed. Longmans, Green & Co., London. 452 pp.
Goodman RM, Hauptli H, Crossway A, Knauf VC. 1987. Gene transfer in crop improvement. *Science* 236:48-54.
Goody JB, Franklin MT, Hooper DJ. 1965. *The Nematode Parasites of Plants Catalogued under their Hosts.* Commonwealth Agricultural Bureaux, Farnham Royal Co., Bucks, U.K. 214 pp.
Goriainov AA. 1914. *The Pests of Agricultural Plants in the District of Riazan.* Riazan Government Printing Office, Riazan, USSR. 67 pp.
Gorske SF, Hopen HJ, Randell R. 1976. Host specificity of the purslane sawfly and its response to selected pesticides. *HortScience* 11:580-582.
Gorske SF, Hopen HJ, Randell R. 1977. Bionomics of the purslane sawfly, *Schizocerella pilicornis. Annals Entomological Society America* 70:104-106.
Goswami BK, Vijayalakshmi K. 1986. Efficacy of some indigenous plant materials and oil cake amended soil on the growth of tomato and root-knot nematode population. *Ann. Agric. Res.* 7(2): 263-266.
Gould F. 1991. The evolutionary potential of crop pests. *American Scientist* 79:496-507.
Gould F. 1998. Sustainability of transgenic insecticidal cultivars: integrating pest genetics and ecology. *Annual Review Entomology* 43:701-726.
Goureau C. 1866. *Les Insectes Nuisibles a l'homme, aux animaux et a l'economie domestiques.* 2nd Supplement. V. Masson, Paris. 258 pp.
Greuter W *et al.* 1994. *International Code of Botanical Nomenclature (Tokyo Code).* Koeltz Scientific Books, Königstein. 389 pp.
Grewal PS. 1989. Effects of leaf-matter incorporation on *Aphelenchoides composticola* (Nematoda), mycofloral composition, mushroom compost quality and yield of *Agaricus bisporus. Annals Applied Biology* 115:299-312.
Grigoryev SV. 1998. Survey of the VIR *Cannabis* collection: resistance of accessions to corn stem borer (*Ostrinia nubilalis* Hb.). *J. International Hemp Association* 5(2):72-74.

Grime JP, Hodgson JG, Hunt R. 1988. *Comparative Plant Ecology*. Unwin Hyman Publishers, London. 742 pp.
Griswold GH. 1926. Notes on some feeding habits of two chalcid parasites. *Annals Entomological Society America* 19:331-334.
Groves WB. 1935. *British Stem and Leaf Fungi*. 2 vol. Cambridge University Press.
Gruyter J de, Noordeloos ME, Boerema GH. 1993. Contributions towards a monograph of *Phoma* (Coelomycetes) - I. 2. Section *Phoma:* additional taxa with very small conidia and taxa with conidia up to 7 µm long. *Persoonia* 15:369-400.
Guirao P, Moya A, Cenis JL. 1995. Optimal use of random amplified polymorphic DNA in estimating the genetic relationship of four major *Meloidogyne* spp. *Pytopathology* 85:547-551.
Gupta RC. 1985. A new leaf spot disease of *Cannabis sativa* caused by *Phomopsis cannabina* Curzi. *Current Science* 54(24):1287.
Gupta SK. 1985. *Plant Mites of India*. Government of India Press, Calcutta. 520 pp.
Gupta RP, Singh A. 1983. Effect of certain plant extracts and chemicals on teliospore germination of *Neovossia indica*. *Indian J. Mycology & Plant Pathology* 13(1):116-117.
Gutberlet V, Karus M. 1995. *Parasitäre Krankheiten und Schädlinge an Hanf* (*Cannabis sativa*). Nova Institut, Köln, Germany. 57pp.
Gutner LS. 1933. [New diseases of hemp, okra and Spanish broom], p. 71 in *Bolyezni i Vredit. nov. lubyan. Kul'tur* [Diseases and Pests of newly cultivated Fibre Plants]. Institut Novogo Lubianogo Syriia [Institute of New Bast Raw Materials], Moscow.
Gzebenyuk NV. 1984. The occurrence of fungi on hemp stems. *Mikologiya i Fitopathologiya* 18(4):322-326.
Hackleman JC, Domingo WE. 1943. Hemp: an Illinois war crop. *University of Illinois Agricultural Experiment Station Circular* No. 547. Urbana, Illinois. 8 pp.
Hajek AE. 1993. "New options for insect control using fungi," pp. 54-62 in *Pest Management: Biologically Based Technologies*. RO Lumsden, JL Vaughn, eds. American Chemical Society, Washington, D.C. 435 pp.
Hamilton E. 1852. "*Cannabis sativa*." pp. 133-142 in *Flora Homoeopathica,* Vol. I. Bailliere Publ. Co., London. 300 pp.
Hammond CT, Mahlberg PG. 1977. Morphogenesis of capitate glandular hairs of *Cannabis sativa* (Cannabaceae). *Amer. J. Bot.* 64:1023-1031.
Hancer JH. 1992. Buds of paradise. *High Times* No. 199: 48-50.
Haney A, Bazzaz FA. 1970. "Some ecological implications of the distribution of hemp (*Cannabis sativa* L.) in the United States of America." pp. 39-48 in *The Botany and Chemistry of Cannabis*. CRB Joyce & SH Curry, eds. J & A Churchill, London. 217 pp.
Haney A, Kutscheid BB. 1973. Quantitative variation in the chemical constituents of marihuana from stands of naturalized *Cannabis sativa* L. in East-Central Illinois. *Economic Botany* 27:193-203.
Haney A, Kutscheid BB. 1975. An ecological study of naturalized hemp (*Cannabis sativa* L.) in east-central Illinois. *American Midland Naturalist* 93:1-24.
Hanson AA. 1990. *CRC Practical Handbook of Agricultural Science*. CRC Press, Boca Raton, FL. 534 pp.
Hanson HC. 1963. *Diseases and Pests of Economic Plants of Vietnam, Laos and Cambodia*. American Institute of Crop Ecology, Washington D.C. 155 pp.
Hanson HC, Kossack CW. 1963. The mourning dove in Illinois. *Ill. Dept. Conserv. Tech. Bull.* No.2. 133 pp.
Hara AH, Kaya HK, Gaugler R, LeBeck LM, Mello CL. 1993. Entomopathogenic nematodes for biologial control of the leafminer *Liriomyza trifolii*. *Entomophaga* 38:359-369.
Harada T. 1930. On the insects injurious to hemp, especially *Rhinocus pericarpius*. *Konchu Sekae [Insect World]* 34:118-123.
Harlan JP, DeWet JM, Stemler AB. 1976. *Origins of African Plant Domestication*. Mouton Publishing Company, The Hague. 498 pp.
Harman GE, Norton JM, Stasz TE. 1987. Nyolate seed treatment of *Brassica* spp. to eradicate or reduce black rot caused by *Xanthomonas campestris* pv. *campestris*. *Plant Disease* 71:27-30.
Hartowicz LE, Knutson H, Paulsen A, *et al.* 1971. Possible biocontrol of wild hemp. *North Central Weed Control Conference, Proceedings* 26:69.
Hartowicz LE, Eaton BJ. 1971. Reducing the impact of wild hemp control on farm game. *North Central Weed Control Conference, Proceedings* 26:70.
Harukawa C, Kondo S. 1930. On *Hylemyia cilicrura* Rondani. Nogaku Kenkyu 14:449-469.
Harvey JV. 1925. A study of the water molds and pythiums occuring in the soils of Chapel Hill. *J. Elisha Mitchell Sci. Soc.* 41:151-164.
Haseeb A, Singh B, Khan AM, Saxena SK. 1978. Evaluation of nematicidal property in certain alkaloid-bearing plants. *Geobios [India]* 5:116-118.
Hawksworth DL, Kirk PM, Sutton BC, Pegler DN. 1995. *Dictionary of the Fungi,* 8th Ed. CAB International, Wallingford, UK. 616 pp.
Hayman DS. 1982. Influence of soils and fertility on activity and survival of VA mycorrhizal fungi. *Phytopathology* 72:1119-1125.
Headrick DH, Bellows TS, Perring TM. 1995. Behaviors of *Eretmocerus* sp. nr. *californicus* (Hymenoptera: Aphelinidae) attacking *Bemisia argentifolii* (Homoptera: Aleyrodidae) on sweet potato. *Environmental Entomology* 24:412-422.
Hector GP. 1931. Bengal Department of Agriculture, *Annual Report for the Year 1930-31,* pp. 35-44.
Heinz KM, Nelson JM. 1996. Interspecific interactions among natural enemies of *Bemisia* in an inundative biological control program. *Biological Control* 6:384-393.
Helle W. 1962. Genetics of resistance to organosphosphorus compounds and its relation to diapause in *Tetranychus urticae* Koch (Acari). *Tijdschrift over Plantenziekten* 68:155-195.
Helle W, Sabelis MW, eds. 1985. *Spider Mites: their Biology, Natural Enemies, and Control*. Elseivier, Amsterdam. Vol 1A, 405 pp.; Vol 1B, 458 pp.
Hembra RL. 1993. Pesticide reregistration may not be completed until 2006. *Government Accounting Office Memo* GAO/RCED-93-94. 37 pp.
Hendel F. 1932. "59. Agromyzidae" pp. 1-570 in *Die Fliegen der Paläearktishen Region,* Band 6(2), E. Lindner, Ed. Stuttgart.
Hendriks H, Malingré TM, Batterman S, Bos R. 1975. Mono- and sesqui-terpene hydrocarbons of the essential oil of *Cannabis sativa*. *Phytochemistry* 14:814-815.
Hendrischke K, Lickfett T, Buttlar H-B von. 1998. Hemp: a ground water protecting crop? Yields and nitrogen dynamics in plant and soil. *J. International Hemp Association* 5(1):24-28.
Hennink S, de Meijer EPM, van Soest LJM, van der Werf HMG. 1993. "Rassenperspectief van hennep ten aanzien van opbrengst, produktkwaliteit en EG-subsidies," pp. 108-120 in *Papier uit hennep van Nederlandse grond,* JM van Berlo, ed. ATO-DLO, Wageningen, The Netherlands.

Herer J. 1985. *The Emperor Wears No Clothes.* Queen of Clubs Publishing Co. (Homestead Book Co.), Seattle. 125 pp.

Herer J. 1991. *The Emperor Wears No Clothes,* Revised Edition. HEMP Press, Van Nuys, CA. 182 pp.

Hering EM. 1951. *Biology of the Leaf Miners.* Junk Publ., Gravenhage, The Netherlands. 420 pp.

Hering M. 1937. *Die Blattminen Mittel- und Nord-Europas.* Verlag Gustav Feller, Neubranddenburg. 631 pp.

Herodotus. 1906 Reprint. *Herodotus IV (Melpomene).* University Press, Cambridge.

Heslop-Harrison J. 1957. The experimental modification of sex expression in flowering plants. *Biol. Review* 32:1-52.

Hewitt HG. 1998. *Fungicides in Crop Protection.* CAB International, Wallingford, UK. 221 pp.

Hilbeck A, Baumgartner M, Fried PM, Bigler F. 1998. Effects of transgenic *Bacillus thuringiensis* corn-fed prey on mortality and development time of immature *Chrysoperla carnea* (Neuroptera: Chrysopidae). *Environmental Entomology* 27:480-487.

Hildebrand DC, McCain AM. 1978. The use of various substrates for large scale production of *Fusarium oxysporum* f. sp. *cannabis* inoculum. *Phytopathology* 68: 1099-1101.

Hill DS. 1983. *Agricultural Insect Pests of the Tropics and their Control,* 2nd Ed. Cambridge University Press, Cambridge. 332 pp.

Hill DS. 1994. *Agricultural Entomology.* Timber Press, 133 SW 2nd Ave., Suite 450, Portland, OR. 635 pp.

Hirata K. 1966. *Host Range and Geographical Distribution of the Powdery Mildews.* Niigata University Press, Niigata, Japan. 472 pp.

Hockey JF. 1927. *Report of the Dominion field laboratory of plant pathology, Kentville, Nova Scotia.* Canada Department of Agriculture, pp. 28-36.

Hoerner GR. 1940. The infection capabilities of hop downy mildew. *J. Agric. Res.* 61:331-334.

Hoffman GM. 1958. Das Auftreten einer Anthraknose des Hanfes in Mecklenburg und Brandenburg. *Nachrichtenbl. Deutsch. Pflanzenschutz-dienst NF* 12:96-99.

Hoffman GM. 1959. Untersuchungen über die Anthraknose des Hanfes (*Cannabis sativa* L.) *Phytopathologische Zeitschrift* 35:31-57.

Hoffmann MP, Frondsham AC. 1993. *Natural Enemies of Vegetable Insect Pests.* Cooperative Extension Bulletin, Cornell University, Ithaca, NY. 63 pp.

Höhnel F. 1915. Fragmente zur mykologie XVII. Sitzungsberichten de Kaiserl. *Akademie der Wissenschaften in Wien* I(124:1) p. 24 [72].

Hokkanen HMT. 1991. Trap cropping in pest management. *Annual Review Entomology* 36:119-138.

Holland WJ. 1937. *The Moth Book.* Doubleday, Doran & Co., Inc. Garden City, NY.

Holliday P. 1980. *Fungus Diseases of Tropical Crops.* Cambridge University Press, London. 607 pp.

Holm L. 1957. Etudes taxonomiques sur les Pléosporacées. *Symb. Bot. Upsal.* 14:1-188.

Holmes FO. 1939. Proposal for extension of the binomial system of nomenclature to include viruses. *Phytopathology* 29:431-436.

Hood LVS, Dames ME, Barry GT. 1973. Headspace volatiles of marijuana. *Nature* 242:402-403.

Horowitz M. 1977. Herbicidal treatments for control of *Cannabis sativa* L. *Bulletin on Narcotics* 29(1):75-84.

Houten YM van, Rijn PCJ van, Tanigoshi LK, Stratum P van, Bruin J. 1995. Preselection of predatory mites to improve year-round biological control of western flower thrips in greenhouse crops. *Entomologia Experimentalis et Applicata* 74:225-234.

Howard AJ. 1943. *An Agricultural Testament.* Oxford University Press, London. 253 pp.

Howard RJ, Garland JA, Seaman WL, eds. 1994. *Diseases and Pests of Vegetable Crops in Canada.* Entomological Society of Canada, Ottawa, Ontario. 554 pp.

Howarth FG. 1991. Environmental impacts of classical biological control. *Annual Review Entomology* 36:485-509.

Howse PE, Stevens IDR, Jones OT. 1998. *Insect Pheromones and their Use in Pest Management.* Chapman & Hall, London. 369 pp.

Huang F, Buschman LL, Higgins RA, McGaughey WH. 1999. Inheritance of resistance to *Bacillus thuringiensis* toxin (Dipel ES) in the European corn borer. *Science* 284:965-967.

Hunter-Fujita FR, Entwistle PF, Evans HF, Crook NE. 1998. *Insect Viruses and Pest Management.* John Wiley & Sons, NY.

Hussey NW, Scopes N. 1985. *Biological Pest Control: the Glasshouse Experience.* Blandford Press, Poole, Dorset, UK. 240 pp.

Indian Hemp Drugs Commission. 1894. *Report of the Indian Hemp drugs commission.* Government Printing Office, Simla, India (reprinted 1969 by Thos. Jefferson Publ. Co., Silver Springs, MD).

Imms AD. 1948. *A General Textbook of Entomology,* 7th Edition. E.P. Dutton & Co., NY. 727 pp.

Israel S. 1981. An in-depth plant companionship chart. *Mother Earth News* 69:94-95.

Jackson D. 1983. Controlling spider mites with predatory mites. *Sinsemilla Tips* 3(4):11-13.

Jackson WR. 1993. *Humic, Fulvic and Microbial Balance: Organic Soil Conditioning.* Jackson Research Centre, Evergreen, CO. 958 pp.

Jaczewski AA. 1927. Mucnisto-rosjanye Griby [Powdery-mildew Fungi]. *Karmannyj Opreditel' Gribov* 5(2):412.

Jagadish V, Robertson J, Gibbs A. 1996. RAPD analysis distinguishes *Cannabis sativa* samples from different sources. *Forensic Science International* 79:113-121.

Jain MC, Arora N. 1988. Ganja (*Cannabis sativa*) refuse as cattle feed. *Indian J. Animal Sci.* 58:865-867.

Jalees S, Sharma SK, Rahman SJ, Verghese T. 1993. Evaluation of insecticidal properties of an indigenous plant, *Cannabis sativa* L., against mosquito larvae under laboratory conditions. *J. Entomol. Res.* 17:117-120.

Jarvis WR. 1990. *Managing Diseases in Greenhouse Crops.* APS Press, St. Paul, MN. 228 pp.

Jeffries P, Young TWK. 1994. *Interfungal Parasitic Relationships.* CAB International Press, Egham, Surrey, UK. 296 pp.

Jensen JD. 1983. The development of *Diaporthe phaseolorum* variety *sojae* in culture. *Mycologia* 75:1074-1091.

Jeyasingam JT. 1988. A summary of special problems and considerations related to non-wood fiber pulping world wide. *Proceedings 1988 Pulping Conference* 3:571-579.

Johnson J. 1937. Relation of water-soaked tissues to infection by *Bacterium angulatum* and *B. tabacum* and other organisms. *J. Agricultural Research* 55(8):599-618.

Johnston A. 1964. Additional records of plant nematodes in the south east Asia and Pacific region. *FAO Plant Protect. Comt. So. East Asia & Pacific Region, Tech. Doc.* 30. 6pp.

Joly P. 1964. *Le Genre Alternaria*. Editions Paul Lechevalier, Paris. 250 pp.

Jones JB. 1998. *Plant Nutrition Manual*. CRC Press, Boca Raton, FL. 149 pp.

Jurkowska H, Rogóz A, Wojciechowicz T. 1990. The content of lithium in some species of plants following differentiated doses of nitrogen. *Polish J. Soil Sci.* 23:195-199.

Kabelík J, Krejcí Z, Santavy F. 1960. *Cannabis* as a medicament. *UN Bulletin on Narcotics* 12 (3):5-23.

Kac M. 1976. Preskusanje akaricidov v hmeljiscih [Testing of acaricides in hemp fields]. *Sodobno Kmetijstvo* 9:301-303.

Kagen S. 1981. *Aspergillus:* an inhalable contaminant of marijuana. *New England J. Med.* 304:483-484.

Kagen S, Kurup VP, Sohnle PG, Fink JN. 1983. Marijuana smoking and fungal sensitization. *J. Allergy Clin. Immunol.* 71:389-393.

Kaltenbach JH. 1874. *Die Pflanzenfeinde aus der Klass der Insekten*. Julius Hoffmann, Stuttgart. 848 pp.

Kaneshima H, Mori M, Mizuno N. 1973. Studies on *Cannabis* in Hokkaido (Part 6). The dependence of *Cannabis* plants on iron nutrition. *Hokkaidoritsu Eisei Kenkyushoho* 23:3-5.

Kapoor LD. 1990. *CRC Handbook of Ayurvedic Medicinal Plants*. CRC Press, Boca Raton, FL. 416 pp.

Kapralov SI. 1974. [*Phytomyza* against broomrape.] *Zernovoe Khozyaistvo* 13(7):43-44.

Karus M. 1997. "Newsflash hemp," pg 11 in *Symposium Magazin*, 2nd Biorohstoff Hanf Technisch-wissenschaftliches Symposium. Nova Institut, Köln, Germany.

Kashyap NP, Bhagat RM, Sharma DC, Suri SM. 1992. Efficacy of some useful plant leaves for the control of potato tuber moth, *Phthorimaea operculella* Zell. in stores. *J. Entomological Research* 16:223-227.

Kashyap RK, Kennedy GG, Farrar RR. 1991. Behavioral response of *Trichogramma pretiosum* and *Telenomus sphingis* to trichome/methyl ketone mediated resistance in tomato. *J. Chemical Ecology* 17:543-556.

Kaushal RP, Paul YS. 1989. Inhibitory effects of some plant extracts on some legume pathogens. *Legume Research* 12:131-132.

Ke LS, Xin JL. 1983. [Notes on three new species of the genus *Typhlodromus* (Acari: Phytoseiidae).] *Entomotaxonomia* 5:185-188.

Kees. 1988. *The Marijuana Seed Catalog, '88-'89*. Postbus 1942, 1000BX, Amsterdam. 15 pp.

Kegler H, Spaar D. 1997. Kurzmitteilung zur Virusanfälligkeit von Hanfsorten (*Cannabis sativa* L.). *Archives Phytopathologie & Pflanzenschutz* 30:457-464.

Keijzer P, Lubberts JH, Stripper H. 1990. Evaluation of the growth retardant triapenthol on hemp. *Annals Applied Biology* 116 (Supplement):72-73.

Kelton LA. 1978. *The Insects and Arachnids of Canada. Part 4: The Anthocoridae of Canada and Alaska*. Research Publication No. 1639, Canadian Department of Agriculture, Ottawa, Canada. 101 pp.

Kendrick B. 1985. *The Fifth Kingdom*. Mycologue Publications, Waterloo, Canada. 364 pp.

Kennedy JS. 1951. A biological approach to plant viruses. *Nature* 168:890-894.

Kennedy JS, Booth CO, Kershaw WJS. 1959. Host finding by aphids in the field. *Annals Applied Biology* 47:424-444.

Khamukov VB, Kolotilina ZM. 1987. We are extending utilization of the biological method. *Zashchita Rasteniî* 4:30-31.

Khare BP, Gupta SB, Chandra S. 1974. Biological efficacy of some plant materials against *Sitophilus oryzae* L. *Indian J. Agric. Res.* 8:243-248.

Killough DT. 1920. Unpublished letter of November 16, from USDA Substation No. 5, Temple, TX to USDA headquarters, Beltsville, MD.

King J. 1854. *The American Eclectic Dispensatory*. Moore, Wilstach & Keys, Cincinnati, OH. 1390 pp.

Kirchner O. 1906. "Hanf, *Cannabis sativa* L." pp. 319-323 in *Die Krankheiten und Befehädigungen uhferer landwirtschaftlichen Kulturpflanzen*. E. Ulmer, Stuttgart. 637 pp.

Kirchner O, Boltshauser H. 1898. Blattflecken und Blattminen an Hanf. *Atlas der Krankheiten und Beschädigungen*. Serie 3, Tafel 20. Verlag von Ulmer, Stuttgart.

Kirchner HA. 1966. *Phytopathologie und Pflanzenschutz*. Vol. 2. Akademie Verlag, Berlin. 617 pp.

Kir'yanova ES, Krall EL. 1971. *Plant-Parasitic Nematodes and their Control*, Vol. II. Academy of Sciences of the USSR, Nauka Publishers, Leningrad.

Kishi K. 1988. *Plant Diseases in Japan*. Zenkoku Noson Kyoiku Kyokai Co., Tokyo. 943 pp.

Klement Z, Kaszonyi S. 1960. *A kender új baktériumos betegsége*. Növénytermelés 9(2):153-158.

Klement Z, Lovrekovich L. 1960. Comparative study of *Pseudomonas* species affecting hemp and the mulberry tree. *Acta Microbiol. Acad. Sci. Hungary* 7(2):113-119.

Klingeren B van, Ham MT. 1976. Antibacterial activity of delta-9-tetrahydrocannabinol and cannabidiol. *Antonie van Leeuwenhoek* 42:9-12.

Knipling EF. 1992. Principles of insect parasitism analyzed from new perspectives. USDA, Washington, D.C.

Koehler B. 1946. Hemp seed treatments in relation to different dosages and conditions of storage. *Phytopathology* 36:937-942.

Kok CJ, Coenen GCM. 1996. Host suitability of alternative oilseed and fiber crops to *Pratylenchus penetrans*. *Fundamental & Applied Nematology* 19:205-206.

Kok CJ, Coenen GCM, de Heij A. 1994. The effect of fibre hemp (*Cannabis sativa* L.) on selected soil-borne pathogens. *J. International Hemp Association* 1(1):6-9.

Koo M. 1940. *Studies on Pyrausta nubilalis Hüber attacking the cotton plant*. Unnumbered report, Yamanashi Agricultural Experiment Station, Kofu. 82 pp.

Koppert BV. 1998. *Koppert Webpage: Biological Systems & Products*. www.kopert.nl/english.

Koppert BV. 1999. *Neveneffectsen Gids/Side Effects Guide*. Koppert Press, Berkel en Rodenrijs, the Netherlands. 50 pp.

Korf RP, editor. 1996. *Fungi of China*, by S.C. Teng. Mycotaxon, Ltd. Ithaca, NY. 586 pp.

Kosslak RM, Bohlool BB. 1983. Prevalence of *Azospirillum* spp. in the rhizosphere of tropical plants. *Canadian J. Microbiology* 29:649-652.

Kotev S, Georgieva M. 1969. Kimicheski sredstva za obezzaraza vane semenata na konopa sreshtu bakteriozata. *Zashchita Rasteniî* 17(1):25-29.

Kovacevic Z. 1929. Ueber die wichtigsten Schädlinge der Kulturpflanzen in Slawonien und Backa. *Verh. deuts. Ges. angew. Ent., 7. Mitgliederversamml. München*, 31. Mai-2. Juni 1928, pp.33-41.

Kozhanchikov IV. 1956. Fauna SSSR. *Nasekomyl Cheshuekrylye* 3(2):444. Zoologicheskii Institut Akademii Nauk SSSR.

Kozlowski R. 1995. Interview with Professor R. Kozlowski, director of the Institute of Natural Fibres. *J. International Hemp Association* 2(2):86-87.

Krebs H, Jäggi W. 1999. Pflanzenextrakte gegen Bakterien-Nassfäule der Kartoffeln. *Agrarforschung* 6(1):17-20.

Krejcí Z. 1950. Doctoral dissertation from the Faculty of Natural Sciences, University of Brno, Czech Republic.

Kreutz C. 1995. *Orobanche: die Sommerwurzarten Europas: ein Bestimmungsbuch. 1: Mittel-und Nordeuropa.* Balkema, Rotterdam. 159 pp.

Krishna A. 1995. Chemical control of root rot of ganja. *Current Research - University of Agricultural Sciences (Bangalore)* 24(6):99-100.

Krishna A. 1996. Resistance in ganja cultivars to *Sclerotium rolfsii. Indian J. Mycology & Plant Pathology* 26:307.

Krustev VP. 1957. A new pest of hemp, *Mordellistena parvula* Gyll., in Bulgaria. *Sofia Bulletin Plant Protection* 5(1):87-88.

Kryachko Z, Ignatenko M, Markin A, Zaets V. 1965. Notes on the hemp tortrix. *Zashchita Rasteniî Vredit. Bolez.* 5:51-54.

Kulagin NM. 1915. Insects injurious to cultivated field plants in European Russia in 1914. *Bulletin Moscow Entomological Society* 1:136-161.

Kuoh CL, Huang CL, Tian LX, Ding JH. 1980. [New species and new genus of Delphacidae from China.] *Acta Entomologica Sinica* 23:413-426.

Kurita N, Miyaji M, Kurane R, Takahara Y. 1981. Antifungal activity of components of essential oils. *Agric. Biol. Chemistry* 45:945-952.

Kurup VP, Resnick R, Kagen SL, Cohen SH, Fink JN. 1983. Allergenic fungi and actinomycetes in smoking materials and their health implications. *Mycopathologia* 82:61-64.

Kyokai NSB. 1965. *Narin byogauchu meikan* [Major pests of Economic Plants in Japan]. Japan Plant Protection Association, Tokyo. 412 pp.

Lago PK, Stanford DF. 1989. Phytophagous insects associated with cultivated marijuana, *Cannabis sativa,* in northern Mississippi. *J. Entomological Science* 24:437-445.

Lai T van. 1985. Effects of inbreeding on some major characteristics of hemp. *Acta Agronomica Academiae Scientiarum Hungaricae* 34:77-84.

Lancz G, Specter S, Brown HK. 1990. Suppressive effect of delta-9-tetrahydrocannabinol on herpes simplex virus infectivity *in vitro. Proceedings Society Experimental Medicine & Biology* 196:401-404.

Lancz G, Specter S, Brown HK, Hackney JF, Friedman H. 1991. Interaction of delta-9-tetrahydrocannabinol with herpesviruses and cultural conditions associated with drug-induced anti-cellular effects. *Advances Experimental Medicine & Biology* 288:287-304.

LaRosa FG, Clarke SH, Lefkowitz JB. 1997. Brodifacoum intoxication with marijuana smoking. *Archives Pathology & Laboratory Medicine* 121:67-69.

Laskowska R. 1961. Influence of the age of pollen and stigmas on sex determination in hemp. *Nature* 192:147-148.

Lassen G. 1988. The great outdoors. *Sinsemilla Tips* 8(1):43-48.

Latimer D. 1996. Will success spoil hemp? A debate over industrial-grade *Cannabis. High Times* 249:26-29, 40.

Latta RP, Eaton BJ. 1975. Seasonal fluctuatons in cannabinoid content of Kansas marijuana. *Economic Botany* 29:153-163.

Lawi-Berger C, Wüest J, Tourmel A, Miège J. 1984. Ontogenèse et morphologie des laticifères de *Cannabis sativa* L. *Comptes Rendus de l'Academie des sciences Paris,* Série III. 229(8):327-332.

Lekic M. 1974. [Investigation of the dipteran *Phytomyza orobanchia* Kaltb. as a controller of parasitic phanerogams of the genus Orobanche.] *Savremena Poljoprivreda* 22:93-99.

Lemeshev N, Rumyantseva L, Clarke RC. 1995. Report on the maintenance of hemp (*Cannabis sativa* L.) germplasm accessioned in the Vavilov Research Institute gene bank. *J. International Hemp Association* 2(1):10-13.

Lenicque PM, Paris MR, Poulot M. Effects of some components of *Cannabis sativa* on the regenerating planarian worm *Dugesia tigrina. Experientia* 28:1399-1400.

Lentz P, Turner CE, Robertson LW, Genter WA. 1974. First North American record for *Cercospora cannabina,* with notes on the identification of *C. cannabina* and *C. cannabis. Plant Disease Reporter* 58:165-168.

Lentz P. 1977. Fungi and diseases of *Cannabis.* Unpublished manuscript, notes and collected literature. National Fungus Collections, USDA Bldg 011A ("the bunker"), Beltsville, MD.

Lesik BV. 1958. *Priemy povysheniia kachestva lubianogo volokna.* Gos. isz-vo khoz., Moskva. 230 pp.

Leslie AR. 1994. *Handbook of Integrated Pest Management for Turfgrass and Ornamentals.* CRC Press, Boca Raton, FL.

Lesne P. 1920. Une ancienne invasion du "Botys du millet" (*Pyrausta nubilalis* Hb.) en France. *Bulletin Société de pathologie Végétale de France* 7(1):15-16.

Lewis WJ, Nordlund DA, Gueldner RC, Teal PEA, Tumlinson JH. 1982. Kairomones and their use for management of entomophagous insects, XIII. *J. Chemical Ecology* 8:1323-1331.ß

Lewis WL, Sheehan W. 1997. "Parasitoid foraging from a multitrophic perspective: significance for biological control," pp. 271-281 in *Ecological Interactions and Biological Control.* DA Andow, DW Ragsdale, RF Nyvall, eds. Westview Press, Boulder, CO. 334 pp.

Levin DA. 1973. The role of trichomes in plant defense. *Quarterly Review Biology* 48:3-15.

Levitz SM, Diamond RD. 1991. Aspergillosis and marijuana. *Annals Internal Medicine* 115:578-579.

Li HL. 1974. An archaeological and historical account of *Cannabis* in China. *Economic Botany* 28:437-448.

Li LY. 1994. "Worldwide use of *Trichogramma* for biological control in different crops," pp. 37-53 in *Biological Control with Egg Parasitoids,* E Wajnberg, SA Hassan, eds. CAB International, Wallingford, UK. 286 pp.

Liebermann J. 1944. Sobre la importancia económica de las especies chilenas del género *Dichroplus* Stål (Orth. Acrid. Cyrtacanth.), con algunas consideraciones acerca de su biogeografia. *Rev. Chil. Hist. Nat.* 47:241-247.

Linacre A, Thorpe J. 1998. Detection and identification of *Cannabis* by DNA. *Forensic Science International* 91:71-76.

Lind J. 1913. *Danish Fungi.* Nordisk Forlag, Copenhagen. 648 pp.

Lindeman K. 1882. *Konopliannyi zhuk* [The hemp beetle]. Ministry of Agriculture and Government Estates, St. Petersburg. 5 pp.

Linderman RG, Paulitz TC, Mosier NJ, Griffiths RP, Loper JE, Caldwell BA, Henkels ME. 1991. "Evaluation of the effects of biocontrol agents on mycorrhizal fungi," pp. 379 in *The Rhizosphere and Plant Growth,* DL Keister, PB Cregan, eds. Beltsville Symposia in Agricultural Research, Symposium No. 14. Academic Publishers, London. 386 pp.

Lindemayr H von, Jäger S. 1980. Beruflich erworbene Typ I-Allergie durch Hanfpollen und Haschisch. *Dermatosen* 28:17-19.

Lindley J. 1838. *Flora Medica.* Spottiswoode Printers, London. 656 pp.

Lindquist EE, Sabelis MW, Bruin J, eds. 1996. *Eriophyoid Mites: Their Biology, Natural Enemies and Control.* Elsevier Science, Amsterdam. 790 pp.

Liskow B, Liss JL, Parker CW. 1971. Allergy to marihuana. *Annals Internal Medicine* 75:571-573.

Lisson SN, Mendham NJ. 1995. Tasmanian hemp research. *J. International Hemp Association* 2(2):82-85.

Lisson SN, Mendham NJ. 1998. Response of fiber hemp (*Cannabis sativa* L.) to varying irrigation regimes. *J. International Hemp Association* 5(1):9-15.

Litzenberger SC, Farr ML, Lip HT. 1963. *A Supplementary List of Cambodian Plant Diseases.* United States Agency for International Development to Cambodia, Special Publication. 18pp.

Liu G, Qian Y, Zhang P, *et al.* 1992. Etiological role of *Alternaria alternata* in human esophageal cancer. *Chinese Medical J.* 105:394-400.

Liu T-X, Stansly PA. 1995. Toxicity and repellency of some biorational insecticides to *Bemisia artgentifolii* on tomato plants. *Entomologica Experimentalis et Applicata* 74:137-143.

Llamas R, Hart DR, Schneider NS. 1978. Allergic bronchopulmonary aspergillosis associated with smoking moldy marijuana. *Chest* 73:871-872.

Llewellyn GC, O'Rear CO. 1977. Examination of fungal growth and aflatoxin production on marijuana. *Mycopathologia* 62:109-112.

Locke JC, Larew HG, Walter JF. 1993. "Efficacy of clarified neem seed oil against foliar fungal pathogens and greenhouse whiteflies," pp. 287-289 in *Pest Management: Biologically Based Technologies.* RO Lumsden, JL Vaughn, eds. American Chemical Society, Washington, D.C. 435 pp.

Loewe S. 1946. Studies on the pharmacology and acute toxicity of compounds with marihuana activity. *J. Pharmacology & Expermental Therapeutics* 88:154-164.

Loomans AJM, Lenteren JC van. 1995. Biological control of thrips pests: a review on thrips parasitoids. *Wagenigen Agricultural University Papers* 95.1:92-193.

Lopatin MI. 1936. Susceptibility of plants to *Bacterium tumefaciens. Mikrobiologiia* 5:716-724.

Lösel PM, Lindemann M, Scherkenback J, Campbell CA, Hardie J, Pickett JA, Wadhams LJ. 1996. Effect of primary-host kairomones on the attractiveness of the hop-aphid sex pheromone to *Phorodon humuli* males and gynoparae. *Entomologia Experimentalis* et Applicata 80:79-82.

Losey JE, Calvin DD. 1995. Quality assessment of four commercially available species of *Trichogramma. Journal Economic Entomology* 88:1243-1250.

Losey JE, Ives AR, Haron J, Ballantyne F, Brown C. 1997. A polymorphism maintained by opposite patterns of parasitism and predation. *Nature* 388:269-272.

Losey JE, Rayor LS, Carder ME. 1999. Transgenic pollen harms monarch larvae. *Nature* 399:214.

Lucas GB. 1975. *Diseases of Tobacco,* 3rd Ed. Biological Consulting Assoc., Raleigh, NC. 621 pp.

Lydon J, Teramura AH, Coffman CB. 1987. UV-B radiation effects on photosynthesis, growth and cannabinoid production of two *Cannabis sativa* chemotypes. *Photochemistry & Photobiology* 46(2):201-206.

MacIndoo NL, Stevers AF. 1924. *Plants tested for or reported to possess insecticidal properties.* USDA Department Bulletin No. 1201. 61 pp.

Mackie DB. 1918. Some aliens we do not want, why we do not want them, and how they may arrive. iii. The European cornstalk borer. *Monthly Bulletin California State Commission of Horticulture (Sacramento)* 7(9):541-544.

Mahr S. 1997. Know your friends: *Lydella thompsoni. Midwest Biological Control News Online,* vol. 6, No. 3.

Mahr S. 1998. Know your friends: *Macrocentrus ancylivorus. Midwest Biological Control News Online,* vol. 7, No. 7.

Mahr S. 1999. Mechanized delievery methods for field release of beneficial insects. *Midwest Biological Control News* 6(9):1-2, 6-7.

Malais M, Ravensberg WJ. 1992. *Knowing and Recognizing: The Biology and Glasshouse Pests and their Natural Enemies.* Koppert B.V., Berkel en Rodenrijs, Netherlands. 109 pp.

Malik P, Singh AB, Babu CR, Gangal SV. 1990. Head-high, airborne pollen grains from different areas of metropolitan Delhi. *Allergy* 45:298-305.

Maloney ES, Brodkey MH. 1940. Hemp pollen sensitivity in Omaha. *Nebraska State Medical J.* 25:190-191.

Mandolino G, Ranalli P. 1999. "Advances in biotechnological approaches for hemp breeding and industry," pp. 185-212 in *Advances in Hemp Research,* P. Ranalli, ed. Haworth Press, NY.

Mann C. 1991. Lynn Margulis: science's unruly earth mother. *Science* 252:378-381.

Margolis JS, Clorfene R. 1975. *A Child's Garden of Grass.* Ballantine Books, NY. 151 pp.

Marin AN. 1979. The biomethod — in the field! *Zashchita Rastenii* 11:24.

Marquart B. 1919. *Der Hanfbau.* Paul Parey Publishers, Berlin.

Marras GF, Spanu A, Attene G. 1982. Aspetti di technica colturale in canapa da cellulosa, scelta varietale. *Annali della Facolta di Agraria dell' Universita Sassari* 28:102-110.

Marsh PB, Bollenbacher K. 1949. The fungi concerned in fiber deterioration. *Textile Research Journal* 19(6):313-324.

Marshall GAK. 1917. *Review of Applied Entomology,* Series A (Agriculture) 5:159.

Marshmann JA, Popham RE, Yawney CD. 1976. A note on the cannabinoid content of Jamaican ganja. *Bulletin on Narcotics* 28(4):63-68.

Martelli M. 1940. "Gli insetti piu' esiziali alle colture di canapa," pp. 111-123 in *Lezioni al Corso di Perfezionamento per la Stima del Tiglio di Canapa.* Faenza-Fratelli Lega Editori, Roma. 332 pp.

Martyn T. 1792. *Flora Rustica,* Vol. I. Nodder Publ. Co., London. N.P.

Marudarajan D. 1950. Note on *Orobanche cernua* Loefl. *Current Science* 19:64-65.

Mateeva A. 1995. Use of unfreindly plants against root knot nematodes. *Acta Horticulturae* 382 (Feb):178-182.

Matthews GA. 1992. *Pesticide Application Methods,* 2nd Ed. John Wiley & Sons, NY. 405 pp.

Matthiessen P. 1978. *The Snow Leopard.* Picador editions, Pan Books, Ltd. London. 312 pp.

Mattson WJ. 1980. Herbivory in relation to plant nitrogen content. *Annual Review Ecology & Systematics* 11:119-161.

Maude RB. 1996. *Seedborne Diseases and their Control.* CAB International, Wallingford, UK. 280 pp.

McCain AH, Noviello C. 1985. Biological control of *Cannabis sativa. Proceedings, 6th International Symposium on Biological Control of Weeds,* pp.635-642.

McClean DK, Zimmerman AM. 1976. Action of delta-9-tetrahydrocannabinol on cell division and macromolecular synthesis in division-synchronized protozoa. *Pharmacology* 14:307-321.

McClure HE. 1943. Ecology and management of the morning dove in Iowa. *Iowa Agr. Exp. Sta. Research Bulletin* 310:353-415.

McCurry JB, Hicks AJ. 1925. Canadian Plant Disease Survey. Canada Department of Agriculture, *Experimental Farms Annual Report* No. 5, pg. 22.

McEno J. 1987. The history of *Cannabis* plant pathology, Part I. *Sinsemilla Tips* 7(3):37-39.

McEno J. 1988. The history of *Cannabis* plant pathology, Part II. *Sinsemilla Tips* 7(4):42-44.
McEno J. 1990. Hydroponic IPM. *The Growing Edge* 1(3):35-39.
McEno J., ed. 1991. *Cannabis Ecology: A Compendium of Diseases and Pests.* AMRITA Press, Middlebury, Vermont. 254 pp.
McPartland JM. 1983a. Fungal pathogens of *Cannabis sativa* in Illinois. *Phytopathology* 72:797.
McPartland JM. 1983b. *Phomopsis ganjae* sp. nov. on *Cannabis sativa. Mycotaxon* 18:527-530.
McPartland JM. 1984. Pathogenicity of *Phomopsis ganjae* on *Cannabis sativa* and the fungistatic effect of cannabinoids produced by the host. *Mycopathologia* 87:149-153.
McPartland JM. 1989. Proposed list of common names for diseases of *Cannabis sativa* L. *Phytopathology News* 23(4):46-47.
McPartland JM. 1991. Common names for diseases of *Cannabis sativa* L. *Plant Disease* 75:226-227.
McPartland JM. 1992. The *Cannabis* pathogen project: report of the second five-year plan. *Mycological Society of America Newsletter* 43(1):43.
McPartland JM. 1994. Microbiological contaminants of marijuana. *J. International Hemp Association* 1:41-44.
McPartland JM. 1995a. *Cannabis* pathogens VIII: misidenfications appearing in the literature. *Mycotaxon* 53:407-416.
McPartland JM. 1995b. *Cannabis* pathogens IX: anamorphs of *Botryosphaeria* species. *Mycotaxon* 53:417-424.
McPartland JM. 1995c. *Cannabis* pathogens X: *Phoma, Ascochyta* and *Didymella* species. *Mycologia* 86:870-878.
McPartland JM. 1995d. *Cannabis* pathogens XI: *Septoria* spp. on *Cannabis sativa, sensu strico. Sydowia* 47:44-53.
McPartland JM. 1995e. *Cannabis* pathogens XII: lumper's row. *Mycotaxon* 54:273-279.
McPartland JM. 1995f. Microbiological contaminants of marijuana. *J. International Hemp Association* 1(2):41-44.
McPartland JM. 1995g. Mycotoxin Aternariol: letter to the editor. *Townsend Letter for Doctors,* No. 139:88.
McPartland JM. 1996a. A review of *Cannabis* diseases. *J. International Hemp Association* 3(1):19-23.
McPartland JM. 1996b. *Cannabis* pests. *J. International Hemp Association* 3(2):49, 52-55.
McPartland JM. 1997a. "Krankheiten und Schädlinge an *Cannabis,*" pp. 37-38 in *Symposium Magazin,* 2nd Biorohstoff Hanf Technisch-wissenschaftliches Symposium. Nova Institut, Köln.
McPartland JM. 1997b. *Cannabis* as a repellent crop and botanical pesticide. *J. International Hemp Association* 4(2):89-94.
McPartland JM. 1998. Diseases and pests of hemp in Canada. *Commercial Hemp Magazine* 2(5):33-34.
McPartland JM. 1999a. "A survey of hemp diseases and pests," pp. 109-131 in *Advances in Hemp Research,* P. Ranalli, ed. Haworth Press, NY. 272 pp.
McPartland JM. 2000. *Dendrophoma marconii*—q'est-ce que c'est. Manuscript to be submitted to *Cryptogamie Mycologie.*
McPartland JM, Common R. 2000. A non-lichenous fungus, *Jahniella bohemica,* producing a lichenan reaction. Manuscript to be submitted to *Mycotaxon.*
McPartland JM, Cubeta MA. 1997. New species, combinations, host associations and location records of fungi associated with hemp (*Cannabis sativa*). *Mycological Research* 101:853-857.
McPartland JM, Hosoya T. 1998. Species of *Colletotrichum* on ginseng (*Panax*). *Mycotaxon* 67:3-8.
McPartland JM, Hughes S. 1994. *Cannabis* pathogens VII: a new species, *Schiffnerula cannabis. Mycologia* 86:867-869.
McPartland JM, Mitchell J. 1997. Caffeine and chronic back pain. *Archives Physical Medicine & Rehabilitation* 78(1):61-63.
McPartland JM, Pruitt PL. 1997. Medical marijuana and its use by the immunocompromised. *Alternative Therapies in Health & Medicine* 3(3):39-45.
McPartland JM, Pruitt PL. 1999. Side effects of pharmaceuticals not elicited by comparable herbal medicines: the case of tetrahydrocannabinol and marijuana. *Alternative Therapies in Health & Medicine* 5(4):57-62.
McPartland JM, Schoeneweiss DF. 1984. Hyphal morphology of *Botryosphaeria dothidea* in vessels of unstressed and drought-stressed *Betula alba. Phytopathology* 74:358-362.
McPartland JM, West D. 1999. Killing *Cannabis* with mycoherbicides. Manuscript submitted to *J. International Hemp Association* 4
Mechoulam R. 1986. *Cannabinoids as Therapeutic Agents.* CRC Press, Boca Raton, FL. 186 pp.
Mechoulam R, Gaoni Y. 1965. Hashish IV: the isolation and structure of cannabinolic, cannabiniolic, and cannabigerolic acids. *Tetrahedron* 21:1223-1229.
Mediavilla V, Steinemann S. 1997. Ätherische Öl- erste Prüfung einiger Hanfsorten. *Biorohstoff Hanf Symposium Magazin,* p. 58. Nova-Institut, Hürth, Germany [reprinted as "Essential oil of *Cannabis sativa* strains," *J. International Hemp Association* 4(2):82-84].
Mediavilla V, Spiess E, Zürcher B, Bassetti P, Strasser HR, Konermann M, Spahr J, Christen S, Mosimann E, Aeby P, Ott A, Meister E. 1997. Erfahrungen aus dem Hanfanbau 1996. *Proceedings of 2nd Biorohstoff Hanf Symposium,* p. 253-262.
Meier C, Mediavilla V. 1998. Factors influencing the yield and the quality of hemp (*Cannabis sativa* L.) essential oil. *J. International Hemp Association* 5(1):16-20.
Meister RT, ed. 1998. *Farm Chemicals Handbook '98.* Meister Publishing Company, Willoughby, OH. ca. 816 pp.
Mendvedev Z. 1969. *The Rise and Fall of T.D. Lysenko.* Columbia University Press, NY. 284 pp.
Menge JA. 1983. Utilization of vesicular-arbuscular mycorrhizal fungi in agriculture. *Canadian J. Botany* 61:1015-1024.
Meriwheter WF. 1969. Acute marijuana toxicity in a dog. *Veterinary Medicine* 64:577-8.
Merlin E, Rama Das VS. 1954. Influence of root-metabolites on the growth of tree mycorrhizal fungi. *Physiologia Plantarum* 7:851-858.
Messer A, McCormick K, Sunjaya-Hagedorn HH, Tumbel F, Meinwald J. 1990. Defensive role of tropical tree resins: antitermitic sesquiterpenes from southeast Asian Dipterocarpaceae. *J. Chemical Ecology* 16:3333-3352.
Metcalf C, Flint W, Metcalf R. 1962. *Destructive and Useful Insects,* 4th Ed. McGraw-Hill Book Co., NY. 1087 pp.
Mesquita JF, Santos Dias JD. 1984. Ultrastructural and cytochemical study of the lactifers of *Cannabis sativa* L. *Bol. Soc. Brot.* (Sér 2) 57:337-356.
Metzger FW, Grant DH. 1932. Repellency to the Japanese beetle of extracts made from plants immune to attack. *USDA Technical Bulletin* No. 299. 21 pp.
Mezzetti A. 1951. *Alcune Alterazioni della Canapa Manifestatesi nella Decorsa Aannata Agraria.* Quaderni del Centro di Studi per le Ricerche sulla Lavorazione Coltivazione ed Economia della Canapa (Laboratorio Sperimentale di Patologia Vegetale di Bologna), No. 11. 18 pp.

Mian IH, Godoy G, Shelby RA, Rodrígues-Kábana, Morgan-Jones G. 1982. Chitin amendments for control of *Meloidogyne arenaria* in infested soil. *Nematropica* 12:71-84.

Miège E. 1921. Action de la chloropicrine sur la faculté germinative des graines. *C. R. Hebdom. Acad. Sci.* 172(3):170-173.

Miller NG. 1970. Genera of the Cannabaceae in the southeastern United States. *J. Arnold Arboretum* 51:185-194.

Miller SA. 1995. "Plant disease diagnosis: biotechnological approaches," pp. 461-473 in *Molecular Methods in Plant Pathology.* RP Singh, US Singh, eds. CRC-Lewis Publishers, Boca Raton, FL. 523 pp.

Miller PR, Weiss F, O'Brien MJ. 1960. *Index of Plant Diseases in the United States.* Agriculture Handbook No. 165. USDA, Washington, D.C. 531 pp.

Miller WE. 1982. *Grapholita delineana* (Walker), a Eurasian hemp moth, discovered in North America. *Annals Entomological Society America* 75(2):184-186.

Mishra D. 1987. Damping off of *Cannabis sativa* caused by *Fusarium solani* and its control by seed treatment. *Indian J. Mycology &Plant Pathology* 17(1):100-102.

Misra SB, Dixit SN. 1979. Antifungal activity of leaf extracts of some higher plants. *Acta Botanica Indica* 7:147-150.

Mohan Ram HY, Sett R. 1982. Modification of growth and sex expression in *Cannabis sativa* by aminoethoxyvinylglycine and ethephon. *Z. Pflanzenphysiol.* 105:165-172.

Mohyuddin AI, Scheibelreiter GK. 1973. *Investigations on the Fauna of Papaver spp. and Cannabis sativa.* Annual Report, Commonwealth Institute of Biological Control, Switzerland Station. p32-33.

Mojumder V, Mishra SD, Haque MM, Goswami BK. 1989. Nematicidal efficacy of some wild plants against pigeon pea cyst nematode, *Heterodera cajani. Int. Nematol. Network Newsletter* 6(2):21-24.

Molvray M. 1988. A glossary of Tibetan medicinal plants. *Tibetan Medicine* 11:1-85.

Moody MM, Wharton RC, Schnaper N, Schimpff SC. 1982. Do water pipes prevent transmission of fungi from contaminated marijuana? *New England J. Medicine* 306:1492-1493.

Morningstar PJ. 1985. *Thandai* and *chilam* : traditional Hindu beliefs about the proper uses of *Cannabis. J. Psychoactive Drugs* 17(3):141-165.

Mordue AJ, Blackwell A. 1993. Azadirachtin: an update. *J. Insect Physiology* 39:903-924.

Moss MO. 1996. Mycotoxins. *Mycological Research* 100: 513-523.

Mosse B. 1961. Experimental techniques for obtaining a pure inoculum of an *Endogone* sp., and some observations of the vesicular-arbuscular infections caused by it and other fungi. *Recent Advances in Botany* 2:1728-1732.

Mostafa AR, Messenger PS. 1972. Insects and mites associated with plants of the genera *Argemone, Cannabis, Glaucium, Erythroxylum, Eschscholtzia, Humulus* and *Papaver.* Department of Entomological Sciences, University of California, Berkeley. Unpublished manuscript, 240 pp.

Mountain Girl. 1998. *Primo Plant: Growing Marijuana Outdoors.* Quick American Archives, Oakland, CA. 96 pp.

Mujica F. 1942. Actividades fundamentales del departamento de sanidad vegetal. *Bol. Sanidad Vegetal (Chile)* 2(2):144-155.

Mujica F. 1943. Hongos Chilenos no mencionados anteriormen te en la literatura. *Bol. Sanidad Vegetal (Chile)* 3(1):33-35.

Müller FP, Karl E. 1976. Beitrag zur Kenntnis der Bionomie und Morphologie der Hanfblattlaus, *Phordon cannabis* Passerini, 1860. *Beitr. Ent., Berlin* 26:455-463.

Muminovic S. 1990. Alelopatski efekti ekstrakta nekih korova na klijavost sjemena usjeva. *Fragmenta Herbologica Jugoslavica* 19:93-102.

Mushtaque M, Baloch GM, Ghani MA. 1973. *Natural enemies of Papaver spp. and Cannabis sativa.* Annual report, Commonwealth Institute of Biological Control, Pakistan station, pp. 54-55.

Musselman LJ. 1994. "Taxonomy and spread of *Orobanche*," pp. 27-35 in *Biology and Management of Orobanche.* AH Pieterse, JAC Verkleij, SJ ter Borg, eds. Royal Tropical Institute, Amsterdam.

Nagy B. 1959. Kukoricamoly okozta elváltozások és károsítási formák kenderen. *Különlenyomat a Kísérletügyi Közlemények (Növénytermesztés)* 52(4):49-66.

Nagy B. 1967. The hemp moth (*Grapholith sinana* Feld., Lepid.:Tortricidae), a new pest of hemp in Hungary. *Acta Phytopathologica Academiae Scientiarum Hungaricae* 2:291-294.

Nagy B. 1976. Host selection of the European corn borer (*Ostrinia nubilalis* Hbn.) populations in Hungary. *Sym. Biol. Hung.* 16:191-195.

Nagy B. 1979. Different aspects of flight activity of the hemp moth, *Grapholitha delineana* Walk., related to intergrated control. *Acta Phytopathologica Academiae Scientiarum Hungaricae* 14:481-488.

Nagy B. 1986. European corn borer: historical background to the chages of the host plant pattern in the Carpathian basin. *Proceedings of the 14th Symposium of the International Working Group on Ostrinia,* pp 174-181.

Nagy B, Gulyás S, Pétchy I, Tárkányi-Szücs S. 1982. A kenderormányos (*Ceutorrhynchus rapae* Gyll.) kártételének magyarországi jelentkezése. *Növényvédelem* 18(7):289-298.

Nagy JG, Steinhoff HW, Ward GM. 1964. Effects of essential oils of sagebrush on deer rumen microbial function. *J. Wildlife Management* 28:785-90.

Nair KR, Ponnappa KM. 1974. *Survey for Natural Enemies of Cannabis sativa and Papaver somniferum.* Commonwealth Institute of Biological Control, India Station Report, pp 39-40.

Nattrass RM. 1941. Dodder. *East African Agricultural Journal* 6:187-188.

Naumann ID. 1993. *CSIRO Handbook of Australian Insect Names.* CSIRO Publications, Victoria. 200 pp.

Nazarova ES. 1933. [Spring observations on the diseases of new cultivated bast-yielding plants under Daghestan conditions], pp. 25-28 in *Bolyezni i Vredit. nov. lubyan. Kul'tur* [Diseases and Pests of newly cultivated Fibre Plants]. Institut Novogo Lubianogo Syriia [Institute of New Bast Raw Materials], Moscow.

Nazirov Kh, Tukhtaeva S. 1981. Application of treflan to hemp. *Len i Konoplya* [Moscow] ii:34.

Nearing H. 1980. *Wise Words on the Good Life: an Anthology of Quotations.* Schocken Books, NY. 178 pp.

Neve RA. 1991. *Hops.* Chapman & Hall, London. 266 pp.

Nicholls PJ, Tuxford AF, Hould B. 1991. Bacterial content of cotton, flax, hemp and jute. *Proceedings, 15th Cotton Dust Research Conference,* pp. 289-292.

Nickle WR. 1991. *Manual of Agricultural Nematology.* Marcel Dekker, NY.

Noble M, Richardson MJ. 1968. *An Annotated List of Seed-borne Diseases.* Commonwealth Mycological Institute, Kew, U.K. 191 pp.
Nok AJ, Ibrahim S, Arowosafe S, *et al.* 1994. The trypanocidal effect of *Cannabis sativa* constituents in experimental animal trypanosomiasis. *Veterinary & Human Toxicology* 36:522-524.
Nordlee JA, Taylor SL, Townsend JA, Thomas LA, Bush RK. 1996. Identification of a brazil-nut allergen in transgenic soybeans. *New England J. Med.* 334:688-92.
Norton D.C. 1966. Additions to the known hosts of *Meloidogyne hapla. Plant Disease Reporter* 50:523-524.
Noviello C. 1957. Segnalazione di *Verticillium* sp. su *Cannabis sativa. Ricerche, Osserv. e Divulg. Fitopatol. per la Campania ed il Mezzogiorno* 13-14:161-163.
Noviello C, Snyder WC. 1962. Fusarium wilt of hemp. *Phytopathology* 52:1315-1317.
Noviello C, McCain AH, Aloj B, Scalcione M, Marziano F. 1990. Lotta biologica contro *Cannabis sativa* mediante l'impiego di *Fusarium oxysporum* f. sp. *cannabis. Annali della Facolta di Scienze Agrarie della Universita degli Studi di Napoli, Portici* 24:33-44.
Ogbonna CIC, Pugh GJF. 1982. Nigerian soil fungi. *Nova Hedwigia* 36:795-808.
Ojala JC, Jarrell WM. 1980. Hydroponic culture systems for mycorrhizal research. *Plant and Soil* 57:297-303.
Okabe N. 1949. *Shokubutsu Saikinbyogaku* [Bacterial Diseases of Plants]. Askura Shoten, Tokyo. 424 pp.
Okabe N, Goto. 1965. *Major Pests of Economic Plants in Japan.* Japanese Plant Protection Association, Tokyo. 412 pp.
Olkowski W, Daar S, Olkowski H. 1991. *Common-Sense Pest Control.* Taunton Press, Newtown, CT. 715 pp.
Orr DB, Such CPC. 2000. "Parasitoids and predators," pp. 3-34 in *Biological and Biotechnological Control of Insect Pests.* JE Rechcigl, NA Rechcigl, eds. Lewis Publishers, Boca Raton, FL. 374 pp.
Ondrej M. 1991. Vyskyt hub na stoncích konopí (*Cannabis sativa* L.). *Len a Konopí* [Sumperk, Czech Rep] 21:51-57.
Osler W. 1918. *The Principles and Practice of Medicine,* 8th Ed. Appleton & Co., NY, p. 1089.
O'Shaugnessy WB. 1839. On the preparation of Indian hemp, or gunjah. *Transactions Medical & Physical Society of Bengal,* 1838-1840:71-102; 1842:421-461 [reprinted in Marijuana: Medical Papers 1839-1972. 1973. TH Mikuriya, ed. Medi-Comp Press, Oakland, CA].
Oudemans CAJA. 1920. *Enumeratio Systematica Fungorum* II:945-947. Comitum, The Hague.
Overland L. 1966. The role of allelopathic substances in the "smother crop" barley. *American J. Botany* 53:423-432.
Paclt J. 1976. Der Hanf (*Cannabis sativa*) — eine unbekannte Futterpflanze von *Aglope infausta* (L.) (Lep., Zygaenidae)? *Anzeiger für Schädlingskunde, Pflanzenschutz, Umweltschutz* 49(8):116.
Pakhomov VI, Potushanskii VA. 1977. The winter grain fly in the Ul'yanov region. *Zashchita Rasteniî* 9:18-19.
Palm ME. 1999. Mycology and world trade: a view from the front line. *Mycologia* 91:1-12.
Palti J. 1981. Cultural Practice and Infectious Crop Diseases, Springer Verlag, Berlin.
Pandey KN. 1982. Antifungal activity of some medicinal plants on stored seeds of *Eleusine coracana. J. Indian Phytopathology* 35:499-501.
Pandey J, Mishra SS. 1982. Effects of *Cannabis sativa* L. on yield of rabi maize (*Zea mays* L.). in *Abstracts of papers, Annual conference of Indian Society of Weed Science.* Bihar, India.
Pandotra VR, Sastry KSM. 1967. Wilt: a new disease of hemp in India. *Indian J. Agricultural Science* 37:520.
Panizzi AR. 1997. Wild hosts of pentatomids: ecological significance and role in their pest status on crops. *Annual Review Entomology* 42:99-122.
Parberry DG. 1967. Studies on graminicolous species of *Phyllachora. Australian J. Botany* 15:271-375.
Paris M, Boucher F, Cosson L. 1975. The constituents of *Cannabis sativa* pollen. *Economic Botany* 29:245-253.
Parker C. 1986. "Scope of the agronomic problems caused by *Orobanche* species," pp. 11-17 in *Biology and Control of Orobanche,* SJ Borg, ed. Landbouwhogeschool Press, Wageningen, The Netherlands.
Parker C, Riches CR. 1993. *Parasitic Weeds of the World.* CAB International, Wallingford UK, 332 pp.
Parker WB. 1910. The life history and control of the hop flea beetle. *USDA Entomology Bulletin* 82(4):32-58.
Parker WB. 1913a. The hop aphis in the Pacific region. *USDA Entomology Bulletin* 111:9-39.
Parker WB. 1913b. The red spider on hops in the Sacramento valley of California. *USDA Entomology Bulletin* 117:1-41.
Parkinson J. 1640. *The Theater of Plants — an Universal and Compleate Herbal.* Coates, London, p. 42.
Parmeter JR. 1970. *Rhizoctonia solani: Biology and Pathology.* University of California Press, Berkeley. 255 pp.
Paskovic F. 1941. *Konoplja.* Zagreb, Yugoslavia.
Pate DW. 1983. Possible role of ultraviolet radiation in evolution of *Cannabis* chemotypes. *Economic Botany* 37:396-405.
Pate DW. 1994. Chemical ecology of *Cannabis. J. International Hemp Association* 1(2):29, 32-37.
Pate DW. 1999. "The phytochemistry of Cannabis: its ecological and evolutionary implications," pp. 21-42 in *Advances in Hemp Research,* P. Ranalli, ed. Haworth Press, NY. 272 pp.
Pate DW. 1999b. 1999. "Hemp seed: a valuable food source," pp. 243-255 in *Advances in Hemp Research,* P. Ranalli, ed. Haworth Press, NY. 272 pp.
Patschke K, Gottwald R, Müller R. 1997. Erste Ergebnisse phytopathologischer Beobachtungen im Hanfanbau im Land Brandenburg. *Nachrichtenblatt des Deutschen Pflanxenschutzdienstes* 49:286-290.
Patton J. 1998. Hemp: from seed to feed. *Lexington Herald-Leader,* Lexington, KY. 2 June 1998.
Paulsen AQ. 1971. Plant diseases affecting marijuana (*Cannabis sativa*). Unpublished manuscript, Kansas State University. 11 pp.
Peck C. 1884. *Septoria cannabina. 35th Report N.Y. State Museum, Botany,* p. 137.
Peglion V. 1897. Eine neue Krankheit des Hanfes. *Zeitschrift für Pflanzenkrankheiten* 7:81-84.
Peglion V. 1917. Observations on hemp mildew (*Peronoplasmopara cannabina*) in Italy. *Rend. Cl. Acad. Lincei* 114 [Ser. 5, 26(11)]:618-620.
Persoon CH. 1807. *Cannabis sativa. Synopsis Plantarum* 2:618.
Peteanu S. 1980. Contributii la studiul combaterii biologice si integrate a moliei cînepii (*Grapholita delineana* Walker). *Productia Vegetalâ—Cereale si Plante Tehnice* 32(2):39-43.
Petrak F. 1921. *Diplodina cannabicola* n. sp. *Ann. Mycol.* 19:122-123.
Petri L. 1942. Rassegna dei casi fitopatologici osservati nel 1941. *Bollettino della R. Stazione di Patologia Vegetale di Roma* N.S. 22(1):1-62.
Phatak HC, Lundsgaard T, Verma VS, Singh S. 1975. Mycoplasma-like bodies associated with *Cannabis* phyllody. *Phytopathologische Zeitschrift* 83:281-284.

Phelan PL, Norris KH, Mason JF. 1996. Soil-management history and host preference by *Ostrinia nubilalis:* evidence for plant mineral balance mediating insect-plant interactions. *Environmental Entomology* 25:1329-1336.

Pickett JA, Wadhams LJ, Woodcock CM, Hardic J. 1992. The chemical ecology of aphids. *Annual Review Entomology* 37:67-90.

Piao XG. 1990. Studies on the mites infesting herb medicine in China and Japan and their derivation. *Japanese J. Sanitary Zoology* 41:1-7.

Pietkiewicz TA. 1958. Mikroflora nasion konopi. Przeglad literatury. *Poczn. Nauk rol., Ser. A* 77(4)577-590.

Pimentel D, McLaughlin L, Zepp A, *et al.* 1991. "Environmental and economic impacts of reducing U.S. agricultural pesticide use," pp. 679-686 in *CRC Handbook of Pest Management in Agriculture,* 2nd Ed., Vol. I. CRC Press, Boca Raton, FL.

Plaats-Niterink AJ van der, 1981. Monograph of the genus *Pythium. Stud. Mycol.* 21:1-242.

Pliny the Elder. 1950. *Natural History* [English translation, W. H. S. Jones]. Vol. 6, Book 20, Chapter 97 [pg. 153]. Harvard University Press, Cambridge, and William Heinemann Ltd., London.

Ponnappa K M. 1977. New records of fungi associated with *Cannabis sativa. Indian J. Mycology & Plant Pathology* 7:139-142.

Poorani J, Ramamurthy, VV. 1997. Weevils of the genus *Lepropus* Schoenherr from the Oriental region. *Oriental Insects* 31:1-82.

Pospelov AG, Zapromatov NG, Domasheva AA. 1957. *Fungal Flora of the Kirghiz SSR.* Vol. 1. Systematic list of species and geographical distribution. Frunze, Kirghiz SSR. 128 pp.

Prakash A, Pasalu IC, Mathur KC. 1982. Evaluation of plant products as paddy grain protectants in storage. *International J. Entomology* 1:75-77.

Prakash A, Rao J, Pasalu IC. 1987. *Studies on Stored Grain Pests of Rice and Methods of Minimising Losses caused by them.* Final Project Report (RPF-III) Ent-6/CRRI/ICAR (India). 33 pp.

Press MC, Whittaker JB. 1993. Exploitation of the xylem stream by parasitic organisms. *Philos. Trans. Royal Society London (Series B)* 341:101-111.

Preston DA, Dosdall L. 1955. *Minnesota Plant Diseases.* USDA Special Publication No. 8, Horticultural Crops Research Branch. 184 pp.

Pringle HL, Bradley SG, Harris LS. 1979. Susceptibility of *Naegleria fowleri* to delta-9-tetrahydrocannabinol. *Antimicrobial Agents & Chemotherapy* 16:674-679.

Punithalingam E. 1982. Conidiation and appendage formation in *Macrophomina phaseolina* (Tassi) Goid. *Nova Hedwigia* 36:249-290.

Puritch G. 1982. An inside look at insecticidal soap. *Sinsemilla Tips* 3(3):34.

Quarles W. 1999. Pheromones for aphid control. *IPM Practitioner* 21(1):1-6.

Quimby PC, Birdsall JL. 1995. "Fungal agents for biological control of weeds: classical and augmentative approaches," pp. 293-308 in *Novel Approaches to Integrated Pest Management.* R. Reuveni, ed. Lewis Publ./ CRC Press, Boca Raton, FL.

Radosevic A, Kupinic M, Grlic L. 1962. Antibiotic activity of various types of *Cannabis* resin. *Nature* 195:1007-1009.

Ragazzi G. 1954. Nemici vegetali ed animali della canapa. *Humus* 10(5):27-29.

Ramírez J. 1990. Acute pulmonary histoplasmosis: newly recognized hazard of marijuana hunters. *American J. Medicine* 88(Supplement 5):60N-62N.

Rao YR. 1928. *Administration Report of the Government Entomologist for 1927-28.* Coimbatore [India] Agricultural Research Institute. 30 pp.

Rataj K. 1957. Skodlivi cinitele pradnych rostlin. *Prameny literatury* 2:1-123.

Raychaudhuri, D.N. 1985. *Food Plant Catalogue of Indian Aphididae.* Grafic Printall Press, Calcutta. 188 pp.

Rayllo AL. 1927. Experiments and observations on *Hypochnus solani* disease of the potato. *Annals State Institute of Experimental Agronomy, Leningrad.* v. 2-3, p. 203.

Regev S, Cone WW. 1975. Chemical differences in hop varieties vs. susceptibility to the twospotted spider mite. *Environmental Entomology* 4:69-700.

Rehner SA, Uecker FA. 1994. Nuclear ribosomal internal transcribed spacer phylogeny and host diversity in the coelomycete *Phomopsis. Canadian J. Botany* 72:1666-1674.

Reid DP. 1987. *Chinese Herbal Medicine.* Shambhala Press, Boston. 174 pp.

Reininger WR. 1946. Hashish. *Ciba Symposia* 8:374-404.

Reuveni R. 1995. *Novel Approaches to Integrated Pest Management.* Lewis Publishers, Boca Raton, FL. 369 pp.

Reznik PA, Imbs YG. 1965. *Zoologicheskii Zhurnal* 44:1861-1864.

Richter L. 1911. In Brasilien beobachtete Pflanzenkrankheiten. *Zeitschrift für pflanzenkrankheiten* 21:49-50.

Riley CV. 1885. *On the Cotton Worm Together with a Chapter on the Boll Worm.* Fourth Report of the US Entomological Commission, USDA. Government Printing Office, Washington, D.C. 399 pp + 147 pp appendix.

Riley CV, Howard LO. 1892. Hemp as a protection against weevils. *Insect Life (USDA)* 4: 223.

Rippon JW. 1988. *Medical Mycology,* 3rd ed. W.B.Saunders Co., Philadelphia. 797 pp.

Riudavets J. 1995. Predators of *Frankliniella occidentalis* and *Thrips tabaci:* a review. *Wagenigen Agricultural University Papers* 95.1:46-87.

Robel RJ. 1969. Food habits, weight dynamics, and fat content of bobwhites in relation to food plantings in Kansas. *J. Wildl. Management* 33:237-294.

Roberts DW, Hajek AE. 1992. "Entomophathogenic fungi as bioinsecticides," pp. 144-159 in *Frontiers in Industrial Mycology,* GF Leatham, ed. Chapman & Hall, NY. 222 pp.

Roberts L. 1988. Is there life after climate change? *Science* 242:1010-1012.

Robertson LW, Lyle MA, Billets S. 1975. Biotransformation of cannabinoids by *Syncephalastrum racemosum. Biomedical Mass Spectroscopy* 2:266-271.

Robinson BB. 1943a. Seed treatment of hemp seeds. *J. American Society Agronomy* 35:910-914.

Robinson BB. 1943b. Hemp. *USDA Farmer's Bulletin* No. 1935. Washington, D.C. 16 pp.

Robinson BB. 1946. Dew retting of hemp uncertain west of longitude 95°. *J. American Society Agronomy* 38:1106-1109.

Robinson BB. 1952. Hemp (Revised Edition). *USDA Farmer's Bulletin* No. 1935. Washington, D.C. 16pp.

Röder K. 1937. *Phyllostica cannabis* (Kirchner?) Speg. eine Nebeufruchtform von *Mycosphaerella cannabis* (Winter) n.c. *Zeitschrift für Pfanzenkrankheiten* 47:526-531.

Röder K. 1939. Über einen neuen Hanfschädiger, *Didymella arcuata* n. sp. und seine Nebenfruchtformen. *Phytopathologische Zeitschrift* 12:321-333.

Röder K. 1941. Einige Untersuchungen über ein an Hanf (*Cannabis sativa* L.) auftretendes Virus. *Faserforschung* 15:77-81.

Rogers DP. 1977. L.D. de Schweinitz and early American mycology. *Mycologia* 69:223-245.

Röling NG, Wagemakers, eds. 1998. *Facilitating Sustainable Agriculture.* Cambridge University Press, Cambridge, UK. 318 pp.

Roonwal ML. 1945. Notes on the bionomics of *Hieroglyphus nigrorepletus* Bolivar at Benares, United Provinces, India. *Bulletin Entomological Research* 36(3):339-341.

Rosenthal E. 1990. "Ask Ed." *High Times* No. 173:53-54, 62, 82.

Rosenthal E. 1998. *Marijuana Grower's Handbook.* Quick American Archives, Oakland, CA.261 pp.

Rosenthal E. 1999. Sixteen years of "Ask Ed." *High Times* No. 290:96-104.

Rosenthal E & McPartland J. 1998. Tobacco Mosaic Virus. *High Times* No. 275:97-98.

Ross SA, ElSohly MA. 1996. The volatile oil composition of fresh and air-dried buds of *Cannabis sativa. J. Natural Products* 59:49-51.

Rothschild M, Rowan MR, Fairbairn JW. 1977. Storage of cannabinoids by *Arctia caja* and *Zonocerus elegans* fed on chemically distinct strains of *Cannabis sativa. Nature* 266:650-651.

Rothschild M, Aplin RT, Cockrum PA, Edgar JA, Fairweather P, Lees R. 1979. Pyrrolizidine alkaloids in arctiid moths (Lep.) with a discussion on host plant relationships and the role of these secondary plant substances in the Arctiidae. *Biological J. Linnean Society* 12:305-326.

Rothschild M, Fairbairn JW. 1980. Ovipositing butterfly (*Pieris brassicae* L.) distinguishes between aqueous extracts of two strains of *Cannabis sativa* L. and THC and CBD. *Nature* 286:56-59.

Royalty RR, Perring TM. 1987. Comparative toxicity of acaricides to *Aculops lycopersici* and *Homeopronematous anconai. J. Economic Entomology* 80:348-351.

Rubin V. 1975. *Cannabis and Culture.* Mouton Publishers, The Hague.

Runia WT. 1986. Disinfestation of substrates used in protected cultivtion. *Soilless Cult.* 2:35-44.

Rutkis E. 1976. *Cannabis sativa* L. y su historia. *Boletin, Sociedad Cenezolana de Ciencias Naturales* 32:461-471.

Ryan MF, Byrne O. 1988. Plant-insect coevolution and inhibition of acetylcholinesterase. *J. Chemical Ecology* 14:1965-1975.

Sabelis MW, Janssen A. 1994. "Evolution and life-history patterns in the phytoseiidae," pp. 70-98 in *Mites.* MA Houck, ed. Chapman & Hall, NY. 357 pp.

Saccardo PA. 1882-1925 [-72]. *Sylloge Fungorum omnium hucusque cognitorum,* Volumes 1-26. [reprints 1944, 1967], Padova, Italy.

Saccardo PA, Roumeguère C. 1883. Reliquiae Libertianae (Series III). *Revue Mycologique* 5:233-239.

Samson RA, Evans HC, Latgé JP. 1988. *Atlas of Entomopathogenic Fungi.* Springer-Verlag, Berlin. 187 pp.

Samuels GJ. 1996. *Trichoderma:* a review of biology and systematics of the genus. *Mycological Research* 100: 923-935.

Sandru I. 1975. Eficacitatea unor insecticide granulate si emulsionabile in combaterea moliei cinepii (*Grapholitha delineana*). *Probleme de Protectia Plantelor* 3(2):137-154.

Sandru I. 1976. Eficacitatea unor noi insectide in combaterea moliei cinepii (*Grapholitha delineana*). *Analele Institutului de Cercetari pentru Protectia Plantelor* 11:247-259.

Sandru ID. 1977. Vestejirea bacteriana a frunzelor—o boala noua la cinepa [Bacterial wilt in leaves—a new disease of hemp]. *Productia Vegetala Cereale si Plante Tehnice* 29(7): 34-36.

Sands D.C. 1991. Interim Report: *Cannabis sativus.* January/February Report, Cooperative Agreement 58-3K47-9-036, Bozeman, MT. 2 pp.

Sands D.C. 1995. MSU Progress Bullets—April 14, 1995. Montana State University, Bozeman, MT. 2 pp.

Sands DC, Kennett G, Knox-Zidack N, Miller RV, Ford E. 1987. Demonstration of potential biocontrol agents against *Cannabis.* Department of Plant Pathology, Montana State University, Bozeman, MT. 10 pp.

Sáringer G, Nagy B. 1971. The effect of photoperiod and temperature on the diapause of the hemp moth (*Grapholita sinana* Feld.) and its relevance to the integrated control. *Proceedings, 13th International Congress of Entomology, Moscow, 1968.* 1:435-436.

Sasaki T, Honda Y. 1985. Control of certain diseases of greehouse vegetables with ultraviolet-absorbing vinyl film. *Plant Disease* 69:530-533.

Sasman M. 1938. *Cannabis indica* in pharmaceuticals. *J. Medical Soc. New Jersey* 34:51-52.

Sasser JN, Carter CC. 1985. *An Advanced Treatise on Meloidogyne. Volume I: Biology and Control.* North Carolina State University Graphics, Raleigh, NC. 122 pp.

Sastra KS. 1973. Studies on virus diseases of medical plants. *Indian J. Horticulture* 30:562-566.

Savile DBO. 1954. The fungi as aids in the taxonomy of the flowering plants. *Science* 120:583-585.

Schaefer PW, Ikebe K, Higashiura Y. 1987a. Gypsy Moth, *Lymantria dispar* (L.) and its natural enemies in the far east (especially Japan). *Delaware Agricultural Experimental Station Bulletin* No. 476:1-160.

Schaefer PW, Dysart RJ, Specht HB. 1987b. North American distribution of *Coccinella septempunctata* and its mass appearance in costal Delaware. *Environmental Entomology* 16:368-373.

Scheibelreiter GK. 1976. *Investigations on the fauna of Papaver spp. and Cannabis sativa.* Annual report, Commonwealth Institute of Biological Control, European station, pp. 34-35.

Scheifele G. 1998. *Final Report: Determining the feasibility and potential of field production of low THC industrial hemp (Cannabis sativa L.) for fibre and seed grain in northern Ontario.* Kemptville College/University of Guelph, Thunder Bay, Ontario, Canada. <www.gov.on.ca:80/ OMAFRA>

Scheifele G, Dragla P, Pinsonneault C, Laprise JM. 1997. *Hemp (Cannabis sativa) research report, Kent County, Ontario, Canada.* WWW publication.

Schilling RSF. 1956. Byssinosis in cotton and other textile workers. *Lancet* 271:261-265, 319-324.

Schlechtendal DFL. 1846. Klotzschii herbarium vivum mycologicum, Centuria undecima cura. *Botanische Zeitung* 4:878.

Schleiffer H. 1979. *Narcotic Plants of the World.* Lubrecht & Cramer, Monticello, NY. 192 pp.

Schmelzer K. 1962. Untersuchungen an Viren der Zier- und Wildgehölze. 1. Mitteilung. Virosen an*Viburnum* und *Ribes. Phytopathologische Zeitschrift* 46:17-52.

Schmidt HE, Karl E. 1969. Untersuchungen über eine Flecken- und Streifenbildung am Hanf (*Cannabis sativa* L.). *Zentralblatt Bakteriologie, Parasitenkunden, Infektionskrankheiten, Hygiene.* Zweite Abteilung, 123:310-314.

Schmidt HE, Karl E. 1970. Ein Beitrag zur Analyse der Virosen des Hanfes under berücksichtigung der Hanfplattlaus als Virusvektor. *Zentralblatt Bakteriologie, Parasitenkunden, Infecktionskrankheiten, Hygiene.* Abt. 2, 125:16-22.

Schmidt M. 1929. Blattlausfliegen (Syrphidae) als Vorratsschädlinge. *Mitt. Ges. Voratsschutz* 5(6):80-81.
Schreck R, Meier B, Männel DN, Dröge W, Baeuerle PA. 1992. Dithiocarbamates as potent inhibitors of nuclear factor kB activation in intact cells. *J. Exp. Med.* 175: 1181-94.
Schultz OE, Haffner G. 1959. Zur Kenntnis eines sedativen und antibakteriellen Wirkstoffes aus dem deutschen Faserhanf (*Cannabis sativa*). *Zeitschrift für Naturforschung* (Sec. B?) 14:98-100.
Schoenmakers N. 1986. *The Seed Bank 1986/1987 Catalogue.* Postbus 5, 6576 ZA, Ooy, The Netherlands. 13 pp.
Schropp W. 1938. Beiträge zur Kenntnis der Kalimangelerscheinungen einigen Öl- und Gespinstpfanzen. *Ernähr. Pfl.* 34(10):165-170 *and* 34(11-12):181-186.
Schultes RE. 1970. "Random thoughts and queries on the botany of *Cannabis*," pp. 11-38 in *The Botany and Chemistry of Cannabis.* CRB Joyce, SH Curry, eds. J & A Churchill, London. 217 pp.
Schultes RE, Klein WM, Plowman T, Lockwood TE. 1974. *Cannabis:* an example of taxonomic neglect. *Bot. Mus. Leaflet. Harv. Univ.* 23:337-367.
Schultz H. 1939. Untersuchungen über die rolle von *Pythium* -arten als erreger der Fusskrankheit der Lupine I. *Phytopathologische Zeitschrift* 12:405-420.
Schwär K. 1972. [Study of the reasons of hemp suppressive effect on cultivated plants.] *Fiziologiya Biokhimiya Osn Vzaimodeistviia Rast Fitotsenozakh* 3:35-38.
Schwartz I. 1985. Marijuana and fungal infection. *Am. J. Clin. Path.* 84:256.
Schwartz IS. 1987. Fungal sinusitis and marijuana. *JAMA* 257:2914-2915.
Schwartz IS. 1992. Non-*Aspergillus* sinusitis and marijuana use. *Am. J. Clin. Path.* 97:601.
Schweinitz LD de. 1836. Remarks on the plants of Europe which have become naturalized in a more or less degree in the United States. *Ann. Lyceum Nat. Hist. N.Y.* 3:148-155.
Scopes NEA, Pickford R. 1985. "Mass production of natural enemies," pp. 197-209 in *Biological Pest Control: the Glasshouse Experience.* NW Hussey, N Scopes, eds. Blandford Press, Poole, Dorset, UK. 240 pp.
Scudder JM. 1875. *Specific Medication and Specific Medicines,* 6th Ed. Wilstach Baldwin & Co., Cincinnati, OH.
Sears ER. 1956. The transfer of leaf-rust resistance from *Aegilops umbellulata* to wheat. *Proc. Brookhaven Symp. Biol.* 9:1-22.
Seczkowska K. 1969. *Thrips tabaci* Lind. (Thysanoptera) jako wektor *Lycopersicum* virus 3 w Lubelskim Okregu Upraw Tytoniu Przemyslowego. *Annls Univ. Mariae Curie-Sklodowska Sect. C Biol* 24:341-354.
Sekhon SS, Sajjan SS, Kanta U. 1979. A note on new host plants of greenpeach aphid, *Myzus persicae* from Punjab and Himachal Pradesh. *Indian J. Plant Protection* 7:106.
Selgnij S. 1982. Sun, seeds, soil and soul. *Sinsemilla Tips* 2(4):20-21, 30.
Semenchenko GV, Tiourebaev KS, Dolgovskaya M, Schultz MT, McCarthy MK, Sands D.C. 1995. Phytopathogenic *Fusarium* strains in biological control of *Cannabis sativa. Phytopathology* 85:1200.
Senchenko GI, Kolyadko IV. 1973. [A method for producing broomrape-resistant varieties of hemp.] *Selektsiya i Semenovodstvo* 23:27-33.
Senchenko GI, Timonina MA. 1978. *Konoplia.* Kolos Press, Moscow. 260 pp.
Serzane M. 1962. "Kanepju - *Cannabis sativa* L. Slimibas," pp. 366-369 in *Augu Slimibas, Praktiskie Darbi.* Riga Latvijas Valsts Izdevnieciba, Lativa USSR. 518 pp.
Seven Turtles, Chief. 1988. "Reciprocal recurrent selection in *Cannabis*," pp. 34-37 in *The Best of Sinsemilla Tips.* T Alexander, ed. Full Moon Publishing, Corvallis, OR. 253 pp.
Seymour AB. 1929. *Host Index of the Fungi of North America.* Harvard University Press, Cambridge, MA. 732 pp.
Shanavaskhan AE, Binu S, Muraleedharan Unnithan C, Santhoshkumar ES, Pushpangadan P. 1997. Detoxification techniques of traditional physicians of Kerala, India on some toxic herbal drugs. *Fitoterapia* 68:69-74.
Shapiro M, Dougherty EM. 1993. "The use of fluorescent brighteners as activity enhancers for insect pathogenic viruses," pp. 40-46 in *Pest Management: Biologically Based Technologies.* RO Lumsden, JL Vaughn, eds. American Chemical Society, Washington, D.C. 435 pp.
Sharma GK. 1983. Altitudinal differentiation in *Cannabis sativa* L. *Science &Culture* 49(2):53-55.
Sharma GK, Mann SK. 1984. Variation in vegetative growth and trichomes in *Cannabis sativa* L. (marihuana) in response to environmental pollution. *J. Tennessee Academy Science* 59:38-40.
Sharma DC, Rani S, Kashyap NP. 1997. Oviposition deterrence and ovicidal properties of some plant extracts against potato tuber moth *Phthorimaea operculella* (Zell.). *Pesticide Research J.* 9:241-246.
Shay R. 1975. Easy-gro fungus kills pot among us. *The Daily Californian,* March 14, pg. 3.
Shchegolev VN. 1929. Owlet moths as pests of technical plants in the North Caucasus. *Plant Protection* 6:399-406.
Sherbakoff CD. 1928. An examination of fusaria in the herbarium of the pathological collections, Bureau of Plant Industry, USDA. *Phytopathology* 18:148.
Shiraki T. 1952. *Catalogue of Injurious Insects in Japan.* General Headquarters, Supreme Comm. Allied Powers, Tokyo. 842 pp.
Shoemaker RA. 1964. Conidial states of some *Botryosphaeria* species on *Vitis* and *Quercus. Canadian J. Botany* 42:1297-1301.
Shukla DD, Pathak VN. 1967. A new species of *Ascochyta* on *Cannabis sativa* L. *Sydowia* 21:277-278.
Shutova NN, Strygina SP. 1969. The hemp moth. *Zashchita Rastenii* 14(11):49-50.
Sidorenko NM, Shcherban II. 1987. Tetraploid forms of monoecious hemp. *Len i Konoplya* [Moscow] 1987(1):41-42.
Siegel BZ, Garnier L, Siegel SM. 1988. Mercury in marijuana. *BioScience* 38:619-623.
Siegel RK. 1989. *Intoxication: Life in Pursuit of Artificial Paradise.* E.P. Dutton, NY. 390 pp.
Silander JA, Trenbath BR, Fox LR. 1983. Chemical interference among plants mediated by grazing insects. *Oecologia* 58:415-417.
Silantyev A. 1897. *Results of Investigations of the Hemp (Hop, Flax) and Beet Flea.* Ministry Agriculture and Government Estates, St. Petersburg. 7 pp.
Singh KV, Pathak RK. 1984. Effects of leaf extracts of some higher plants on spore germination of *Ustilago maydes* and *U. nuda. Fitoterapia* 55:318-320.
Sinha RP, Singh BN, Zafar IM, Choudhary RS. 1979. Ganja, *Cannabis sativa* L., a new host for cutworms in Bihar. *Indian J. Entomology* 41:296.
Siniscalco Gigliano G, Caputo P, Cozzolino S. 1997. Ribosomal DNA analysis as a tool for the identification of *Cannabis sativa* specimens of forensic interest. *Science & Justice* 37:171-174.

Sitnik VP. 1981. Inheritance of charcters controlled by the pleiotropic effect of the gene for yellow stem in hemp. *Selektsiya i Semenovodstvo* No. 47, pp. 46-49.

Sivanesan A. 1984. *The Bitunicate Ascomycetes and their Anamorphs.* J. Cramer, Vaduz. 701 pp.

Slonov LKh, Petinov NS. 1980. The content of nucleotides and the ATPase activity in hemp leaves in relation to the water supply. *Fiziologiya Rastenii* 27(5):1095-1100.

Slembrouck I. 1994. Anbau von Hanf: Ertragsbildung unter verschiedenin klimatischen Bedingungen. PhD Thesis, Institute für Pflanzenwissenschaften, Zürich, Switzerland.

Small E. 1975. The case of the curious *Cannabis. Economic Botany* 29:254.

Small E, Cronquist A. 1976. A practical and natural taxonomy for *Cannabis. Taxon* 25:405-435.

Small E, Beckstead HD, Chan A. 1975. The evolution of cannabinoid phenotypes in *Cannabis. Economic Botany* 29:219-232.

Smith RA. 1988. Coma in a ferret after ingestion of *Cannabis. Veterinary & Human Toxicology* 30:486.

Smith SM. 1996. Biological control with *Trichogramma. Annual Review Entomology* 41:375-406.

Smith GE, Haney A. 1973. *Grapholitha tristrigana* (Lepidoptera:Torttricidae) on naturalized hemp (*Cannabis sativa* L.) in east-central Illinois. *Transactions Illinois State Academy Science* 66:38-41.

Smith IM, Dunez J, Phillips DH, Lelliott RA, Archer SA. 1988. *European Handbook of Plant Diseases.* Blackwell Scientific Publications, Oxford. 583 pp.

Smith RL, Olson CA. 1982. Confused flour beetle and other coleoptera in stored marijuana. *Pan-Pacific Entomologist* 58:79-80.

Sohi AS. 1977. New genera and species of Typhlocybinae (Homoptera: Cicadellidae) from north-western India. *Oriental Insects* 11:347-362.

Sohi HS, Nayar SK. 1971. New records of fungi from Himachal Pradesh-III. *Research Bulletin of Punjab University* 22:243-245.

Solomon WR. 1976. Volumetric studies of aeroallergen prevalence. Pollens of weedy forbs at a midwestern station. *J. Allergy Clin. Immunol.* 57:318-327.

Sonan J. 1940. On the life history of the Citrus locust (*Chondracris rosea* DeGeer) in Formosa. *Formosan Agricultural Review* 36(9):839-842.

Sorauer P. 1958. *Handbuch der Pflanzenkrankheiten* (Band 5). 26 Volumes. Paul Parey, Berlin.

Southwood TRE. 1959. *Land and Water Bugs of the British Isles.* Frederick Warne & Co., London. 436 pp.

Spaar D, Kleinhempel H, Fritzsche R. 1990. *Öl- und Faserpflanzen.* Springer-Verlag, Berlin. 248 pp.

Sparrow FK. 1936. A contribution to our knowledge of the aquatic Phycomycetes of Great Britain. *J. Linn. Soc. London (Botany)* 50:417-478.

Sparrow FK. 1960. *Aquatic Phycomycetes.* 2nd Edition. University of Michigan Press, Brill, Leiden. 1187 pp.

Spector I. 1985. AMP: a new form of marijuana. *J. Clin. Psychiatry* 46:498-499.

Srivastara PP, Das LL. 1974. Effect of certain aqueous plant extracts on the germination of *Cyperus rotundus* L. *Science & Culture* 40:318-319.

Srivastara SL, Naithani SC. 1979. *Cannabis sativa* Linn., a new host for *Phoma* sp. *Current Science* 48(22):1004-1005.

Stannard LJ, DeWitt JR, Vance TC. 1970. The marijuana thrips, *Oxythrips cannabensis,* a new record for Illinois and North America. *Transactions Illinois Academy Science* 63:152-6.

Steenis MJ van. 1995. Evaluation of four aphidiine parasitoids for biological control of *Aphis gossypii. Entomologia Experimentalis et Applicata* 75:151-157.

Stefano GB, Rialas CM, Deutsch DG, Salzet M. 1998. Anandamide amidase inhibition enhances anadamide-stimulated nitric oxide release in invertebrate neural tissues. *Brain Research* 793:341-5.

Steiner R. 1924. *Geisteswessenschaftliche Grundlagen zum Gedeihen der Landwirtschaft.* R. Steiner Verlag, Dornach.

Stepanova IV. 1975. [Microflora of hemp seeds and its change during storage.] *Doklady Timiriazevskaia S-Kh. Akad.* 209:117-120.

Stern VM, Smith RF, van den Bosch R, Hagen KS. 1959. The integration of chemical and biological control of the spotted alfalfa aphid: the integrated control concept. *Hilgardia* 29:81-101.

Stevens M. 1975. *How to Grow Marijuana Indoors Under Lights,* 3rd Ed. Sun Magic, Seattle. 73 pp.

Stevens NE. 1933. Two apple black rot fungi in the United States. *Mycologia* 25:536-548.

Stockberger WW. 1915. Drug plants under cultivation. *USDA Farmer's Bulletin No.* 663. Washington, D.C. 39 pp.

Stojanovic D. 1959. A contribution to the knowledge of species and varieties of *Cuscuta* in the territory of North Serbia. *Zashtita Bilja* 54:21-27.

Storm D. 1987. *Marijuana Hydroponics.* And/Or Books, Berkeley CA. 118 pp.

Stratii YI. 1976. Hemp and the Colorado beetle. *Zashchita Rasteniî* 5:61.

Strauss JH, Strauss EG. 1988. Evolution of RNA viruses. *Annual Review Microbiology* 42:657-83.

Strong LA. 1921. Quarantine division: reports for the months of March and April 1921. *Monthly Bulletin, California Department of Agriculture, Sacramento* 10(5-6): 210-215.

Strong WB, Croft BA. 1996. Release strategies and cultural modifications for biological control of twospotted spider mite by *Neosieulus fallicis* (Acari: Tetranychidae, Phytoseiidae) on hops. *Environmental Entomology* 25:529-535.

Stupnicka-Rodzynkiewicz E. 1970. Phenomena of allelopathy between some crop plants and weeds. *Acta Agraria et Silvestria (Series Agraria)* 10(2):75-105.

Sukul NC, Bala SK, Bhattacharyya B. 1986. Prolonged cataleptogenic effects of potentized homoeopathic drugs. *Psychopharmacology* 89:338-339.

Surina LN, Stolbov NM. 1981. [Hemp for controlling *Varroa jacobsoni.*] *Pchelovodstvo* 1981 August (8):21-22.

Sutic D, Dowson WJ. 1959. An investigation of a serious disease of hemp (*Cannabis sativa* L.) in Yugoslavia. *Phytopathologische Zeitschrift* 34:307-314.

Sutton BC. 1977. Coelomycetes VI. Nomenclature of generic names proposed for Coelomycetes. *Commonwealth Mycological Institute Mycol. Pap.* 141:1-253.

Sutton BC. 1980. *The Coelomycetes—Fungi Imperfecti with Pycnidia, Acervuli, and Stromata.* Commonwealth Agricultural Bureau Publications, London. 696 pp.

Sutton BC, Waterston JM. 1966. *Ascochyta phaseolorum. Commonwealth Mycological Institute Descriptions of Pathogenic Fungi and Bacteria* No. 81.

Sutton JC, Li DW, Yu GP, Zhang P, Valdebenito-Sanhueza RM. 1997. *Gliocladium roseum:* a versatile adversary of *Botrytis cinerea* in crops. *Plant Disease* 81:316-328.

Sutula CL. 1996. Quality control and cost effectiveness of indexing proceedures. *Advances in Botanical Research* 23:279-292.

Sydow P, Sydow H. 1924. *Monographia Uredinearum*, Vol. 4. Lipsiae Fratres, Borntraeger. 671 pp.

Synnott K., Editor. 1941. Plant diseases. Notes contributed by the biological branch. *Agricultural Gazette New South Wales* 52(7):369-371, 384 and 52(8):435-438.

Szembel SJ. 1927. A new species of rust on hemp. *Commentarii Instituti Astrachanensis ad Defensionem Plantarum* i(5-6):59-60.

Tai FL. 1936. Notes on Chinese fungi VII. *Chinese Botonical Society Bulletin* 2(2):45-66.

Takahashi S. 1919. Daima no gaichu to daima-iengyu ni tsukite. [Notes on insects injurious to hemp.] *Konchu Sekae [Insect World], Gifu* 23(1):20-24.

Takeuchi K. 1949. A list of the food-plants of Japanese sawflies. *Transactions Kansas Entomological Society* 14(2):47-50.

Tarasov AV. 1971. Herbicides in hemp. *Len i Konoplya* (4)20-21. Inst. lub. Kul'tur, Sumskaya Obl., Glukhov, Ukrainian SSR. (abstracted in *Weed Abstr.* 21:2249)

Tarchokov KH. 1975. Herbicides in hemp. *Len i Konoplya* (4)22-23. Kabardino-Balkarskaya, ASSR. (abstracted in *Weed Abstr.* 25:2907)

Taubenhaus J, Killough DT. 1923. Texas root rot of cotton and methods of its control. *Texas Agricultural Experiment Station Bulletin* No. 307. 98 pp.

Tauber MJ, Tauber CA. 1993. "Adaptation to temporal variations in habitats," pp. 103-127 in *Evolution of Insect Pests*. KC Kim, BA McPheron, eds. John Wiley & Sons, Inc., NY. 479 pp.

Taylor DN *et al.* 1982. Salmonellosis associated with marijuana. *New England J. Medicine* 306:1249-1253.

Tedford EC, Jaffee BA, Muldoon AE. 1995. Suppression of the nematode *Heterodera schachtii* by the fungus *Hirsutella rhossiliensis* as affected by fungus population density and nematode movement. *Phytopathology* 85:613-617.

Tehon LR. 1951. The Drug Plants of Illinois. *Illinois Natural History Survey Circular* 44, Urbana, IL. 135 pp.

Tehon LR, Boewe GH. 1939. Charcoal rot in Illinois. *Plant Disease Reporter* 23:312-321.

Tehon LR, Stout GL. 1930. Epidemic diseases of fruit trees in Illinois 1922-1928. *Bulletin Illinois Natural History Survey* 18(3):415-503.

Teng SC. 1936. Additional fungi from China V. *Sinensia* 7:752-822.

Teng SC. 1939. *A Contribution to our Knowledge of the Higher Fungi of China*. National Institute of Zoology & Botany, Peking.

Termorshuizen AJ. 1991. *Literatuuronderzoek over Ziekten bij Nieuwe Potentiële Gewassen*. IPO-DLO Rapport No. 91-08, Instituut voor Planteziektenkundig Onderzoek, Wageningen. 18 pp.

Thacker JRM, Young RDF, Stevenson S, Curtis DJ. 1995. Effects of a change in pesticide droplet size on topical toxicity of chlorpyrifos and deltamethrin to *Myzus persicae* and *Nebria brevicollis*. *J. Economic Entomology* 88:1560-1565.

Thaysen AC, Bunker HJ. 1927. *The Microbiology of Cellulose, Hemicelluloses, Pectin and Gums*. Oxford University Press, London. 363 pp.

Theiling KM, Croft BA. 1988. Pesticide side-effects on arthropod natural enemies: a database summary. *Agriculture Ecosystems & Environment* 21:191-218.

Theissen F, Sydow H. 1915. Die Dothideales. *Ann. Mycol.* 13:11-736.

Thomas HF, Elwood JH, Elwood PC. 1988. Byssinosis in Belfast ropeworks: an historical note. *Ann. Occup. Hyg.* 32:249-251.

Thompson GR, Rosenkrantz H, Schaeppi UH, Braude MC. 1973. Comparison of acute oral toxicity of cannabinoids in rats, dogs and monkeys. *Toxicology & Applied Pharmacology* 254:363-372.

Thompson LM, Troeh FR . 1973. *Soils and Soil Fertility*, 3rd Ed. McGraw-Hill, NY. 498pp.

Thomson WT. 1992. *A Worldwide Guide to Beneficial Animals used for Pest Control Purposes*. Thomson Publications, Fresno, CA. 91 pp.

Thomson WT. 1998. *Agricultural Chemicals Book I: Insecticides, Acaricides and Ovicides*, 14th Edition. Thomson Publications, Fresno, CA. 269 pp.

Thomson WT. 1997. *Agricultural Chemicals Book IV: Fungicides*, 12th Edition. Thomson Publications, Fresno, CA. 236 pp.

Thurston JM. 1900. *The Philosophy of Physiomedicalism*. Nicholson Printing & Mfg. Co., Richmond, IN. 398 pp.

Tichomirov, W. 1868.*Peziza Kauffmanniana*, eine neue, aus Sclerotium stammende und auf Hanf schmarotzende becherpilz-species. *Bullétin Société Imperiale Naturalistes Moscou* 41(2):295-342.

Tietz HM. 1972. *An Index to the Described Life Histories, Early Stages and Hosts of the Macrolepidoptera of the Continental United States and Canada*. AC Allyn, Sarasota, FL. 2 Vol., 1041 pp.

Tiourebaev KS, Pilgeram AL, Anderson TW, Sands D.C. 1997. Soil penetration of a mycoherbicide facilitated by carrier seedlings. *Phytopathology* 87(6 Supplement):S 97.

Tiourebaev KS, Pilgeram AL, Anderson TW, Baizhanov MK, Sands D.C. 1998. *Fusarium oxysporum* f. sp. *cannabina* [sic] as promising candidate for biocontrol of *Cannabis* sp. *Phytopathology* 88(9 Supplement):S 89.

Tkalich PP. 1967. [The occurrence and use of entomophagous insects against the stem borer on hemp.] *Vozdelÿvanie i Pervichnaya Obrabotka Konopli*, pp 143-146. Urozhaï, Kiev. [abstracted in *Review Applied Entomology* 58:3017].

Toole EH, Toole VK, Nelson EG. 1960. Preservation of hemp and kenaf seed. *USDA-ARS Technical Bulletin* No. 1215. 16 pp.

Toussoun TA, Nelson PA . 1975. Variation and speciation in the fusaria. *Ann. Rev. Phytopathology* 13:71-82.

Transhel V, Gutner L, Khokhryakov M. 1933. A list of fungi found on new cultivated textile plants. *Moscow Inst. Nov. Lubian. Syr'ia, Trudy* 4:127-140.

Traversi BA. 1949. Estudio inicial sobre una enfermeded del girasol (*Helianthus annuus* L.) en Argentina. *Lilloa* 21:271-278.

Tremblay E. 1968. Observations of hemp weevils. Notes on morphology, biology, and chemical control. *Bollettino Portici Lab. Ent. Agr. Filippo Silvestri* 26:139-190.

Tremblay E, Bianco M. 1978. *I punteruoli del cavolfiore in Campania*. Note Divulgative No. 12., Istituto di Entomologia Agraria della Università di Napoli, Portici. 26 pp.

Trunoff GA. 1936. "Contributions to the phytopathological study of hemp," pp. 69-113 in *Collection of Scientific Papers on Plant Protection*. Ukr. St. Publ. Off. Collect. Co-op., Kiev.

Tumosa CS. 1984. A lectin in the pollen of marihuana, *Cannabis sativa* L. *Experientia* 40:718-9.

Turlings TC, Tumlinson JH, Lewis WJ. 1990. Exploitation of herbivore-induced plant odors by host-seeking parasitic wasps. *Science* 250:1251-1253.

Turner CE, Elsohly MA. 1981. Biological activity of cannabichromene, its homologs and isomers. *J. Clin. Pharmacol.* 21(supplement): 283S-291S.

Turner CE, Elsohly MA, Boeren EG. 1980. Constituents of *Cannabis sativa* L. XVII. A review of the natural constituents. *J. Natural Products* 43:169-234.

Ungerleider JT, Andrysiak T, Tashkin DP, Gale RP. 1982. Contamination of marijuana cigarettes with pathogenic bacteria. *Cancer Treatment Reports* 66(3):589-590.

Upadhyaya ML, Gupta RC. 1990. Effect of extracts of some medicinal plants on the growth of *Curvularia lunata. Indian J. Mycology & Plant Pathology* 20:144-145.

Uppal BN. 1933. India: diseases in the Bombay presidency. *International Bulletin Plant Protection* 6(5):103-104.

Urban L. 1979. Fagyhatás és fagyvédelem. *Növénternelés* 28(5): 473-478.

Vakhrusheva TE. 1979. *Methodological instructions for categorizing diseases and cultures of flax and hemp.* All-Union N. I. Vavilov Scientific Research Institute of Horticulture, Leningrad. 195 pp.

Vance JM. 1971. Marijuana is for the birds. *Outdoor Life* 147(6):53-55, 96-100.

Van der Werf HMG. 1994. Crop physiology of fibre hemp (*Cannabis sativa* L.) Doctoral thesis, Wageningen Agricultural University, Wageningen, the Netherlands. 152 pp.

Van der Werf HMG, Van Geel WCA, Wijhuizen M. 1995. Agronomic research on hemp (*Cannabis sativa* L.) in the Netherlands, 1987-1993. *J. International Hemp Association* 2:14-17.

Van Driesche RG, Bellows TS. 1996. *Biological Control.* Chapman & Hall, NY. 539 pp.

Van Lenteren JC. 1995. "Integrated pest management in protected crops," pp. 311-343 in *Integrated Pest Management.* D Dent, ed. Chapman & Hall, London.

Vassilieff AA. 1933. [Wilt of cultivated bast-yielding plants under central Asian conditions], pp. 22-24 in *Bolyezni i Vredit. nov. lubyan. Kul'tur* [Diseases and Pests of Newly Cultivated Fibre Plants]. Institut Novogo Lubianogo Syriia [Institute of New Bast Raw Materials], Moscow.

Vasudeva RS. 1961. *Indian Cercosporae.* Indian Council of Agricultural Research, New Delhi.

Vavilov NI. 1926. "The origin of the cultivation of "primary" crops, in particular of cultivated hemp," pp. 221-233 in *Studies on the Origin of Cultivated Plants.* Institute of Applied Botany and Plant Breeding, Leningrad. 248 pp.

Vavilov NI. 1957. *Agroecological Survey of the Main Field Crops.* Academy of Sciences of the USSR, Moscow. 427 pp.

Vavilov NI, Bukinich DD. 1929. *Zemledelcheskii Afghanistan. Trudy po Prikladnoi Botanike, Genetike i Selektsii* (reissued 1959 by Izdatel'stuo Akademii Nauk SSSR, Moskva-Leningrad). 610 pp.

Vegter IH. 1983. *Index Herbariorum: Collectors, Part II (5).* W. Junk, Publishers, The Hague.

Veliky IA, Genest K. 1972. Growth and metabolites of *Cannabis sativa* cell suspension cultures. *Lloydia* 35:450-456.

Veliky IA, Latta RK. 1974. Antimicrobial activity of cultured plant cells and tissues. *Lloydia* 37:611-620.

Vetter C. 1985. Early harvest: against the wind. *Playboy Magazine* 32(7):47.

Viégas AP. 1961. *Indices de Fungos da America do Sul.* Seçao de Fitopathologia. Campinas, Brazil. 921 pp.

Vijai P, Jalali I, Parashar RD. 1993. Suppression of bacterial soft rot of potato by common weed extracts. *J. Indian Potato Association* 20:206-209.

Vijayalakshmi K, Mishra SD, Prasad SK. 1979. Nematicidal properties of some indigenous plant materials against second stage juveniles of *Meloidogyne incognita. Indian J. Entomology* 41:326-331.

Vinokurov GM. 1927. Larvae of the lucerne or flax noctuid (*Chloridea dipsacea* L.) as a pest of grain in the ear. *Bull. Irkutsk Pl. Prot. Sta.* 1:67-79.

Virovets VG, Lepskaya LA. 1983. "Varietal resistance in hemp to European corn borers (*Ostrinia nubilalis* Hb.)," pp. 53-58 in *Biologicheskie osobennosti, tekhnologiya vozdelyvaniya i pervichnaya obrabotka lubyanykh kul'tur.* Glukhov, Ukrainia.

Vismal OP, Shukla GC. 1970. Chemical growth inhibitors liberated during decomposition of submerged weeds and their effect on the growth of rice crop. *Indian J. Agricul. Sci.* 40:535-545.

Voglino P. 1912. Pflanzenkrankheiten im Piemont. *Zeitschrift für Pflanzenkrankheiten* 22:153-155.

Vysots'kyi GA. 1962. Zastosuvannya fitontsÿdiv Konopel' u borot'bi z fuzarizom Sosnÿ. *Mikrobiolohichnyi Zhurnal (Kiev)* 24(2):65-66.

Wahlin B. 1944. Några fall av manganbrist sommaren 1943. *Växtskyddsanst Stockholm* 8(1):11-15.

Wang B, Ferro DN, Hosmer DW. 1997. Importance of plant size, distribution of egg masses, and weather conditions on egg parasitism of the European corn borer, *Ostrinia nubilalis* by *Trichogramma ostriniae* in sweet corn. *Entomologia Experimentalis et Applicata* 83:337-345.

Wang T, Cui L, Wan Z. 1987. A study on the hemp sawfly. *Acta Entomologica Sinica* 30:407-413.

Wang T, Shen G, Feng X, Guo D, Lui L. 1995. A study on the bionomics of the hemp weevil *Rhinoncus pericarpius. China's Fiber Crops* 17(1):37-39.

Watanabe T, Takesawa M. 1936. Studies on the leaf-spot disease of the hemp. *Ann. Phytopath. Soc. Japan* 6:30-47.

Waterhouse GM, Brothers MP. 1981. The taxonomy of *Pseudoperonospora. Commonwealth Mycological Institute Mycological Paper* No. 148. 28pp.

Watling RJ. 1998. Sourcing the provenance of *Cannabis* crops using inter-element association patterns "fingerprinting" and laser ablation inductively coupled plasma mass spectrometry. *J. Analytical Atomic Spectrometry* 13:917-926.

Watson ES, Murphy JC, Turner CE. 1983. Allergenic properties of naturally occurring cannabinoids. *J. Pharmaceutical Sciences* 72:954-5.

Weber H. 1930. *Biologie der Hepipteren.* Verlag Von Julius Springer, Berlin. 543 pp.

Wehmer CW. 1903. Der *Mucor* der Hanfrötte, *M. hiemalis* nov. spec. *Ann. Mycologici* 1:39-41.

Wei CS, Cai P. 1998. [Two new species of the genus *Macropsis* Lewis (Homptera: Cicadellidae) from China.] *Entomotaxonomia* 20 (2):119-122.

Weigert J, Fürst F. 1939. Die wirkung von spurenelementen in dem randgeibieten südbayerischer moore. *Prakt. Bl. Pflanzenb.* 17:117-140.

Westcott C. 1964. *The Gardener's Bug Book,* 3rd Ed. Doubleday, Garden City NY. 625 pp.

Westcott C. 1990. *Wescott's Plant Disease Handbook,* 5th Ed., revised by R.K. Horst. Van Nostrand Reinhold, NY. 953 pp.

Westendorp GD. 1854. *Les Cryptogames.* I.S. Van Doosselaere. Gand, Belgium. 301 pp.

Wheeler QD. 1997. "The role of taxonomy in genetic resource management," pp. 59-70 in *Global Genetic Resources: Access, Ownership, and Intellectual Property Rights,* KE Hoagland, AY Rossman, eds. Association of Systematics Collections, Washington, D.C.

Whipps JM, Gerlagh M. 1992. Biology of *Coniothyrium minitans* and its potential for use in disease biocontrol. *Mycological Research* 96(11):897-907.

Windels CE. 1997. "Altering community balance: organic amendments, selection pressures, and biocontrols," pp. 282-300 in *Ecological Interactions and Biological Control*. DA Andow, DW Ragsdale, RF Nyvall, eds. Westview Press, Boulder, CO. 334 pp.

Williams R J, Ayanaba A. 1975. Increased incidence of *Pythium* stem rot in cowpeas treated with benomyl and related fungicides. *Phytopathology* 65:217-218.

Wilsie CP, Reddy CS. 1946. Seed treatment experiments with hemp. *J. American Society Agronomy* 38:693-701.

Wilson CL, ElGhaouth A, Wisniewski ME. 1997. Rapid evaluation of plant extracts and essential oils for antifungal activity against *Botrytis cinerea*. *Plant Disease* 81:204-210.

Wilson GF. 1938. Contributions from the Wisley Laboratory No. 85: the glasshouse leaf-hopper, *Erythroneura pallidifrons*. *J. Royal Horticultural Society* 1938:481-484.

Wilson CL, Ehalutz E. 1991. *Biological Control of Postharvest Diseases of Fruits and Vegetables*. USDA-ARS No. 92, Washington, D.C. 324 pp.

Wilson GWT, Daniels-Hetrick BA, Gerschefske-Kitt B. 1988. Suppression of mycorrhizal growth response of big bluestem by non-sterile soil. *Mycologia* 80:338-343.

Winslow RD. 1954. Provisional lists of host plants of some root eelworms (*Heterodera* spp.). *Annals Applied Biology* 41:591-605.

Wirstshafter D. 1997. "Nutritional value of hemp seed and hemp seed oil, pp. 181-191 in *Cannabis in Medical Practice*. ML Mathre, ed. McFarland & Co., Jefferson, NC. 239 pp.

Wodehouse RP. 1945. *Hayfever Plants*. Chronica Botanica Co., Waltham, MA. 245 pp.

Wolf B. 1999. *The Fertile Triangle: The Interrelationship of Air, Water, and Nutrients in Maximizing Soil Productivity*. Haworth Press, New York. 463 pp.

Wollenweber HW. 1926. *Hypomyces cancri. Fusaria Autographice Delineata* No. 55. Berolini, Italy. 509 tabulis.

Wollenweber HW, Reinking OA. 1935. *Die Fusarien*. Verlag Paul Parey, Berlin. 355 pp.

Wood A. 1989. *Insects of Economic Importance: a checklist of preferred names*. CABI, Wallingford, UK. 150 pp.

Wu FF, Dobberstein RH, Morris RW. 1978. The effects of selected fungicides and insecticides on growth and thebaine production of *Papaver bracteatum*. *Lloydia* 41:355-360.

Yarwood CE. 1973. "Erysiphales," pp. 71-86 in *The Fungi: An Advanced Treatise*, Vol. 4A. Ainsworth GC, Sparrow FK, Sussman AS, eds. Academic Press, NY. 621 pp.

Yepsen RB. 1976. *Organic Plant Protection*. Rodale Press, Inc. Emmaus, PA. 688 pp.

Young M. 1997. *The Natural History of Moths*. T & AD Poyser, Ltd., London. 271 pp.

Yu S-L. 1973. Fungal pathogens of *Cannabis sativa* grown in the USDA garden. Unpublished manuscript, University of Mississippi, University, Mississippi. 11 pp.

Yuasa H. 1927. Notes on the Japanese Chrysomelidae. 1. On the food plants of several species. *Kontyu* 2(2): 130-133.

Yunker TG. 1932. *Cuscuta* monograph. *Mem. Torrey Bot. Club* 18:113-331.

Zabrin R. 1981. The fungus that destroys pot. *War on Drugs Action Reporter*, June 1981:61-62.

Zaborski E. 1998. Earthworms: indicators of soil health. *Illinois Natural History Survey Reports* 351:4-5.

Zabrodskii AE. 1968. Herbicides in hemp crops. *Len i Konoplia* [Moscow]13(4):27-29. Inst. lubyan. Kul't., Sumskaya obl., Glukhov, ul. Lenina 35, USSR. (abstracted in *Weed Abstr.* 18:2581)

Zander A. 1928. Über Verlauf und Entstehung der Michröhren des Hanfes (*Cannabis sativa*). *Flora* 23:195-218.

Zarzycka H, Jaranowska B. 1977. Alternarioza konopi. *Acta Agrobotanica* 30:419-421.

Zelenay A. 1960. Fungi of the genus *Fusarium* occuring on seeds and seedlings of hemp and their pathogenicity. *Prace nauk. Inst. Ochr. Rosl., Pozan* 2(2):248-249.

Zelepukha SI. 1960. [The third conference on the problem of phytoncides.] *Mikrobiolohichnyi Khurnal (Kiev)* 22(1):68-71.

Zelepukha SI, Rabinovich AS, Pochinok PY, Negrash AK, Kudryavtsev VA. 1963. [Antimicrobial properties of preparations from hemp—cansantin.] *Mikrobiolohichnyi Khurnal (Kiev)* 25(2):42-46.

Zhalnina LS. 1969. Porazhennost' konopli fuzariozom pri dlitel' nom primenenii udobrenii. *Khimiya sel'. Khoz.* 7(11):33-34.

Zottini M, Mandolino G, Ranalli P. 1997. Effects of g-ray treatment on *Cannabis sativa* pollen viability. *Plant Cell, Tissue & Organ Culture* 47:189-194.

Zubrin R. 1981. The fungus that destroys pot. *War on Drugs Action Reporter*, June 1981:61-62. LaRouche Publ. Co.

Zuskin E, Valic F, Bouhuys A. 1976. Byssinosis and airway responses due to exposure to textile dust. *Lung* 154:17-24.

Zycha H, Siepmann R, Linnemann G. 1969. *Mucorales*. J. Cramer, Lehre, Germany. 355 pp.

Zylinski T. 1958. *Nauka o Włóknie*. Centralny Instytut Informacji Naukowo-technicznej i Ekonomicznej. Warszawa. 681 pp.

Index

Note: page numbers in *italics* refer to figures and tables